Lecture Notes in Computer Science 13811

The series Lecture Notes in Computer Science (LNCS), including its subseries Lecture Notes in Artificial Intelligence (LNAI) and Lecture Notes in Bioinformatics (LNBI), has established itself as a medium for the publication of new developments in computer science and information technology research, teaching, and education.

LNCS enjoys close cooperation with the computer science R & D community, the series counts many renowned academics among its volume editors and paper authors, and collaborates with prestigious societies. Its mission is to serve this international community by providing an invaluable service, mainly focused on the publication of conference and workshop proceedings and postproceedings. LNCS commenced publication in 1973.

Giuseppe Nicosia · Varun Ojha ·
Emanuele La Malfa · Gabriele La Malfa ·
Panos Pardalos · Giuseppe Di Fatta ·
Giovanni Giuffrida · Renato Umeton
Editors

Machine Learning, Optimization, and Data Science

8th International Conference, LOD 2022
Certosa di Pontignano, Italy, September 18–22, 2022
Revised Selected Papers, Part II

Editors
Giuseppe Nicosia
University of Catania
Catania, Italy

Varun Ojha
University of Reading
Reading, UK

Emanuele La Malfa
University of Oxford
Oxford, UK

Gabriele La Malfa
University of Cambridge
Cambridge, UK

Panos Pardalos
University of Florida
Gainesville, FL, USA

Giuseppe Di Fatta
Free University of Bozen-Bolzano
Bolzano, Italy

Giovanni Giuffrida
University of Catania
Catania, Italy

Renato Umeton
Dana-Farber Cancer Institute
Boston, MA, USA

ISSN 0302-9743 ISSN 1611-3349 (electronic)
Lecture Notes in Computer Science
ISBN 978-3-031-25890-9 ISBN 978-3-031-25891-6 (eBook)
https://doi.org/10.1007/978-3-031-25891-6

This Springer imprint is published by the registered company Springer Nature Switzerland AG
The registered company address is: Gewerbestrasse 11, 6330 Cham, Switzerland

Preface

LOD is the international conference embracing the fields of machine learning, deep learning, optimization, and data science. The eighth edition, LOD 2022, was organized on September 18–22, 2022, in Castelnuovo Berardenga, Italy. LOD 2022 was held successfully online and onsite to meet challenges posed by the worldwide outbreak of COVID-19. As in the previous edition, LOD 2022 hosted the second edition of the Advanced Course and Symposium on Artificial Intelligence & Neuroscience – ACAIN 2022. In fact, this year, in the LOD proceedings we decided to also include the papers of the second edition of the Symposium on Artificial Intelligence and Neuroscience (ACAIN 2022). The ACAIN 2022 chairs were:

Giuseppe Nicosia, University of Catania, Italy, and University of Cambridge, UK
Varun Ojha, University of Reading, UK
Panos Pardalos, University of Florida, USA

The review process of papers submitted to ACAIN 2022 was double blind, performed rigorously by an international program committee consisting of leading experts in the field. Therefore, the last thirteen articles in the Table of Contents are the articles accepted at ACAIN 2022.

Since 2015, the LOD conference has brought together academics, researchers and industrial researchers in a unique *pandisciplinary community* to discuss the state of the art and the latest advances in the integration of machine learning, deep learning, nonlinear optimization and data science to provide and support the scientific and technological foundations for interpretable, explainable and trustworthy AI. Since 2017, LOD has adopted the *Asilomar AI Principles.*

The annual conference on machine Learning, Optimization and Data science (LOD) is an international conference on machine learning, computational optimization and big data that includes invited talks, tutorial talks, special sessions, industrial tracks, demonstrations and oral and poster presentations of refereed papers.

LOD has established itself as a premier multidisciplinary conference in machine learning, computational optimization and data science. It provides an international forum for presentation of original multidisciplinary research results, as well as exchange and dissemination of innovative and practical development experiences.

The manifesto of the LOD conference is:

"The problem of understanding intelligence is said to be the greatest problem in science today and "the" problem for this century – as deciphering the genetic code was for the second half of the last one. Arguably, the problem of learning represents a gateway to understanding intelligence in brains and machines, to discovering how the human brain works, and to making intelligent machines that learn from experience and improve their competences as children do. In engineering, learning techniques would make it possible to develop software that can be quickly customized to deal with the increasing amount of information and the flood of data around us."

The Mathematics of Learning: Dealing with Data
Tomaso Poggio (MOD 2015 & LOD 2020 Keynote Speaker) and Steve Smale

"Artificial Intelligence has already provided beneficial tools that are used every day by people around the world. Its continued development, guided by the Asilomar principles of AI, will offer amazing opportunities to help and empower people in the decades and centuries ahead."
The Asilomar AI Principles

LOD 2022 attracted leading experts from industry and the academic world with the aim of strengthening the connection between these institutions. The 2022 edition of LOD represented a great opportunity for professors, scientists, industry experts, and research students to learn about recent developments in their own research areas and to learn about research in contiguous research areas, with the aim of creating an environment to share ideas and trigger new collaborations.

As chairs, it was an honour to organize a premier conference in these areas and to have received a large variety of innovative and original scientific contributions.

During LOD 2022, 11 plenary talks were presented by leading experts:

LOD 2022 Keynote Speakers:

Jürgen Bajorath, University of Bonn, Germany
Pierre Baldi, University of California Irvine, USA
Ross King, University of Cambridge, UK and The Alan Turing Institute, UK
Rema Padman, Carnegie Mellon University, USA
Panos Pardalos, University of Florida, USA

ACAIN 2022 Keynote Lecturers:

Karl Friston, University College London, UK
Wulfram Gerstner, EPFL, Switzerland
Máté Lengyel, Cambridge University, UK
Christos Papadimitriou, Columbia Engineering, Columbia University, USA
Panos Pardalos, University of Florida, USA
Michail Tsodyks, Institute for Advanced Study, USA

LOD 2022 received 226 submissions from 69 countries in five continents, and each manuscript was independently reviewed by a committee formed by at least five members. These proceedings contain 85 research articles written by leading scientists in the fields of machine learning, artificial intelligence, reinforcement learning, computational optimization, neuroscience, and data science, presenting a substantial array of ideas, technologies, algorithms, methods, and applications.

At LOD 2022, Springer LNCS generously sponsored the LOD Best Paper Award. This year, the paper by *Pedro Henrique da Costa Avelar, Roman Laddach, Sophia Karagiannis, Min Wu and Sophia Tsoka* titled "*Multi-Omic Data Integration and Feature Selection for Survival-based Patient Stratification via Supervised Concrete Autoencoders*", received the LOD 2022 Best Paper Award.

This conference could not have been organized without the contributions of exceptional researchers and visionary industry experts, so we thank them all for participating.

A sincere thank you goes also to the 35 sub-reviewers and to the Program Committee, of more than 170 scientists from academia and industry, for their valuable and essential work in selecting the scientific contributions.

Finally, we would like to express our appreciation to the keynote speakers who accepted our invitation, and to all the authors who submitted their research papers to LOD 2022.

October 2022

Giuseppe Nicosia
Varun Ojha
Emanuele La Malfa
Gabriele La Malfa
Panos Pardalos
Giuseppe Di Fatta
Giovanni Giuffrida
Renato Umeton

Organization

General Chairs

Giovanni Giuffrida	University of Catania and NeoData Group, Italy
Renato Umeton	Dana-Farber Cancer Institute, MIT, Harvard T.H. Chan School of Public Health & Weill Cornell Medicine, USA

Conference and Technical Program Committee Co-chairs

Giuseppe Di Fatta	Free University of Bozen-Bolzano, Bolzano, Italy
Varun Ojha	University of Reading, UK
Panos Pardalos	University of Florida, USA

Special Sessions Chairs

Gabriele La Malfa	University of Cambridge, UK
Emanuele La Malfa	University of Oxford, UK

Tutorial Chair

Giorgio Jansen	University of Cambridge, UK

Steering Committee

Giuseppe Nicosia	University of Catania, Italy
Panos Pardalos	University of Florida, USA

Program Committee Members

Jason Adair	University of Stirling, UK
Agostinho Agra	Universidade de Aveiro, Portugal
Richard Allmendinger	University of Manchester, UK

Paula Amaral	University NOVA of Lisbon, Portugal
Hoai An Le Thi	Université de Lorraine, France
Adam Arany	University of Leuven, Belgium
Alberto Archetti	Politecnico di Milano, Italy
Roberto Aringhieri	University of Torino, Italy
Heder Bernardino	Universidade Federal de Juiz de Fora, Brazil
Daniel Berrar	Tokyo Institute of Technology, Japan
Martin Berzins	SCI Institute, University of Utah, USA
Hans-Georg Beyer	FH Vorarlberg University of Applied Sciences, Austria
Martin Boldt	Blekinge Institute of Technology, Sweden
Fabio Bonassi	Politecnico di Milano, Italy
Anton Borg	Blekinge Institute of Technology, Sweden
Michele Braccini	Università di Bologna, Italy
Will Browne	Victoria University of Wellington, New Zealand
Sergiy I. Butenko	Texas A&M University, USA
Luca Cagliero	Politecnico di Torino, Italy
Timoteo Carletti	University of Namur, Belgium
Michelangelo Ceci	University of Bari, Italy
Adelaide Cerveira	INESC-TEC & Universidade de Trás-os-Montes e Alto Douro, Portugal
Uday Chakraborty	University of Missouri, St. Louis, USA
Keke Chen	Wright State University, USA
John W. Chinneck	Carleton University, Canada
Miroslav Chlebik	University of Sussex, UK
Sung-Bae Cho	Yonsei University, South Korea
Stephane Chretien	ASSP/ERIC, Université Lyon 2, France
Andre Augusto Cire	University of Toronto Scarborough, Canada
Philippe Codognet	JFLI - CNRS/Sorbonne University/University of Tokyo, France/Japan
Sergio Consoli	European Commission, Joint Research Centre (DG-JRC), Belgium
Chiara Damiani	University of Milan Bicocca, Italy
Thomas Dandekar	University of Wuerzburg, Germany
Renato De Leone	Università di Camerino, Italy
Nicoletta Del Buono	University of Bari, Italy
Mauro Dell'Amico	Università degli Studi di Modena e Reggio Emilia, Italy
Brian Denton	University of Michigan, USA
Ralf Der	MPG, Germany
Clarisse Dhaenens	University of Lille, France
Giuseppe Di Fatta	Free University of Bozen-Bolzano, Bolzano, Italy

Valery Kalyagin	National Research University Higher School of Economics, Russia
George Karakostas	McMaster University, Canada
Branko Kavšek	University of Primorska, Slovenia
Michael Khachay	Ural Federal University Ekaterinburg, Russia
Zeynep Kiziltan	University of Bologna, Italy
Yury Kochetov	Novosibirsk State University, Russia
Hennie Kruger	North-West University, South Africa
Vinod Kumar Chauhan	University of Cambridge, UK
Dmitri Kvasov	University of Calabria, Italy
Niklas Lavesson	Jönköping University, Sweden
Eva K. Lee	Georgia Tech, USA
Carson Leung	University of Manitoba, Canada
Gianfranco Lombardo	University of Parma, Italy
Paul Lu	University of Alberta, Canada
Angelo Lucia	University of Rhode Island, USA
Pasi Luukka	Lappeenranta-Lahti University of Technology, Finland
Anthony Man-Cho So	The Chinese University of Hong Kong, Hong Kong, China
Luca Manzoni	University of Trieste, Italy
Marco Maratea	University of Genova, Italy
Magdalene Marinaki	Technical University of Crete, Greece
Yannis Marinakis	Technical University of Crete, Greece
Ivan Martino	Royal Institute of Technology, Sweden
Petr Masa	University of Economics, Prague, Czech Republic
Joana Matos Dias	Universidade de Coimbra, Portugal
Nikolaos Matsatsinis	Technical University of Crete, Greece
Stefano Mauceri	University College Dublin, Ireland
George Michailidis	University of Florida, USA
Kaisa Miettinen	University of Jyväskylä, Finland
Shokoufeh Monjezi Kouchak	Arizona State University, USA
Rafael M. Moraes	Viasat Inc., USA
Mohamed Nadif	University of Paris, France
Mirco Nanni	CNR - ISTI, Italy
Giuseppe Nicosia	University of Catania, Italy
Wieslaw Nowak	Nicolaus Copernicus University, Poland
Varun Ojha	University of Reading, UK
Marcin Orchel	AGH University of Science and Technology, Poland
Mathias Pacher	Goethe Universität Frankfurt am Main, Germany
Pramudita Satria Palar	Bandung Institute of Technology, Indonesia

Panos Pardalos	University of Florida, USA
Konstantinos Parsopoulos	University of Ioannina, Greece
Andrea Patanè	University of Oxford, UK
David A. Pelta	Universidad de Granada, Spain
Dimitri Perrin	Queensland University of Technology, Australia
Milena Petkovic	Zuse Institute Berlin, Germany
Koumoutsakos Petros	ETH, Switzerland
Nikolaos Ploskas	University of Western Macedonia, Greece
Alessandro S. Podda	University of Cagliari, Italy
Buyue Qian	IBM T. J. Watson, USA
Michela Quadrini	University of Padova, Italy
Günther Raidl	TU Wien, Austria
Jan Rauch	University of Economics, Prague, Czech Republic
Steffen Rebennack	Karlsruhe Institute of Technology (KIT), Germany
Wolfgang Reif	University of Augsburg, Germany
Cristina Requejo	Universidade de Aveiro, Portugal
Francesco Rinaldi	University of Padova, Italy
Laura Anna Ripamonti	Università degli Studi di Milano, Italy
Humberto Rocha	University of Coimbra, Portugal
Vittorio Romano	University of Catania, Italy
Andre Rosendo	University of Cambridge, UK
Arnab Roy	Fujitsu Laboratories of America, USA
Andrea Santoro	Queen Mary University London, UK
Giorgio Sartor	SINTEF, Norway
Claudio Sartori	University of Bologna, Italy
Fréderic Saubion	Université d'Angers, France
Robert Schaefer	AGH University of Science and Technology, Kraków, Poland
Andrea Schaerf	University of Udine, Italy
Oliver Schuetze	CINVESTAV-IPN, Mexico
Bryan Scotney	Ulster University, UK
Andrea Serani	CNR-INM, National Research Council-Institute of Marine Engineering, Rome, Italy
Marc Sevaux	Université de Bretagne-Sud, France
Vladimir Shenmaier	Sobolev Institute of Mathematics, Russia
Zeren Shui	University of Minnesota, USA
Patrick Siarry	Université Paris-Est Creteil, France
Konstantina Skouri	University of Ioannina, Greece
Cole Smith	New York University, USA
Antoine Spicher	LACL University Paris Est Creteil, France
Catalin Stoean	University of Craiova, Romania

Johan Suykens	KU Leuven, Belgium
Tatiana Tchemisova Cordeiro	University of Aveiro, Portugal
Gabriele Tolomei	Sapienza University of Rome, Italy
Elio Tuci	University of Namur, Belgium
Gabriel Turinici	Université Paris Dauphine - PSL, France
Gregor Ulm	Fraunhofer-Chalmers Research Centre for Industrial Mathematics, Sweden
Renato Umeton	Dana-Farber Cancer Institute, MIT & Harvard T.H. Chan School of Public Health, USA
Werner Van Geit	Blue Brain Project, EPFL, Switzerland
Carlos Varela	Rensselaer Polytechnic Institute, USA
Herna L. Viktor	University of Ottawa, Canada
Marco Villani	University of Modena and Reggio Emilia, Italy
Mirko Viroli	Università di Bologna, Italy
Dachuan Xu	Beijing University of Technology, China
Xin-She Yang	Middlesex University London, UK
Shiu Yin Yuen	City University of Hong Kong, China
Qi Yu	Rochester Institute of Technology, USA
Zelda Zabinsky	University of Washington, USA

Best Paper Awards

LOD 2022 Best Paper Award

"Multi-Omic Data Integration and Feature Selection for Survival-based Patient Stratification via Supervised Concrete Autoencoders"
Pedro Henrique da Costa Avelar[1], Roman Laddach[2], Sophia Karagiannis[2], Min Wu[3] and Sophia Tsoka[1]
[1] Department of Informatics, Faculty of Natural, Mathematical and Engineering Sciences, King's College London, UK
[2] St John's Institute of Dermatology, School of Basic and Medical Biosciences, King's College London, UK
[3] Machine Intellection Department, Institute for Infocomm Research, A*STAR, Singapore
Springer sponsored the LOD 2022 Best Paper Award with a cash prize of EUR 1,000.

Special Mention

"A Two-Country Study of Default Risk Prediction using Bayesian Machine-Learning"
Fabio Incerti[1], Falco Joannes Bargagli-Stoffi[2] and Massimo Riccaboni[1]
[1] IMT School for Advanced Studies of Lucca, Italy
[2] Harvard University, USA

"Helping the Oracle: Vector Sign Constraints in Model Shrinkage Methodologies"
Ana Boskovic[1] and Marco Gross[2]
[1] ETH Zurich, Switzerland
[2] International Monetary Fund, USA

"Parallel Bayesian Optimization of Agent-based Transportation Simulation"
Kiran Chhatre[1,2], Sidney Feygin[3], Colin Sheppard[1] and Rashid Waraich[1]
[1] Lawrence Berkeley National Laboratory, USA
[2] KTH Royal Institute of Technolgy, Sweden
[3] Marain Inc., Palo Alto, USA

"Source Attribution and Leak Quantification for Methane Emissions"
Mirco Milletarí[1], Sara Malvar[2], Yagna Oruganti[3], Leonardo Nunes[2], Yazeed Alaudah[3] and Anirudh Badam[3]
[1] Microsoft, Singapore
[2] Microsoft, Brazil
[3] Microsoft, USA

ACAIN 2022 Special Mention

"*Brain-like combination of feedforward and recurrent network components achieves prototype extraction and robust pattern recognition*"
Naresh Balaji Ravichandran, Anders Lansner and Pawel Herman
KTH Royal Institute of Technology, Sweden

"*Topology-based Comparison of Neural Population Responses via Persistent Homology and p-Wasserstein Distance*"
Liu Zhang[1], Fei Han[2] and Kelin Xia[3]
[1] Princeton University, USA
[2] National University of Singapore, Singapore
[3] Nanyang Technological University, Singapore

Contents – Part II

Contents – Part I

Pooling Graph Convolutional Networks for Structural Performance Prediction

Lorenz Wendlinger(✉), Michael Granitzer, and Christofer Fellicious

Chair of Data Science, Universität Passau, Passau, Germany
{lorenz.wendlinger,michael.granitzer,christofer.fellicious}@uni-passau.de

Abstract. Neural Architecture Search can help in finding high-performance task specific neural network architectures. However, the training of architectures that is required for fitness computation can be prohibitively expensive. Employing surrogate models as performance predictors can reduce or remove the need for these costly evaluations. We present a deep graph learning approach that achieves state-of-the-art performance in multiple NAS performance prediction benchmarks. In contrast to other methods, this model is purely supervised, which can be a methodologic advantage, as it does not rely on unlabeled instances sampled from complex search spaces.

Keywords: Performance prediction · Neural architecture search · Graph neural networks

1 Background

Convolutional Neural Networks (**CNNs**) have emerged as the de facto default approach for many computer vision tasks. They achieve human equivalent or better performance in image recognition, c.f. [5,8,11]. While there are many sophisticated models trained on large corpora that transfer well to other tasks, they can be inefficient when used in less demanding tasks. Developing task-specific architectures can be difficult as it requires expert knowledge. Furthermore, the computational complexity of training and evaluating these models introduces idle periods for developers, making manual search even more labor-intensive.

Neural Architecture Search (**NAS**) is a research field dedicated to methods that automatically find high-performance neural architectures. In NAS for image recognition, an individuals fitness on a large image dataset can be used as the reinforcement signal. However, this requires a full training pass with back-propagation for each architecture sampled from the search space. It is prudent to also consider the environmental cost of running extensive searches on power-hungry hardware. E.g. a full search using the method of [29] uses over 500 000 GPU hours, consuming at least 100 MWh of electrical energy. This results in more than 70 metric tons of CO_2 being released, equivalent to circumnavigating the globe more than 7 times in an average US car, according to [24].

Performance prediction can be employed to mitigate the high computational cost of NAS. Performance prediction either utilizes patterns in the network's

G. Nicosia et al. (Eds.): LOD 2022, LNCS 13811, pp. 1–16, 2023.
https://doi.org/10.1007/978-3-031-25891-6_1

architecture itself or analyses model behavior in the initial training epochs such that training can be cut short or foregone entirely. We refer to the former as structure based performance prediction and the latter as training curve analysis. Purely structure based prediction is the preferable option, as it does not necessitate partial training for inference. It only requires full back-propagation to create an initial training corpus during a warm-up phase. Consequently, methods that handle data sparsity well, such as semi-supervised algorithms, can further accelerate the NAS process by reducing the number of such training points needed.

Most NAS methods represent a networks architecture as heterogeneous Directed Acyclic Graphs (**DAG**s) comprised of high-level computational blocks, with a search space spanning over all heterogeneous DAGs fulfilling certain architectural constraints. These computational blocks represent CNN-specific functions such as convolutions or pooling, while the edges between them express data flow. We call these operation-on-node (**OON**) spaces. An alternative representation of neural architectures are operation-on-edge (**OOE**) spaces, in which edges represent the available computations.

The high dimensional and heterogeneous nature of these graphs makes the identification of performance predicting patterns difficult. Furthermore, as nodes are unordered, application of traditional methods developed for Euclidean data, such as those used for image processing, is not possible. More specifically, patterns that fulfill the same function in the fully trained neural network can be expressed differently in the architecture DAG as graph isomorphisms. This means that structural performance prediction needs to respect architecture isomorphisms by assigning them equal internal representations.

The field of network science is large and offers solutions for many tasks with graph data. However, it is deeply rooted in the analysis of large social networks, biological networks and other similar data. As a result, many methods are intended for graphs with similar properties. Unfortunately, NAS DAGs do not share all of these properties, most prominently they are much smaller in size. Also, as per the survey of [27], both heterogeneous as well as directed graphs are seldom considered. The disregard of those properties can incur a large loss of information when applied to architecture DAGs. Therefore most established methods are not easily applicable to the many smaller heterogeneous DAGs encountered in NAS.

However, graph neural networks (**GNN**s) have emerged as a technique for generating salient dense representation from arbitrary graph data in recent years. Some methods that work well on search spaces developed as NAS benchmarks use adapted GNNs. However, they do not generally retain the capability to handle graphs of varying sizes. This, in turn, makes them unsuitable for larger search spaces with fewer restrictions, such as CGP-CNN from [22].

Further, most state-of-the-art approaches are semi-supervised. The resulting need for additional unlabeled training data can become a problem in settings where the search space cannot easily be sampled. For most NAS search spaces extensive sampling is not tractable, excluding smaller NAS performance prediction benchmark datasets. Especially in scenarios with few available labeled data

- in which semi-supervised methods usually excel - the exact sampling parameters might be unknown, prohibiting the generation of unlabeled data from the same manifold. This can complicate adaptation of such methods to new tasks and require additional hyper-parameter tuning.

In this work, we propose a model that can predict performance for various NAS search spaces with variable-size architectures in a fully supervised manner. Specifically, our contributions comprise:

1. We develop a multi-objective residual pooling graph convolutional model (**MORP-GCN**) that can produce whole-graph predictions. It can natively handle variable-length data and does not impose restrictions on the structure of a graph.
2. We show that this model achieves state-of-the-art performance in a NAS prediction task, especially with limited training data available.
3. We illustrate that our method can be adapted to OOE search spaces with minimal loss in prediction quality.
4. We demonstrate that our method can compete with learning curve based prediction in a complex search space.
5. We show that the explicit residual computation in our model improves predictions in various scenarios.

We start off by describing related work in the field of performance prediction for NAS and go into some of their limitations. We then introduce our MORP-GCN model that addresses several of these issues. A description of the three datasets that we use to show the performance of MORP-GCN in three diverse scenarios follows. Experimental results for performance prediction on those datasets are presented and compared with related work. Furthermore, we present ablation studies into the effect of residual computation as well as input graph data structure.

2 Related Work

Various recent advances in graph representation learning have helped in facilitating the use of graph neural networks for NAS performance prediction. We give a brief overview of the most relevant related work below.

The Graph Convolutions of [15] capture the neighborhood of a node through convolutional filters, similar to those used in Convolutional Neural Networks for image classification. The resulting node embeddings can be used for semi-supervised tasks or as latent feature vectors.

Shi et al. use a GCN based neural network assessor with an artificial global node, that collects a representation for a whole graph in [21]. Therefore, this model can handle variable-sized input.

Tang et al. [23] use representations generated by an auto-encoder trained on unlabeled data to obtain a relational graph of similarities between graphs. This information is then used to train a GCN regressor that can be optimized jointly with the en- and decoder in a semi-supervised manner. This method requires a large sampling of unlabeled data from a search space, which is not always trivial to obtain.

The smooth variational graph embeddings of [18] incorporate the fact that information travels forward in a neural network while gradients travel backward. This is done by using an auto-encoder, in which decoder and encoder both use the original graph as well as its reverse. They also use learned node type embeddings as input node features that can express fine-grained information about node functions. Zero-padding to a fixed length allows this model to compute a whole graph representation via a gated sum, which, however limits applicability to graphs of varying sizes.

GATES, as proposed by [19] is a generic encoding scheme for neural architecture that works for structural performance prediction. The capabilities of this model are derived from modeling the data processing that is performed by the node operations in the encoding scheme, which is then used as an unary soft attention mask. GATES requires a singular sink node, i.e. the output of an architecture, to obtain the graph representation. Their method achieves state-of-the-art results in performance prediction on multiple benchmark datasets. They furthermore suggest a Line-Graph GCN for OOE spaces that concatenates the node embeddings generate by a GCN for fixed-size graphs.

3 Multi-objective Residual Pooling Graph Convolutional Networks

Here we present the Multi-objective Residual Pooling Graph Convolutional Network (MORP-GCN), that forms the core of our work, as well as the concept of graph convolutions and related techniques from the realm of deep learning.

Graph Convolutional Networks (**GCNs**), introduced by [15], compute node-level functions based on graph structures and input node features. Akin to convolutions in image recognition, features are computed from neighboring nodes' features via linear combination in multiple parallel filters. The weights and biases for these filters are learned through back-propagation, c.f. [20].

The representation of neural network architectures as heterogeneous directed acyclic graph is obvious and common. More precisely, a graph $\mathcal{G} = (\mathcal{V}, \mathcal{E}, \lambda_l)$ is composed of nodes (or vertices) $\mathcal{V}$ and edges $\mathcal{E} \subseteq \{(u, v) : u, v \in \mathcal{V} \wedge u \neq v\}$. The one-hot-encoded node type is given by $\lambda : \mathcal{V} \rightarrow \{0, 1\}^{n_l}$, while the edges $\mathcal{E}$ express the directional data flow between nodes.

The node labeling function λ can be extended to additional real-valued node features. Node hyper-parameters are a prime candidate, though they can be very sparse if only applicable for some node types. It is further possible to extend λ with node properties, such as centrality measures. This can help emphasizing the roles of nodes in $\mathcal{G}$ and create more meaningful input node features. We use the harmonic centrality described by [2] for this purpose. To include the backward direction the gradients travel in during neural network training, we transform each graph to the union of itself and its reverse. Both the backward as well as the forward representations of each node are later pooled to obtain a joint representation.

We use the topology adaptive graph convolution of [7] that uses multiple receptive field sizes. This is done with k filters that use neighborhoods of $1..k$ hops. The result of these filters are aggregated via a linear combination with learned weights. We adopt this method with $k = 2$ for its flexibility and improved expressive power over graph convolutions with only one receptive field size.

We further add skip connections as defined in [9] between same size layers so residuals can be learned explicitly. They allow for the stacking of multiple GCN layers without the problematic vanishing gradients that plague back-propagation in deep architectures. This can help more expressive models converge and generally improve performance according to [3].

Graph-Level Pooling. The GCN described so far can generate meaningful node-level representations, i.e. a F_L sized representation for each $v \in \mathcal{V}$., but no graph representations. However, in the context of NAS, we may want a single joint embedding that encodes the structure of a whole graph, to facilitate comparison between architectures.

Related work proposes multiple solutions to this problem, chief among which is the *global node* approach. Thereby a node, that is connected to every other node, is added to the graph. The representation of this node is consequently a linear combination of all other node representations and can be used vicariously. In many NAS scenarios, the output node is a successor to all other nodes, and thereby performs a similar function already. GATES [19] therefore use its representation for the whole graph. They compare this method to mean pooling across all nodes as well as concatenation, but find results lacking compared to their method. However, clearly defined output nodes are not always available.

So we introduce global pooling, specifically a function $p_i : \mathbb{R}^{|V| \times F_L} \rightarrow \mathbb{R}^{F_L}$ to obtain a fixed-size representation regardless of graph size. We can use a set P of functionals to pool across each embedding dimension. If they are chosen so that they sufficiently describe the node distribution, information loss is minimal. We therefore choose the mean, standard deviation, minimum and maximum to preserve the center, spread and outliers, respectively. This produces an output of size $|P| \times F_L$ for each graph. These pooling results are then scaled via batch normalization as per [12] to reduce the impact of different scales induced by the dissimilar pooling operations. To obtain an output of size F_L, they are aggregated via a learned linear combination.

Thereby the model can handle architectures of varying size and natively respects isomorphisms between them. Further, this method does not incur the penalty of computing representations for an additional global node. As it does not rely on a single output node being present in every graph, it can also easily be adapted to tasks in which the presence of such a sink node is not guaranteed.

For unsupervised learning, such as generating latent representations, this is the final output. The model can also be adapted to obtain an end-to-end model for supervised tasks by adding dense layers as an integrated MLP estimator. For classification, a *softmax*-activated dense layer with one output per encountered class can be added. For regression, a dense layer with one output is used as the last layer. The main components of this MORP-GCN are shown in Fig. 1.

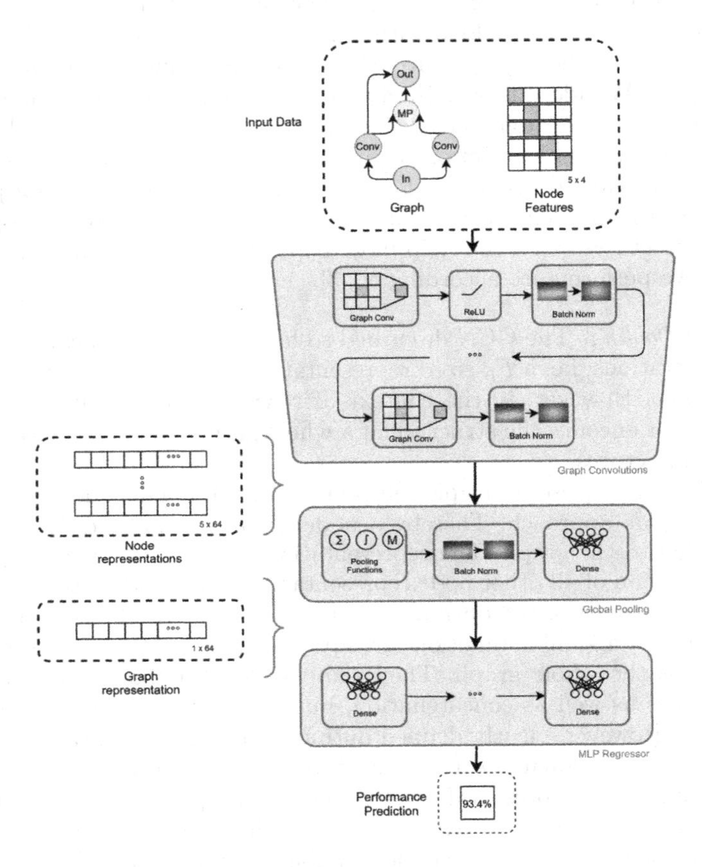

Fig. 1. MORP-GCN architecture for NAS performance prediction.

Training Procedure. The adaptive momentum gradient descent method of [14] is used to train MORP-GCN in mini-batches. We use an exponential learning rate decay with a factor γ at each epoch to encourage exploitation in the later stages of training. For improved convergence behavior and added regularization, batch normalization is also applied to each graph convolution layer's output before rectification. In the same vein, dropout as proposed in [10] is used at the last graph convolution during training to encourage the learning of more robust representations. Furthermore, we apply decoupled weight decay, described in [17], with a factor of 0.01.

We optimize a linear combination of MSE loss and the pairwise hinge ranking loss as defined in [19]. For true performance y and predicted performances $\hat{y}$ of length N and with margin $m = 0.05$ it is defined as:

$$L_c(y,\hat{y}) = w_1 \cdot MSE(y,\hat{y}) + w_2 \cdot L_r(y,\hat{y})$$
$$L_r(y,\hat{y}) = \sum_{j=1}^{N} \sum_{i:\ y_i > y_j} \max\left(0, m - (\hat{y}_i - \hat{y}_j)\right) \tag{1}$$

Through this multi-objective loss, emphasis can be placed on optimizing for high ranking correlation or low squared error by varying the respective weights w_2 and w_1. This allows for adaptation to different NAS selection strategies, e.g. by using a high w_2 for rank-based methods.

4 Datasets

We use three datasets to confirm the suitability of MORP-GCN for NAS performance prediction. In this way we can compare our approach to the state of the art and also evaluate how it performs on different search spaces or datasets created with different sampling strategies.

Both NAS-Bench-101 of [28] as well as NAS-Bench-201, devised in [6], are relatively small search spaces that can be fully sampled and evaluated. Also both search spaces do not directly model CNN architectures but rather cells that are stacked to construct networks. The CGP-CNN search space defined in [22] that evofficient[1] is based on is considerable larger and less restrictive.

Both NAS-Bench-101 and evofficient are operation-on-node (OON) search spaces, in which data processing steps are represented as nodes. NAS-Bench-201 models CNNs as operation-on-edge (OOE) graphs. They have a fixed fully-connected layout and special edges that model operations. These operations encompass CNN functions such as convolutions as well as skip-connections or dead ends through `zeroize` edges.

All three datasets include evaluations on CIFAR10 [16] as targets. NAS-Bench-201 further contains per architecture accuracies on CIFAR100, also from [16], as well as ImageNet-16-120, introduced in [4].

4.1 Pre-processing

For NAS-Bench-101 no pre-processing is necessary. As NAS-Bench-201 is a OOE search space, its architectures need to be mapped into an OON space first. We use the same schema as [19], i.e. transformation to line graphs. Furthermore, `zeroize` edges are removed along with resulting isolated nodes. It should be mentioned that this pre-processing schema would not work with some methods from related work because it generates variable-size representations from the fixed-size OOE graphs of NAS-Bench-201 (Table 1).

[1] We use auxiliary experiment data published with [26] to create the evofficient dataset. It is available at https://github.com/wendli01/morp_gcn/tree/main/experiments/datasets/evofficient.

Table 1. Statistics for datasets used for performance prediction. † 7 unique node types plus the 3 different configurations for convolutional layers

	NAS-Bench	NAS-Bench	Evofficient
Statistic	−101	−201	
Architectures	423k	15.6k	3k
Evaluations	1.27M	46.9k	3k
Node types	5	6	7 + 3†
Graph size	$8.73_{\pm 0.55}$	$6.4_{\pm 1.5}$	$14.2_{\pm 4.1}$

As evofficient instances represent full neural network architectures, we can use certain phenotypic characteristics as node features for MORP-GCN. These are the output shapes, i.e. the dimensions of the tensor volume, after the processing of a node. Furthermore, node configurations, i.e. kernel size, number of filter and strides, are used as well. For evofficient we only retain the 2 969 architectures that achieve a final accuracy > 0.1 in order to remove non-deterministic outliers.

5 Experiments

We conduct multiple experiments to validate the suitability of MORP-GCN for NAS performance prediction on the datasets described above.

5.1 Evaluation Methodology

To obtain a realistic evaluation, the mean and standard deviation across 5 trials with different dataset splits is reported. The seed for the pseudo-random number generator is varied for each split, resulting in different network initializations and batch compositions.

NAS performance prediction differs from traditional regression in that the absolute difference or linear correlation are of lower importance. This is because most accelerated NAS methods only compare performance predictions with other predicted values, i.e. used as a full substitute. The aim is therefore to generate predictions that show high correlation with the target, rather than low error. Furthermore, as most NAS methods rank architectures by performance for selection, ranking correlation is more important than linear correlation. In concordance with most related work, we report the Kendall ranking coefficient [13] τ_k as the main measure of predictive performance.

The lack of published and usable code makes reproduction difficult for many methods from related work. This limits comparability between approaches, as it needs to account for different metrics and missing error estimates in reporting. Consequently, for comparability to other work, we also list mean squared error and Pearson correlation ρ_p where necessary.

The default hyper-parameters for MORP-GCN are listed in Table 2. They are identical for all three datasets.

Table 2. Parameter Setting for the MORP-GCN for NAS performance prediction, for training set size N_l.

Parameter name	Default value
GCN layer sizes	(128, 128, 128, 128, 128, 64)
Pooling operations	(max, min, mean, stdev)
Epochs	150
Dropout probability	0.05
Initial learning rate	0.005
Dense layer sizes	(64, 64)
MSE loss weight w_1	0.25
Ranking loss weight w_2	0.75
Learning rate decay	0.95
Additional features	(harmonic centrality)
Batch Size	$\lceil N_l \div 10 \rceil$

5.2 Performance Prediction on NAS-Bench-101

MORP-GCN outperforms the state of the art in performance prediction on NAS-Bench-101 w.r.t. to the τ_k, c.f. Table 3a. The margin is largest for smaller training corpora, whereas most approaches converge to $\tau_k \approx 0.9$ for $N_l \geq 100\,000$.

Table 3. Performance prediction results on NAS-Bench-101 with N_l training instances. Mean and standard deviation over 5 random trials for multiple evaluation criteria. Results taken from the corresponding paper unless indicated otherwise. For some methods, results for some metrics and standard deviations are missing as they were not reported and reproduction infeasible. *results from [21]. †results from [19].

(a) Ranking correlation τ_k

N_l	GCN†	Tang et al.	GATES	MORP-GCN (Ours)
381	0.5790		0.7789	**0.8061**$_{\pm 0.008}$
1k		0.6541$_{\pm 0.008}$		**0.841**$_{\pm 0.004}$
3813	0.8277		0.8681	**0.8708**$_{\pm 0.002}$
10k		0.7814$_{\pm 0.004}$		**0.884**$_{\pm 0.0007}$
19063	0.8641		0.8841	**0.8916**$_{\pm 0.001}$
38126	0.8747		0.8922	**0.8971**$_{\pm 0.02}$
100k		0.8456$_{\pm 0.003}$		**0.9018**$_{\pm 0.0008}$

(b) MSE

N_l	SVGe	MORP-GCN (Ours)
1k	0.0084$_{\pm 0.00002}$	0.0172$_{\pm 0.0008}$
10k	0.0023$_{\pm 0.00004}$	0.0174$_{\pm 0.0003}$
100k	0.002$_{\pm 0.00003}$	0.0176$_{\pm 0.0004}$

(c) Linear correlation ρ_p

N_l	GCN*	MORP-GCN (Ours)
1k	0.819	0.5412$_{\pm 0.005}$

Furthermore, it can be seen that MORP-GCN outperforms the global node GCN approach of [19] by a significant margin. Even a MORP-GCN that has only seen 50 labeled instances achieves a higher τ_k than the global node GCN trained on 381 samples. This illustrates the advantage of the pooling component in MORP-GCN over using a global node.

Figure 2 shows predictions for a single fold in detail. It can be observed that, while there is a linear relationship between prediction and target, it is far from perfect. However, the rankings are much more closely correlated, except for some high-performing instances that the model underestimates the performance of.

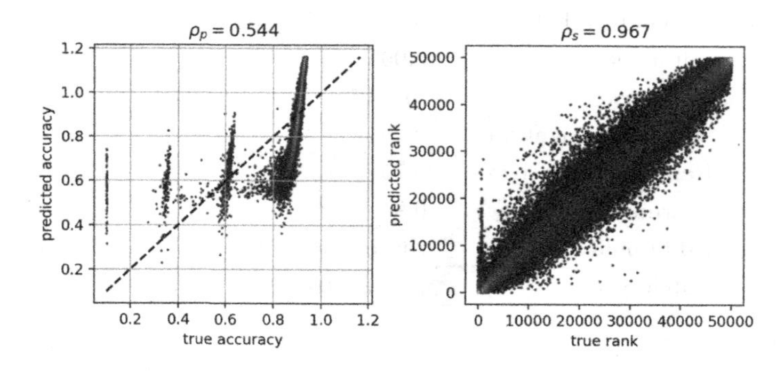

Fig. 2. Performance predictions for 50 000 unseen instances from the NAS-Bench-101 dataset generated by a MORP-GCN trained on 1k samples. Accuracy values (left) and rankings (right) with Pearson and Spearman correlation.

5.3 Performance Prediction on Evofficient

As the evofficient dataset is comprised of variable-size DAGs, not all performance prediction methods from related work are applicable. It is also generally more difficult to generate performance predictions for due to its more complex and varied structure. The dataset also contains learning curves for the 50 epochs of training with back-propagation on CIFAR10, so we can use learning curve methods for performance prediction. Sequential Regression Models (**SRMs**), introduced in [1] and successfully used for acceleration on the evofficient dataset in [26], heavily depend on the available portion of the learning curve τ. However, a large τ reduces the potential speed-up as it results in longer partial training times. As can be seen in Fig. 4, learning curve methods also benefit less from a larger training corpus, so their predictive performance cannot be improved by re-training on additional data labeled through full evaluations in the NAS. Higher training data availability does however increase consistency for SRM.

Structure-based performance prediction removes the requirement of training inference altogether. This can achieve further speed-up and allow for more complex search methods that rely on sampling more fitness value points. Even with only 100 training instances, MORP-GCN performs on par with a $\tau = 0.1$ SRM, and trained on 1000 samples it comes close to the $\tau = 0.2$ SRM performance.

Table 4. Performance prediction results on the evofficient dataset with N_l training instances. Mean and standard deviation of ranking correlation τ_k over 5 random trials. Proportion τ of learning curve available given for SRM.

N_l	Learning curve methods		MORP-GCN
	SRM $\tau = .1$	SRM $\tau = .2$	
100	$0.352_{\pm 0.015}$	$0.509_{\pm 0.026}$	$0.3558_{\pm 0.02}$
200	$0.369_{\pm 0.015}$	$0.524_{\pm 0.016}$	$0.4137_{\pm 0.018}$
1k	$0.377_{\pm 0.0076}$	$0.528_{\pm 0.0058}$	$0.4811_{\pm 0.013}$

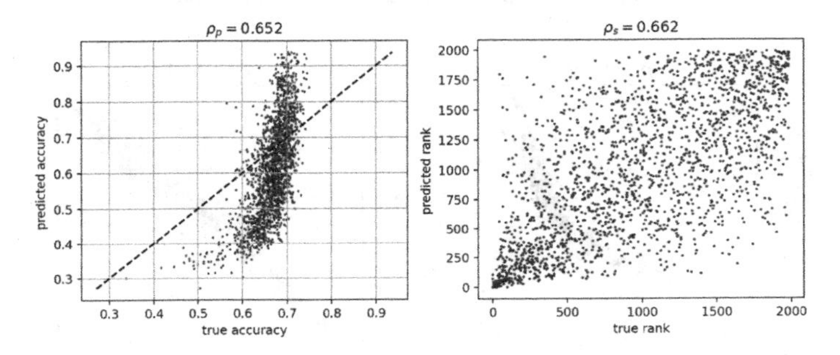

Fig. 3. Performance predictions for 1 969 unseen instances from the evofficient dataset generated by a MORP-GCN trained on 1k samples. Accuracy values (left) and rankings (right) with Pearson and Spearman correlation.

As outliers that could not be trained were considered as noise and removed from the evofficient dataset in a pre-processing step, the predictive quality does not suffer in low performance regions (c.f. Fig. 3).

5.4 Performance Prediction on NAS-Bench-201

MORP-GCN outperforms other supervised methods on NAS-Bench-201 by some margin, c.f. Table 5. It achieves performance slightly below GATES in scenarios with medium or high data availability. Prediction results for larger training corpus sizes are roughly on par with GATES. However, the performance in high sparsity is significantly worse.

This shows that MORP-GCN can handle supervised learning in OOE spaces, which can be attributed to its design that makes it compatible with variable-size data and allows for it to use a more concise data representation. It also confirms that for NAS-Bench-201 conversion to line graphs and removal of `zeroize` edges is a valid pre-processing step that does not incur information loss.

Figure 4 shows that MORP-GCN predictions exhibit very high linear and ranking correlation in this scenario. Some erroneous predictions originate from low performance outliers that are not consistently placed at zero accuracy.

Table 5. MORP-GCN Performance prediction results on the NAS-Bench-201 dataset with N_l training instances compared to related work. Mean and standard deviation of τ_k over 5 random trials. *results from [25]. †results from [19].

N_l	LSTM*	GCN†	GATES†	MORP-GCN
78	0.5550	0.5063	**0.7401**	$0.639_{\pm 0.018}$
390	0.6407	0.6822	**0.8628**	$0.8513_{\pm 0.0088}$
781	0.7268	0.7567	0.8802	$\mathbf{0.8826}_{\pm 0.0062}$
3906	0.8791	0.8676	**0.9192**	$0.9147_{\pm 0.0021}$
7813	0.9002	0.9002	**0.9259**	$0.9198_{\pm 0.0023}$

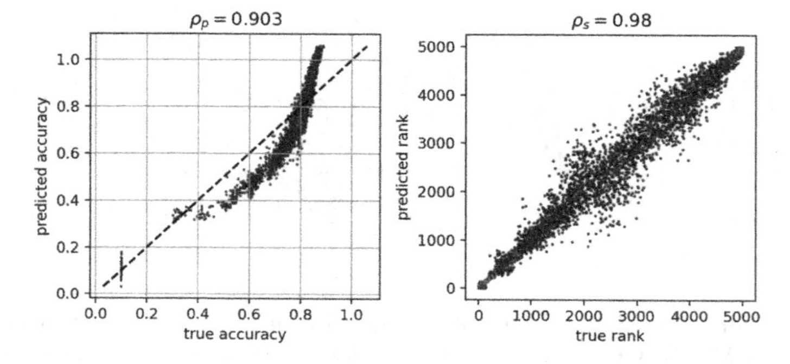

Fig. 4. Performance predictions for 14 625 unseen instances from the NAS-Bench-201 dataset generated by a MORP-GCN trained on 781 labeled instances. Comparison of accuracy values (left) and rankings (right) with Pearson and Spearman coefficient given.

Predictive performance for alternative targets is similar, with $\tau_k = 0.886_{\pm 0.0071}$ for CIFAR100 and $\tau_k = 0.868_{\pm 0.008}$ on ImageNet-16-120 labels also using 781 training instances.

5.5 Ablation Studies

We perform several ablation studies to both motivate MORP-GCN design choices as well as further investigate its capabilities.

Explicit Residual Computation. Adding explicit residual computation in the form of skip connections consistently improves prediction results. This behavior can be observed for various different GCN layer configurations, c.f. Fig. 5.

It shows the largest improvements for deeper and wider configurations. We assume that this is because the skip connections allow for the learning of less complex features for basic patterns in the input data. They do this while still maintaining the ability to generate more sophisticated features without bypass. The addition of skip connections also reduces the influence of GCN layer config-

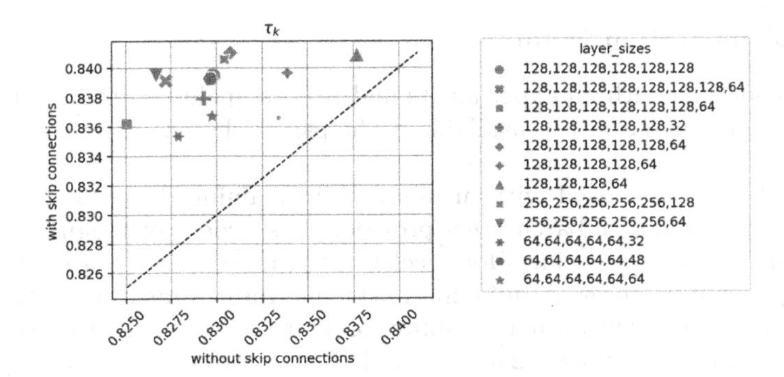

Fig. 5. Effect of explicit residual computations between GCN layers for different combinations of graph convolution layer sizes. Mean τ_k over 5 folds on 1000 instances of NAS-Bench-101.

Table 6. Effect of input data structure on MORP-GCN predictive performance on NAS-bench-101 trained on 1000 instances. Mean and standard deviation over 5 random trials for multiple evaluation criteria.

Graph structure	τ_k	MSE
Default	$\mathbf{0.841}_{\pm 0.0083}$	$0.0172_{\pm 0.00075}$
Forward Pass only	$0.8333_{\pm 0.0044}$	$0.018_{\pm 0.00043}$
Undirected	$0.7591_{\pm 0.0066}$	$0.0176_{\pm 0.00092}$
Homogeneous	$0.2466_{\pm 0.017}$	$\mathbf{0.00857}_{\pm 0.00069}$

uration overall, making MORP-GCN more robust to hyper-parameter changes and potentially easier to transfer to new datasets or tasks.

Graph Data Structure. As MORP-GCN was explicitly designed to capture the directed as well as the heterogeneous property of NAS DAGs, we seek to quantify the performance advantage that this effects. Table 6 shows how changing the input data structure affects prediction quality on NAS-Bench-101.

It can be observed that the consideration of the backward pass results in a small yet tangible improvement that also increases repeatability. The removal of edge direction information significantly decreases correlation with the target variable. This shows that the direction is indeed important information that can be used effectively by MORP-GCN. Data without node type information shows very low ranking correlation, making it and other methods that ignore node heterogeneity unsuitable for NAS acceleration. However, MORP-GCN still produces predictions with low MSE for homogeneous graphs.

5.6 Implementation Details

The used codebase with all experiments and results, including the used dataset splits and random seeds, are available at https://github.com/wendli01/morp_gcn.

MORP-GCN scales well with an increase in training data due to the efficiency boost that larger batch sizes provide. It is orders of magnitude faster than a NAS run, making it ideal for accelerating the search because of the short warm-up phase it requires. A full experiment, i.e. full training on N_l instances, averaged over 5 runs with different splits for 1 000 NAS-Bench-101 takes 34.6 s $_{\pm 0.2}$, wile 10k instances take 112.4 s $_{\pm 1.2}$ and 10k samples take 1097.3 s $_{\pm 3.2}$. This includes some data processing steps as well as the extraction of any additional node features. Depending on the number of node features, MORP-GCN with hyper-parameters set as per Table 2 has about 250k trainable parameters. All experiments were run on an Nvidia GTX 1080 with 8 GB of video memory.

6 Conclusion

We present a GCN based alternative to semi-supervised structural performance prediction that is natively compatible with variable size graphs. It achieves similar or superior results while offering methodological advantages. The chosen multi-pooling method is computationally inexpensive and flexible. The model also adapts well to operation-on-edge search spaces, which is not possible with other GNN approaches. The multi-objective loss allows for easy tuning to NAS tasks with different selection methods. This shows the power of residual pooling GCNs in extracting useful features from heterogeneous directed acyclic graphs.

In summary, we propose MORP-GCN as a simple yet powerful and extensible variant of graph convolutions. Especially future work that sees it applied to different tasks and domains that present difficult to represent and analyze heterogeneous DAGs can be of interest. In the domain of NAS performance prediction, we view MORP-GCN as a suitable approach for structure based acceleration in larger and more complex search spaces than are currently considered.

Acknowledgments. The research reported in this paper has been supported by the FFG BRIDGE project KnoP-2D (grant no. 871299).

References

1. Baker, B., Gupta, O., Raskar, R., Naik, N.: Accelerating neural architecture search using performance prediction. arXiv preprint arXiv:1705.10823 (2017)
2. Boldi, P., Vigna, S.: Axioms for centrality. Internet Math. **10**(3–4), 222–262 (2014)
3. Bresson, X., Laurent, T.: Residual gated graph convnets. arXiv preprint arXiv:1711.07553 (2017)
4. Chrabaszcz, P., Loshchilov, I., Hutter, F.: A downsampled variant of imagenet as an alternative to the CIFAR datasets. arXiv preprint arXiv:1707.08819 (2017)

5. Ciregan, D., Meier, U., Schmidhuber, J.: Multi-column deep neural networks for image classification. In: 2012 IEEE Conference on Computer Vision and Pattern Recognition, pp. 3642–3649. IEEE (2012)
6. Dong, X., Yang, Y.: NAS-bench-201: extending the scope of reproducible neural architecture search. arXiv preprint arXiv:2001.00326 (2020)
7. Du, J., Zhang, S., Wu, G., Moura, J.M., Kar, S.: Topology adaptive graph convolutional networks. arXiv preprint arXiv:1710.10370 (2017)
8. He, K., Zhang, X., Ren, S., Sun, J.: Delving deep into rectifiers: surpassing human-level performance on imagenet classification. In: Proceedings of the IEEE International Conference on Computer Vision, pp. 1026–1034 (2015)
9. He, K., Zhang, X., Ren, S., Sun, J.: Deep residual learning for image recognition. In: Proceedings of the IEEE Conference on Computer Vision and Pattern Recognition, pp. 770–778 (2016)
10. Hinton, G.E., Srivastava, N., Krizhevsky, A., Sutskever, I., Salakhutdinov, R.R.: Improving neural networks by preventing co-adaptation of feature detectors. arXiv preprint arXiv:1207.0580 (2012)
11. Ho-Phuoc, T.: CIFAR10 to compare visual recognition performance between deep neural networks and humans. arXiv preprint arXiv:1811.07270 (2018)
12. Ioffe, S., Szegedy, C.: Batch normalization: accelerating deep network training by reducing internal covariate shift. arXiv preprint arXiv:1502.03167 (2015)
13. Kendall, M.G.: A new measure of rank correlation. Biometrika **30**(1/2), 81–93 (1938)
14. Kingma, D.P., Ba, J.: Adam: a method for stochastic optimization. arXiv preprint arXiv:1412.6980 (2014)
15. Kipf, T.N., Welling, M.: Semi-supervised classification with graph convolutional networks. arXiv preprint arXiv:1609.02907 (2016)
16. Krizhevsky, A., Hinton, G., et al.: Learning multiple layers of features from tiny images (2009)
17. Loshchilov, I., Hutter, F.: Decoupled weight decay regularization. arXiv preprint arXiv:1711.05101 (2017)
18. Lukasik, J., Friede, D., Zela, A., Stuckenschmidt, H., Hutter, F., Keuper, M.: Smooth variational graph embeddings for efficient neural architecture search. arXiv preprint arXiv:2010.04683 (2020)
19. Ning, X., Zheng, Y., Zhao, T., Wang, Y., Yang, H.: A generic graph-based neural architecture encoding scheme for predictor-based NAS. In: Vedaldi, A., Bischof, H., Brox, T., Frahm, J.-M. (eds.) ECCV 2020. LNCS, vol. 12358, pp. 189–204. Springer, Cham (2020). https://doi.org/10.1007/978-3-030-58601-0_12
20. Rumelhart, D.E., Hinton, G.E., Williams, R.J.: Learning representations by back-propagating errors. Nature **323**(6088), 533–536 (1986)
21. Shi, H., Pi, R., Xu, H., Li, Z., Kwok, J.T., Zhang, T.: Multi-objective neural architecture search via predictive network performance optimization (2019)
22. Suganuma, M., Shirakawa, S., Nagao, T.: A genetic programming approach to designing convolutional neural network architectures. In: Proceedings of the Genetic and Evolutionary Computation Conference, pp. 497–504 (2017)
23. Tang, Y., et al.: A semi-supervised assessor of neural architectures. In: Proceedings of the IEEE/CVF Conference on Computer Vision and Pattern Recognition, pp. 1810–1819 (2020)
24. United States Environmental Protection Agency: Greenhouse gas equivalencies calculator (2021). https://www.epa.gov/energy/greenhouse-gas-equivalencies-calculator

25. Wang, L., Zhao, Y., Jinnai, Y., Tian, Y., Fonseca, R.: AlphaX: exploring neural architectures with deep neural networks and Monte Carlo tree search. arXiv preprint arXiv:1903.11059 (2019)
26. Wendlinger, L., Stier, J., Granitzer, M.: Evofficient: reproducing a cartesian genetic programming method. In: Hu, T., Lourenço, N., Medvet, E. (eds.) EuroGP 2021. LNCS, vol. 12691, pp. 162–178. Springer, Cham (2021). https://doi.org/10.1007/978-3-030-72812-0_11
27. Wu, Z., Pan, S., Chen, F., Long, G., Zhang, C., Philip, S.Y.: A comprehensive survey on graph neural networks. IEEE Trans. Neural Netw. Learn. Syst. **32**, 4–24 (2020)
28. Ying, C., Klein, A., Christiansen, E., Real, E., Murphy, K., Hutter, F.: NAS-bench-101: towards reproducible neural architecture search. In: International Conference on Machine Learning, pp. 7105–7114. PMLR (2019)
29. Zoph, B., Le, Q.V.: Neural architecture search with reinforcement learning. arXiv preprint arXiv:1611.01578 (2016)

A Game Theoretic Flavoured Decision Tree for Classification

Mihai-Alexandru Suciu and Rodica-Ioana Lung(✉)

Center for the Study of Complexity, Babeş-Bolyai University, Cluj Napoca, Romania
{mihai.suciu,rodica.lung}@ubbcluj.ro

Abstract. A game theoretic flavoured decision tree is designed for multi-class classification. Node data is split by using a game between sub-nodes that try to minimize their entropy. The splitting parameter is approximated by a naive approach that explores the deviations of players that can improve payoffs by unilateral deviations in order to imitate the behavior of the Nash equilibrium of the game. The potential of the approach is illustrated by comparing its performance with other decision tree-based approaches on a set of synthetic data.

Keywords: Oblique decision trees · Game theory · Classification

1 Introduction

Classification problems have become ubiquitous as increasingly more fields of science begin to rely on data science as a tool for discovery and advancement. Applications range across a variety of fields, with examples from economic decision-making [6], drug discovery [13], biology [2], chemistry [3], security [8] and many more.

Decision trees (DTs) are classification methods that aim to find rules for splitting given data into regions as pure as possible. Various indicators can be used to evaluate the purity, most of them relying on the proportion of data labels in that region. They are versatile, adaptive, and, if small enough, provide an intuitive interpretation of the classification.

DTs are built recursively: at each node level, data is split according to a specific rule. When a maximum depth for the tree is reached or if the data in the node is pure enough, the splitting process stops. The first variants of DTs used axis-parallel splits [1]. Inherent limitations have led to the proposal of oblique [10,15] and nonlinear trees [7]. Even in the simplest form, their efficiency has been exploited by creating random forests that use many DTs, usually built on bootstrapped data on a subset of the attributes.

While DTs are highly regarded as top classification methods [16] they also present some disadvantages. First, they can easily overfit data when grown too

This work was supported by a grant of the Romanian Ministry of Education and Research, CNCS - UEFISCDI, project number PN-III-P4-ID-PCE-2020-2360, within PNCDI III.

G. Nicosia et al. (Eds.): LOD 2022, LNCS 13811, pp. 17–26, 2023.
https://doi.org/10.1007/978-3-031-25891-6_2

large, so a trade-off has to be found between the size of the tree and training accuracies. Optimal complex trees such as oblique or nonlinear are computationally expensive to build. The trade-off here is usually achieved by using search heuristics or statistical methods to approximate optimal solutions [14].

This paper proposes a game-theoretic inspired splitting mechanism for decision trees node data by oblique hyper-planes. Instead of searching for a hyperplane that optimizes a purity criterion, a game is devised between the two subnodes, in which each one of them tries to find a hyper-plane that improves its entropy. An equilibrium of this game consists of a split such that none of the subnodes can propose another hyperplane to decrease its entropy further. Numerical experiments are used to illustrate the efficiency of such an approach compared to other decision tree models.

2 A Game Theoretic Splitting Mechanism for Decision Trees

Let $\mathcal{X} \subset \mathbb{R}^{N \times p}$ be a data set containing N instances $x_i \in \mathbb{R}^p$, $i = 1, \ldots, N$ and $\mathcal{Y} = \{c_1, \ldots, c_N\}$ their corresponding labels (or classes); $c_i \in C$ is the label of instance x_i, and $C = \{0, \ldots, k\}$ is the set of labels. The goal of a classifier is to find a rule to assign labels to instances in $\mathcal{X}$ as accurate as possible, i.e. as closest to $\mathcal{Y}$ as possible.

A decision tree splits the data in $\mathcal{X}$ into subsets or regions. It assigns instances in regions labels with probabilities proportional to the number of instances in each class in that region. As splitting is performed recursively, DT are represented as trees, and the regions they define correspond to the leaves of the trees. At each node level, data is split by defining a splitting rule that optimizes some criterion for the data in the nodes. The rule can be applied either on one attribute of the data, in which case axis parallel hyper-planes are defined. When all attributes or a subset of the attributes are considered, oblique hyper-planes are devised.

If X, Y is the data corresponding to a node, then a splitting hyperplane for all attributes can be expressed by using a parameter $\beta \in \mathbb{R}^p$:

$$\begin{aligned} X_L &= \{x \in X | \sum_{j=1}^{p} x_j \beta_j \leq 0\} \\ X_R &= \{x \in X | \sum_{j=1}^{p} x_j \beta_j > 0\} \end{aligned} \tag{1}$$

where X_L and X_R are the data corresponding to the left and right sub-nodes respectively.

Different decision trees use different criteria and methods to compute parameter β. In the present approach, the following node game $\Gamma(X, Y)$ is devised in order to decide how to split node data:

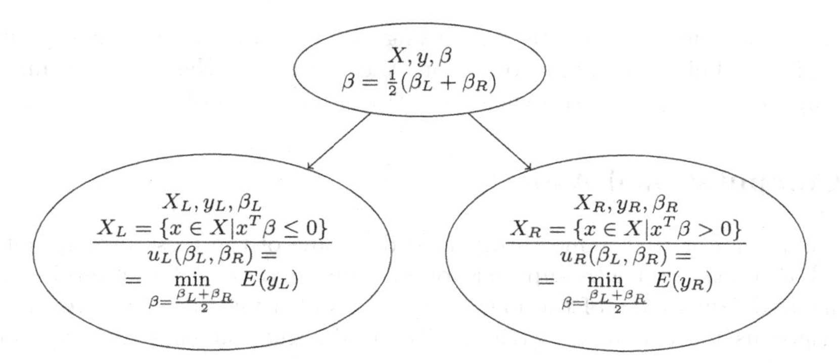

Fig. 1. Game-based node splitting procedure. Sub-nodes to the left and the right choose parameters β_L and β_R to minimize their entropy. The parent node is split using β, the average of the two, thus ensuring that the players' payoffs depend on both their strategies.

- players are represented by the two sub-nodes, L-left and R-right;
- the strategy of each player is to propose a parameter: β_L - for the left sub-node and β_R - for the right sub-node; $\beta_L, \beta_R \in \mathbb{R}^p$;
- the players' payoffs, $u_L(\beta_L, \beta_R)$ and $u_R(\beta_L, \beta_R)$ are computed as the entropy of the data corresponding to the split performed by using:

$$\beta = \frac{1}{2}(\beta_L + \beta_R) \tag{2}$$

Figure 1 illustrates the splitting mechanism based on game payoffs.

A possible and appealing solution to a game is the Nash equilibrium (NE), a strategy profile in which none of the players has an incentive for unilateral deviation. The Nash equilibrium can be used as an alternative solution to complex optimization problems in which trade-offs cannot be avoided. For game $\Gamma(X, Y)$, the NE is a strategy profile (β_L, β_R) such that none of the players can improve its payoff by changing only its strategy. In other words, there is no "movement" of the hyperplane "towards" any of the players that would decrease the entropy of either sub-node.

One advantage of using the Nash equilibrium instead of maximizing the information gain is decentralization. Each node minimizes its entropy resulting in a minimization of their sum, instead of minimizing their sum (which is what happens in maximizing information gain) in order to minimize them individually. Thus, for situations with equal information gain, the equilibrium would provide additional properties that can be exploited.

One way overfitting occurs is by defining regions that are too narrow around node data to be able to make reliable predictions. This may happen by maximizing information gain by minimizing the entropy of one node at the expense of the entropy of the other while still resulting in an optimal value for the information gain. The "pure" node may lead to overfitting, while the other may contain messy, inseparable data. In this case, using an alternate solution concept may be beneficial. This work aims to explore the use of the Nash equilibrium for splitting node data to assess if such an assertion may be sustained.

3 Computational Aspects

There is no doubt that game theory provides some of the most useful, mathematical, theoretically backed-up tools for modeling strategic interactions in many settings and forms. One of the main advantages of using them lies precisely in their rigorous mathematical setting, with results guaranteeing their existence and appealing properties.

However, the same robust mathematical framework limits the use of the game-theoretic concepts outside of it, creating a large gap between theory and practice, as many applications would benefit from using the solutions concepts proposed by game theory. However, as they do not fit the framework, they cannot compute or envisage how such solutions would look and if they are worth exploring.

A possible solution for bridging this gap is using a heuristic search for equilibria. Solutions provided by heuristics may not be exact and, in the case of equilibria, maybe even difficult to validate, but they may prove interesting from a practical point of view. In this case, one would know that they are worth exploring and investing in building a theoretical framework for a particular application.

Thus, in this paper, we are not concerned with the theoretical aspects related to the equilibrium of the node game $\Gamma(X, Y)$, but with practical approaches to approximate it in order to find if there are indications that it may be indeed considered a possible solution for the classification task, better than those provided by classical approaches.

Our endeavor starts from the definition of NE: there is no unilateral deviation that can improve a player's payoff. For normal form games, the sum of all improvements in unilateral deviations from a strategy can be used as a quality measure for it [9] and minimized in order to compute NEs in mixed form. Since our search space is continuous, we cannot replicate the sum for *all* unilateral deviations, but, since we are in the realm of heuristics, we can replace it with an approximation of it: generate a number of uniformly distributed unilateral deviations for each player and sum the squared deviations that lead to improvements in the players' payoffs. Thus we can define function $\nu(\beta_{\mathcal{L}}, \beta_{\mathcal{R}}|X, y, d)$:

$$\begin{aligned}\nu(\beta_L, \beta_R|X, y, d) = \sum_{k=1}^{d} \big(&\max\{0, E(\beta_L, \beta_R) - E(\beta_L + U([a,b],p), \beta_R)\}^2 \\ &+ \max\{0, E(\beta_L, \beta_R) - E(\beta_L, \beta_R + U([a,b],p))\}^2\big)\end{aligned} \quad (3)$$

where $U([a,b],p))$ denotes p uniformly distributed values in the interval $[a,b]$ and d the number of deviations used for the evaluation.

Function $\nu(\cdot)$ can be used as an objective function to be minimized by an optimization heuristic. Intuitively, we know that as $d \to \infty$ the value of $\nu(\cdot) \to 0$ for NEs of the game. Population-based heuristics for optimization have shown that randomization works: we do not need to consider *all* possible deviations to assess the quality of a solution with respect to the equilibrium. Furthermore, its evaluation is computationally expensive when considering a population-based search heuristic, especially when needed to be applied to each node of the tree. However, evaluating $\nu(\cdot)$ for a given strategy provides a lot of useful information regarding possible improvements in players' payoffs, and this information can be used within the search.

The following naive search heuristic for solving node games is proposed: starting with a random strategy profile (β_L, β_R) the value of $\nu(\cdot)$ is evaluated and the deviations $d_L, d_R \in \mathbb{R}^p$ providing the greatest improvement in each players' payoff are preserved. Then we set:

$$\beta_L = \beta_L + d_L, \quad \beta_R = \beta_R + d_R \tag{4}$$

and re-evaluate $\nu()$ for the deviated values. This process is repeated for a number of iterations. By replacing each iteration players' strategies with their largest possible deviation successively, we perform a greedy search for the equilibrium. In order to check if this naive approach works, we can look at the values of ν at each iteration: decreasing values indicate a possible success, reaching a 0 shows that a solution that may be close to the game equilibrium has been reached. Algorithm 1 outlines the steps described above.

Algorithm 1. Naive equilibrium approximation

1: **Input**: node data: X, y, number of deviations d, number of iterations n_{it};
2: **Output**: X_L, y_L, X_R, y_R, and β to define the split rule for the node;
3: Generate β_L, β_R at random following the standard normal distribution;
4: **for** n_{it} iterations **do**
5: Generate a number of d deviations D_L, D_R from $U([a, b], p)$ for each player;
6: Evaluate $\nu(\beta_L, \beta_R)$ using deviations D_L and D_R;
7: Set

$$d_L = \max_{d \in D_L} \max\{0, E(\beta_L, \beta_R) - E(\beta_L + d, \beta_R)\}^2$$

and

$$d_R = \max_{d \in D_R} \max\{0, E(\beta_L, \beta_R) - E(\beta_L, \beta_R + d)\}^2;$$

8: Set $\beta_L = \beta_L + d_L, \quad \beta_R = \beta_R + d_R$;
9: **end for**
10: Set $\beta = \frac{1}{2}(\beta_L + \beta_R)$
11: Set $X_L = \{x \in X | x^T\beta \leq 0\}$ and
$y_L = \{y \in Y | x \in X_L\}$;
12: Set $X_R = \{x \in X | x^T\beta > 0\}$ and
$y_R = \{y \in Y | x \in X_R\}$
13: **Return:** $X_L, y_L, X_R, y_R, \beta = \frac{1}{2}(\beta_L + \beta_R)$;

The decision tree constructed by using Algorithm 1 to split data at each node is called Naive Equilibrium Decision Tree (NEDT). NEDT starts from the root with the entire data set $\mathcal{X}, \mathcal{Y}$ and recursively splits data using Algorithm 1 until a maximum tree depth is reached or data in a node can be considered pure.

4 Numerical Experiments

Numerical experiments are performed to show the potential of the proposed approach. We compared our approach to other decision trees on synthetic and real world data sets with various degree of difficulty.

Data. To ensure reproducibility we generate different synthetic data sets using the `make_classification` function from the `scikit-learn` Python library [11]. The parameters used to generate data sets are combinations of: number of instances $(200, 500, 1000, 1500)$, number of classes $(6, 9, 12)$, number of attributes $(6, 9, 12)$, class separator $(0.1, 1)$, and the random state (50).

We also use real world data sets with different characteristics[1]. We test our approach on: the wine data set (D1) with 178 instances, 13 attributes and three classes, iris data set (D2) from which we removed the *setosa* instances in order to obtain a linear non separable binary classification problem, and Haberman's survival data set (D3) with 306 instances, three attributes, and two classes.

Comparisons with Other Methods. We compare our approach with other Decision Tree based classifiers by using their implementations from *scikit-learn* for reproducibility. The parameters for the compared models are: split criterion (*gini* and *entropy*), maximum depth (0 - split until the leafs are pure, 5 and 10).

Performance Evaluation. To evaluate the performance and compare with other classification models we use as metric the AUC - area under the ROC curve [4,12], for binary and multi-class classification. The AUC takes values in the interval $[0, 1]$, a higher value indicates higher probability to correctly classify instances. We use k-fold cross validation: we split the data in 10 folds using the `StratifiedKFold` method from `scikit-learn` (with seeds $60, 1, 2, 3, 4$) [5].

We report the AUC for each test fold from the 10 folds, resulting in 50 (fold) AUC values for each data set and each method. The 50 AUC values are compared using a paired t-test to test the null hypothesis that NEDT AUC values are smaller than the compared method. Corresponding p values are reported. Values smaller than 0.05 indicate that the null hypothesis can be rejected, and NEDT results may be considered significantly better than the other one for the tested data.

Results. Synthetic datasets represent a reliable test-bed for classification methods as their degree of difficulty can be controlled. Figures 2, 3 and 4 illustrate comparisons results between NEDT and the other methods for the synthetic datasets considered. Out of the total 576 comparisons performed, in 289 cases,

[1] UCI Machine Learning Repository https://archive.ics.uci.edu/ml/index.php, accessed March 2022.

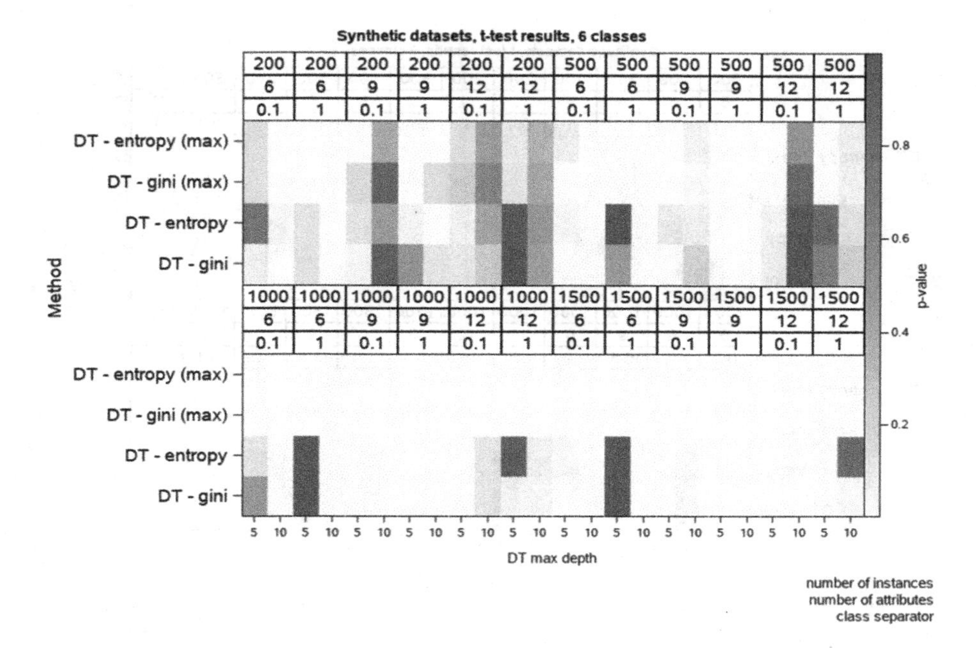

Fig. 2. Comparisons with other DT based, either using the same maximum depth or without a limit (max). The data set has 6 classes. p values indicate the probability that better results reported by NEDT are due to chance. Headers contain: number of instances, number of attributes and class separator values.

i.e. 50.14%, AUC values reported by NEDT can be considered significantly better than the others, based on t-test p-values.

The number of classes does not influence the performance of NEDT compared with the other methods significantly: 47.91% for 6 classes, 55.20% for 9 classes and 47.39% for 12 classes are the proportions NEDT results can be considered significantly better for each case.

When looking at the number of instances, we have the following values: out of 144 comparisons performed for each number of instances, NEDT results were significantly better in 29.86% of cases for data with 200 instances, 52.08% for 500, 56.25% for 1000, and 62.5% for 1500, indicating a possible better performance for large data sets for NEDT.

The number of attributes tested does not seem to influence results either: 54.16% of comparisons performed on synthetic data sets with six attributes can be considered in favor of NEDT from a statistical significance point of view. Corresponding values for 9 and 12 classes are 51.56% and 44.79%, respectively.

A similar situation occurs with the class separator values: for 0.1 - the most challenging case - in 49.30% of the cases, NEDT reported better results, and in 51.04% of cases for 1. For the two values of maximum tree depth tested, 5 and

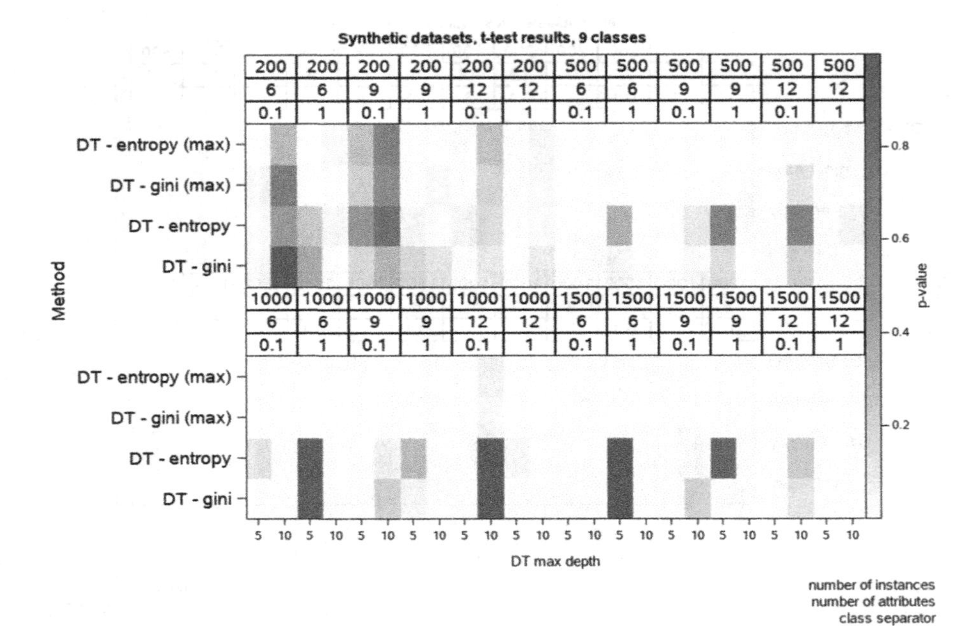

Fig. 3. Same as Fig. 2, when data contain 9 classes.

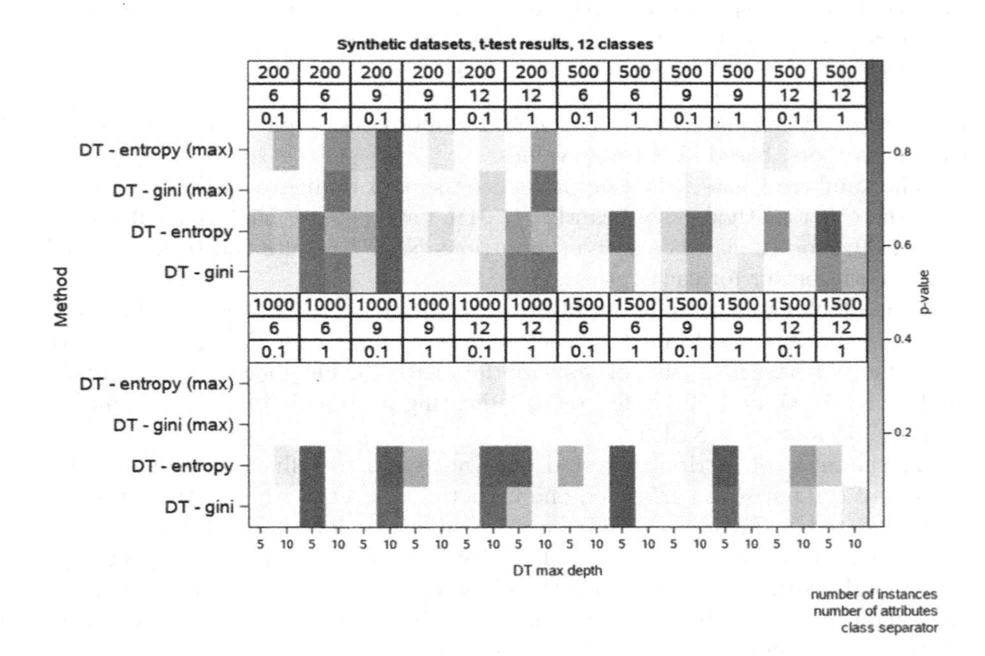

Fig. 4. Same as Fig. 2, 12 classes.

Table 1. Results for real data set, maximum tree depth of 5, for NEDT and compared models. Mean and standard deviation of the AUC indicator for all models over all test folds. A ◇ indicates a p value smaller that 0.05, i.e. NEDT results can be considered significantly better and a ∘ indicates a p value greater than 0.95 when comparing the results of NEDT with the other models.

Method	D1	D2	D3
NEDT	0.94(±0.05)	0.94(±0.08)	0.60(±0.12)
DT - gini	0.92(±0.06)◇	0.91(±0.09)◇	0.63(±0.12)
DT - entropy	0.95(±0.04)	0.92(±0.08)◇	0.62(±0.11)
DT - gini (max)	0.92(±0.05)◇	0.91(±0.09)◇	0.55(±0.08)◇
DT - entropy (max)	0.95(±0.04)	0.92(±0.08)◇	0.55(±0.08)◇

Table 2. Same as Table 1, max depth 10 for NEDT and compared models.

Method	D1	D2	D3
NEDT	0.93(±0.05)	0.93(±0.07)	0.61(±0.10)
DT - gini	0.92(±0.05)◇	0.92(±0.08)◇	0.57(±0.09)◇
DT - entropy	0.96(±0.04)∘	0.92(±0.08)	0.57(±0.09)◇
DT - gini (max)	0.92(±0.05)◇	0.91(±0.09)◇	0.55(±0.08)◇
DT - entropy (max)	0.95(±0.04)∘	0.92(±0.08)	0.55(±0.08)◇

10, corresponding values are 50.59% and 49.65%. These results are promising, indicating that this concept can be further explored.

Tables 1 and 2 present numerical results obtained by all models on the real world data sets. We find that for real data sets, NEDT results are comparable with the other methods, in some situations significantly better, but also outperformed in some settings.

5 Conclusions

A decision tree that naively uses the Nash equilibrium concept to split node data is proposed. The attempt is exploratory to verify if such an approach may be further considered from a theoretical point of view. Results reported on synthetic benchmarks indicate that, indeed, trees that use the approximation of Nash equilibria may provide better solutions to the classification problem than standard ones, indicating that further endeavors may lead to more interesting results.

References

1. Breiman, L., Friedman, J.H., Olshen, R.A., Stone, C.J.: Classification and Regression Trees. Wadsworth and Brooks, Monterey (1984)

2. Chen, Y., He, X., Xu, J., Guo, L., Lu, Y., Zhang, R.: Decision tree-based classification in coastal area integrating polarimetric SAR and optical data. Data Technol. Appl. **56**, 342–357 (2022). https://doi.org/10.1108/DTA-08-2019-0149
3. Chioson, F.B., Munsayac, F.E.T.J., III., Luta, R.B.G., Baldovino, R.G., Bugtai, N.T.: Classification and determination of pH value: a decision tree learning approach. In: 2018 IEEE 10th International Conference on Humanoid, Nanotechnology, Information Technology, Communication and Control, Environment and Management (HNICEM), Baguio City, Philippines. IEEE (2018)
4. Fawcett, T.: An introduction to ROC analysis. Pattern Recogn. Lett. **27**(8), 861–874 (2006). https://doi.org/10.1016/j.patrec.2005.10.010. ROC Analysis in Pattern Recognition
5. Hastie, T., Tibshirani, R., Friedman, J.: The Elements of Statistical Learning: Data Mining, Inference and Prediction, 2nd edn. Springer, Heidelberg (2009)
6. Krupka, J., Kasparova, M., Jirava, P.: Quality of life modelling based on decision trees. E & M Ekonomie Manag. **13**(3), 130–146 (2010)
7. Li, Y., Dong, M., Kothari, R.: Classifiability-based omnivariate decision trees. IEEE Trans. Neural Netw. **16**(6), 1547–1560 (2005)
8. Liang, J., Qin, Z., Xiao, S., Ou, L., Lin, X.: Efficient and secure decision tree classification for cloud-assisted online diagnosis services. IEEE Trans. Dependable Secure Comput. **18**(4), 1632–1644 (2021). https://doi.org/10.1109/TDSC.2019.2922958
9. McKelvey, R.D., McLennan, A.: Computation of equilibria in finite games. Handb. Comput. Econ. **1**, 87–142 (1996)
10. Murthy, S.K., Kasif, S., Salzberg, S.: A system for induction of oblique decision trees. J. Artif. Intell. Res. **2**, 1–32 (1994)
11. Pedregosa, F., et al.: Scikit-learn: machine learning in Python. J. Mach. Learn. Res. **12**, 2825–2830 (2011)
12. Rosset, S.: Model selection via the AUC. In: Proceedings of the Twenty-First International Conference on Machine Learning, ICML 2004, p. 89. Association for Computing Machinery, New York (2004). https://doi.org/10.1145/1015330.1015400
13. Singh, M., Wadhwa, P.K., Kaur, S.: Predicting protein function using decision tree. In: Ardil, C. (ed.) Proceedings of World Academy of Science, Engineering and Technology, Bangkok, Thailand, 21–23 May 2008, vol. 29, p. 350+ (2008)
14. Song, Y.Y., Lu, Y.: Decision tree methods: applications for classification and prediction. Shanghai Arch. Psychiatry **27**, 130–135 (2015)
15. Wickramarachchi, D., Robertson, B., Reale, M., Price, C., Brown, J.: HHCART: an oblique decision tree. Comput. Stat. Data Anal. **96**, 12–23 (2016). https://doi.org/10.1016/j.csda.2015.11.006
16. Wu, X., et al.: Top 10 algorithms in data mining. Knowl. Inf. Syst. **14**(1), 1–37 (2008). https://doi.org/10.1007/s10115-007-0114-2

Human-Centric Machine Learning Approach for Injection Mold Design: Towards Automated Ejector Pin Placement

Robert Jungnickel(✉), Johanna Lauwigi, Vladimir Samsonov, and Daniel Lütticke

Information Management in Mechanical Engineering, RWTH Aachen University, Aachen, Germany
{robert.jungnickel,johanna.lauwigi,vladimir.samsonov, daniel.lutticke}@ima.rwth-aachen.de
https://cybernetics-lab.de/en/

Abstract. Nowadays, injection molds are manually designed by humans using computer-aided design (CAD) systems. The placement of ejector pins is a critical step in the injection mold design to enable demolding complex parts in the production. Since each injection mold is unique, designers are limited in using standard ejector layouts or previous mold designs, which results in high design time so that an automation of the design process is needed. For such a system, human knowledge is essential. Therefore, we propose a human-centric machine learning (HCML) approach for the automatic placement of ejector pins for injection molds. In this work, we extract mental models of injection mold designers to obtain machine-readable fundamental design rules and train a machine learning model using an ongoing human-machine learning approach.

Keywords: Human-centric machine learning · Mental models · Injection molding · Ejector pins · Expert system

1 Introduction

Injection molding is the most important manufacturing process in plastics technology. In the last decades, automation solutions have enabled significant productivity and efficiency improvement in the mechanical production of injection molding. However, the design of injection molds is still carried out manually. As a result, over 70% of the total cost of an injection molded part is determined by mold design only [1], whereby the precise ejector pin placement is a crucial step in ejecting the molded part without damage [2]. Hence, an automation of the

The work of RWTH Aachen University within the AutoEnSys joint research project is funded by the German Federal Ministry of Education and Research (BMBF) under the grant number 01IS20081C.

G. Nicosia et al. (Eds.): LOD 2022, LNCS 13811, pp. 27–31, 2023.
https://doi.org/10.1007/978-3-031-25891-6_3

ejector pin placement for injection molds would allow a reduction of time and cost. However, the ejector placement depends highly on experience of injection mold designers and is therefore a human knowledge-driven process [3]. The experiences are stored in mental models as a set of rules of injection mold designers. Extracting these mental models is a complex task as humans are often not able to describe the rules they use explicitly [4]. Hence, the challenge is to extract these rules in a systematic and quantified way to make sure that all implicit knowledge is part of the automated design system.

In this work, we aim to develop a novel human-centric machine learning approach to obtain fundamental design rules from injection mold designers. This allows for ongoing human-machine learning in which the human knowledge is integrated into the machine learning process. Motivated by this, we focus on the use of implicit knowledge from injection mold designers by including rules into an expert system and continuously adjusting these rules based on the human feedback.

This work provides the following contributions:

- We present a novel HCML approach for automatic ejector pin placement in injection mold design.
- We integrate human mental models as design rules into the design process to enable ongoing human-machine learning to optimize the ejector pin placement.
- We exemplify our approach with the use of a box part geometry.

2 Related Work

Investigation for ejection pin placement focus mainly on: the numerical layout design and size of the ejector pins [2], the engineering of ejector pins using a balancing algorithm [5], the development of a multidisciplinary framework that takes customer preferences and their impositions into account [6], multi-objective optimization of injection molding process [7,8], and an expert system to estimate injection mold cost design [9].

The optimization of injection molding solutions and the reduction of design costs are one of the key aspects of current injection molding research and is considered in multiple works [1–8]. While most of these approaches are based on machine-generated numerical simulations, only Chin and Wong [9] apply a set of design rules, but do not consider ongoing human feedback in an automated design process, as proposed herein.

Human-Centric Machine Learning (HCML) is a research field that includes the human and the machine in the overall model development process [10]. However, the integration of the human input into the machine learning process remains a challenge [11]. In the proposed approach, we extract human mental models [12] of injection mold designers to obtain fundamental design rules for the ejector pin placement. Based on these models, we build an expert system, which can be trained in the following process. This optimization of the expert system is based on online learning [13].

3 Human-Centric Machine Learning Framework

We consider a two-step learning approach, depicted in Fig. 1. First, during the one-time extraction step we obtain basic design rules from human mental models and the features of a machine generated 3D-CAD model.

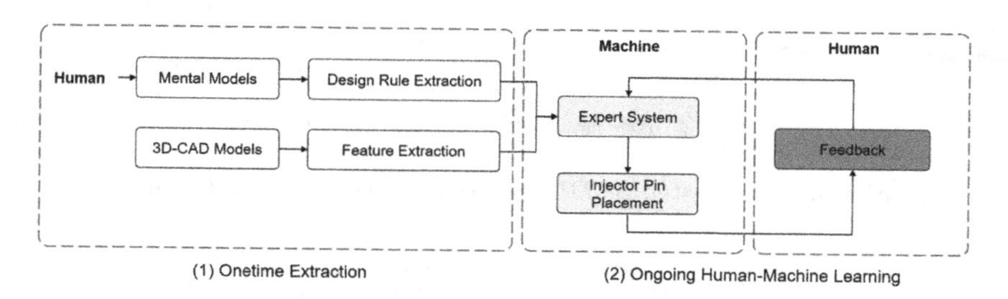

Fig. 1. Human-centric machine learning framework for injection mold design.

Second, the extracted models are integrated in an expert system in the ongoing human-machine learning step. The expert system places the ejector pins automatically based on the fundamental design rules saved in the expert system. It proposes the ejector pin placement to the injection mold designer and continuously learns from the designer's feedback to the system to optimise the suggestions over time.

Figure 2 details the process of the one-time extraction and ongoing human-machine learning steps. The one-time extraction step consist of two fundamentals. First, human design rules for ejector pin placement are extracted from different injection mold part geometries at time steps t_{e1}^{H} to t_{ei}^{H} to obtain the machine learning constraints for the expert system. Each geometry $1 - i$ represents a different kind of injection molded part geometry and build its own rule set. Second, 3D-CAD model features are extracted from injection molded part geometries 1-i at time t_{e1}^{M} to t_{ei}^{M}. These features are necessary to execute the rules from the rule set so that the position of the ejector pin placement can be determined by the expert system at time t_{e1}^{HM} to t_{ei}^{HM}.

The ongoing human-machine learning step starts with a proposition of the expert system about the ejector pin placement at t_{l1}^{HM} to the human, who validates the results at time t_{l1}^{H}. This feedback is sent back to the expert system at time t_{l2}^{H} to retrain the ejector pin placement at t_{l2}^{HM}. These iterative steps proceed until the expert system is able to recommend ejector pin placements which completely satisfies the expert's requirements.

4 Example: Box Geometry

To showcase the capability of the framework we test the approach on a box geometry, shown in Fig. 3(a). The given geometry represents a category of injection

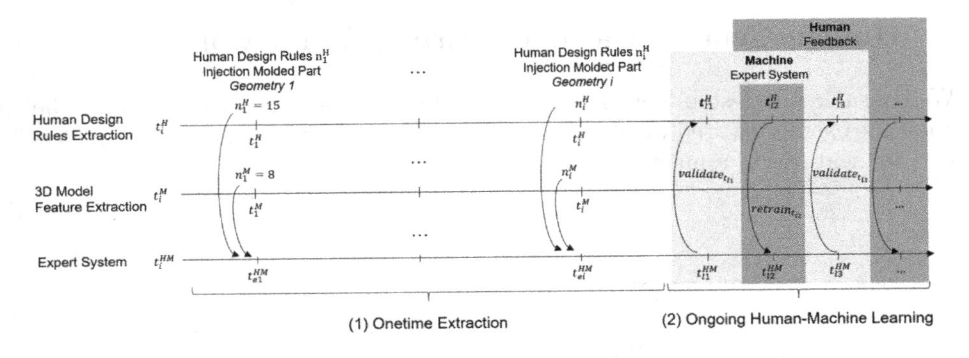

Fig. 2. The two-step HCML process for injection mold design.

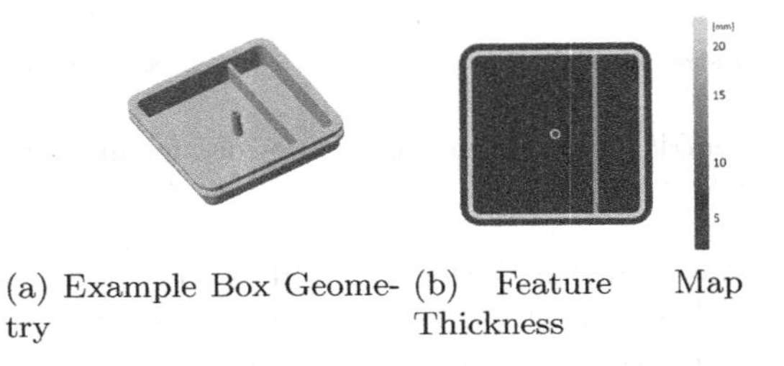

(a) Example Box Geometry (b) Feature Map Thickness

Fig. 3. A box geometry is used as a test geometry for the proposed concept.

molded parts with similar box geometries. Based on the experience of injection mold designers, 15 geometry-specific placement rules for ejector pins were extracted at time t_{e1}^{HM}, for example *"Place ejector pins at corners"* or *"Do not place ejector pins on thin material (thin equals less than 3/4 of mean thickness)"*. From these rules, the relevant features $n_{e1}^{M} = 8$ of the 3D-model are derived: material thickness, height, corners, ribs, walls, tubes, slopes, and position of cooling system. An example of the extracted thickness is shown in Fig. 3(b).

The extracted knowledge and features build the basis for the expert system. That will be trained by the expert's feedback, giving the expert three different actions: 1. move a position; 2. delete a position; 3. add a position. In an online learning process, the algorithm will take the feedback and update parameters in rules and prioritize the rules depended on the feedback.

5 Conclusion

We propose an approach to integrate human knowledge into the development of an automated ejector pins placement for injection mold design. Therefore, we formulate fundamental rules for injection mold design based on the mental models from human experts. We demonstrate with a box part geometry that the one-time extraction of the human mental model leads to a set of rules that can automate the placement of ejector pins. However, the execution of the rules still needs to be trained to enable the automated design process. Therefore, the approach establishes an human-centric machine learning to allow an ongoing online human-machine learning process.

References

1. Ehrlenspiel, K., Kiewert, A., Mörtl, M., Lindemann, U.: Kostengünstig Entwickeln und Konstruieren. Kostenmanagement bei der integrierten Produktentwicklung, 8th edn. Springer, Heidelberg (2020)
2. Kwak, S., Kim, T., Park, S., Lee, K.: Layout and sizing of ejector pins for injection mould desian using the wavelet transform. Proc. Inst. Mech. Eng. Part B: J. Eng. Manuf. **217**(4), 463–473 (2003). https://doi.org/10.1243/095440503321628143
3. Chin, K.-S., Wong, T.N.: An expert system for injection mold cost estimation. Adv. Polym. Technol. **14**(4), 303–314 (1995). https://doi.org/10.1002/ADV.1995.060140404
4. Sterman, J.D.: Business Dynamics. Systems Thinking and Modeling for a Complex World. Irwin/McGraw-Hill, Boston (2000)
5. Wang, Z., et al.: Optimum ejector system design for plastic injection moulds. Int. J. Comput. Appl. Technol. **9**(4), 211–218. (1996). ScholarBank@NUS Repository. https://doi.org/10.1504/IJMPT.1996.036339
6. Ferreira, I., Cabral, J.A., Saraiva, P., Oliveira, M.C.: A multidisciplinary framework to support the design of injection mold tools. Struct. Multidisc. Optim. **49**(3), 501–521 (2014). https://doi.org/10.1007/s00158-013-0990-x
7. Alvarado-Iniesta, A., Cuate, O., Schütze, O.: Multi-objective and many objective design of plastic injection molding process. Int. J. Adv. Manuf. Technol. **102**(9–12), 3165–3180 (2019). https://doi.org/10.1007/s00170-019-03432-8
8. Li, K., Yan, S., Zhong, Y., Pan, W., Zhao, G.: Multi-objective optimization of the fiber-reinforced composite injection molding process using Taguchi method, RSM, and NSGA-II. Simul. Model. Pract. Theory **91**, 69–82 (2019). https://doi.org/10.1016/j.simpat.2018.09.003
9. Chin, K.-S., Wong, T.N.: An expert system for injection mold cost estimation. Adv. Polym. Technol. **9**(4), 303–314 (1995). https://doi.org/10.1002/ADV.1995.060140404
10. Kaluarachchi, T., Reis, A., Nanayakkara, S.: A review of recent deep learning approaches in human-centered machine learning. **21**(7) (2021). https://doi.org/10.3390/s21072514
11. Sperrle, F., et al.: A survey of human-centered evaluations in human-centered machine learning. Comput. Graph. Forum **40**(3), 543–568 (2021). https://doi.org/10.1111/cgf.14329
12. Gary, M.S., Wood, R.E.: Unpacking mental models through laboratory experiments. Syst. Dyn. Rev. **32**(2), 101–129 (2016). https://doi.org/10.1002/sdr.1560
13. Bubeck, S.: Convex Optimization. Algorithms and Complexity. Foundations and Trends® in Machine Learning Ser, 26th. Now Publishers, Norwell (2015)

A Generative Adversarial Network Based Approach to Malware Generation Based on Behavioural Graphs

Ross A. J. McLaren, Kehinde Oluwatoyin Babaagba(✉), and Zhiyuan Tan

School of Computing, Edinburgh Napier University, Edinburgh EH10 5DT, UK
40174116@live.napier.ac.uk, {K.Babaagba,Z.Tan}@napier.ac.uk

Abstract. As the field of malware detection continues to grow, a shift in focus is occurring from feature vectors and other common, but easily obfuscated elements to a semantics based approach. This is due to the emergence of more complex malware families that use obfuscation techniques to evade detection. Whilst many different methods for developing adversarial examples have been presented against older, non semantics based approaches to malware detection, currently only few seek to generate adversarial examples for the testing of these new semantics based approaches. The model defined in this paper is a step towards such a generator, building on the work of the successful Malware Generative Adversarial Network (MalGAN) to incorporate behavioural graphs in order to build adversarial examples which obfuscate at the semantics level. This work provides initial results showing the viability of the Graph based MalGAN and provides preliminary steps regarding instantiating the model.

Keywords: Malware · Malware detection · Adversarial examples · Generative Adversarial Network (GAN) · Behavioural graphs

1 Introduction

Malware attack landscape has evolved over time as malware authors and attackers now employ a number of sophisticated techniques in launching attacks. Some of these include the use of encryption [9], polymorphism [16] among others to evade detection. Consequently, most traditional detection mechanisms are vulnerable in defending against these new intrusive techniques.

As more and more of the world becomes connected through the internet, or automated through technological advancements, defence against attacks on these systems becomes a growing priority. As attackers become more sophisticated it can be a struggle for defence vendors to keep up as it is not enough if they can fix the issue after it occurs, preemptive defence of systems will always be superior. Due to this, a crucial area in software security research is trying to anticipate what attackers will do to keep systems safe. One such way to do this

G. Nicosia et al. (Eds.): LOD 2022, LNCS 13811, pp. 32–46, 2023.
https://doi.org/10.1007/978-3-031-25891-6_4

is to generate adversarial examples of malware in a secure environment with the sole purpose of training automated detectors to be able to detect a wide range of new, never before seen samples.

Machine learning has shown encouraging results as a useful tool in the detection of malicious software. It has been used to learn patterns within header files, raw bytes and instruction sequences which identify a piece of software as malicious or benign [18]. Commonly, these machine learning solutions are created as a black box in an effort to increase the security of the underlying system. This follows an idea that if the attacker cannot access the underlying machine learning algorithm then it is far more difficult for them to exploit it for their own ends. This, however, has presented a weakness in such thinking as attackers are able to probe the network and from that, determine which design features will be flagged as malicious allowing them to design software that evades detection by the machine learning models [13].

In addressing the aforementioned weakness, adversarial learning [3,4,11] was introduced. The aim of an adversarial method is to create malicious software which exploits the loopholes in other machine learning models. To this effect, they collect data using models such as deep, convolutional neural networks used in malware analysis in order to categorize what they will allow as a benign piece of software. Another network is then trained to create malware which should be considered benign by the first network, initially this yields poor results but over numerous iterations the generated malware signature should successfully evade detection. This is known as gradient based adversarial generation [10] and has been done to great success with minuscule amounts of the malicious software needing to be changed to evade detection.

In recent years, attackers have been able to stay ahead of defenders by utilising numerous obfuscation techniques. To combat this, steps are being taken to move towards a broader semantics based, higher level approach to detection [1]. Malware behavioural graphs are a step towards this, as they utilise a Control Flow Graph (CFG) as a signature. CFGs are used because the domain of a CFG is at least as complex as the domain of strings in the same function [5]. A CFG consists of linked nodes with each node being a different element of the program. In assembly, these could be the different calls such as jmp, call or end [7]. This allows a mapping of the flow of data and functions within a greater piece of malware [8]. This not only provides the information on the number of times a function is called, but the order in which they were called thus mapping out the actual behaviour of the malware [1]. The structures of the created graphs can be compared to identify crossovers and similarities in the behaviour of two programs, allowing the classification of new malware into existing families based on the behavioural patterns they exhibit [5]. Unlike other classical detection techniques, CFG detection improves its detection accuracy as the size of the program increases [8]. This is due to a larger program creating a more detailed end graph. It is also resistant to common obfuscation techniques due to not being based on checks of specific vector or string details of a signature [7].

Generative Adversarial Networks (GANs) [15] have already been used to much success in the area of adversarial example generation, however, these attempts suffer from a lack of diversity within the malware used and the limited features they use for generation. This paper seeks to outline a solution to these problems by building the examples based on behavioural graphs rather than simple feature semantics. Two research questions are addressed in this work and they include:

1. To what extent can a Graph based Malware Generative Adversarial Network (MalGAN) be used in creating adversarial samples?
2. How does the Graph based MalGAN compare with the original MalGAN?

The rest of the paper is structured as follows. The second section presents related work. Section 3 describes the research methodology. The results and evaluation are discussed in Sect. 4 and we present the conclusion and future work in Sect. 5.

2 Related Work

Adversarial examples of malware have been utilized to show weaknesses in machine learning based malware detection systems [13]. These examples are capable of bypassing the traditional black-box malware detection systems by allowing attackers to infer the features likely to be flagged as malware [14]. Till very recently, neural network based models for generating adversarial examples have been primarily gradient based. These have had some success but struggle to reach a detection rate of zero. They are also able to be quickly retrained against by most defensive methods [17].

MalGAN has been introduced as a new model for generating adversarial examples [13]. It is based on the GAN model for generators and uses a system which comprises a generator and discriminator. It takes API features as an example of how to represent a program. The main difference between this given model and current models is the fact that the generator can update dynamically in relation to the feedback it receives from the black-box detector. Currently, most models use a static gradient to produce examples instead. The generator transforms an API feature vector into an adversarial malware example. It creates a binary version of the feature vector of a piece of malware, showing a zero if the API feature does not exist and a one if it does. This is concatenated with a random noise function, altered to return values between 0 and 1 in an effort to add non-malicious features to the example. This is fed into a multi-layer feed-forward perceptron with the last layer's activation function being sigmoid with a restricted output range of 0 to 1 [13]. During the generation, only irrelevant features are considered when adding features to the malware. The removal of features is not considered as this could result in a cracked malware. The black-box detector used is also a multi-layered feed-forward neural network that takes in a feature vector as an input and outputs if the vector is malicious or benign [13].

The training data used for the black box detector consists of a mix of the generator's examples, and benign pieces of software. MalGAN is also significantly more dynamic than a gradient based approach which allows it to keep up with advances in security. This is because MalGAN only needs to be retrained on the new detectors in order to be able to create adversarial examples against it.

It has a limitation of only currently generating feature vector examples, which could make it difficult to produce examples which can fool higher-level, broader semantics based detectors which utilize the behaviour of the malware rather than its API features [14]. That is why this paper seeks to present a new generator based on MalGAN that generates graph-based examples rather than feature vectors and evaluate how this affects the detection rate against its examples. MalGAN has also been suggested to have some issues which limit its functionality in a real-world application [14]. For example, by using a set of the features in a piece of malware rather than the entirety of the feature list, it limits to what degree the adversarial examples it generates can be properly used to actually harden machine learning based malware detection approaches. Also, by having the generator and discriminator built and trained within the same process as the generator, it creates an unrealistic advantage that traditionally attackers would not have [14]. In the case of both of these versions of MalGAN, they are able to reach under three percent detection rates based on the True Positive Rate (TPR).

3 Methodology

3.1 Overview

The implementation of the Graph based MalGAN comprises of multiple components. It required the collection of API features from analysed examples, the connection of dependencies between these calls and the encoding into a format a neural network can utilise. These steps led on to the model construction and neural network training and testing. This section gives a brief overview of these parts and presents an explanation of how the model works.

Dynamic Analysis. This was used in order to get the correct API calls that were used at run time for each program, rather than relying on all possible API calls as is present in static analysis. To facilitate this, a Windows Operating System was procured and loaded onto a Virtual Box Machine. This machine was then hardened and set up for malware analysis, being completely cut off from being able to connect to the host or the Internet and had all of the Windows Defender settings turned off. The host machine was also prepared as per the specifications listed by the Cuckoo[1] sandbox website and finally the Cuckoo agent was loaded onto the Virtual Machine to allow for secure communication between the host and machine. After the malware binaries were analysed on the

[1] Cuckoo - https://cuckoosandbox.org/.

Virtual Machine, several python scripts were developed which first extracted the API calls from the Cuckoo log which consists of a JSON file, then removed any exact duplicate calls before assembling them into a behavioural graph that represents the initial program. One-hot encoding was then used to have valid inputs for the neural network.

The Neural Network (NN) Model. The NN model put forward for Graph based MalGAN consists of a generator and discriminator as in a typical GAN architecture. The generator takes a concatenated input consisting of a malware example and a noise vector and outputs an adversarial example that the discriminator takes as an input. The discriminator then determines whether or not the example is malicious based on the classification given to it from an outside detector, the detector an attacker is trying to bypass in a real world scenario. The generator and discriminator train with the goal being to minimise the number of samples correctly identified by the external detector as shown in Fig. 1.

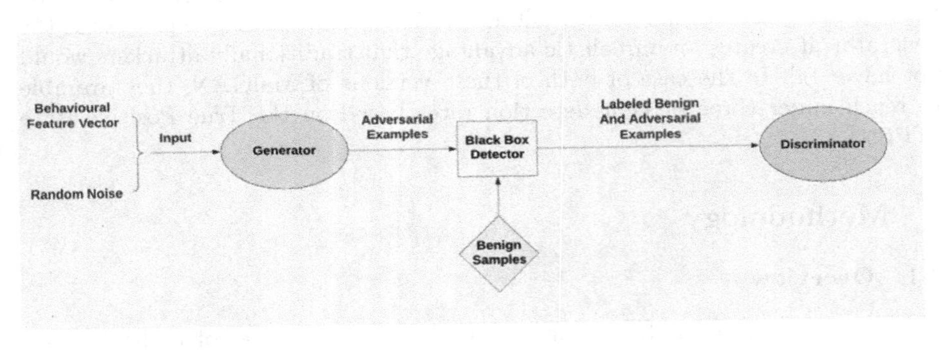

Fig. 1. A simple overview of the entire network model

3.2 Input Formatting and Encoding

Behaviour-based malware detection focuses on security critical operations, and this was taken into consideration when deciding what operations and API calls to focus on for the API graphs. Unlike in the original MalGAN paper, none of the API call graphs were capped. The resulting behavioural graph for each program only counts unique API call graphs, as duplicate behaviours are rare and overall unlikely to affect the behavioural obfuscation displayed in this paper.

Whilst all of the calls selected appear in benign samples, the combination of them can lead to intentional malicious design. From this, the generator can learn what combination of behaviours are likely to be assessed as malicious and how to alter them to avoid detection without cracking the underlying malware.

Behavioural graphs are an un-ordered set of API call graphs for a given program and are used as the basis of behaviour-based malware detection. By cleaning and processing a set of API calls from a given program, dependencies can be built between calls which act upon the same Operating System resource, this lets a user view the API calls in a graph structure. This graph structure presents the function of a piece of code within the program in a human friendly format, allowing for better understanding of the behaviour of the code. A behavioural graph is the next step up, and instead of displaying the behaviour of a particular section of code, it combines all of them into an un-ordered graph which describes the behaviour of the whole program.

Both the API call and its relevant Operating System resources are extracted from an analysed program and used to construct the API call graph. The API calls are assigned a numerical value in order to aid in the labelling during the encoding stage. The graph is directed and acylic with the nodes being defined as the API calls and the edges between them being their connected Operating System resource and the dependence between them. It can be defined as:

$$G = (V, E, f) \tag{1}$$

wherein, API call graph G has a set of nodes, V, and a set of edges represented by E which is a member of $V \times V$ and an incidence function f, mapping E to V.

In order to allow for further classification of API call graphs within the overarching behavioural graph, different commands which serve the same purpose are treated as the same call. An example of this is seen in samples with multiple calls which all serve the purpose of creating a new key for the registry -

1. RegCreateKeyExW
2. RegCreateKeyExA
3. RegCreateKeyW

For the API call graph, all of these calls have been transformed into RegCreateKey.

Firstly, a dataset of program samples was constructed by analysing both malicious and benign samples within the Cuckoo sandbox. This dataset ΔB_G contained sixteen-hundred malicious examples and five-hundred-and-fifty benign examples. After being analysed, the API calls, their Operating System resource arguments and the time they were called are all extracted from the Cuckoo reports and held in a JSON file. At this point, exact duplicate calls which happen in sequence are removed as during dynamic analysis it is possible for calls to be logged which have not actually occurred and are as such, simply copies of the previous call.

From here, API calls are linked together through the dependencies present in both their Operating System arguments using both the addresses in memory and the names of the processes, unlike in relevant studies using behavioural graphs which only use the memory addresses. Whilst it should be noted that the malware may use a different name for a process than the actual process' name, it was decided to still be an important factor in correctly linking API calls within this study. These graphs are then ordered based on the time at which the call was made in order to appropriately structure the order they happen in. This is crucial for identifying the correct patterns because the transference of data

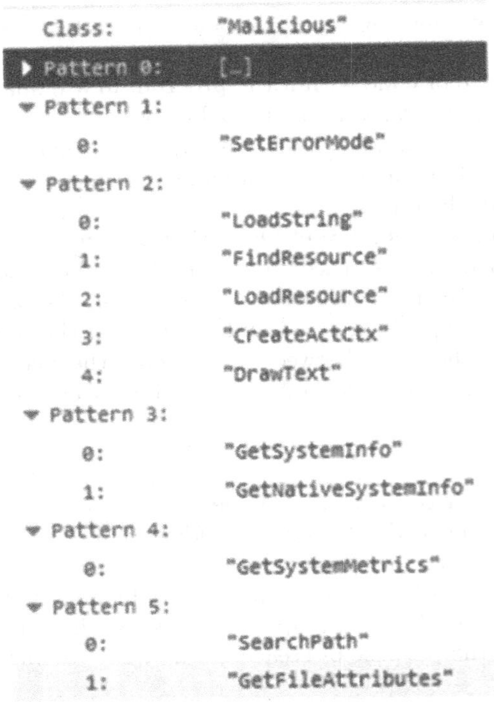

```
Class:          "Malicious"
▸ Pattern 0:    [...]
▾ Pattern 1:
    0:          "SetErrorMode"
▾ Pattern 2:
    0:          "LoadString"
    1:          "FindResource"
    2:          "LoadResource"
    3:          "CreateActCtx"
    4:          "DrawText"
▾ Pattern 3:
    0:          "GetSystemInfo"
    1:          "GetNativeSystemInfo"
▾ Pattern 4:
    0:          "GetSystemMetrics"
▾ Pattern 5:
    0:          "SearchPath"
    1:          "GetFileAttributes"
```

Fig. 2. Excerpt from a Behavioural Graph in JSON format

between different calls within the API graphs is important for their underlying functionality (Fig. 2).

Once all of the patterns for a program have been successfully identified, duplicates within the same program are removed. This is because, whereas in MalGAN the objective is the generation of adversarial examples for specific malware, the objective here is training a generator which can successfully obfuscate these patterns to generate adversarial examples. At this point the dataset of behavioural graphs ΔB_G is created by iterating through each program's behavioural graph and creating a list of every unique occurrence.

By utilising one hot encoding, each program's behavioural graph is compared to the dataset and at each pattern, given a one or zero if the pattern exists within the program.

$$T_e = (V, C)T_{BG} \in \Delta B_G T_{eVn} = \{0|1\}T_{eV1} = 1 \iff B_{G1} \in T_{BG}$$

where T_e is equal to an encoded program and contains a binary vector V and a class definition C which defines if T is benign or malicious when training the discriminator. T_{BG} refers to the behavioural graph for program T, ΔB_G is the dataset of all behavioural graphs used in this work. As with one hot encoding T_{eVn} equals a one if the nth value of the greater dataset exists within the behavioural graph for program T. This results in a binary vector which can be used as the inputs for a neural network.

3.3 Dataset

The dataset used for this work was the MC-binary-dataset [2]. Sixteen-hundred malicious samples and five-hundred-and-fifty benign samples were used to create the behavioural graph dataset and used for training the black box detector and generator system. The malicious dataset was split with sixty percent going to the generator system and forty percent going to the black box detector. This results in the generator having nine-hundred-and-sixty malicious examples to utilise and the black box detector having six-hundred-and-forty to train on. This split is for two reasons:-

1. It better simulates a real world scenario where an attacker is unlikely to have the exact malicious examples used to train a detection system.
2. It forces the generator to obfuscate the malicious patterns, rather than obfuscating specific malware as in both the original and improved MalGAN papers.

3.4 Network Architecture

Generator. The generator is a dense, multi-layer, feed-forward network with weights classified as θ_g. The output layer is the same shape as the number of patterns within the dataset. The weighting of the graph is informed from the feedback of the discriminator and effects the distribution of the noise throughout the vector in an attempt to favour patterns which allow the generated example to appear benign. The final layer uses a sigmoid activation function for the express purpose of limiting the outputs to the range [O, 1], and in order to ensure that the malware remains functionally malicious, the generated adversarial example is combined with the original sample using the OR function, allowing the non-zero values in the adversarial example to act as pattern obfuscation. This resulting vector is transformed into a binary by affixing a one in every position where the value is greater than 0.5 and a zero when below this threshold (Table 1).

This obfuscated behavioural pattern vector is then able to be used as the input for the black box detector to be labelled, and then based on this labelling receives gradient information from the discriminator in order to improve its weighting and results. The goal of the generator is to have its adversarial examples mislabelled as benign by the detector.

Discriminator. The discriminator in this instance is a substitute detector. It acts as a stand-in for the black box detector system to enforce the real world scenario where attackers would be unable to access the underlying system within the detector. It is fed classifications from the black box detector based on the adversarial examples and benign code given to the black box which can then be used to train the generator. It does this by learning the classification rules

Table 1. The model used as the generator in graph based MalGAN

Layer	Activation	Size
Input		
Dense	ReLU	256
Batch Normalization		
Dense	ReLu	128
Dense	ReLU	64
Batch Normalization		
Dropout(0.5)		
Output Layer	Sigmoid	

involved in the black box detector. Like the generator it is a dense, multi-layer, feed forward network (Table 2).

As input it takes in either a benign example or an adversarial one that has been classified and labelled by the black box detector. As its purpose is as a stand-in for the black box detector it is only able to see the samples that the detector itself has labelled as malicious, rather than the data set of actual malicious examples. This ensures that it trains the generator to successfully fool the detector. As with the generator, the goal and associated loss function of the discriminator relates to lowering the number of examples correctly identified as malware by the black box detector.

Table 2. The model used as the discriminator in graph based MalGAN

Layer	Activation	Size
Input		
Dense	ReLU	256
Dense	ReLu	128
Dropout(0.5)		
Dense	ReLU	64
Output Layer	Sigmoid	

3.5 Machine Learning Models Used for Comparison

Multilayer Perceptron (MLP). MLP is a deep Artificial Neural Network (ANN) which has different layers consisting of at least an input layer, a hidden layer and an output layer. Input reception is handled by the input layer, the hidden layer is termed the computation engine and decision making or predictive analysis is carried out by the output layer [19].

Random Forest (RF). RF is also referred to as random decision forest. This is a machine learning ensemble that combines several algorithms to derive better learning results. RF as the name suggests, comprises of several separate decision trees that form an ensemble with every of these trees spitting out a prediction for a class and the class that has the highest vote then becomes the prediction of the model [6].

Logistic Regression (LR). LR is commonly employed for binary data and for categorical target variables. An example would be in the prediction of an email as being either benign (0) or spam (1). A logit transformation is used to force the Y value to take on differing values between 0 and 1. The probability $P = 1/(1 + e - (c + bX))$ is initially calculated after that X is then linearly correlated to $log_n P/(1 - P)$ [12].

4 Results Discussion and Evaluation

Here, in answering the two research questions, (1) To what extent can a Graph based MalGAN be used in creating adversarial samples? and (2) How does the Graph based MalGAN compare with the original MalGAN?, we present an overview of the performance of the Graph based MalGAN model which is provided in Table 3 which compares MalGAN's True Positive Rate (TPR) against different machine learning algorithms. The TPR is the percentage of the samples that the black box has correctly classified and the lower the value is, the better the model has performed. We present a graph for each learning algorithm, visually showing the change in the TPR throughout the training and testing phases of each epoch.

4.1 Overall Performance

Table 3 presents the TPR of malware detection for the blackbox detector showing how many of the original malware samples the different algorithms were able to classify, how many of the original MalGAN samples the algorithms were able to classify and finally, how many of the Graph based MalGAN samples that they were able to classify. It is important to also include the TPR figures for the malware examples before they are obfuscated as high rates of correct classification at this stage add validity to the behavioural graph method for presenting malware. If the detector was incapable of classifying malware correctly even before the adversarial examples were generated then it could be determined that the behavioural graph method was flawed and as such, results for the classification of their adversarial examples would be invalid. The results for the Graph based MalGAN are also compared to the original MalGAN in order to determine if any improvements to the detection rate has been observed. In this instance, as the objective is successfully generating examples which fool the detector, a lower true positive rate is desired.

It can be observed that against the unaltered malware examples, the blackbox detector was able to successfully classify the behavioural graph's patterns as malicious. This adds validity that the behavioural graph method for malware detection is a valid and good method of detection and therefore a system which can successfully obfuscate these patterns is useful and results gathered from such obfuscation are valid. The Graph based MalGAN was able to successfully reduce the TPR for the blackbox detector consistently against multiple different algorithms across both the training and testing set. From this, it can be observed that neither the discriminator nor the generator became over fitted against the training set. This infers that the Graph based MalGAN would consistently work against new datasets and against multiple algorithms implying robustness against change in the model.

Table 3. TPR for each algorithm

	TPR for Blackbox Detector					
	Original malware		Original MalGAN		Graph based MalGAN	
	Training set	Testing set	Training set	Testing set	Training set	Testing set
MLP	93.00%	94.00%	0.00%	0.00%	1.90%	2.70%
RF	98.00%	100.00%	0.20%	0.19%	0.00%	1.00%
LR	97.00%	96.00%	0.00%	0.00%	0.80%	2.00%

MLP. The progress for how the TPR changed over training was mapped to a graph for easy visualisation. It can be observed that for the algorithm MLP, the TPR had some spikes where it increased, but overall trended towards zero. The spiked increases may be caused by low variance within the dataset, and the randomness at which the noise is applied (Fig. 3).

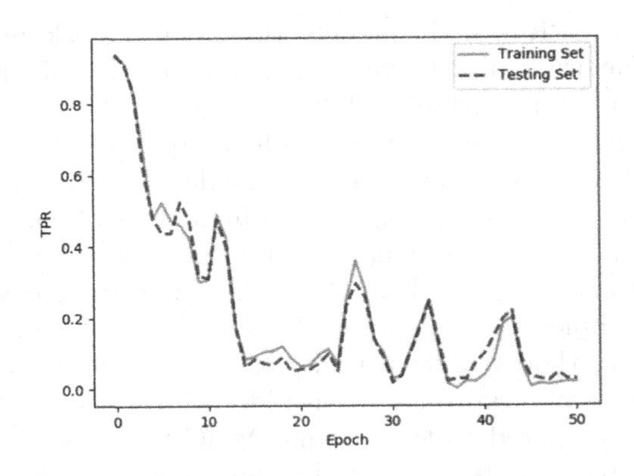

Fig. 3. TPR progress Graph for MLP algorithm

RF. The progress for how the TPR changed over training was mapped to a graph for easy visualisation. For algorithm RF, the TPR trended quickly towards zero before spiking and then had a gradual trend towards zero (Fig. 4).

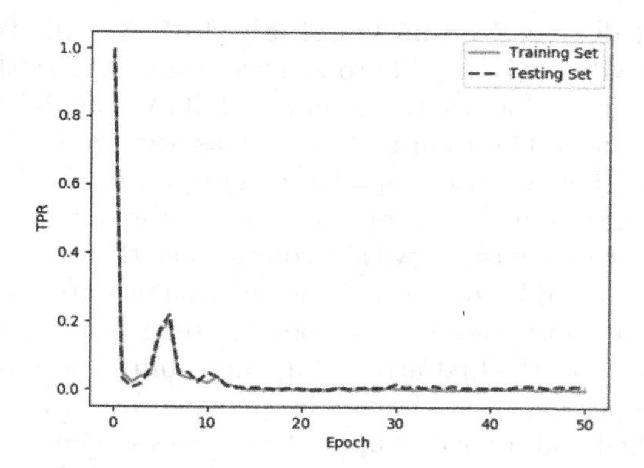

Fig. 4. TPR progress Graph for RF algorithm

LR. The progress for how the TPR changed over training was mapped to a graph for easy visualisation. For algorithm LR, the TPR had a spiky descent before evening out and trending towards zero (Fig. 5).

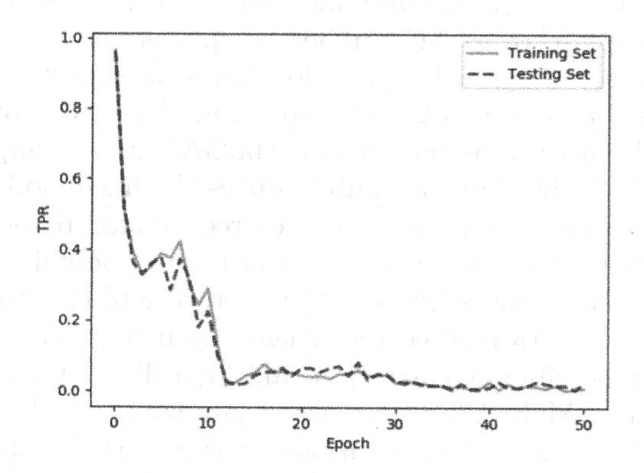

Fig. 5. TPR progress Graph for LR algorithm

4.2 Evaluation

Whilst the model performed well, the input range given by behavioural graphs is very large and makes the generation of adversarial examples challenging as a balance must be struck between generating enough noise to create adversarial examples and not generating so much as to create patterns which belong in neither dataset.

Graph Based MalGAN Versus Original MalGAN in Terms of TPR. Whilst the presented model was able to achieve results of less than two percent detection, this is not as low as the original MalGAN model which produced results all of which were less than one percent as seen in Table 3. It should be noted, however, that behavioural graph based interpretations of malware capture more semantic information than simple feature vectors as they rely on specific patterns of API features existing within code rather than simply the existence of a specific call. Being able to obfuscate adversarial examples against detectors which work at the semantics level of the code is a progressive step forward in the field and as this is one of the first attempt at such obfuscation, the low detection rates of under two percent shows its validity moving forward and that it should be further researched and improved upon. Graph based MalGAN also continues the same robustness as its predecessor as demonstrated by their comparable differences in between algorithms.

Graph Based MalGAN Versus Original MalGAN in Terms of Graphs Mapping TPR over Epochs. Whilst the original MalGAN paper does not show all of their graphs mapping the change in TPR over the number of epochs, those that they do show demonstrate a similar gradient as those presented by the Graph based MalGAN. One of the few differences between them is the rate at which the TPR lowers with the original MalGAN presenting smoother graphs with a gradual decrease over the number of epochs versus the spiky graphs presented in this paper. A possible reason for this is the size of the patterns being obfuscated. Graph based MalGAN uses an input layer size of one-thousand-eight-hundred-and-ten whereas the original MalGAN uses an input layer of one-hundred-and-twenty-eight. This not only reduces the likelihood that the Graph based MalGAN will randomly assign the correct vector to obfuscate a given piece of malware but increases the time it will take to be able to obfuscate the example. It also makes it more likely for the detector to classify an example as malicious unless the generator is correctly learning benign features to add to its examples. Another possible explanation could be a difference in the size of the datasets. The original MalGAN uses a larger data set for both their benign and malicious examples as one of the limitations for this work was finding an equally sized dataset to use. This could result in the detector simply learning the entire data set and classifying any example that does not exist in its original benign set as malicious.

5 Conclusion and Future Work

The work presented in this paper sought to build on the existing work of MalGAN and present a modern take on adversarial malware generation based on similar research found in the field of malware detection. It presented code not only for the feature extraction, ordering, graphing and encoding of malware into a behavioural pattern based model but also a neural network capable of generating similar but deceiving models. A wider range of both malware and benign samples could have been collected for the training of this model and it is impossible to determine if that has impacted the results gathered here.

Moving forward the work outlined in this paper could be utilised to generate completely new malware based on associated behaviours. In theory, if the behavioural graph approach were to be combined with the original MalGAN's feature vector approach and applied to a semi-random walking function it would be possible to define new malicious patterns for code. Undertaking such a task would potentially involve the combination of both a GAN as presented here and some form of auto-encoder system to handle the wide array of potential variables involved in creating such a pattern generator. Future work could also be undertaken to combine sparse auto encoders within the generator model as this could also potentially lead to a generator capable of creating malware with undocumented behaviour and patterns.

References

1. Anderson, B., Quist, D., Neil, J., Storlie, C., Lane, T.: Graph-based malware detection using dynamic analysis. J. Comput. Virol. **7**(4), 247–258 (2011)
2. Andrade, E.D.O.: MC-dataset-binary (2018). https://figshare.com/articles/MC-dataset-binary/5995408/1
3. Babaagba, K.O., Tan, Z., Hart, E.: Nowhere metamorphic malware can hide - a biological evolution inspired detection scheme. In: Wang, G., Bhuiyan, M.Z.A., De Capitani di Vimercati, S., Ren, Y. (eds.) DependSys 2019. CCIS, vol. 1123, pp. 369–382. Springer, Singapore (2019). https://doi.org/10.1007/978-981-15-1304-6_29
4. Babaagba, K.O., Tan, Z., Hart, E.: Automatic generation of adversarial metamorphic malware using MAP-elites. In: Castillo, P.A., Jiménez Laredo, J.L., Fernández de Vega, F. (eds.) EvoApplications 2020. LNCS, vol. 12104, pp. 117–132. Springer, Cham (2020). https://doi.org/10.1007/978-3-030-43722-0_8
5. Bonfante, G., Kaczmarek, M., Marion, J.Y.: Control flow graphs as malware signatures (2007)
6. Breiman, L.: Random forests. Mach. Learn. **45**(1), 5–32 (2001)
7. Cesare, S., Xiang, Y.: Malware variant detection using similarity search over sets of control flow graphs. In: 2011 IEEE 10th International Conference on Trust, Security and Privacy in Computing and Communications, pp. 181–189. IEEE (2011)
8. Cesare, S., Xiang, Y., Zhou, W.: Control flow-based malware variantdetection. IEEE Trans. Dependable Secure Comput. **11**(4), 307–317 (2013)
9. Chuman, T., Sirichotedumrong, W., Kiya, H.: Encryption-then-compression systems using grayscale-based image encryption for jpeg images. IEEE Trans. Inf. Forensics Secur. **14**(6), 1515–1525 (2018)
10. Guo, C., Sablayrolles, A., Jégou, H., Kiela, D.: Gradient-based adversarial attacks against text transformers. arXiv preprint arXiv:2104.13733 (2021)
11. He, R., Li, Y., Wu, X., Song, L., Chai, Z., Wei, X.: Coupled adversarial learning for semi-supervised heterogeneous face recognition. Pattern Recogn. **110**, 107618 (2021)

12. Hoffman, J.I.: Logistic regression, chapter 33. In: Hoffman, J.I. (ed.) Basic Biostatistics for Medical and Biomedical Practitioners, 2nd edn., pp. 581–589. Academic Press (2019)
13. Hu, W., Tan, Y.: Generating adversarial malware examples for black-box attacks based on GAN. arXiv preprint arXiv:1702.05983 (2017)
14. Kawai, M., Ota, K., Dong, M.: Improved MalGAN: avoiding malware detector by leaning cleanware features. In: 2019 International Conference on Artificial Intelligence in Information and Communication (ICAIIC), pp. 040–045 (2019)
15. Maeda, H., Kashiyama, T., Sekimoto, Y., Seto, T., Omata, H.: Generative adversarial network for road damage detection. Comput.-Aided Civil Infrastruct. Eng. **36**(1), 47–60 (2021)
16. Popli, N.K., Girdhar, A.: Behavioural analysis of recent ransomwares and prediction of future attacks by polymorphic and metamorphic ransomware. In: Verma, N., Ghosh, A. (eds.) Computational Intelligence: Theories, Applications and Future Directions-Volume II. AISC, vol. 799, pp. 65–80. Springer, Cham (2019). https://doi.org/10.1007/978-981-13-1135-2_6
17. Saxe, J., Berlin, K.: Deep neural network based malware detection using two dimensional binary program features. In: 2015 10th International Conference on Malicious and Unwanted Software (MALWARE), pp. 11–20 (2015)
18. Singh, J., Singh, J.: A survey on machine learning-based malware detection in executable files. J. Syst. Architect. 101861 (2020)
19. Taud, H., Mas, J.F.: Multilayer perceptron (MLP). In: Camacho Olmedo, M.T., Paegelow, M., Mas, J.-F., Escobar, F. (eds.) Geomatic Approaches for Modeling Land Change Scenarios. LNGC, pp. 451–455. Springer, Cham (2018). https://doi.org/10.1007/978-3-319-60801-3_27

Multi-omic Data Integration and Feature Selection for Survival-Based Patient Stratification via Supervised Concrete Autoencoders

Pedro Henrique da Costa Avelar[1,2(✉)], Roman Laddach[1], Sophia N. Karagiannis[1], Min Wu[2], and Sophia Tsoka[1]

[1] King's College London, London, UK
{pedro_henrique.da_costa_avelar,roman.laddach,sophia.karagiannis, sophia.tsoka}@kcl.ac.uk

[2] Institute for Infocomm Research, A*STAR, Singapore, Singapore
wumin@i2r.a-star.edu.sg

Abstract. Cancer is a complex disease with significant social and economic impact. Advancements in high-throughput molecular assays and the reduced cost for performing high-quality multi-omic measurements have fuelled insights through machine learning. Previous studies have shown promise on using multiple omic layers to predict survival and stratify cancer patients. In this paper, we develop and report a Supervised Autoencoder (SAE) model for survival-based multi-omic integration, which improves upon previous work, as well as a Concrete Supervised Autoencoder model (CSAE) which uses feature selection to jointly reconstruct the input features as well as to predict survival. Our results show that our models either outperform or are on par with some of the most commonly used baselines, while either providing a better survival separation (SAE) or being more interpretable (CSAE). Feature selection stability analysis on our models shows a power-law relationship with features commonly associated with survival. The code for this project is available at: https://github.com/phcavelar/coxae.

Keywords: Supervised autoencoders · Concrete autoencoders · Multi-omic integration · Survival prediction · Survival stratification

1 Introduction

Given the rapid advance of high-throughput molecular assays, the reduction in cost for performing such experiments and the joint efforts by the community in producing high-quality datasets with multi-omic measurements available, the integration of these multiple omic layers has become a major focus for precision medicine [4,16]. Such integration involves analysis of clinical data across multiple modalities for each patient, providing a holistic view of underlying mechanisms during development of disease or patient response to treatment.

Methods for multi-omic data integration can be classified as sequential, late, or joint integration approaches, depending on the order of implemented tasks and at what

G. Nicosia et al. (Eds.): LOD 2022, LNCS 13811, pp. 47–61, 2023.
https://doi.org/10.1007/978-3-031-25891-6_5

point multi-omic data is integrated [21]. Specifically, in *sequential* integration each omic layer is analysed in sequence, i.e. one after the other, and has been the method of choice in various early approaches. In *late* integration methods [17,20,22], each omic layer is analysed separately and then results are integrated, which helps capture patterns that are reproducible between different layers, but making the method blind to cross-modal patterns. Finally, in *joint* integration methods [1,6,14,19,21–23] all omics layers are analysed simultaneously, often employing a dimensionality reduction method that maps all layers into a joint latent space that represents all layers [5]. Such methods enable cross-modal pattern analysis to identify how multiple layers interact to affect a biological process.

All of the aforementioned approaches can be linked to particular challenges in the machine learning task. First, many of the datasets are sparse, often with some omics features missing between samples or studies, or even omics layers being unavailable for some patients. Second, molecular assay data are often highly complex, comprising thousands to tens-of-thousands of different features for each modality. Third, even with reduced cost of profiling, availability of data may still be prohibitively expensive and specialised, limiting the number of publicly-available datasets for analysis. Fourth, although experiments might be performed using the same assaying technology and be collected with the same system in mind, there can be varying experimental conditions between datasets [12] and batch effects may be present [11]. Finally, current interest in profiling at the single-cell level has resulted in dramatic dataset size increase, posing challenges to available methodologies.

The benefits of integrated datasets include more accurate patient stratification (e.g. high/low risk), disease classification or prediction of disease progression. All of those may suggest better treatment strategies, resulting in better patient outcomes. Additionally, the combined information may also be used for biomarker identification supporting further research.

In this paper we provide several contributions to the field of survival-based autoencoder (AE) integration methods, which can be summarized in the following points: 1. In Subsect. 3.4, we develop a simpler Supervised Autoencoder (SAE) as an alternative to the HierSAE model [22] for data integration - a method which provides stable and efficient survival separation - and use it as an upper-bound baseline for testing the performance of our Concrete Supervised Autoencoder. 2. In Subsect. 3.5, we propose the Concrete Supervised Autoencoder (CSAE), building on Concrete Autoencoders [2], a method for supervised feature selection, which we apply to survival-based feature selection. 3. In Subsect. 3.3, we provide a more stringent testing framework, compared to what was used in some previous work. Our results are also compared with a standard PCA pipeline as well as the more advanced Maui method [19]. 4. With our testing framework, in Sect. 4, we show that the Concrete Supervised Autoencoder has achieved performance on par with that of more complex baselines, while simultaneously being more interpretable. Also, we provide, to the best of our knowledge, the first feature importance analysis with multiple runs of a model of this type.

2 Related Work

Autoencoders for Multi-omic. After an initial publication in 2018 showing the use of Autoencoders (AEs) for dimensionality reduction in multi-omic datasets [6], there has been significant interest in re-application of this technique in cancer risk stratification, prognostication, and biomarker identification [1,6,14,17,23]. All of the applications of these methods share similar pipelines, and most [1,6,17,23] use the same techniques up to the AE optimisation stage, with only minor differences in hyperparameters and loss functions. Only [14] has a major difference in their AE model, using Adam instead of SGD as the optimiser. Another publication in this vein is [17], which performs late-integration, instead opting to project the input for each omics layer separately, concatenating features from different layers after Cox-PH selection, and using a boosting approach to train and merge multiple models into a single predictor. In these methods AEs are used to perform risk subgroup separation on the whole dataset, which is then used as ground truth for another classification model, whereas in our work the pipeline is cross-validated in its entirety.

In [19] a method was proposed using a Variational Autoencoder (VAE), which is named Maui, to learn reduced-dimensionality fingerprints of multiple omics layers for colorectal cancer types, showing that their method both mapped most samples into the existing subtypes correctly and identified more nuanced subtypes, while still keeping a level of interpretability by correlating input features with embedding features. The same group expanded their analysis on a pan-cancer study [21], changing their interpretability approach to consider the absolute value of the multiplication of the neural path weights for each input-fingerprint feature pair. This interpretability is one of the many methodological differences that sets these works apart from the aforementioned AE approaches based on the work reported in [6]. These VAE-based works also use the fingerprints to cluster samples into different risk subgroups and hazard regression. The main differences with our proposed framework are the type of AE used (VAE instead of AE) and that our models are supervised with a Cox loss. Furthermore, we also provide a different type of encoding function, used by our Concrete Supervised Autoencoder.

Supervised Autoencoders. One can also perform Cox regression on neural networks [7–9], and this implies that one can add a hazard-predicting neural network block on other neural network architectures, including on an AE fingerprints. Independently from our Cox-SAE model, presented in Subsect. 3.4, two other works developed similar techniques. In [20], the same principles of performing Cox-PH regression on the fingerprints generated by the autoencoder were used. However, the main difference is that the integration is done on the fingerprint-level, i.e. dimensionality reduction is performed through the autoencoder as normal, and then the fingerprints are concatenated to perform Cox-PH regression, or cross-omics decoding is implemented with the Cox-PH loss being calculated on the average of both generated fingerprints. However, these cases are limited to 2-omics integration. A recent paper also improved on this idea by proposing an autoencoder which reconstructs the concatenation of the fingerprints generated through other encoders [22] while performing integration on 6 omics layers and clinical data.

Concrete Autoencoders. Recently, the efficacy of using a concrete selection layer as the encoder of an autoencoder was reported [2], referred there as Concrete Autoencoder. They provided tests with different feature types including gene expression, using the Concrete AE as an alternative to the "943 landmark genes" [13], as well as mouse model protein expression levels. A concrete selection layer [15] is an end-to-end differentiable feature selection method, that uses the reparametrisation trick [10] to provide a continuous approximation of discrete random variables, which Balin et al. used in its autoencoder model with an exponentially decreasing temperature during training to provide a smooth transition from random feature selection to discrete feature selection [2].

3 Methods

3.1 Datasets

Our models were tested on open high-quality cancer data derived from TCGA with multiple omics layers and survival information. This data has been used as a baseline testing dataset in various recent studies [6, 17, 19–22]. Here, we employ datasets as preprocessed by [22], described in Table 1. However, we use our own set of splits for cross validation, since we perform 10-fold cross validation with 10 repeats, as compared to 2 repeats of 5-fold cross validation in the previous study.

Table 1. Number of features in each of the used TCGA datasets. The "Used" column indicates how many features we expect the models to use after the second pipeline step, which involved selecting the top 1000 features for each omics layer. All datasets were used as preprocessed and made available by [22].

Dataset	Samples	Clinical	GEx	CNV	Methylation	μRNA	Mutation	RPPA	Total	Used
BLCA	325	9	20225	24776	22124	740	16317	189	84380	4938
BRCA	765	9	20227	24776	19371	737	15358	190	80668	4936
COAD	284	16	17507	24776	21424	740	17569	189	82221	4945
ESCA	118	17	19076	24776	21941	737	9012	193	75752	4947
HNSC	201	16	20169	24776	21647	735	11752	191	79286	4942
KIRC	309	14	20230	24776	19456	735	9252	189	74652	4938
KIRP	199	5	20178	24776	21921	738	8486	190	76294	4933
LGG	395	15	20209	24776	21564	740	10760	190	78254	4945
LIHC	157	3	20078	24776	21739	742	8719	190	76247	4935
LUAD	338	11	20165	24776	21059	739	16060	189	82999	4939
LUSC	280	20	20232	24776	20659	739	15510	189	82125	4948
OV	161	17	19064	24776	19639	731	8347	189	72763	4937
PAAD	100	26	19932	24776	21586	732	9412	190	76654	4948
SARC	190	45	20206	24776	21724	739	8385	193	76068	4977
SKCM	238	3	20179	24776	21635	741	17731	189	85254	4933
STAD	304	7	16765	24776	21506	743	16870	193	80860	4943
UCEC	392	24	17507	24776	21692	743	19199	189	84130	4956

3.2 Evaluation and Metrics

Concordance Index. The most commonly used quantitative metric for both survival regression and stratification is the concordance index (C-index), which can be seen as a generalisation of the AUC metric for regression, with a C-index of 0 representing perfect anti-concordance, 1 representing perfect concordance, and 0.5 being the expected result from random predictions. The metric is calculated by analysing the number of times a set of model predictions $f(x_i) > f(x_j)$ given that $y_i > y_j$, while also handling censored data, since if y_j is censored one cannot be certain that y_i is in fact greater than y_j. That is, given a set of features X to which a function f is applied to, and the ground-truth consisting of both the set event occurrences E as well as the drop-out times Y, we would define the metric as:

$$\text{concordance\_index}(f(X),Y,E) = \frac{\text{correct\_pairs}(f(X),Y,E) + \frac{\text{tied\_pairs}(f(X),Y,E)}{2}}{\text{admissable\_pairs}(f(X),Y,E)} \tag{1}$$

where $\text{CI}(f(X),Y,E)$ is the concordance index, $\text{CP}(f(X),Y,E)$ the number of correct pairs, $\text{TP}(f(X),Y,E)$ the number of tied pairs, and $\text{AP}(f(X),Y,E)$ the number of admissible pairs. An admissible pair is one that both events were observed or where a single event e_i was observed and $y_i \leq y_j$. The number of correct and tied pairs are taken only from the admissible pairs. To the best of our knowledge, none of the related work has the entire pipeline validated as we show here, with most related work normalising the entire dataset prior to application of the pipeline [6,19,20,22]

Qualitative Analysis. Another way to qualitatively validate the results of the model is by analysing whether patient stratification yields well-separated survival curves. This task entails fitting Kaplan-Meier (KM) curves for each subgroup identified by the model and has been done in many recent studies, where KM curves are reported, accompanied by the logrank p values for the subgroup separation [6,17,19,21]. It is noted that instead of using all samples (training and testing) to build the KM curves, for a quantitative analysis it would be more appropriate if only testing samples are included.

3.3 Main Testing Pipeline

We followed the common and well-established practice of cross validation of the entire pipeline. During preliminary testing, performing scaling only on training samples versus on the entire dataset resulted in drastic performance change in terms of c-index. Also, analysing logrank p-values and concordance indexes on the whole dataset generally means that the logrank p values will be significant simply due to the training dataset being larger than the test dataset, thus skewing the results towards already-seen data. The testing framework we adopt solves both these issues, and our testing framework is one of the key methodological contributions of our model compared to previous work [1,6,14,17,19,21,23], which either provide external validation through other cohorts, validate piecewise, or both.

Our methodology is comprised of 6 steps, shown with letters A to F in Fig. 1: (A) First feature selection is applied to each omic layer by selecting the k most variable features, as was done in previous work [17,19–21]; (B) Feature scaling is then performed by calculating the z-score of the selected features with, different to most related work,

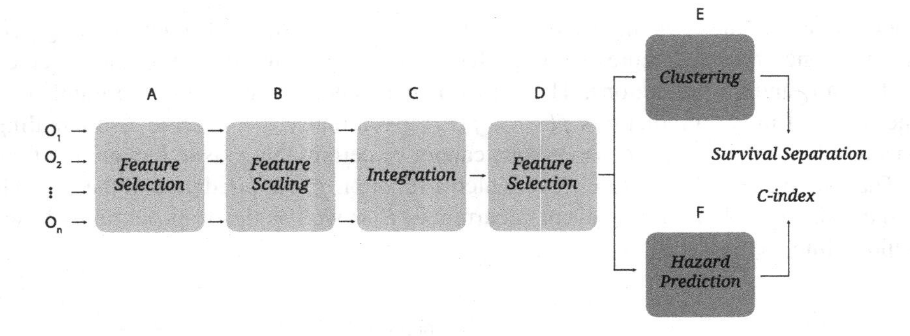

Fig. 1. A diagram showing the pipeline used in our testing framework.

only the samples available during training; (C) Then, feature compression (i.e., integration, selection, or linear combination) is performed through the the relevant model; (D) The following step involves performing Cox-PH univariate feature selection to select those features that are considered relevant for survival; (E) 2-clustering (with a clustering algorithm that allows for clustering new inputs) on this integrated dataset for survival subgroup separation, which can be used for differential expression analyses; (F) And performing hazard prediction, which can be used to rank patients with regards to survival probability, and is also the endpoint from which we calculate C-indexes and evaluate the stratification outcome.

Experiments were run on a HPC cluster, creating a single process for each dataset and using the same consistent seeds across algorithm repetitions to ensure the same cross validation splits were used for all models. We used 10 repetitions of 10-fold cross validation for all datasets except for BRCA and STAD, where 5 repetitions of 10-fold cross validation were applied due to limitations in our compute budget. For the BRCA, STAD and KIRP datasets the experiment used 16000MB of available RAM memory, and for all other datasets 8000MB of RAM was made available. All experiments were run with 16 cores available to the program.

All models use the following: Fig. 1: (A) $k \leq 1000$ for feature selection on each omics layer (see "Used" column in Table 1 for the total number of features); (B) feature scaling using the mean and standard deviation in the training dataset for each fold; (C) A default of 128 target "fingerprint" features for all models and, on traditional autoencoder-based models, 512 neurons on the hidden layer for roughly 10x, and then a further 5x compression on the input feature size. The Maui model is trained for 400 epochs (taken as a default from their codebase) whereas our models are trained for 256 epochs, with the Adam optimiser using 0.01 as a learning rate and 0.001 as the l2 normalisation weight. All AE models implemented by us use 0.3 dropout rate and a gaussian noise with zero mean and 0.2 standard deviation added to input features during training. We use rectified linear units as a nonlinearity on all intermediate layers. We use the same temperature settings provided in [2] for our concrete selection layers, starting with a temperature of 10 and ending with a temperature of 0.1; (D) Cox-PH univariate "fingerprint" feature selection, with a significance threshold of $p < 0.05$, using all the of the fingerprints if no fingerprint is identified as significant for survival; (E) 2-clustering with KMeans with 10 initialisations, using the best in terms of inertia, with a maximum of 300 iterations and a tolerance of 0.001; (F) Non-penalised Cox-Regression, unless the model failed to converge, in which case we use Cox-Regression with 0.1 penalisation. Furthermore, if a model fails to run on any fold, that value was dropped. Since we perform 10 repetitions of the 10-fold cross validation, this means that models have at least some results for each dataset, except for the PCA model, which failed to produce any results on two datasets due to convergence problems.

3.4 Cox-Supervised Autoencoder

Many methods available in the literature report using autoencoders to perform dimensionality reduction and then select survival-relevant features from the autoencoder fingerprints through Cox-PH regression [1,6,14,17,19,21,23]. We argue that the assumption contained within an autoencoder's loss function is insufficient to provide features relevant for survival, since feature combinations that are good at predicting the input features might not necessarily be good for survival prediction. This has support on our preliminary tests where non-trained models performed just as well as trained models. To solve this issue, we propose our independently developed methodology of a Cox-Supervised Autoencoder (SAE).

To address the lack of inductive bias towards survival, we introduce a Cox-PH model within the neural network as a normalising loss, so that the model learns not only to create codes which are efficient at reconstructing the input, but also codes that are indicative of survival. Since a Cox-PH model is end-to-end differentiable, this is easily done by simply adding a Cox-PH model on the generated fingerprints, and performing Cox-PH regression with regards to the input survival times and observed events. A schematic overview of such a model is shown in Fig. 2.

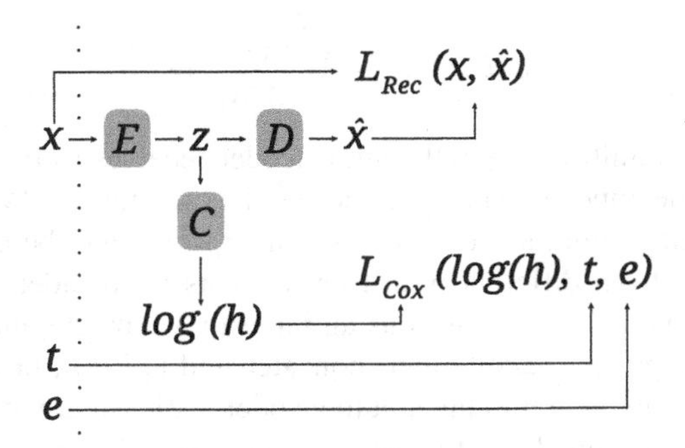

Fig. 2. Diagrammatic overview of our Cox-Supervised Autoencoder (SAE) model. The multi-omic input x generates the encoding z, which is used to both provide the reconstruction $\hat{x}$ as well as to calculate the log hazards $log(h)$ using another neural module, which is trained using a Cox-PH loss alongside the time and event information, t and e.

In mathematical terms, then, we would have a model composed of the three neural network blocks shown in Fig. 2: An encoder E, which takes the input x and produces a set of fingerprints z from which a decoder D produces a reconstruction of the input $\hat{x}$, and finally a linear layer C which uses the same fingerprints z to produce a log-hazard estimate $log(h)$ for each patient. We then perform gradient descent on a loss L which is composed not only of a reconstruction loss $L_{Rec}(x,\hat{x})$ and a normalisation loss $L_{norm}(E,D,C)$ on the weights of E, D, and C, but also the Cox-PH regression loss $L_{Cox}(log(h),t,e)$ using the information available about the survival time t and the binary information of whether the event was censored or not e, yielding the equation below:

$$L(x,t,e,E,D,C) = L_{rec}(x,D(E(x))) + L_{cox}(C(E(x)),t,e) + L_{norm}(E,D,C) \quad (2)$$

Here we use the Cox-PH loss from [9] as implemented by the pycox library (https://github.com/havakv/pycox) version 0.2.3, which, assuming that x, t, and e are sorted on t, and that we have k examples, is defined as:

$$\begin{aligned} L_{cox}(log(h)),t,e) &= -\frac{\sum_{1\leq i\leq k} e_i \cdot (log(h_i) - log(g_i) + \gamma)}{\sum_{1\leq i\leq k} e_i}, \\ g_i &= \varepsilon + \sum_{1\leq j\leq i} e^{log(h_j)-\gamma}, \\ \gamma &= max(log(h)) \end{aligned} \quad (3)$$

For the reconstruction, the Mean Square Error (MSE) loss is employed due to its symmetry, as below:

$$L_{rec}(x,\hat{x}) = \frac{\sum 1 \leq i \leq k \sum 1 \leq j \leq d (x_{i,j} - \hat{x_{i,j}})^2}{k} \quad (4)$$

And, finally, we use the L2 norm of the model weights as our normalisation loss, using the Frobenius norm $||\cdot||_F$ and a hyperparameter λ to control the how much of the norm is applied:

$$L_{norm}(E,D,C) = \lambda \sum_{w\in E,D,C} ||w||_F^2 \quad (5)$$

Note that in our definitions here, the autoencoder receives as input all of the omics layers at the same time, much like many models in the literature [6, 19, 21], and we argue that this provides integration, since all of the fingerprints may be composed of combinations of features from different omics levels. This is considerably different from models where each omics layer is used as an input to a separate autoencoder, and the autoencoders' "fingerprints" being either: concatenated [17,20]; or combined through pooling [22]; or through an hierarchical autoencoder [22]. We argue that only the hierarchical autoencoder would de-facto integrate the omics layers, since it is combining the separately-compressed layer fingerprints, albeit through a much more complex procedure.

3.5 Concrete Supervised Autoencoder

Another possible point of concern for many multi-omic analysis pipelines is that using neural-network-based models can lead to less interpretable results. To address this, we use the Concrete Autoencoder [2] to build a Concrete Supervised Autoencoder (CSAE), using the same concrete selection layer and reparametrization tricks as described previously [2, 10, 15]. A diagramatic representation of a concrete selection layer and the Gumbel distribution used in its training is shown in Fig. 3.

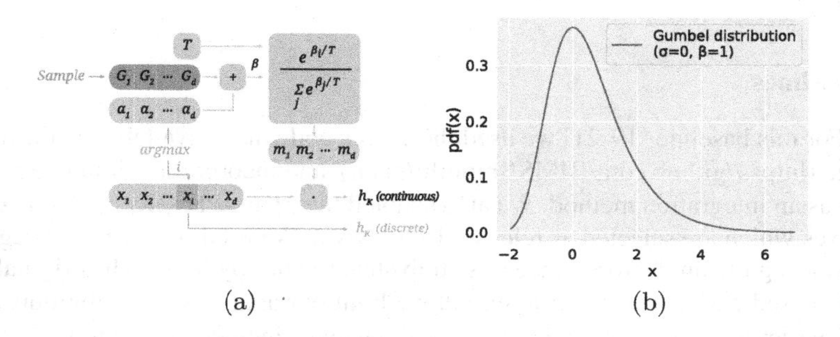

Fig. 3. A diagrammatic Concrete Selection Layer's neuron (a), where blue elements represent continuous stochastic nodes, yellow elements are continuous and deterministic, and brown elements show the discrete selection path, generally followed after training. The values of G_i are sampled from the Gumbel distribution, whose PDF we can see on (b). (Color figure online)

Thus, our encoder follows the same exponential decay rate and model reparametrisation described previously [2], to give us a sampling in d dimensions with parameters $\alpha \in \mathscr{R}^d, \alpha_i > 0 \forall \alpha_i$, regulated by a temperature $T(b)$, and with d values $g \in \mathscr{R}^d$ sampled from a Gumbel distribution, giving us a probability distribution $m \in \mathscr{R}^d$:

$$g_j = \frac{e^{log(\alpha_j + g_j)/T(b)}}{\sum_{k=1}^{d} e^{log(\alpha_k + g_k)/T(b)}} \tag{6}$$

During training, each element outputted by a concrete selection "neuron" i, would be a linear combination of the input $x \in \mathscr{R}^d$:

$$E(x)_i = x_i * g_j \tag{7}$$

But as the temperature $T(b) \to 0$, we have that each node will select only a single input, at which point we can switch from the linear combination to simple indexing:

$$E(x)_i = \arg\max_j \alpha_j \tag{8}$$

This allows us to smoothly transition for feature combination and also to reduce evaluation-time memory requirements, since the indexing takes $O(1)$ space per neuron instead of $O(d)$ of the linear approximation. To perform such smooth transition, we use the previously defined exponential temperature decay schedule, with T_0 being the initial temperature, T_B the minimum temperature, b being the current epoch, and B the maximum number of epochs:

$$T(b) = T_0(T_B/T_0)^{b/B} \tag{9}$$

Having our encoder thus defined, we simply apply the same optimisation as in Eq. 2, replacing the traditional pyramidal MLP used as the encoder for the Autoencoder with our concrete selection layer, and performing both input reconstruction with the decoder reverse-pyramidal decoder D as well as hazard prediction through the hazard prediction network C.

3.6 Baselines

Maui. For this baseline [19,21] we used the source code made available by the authors on github (https://github.com/BIMSBbioinfo/maui), and incorporated it into our testing pipeline as an integration method. We also adapted their Cox-PH selection code to save the indexes which are selected as relevant for survival. We used our own Cox-regressor class on the significant factors, since it is equivalent to ones by Maui. The original Maui paper also used KMeans as a clusterer, but we limit our analyses to 2 subgroups, since this is a harder test for the model. Other small changes include changing their code to work with a more recent version of Keras and Tensorflow. Note that, although we are using the Maui original code or adaptations of their code (where the original code does not store a model for later use), our pipeline is drastically different to Maui and more stringent, which might lead to the different performance reported here.

Autoencoder + Cox-PH. Our AE baseline could be seen as a rough equivalent of [14], and is the base on which our Supervised Autoencoder model was built. The model that we've first seen following a similar approach [6] uses a SGD optimiser instead of an Adam optimiser, and does so for a small amount of training epochs, which in our initial testing proved to be equivalent to not training the model at all, and might be seen as a form of random projections, as in [3]. Another approach used both omics-specific autoencoders (thus not performing cross-omics combinations in the fingerprints) and makes heavy use of boosting to improve the models' joint performance, while still using the SGD optimiser [17]. Using code available on the model's github page (https://github.com/lanagarmire/DeepProg), we attested to the model providing similar outputs when not trained and when trained with the default number of epochs, most likely due to the use of SGD as an optimiser, which again makes the model interpretable as a form of random projection [3]. These models were not included in our final comparison due to the abovementioned methodological differences.

PCA + Cox-PH. Our main baseline is PCA, which has been thoroughly used as a baseline in other papers, including [2] as an upper-bound for the reconstruction loss of their CAE model. For the PCA baseline the first d principal components are used as a drop-in equivalent for the d fingerprint nodes in any of the autoencoder-based models.

4 Results

Concordance Index Analysis. Using our stringent testing pipeline we evaluated all the aforementioned models in predicting same-cancer, out-of-sample instances through the cross-validation scheme. This is due to the fact that normalisation was calibrated using only training data, which is different, for example, to the methodology previously used for MAUI [19], where normalisation is done on all samples before validation, and the impact for this can be clearly seen on Fig. 4, where we see that the concordance indexes reported in the original paper are higher than the ones we've encountered with this difference, something which was also noticed during our preliminary studies.

Also in Fig. 4, we can see that our models either perform better than a PCA-based model or are not significantly different from the traditional PCA pipeline. In fact, Maui was the model with the worst average rank (4.88), whereas the CSAE ranked equal to the PCA model at 2.94, only slightly worse than an AE model without Cox supervision with 2.53 average rank, with the Cox-supervised autoencoder having 1.82 average rank.

With regards to overall Concordance-index, the previous ordering remains very much the same, with the SAE being the best overall with an average test score of 0.632 (all p-values statistically significant $< 10^{-4}$ through an independent two-sided t-test), with the AE model having an average score of 0.610 (statistically significant difference $p < 0.007$ to the Maui and PCA models), the CSAE with an average score of 0.603 (statistically significant difference $p < 10^{-46}$ to the Maui model), and the Maui model having an average test C-index of 0.526.

These results lead to the conclusion that joint supervision on both the Cox and reconstruction objectives improves out-of-sample performance in autoencoder based models. This was already somewhat attested to previously [20,22], but none of these studies performed joint integration directly, using one autoencoder per omics layer [20], which can be argued to not amount to integration at all, since the omics layers do not cross-contribute to the generated fingerprints, or having the generated fingerprints be de-facto integrated only through a two-step hierarchical process [22]. Here we attest that models that do perform this direct multi-omic integration, such as [1,6,14,19,21,23] could greatly benefit from Cox Supervision as a joint optimisation loss.

SKCM Analysis. We can see that the Cox-Supervised Autoencoder fingerprints for the SKCM dataset provide a very clear separation in terms of survival outcomes. In Fig. 5a, we see the 2-clustering results on the survival-relevant fingerprints for this model, which clearly separates a group of high-risk patients and a group of low-risk patients, despite it being 3rd place in terms of average C-index (only statistically significantly worse to PCA $p = 0.02333$) provided a better separation on the Kaplan-Meier analysis than what was previously reported with whole-dataset Kaplan-Meier plots in previous works (e.g., [17]). The Concrete-SAE model, which was 4th place (but also only statistically significantly worse than PCA, $p = 0.001685$), still manages an adequate survival stratification despite its "fingerprints" consisting of the original features, showing that it can still provide a good survival separation only on the basis of highly-interpretable feature selection (Fig. 5b).

With regards to layer selections, we ran our models 32 times each and analysed the most important feature for each of their fingerprint features. In the case of the SAE

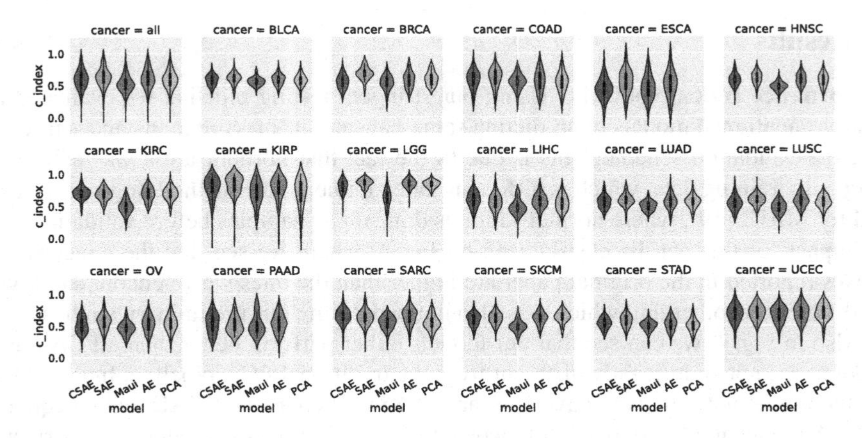

Fig. 4. Violin plot of the Concordance Index (c_index) cross-validation test performance of methods tested across various cancer types, as well as the Violin plot over all results. Even though our CSAE model relies on simpler feature selection instead of multiple nonlinear combinations, it manages to perform on-par with most other baselines.

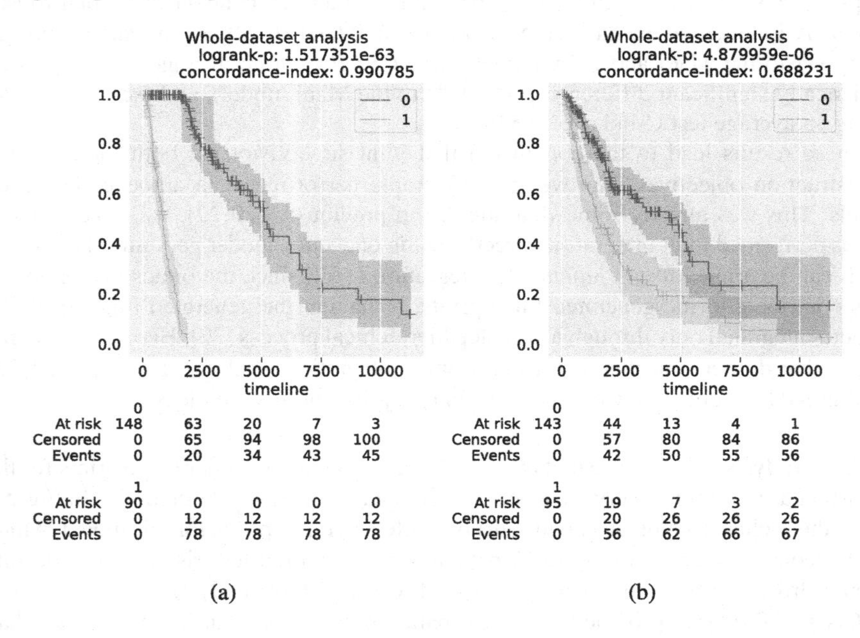

Fig. 5. Survival curve separations for the SKCM dataset using our Supervised Autoencoder model a and Concrete Supervised Autoencoder b

model, possibly due to the fact that each fingerprint consists of a combination of combination of features, one feature consistently outranked the others and thus was used as the most important one with regards to absolute neural path weight [21]. The Concrete SAE model, however, had a richer feature set selection, due to the fact that each fingerprint feature maps directly to an input feature. The feature distributions of the CSAE model can be seen in Fig. 6 and they show a varied selection from multiple omics layers as well as a strong preference for a single clinical factor which is very relevant for survival.

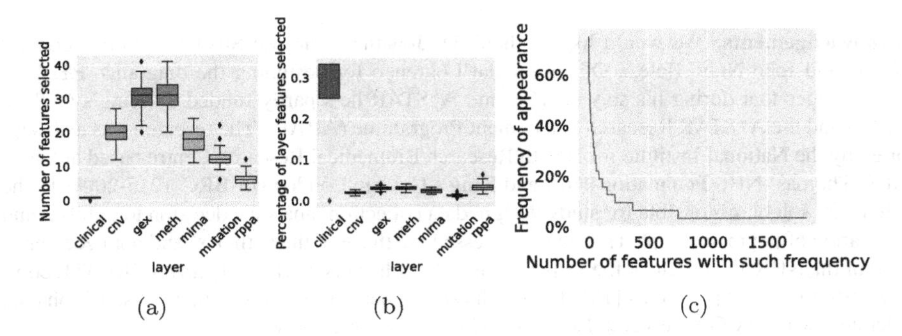

Fig. 6. Layer feature selection distribution for the SKCM dataset. In a we can see how many features were selected per layer, with b showing the same information normalised by omics layer size, and c showing the distribution of features that appear over multiple random initialisations of the algorithm.

5 Discussion

In this paper we proposed two models for multi-omic analysis. Our independently developed Cox-Supervised Autoencoder model, which is conceptually simpler than previously described models which also attempt survival-based multi-omic analysis [20,22], proved to be very efficient, while providing true integration in the same sense adopted in previous work [1,6,14,19,21,23]. Our Concrete Cox-Supervised Autoencoder model, which performs multi-omic feature selection, instead, also proved to be a strong alternative for cases where interpretability is more favourable than expressive power. It is more interpretable than and as powerful as the PCA baseline, whilst not straying too far from its theoretical maximal baseline, our Cox-Supervised Autoencoder model.

Our proposed models, however, are not a one-size-fits-all solution to survival-based multi-omic integration and feature selection challenges. Although one of our models ranked at least first or second with regards to survival separation on all but one dataset, the Concrete Supervised Autoencoder model is not expressive enough to capture cross-omics relationships due to its simple feature selection method, and our Supervised Autoencoder model might still be less expressive than its more complicated counterpart, the Hierarchical Supervised Autoencoder [22], a comparison which we left for future work.

We believe that our Cox-Supervised Autoencoder model presented here provides a clear path forward, with a simple method for survival-based multi-omic integration, which can be further enriched with multitasking in its supervision, possibly also integrating drug responses, which can then lead to further applications in drug discovery.

Our Concrete Cox-Supervised Autoencoder model also makes use of recent advances [2, 15] to provide an end-to-end differentiable feature selection, whose ramifications can range from finding specific sets of omics features that map to tasks other than survival, allowing us to leverage the power of differentiable programming techniques to discover new relationships in molecular assay datasets.

Acknowledgements. We would like to thank Dr Jonathan Cardoso-Silva for fruitful conversations, and João Nuno Beleza Oliveira Vidal Lourenço for designing the diagrams. P.H.C.A. acknowledges that during his stay at KCL and A*STAR he's partly funded by King's College London and the A*STAR Research Attachment Programme (ARAP). The research was also supported by the National Institute for Health Research Biomedical Research Centre based at Guy's and St Thomas' NHS Foundation Trust and King's College London (IS-BRC-1215-20006). The authors are solely responsible for study design, data collection, analysis, decision to publish, and preparation of the manuscript. The views expressed are those of the author(s) and not necessarily those of the NHS, the NIHR or the Department of Health. This work used King's CREATE compute cluster for its experiments [18]. The results shown here are in whole or part based upon data generated by the TCGA Research Network: https://www.cancer.gov/tcga.

References

1. Asada, K., et al.: Uncovering prognosis-related genes and pathways by multi-omics analysis in lung cancer. Biomolecules **10**(4), 524 (2020)
2. Balın, M.F., Abid, A., Zou, J.: Concrete autoencoders: differentiable feature selection and reconstruction. In: International Conference on Machine Learning, pp. 444–453. PMLR (2019)
3. Bingham, E., Mannila, H.: Random projection in dimensionality reduction: applications to image and text data. In: Proceedings of the 7th ACM SIGKDD, KDD 2001, pp. 245–250. Association for Computing Machinery, New York (2001). https://doi.org/10.1145/502512.502546
4. Bode, A.M., Dong, Z.: Precision oncology-the future of personalized cancer medicine? NPJ Precis. Oncol. **1**(1), 1–2 (2017). https://doi.org/10.1038/s41698-017-0010-5
5. Cantini, L., et al.: Benchmarking joint multi-omics dimensionality reduction approaches for the study of cancer. Nat. Commun. **12**(1), 1–12 (2021)
6. Chaudhary, K., Poirion, O.B., Lu, L., Garmire, L.X.: Deep learning-based multi-omics integration robustly predicts survival in liver cancer. Clin. Can. Res. **24**(6), 1248–1259 (2018)
7. Ching, T., Zhu, X., Garmire, L.X.: Cox-nnet: an artificial neural network method for prognosis prediction of high-throughput omics data. PLoS Comput. Biol. **14**(4), e1006076 (2018)
8. Huang, Z., et al.: SALMON: survival analysis learning with multi-omics neural networks on breast cancer. Front. Genet. **10**, 166 (2019)
9. Katzman, J.L., Shaham, U., Cloninger, A., Bates, J., Jiang, T., Kluger, Y.: DeepSurv: personalized treatment recommender system using a cox proportional hazards deep neural network. BMC Med. Res. Methodol. **18**(1), 1–12 (2018)
10. Kingma, D.P., Welling, M.: Auto-encoding variational bayes. In: Bengio, Y., LeCun, Y. (eds.) 2nd ICLR, Banff, AB, Canada, 14–16 April 2014, Conference Track Proceedings (2014)
11. Koch, C.M., et al.: A beginner's guide to analysis of RNA sequencing data. Am. J. Respir. Cell Mol. Biol. **59**(2), 145–157 (2018)
12. Korsunsky, I., et al.: Fast, sensitive and accurate integration of single-cell data with harmony. Nat. Methods **16**(12), 1289–1296 (2019)

13. Lamb, J., et al.: The connectivity map: using gene-expression signatures to connect small molecules, genes, and disease. Science **313**(5795), 1929–1935 (2006). https://doi.org/10.1126/science.1132939
14. Lee, T.Y., Huang, K.Y., Chuang, C.H., Lee, C.Y., Chang, T.H.: Incorporating deep learning and multi-omics autoencoding for analysis of lung adenocarcinoma prognostication. Comput. Biol. Chem. **87**, 107277 (2020)
15. Maddison, C.J., Mnih, A., Teh, Y.W.: The concrete distribution: a continuous relaxation of discrete random variables. arXiv:1611.00712 [cs, stat] (2017)
16. Nicora, G., Vitali, F., Dagliati, A., Geifman, N., Bellazzi, R.: Integrated multi-omics analyses in oncology: a review of machine learning methods and tools. Front. Oncol. **10**, 1030 (2020)
17. Poirion, O.B., Jing, Z., Chaudhary, K., Huang, S., Garmire, L.X.: DeepProg: an ensemble of deep-learning and machine-learning models for prognosis prediction using multi-omics data. Genome Med. **13**(1), 1–15 (2021)
18. King's College London e Research Team: King's Computational Research, Engineering and Technology Environment (CREATE) (2022). https://doi.org/10.18742/RNVF-M076. https://docs.er.kcl.ac.uk/
19. Ronen, J., Hayat, S., Akalin, A.: Evaluation of colorectal cancer subtypes and cell lines using deep learning. Life Sci. Alliance **2**(6) (2019)
20. Tong, L., Mitchel, J., Chatlin, K., Wang, M.D.: Deep learning based feature-level integration of multi-omics data for breast cancer patients survival analysis. BMC Med. Inform. Decis. Mak. **20**(1), 225 (2020). https://doi.org/10.1186/s12911-020-01225-8
21. Uyar, B., Ronen, J., Franke, V., Gargiulo, G., Akalin, A.: Multi-omics and deep learning provide a multifaceted view of cancer. bioRxiv (2021)
22. Wissel, D., Rowson, D., Boeva, V.: Hierarchical autoencoder-based integration improves performance in multi-omics cancer survival models through soft modality selection. Technical report, bioRxiv (2022). https://doi.org/10.1101/2021.09.16.460589. Section: New Results Type: article
23. Zhang, L., et al.: Deep learning-based multi-omics data integration reveals two prognostic subtypes in high-risk neuroblastoma. Front. Genet. **9**, 477 (2018)

Smart Pruning of Deep Neural Networks Using Curve Fitting and Evolution of Weights

Ashhadul Islam(✉) and Samir Brahim Belhaouari

Hamad Bin Khalifa University, Doha, Qatar
{aislam,sbelhaouari}@hbku.edu.qa
https://www.hbku.edu.qa/en

Abstract. Compression of the deep neural networks is a critical problem area when it comes to enhancing the capability of embedded devices. As deep neural networks are space and compute-intensive, they are generally unsuitable for use in edge devices and thereby lose their ubiquity. This paper discusses novel methods of neural network pruning, making them lighter, faster, and immune to noise and over-fitting without compromising the accuracy of the models. It poses two questions about the accepted methods of pruning and proffers two new strategies - evolution of weights and smart pruning to compress the deep neural networks better. These methods are then compared with the standard pruning mechanism on benchmark data sets to establish their efficiency. The code is made available online for public use.

Keywords: Pruning · Deep neural networks · Forward optimization · Explainability · Weight manipulation

1 Introduction

Over time, Neural networks have evolved from a single neuron(perceptron [15]) to Artificial Neural Networks(multi layer perceptron) [9,17], followed by Convolution Neural Networks(CNN) [8,13,16,19] and then moving on to Recurrent Neural Networks [14]. The elementary unit of a neural network is a neuron (filter in case of CNN) which consists of weights. Considering a deep neural network tasked with classification among different categories of data, as the depth and density of the network increases, so does the number of neurons and consequently the number of weights in the network. The training process is nothing but setting values to these weights to create a good discriminator between the data points of different classes. Once the training is complete, a standard method of reducing the network size is by pruning the weights with low values to 0 [10]. The technique consists of repeated pruning attempts followed by a retraining step to ensure that the model's accuracy does not deteriorate. This method is widely acclaimed and has resulted in considerable reduction of the size of the networks,

G. Nicosia et al. (Eds.): LOD 2022, LNCS 13811, pp. 62–76, 2023.
https://doi.org/10.1007/978-3-031-25891-6_6

often surpassing the accuracy level of an un-pruned network. This method is the basis for many different kinds of weight-based pruning algorithms. This paper takes a critical look at the established methods of pruning, putting across some fundamental questions and proposing a modified approach to better prune neural networks.

2 Related Work

Some of the common techniques of neural network pruning include applying optimal brain damage [12] and optimal brain surgery [2] on individual layers. As this involves 2_{nd} order differential, this method is applied layer-wise, keeping the computation feasible [6]. Another method is the Net-Trim, where an optimization scheme is proposed for each layer of the network [1]. This scheme enforces weight sparsity while maintaining the consistency of results between the layers. A generic algorithm is proposed by [4] which alternates between optimizing the weights and pruning them in a data-independent manner. The algorithm explores subsets of weights instead of considering the entire network, so as to reduce the cost of computation. The paper by [22] attempts to combat the uncertainty involved in compressing a neural network. As there is no prior knowledge about the number of iterations required to compress the model they break the non-convex optimization problem into two sub-problems that are solved in an iterative manner either using stochastic gradient descent or analytically.

The work by [23] in ascertaining the efficacy of the pruning method has shown that compressing deep neural networks by removing weights can improve the neural networks' performance by reducing size, decreasing the computation time, and alleviating the problem of overfitting in highly parameterized networks. These results have encouraged researchers to undertake different approaches to optimizing neural networks. Besides weights in fully connected layers, filters in CNN layers have also been subjected to pruning [21]. Another noteworthy innovation is the Lottery ticket hypothesis [7] which claims that the essence of a vast neural network could be found to be a sub network that provides the same accuracy given by the complete network. With an increase in the number of CNNs, recent work has focused more on reducing the filter size to reduce the network size. The methods proposed in this paper is applicable to weights of a neuron in a fully connected layer as well as the filters in a CNN layer. However the crux of the method is grasped by re-thinking how pruning is done and by questioning the methodologies involved.

3 Re-thinking Pruning

The standard pruning approach takes a snapshot of the neurons' weights in a neural network at the end of the training process and applies a threshold on the absolute values of the weights, preserving the ones with higher magnitudes. This paper raises two questions regarding the method.

- Is viewing the latest values of the weights the best way to ascertain their importance? Can a weight with a high value become lower with more training epochs and vice versa? Would it be better then, to watch the trend of progression of weight values or the **evolution of weights** as its called in this paper, to get a more holistic idea about the importance of the individual weights?
- During standard pruning, some weights are nullified to 0 and maintained at that value even during re training to ensure fixed rate of compression. We call this hard pruning and ponder if a chance can be given to the weights to attain non zero magnitude even after they have been nullified. We call this method **smart pruning** and discuss it in detail in the following sections.

3.1 Evolution of Weights

Generally the weights of a neural network are randomly set at the beginning. With the progress of training, the values change, often increasing and decreasing at times. Figure 1a shows the change in values of some randomly selected weights over 100 epochs for CIFAR-10 dataset [11]. For some weights, the values rise and then fall again. For some, the values keep on increasing till the end of training. The conventional pruning methods consider the weight values only at the end of the training. The pruning logic and algorithms are applied giving preference to weights that have a higher value at the end of the training process and generally disregarding the weights that have a lower value at the end of all epochs [10].

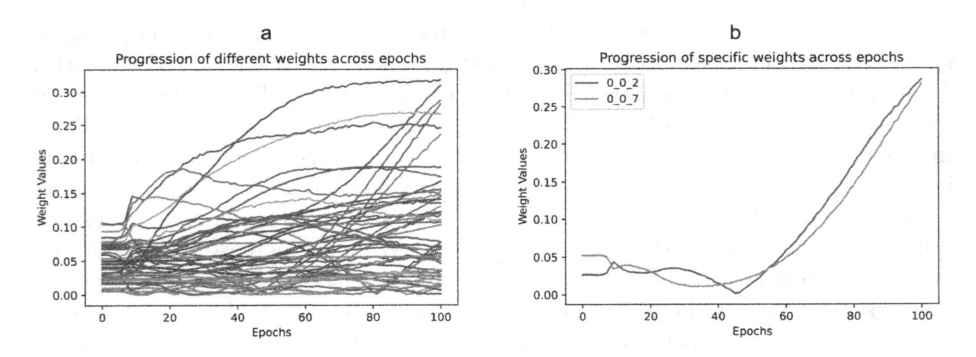

Fig. 1. a. The plot of randomly chosen weights across 100 epochs of training. b. The plot of specific weights across 100 epochs of training. (Color figure online)

Figure 1b shows a specific case where the values of two weights have a similar trajectory. The orange line represents values of the 2nd weight of the 1st neuron in the 1st layer of a Neural network while the blue line represents the 7th weight in the 1st neuron of the 1st layer of the same neural network. They end up with almost the same magnitude, but if we had to compare the two weights on the basis of the traversal of their values, we might give preference to one over the other based on the rise and fall of their magnitudes.

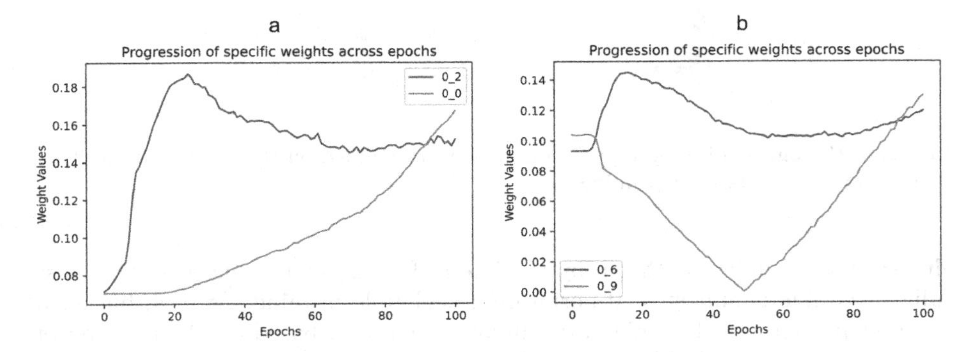

Fig. 2. a. Two weights having similar magnitudes at the end but different rise. b. Two weights having similar magnitudes at the beginning but different pattern of rise (Color figure online)

Figure 2a depicts the magnitudes of the 2nd weight in the 1st neuron and the 1st weight of the 1st neuron of the same network. Although the values are similar at the end the weight marked by the orange line has shown a steady rise while the weight illustrated by the blue line has shown a rise followed by a steady fall with intermittent upward peaks. In this case also, a judicial comparison of the weight evolution may produce a different selection when compared with selection by the final magnitudes. The point is driven home by the final example in Fig. 2b. Two weights are initialized with similar values, follow a radically different trajectory. The weight marked in orange faces a considerable fall before it rises meteorically and surpasses the magnitude of the weight in blue. The weight marked in blue on the other hand, experiences an early rise in value and then a considerable fall. All these examples provide evidence to substantiate the need to consider the evolution of weights across the epochs, instead of considering their final values.

3.2 Smart Pruning

In case of neural network pruning, the standard method prunes or nullifies all weights below a certain threshold by making them 0. It does not modify the values with higher weights. The process is shown in Fig. 3. Figure 3a shows the weights before pruning, post-pruning a certain percentage of weights become 0 while the rest are not modified. In the subsequent fine-tuning or training, the weights which were made 0 are kept nullified by use of masks while the rest are allowed to be modified. This paper proposes an alternate method of pruning called soft pruning. Figure 4 illustrates the proposed method. After dividing the weights into buckets we perform a sweeping pruning across all the buckets at varying proportions. For the weight buckets closer to magnitude 0 we perform

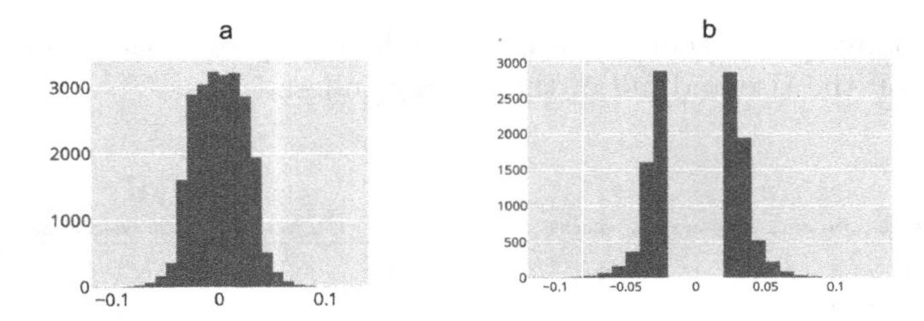

Fig. 3. a. Histogram of weights in a neural network trained on the MNIST dataset. b. Histogram of weights after pruning

almost 100% pruning as shown by the line in Fig. 4. For the buckets at highest values, the line is almost at 0 performing negligible pruning. At each bucket a different percentage of weights are randomly selected, based on the position of the line to be pruned. This results in pruning judiciously with ample chance given to the weights to bounce back into relevance. The selection of the line plays an important role in pruning and is discussed in the following sections.

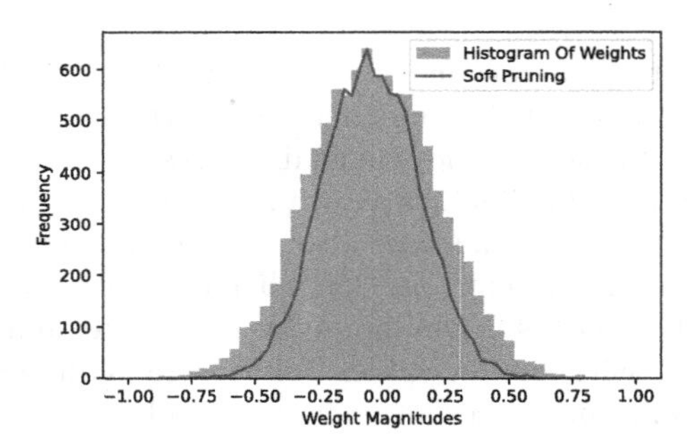

Fig. 4. Sample histogram of weights in a neural network. Line shows the degree of pruning applied at each bucket

4 Proposed Approach

This paper proposes a novel method of pruning by considering the progression of weights over the last few epochs. Instead of considering the final value of the weights, the method calculates a weighted aggregate of the magnitudes giving higher priority to weight values in epochs closer to completion. An example is given in Table 1 where 4 weight values are shown across 3 epochs. The final values of the weights after three epochs are 6, 9, 1 and 3 respectively. If we just go by the magnitude at the end of all epochs, the top 3 weights would be Weight 2 (final magnitude 9), followed by Weight 1 (final magnitude of 6) and weight 4 (final magnitude of 3). Here we recommend an alternate way of looking at the magnitudes. Instead of only considering the final values of the weights we perform a weighted sum of the magnitudes. The weights at 3rd epoch are multiplied by 3, the values at 2nd epoch are multiplied by 2 and so on. In this way, we give priority to weights that are closer to the final epoch without absolutely neglecting the values at the previous epochs. We then aggregate this product and calculate an average importance score. Using this method, the top 3 weights are Weight 1 (Importance score of 14), followed by Weight 2 (Importance score of 13) and Weight 3 (Importance score of 6.66). The calculation is generalized as follows: for each weight or filter i, we persist a vector $val_i = [val_{i1}, val_{i2}, val_{i3}, ...val_{in}]$ where each value in the vector is the magnitude of the weight or filter i at every epoch for n epochs of training. This array is utilized to calculated the weighted importance of the magnitude of any weight or filter i. For example, considering the last k epochs, the weighted importance is calculated using the equation below

$$Imp_i = \frac{\sum_{L=0}^{k} val_{i(n-L)} * (n-L)}{\sum_{L=0}^{k} (n-L)} \tag{1}$$

where value of L ranges from 0 to k, 0 being the immediate final epoch of training and k being the kth epoch going backwards from the end of training. This importance matrix generated from an extended analysis of the value of weights across epochs is then used as a primary reference to understand the importance of weights and prune them accordingly.

Table 1. Aggregation of weight magnitudes across epochs. For each weight we record the value at the end of each epoch. Aggregated importance is the weighted sum of the magnitudes. The last row, Multiplier, shows the value with which the weight magnitudes are multiplied.

Weight number	Epoch 1	Epoch 2	Epoch 3	Aggregated importance
1	8	8	6	14
2	6	3	9	13
3	7	5	1	6.66
4	1	2	3	4.66
Multiplier	X1	X2	X3	

4.1 Pruning Distributions

While the importance scores of the weights discussed in the previous section serve as an analytics tool to measure the significance of individual weights and filters, this paper also proposes a novel method of the pruning operation. The popular methods of network pruning look at the magnitudes of weights of a layer or a neuron and reduce the weights below a specific threshold to 0 [10]. Figure 3a shows a typical weight distribution in the neurons of a neural network. As can be seen the weights generally follow a normal distribution with a high number of weights clustering around the value 0. These are the not-so-important weights, while the weights farther away from 0 in both directions are considered to be crucial. A standard pruning algorithm reduces the negligible weights to 0, retaining the ones with high values. Figure 3b shows the values of the weights after pruning. The weights near 0 have been removed, retaining the higher valued ones. The approach described in this paper deviates from the conventional pruning method by proposing a filter method to weed out weights randomly, allowing them to re-appear even after being pruned at the initial phase. This is performed by superimposing different distributions on the weight histograms and randomly removing the weights that fall under the curve of the filter.

The technique applied in this paper is derived from the Bezier Curve. It is defined by

$$P(t) = \sum_{i=0}^{n} B_i^n(t) \cdot P_i, \quad t \in [0, 1] \tag{2}$$

where P_0 ... P_n are n+1 control points and B(t) is the Bernstein polynomial defined by

$$B_i^n(t) = \binom{n}{i} t^i (1-t)^{n-i} \tag{3}$$

The curves generated using this method can be made to have varying shapes and sizes depending on the values of t. Figure 5 shows the different curves made using different values of t.

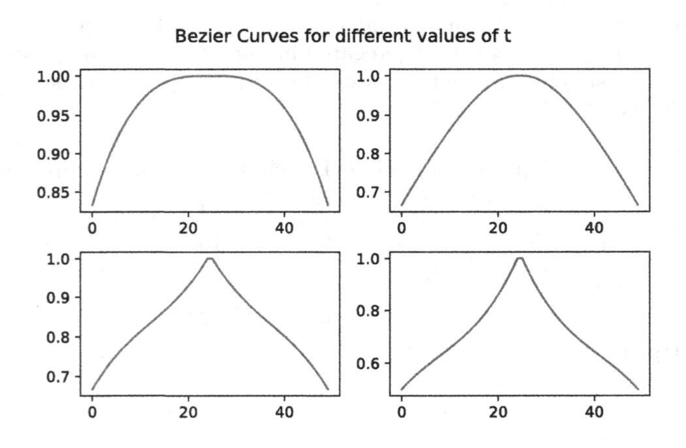

Fig. 5. Curves generated for different values of t

4.2 Smart Pruning

The Bezier curve described in the previous section is divided into a number of buckets. The number of buckets is taken to be the square root of the number of neurons. The weights of each neuron are also segregated into an equal number of bins. The values of the Bezier curve, normalized from 0 to 1, gives the proportion of weights to be pruned. Figure 6b shows a single Bezier curve with 50 buckets.

The values are marked with orange circles. For example, the value at bucket number 10 is approximately 81%. This means that 81% of the weights at bucket number 10 for the histogram of weights (Fig. 3a) will be randomly removed. Referring back to Fig. 3a the same Bezier Curve could be applied to the weights of the neural netowrk and weights falling within the curve would be pruned. Figure 6a illustrates the idea by overlapping the Bezier curve on histogram of weights. The orange curve represents the Bezier Curve while the blue curve represents the top points of the weight histogram. The effect of this pruning method is seen in Fig. 6c which is quite different from the method of pruning by using a certain value as threshold (Fig. 3b). The difference is further highlighted in Fig. 7 where weights remaining after pruning with different Bezier curves are shown. The figures on the left show the bezier curve superimposed on the weights while the figures on the right show the weights remaining after the pruning process, using the bezier curve as a filtering criteria.

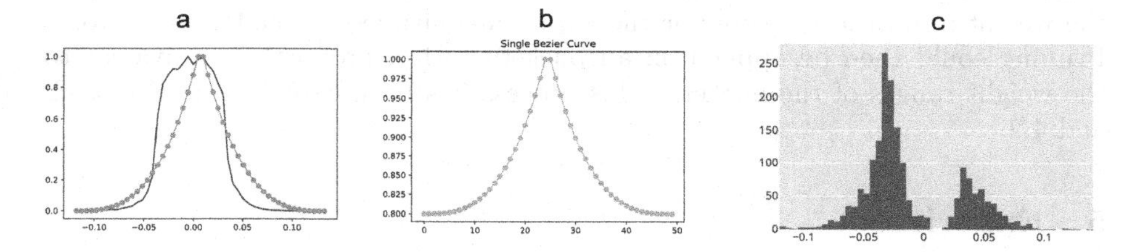

Fig. 6. Effect of Bezier Curves on neural network weights. Figure a shows Bezier Curve super imposed on weight histogram. Figure b displays an isolated bezier curve and c shows the pruned weights histogram after applying Bezier Curve on the weights. The histogram shows all non-zero weights. (Color figure online)

4.3 Formulation

This paper aims to arm the pruning research body with two new tools

- A novel method of generating importance of filters and weights (called weight importance vectors) explained in Sect. 4.
- Smart pruning applying Bezier Curves.

The two tools could be combined in three different ways to generate distinct pruning strategies. They are explained as follows along with pseudo codes to illustrate the approaches.

- **Applying threshold on weight importance vector:** This method generates the weight importance vectors and applies hard pruning on the neuron weights using this vector as a metric.

- **Applying Bezier Curve on weight importance vectors:** This process generates the weight importance vector and then applies smart pruning on the neuron weights using the weight importance vector as metric.
- **Applying soft pruning on weight magnitudes:** This approach applies soft pruning directly on the weights of the neurons.

The process can be further explained as below. First, usual methods are used to train the Neural network. As the training progresses, the magnitudes of the weights are stored at each epoch. On completion of training, the evolution of weights is analyzed using the methods in Sect. 4. The output of the analysis is a vector of weight importance values with dimensions same as that of the network's neurons. The dimensional mapping makes applying the pruning logic to the neurons easy. We could directly apply the standard pruning mechanisms to the neurons, using the weight importance vector instead of the individual weight magnitudes. The second method proposed in the paper - soft pruning, could be applied next. The weights would be aggregated into buckets based on the weight importance vectors or the actual magnitudes of weights in neurons. Pruning could then be applied on all buckets with a proportion equivalent to the weight ranges of the buckets. This process has been explained in Sects. 3.2 and 4.2.

5 Experiment

5.1 Settings

The efficacy of the novel pruning methods is established by comparing the accuracy of models that are pruned using conventional method versus methods proposed in this paper. We use 3 representative benchmark datasets, including the CIFAR-10 [11], MNIST [5] and FashionMNIST [20]. The CIFAR-10 dataset consists of 60000 32×32 colour images divided into 50k training and 10k test images. The data is categorized into 10 classes. Each class is represented by 6000 images. The MNIST dataset consists of 60,000 images of size 28×28. Each image is a handwritten single digit between 0 and 9. It also consists of 10,000 test images. The Fashion-MNIST dataset has a distribution similar to MNIST consisting of 28×28 grayscale images of garments and fashion accessories divided into 10 classes. It also has 60,000 images for training and 10,000 for testing.

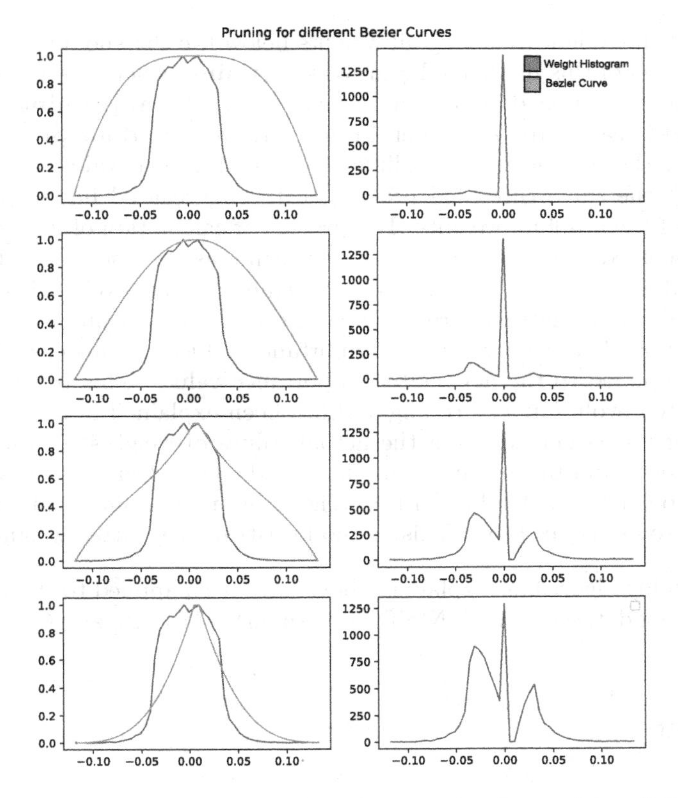

Fig. 7. Pruned weights after removing random weights by applying different Bezier Curves. Orange line in the graphs to the left show the different Bezier Curves. The blue lines represent histogram of weights. Every row represents the transformation of weights after applying a different Bezier Curve (Color figure online)

Neural Networks and Training In order to apply our technique to both weights as well as filters, we have used both Convolution based neural networks, containing filters and weights as well as Fully Connected Layer based neural networks containing only weights. The CIFAR-10 dataset is trained on a Neural Network built from scratch with 2 convolution layers followed by 3 fully connected layers. The Fashion-MNIST and MNIST datasets are trained on a Neural Network built with 4 fully connected layers. All the networks are trained using Stochastic Gradient Descent (SGD) with a learning rate of 0.01 and batch size of 128. As the training progresses, the magnitudes of the weights and filters at each epoch are persisted into the disk for use during pruning. Our implementation is done on the Pytorch framework [18]. The experiments are thus divided into four parts.

- Applying the standard method of pruning by using a threshold weight. This is the established practice where different thresholds are applied on the mag-

nitude of the weight, reducing all weights below the threshold to 0. We apply three flavors of this method - **layer-wise pruning**, where the thresholds are calculated and applied on a layer-by-layer basis, **global pruning**, where the threshold is calculated on the entire network and **random pruning** where randomly chosen weights are nullified by changing their values to 0 [3].
- Applying the same threshold process on the aggregated importance values instead of the absolute weights. The process of calculation of importance has been discussed in Sect. 4. The process is same as previous, except that the threshold is applied on the value of importance of each weight instead of its magnitude. This method is referred to as evolutionary pruning.
- Applying the bezier curve on the importance values. In this case, the bezier curve is applied to the aggregated importance values. This is referred to as curve fitted evolutionary pruning and has been explained in Sect. 4.2.
- Applying the bezier curve on the actual values of weights. In this case the aggregated importance values are not used. The Bezier curve is applied directly on the actual values of the weights in order to filter out the weights that are to be reduced to 0. This method is titled curve fitted weight pruning.

The pruning algorithms explained above have been applied to three standard Deep Learning datasets - the MNIST [5], FashionMNIST [20] and CIFAR-10 [11] dataset.

6 Results

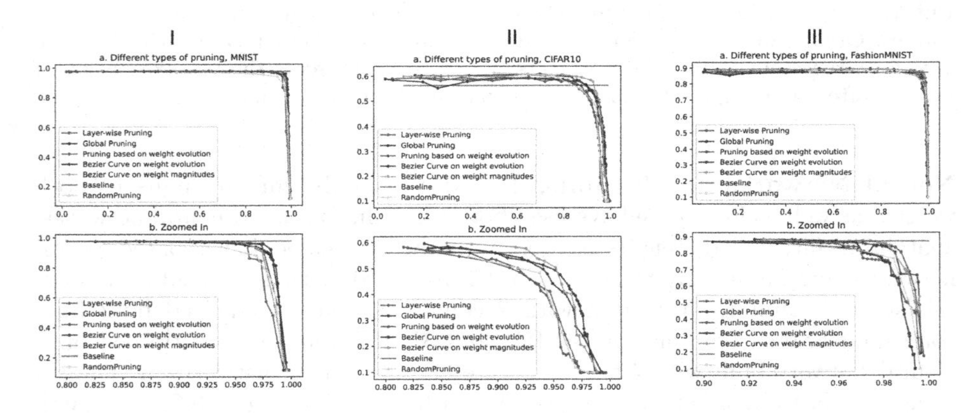

Fig. 8. Accuracy against compression for the different techniques on different datasets (Color figure online)

Figure 8 shows the accuracy of different compression methods for CIFAR-10, MNIST and FashionMNIST datasets respectively. Figures a in each case show the overall change in accuracy with increasing compression while Figures b specifies the accuracy over selected compression ranges. The red line depicts the layer-wise method of pruning using different thresholds to nullify the weights. The purple line shows the global pruning method and the yellow line illustrates the random pruning process. As the threshold increases, the compression also increases. After a certain point this deteriorates the accuracy of the system. In the case depicted with the green line, the threshold is applied on the aggregated importance score of every weight accrued over the last 90 epochs of training instead of the actual magnitude of weights at the end of training. The blue line shows the second experiment where we apply the Bezier Curves on the importance of the weights and prune randomly chosen weights that fall under the curve. This does not involve any thresholding, rather it creates buckets of weights at different bands of importance and removes some of the weights, depending on the value of the bezier curve at that band. The process has been explained in Sect. 4.2. The third experiment is illustrated by the yellow line where we apply the Bezier Curve on absolute value of the weights, instead of using the aggregated value.

Figure 8Ib shows the zoomed in accuracy over compression ranging from 80% to 99% for the MNIST dataset. Figure 8IIb shows the shows the zoomed in accuracy over compression ranging from 82% to 99% for the CIFAR-10 dataset. Figure 8IIIb shows the zoomed in accuracy over compression ranging from 90% to 99% for the FashionMNIST dataset.

The accuracy of the second and third method of experiment is considerably higher than that obtained in the layer-wise method. These methods provide a higher tolerance for compression in comparison to the standard methods of pruning. The superiority of the methods is also borne out by Table 2 which shows the different accuracy scores for the same compression ratio and Table 3 depicting the different compression values for the same accuracy. The second and third experiments viz. applying Bezier Curve on importance of weights and applying Bezier Curve on actual magnitude of weights perform better than the other methods.These methods seem to sustain greater amount of pruning before the accuracy reduces drastically. As can be seen in Table 2 in case of the MNIST dataset, the model pruned by applying Bezier Curve on the Importance values (Curve On Evolution) and the one pruned by applying Bezier Curve on weight magnitudes (Curve On Weights) achieve 10 and 7% more accuracy than the layer-wise method at 97% compression. Similarly in Table 3 in case of the CIFAR-10 dataset, the same pruning methods achieve 4 and 9% more compression for the same accuracy attained by the layer-wise method. The comparison is done with the layer-wise pruning method as the approaches proposed in the paper are also implemented in a layer-by-layer manner. The pruning methods are further tested by choosing neural networks at similar compression level and then applying them on noisy data.

Table 2. Accuracy at different compression rates

Dataset	Compression rate	Layer-wise pruning	Compression evolution	Curve on evolution	Curve on weights
CIFAR10	94–94.5	39.7	42.05	49.84	**52.86**
	86–88	54.77	56.09	58.02	**59.4**
MNIST	97–97.5	85.38	95.12	**95.83**	92.15
	96–97	86.3	**96.16**	96	92.75
FashionMNIST	97.9	75.67	84.17	84.84	**85.85**
	98–98.3	75.54	84.13	84.11	**85.62**
	98.6–99	60.69	68.71	67.36	**70.21**

Table 3. Compression rates at different levels of accuracy

Dataset	Accuracy level	Layer-wise pruning	Compression evolution	Curve on evolution	Curve on weights
CIFAR10	43–44	93.33	93.5	94.9	**95.7**
	58–58.3	82.03	81.65	86.72	**91.6**
MNIST	92–96	96.05	96.36	**97.48**	97.26
FashionMNIST	86–87.5	95.89	97.07	96.78	**97.11**
	83.5–84.8	96.88	97.93	98.04	**98.25**

6.1 Test on Noisy Data

A Gaussian noise was applied to all images in the test dataset of CIFAR-10, MNIST and FashionMNIST datasets. The probability density function p of a Gaussian random variable is given by

$$p_G(z) = \frac{1}{\sigma\sqrt{2\pi}} e^{-\frac{(z-\mu)^2}{2\sigma^2}} \tag{4}$$

where z represents the data, μ and σ represnt the mean and standard deviation of the data respectively. We vary the value of σ from 0 to 1 and track the accuracy of each type of compressed network while keeping μ fixed at 0. Figure 9 shows the accuracy levels of the pruned networks for noisy data stretching from a standard deviation of 0 to 1. Figure 9a shows the performance of networks pruned at 85% for the noisy FashionMNIST Dataset. Figure 9b shows the performance of networks pruned at 75% for the noisy CIFAR10 dataset while Fig. 9c shows the performance of networks pruned at 85% for the noisy MNIST dataset. The pruned models have same compression for each of the dataset and start from comparable accuracy scores. As the noise is increased, the performance of the models deteriorate and accuracy drops. In spite of starting at a lower accuracy, the proposed approaches achieve a lower rate of decline than the layer-wise pruning method.

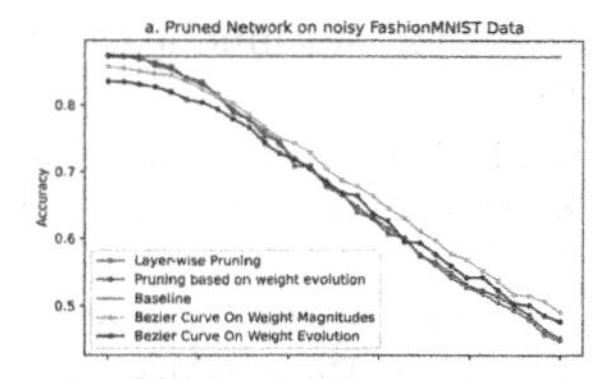

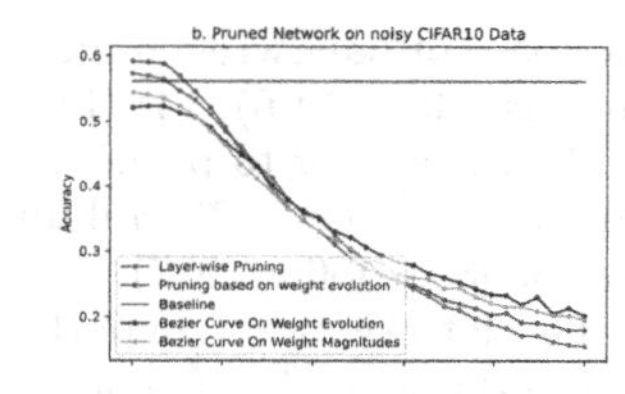

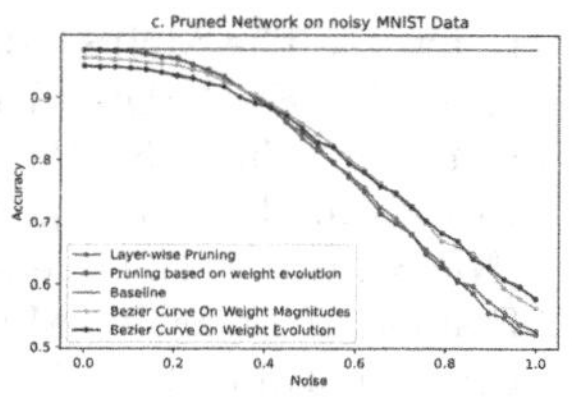

Fig. 9. Accuracy against noisy data for different compression techniques. Figures show the accuracy of pruned network models using the 4 techniques on the 3 different datasets as the data is perturbed by adding noise using the Gaussian distribution with mean fixed at 0 and standard deviation increasing from 0 to 1. The networks begin at same compression with similar accuracies and deteriorate in performance as noise is increased.

7 Discussion

The approaches described in this paper stem from two core concepts - firstly, the importance of any weight in a neural network should not be measured by its final magnitude at the end of the training. Rather, the progression of its values need to be monitored to have a holistic view of its significance across the entire training period. The second idea is that, at any point during pruning and fine-tuning, even if a weight is nullified, it should have a finite probability of coming back into prominence at a later stage. This covers the case where weights with lower magnitude prove important to the performance of the network and those with higher magnitude prove detrimental. The three approaches described in the paper reach higher compression with more accuracy in comparison to the standard method of pruning and the code is made available at https://github.com/ashhadulislam/SmartPruningDeepNN. They also prove to be more robust against noise. In the future, these methods can be tested to provide resistance against overfitting and adversarial inputs. The Bezier Curve used in this paper is a precursor to different methods of random weight selection that can be employed in weight selection.

References

1. Aghasi, A., Abdi, A., Nguyen, N., Romberg, J.: Net-trim: convex pruning of deep neural networks with performance guarantee. In: Advances in Neural Information Processing Systems, vol. 2017-Decem, pp. 3178–3187 (2017)
2. Babak Hassibi, D.G.S.: Second order derivatives for network pruning: optimal brain surgeon (October), pp. 1–8 (2014). https://authors.library.caltech.edu/54983/3/647-second-order-derivatives-for-network-pruning-optimal-brain-surgeon(1).pdf
3. Blalock, D., Ortiz, J.J.G., Frankle, J., Guttag, J.: What is the state of neural network pruning? (2020). https://arxiv.org/abs/2003.03033
4. Carreira-Perpiñán, M.A., Idelbayev, Y.: 'Learning-compression' algorithms for neural net pruning. In: Proceedings of the IEEE Computer Society Conference on Computer Vision and Pattern Recognition (LC), pp. 8532–8541 (2018). https://doi.org/10.1109/CVPR.2018.00890

5. Deng, L.: The MNIST database of handwritten digit images for machine learning research. IEEE Signal Process. Mag. **29**(6), 141–142 (2012)
6. Dong, X., Chen, S., Pan, S.J.: Learning to prune deep neural networks via layer-wise optimal brain surgeon. In: Advances in Neural Information Processing Systems (NIPS), vol. 2017-Decem, pp. 4858–4868 (2017)
7. Frankle, J., Carbin, M.: The lottery ticket hypothesis: finding sparse, trainable neural networks. In: 7th International Conference on Learning Representations, ICLR 2019, pp. 1–42 (2019)
8. Fukushima, K.: Neocognitron: a self-organizing neural network model for a mechanism of pattern recognition unaffected by shift in position. Biol. Cybern. **36**(4), 193–202 (1980). https://doi.org/10.1007/BF00344251
9. Guo, Y., Liu, Y., Georgiou, T., Lew, M.S.: A review of semantic segmentation using deep neural networks. Int. J. Multimed. Inf. Retrieval **7**(2), 87–93 (2018)
10. Han, S., Mao, H., Dally, W.J.: Deep compression: compressing deep neural networks with pruning, trained quantization and Huffman coding. In: 4th International Conference on Learning Representations, ICLR 2016 - Conference Track Proceedings, pp. 1–14 (2016)
11. Krizhevsky, A.: Learning multiple layers of features from tiny images (2009)
12. Le Cun, Y., Denker, J.S., Solla, S.A.: Optimal brain damage, vol. 1, pp. 4–11 (1989)
13. LeCun, Y., et al.: Backpropagation applied to digit recognition (1989). https://www.ics.uci.edu/~welling/teaching/273ASpring09/lecun-89e.pdf
14. Robinson, A., Fallside, F.: The utility driven dynamic error propagation network, vol. 1. IEEE (1987)
15. Rosenblatt, F.: The perceptron: a probabilistic model for information storage and organization in the brain. Psychol. Rev. **65**(6), 386–408 (1958)
16. Salman, H., Grover, J., Shankar, T.: Hierarchical reinforcement learning for sequencing behaviors, vol. 2733, no. March, pp. 2709–2733 (2018)
17. Sharma, P., Singh, A.: Era of deep neural networks: a review. In: 8th International Conference on Computing, Communications and Networking Technologies, ICCCNT 2017 (2017). https://doi.org/10.1109/ICCCNT.2017.8203938
18. Shen, S., Li, R., Zhao, Z., Zhang, H., Zhou, Y.: Learning to prune in training via dynamic channel propagation. In: Proceedings - International Conference on Pattern Recognition, pp. 939–945 (2020). https://doi.org/10.1109/ICPR48806.2021.9412191
19. Weng, J.J., Ahuja, N., Huang, T.S.: Learning recognition and segmentation of 3-D objects from 2-D images. In: 1993 IEEE 4th International Conference on Computer Vision, pp. 121–127 (1993). https://doi.org/10.1109/iccv.1993.378228
20. Xiao, H., Rasul, K., Vollgraf, R.: Fashion-MNIST: a novel image dataset for benchmarking machine learning algorithms, pp. 1–6 (2017). http://arxiv.org/abs/1708.07747
21. Ye, J., Lu, X., Lin, Z., Wang, J.Z.: Rethinking the smaller-norm-less-informative assumption in channel pruning of convolution layers. In: 6th International Conference on Learning Representations, ICLR 2018 - Conference Track Proceedings 2017, pp. 1–11 (2018)
22. Zhang, T., et al.: A systematic DNN weight pruning framework using alternating direction method of multipliers. In: Ferrari, V., Hebert, M., Sminchisescu, C., Weiss, Y. (eds.) ECCV 2018. LNCS, vol. 11212, pp. 191–207. Springer, Cham (2018). https://doi.org/10.1007/978-3-030-01237-3_12
23. Zhu, M.H., Gupta, S.: To prune, or not to prune: exploring the efficacy of pruning for model compression (2017)

Biologically Inspired Unified Artificial Immune System for Industrial Equipment Diagnostic

Galina Samigulina[1] and Zarina Samigulina[2](✉)

[1] Institute of Information and Computing Technologies, Almaty, Kazakhstan
kz.galinasamigulina@gmail.com
[2] Kazakh-British Technical University, Almaty, Kazakhstan
z.samigulina@kbtu.kz

Abstract. Modern high-tech industrial enterprises are equipped with sophisticated equipment and microprocessor technology. The maintenance of such production facilities is an expensive procedure, and the replacement of equipment due to breakdown or wear and tear leads to significant financial costs. In the oil and gas industry, the idle operation of an enterprise for several hours or a working day can cause serious losses to the company. In this regard, it is relevant to develop intelligent diagnostic systems aimed at timely detection of faults, assessing the degree of their criticality and predicting possible breakdowns in the future. The use of bioinspired machine learning methods for diagnosing industrial equipment in real industrial production is a promising area of research. The article presents the developed diagnostic system for industrial equipment based on the methodology of analysis of modes, failures of their influence, degree of criticality (Failure Mode and Effects Analysis, FMEA) and a unified artificial immune system (UAIS), created on the basis of systematization and classification of modified algorithms of artificial immune systems (AIS). Unification is used to select the most efficient modified AIS algorithm based on the theories of clonal selection, negative selection and the immune network for processing heterogeneous data. UAIS is especially effective in the analysis of dynamically changing production data and a small number of training samples corresponding to equipment failures. Simulation results obtained on real data of TengizChevroil refinery.

Keywords: Bioinspired algorithms · Unified artificial immune system · Diagnostic of industrial equipment · Fault detection

1 Introduction

Currently, industrial plants are created on the basis of modern microprocessor technology using sophisticated software, which, according to international standards (for example, IEC 61131-3) of the independent organization PLCopen, provides automation efficiency based on user needs. Such systems allow real-time monitoring, control, diagnostics and operational management of technological processes and generate a huge flow of production information, which, as a rule, is not fully analyzed. For example, an industrial controller's sensors are polled on average once every 20 ms, if 30 parameters are adjusted

G. Nicosia et al. (Eds.): LOD 2022, LNCS 13811, pp. 77–92, 2023.
https://doi.org/10.1007/978-3-031-25891-6_7

on one node, then 90 indicators are recorded per minute or 5,400 per hour, and in conditions of a large technological process, the number of production data increases many times. Timely analysis of data on production line failures and forecasting of equipment condition helps to prevent serious accidents at the enterprise.

Currently, the world's largest manufacturers of microprocessor technology are developing specialized intelligent systems for solving production problems. Schneider Electric (France) and Microsoft launched a large-scale project in 2019 with Inria, Sigfox, Elaia, Energize and France Digital to help start-ups (Accenta, BeeBryte, Craft AI) engaged in AI for Europe transition to efficient renewable sources [1]. Siemens (Germany) has developed a TM NPU neuroprocessor module with a neural network for processing incomplete, partially noisy or distorted data [2]. Omron (Japan) has created a Sysmac AI controller with a machine learning function and the "AI Predictive Maintenance" library. Artificial intelligence is used to increase the life of equipment, detect faults and prevent failures [3, 4].

However, the production environment imposes its own limitations on the use of intelligent models for solving various engineering problems. For example, efficient AI algorithms cannot be implemented directly on controllers due to the small amount of memory and operational capabilities of processors. In this regard, it is promising to study bioinspired machine learning algorithms that can adapt to the requirements of the production environment with the ability to operate in close to real time.

The approach based on artificial immune systems (AIS) [5, 6] and its modified algorithms for solving engineering problems [7, 8] has proven itself well. The following approaches of AIS are widely used: clonal selection [9, 10], negative selection, immune network modelling. For example, in [11], there is considered a method for detecting multiclass data anomalies based on a combination of negative selection and clonal selection algorithms. The proposed algorithm was tested using a data set of vacuum deposition bearings using the developed intelligent equipment maintenance system. Studies [12] are devoted to the use of AIS for predicting software failures. The effectiveness of 8 immunological systems for detecting program defects was evaluated against three different control indicators. The simulation results are based on the analysis of 41 data sets from the Promise repository. In [13], AIS is used to detect and classify faults in semiconductor equipment. Equipment health data monitoring, wafer metrology and wafer inspection are used to identify any anomalies in the manufacturing process that could affect the quality of the final chip. The article [14] considers the developed hybrid system based on the multi-agent approach and AIS for diagnostics, forecasting and maintenance of geographically dispersed industrial facilities. Using a multi-agent approach allows to combine autonomy and distributed processing to solve the problem of a small number of training data. The work [15] is devoted to the AIS-based diagnostics of failures in the operation of robot swarms. Research has shown that robot swarms are not always tolerant of failures by individual robots that are partially out of order but continue to influence collective behavior. In this regard, the development of fault-tolerant robotic systems, where the swarm can identify and eliminate faults during operation, is relevant. The results of the experiments were obtained on the basis of a real robot and showed that the proposed system for diagnosing failures is flexible, scalable and increases the stability of the swarm to various electromechanical failures. The article [16] considers fault diagnostics in a wireless sensor network using the principle of clonal selection and a probabilistic neural network approach. An algorithm for detecting faulty sensory nodes based on clonal selection is proposed. Faults were classified into permanent, intermittent and temporary using a probabilistic artificial neural network approach. The evaluation of the algorithm efficiency was carried out on the basis of such indicators as: detection accuracy, false positive rate, false classification rate, diagnostic delay and energy consumption. The proposed algorithm provides lower diagnostic latency and consumes less power than existing algorithms developed by Mohapatra et al., Panda et al. and Elhadef et al, for a network of wireless sensors. The work [17] is devoted to an intelligent diagnostic method based on classification by clonal selection using a pool of B cells. Since there are difficulties in obtaining samples of mechanical damage, then intelligent diagnostic methods should be adapted to training small samples and have the interpretability of a "white box" model. A method is proposed for creating unique pools of B cells corresponding to a specific antigen. A greedy strategy is used to obtain pools of B-memory cells. The simulation results were based on a dataset of bearing defects at Case Western Reserve University. In [18], a deep learning (DL) method is considered for diagnosing faults in rolling bearings using AIS. An adaptive enhanced deep convolution algorithm is applied to extract the feature set. Experimental results show that the proposed deep convolutional neural network model can extract several signs of rolling bearing failures, improve classification and detection accuracy by reducing the number of false positives. Research [19] is devoted to the analysis of the hybrid method CSA-DEA (Clonal Selection Algorithm with a Differential Evolution Algorithm) for flaw detection of structures with cracks. The formation of damage in a structural element often causes failures. Determining the location and severity of damage can help in taking the necessary steps to mitigate structural catastrophic failures. In this regard, methods for detecting damage without destroying the structure have become popular. The input data of the hybrid system are the relative frequencies of the damaged structure, and the output data are the relative crack locations and depth. The results of the experiments showed the effectiveness of the proposed algorithm. In [20], a bioinspired anomaly detection algorithm based on the Cursory Dendritic Cell Algorithm (CDCA) is considered. Autonomous and adaptive anomaly identification in industrial systems such as Cyber-Physical Production Systems (CPPS) requires new technologies to correctly identify failures. Most studies do not take into account the dynamics of such systems. Due to the complexity and multidimensionality of CPPS, a scalable, adaptive and fast anomaly detection system is needed, taking into account new technologies developed as part of the Industry 4.0 concept. Immune models such as the Dendritic Cell Algorithm (DCA) is a promising area of research, since the problem of detecting anomalies in CPPS resembles the work of dendritic cells in protecting the human body from dangerous pathogens. A new version of DCA, the CDCA algorithm, is proposed for continuous monitoring of industrial processes and online anomaly detection. The results of the experiments proved to be effective and were carried out on the basis of two industrial data sets for detecting physical anomalies and network intrusions (Skoltech Anomaly Benchmark (SKAB) and M2M using OPC UA).

An analysis of the literature showed the prospects and possibilities of using the AIS approach to develop effective industrial diagnostic systems, taking into account the requirements of real industries.

The following structure of the article is proposed: the second section presents the formulation of the research problem, section three is devoted to the methods used in the work, section four presents the developed system for diagnosing industrial equipment based on the FMEA and UAIS approaches, the fifth section contains the simulation results based on the production data of the TengizChevroil enterprise, the sixth section describes the main conclusions and prospects for future research.

2 Statement of the Problem

Maintenance of expensive production equipment in working order, timely diagnostics and forecast of critical conditions is a priority task in the maintenance of production lines. Currently, a large number of models have been developed for equipment failure analysis, such as: FMEA [21], HAZOP (Hazard and Operability Study) [22], FTA (Fault Tree Analysis) [23], etc. These methods allow errors to be eliminated at an early stage, are simple and intuitive, and do not contain complex calculations, but can be time consuming for complex systems. Errors can accumulate in a redundancy system when it is necessary to track several levels of a hierarchical system, and a big problem is the inability to assess the reliability of the entire complex system. The problem can be solved using an extended FMEA equipment failure model using modern advanced biospirited AIS algorithms.

The statement of the research problem is formulated as follows, it is necessary to develop a system for diagnosing industrial equipment based on the methodology for analyzing modes, failures, their influence, the degree of criticality of FMEA and a unified artificial immune system using the example of real production data of the TengizChevroil enterprise.

Definition 1. A unified artificial immune system is understood as a system created on the basis of systematization and classification of modified AIS algorithms, taking into account weaknesses and strengths, in order to select the most effective immune response mechanisms for analyzing and predicting multidimensional data (different in structure, type, etc.).

Definition 2. The modified algorithm of artificial immune systems is an improved prediction algorithm with an optimal data set obtained by extracting informative features based on bioinspired artificial intelligence methods or statistical data analysis. A modified AIS algorithm is also understood as an algorithm with an improved structure, the efficiency of which exceeds the classical analogue due to the introduced modifications.

3 Approaches and Methods

3.1 Failure Mode and Effects Analysis

An effective methodology for analyzing failures, their impact on the system and assessing the degree of criticality is FMEA. This approach was first developed in the USA (NASA) and was applied in the following areas: aerospace, automotive, standards development (Germany), medicine and pharmacology [21]. Figure 1 presents a risk assessment table using FMEA. The table describes potential failures (Fig. 1, columns 1–3) or system defects and their causes to prevent errors.

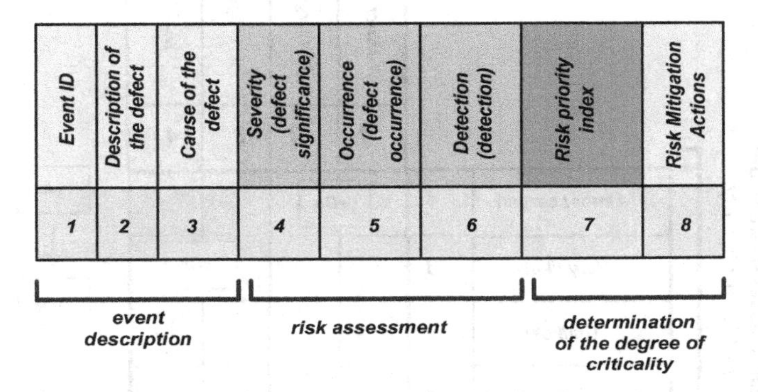

Fig. 1. FMEA methodology

The first column contains the equipment failure ID number, as well as the possible location of the failure at the enterprise. The second column describes the nature of the defect. The next column lists the causes of the failure. Columns 4–6 are filled in by experts in accordance with the following indicators:

1. «Severity» - the significance of the defect, which shows the degree of failure consequences and is ranked in the range: $S = \{1..4\}$.
2. «Occurrence» - the occurrence of a defect is an estimate representing the probability of occurrence of the cause of the defect. The criterion is set in the range: O = {1..4}.
3. «Detection» - defect detection is an indicator of the probability of detecting defects, their causes and their consequences for the entire system. The criterion is ranked in the range: $D = \{1..4\}$.

Next, the Risk Priority Index is calculated, which is a quantitative assessment of the complex risk (Fig. 1, column 7). If the risk value for several failures will have the same value, the most serious risk is considered to be the risk for which the "Severity" criterion has the highest value. This coefficient is calculated by the formula:

$$RPI = Severity \cdot Occurence \cdot Detection \quad (1)$$

At the next stage of the FMEA methodology, a risk matrix is compiled, which is a mechanism for assessing the degree of a failure criticality. This assessment is effective in case of incorrect prioritization of risks by experts (Fig. 1, columns 4–6) and allows assessing the criticality of defects from different points of view. Figure 2 shows the visualization of the risk matrix, taking into account the ranking of criteria {S, O, D} in the range from 1 to 4.

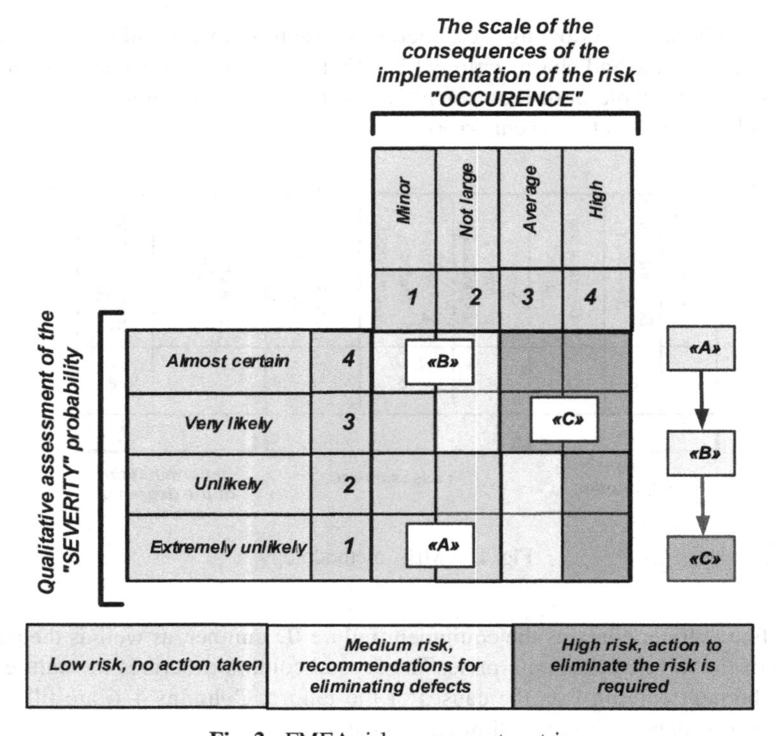

Fig. 2. FMEA risk assessment matrix

The matrix is divided into three zones: "ZONE A" - low risk; "ZONE B" - medium degree of risk, "ZONE C" - high degree of risk. Depending on the zone in which the failure is located, decisions are made regarding the degree of criticality and the need for prompt repair.

The advantages of this approach are: easy interpretation of the results, the absence of complex mathematical calculations. As disadvantages, we can indicate the complexity of filling in case of complex industrial complexes.

3.2 Unified Artificial Immune System

The study of the complex mechanisms of the human biological immune system makes it possible to develop effective and reliable approaches for processing multidimensional information to solve complex problems.

Artificial immune systems can analyze, memorize, adapt previously received information, learn and perform pattern recognition [24]. The field of AIS includes two areas of research: the use of mathematical and computational methods in modeling immunology and the inclusion of the principles of the functioning of the immune system in relation to the development of technical applications [25]. Currently, AIS uses separate disparate mechanisms of the immune system (molecular recognition, clonal and negative selection) to create various algorithms and applications based on them. However, the immune system functions as a holistic system to protect the human body. Therefore, the first attempts to create a single unified AIS, that is, to integrate independent AIS algorithms to create a holistic AIS in order to raise this approach to a higher level, began a long time ago. Separate applications were developed for solving narrow problems [26]. However, this complex problem has not yet been fully solved. Unfortunately, one of the obvious reasons is that the principles of operation of numerous modern AIS algorithms differ significantly from their biological prototypes. It has been proven that proteins play an important role in information processing in the human body. Awareness and application of the basic biological principles, development of the theoretical foundations of AIS is a necessary condition for the further successful development of this area and has great prospects. Another factor complicating the solution of the problem is the interdisciplinary nature of research, affecting many areas of science and the lack of specialists with sufficient knowledge of the necessary knowledge.

Figure 3 shows the architecture of UAIS, based on the systematization and classification of modified AIS algorithms.

UAIS includes the following types of algorithms based on clonal selection, negative selection and immune network. Each type contains modified AIS algorithms (Fig. 3) for solving the classification problem. Next, the effectiveness of the modified algorithm is evaluated on a specific data set according to the metrics {Score 1. Score N}: Accuracy, Classification Error, Precision, Recall, F-measure, etc. As a result, a matrix of estimates is compiled, which allows to conclude how effective this algorithm is with respect to the data set under consideration. Next, the evaluation matrices of the modified algorithms are analyzed and decisions are made based on the results obtained.

The scientific novelty of the proposed UAIS lies in the use of not classical, but modified AIS algorithms based on clonal selection, negative selection, and immune networks. Modifications are possible both in the AIS algorithms themselves and in the data preprocessing algorithms. The unification is aimed at reducing the variety of modified AIS algorithms in order to identify the most effective ones for a particular data set. For example, the authors developed a modified CPSOIW-AIS algorithm based on the cooperative particle swarm algorithm with inertia weight [27] and the AIS immune network algorithm. By modifying the particle swarm algorithm with the inertia weight, the process of extracting informative features takes less time, and by modifying the immune network algorithm based on homologues, pattern recognition at class boundaries is increased. A modified RF-AIS algorithm was also developed, the predictive ability of this model showed high accuracy due to the reduction of non-informative features after data processing by a random forest [28].

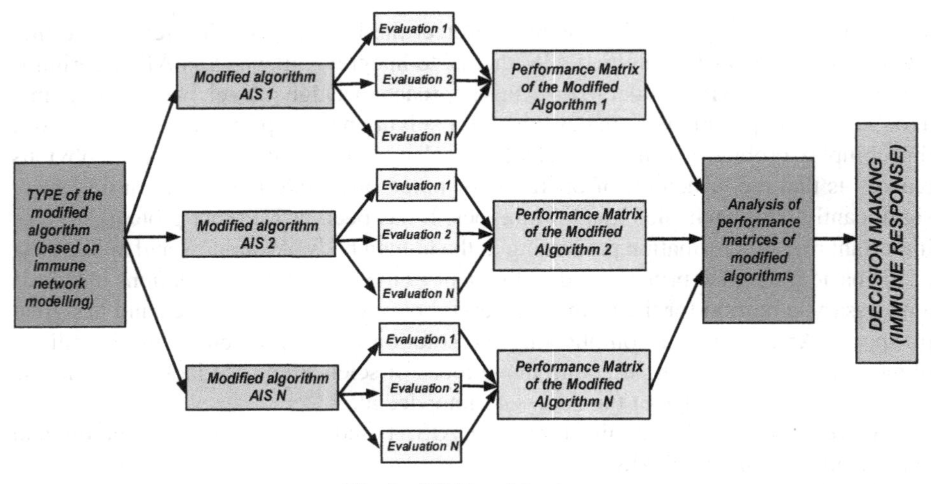

Fig. 3. UAIS architecture

The combination of the advantages of UAIS and the basic capabilities of the FMEA model will allow developing an intelligent system for diagnosing industrial equipment, adapted to the requirements of real production.

4 Development of Equipment Diagnostics System Based on FMEA and UAIS

Let us consider the principles of equipment diagnostics based on FMEA and UAIS. Figure 4 shows a modification of the classical methodology.

CLASSIC FMEA MODEL								UAIS-BASED INTELLIGENT DATA PROCESSING					
Event ID	Defect Description	Cause of the defect	Severity (defect significance)	Occurrence (defect occurrence)	Detection (detection)	Risk priority index	Risk Mitigation Actions	Modified Algorithm 1	Modified Algorithm 2	....	Modified Algorithm N	Failure criticality	Equipment wear percentage
1	2	3	4	5	6	7	8	9	10	11	12	13	14
event description			risk assessment			determination of the degree of criticality		Equipment diagnostics based on UAIS					

Fig. 4. Intelligent equipment diagnostics based on FMEA and UAIS

The following FMEA_UAIS algorithm for detecting failures and assessing the degree of their criticality is presented:

1. Creation of a database of industrial equipment failures based on polling sensors, alarm information and internal diagnostic capabilities of microprocessor technology.

2. Creation of a database of typical equipment breakdowns with possible troubleshooting options based on the technical regulations of production lines.
3. Development of the FMEA table according to Fig. 1. Description of the complex object of the production line where the failure occurred, indication of the type of equipment, description of the failure, possible causes and consequences of the failure.
4. Expert evaluation by categories {Severity, Occurence, Detection} according to the FMEA methodology (Fig. 1, 2).
5. Calculation of the risk priority index according to the formula (1).
6. Description of actions aimed at reducing the risk.
7. Processing the database of equipment failures based on the UAIS methodology (Sect. 3.2).
8. Evaluation of the degree of failure criticality based on UAIS.
9. Calculation of the percentage of equipment wear.

The proposed system of equipment diagnostics allows to reduce the risks associated with an erroneous expert assessment of the degree of criticality of an individual failure, as well as to consider not only one failure, but also its impact on the entire system as a whole.

5 Simulation and Experiment Results

The study uses information about equipment failures at the TengizChevroil LLP enterprise. Unit 700 for cooling, separation, fractionation of sweet gas and commercial gas compression is considered. Figure 5 shows a simplified diagram of Unit 700 [29], the products of which are: commercial gas, liquefied propane, liquefied butane, etc.

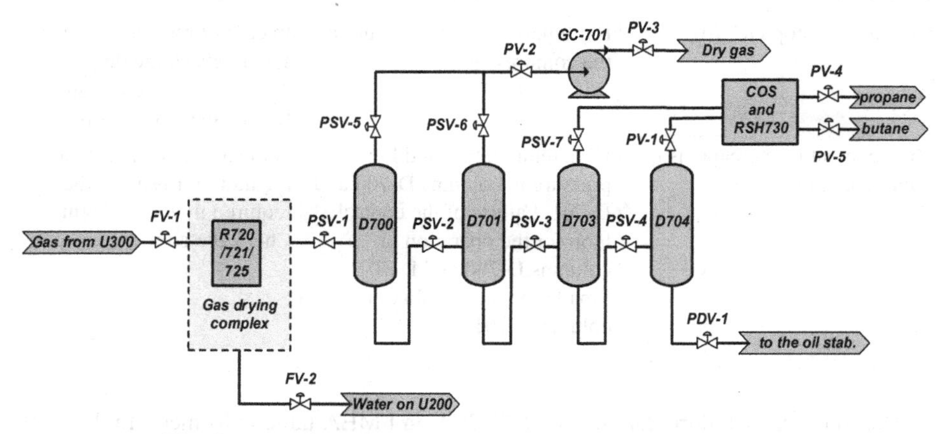

Fig. 5. Simplified diagram of U700 operation at TengizChevroil

Table 1 describes the specification of the main Unit 700 equipment. Column parameters are controlled using the following sensors: TRC-1 – column top temperature sensor D700, LICAHL-1/LICAHL-2 – column level sensor D700, PRCHL-1 – column pressure sensor D700; TI-1/TI-2 – column pressure sensor D701, LICAHL-3/LICAHL-4 – column level sensor D701, PIC-1 – column pressure sensor D701; TI-3 - column temperature sensor D703, LAH-5/LAL-7 - column level sensor D703, PAH-1/PAL-2 - column pressure sensor D703, TRC-1 - column temperature sensor D704, LAH-8/LAL-9 - column level sensor D704, PAH-2 - column pressure sensor D704, etc.

Table 1. Unit 700 Equipment specification

Aggregate designation	Function
R720/721/725	Gas drying reactors
D700	Demethanization column
D701	Deethanization column
D703	Depropanization column
D704	Debutanization column
GC-701	Commercial gas compressor

The technical documentation of TengizChevroil contains information about possible deviations of the technological process, as well as their causes and ways to eliminate them [30]. A fragment of the specification of typical failures is presented in Table 2.

Table 2. Fragment of the specification of Unit 700 failures

Failure type	Reasons for the occurrence	Solutions
Compressor stop GC-701	Extremely high temperature of the compressor	In case of receipt of one signal, check the devices, 2, 3 signals, take measures for an emergency stop
High content of mercaptans in commercial gas	High temperature and low pressure in columns D-700 and D-701. Failure of the control loop for the operation of columns D-700 and D-701. Small amount of reflux on column D-701	Temperature and pressure regulation. Ensuring the required flow of phlegm. Check the control loops

Based on daily failure data from UNIT 700, an FMEA table is formed (Table 3). If the RPI index (formula 1) has a maximum value, then the failure is critical and prompt troubleshooting is necessary. This FMEA model is extended with UAIS.

Table 4 shows a fragment of the database of equipment failures, where Unit 1 is the R 720/721/722 unit, Sensor 1 - FRC 20002 (regeneration gas flow), Sensor 2 - PDT-720007 (pressure drop 721), Sensor 3 - PDT- 720016 (differential pressure R720), etc.

[29–31]. For example, 2 classes are allocated: class 1 - equipment failure, class 2 - normal operation mode (column "Failure"). Failure criticality is evaluated in a scoring system ranging from 1 to 4: "1" - minimum criticality, "2" - pre-emergency condition, "3" - emergency condition, "4" – critical condition.

Table 3. FMEA diagnostic example

Complex facilities	Type of equipment	Failure	Reasons for failure	Consequences of failure	Severity	Occurence	Detection	Risk mitigation actions	RPI INDEX
1	2	3	4	5	6	7	8	9	10
UNIT 700	GC-701	Overheat	Uneven load	Breaking	3	2	2	Unit load planning	**12**
	T1G703B	Noise, vibration	Wear	Breaking	4	3	2	Replacement of parts of the unit	**24**
	T3 G301	End seal leak	Seal wear	Breaking	4	3	3	Repair	**36**
	T5 EA 33	Overload stop	Overload	Breaking	4	4	3	Repair	**48**

The dimension of the database (DB) is 8460 data instances.

Table 4. Fragment of the database of Unit 700equipment failures

Unit ID	Sensor 1	Sensor 2	Sensor 3	Sensor N	Failure
Unit 1	11 700, 869	6,6	3,2	400, 347	Yes
Unit 2	2,5	256,454	453,342	423,987	No
Unit 3	25,234	85,643	98	187,356	No
Unit 4	75	0,333	125	7	No
...	...	...	...	...	...
Unit M	98,763	184,456	98,718	45,565	Yes

The equipment failure database analysis for Unit 700 was carried out using modified UAIS algorithms: RF-AIS [28], FPA-AIS [29], GWO-AIS [29], CPSOIW-AIS [27]. In the RF-AIS model, informative features are extracted based on the random forest algorithm, for FPA-AIS, the flower pollination method is used, in the GWO-AIS model, the grey wolf optimization method is used, and in CPSOIW-AIS, informative features are extracted using a cooperative algorithm with inertia weight (Fig. 6).

Parameters that have the least weight are subject to reduction. After solving the problem of extracting informative features, the optimal immune network model is created. Then the problem of pattern recognition is solved.

Evaluation of the effectiveness of the modified UAIS algorithms is presented in Table 5.

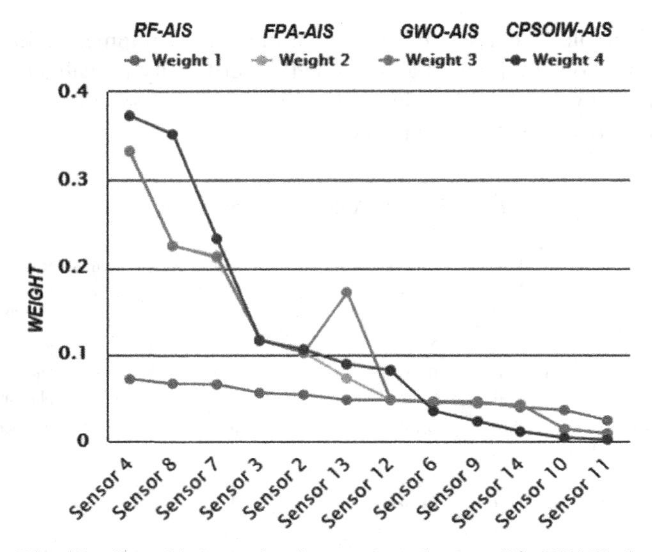

Fig. 6. Identification of informative features based on modified UAIS algorithms

Table 5. Evaluation of the effectiveness of modified UAIS algorithms

Algorithm of UAIS	Accuracy	Classification error	AUC	Precision	Recall	F-measure	Sensitivity	Specificity
RF-AIS	97,8%	2,2%	0,98	100%	96%	97,8%	96%	100%
FPA-AIS	93,1%	6,9%	1	92%	96%	93,3%	96%	91%
GWO-AIS	95,6%	4,4%	1	96%	96%	95,6%	96%	96%
CPSOIW-AIS	95,3%	4,7%	0,955	96%	96%	95,6%	96%	95%

Since the analysis of real production data is carried out, temporal characteristics play an important role. For example, the most efficient modified RF-AIS algorithm requires the most data processing time, which is undesirable in systems operating close to real time (Table 6).

Table 6. Temporal characteristics of UAIS algorithms

Algorithm of UAIS	Total time	Training time	Scoring time
RF_AIS	9 s	302 ms	130 ms
FPA_AIS	2 s	59 ms	106 ms
GWO_AIS	1 s	50 ms	56 ms
CPSOIW-AIS	2 s	562 ms	130 ms

Based on the results of the obtained data, the FMEA model is supplemented in accordance with Fig. 4 (Table 7).

Table 7. Extended FMEA model based on UAIS

Objects of the complex	Type of equipment		Severity	Occurence	Detection	Risk Mitigation Actions	RPI INDEX	RF-AIS	FPA-AIS	GWO-AIS	CPSOIW-AIS	Criticality of failure	% wear and tear of equipment
1	2	...	6	7	8	9	10	11	12	13	14	15	16
UNIT 700	GC-701	...	3	2	2	unit load plannin g	**12**	1	2	1	1	1,25	**31,25%**
	T1G 703B	...	4	3	2	replace ment of parts of the unit	**24**	2	1	1	2	1,5	**37,5%**
	T3 G301	...	4	3	3	repair	**36**	1	3	1	3	2	**50%**
	T5 EA 33	...	4	4	3	repair	**48**	3	4	4	3	**3,5**	**87,5%**

Equipment wear rate A_{UAIS} as a percentage based on UAIS is calculated using the following formula [23]:

$$A_{UAIS} = \frac{\sum K_{MAIS}}{N} \cdot k \cdot 100\% \tag{2}$$

where K_{MAIS} – the value of the criticality coefficients obtained from the results of data analysis based on modified AIS algorithms; N–number of considered modified AIS algorithms; k – coefficient for calculating equipment wear ($k = 0, 25$).

6 Conclusion

The creation of intelligent diagnostic systems using bioinspired modified UAIS algorithms and the FMEA methodology makes it possible to timely detect equipment failures and carry out an automated assessment of their criticality based on the analysis of multidimensional production data obtained using modern microprocessor technology. The use of a unified artificial immune system created on the basis of systematization and classification of modified AIS algorithms solves a number of serious problems of existing diagnostic systems, such as the impossibility of assessing the impact of a failure on the entire system in the case of a complex hierarchical structure; incorrect ranking of coefficients by an expert; the inability to assess the impact of several failures when they occur together, which can have significant consequences, while each of them individually has a low probability of a critical situation. With the help of UAIS, the problem of a small training sample containing equipment failures is successfully solved. The use of various modified UAIS algorithms allows choosing the best method for a specific data set, which differs in size and structure depending on the type of aggregate. Thus, the new bioinspired modified UAIS algorithms are successfully used in industrial environments and adapt to the limitations associated with the capabilities of modern microprocessor technology.

Acknowledgements. This research has been funded by the Science Committee of the Ministry of Education and Science of the Republic Kazakhstan (Grant No. AP09258508) 2021–2023.

References

1. Rueil-Malmaison. Start-ups from Schneider Electric and Microsoft's joint accelerator are transforming the energy sector in Europe with artificial intelligence. Press release Schneider Electric, pp. 1–3 (2020)
2. Sundaram, K., Natarajan, N.: Artificial intelligence in the shop floor envisioning the future of intelligent automation and its impact on manufacturing. A Frost & Sullivan White Paper, pp. 1–17 (2018)
3. Torres, P.: Improve OEE with Artificial Intelligence at the Edge in Food Manufacturing (2019)
4. https://industrial.omron.ru/
5. Cutello, V., Nicosia, G.: Multiple learning using immune algorithms. In: Proceedings of 4th International Conference on Recent Advances in Soft Computing, RASC, pp. 102–107 (2022)
6. Cutello, V., Lee, D., Nicosia, G., Pavone, M., Prizzi, I.: Aligning multiple protein sequences by hybrid clonal selection algorithm with insert-remove-gaps and blockshuffling operators. In: Bersini, H., Carneiro, J. (eds.) ICARIS 2006. LNCS, vol. 4163, pp. 321–334. Springer, Heidelberg (2006). https://doi.org/10.1007/11823940_25
7. Ciccazzo, A., Conca, P., Nicosia, G., Stracquadanio, G.: An advanced clonal selection algorithm with ad-hoc network-based hypermutation operators for synthesis of topology and sizing of analog electrical circuits. In: Bentley, P.J., Lee, D., Jung, S. (eds.) ICARIS 2008. LNCS, vol. 5132, pp. 60–70. Springer, Heidelberg (2008). https://doi.org/10.1007/978-3-540-85072-4_6
8. Conca, P., Nicosia, G., Stracquadanio, G., Timmis, J.: Nominal-yield-area tradeoff in automatic synthesis of analog circuits: a genetic programming approach using immune-inspired operators. In: NASA/ESA Conference on Adaptive Hardware and Systems, pp. 399–406 (2009)
9. Cutello, V., Narzisi, G., Nicosia, G., Pavone, M.: Clonal selection algorithms: a comparative case study using effective mutation potentials. In: Jacob, C., Pilat, M.L., Bentley, P.J., Timmis, J.I. (eds.) ICARIS 2005. LNCS, vol. 3627, pp. 13–28. Springer, Heidelberg (2005). https://doi.org/10.1007/11536444_2
10. Cutello, V., Nicosia, G.: A clonal selection algorithm for coloring, hitting set and satisfiability problems. In: Apolloni, B., Marinaro, M., Nicosia, G., Tagliaferri, R. (eds.) NAIS/WIRN - 2005. LNCS, vol. 3931, pp. 324–337. Springer, Heidelberg (2006). https://doi.org/10.1007/11731177_39
11. Kim, Y., Nam, W., Lee, J.: Multiclass anomaly detection for unsupervised and semi-supervised data based on a combination of negative selection and clonal selection algorithms. Appl. Soft Comput. **122**(108838), 1–12 (2022). https://doi.org/10.1016/j.asoc.2022.108838

12. Haouari, A.T., Souici-Meslati, L., Atil, F., Meslati, D.: Empirical comparison and evaluation of artificial immune systems in inter-release software fault prediction. Appl. Soft Comput. **96**, 1–18 (2020)
13. Park, H., Choi, J., Kim, D., Hong, S.J.: Artifical immune system for fault detection and classification of semiconductor equipment. Electronics **10**(8), 944 (2021). https://doi.org/10.3390/electronics10080944
14. Fasanotti, L., Cavalieri, S., Dovere, E., Gaiardelli, P., Pereira, C.E.: An artificial immune intelligent maintenance system for distributed industrial environments. Proc. Inst. Mech. Eng. Part O J. Risk Reliab. **232**(4), 401–414 (2018). https://doi.org/10.1177/1748006x18769208
15. O'Keeffe, J.: Immune-Inspired Fault Diagnosis for Robot Swarms. University of York. Electronic Engineering, 127 p. (2019)
16. Mohapatra, S., Khilar, P.M., Swain, R.: Fault diagnosis in wireless sensor network using clonal selection principle and probabilistic neural network approach. Int. J. Commun. Syst. **32**(16), 1–20 (2019). https://doi.org/10.1002/dac.4138
17. Lan, C., Zhang, H., Sun, X., Ren, Z.: An intelligent diagnostic method based on optimizing B-cell pool clonal selection classification algorithm. Turk. J. Electr. Eng. Comput. Sci. **28**, 3270–3284 (2020)
18. Tian, Y., Liu, X.: A deep adaptive learning method for rolling bearing fault diagnosis using immunity. Tsinghua Sci. Technol. **24**(6), 1–14 (2019). https://doi.org/10.26599/TST.2018.9010144
19. Sahu, S., Kumar, P.B., Parhi, D.R.: Analysis of hybrid CSA-DEA method for fault detection of cracked structures. J. Theor. Appl. Mech. **57**(2), 369–382 (2019). https://doi.org/10.15632/jtam-pl/104590
20. Pinto, C., Pinto, R., Gonçalves, G.: Towards bio-inspired anomaly detection using the cursory dendritic cell algorithm. Algorithms **15**(1), 1–28 (2022). https://doi.org/10.3390/a15010001
21. Häring, I.: Failure modes and effects analysis. In: Häring, I. (ed.) Technical Safety, Reliability and Resilience, pp. 101–126. Springer, Singapore (2021). https://doi.org/10.1007/978-981-33-4272-9_7
22. Signoret, J.P., Leroy, A.: Hazard and operability study (HAZOP). In: Signoret, J.P., Leroy, A. (eds.) Reliability Assessment of Safety and Production Systems. Springer Series in Reliability Engineering, pp. 157–164. Springer, Cham (2021). https://doi.org/10.1007/978-3-030-64708-7_9
23. Fuentes-Bargue, J.L., González-Cruz, M., González-Gaya, C., Baixauli-Pérez, M.P.: Risk analysis of a fuel storage terminal using HAZOP and FTA. Int. J. Environ. Res. Public Health **14**(705), 1–26 (2017). https://doi.org/10.3390/ijerph14070705
24. Schaust, S., Szczerbicka, H.: Artificial immune systems in the context of misbehavior detection. Cybern. Syst. **39**(2), 136–154 (2008). https://doi.org/10.1080/01969720701853434
25. Read, M., Andrews, P.S., Timmis, J.: An introduction to artificial immune systems. In: Rozenberg, G., Bäck, T., Kok, J.N. (eds.) Handbook of Natural Computing, pp. 1575–1597. Springer, Heidelberg (2012). https://doi.org/10.1007/978-3-540-92910-9_47
26. Chen, Y., Wang, X., Zhang, Q., Tang, C.: Unified artificial immune system. In: Proceedings of 5th International Conference on Computational Intelligence and Communication Networks. Mathura, pp. 617–621 (2013). doi:https://doi.org/10.1109/CICN.2013.135
27. Samigulina, G.A., Massimkanova, Z.: Development of modified cooperative particle swarm optimization with inertia weight for feature selection. Cogent Eng. **7**(1), 1–10 (2020)
28. Samigulina, G.A., Samigulina, Z.I.: Modified immune network algorithm based on the Random Forest approach for the complex objects control. Artif. Intell. Rev. **52**(4), 2457–2473 (2018). https://doi.org/10.1007/s10462-018-9621-7

29. Samigulina, G., Samigulina, Z.: Diagnostics of industrial equipment and faults prediction based on modified algorithms of artificial immune systems. J. Intell. Manuf. **33**, 1–18 (2021). https://doi.org/10.1007/s10845-020-01732-5
30. Permanent technological regulations for the process of extracting LPG at U-700. TengizChevroil, TP-ZVP-700-11 (2017)
31. KTL TCO COMPLEX COORDINATOR DAILY REPORT as of 11-January-2017

A Nonparametric Pooling Operator Capable of Texture Extraction

V. Vigneron(✉) and H. Maaref

Univ Evry, Université Paris-Saclay, IBISC EA 4526, Evry, France
{vincent.vigneron,hichem.maaref}@univ-evry.fr

Abstract. Much of Convolutional neural networks (CNNs)'s profound success lies in translation invariance. The other part lies in the almost infinite ways of arranging the layers of the neural network to make decisions in particular in computer vision problems, taking into account the whole image. This work proposes an alternative way to extend the pooling function, we named rank-order pooling, capable of extracting texture descriptors from images. Efforts to improve pooling layers or replace-add their functionality to other CNN layers is still an active area of research despite already a quite long history of architecture. Rank-order clustering is non-parametric, independent of geometric layout or image region sizes, and can therefore better tolerate rotations. Many related metrics are available for rank aggregation. In this article we present the properties of some of these metrics, their concordance indices and how they contribute to the efficiency of this new pooling operator.

Keywords: Deep CNN · Pooling function · Rank aggregation · LBP · Optimization · Linear programming · Rank-order · Contour extraction · Segmentation

1 Introduction

A deep CNN stacks four different processing layers: convolution, pooling, ReLU and fully-connected [5]. CNNs architecture is augmented by multi-resolution (*pyramidal*) structures which come from the idea that the network needs to see different levels of resolutions to produce good results.

Placed between two convolutional layers, the pooling layer receives several input feature maps. Pooling [4] (a) reduces the number of parameters in the model (subsampling) and computations in the network while preserving their important characteristics (b) improves the efficiency of the network (c) prevents overtraining. Even though pooling layers do not have parameters, they affect the backpropagation (derivatives) calculation. Back-propagating through the max-pooling layer simply selects the maximum neuron from the previous

This research was supported by the program *Cátedras Franco-Brasileiras no Estado de São Paulo*, an initiative of the French consulate and the state of São Paulo (Brazil). We thank our colleagues Prof. João M. T. Romano, Dr. Kenji Nose and Dr. Michele Costa, who provided insights that greatly assisted this work.

G. Nicosia et al. (Eds.): LOD 2022, LNCS 13811, pp. 93–107, 2023.
https://doi.org/10.1007/978-3-031-25891-6_8

layer (on which the max-pooling was performed) and continues backpropagation only through it. The max function is locally linear for the activation that obtained the max, so its derivative is 1, and 0 for the activation which did not succeed. This is conceptually very similar to the differentiation of the activation function $\text{ReLU}(x) = \max(0, x)$.

Suppose a layer H_ℓ comes on top of a layer $H_{\ell-1}$. Then the activation of the ith neuron of the layer H_ℓ is

$$H_{\ell i} = f(\sum_j w_{ij} H_{(\ell-1)i}), \tag{1}$$

where f is the activation function and $W = \{w_{ij}\}$ are the weights. The derivation of Eq. (1) by the chain-rule gives the gradient flows as follows

$$\text{grad}(H_{(\ell-1)i}) = \sum_i \text{grad}(H_{\ell i}) f' w_{ij}. \tag{2}$$

In the case of max-pooling, $f = \text{id}$ for the max neuron and $f = 0$ for all other neurons, so $f' = 1$ for the max neuron of the previous layer and $f' = 0$ for all other neurons. Back-propagating through the max-pooling layer simply selects the max neuron from the previous layer (on which the max-pooling was done) and continues back-propagation only through that.

The max-pooling function downsamples the input representation (image, hidden layer output matrix, etc.) making it less sensitive to re-cropping, rotation, shifting, and other minor changes.

Weaknesses of pooling functions are well identified [21]: (a) they don't preserve all spatial information (b) the maximum chosen by the max-pooling in the pixel grid is not the true maximum (c) average pooling assumes a single mode with a single centroïd. How optimally take into account the characteristics of the input image grouped in the pooling operation ? Part of the answer lies in the work of Lazebnik's who demonstrated the importance of the spatial structure of close neighborhoods [8]: indeed, local spatial variations of image pixel intensities (also called textures) characterize an "organized area phenomenon" [11] which cannot be captured in pooling layers.

This paper proposes a new pooling operation, independent of the geometric arrangement or sizes of image regions, and can therefore better tolerate rotations [1].

Notations. Throughout this paper small Latin letters $a, b, \dots$ represent integers. Small bold letters $\mathbf{a}, \mathbf{b}$ are put for vectors and capital letters A, B for matrices or tensor depending of the context. The dot product between two vectors is denoted $<\mathbf{a}, \mathbf{b}>$. We denote by $\|\mathbf{a}\| = \sqrt{<\mathbf{a}, \mathbf{a}>}$, the ℓ_2 norm of a vector. $X_1, \dots, X_n$ are non ordered variates, $x_1, \dots, x_n$ non ordered observations. "Ordered statistics" means either $p_{(1)} \leq \dots \leq p_{(n)}$ (ordered variates) and $p_{(1)} \leq \dots \leq p_{(n)}$ (ordered observations). The extreme order statistics are $p_{(1)} = \min\{x_1, x_2 \dots, x_n\}$, $p_{(n)} = \max\{x_1, x_2, \dots, x_n\}$. The sample range is $p_{(n)} - p_{(1)}$. The $p_{(i)}$ are necessarily *dependent* because of the inequality relations among them.

Definition 1 (Savage [12]). *The rank order corresponding to the n distinct numbers $x_1, \ldots, x_n$ is the vector $\boldsymbol{r} = (r_1, \ldots, r_n)^T$ where r_i is the number of x_j's$\leq x_i$ and $i \neq j$.*

The rank order $\boldsymbol{r}$ is always unambiguously defined as a *permutation* of the first n integers.

2 Texture Encoding

Local image descriptors are compact and rich descriptions capable of encoding local patterns *e.g.* local binary patterns (LBP) and its variants [2]. Since imagery data is noisy, locally correlated, and usually with too many data points per "unit" of useful information, these intermediate representations lead to an understanding of the scene in an image without the noisy and devoid influence sense of very many pixels that carry little inferable information. These representations typically encode either structural (texture) information in the form of a set of repeated primitive *textons*, or statistical information indicating how different pixel intensities are arranged in a local neighborhood.

The amount of information extracted from different regions of an image depends on (a) the size of the neighborhood (b) the reading order of the neighbors (c) and the mathematical function that is used to extract the relationship between two neighboring pixels. Texture gives information on the spatial arrangement of the grey levels in an image.

For instance, given a monochromatic image I, LBP generates 8-bit string for a 3×3 neighborhood by computing the Heaviside function $t(x)$ of the difference of neighboring pixel $\{g_i | i = 0, \ldots, p-1\}$ and the central pixel g_c *i.e.* $(g_i - g_c)$ (see Fig. 1), but $(g_c - g_i)$ in case of Census transform. The only difference between these two descriptors is the reading order of neighboring pixels and the sign of the difference which results in 2 different bit patterns. Given the 8-bit string, the LBP code is calculated in the range of 0 to 255 as:

$$L_{p,r} = \sum_{i=0}^{p-1} 2^p \cdot t(g_i - g_c) \text{with } t(x) = \begin{cases} 1 & \text{if } x \geq 0 \\ 0 & \text{otherwise} \end{cases}, \tag{3}$$

where p counts the number of pixels in the neighborhood of g_c, considering the distance R between central pixel g_c and the neighboring pixel g_i.

example

121	201	200
190	100	164
78	77	65

→ thresholded

1	1	1
1		1
0	0	0

→ LBP weights

1	2	4
128		8
64	32	16

Pattern $(10001111)_2$
LBP 128+8+4+2+1 = 143

Fig. 1. Example of 3×3 image neighborhood ($p = 8$ and $R = 1$).

Invariance w.r.t. any monotone transformation of the gray scale is obtained by considering in (3) the signs of the differences $t(g_i - g_c), i = 0, \ldots, P - 1$.

But the *independence* of g_c and $\{|g_0 - g_c|, \ldots, |g_{P-1} - g_c|\}$ is not warranted in practice. Moreover, under certain circumstances, LBP misses the local structure as it does not consider the central pixel. The binary data produced by these descriptors are sensitive to noise mainly in uniform regions of an image. To reduce the noise sensitivity [15] have proposed a 3-level operator which describes a pixel relationship with its neighbor by a ternary code *i.e.* $-1, 0, 1$ rather than a binary code *i.e.* 0, 1. These methods ignore the relationship between neighboring pixels themselves and fail to get robust results [10]. It is inspired by the traditional image feature extraction method HOG [19] which extracts the relative orders on small local areas.

In this paper, a new encoding function is used to decide the relative order between the pixels in each local area of a given pixel. This cost function is minimized by linear programming and generates rank orders (including ex-aequos positions) which could be used in contour detection, segmentation or adaptive image quantization.

3 Total Rank Order for Image Texture Encoding

Let $A = \{\mathsf{a}_1, \mathsf{a}_2, \ldots, \mathsf{a}_n\}$ be a set of alternatives, candidates, individuals, etc. with cardinality $|A| = n$ and let V be a set of voters, judges, criteria, etc. with $|V| = m$. The data is collected in a $(n \times m)$ table T of general term $\{t_{ij}\}$ crossing the sets A and V (Table 1a). t_{ij} can be marks ($t_{ij} \in \mathbb{N}$), value scales ($t_{ij} \in \mathbb{R}$), ranks of notes or binary ($t_{ij} \in \{0, 1\}$ such as opinion yes/no).

In the following, m is the number of alternatives and m the number of voters.

Table 1. The data are collected in a $(n \times m)$ table T.

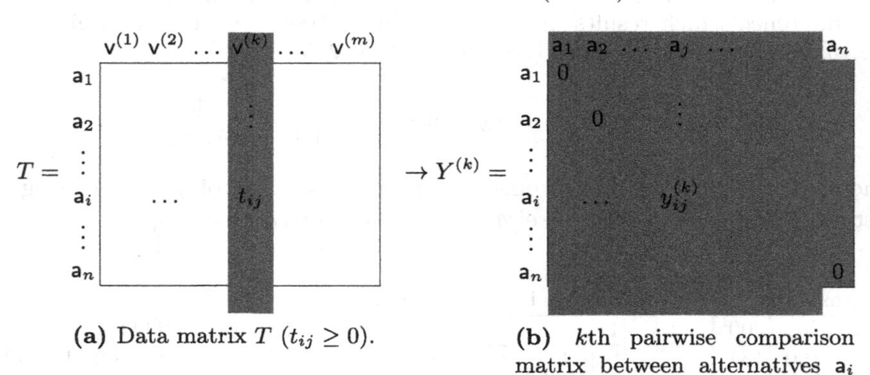

(a) Data matrix T ($t_{ij} \geq 0$).

(b) kth pairwise comparison matrix between alternatives a_i and a_j.

The objective is to find a distribution of values $\boldsymbol{x}^*$ attributed by a virtual judge to the n individuals of the studied population by minimizing the disagreements of opinions of the m judges, *i.e.*

$$\boldsymbol{x}^* = \arg\min_{\boldsymbol{t}} \sum_{k=1}^{m} \mathrm{d}(\boldsymbol{t}, \boldsymbol{t}^{(k)}), \quad \text{s.t.} \quad \boldsymbol{x} \geq 0, \tag{4}$$

$\mathrm{d}(\boldsymbol{t}, \boldsymbol{t}^{(k)})$ being a metric measuring the proximity between $\boldsymbol{t}$ and $\boldsymbol{t}^{(k)}$, chosen a priori, and $\boldsymbol{t}^{(k)}$ is the kth column of the table T. Depending on the nature of d, we will see that we will be dealing with a nonlinear optimization program with an explicit or implicit solution.

One could also stand the dual problem of the previous one, *i.e.*: is there a distribution of ratings/values that could have been attributed by the m voters to a virtual alternative 'a' summarizing the behaviour of the set of individuals A [20]? The first problem is linked to the idea of aggregating of points of view, the second to the idea of summarizing behaviors.

3.1 Explicit Resolution

The distance $\mathrm{d}(\boldsymbol{t}^{(k)}, \boldsymbol{t}^{(k')})$ between the voter k and the voter k' can be chosen for instance as the *disagreement* distance $\sum_{i=1}^{n} \mathrm{sgn}\,|t_{ik} - t_{ik'}|$ where $\mathrm{sgn}(a) = a/|a|$, the holder distance $(\sum_{i=1}^{n} |t_{ik} - t_{ik'}|^m)^{1/m}, m > 1$, the d_∞ distance defined as $\max_i |t_{ik} - t_{ik'}|$ which is also the limit of the Holder distance when $m \to \infty$, la distance du χ^2, etc.

Surprisingly, the explicit resolution with linear programming optimization explicitly leads to central tendency statistics, *e.g.* the mode for the disagreement distance, the median for the absolute deviations, the mean for the Holder distance ($m = 2$), etc. [3].

Let's take an example. Assume the t_{ik} represent the scores obtained by a neurologist i in m services: emergency, surgery, palliative care, rehabilitation, etc. and suppose that n neurologists are candidates for a position in a neurovascular unit (NVU). What can be the hiring technique of the chief of the NVU? If he is careful and wants his staff to be interchangeable throughout the year, he can take either of these "medium" measures. If he wants to be efficient, and accepts that the neurologist is not interchangeable, he will take the disagreement metric, and choose the neurologist whose mode is the strongest. If the chief of the NVU wants to satisfy neurologists with the breadth of their capabilities, he will use the d_∞ metric.

In this example it is asked *in fine* to compare different individuals using a common scale.

On the opposite, ranked variables are (a) easily collated (b) easily categorized (c) easy to analyze (d) they have an intuitive and plausible interpretation (e) they provide the best possible description of the process of ranking items as performed by a human (f) provide very good concordance indicator.

We still have n voters, and m alternatives but in this case, the m voters give a ranking of the candidates with or without ex-aequo in the form of the table T or $t_{ik} = r_{ik}$ (rank given by voter k to individual i). Each ranking V^k is a permutation P of the first n integers identifying the candidates in case of strict preference.

3.2 Order Disagreement Distance

When looking for the optimal consensus $\boldsymbol{r}^*$ of m voters who attributed the votes $\boldsymbol{r}^{(1)}, \boldsymbol{r}^{(2)}, \ldots, \boldsymbol{r}^{(m)}$ to the n candidates $\{\mathsf{a}_1, \mathsf{a}_2, \ldots, \mathsf{a}_n\}$, we minimize the absolute deviation distance

$$\sum_{k=1}^{m} \sum_{i=1}^{n} |r_i^* - r_{ik}|. \tag{5}$$

where, for ease of writing, $r_{ik} = r_i^{(k)}$. A voter can give *ex-aequo* positions. Note that $\boldsymbol{r}^{(k)} \notin \mathcal{S}_n$, with $\mathcal{S}_n$ the symmetric group of the $n!$ permutations [6] because of the ex-aequo. Hence $\boldsymbol{r}^* \notin \mathcal{S}_n$.

To define this distance, we define a new set of tables $\{Y^{(1)}, \ldots, Y^{(m)}\}$, where $Y_{ij}^{(k)} = \mathbb{1}_{i<j}$ denotes the indicator matrix for which $y_{ij}^{(k)} = 1$ if the rank of the alternative a_i is *less* than the alternative a_j and 0 otherwise (see Table 1b).

Using the tables $Y^{(k)}$

$$\sum_{k=1}^{m} \sum_{i=1}^{n} |r_i^* - r_{ik}| = \frac{1}{2} \sum_{i} \sum_{j} |y_{ij}^{(k)} - y_{ij}^{(k')}| \tag{6}$$

or $\frac{1}{2} \sum_i \sum_j (y_{ik}^{(k)} - y_{ik}^{(k')})^2$. Which can be simplified in the case of total order as

$$\sum_{i} \sum_{j} y_{ij}^{(k)} y_{ji}^{(k')}. \tag{7}$$

The $y_{ij}^{(k)}$ verify *(a)* the transitivity relationship: if $y_{ij}^{(k)} = 1$ and $y_{j\ell}^{(k)} = 1$ then $y_{i\ell}^{(k)} = 1$, equivalent to $y_{ij} + y_{ji} - y_{ik} \leq 1, i \neq j \neq k,\ y_{ij} \in \{0, 1\}$ *(b)* $y_{ij} + y_{ji} \leq 1$, with equality only when $\boldsymbol{r}^{(k)}$ is a total order.

As $y_{ij}^2 = y_{ij} = {y_{ij}^{(k)}}^2 = y_{ij}^{(k)} = 0$ or 1, the distance function associated to Eq. (6) is given by

$$\frac{1}{2} \left[\sum_{i=1}^{n} \sum_{j=1}^{n} m y_{ij} + \sum_{i=1}^{n} \sum_{j=1}^{n} \left(\sum_{k=1}^{m} y_{ij} \right) - 2 \sum_{i=1}^{n} \sum_{j=1}^{n} y_{ij} \sum_{k=1}^{m} y_{ij}^{(k)} \right]. \tag{8}$$

Let $\alpha_{ij} = \sum_{k=1}^{p} y_{ij}^{(k)}$ the total number of voters preferring alternative a_i to a_j and define a matrix $\mathcal{A} = \{\alpha_{ij}\}$, summing the m matrices $Y^{(k)}$ associated to the rankings $\boldsymbol{r}^{(k)}$ of the voters $V^{(k)}$. Equation (8) can be rewritten using the α_{ij}:

$$\frac{1}{2} \left[\sum_{i=1}^{n} \sum_{j=1}^{n} m y_{ij} + \sum_{i=1}^{n} \sum_{j=1}^{n} \alpha_{ij} - 2 \sum_{i=1}^{n} \sum_{j=1}^{n} \alpha_{ij} y_{ij} \right] \tag{9}$$

As $\sum_{i=1}^{n} \sum_{j=1}^{n} \alpha_{ij} < \dfrac{n(n-1)}{2}$, and let $K = \sum_{i=1}^{n} \sum_{j=1}^{n} \alpha_{ij}$ a constant. Then (9) is:

$$K - \sum_{i=1}^{n} \sum_{j=1}^{n} (\alpha_{ij} - m/2) y_{ij}. \tag{10}$$

Finally the search of a order given by a matrix Y is the optimal solution of the following *linear program*

$$\max_Y \sum_{i=1}^{n}\sum_{j=1}^{n}(\alpha_{ij} - m/2)y_{ij} \quad \text{s.t.} \quad \alpha_{ij} = \sum_{k=1}^{m} y_{ij}^{(k)},$$
$$y_{ij} + y_{ji} = 1, i < j, y_{ii} = 0$$
$$y_{ij} + y_{ji} - y_{ik} \leq 1, i \neq j \neq k, \; y_{ij} \in \{0, 1\}. \tag{11}$$

From a machine learning perspective, Eq. (11) is remarkably simple and provide an exact solution using a linear programming solver [7].

4 Rank-Order Principal Components

4.1 Importance Ranking

Several strategies have been proposed in the literature to extract important variables or develop parsimonious models and deal with the dimensionality. The dimension of observed data being generally higher than their intrinsic dimension, it is theoretically possible to reduce the dimension without loosing information.

Among the unsupervised tools, principal component analysis (PCA) or factor analysis (FA) are certainly the most used techniques to optimize the understanding insight into of a data set. They aim to project the data onto a lower dimensional subspace in which axes are constructed either by maximizing the variance of the projected data or by explaining the overall covariance structure.

PCA and FA are both linear tools. This means that nonlinear dependencies are not taken into account.

The question is simply: can we extract a set of the most decorrelated rank-order variables to each other capable of capturing distinct information? The overall framework for this objective suggests a rank-order decomposition. The principle remains remarkably simple: it consists into a re-distributive effect of the rank variables – similar to PCA – on a Hilbert space.

Lemma 1 (Vigneron and Duarte [16]). *Consider a collection of rank-orders (with ex-aequo or not) $R = \{\boldsymbol{r}_1, \boldsymbol{r}_2, \ldots, \boldsymbol{r}_m\}$ (data). It is always possible to extract a total rank-order component $\boldsymbol{g}_\ell$ minimizing its proximity to the data $\{\boldsymbol{r}_1, \boldsymbol{r}_2, \ldots, \boldsymbol{r}_m\}$ and simultaneously maximizing the distance to the collection of previously calculated ranks $\{\boldsymbol{g}_1, \ldots, \boldsymbol{g}_{\ell-1}\}$.*

The algorithm is as follows:

Algorithm 1. Rank-order decomposition Algorithm.

Require: $Y^{(1)}, \ldots, Y^{(m)} \leftarrow \{\boldsymbol{r}_1, \boldsymbol{r}_2, \ldots, \boldsymbol{r}_m\}$ {order disagreement matrices $Y^{(k)} = \{y_{ij}^{(k)}\}\}\vee$ stack $A = \emptyset$ {contain the reranked components}
Ensure: $\{\boldsymbol{g}_1, \ldots, \boldsymbol{g}_m\}$ {Postcondition}
1: **for** $\ell = 1$ **to** m **do**
2: Compute $\alpha_{ij} = \sum_{k=1}^{m} y_{ij}^{(k)}$, $\beta_{ij} = \sum_{k=1}^{\ell-1} z_{ij}^{(k)}$
3: $\mathcal{A} = \{\alpha_{ij}\}, \mathcal{B} = \{\beta_{ij}\}$
4: **LP**$(\mathcal{A}, \mathcal{B}, Z^{(\ell)})$ under constraints (12) {solve linear program}
5: $\boldsymbol{g}_\ell \leftarrow Z^{(\ell)}$
6: **end for**
7: **return** $\{\boldsymbol{g}_1, \boldsymbol{g}_2, \ldots, \boldsymbol{g}_m\}$

At stage ℓ, the search of the ℓth total order $\boldsymbol{g}_\ell$ represented by the matrix $Z^{(\ell)}$ in the case of the order disagreement distance (see Sect. 3.2) reduced to

$$\begin{aligned} &\max_{Z^{(\ell)}} \sum_{i=1}^{n}\sum_{j=1}^{n}(\alpha_{ij} - m/2)z_{ij}^{(\ell)} - \max_{Z^{(\ell)}} \sum_{i=1}^{n}\sum_{j=1}^{n}\beta_{ij} z_{ij}^{(\ell)} \quad \text{s.t.} \\ &\alpha_{ij} = \sum_{k=1}^{m} y_{ij}^{(k)}, \ \beta_{ij} = \sum_{k=1}^{\ell-1} z_{ij}^{(k)}, \ z_{ij}^{(\ell)} + z_{ji}^{(\ell)} = 1, i < j, \\ &z_{ii} = 0 \ \ z_{ij}^{(\ell)} + z_{ji}^{(\ell)} - z_{ik}^{(\ell)} \leq 1, i \neq j \neq k, \ z_{ij}^{(\ell)} \in \{0, 1\}. \end{aligned} \tag{12}$$

Algorithm 1 stops when $\ell = m$ and provide $\{\boldsymbol{g}_1, \ldots, \boldsymbol{g}_m\}$ such that $\boldsymbol{g}_\ell$ is the most decorrelated to the previous ranks $\{\boldsymbol{g}_1, \ldots, \boldsymbol{g}_{\ell-1}\}$. Until now, on the contrary to PCA, there is no index capable of indicating the quantity of information captured by each vector $\boldsymbol{g}_\ell$.

4.2 Experiment: Application of Rank-Order Pooling (RO) Principal Components to Textured Image

Now consider the "neighborhood" of a pixel 'p' in an image I, *i.e.* the set of pixels touching it (a maximum of 8 pixels) as shown in Fig. 1.

The 4×4 image I in Fig. 2a can be transformed using the 8-connectivity into the 4×8 matrix C (Fig. 2b) where column 0 refers to the 1st pixel (clockwise), column 1 refers to the 2nd pixel, etc. The 8 neighboring pixels around 'p' can be seen as "voters" from which a pseudo-rank is expected. The matrix C is then transformed into the rank-matrix $R = \{\boldsymbol{r}_1, \boldsymbol{r}_2, \ldots, \boldsymbol{r}_8\}$ (Fig. 2c) by simple ordering on which Algorithm 1 can be applied. Note that usually the peripheral zone is filled with zero, *i.e.* added with borders of zeros, to preserve the size of the image.

As an illustration LBP is applied to original picture Fig. 3a of Lena which provides the texture representation given in Fig. 3b. From the Fig. 3a, Algorithm 1

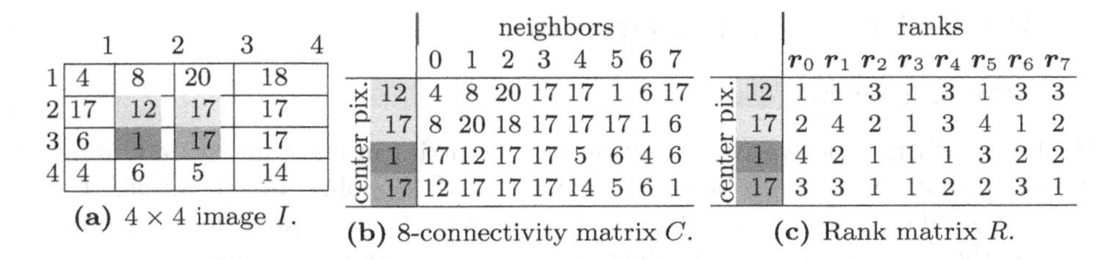

	1	2	3	4
1	4	8	20	18
2	17	12	17	17
3	6	1	17	17
4	4	6	5	14

(a) 4×4 image I.

center pix.	neighbors 0	1	2	3	4	5	6	7
12	4	8	20	17	17	1	6	17
17	8	20	18	17	17	17	1	6
1	17	12	17	17	5	6	4	6
17	12	17	17	17	14	5	6	1

(b) 8-connectivity matrix C.

center pix.	ranks r_0	r_1	r_2	r_3	r_4	r_5	r_6	r_7
12	1	1	3	1	3	1	3	3
17	2	4	2	1	3	4	1	2
1	4	2	1	1	1	3	2	2
17	3	3	1	1	2	2	3	1

(c) Rank matrix R.

Fig. 2. The colored pixels are 8-connected. The 8 neighbor pixels around the central pixel 'p' can be seen as "voters" from which we expect a pseudo-rank. (Color figure online)

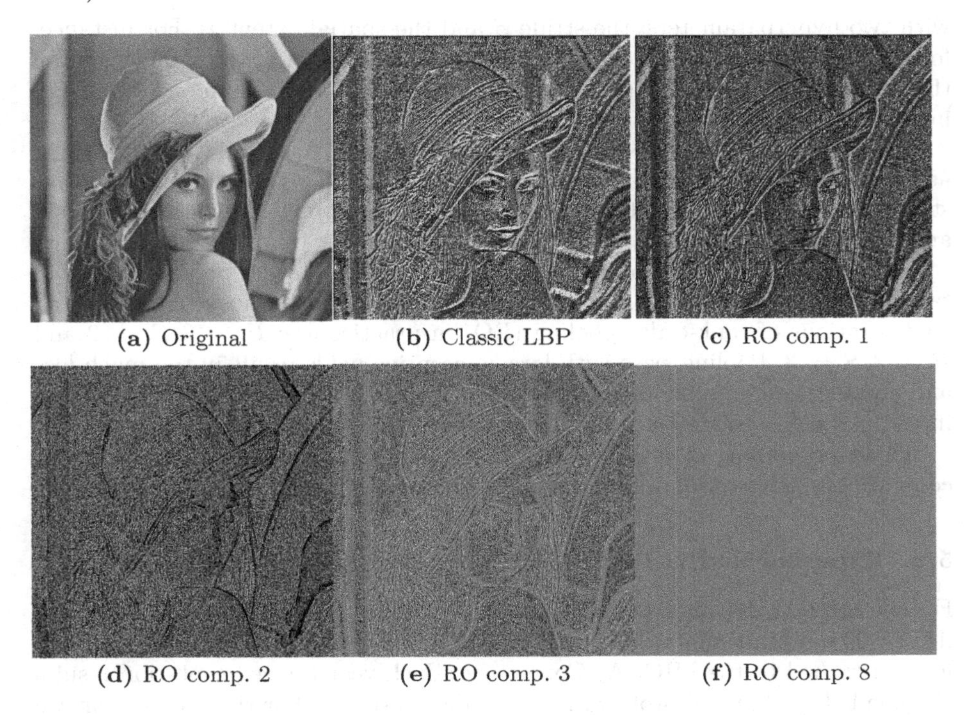

(a) Original **(b)** Classic LBP **(c)** RO comp. 1

(d) RO comp. 2 **(e)** RO comp. 3 **(f)** RO comp. 8

Fig. 3. Lena' original (a) is compared to classic LBP representation (b) and with the 1st, 2nd, 3rd and eighth rank-order components obtained from Algorithm 1. The eighth component is apparently the less informative component.

provides a new set of images (Figs. 3c–3f). Usually, the 1st plot is more informative the 2nd one, which itself is more informative than the 3rd, etc. Vigneron and Duarte have shown [17] that this decomposition is sensitive to the chosen metric.

5 Rank-Order Pooling Operator

5.1 Pooling Layer

Pooling perform a derivative operation of the entries in the window (*e.g.* max, mean, median, etc.) to "collapse" over values. Max-pooling down-samples the convolutional output in a discrete way. The method of calculating the 2d- pooling layer is the same as for the convolutional layer and produces a image of size:

$$\left(\left\lfloor\frac{I_x - P}{S}\right\rfloor + 1, \left\lfloor\frac{I_y - P}{S}\right\rfloor + 1\right) \tag{13}$$

with two hyperparameters: the stride S and the spatial extent P. For instance, for a 13×13 image, 2×2 pooling window and a stride of 2, the x-dimension of the resulting image is floor$((13 - 2)/2) + 1 = 6$. It is not common to pad the input using zero-padding for the pooling.

The rules presented in Sect. 1 apply to all types of pooling layers with some adjustments. Forward propagation for max pooling means creating a mask that stores the position of the retained maximum values through which the gradients are transferred.

Instead of taking the maxima in each filter, the rank-order pooling approach takes the consensus rank in a neighborhood. Interestingly, the most common tested configurations of rank-order pooling (RO) in practice are: $P = 3, S = 2$ and $P = 3, S = 3$. Pooling sizes with larger receptive fields result in too much loss and are destructive. It should also be noted that as max-pooling, the RO pooling introduces zero parameters since it computes a fixed function of the input.

Different pooling operations were performed in a categorization context to compare the behavior of different pooling operations in a categorization context.

5.2 Experiment: RO Pooling for Automatic Tumor Segmentation

For the second experiment we chose a real data set which is renowned for its difficulty. The brain tumor segmentation (BraTS) [9] challenge is a recurring challenge attached to the MICCAI Conference. Each year the segmentation results become better, but the problem is an ongoing research. For this experiment we use the high grade glioma part of the BraTS 2017 data-set[1]. It contains multimodal MRI of 210 patients which were manually segmented my experts, *i.e.* a ground truth is available. On these image three different classes have to be segmented from the background. The enhancing tumor, the necrotic and non-enhancing tumor and as a third class the peritumoral edema. This makes it an ideal real data-set for supervised learning of a multi-class segmentation task.

Architectures. We compare architectures containing the RO pooling to model without [18]. The U-Net was chosen because it is the archetype of modern convolutional networks used for bio-medical image segmentation tasks and achieved

[1] www.med.upenn.edu/sbia/brats2017.html.

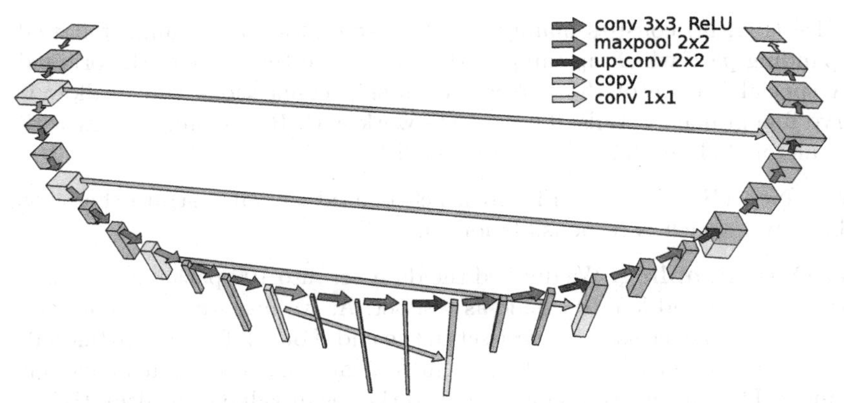

(a) U-Net. Arrows represent operations and cubes represent feature maps where the height of the cube stands for the number of feature maps and the width and depth of the cubes for the size of the feature maps. It is clearly visible that the U-Net uses a pyramidal structure for feature extraction.

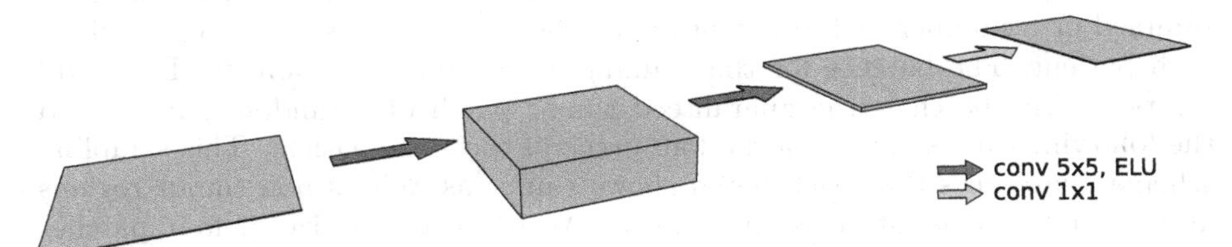

(b) Reference-Net.The most basic CNN. It suffers greatly from a small receptive field. As expected our first experiment showed that it has trouble with objects that are larger than the convolution kernels. This weakness can be overcome either by adding a single transfer layer to the architecture (Transfer-Net), or by adding so many convolutional layers that the chained convolution kernels extend the receptive field to the whole input image (U-Net).

Fig. 4. DNN architectures.

good performance in many applications. To prove that the tasks are not too simple and that the RO layer is responsible for the improved results, a simple reference network with max-pooling layers is included. For evaluating the results we not only use the loss but also the Dice coefficient [14] which is the standard measure for segmentation quality.

All of the following network architectures are implemented in Pytorch[2] and the implementations will be freely accessible. Convolution layers are chosen to keep the output size the same as the input size by padding the input image with zeros. Also all architectures are followed by a softmax layer. And cross entropy is used as loss function. The loss is minimised using Adam [13]. The best learning rate for each architecture has been determined experimentally.

[2] pytorch.org.

U-Net: [18] (Fig. 4a) consists mainly of a feature extraction pyramid followed by an expanding path which up-samples the features to the space of the original image. A special feature of the U-Net are its skip connections which allow it to preserve fine grained details. The U-Nets work with RO pooling layers with hyperparameters 3,3 and 3,2 (F, S), see Sect. 5.1.

Reference Net (R-Net): See Fig. 4b which is used to demonstrate that it is our pooling layer which is responsible for our results.

Training Data Sampling: We divided the data-set into 100 patients for learning, 4 for validation and left 106 aside as test set. As we just wanted to demonstrate a concept we never used the test set in the end. Four different MRI modalities (compare Fig. 5a)–d)) are available, which means that the input image has four channels. The images were normalized to the mean value of healthy tissue, *i.e.* the intensity values were divided by the intensity value of the highest peak in the histogram of brain tissues.

As the whole MRI is too large to process in one step it was broken down into patches of size $64 \times 64 \times 10$. Patches are generated until the whole brain is sampled or a number of 100 patches is reached. This procedure is repeated for each patient. The batches for the training are generated as follows: Each odd sample of the batch is the guaranteed tumor patch of a random patient and the following even sample is a random patch of the same patient. This sampling scheme guarantees that each batch shows tumor as well as non tumor regions and effectively combats class imbalance. We use a batch size of four patches per batch, *i.e.* after 50 patches, the guaranteed tumor patch of all 100 training patients have been seen. This may be considered an epoch. All patches are not guaranteed to be seen during training.

We forego any advanced data augmentation. Only a small random constant is added to the entire patch.

Table 2. Network configurations for the BraTS experiment and their final dice score on the training and validation set. $m \in \mathbb{N}$ is the multiplicity, *i.e.* the number of input channels.

Model	U-Net Maxpool	U-Net 3.3	U-Net 3.2	R-Net 3.2	R-Net maxPool
Kernel size	3	3	5	5	5
Parameters	31032516	7698116	141929	19359	19359
Learning rate	0.001	0.002	0.0005	0.002	0.002
m	–	–	10	10	–
Training speed τ	10737	15032	13322	28761	48767
Training dice	0.96	0.82	0.87	0.80	0.64
Validation dice	0.89	0.51	0.77	0.51	0.49

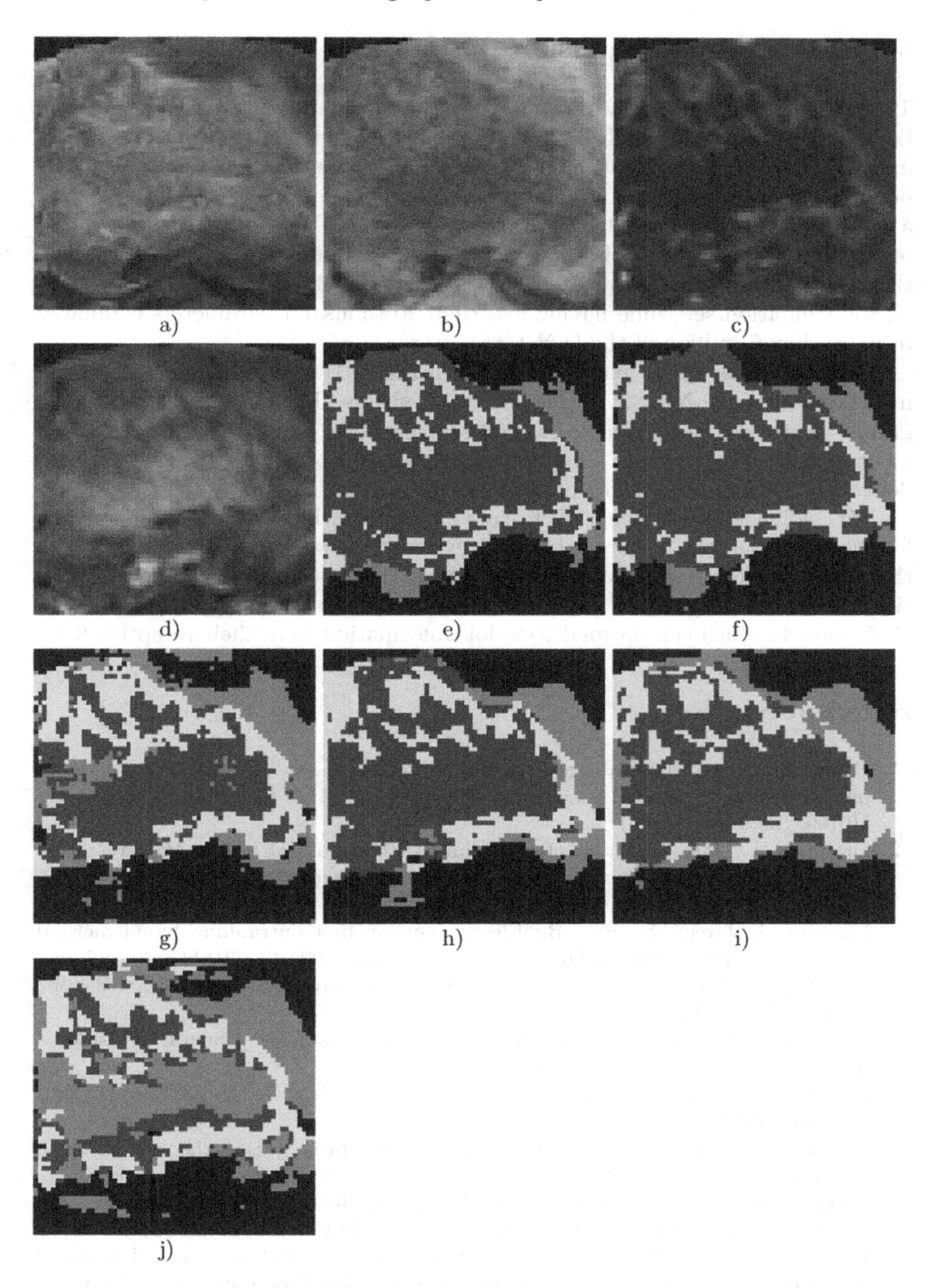

Fig. 5. One patch out of the BraTs validation-set. The input image a)–d) consists of the MRI modalities Flair a), T1 b), contrast enhanced T1 c) and T2 d). The ground truth is seen in e). The following are the predictions of the five networks: U-Net 3.2 f), U-Net 3.3g), U-Net max-pooling h), R-net 3.2 i) and the R-Net max-pooling).

6 Results and Conclusion

This real data-set is a way more difficult task than texture characterization. Looking at the dice score (Table 2) shows a lower loss value does not necessarily indicate a better segmentation performance. The networks based on the RO pooling perform well. Of course we cannot expect to beat the U-Net 3.2 with a network which has only 0.44% of the parameters. But the R-Net 3.2 comes remarkably close in terms of performance and clearly beats the U-Net 3.3 which still has 56 times more parameters. Even the R-Net 3.2 is close to the U-Net 4 on the validation set while having less than 20 thousand parameters compared to more than 7 million of the U-Net 4.

Segmentations of one patch are given in Fig. 5. Here the quality of the segmentation of the U-Net 3.2 and R-Net 3.2 is visible. They are very close to the ground truth and the U-Net MaxPool segmentation while the U-Net 3.3 segmentation is clearly worse. This proves that theRO pooling works, and that it delivers performance at a low parameter cost.

We challenged the concept for feature extraction which has been uncontested for three decades, the feature extraction pyramid. In two experiments we showed that our RO pooling can master the same tasks as a feature extraction in emphasizing low/high frequencies, contours, etc. We shows rank-order pooling leads to CNN models which can optimally exploit information from their receptive field.

References

1. Baker, R.: New order-statistics-based ranking models and faster computation of outcome probabilities. IMA J. Manag. Math. **31**(1), 33–48 (2019). https://doi.org/10.1093/imaman/dpz001
2. Brahnam, S., Jain, L., Nanni, L., Lumini, A. (eds.): Local Binary Patterns: New Variants and Applications. SCI, vol. 506. Springer, Heidelberg (2014). https://doi.org/10.1007/978-3-642-39289-4
3. Figueira, J., Greco, S., Roy, B.: Electre-score: a first outranking based method for scoring actions. Eur. J. Oper. Res. **297**(3), 986–1005 (2022). https://doi.org/10.1016/j.ejor.2021.05.017. https://www.sciencedirect.com/science/article/pii/S0377221721004318
4. Gholamalinezhad, H., Khosravi, H.: Pooling methods in deep neural networks, a review. arXiv abs/2009.07485 (2020)
5. Goodfellow, I., Bengio, Y., Courville, A.: Deep Learning, vol. 1. MIT Press, Cambridge (2016)
6. Gurevich, S., Howe, R.E.: Rank and duality in representation theory. Japan. J. Math. **15**, 223–309 (2020)
7. Iatan, I.F.: Issues in the Use of Neural Networks in Information Retrieval. SCI, vol. 661. Springer, Cham (2017). https://doi.org/10.1007/978-3-319-43871-9
8. Lazebnik, S., Schmid, C., Ponce, J.: Beyond bags of features: spatial pyramid matching for recognizing natural scene categories. In: Proceedings of IEEE Computer Society Conference on Computer Vision and Pattern Recognition, vol. 2, pp. 2169–2178 (2006). https://doi.org/10.1109/CVPR.2006.68

9. Menze, H.: The multimodal brain tumor image segmentation benchmark (brats). IEEE Trans. Med. Imaging **34**(10), 1993–2024 (2015). https://doi.org/10.1109/TMI.2014.2377694
10. Ouslimani, F., Ouslimani, A., Ameur, Z.: Rotation-invariant features based on directional coding for texture classification. Neural Comput. Appl. **31**, 1–8 (2019). https://doi.org/10.1007/s00521-018-3462-9
11. Satoh, Y., Hirata, K., Tamada, D., Funayama, S., Onishi, H.: Texture analysis in the diagnosis of primary breast cancer: comparison of high-resolution dedicated breast positron emission tomography (dbPET) and whole-body PET/CT. Front. Med. **7** (2020). https://doi.org/10.3389/fmed.2020.603303. https://www.frontiersin.org/article/10.3389/fmed.2020.603303
12. Savage, R.: Contributions to the theory of rank-order statistics - the trend case. Ann. Math. Stat. **27**(3), 590–615 (1956)
13. Sen, S., Ozkurt, N.: Convolutional neural network hyperparameter tuning with adam optimizer for ECG classification. In: 2020 Innovations in Intelligent Systems and Applications Conference (ASYU), pp. 1–6 (2020). https://doi.org/10.1109/ASYU50717.2020.9259896
14. Shamir, R., Duchin, Y., Kim, J., Sapiro, G., Harel, N.: Continuous dice coefficient: a method for evaluating probabilistic segmentations. bioRxiv (2018). https://doi.org/10.1101/306977. https://www.biorxiv.org/content/early/2018/04/25/306977
15. Tan, X., Triggs, B.: Enhanced local texture feature sets for face recognition under difficult lighting conditions. In: Zhou, S.K., Zhao, W., Tang, X., Gong, S. (eds.) AMFG 2007. LNCS, vol. 4778, pp. 168–182. Springer, Heidelberg (2007). https://doi.org/10.1007/978-3-540-75690-3_13
16. Vigneron, V., Duarte, L.T.: Toward rank disaggregation: an approach based on linear programming and latent variable analysis. In: Tichavský, P., Babaie-Zadeh, M., Michel, O.J.J., Thirion-Moreau, N. (eds.) LVA/ICA 2017. LNCS, vol. 10169, pp. 192–200. Springer, Cham (2017). https://doi.org/10.1007/978-3-319-53547-0_19
17. Vigneron, V., Tomazeli Duarte, L.: Rank-order principal components. A separation algorithm for ordinal data exploration. In: 2018 International Joint Conference on Neural Networks, IJCNN 2018, Rio de Janeiro, Brazil, 8–13 July 2018, pp. 1–6 (2018)
18. Wang, F., Jiang, R., Zheng, L., Meng, C., Biswal, B.: 3D U-Net based brain tumor segmentation and survival days prediction. In: Crimi, A., Bakas, S. (eds.) BrainLes 2019. LNCS, vol. 11992, pp. 131–141. Springer, Cham (2020). https://doi.org/10.1007/978-3-030-46640-4_13
19. Wang, L., Li, T.: Research on image feature extraction method fusing hog and canny algorithm. In: 2021 4th International Conference on Data Science and Information Technology, DSIT 2021, pp. 208–211. Association for Computing Machinery, New York (2021). https://doi.org/10.1145/3478905.3478947
20. Yadav, N., Yadav, A., Kumar, M.: An Introduction to Neural Network Methods for Differential Equations. SAST, Springer, Dordrecht (2015). https://doi.org/10.1007/978-94-017-9816-7
21. Yu, D., Wang, H., Chen, P., Wei, Z.: Mixed pooling for convolutional neural networks. In: Miao, D., Pedrycz, W., Ślęzak, D., Peters, G., Hu, Q., Wang, R. (eds.) RSKT 2014. LNCS (LNAI), vol. 8818, pp. 364–375. Springer, Cham (2014). https://doi.org/10.1007/978-3-319-11740-9_34

Strategic Workforce Planning with Deep Reinforcement Learning

Yannick Smit[1], Floris den Hengst[2,3](✉), Sandjai Bhulai[2], and Ehsan Mehdad[3]

[1] Universiteit van Amsterdam, Amsterdam, The Netherlands
yannick.smit@xs4all.nl
[2] Vrije Universiteit Amsterdam, Amsterdam, The Netherlands
s.bhulai@vu.nl
[3] ING Bank N.V., Amsterdam, The Netherlands
{Floris.den.Hengst,Ehsan.Mehdad1}@ing.com

Abstract. This paper presents a simulation-optimization approach to strategic workforce planning based on deep reinforcement learning. A domain expert expresses the organization's high-level, strategic workforce goals over the workforce composition. A policy that optimizes these goals is then learned in a simulation-optimization loop. Any suitable simulator can be used, and we describe how a simulator can be derived from historical data. The optimizer is driven by deep reinforcement learning and directly optimizes for the high-level strategic goals as a result. We compare the proposed approach with a linear programming-based approach on two types of workforce goals. The first type of goal, consisting of a target workforce, is relatively easy to optimize for but hard to specify in practice and is called *operational* in this work. The second, *strategic*, type of goal is a possibly non-linear combination of high-level workforce metrics. These goals can easily be specified by domain experts but may be hard to optimize for with existing approaches. The proposed approach performs significantly better on the strategic goal while performing comparably on the operational goal for both a synthetic and a real-world organization. Our novel approach based on deep reinforcement learning and simulation-optimization has a large potential for impact in the workforce planning domain. It directly optimizes for an organization's workforce goals that may be non-linear in the workforce composition and composed of arbitrary workforce composition metrics.

Keywords: Deep reinforcement learning · Optimization · Simulation · Strategic workforce planning

1 Introduction

In order to achieve their strategic goals, organizations need to have the right people in the right place at the right time. *Strategic workforce planning* (SWP) is the business process in which the required actions to meet an organization's

Y. Smit and F. den Hengst—Authors contributed equally.

G. Nicosia et al. (Eds.): LOD 2022, LNCS 13811, pp. 108–122, 2023.
https://doi.org/10.1007/978-3-031-25891-6_9

workforce needs are identified [1]. SWP has been recognized as an important problem across sectors [3–5] and is expected to grow in importance with knowledge and human capital becoming increasingly important drivers of economic growth [18]. Workforce planning helps organizations with forecasting their workforce needs given a range of possible business scenarios and includes predicting the impact of various programs and policies on talent attraction and retention, showing how the impact varies across different segments of the workforce, modeling the impact of employee attrition and movements within the organization, and quantifying the financial impact of workforce decisions [1].

SWP problems are challenging since they require a deep understanding of the organization's high-level strategic goals and constraints on the one hand and technical knowledge to express these as an optimization problem on the other. The problem formulation should correctly capture the organization's workforce goals and constraints into its objective, address the aforementioned aspects of uncertainty, and be both actionable and computationally tractable. As a result, achieving impact with SWP typically requires careful collaboration between experts from the HR and analytics domains.

The SWP problem has attracted substantial interest from researchers as a result. Historically, these have focused on relatively simple and specific settings, e.g., problems of a relatively small scale [16], with a homogeneous workforce [4,20], and an objective function linear in the workforce composition [9,10]. Recently, researchers have addressed some of these limitations with more advanced techniques that explicitly include uncertainty of the workforce dynamics [13], that include employee attributes, such as age, skill, and position [3,6], and that use a piece-wise linear objective [7]. Although more general than previous methods, these still rely on problem specifics to cast the organizations' goals and constraints into a tractable optimization problem. This limits their applicability and comes at a significant analysis and modeling burden.

In this work, we propose a generic and widely applicable approach. In our approach, a policy that optimizes a strategic workforce objective is derived with deep reinforcement learning (DRL). Since DRL does not depend on the specifics of the objective, it can be defined as a non-linear combination of high-level workforce metrics. The optimal policy is determined with DRL in a simulation-optimization loop. The optimization step in this loop does not depend on the internals of the simulator, so that the approach can be applied to a wide range of simulators. We also describe how a simulator can be estimated from data on historical workforce compositions so that only the objective and a data set are required as inputs. Additionally, our approach is capable of handling large problems and fine-grained decision-making as a result of the usage of neural networks in estimating the optimal policy. Our approach improves the usability, granularity, and quality of SWP decision support.

1.1 Related Work

The application of different simulation paradigms in finding the optimal workforce planning decisions is very popular; see [2,14,15] and also see [1] for a discussion of simulation in workforce planning in industry. The adoption of deep reinforcement learning for simulation-optimization has recently become popular in academia and industry; see [11], Pathmind and project Bonsai by Microsoft. To the best of our knowledge, however, no studies have proposed to address SWP with DRL, which brings various benefits to this domain: it does not require any specific domain knowledge, scales well to large problems, and makes no assumptions on, e.g., linearity of the objective function.

This work is organized as follows. We first introduce SWP as an optimization problem, including the modeling of the workforce dynamic and the formulation of optimization objectives. We then introduce the simulation-optimization loop and detail the DRL optimizer. We describe the experimental setup and results, which show that our approach finds suitable policies for high-level objectives for both a synthetic and real-life organization. We conclude that our approach enables direct optimization of strategic workforce goals.

2 Strategic Workforce Planning as Optimization

In this section, we present a quantitative framework for SWP. We first detail a descriptive model of the workforce. This model factors the total workforce into groups of individuals with similar attributes of interest called *cohorts*. Attributes, such as productivity, skills, and manager status, can be included based on the goals and constraints of the organization. We then detail how the dynamics are modeled. Finally, we describe how strategic workforce goals and constraints can be formulated as optimization objectives.

2.1 Cohort Model

We define employee attributes as a set of variables $Y = (Y_1, \dots, Y_m)$ so that each employee with attributes $(Y_1 = y_1, \dots, Y_m = y_m)$ can be described by values $(y_1, \dots, y_m)$ and all employees with the same values can be grouped into the same cohort $C_i \in \mathcal{C} = \{C_1, \dots, C_n\}$. The number of cohorts n depends on the number of attributes m and the cardinalities $|Y_i|$ of these attributes, i.e., $n = |\mathcal{C}| = \prod_{i=1}^{m} |Y_i|$. Note that n grows as a combination of attributes, so that more fine-grained modeling results in a larger number of cohorts quickly.

We now turn to a model of the evolution of a workforce over time. Specifically, we consider discrete time steps of an arbitrary fixed length (e.g., monthly, quarterly, or yearly) $0 < t \leq T$ for some finite horizon $T < \infty$. At each time point t, the number of employees for a particular cohort C_i is defined as a random variable (r.v.) $X_{i,t} \in \mathbb{N}_{\geq 0}$ and the total workforce as a combination of all cohorts $X_t = \left(X_{(1,t)}, \dots, X_{(n,t)}\right) \in \mathbb{N}^n_{\geq 0}$. The dynamics of these so-called *headcounts* can now be modeled as a Markov chain. Its state space consists of all possible headcount compositions. We assume a scalar $X_{\max} < \infty$ for the maximum number of employees per cohort and define the state space of the Markov chain $\mathcal{S} = \{s \in \mathbb{N}^n_{\geq 0} | s \leq (X_{\max})^n\}$.

For any organization and for any time step, we know that an individual can either (i) leave the organization organically due to, e.g., retirement, voluntarily leaving etc. (ii) leave the organization as a result of a management decision, (iii) move from one cohort to another cohort organically, (iv) be moved from one cohort to another cohort by the organization and (v) enter the organization. With this knowledge, the transition function can be factorized into components, so that for every t:

$$X_{t+1} = X_t - O_t - L_t + \mathbb{1}^n M_t - \mathbb{1}^n M_t' + \mathbb{1}^n N_t - \mathbb{1}^n N_t' + H_t, \tag{1}$$

where $\mathbb{1}^n$ is an n-dimensional vector of ones and (i) O_t an n-dimensional r.v. representing organic leavers per cohort, (ii) L_t an n-dimensional r.v. representing organization-initiated leavers, (iii) M_t an $n \times n$ random matrix of employees moving between cohorts organically, (iv) N_t an $n \times n$ random matrix of moves between cohorts initiated by the organization, and (v) H_t an n-dimensional r.v. of new hires. This model describes how the workforce changes over time and it allows to easily formalize strategic workforce goals as optimization objectives as described in the next section.

2.2 Optimizing the Cohort Model

In this section, we cast the SWP problem as an optimization problem. The first step is to identify the actions available to the organization. We assume that these are direct and indirect controls on the Markov chain in Eq. 1. In general, the transitions L_t, N_t, and H_t are controlled by the organization directly. Additional controls may be in place to affect the other r.v.'s indirectly. For example, an employee retention plan can be included to affect the attrition O_t. The cohort model supports both direct and indirect controls, and these can be included based on the organization's needs.

The organization should take those actions that result in the most suitable workforce at every time step. We here formalize the organization's actions as some set $\mathcal{A}$ and a particular action at time t as $A_t \in \mathcal{A}$ and refer the reader to Sect. 4 for examples. We assume that each workforce composition X_t and A_t can be assigned a numerical value corresponding to the particular SWP goal of the organization with some function $r : \mathcal{S} \times \mathcal{A} \to \mathbb{R}$. The objective, now, is to maximize this value over time by sampling appropriate states and actions in the system in Eq. 1 until some horizon T:

$$\begin{aligned} A^* = &\operatorname*{arg\,max}_{A_0,\ldots,A_T} \mathbb{E}\left[\sum_{t=0}^{T} r(X_{t+1}, A_t)\right] \\ \text{s.t. } &X_{t+1} = X_t - O_t - L_t + \mathbb{1}^n M_t - \mathbb{1}^n M_t' + \mathbb{1}^n N_t - \mathbb{1}^n N_t' + H_t, \\ &\text{and } O_t, L_t, M_t, N_t, H_t \text{ dependent on } A_t \\ &\text{for } t = 0, \ldots, T, \text{ and a given } X_0. \end{aligned} \tag{2}$$

Having defined the general optimization objective, we now turn to examples of suitable reward functions. A reward function should reflect the strategic workforce goals of the organization accurately. Because of the strategic nature of SWP, a goal is usually composed of multiple terms. General terms such as headcount and budget, SWP-specific terms such as average span of control[1], job level[2] and manager status, and finally, organization-specific metrics such as productivity, skills, and diversity may all be included.

Strategic Workforce Goals. The example strategic workforce goal is composed of three components, here presented by decreasing importance. The primary component consists of bounds for headcounts for each cohort. The second component contains a target average span of control across the organization. In general, such a target span of control is attained by multiple workforce compositions. The third component, therefore, specifies that minimal salary costs are preferred. We formalize this strategic goal by formalizing each component and then combining the components in an overall objective.

To formalize the objective based on headcount bounds, we penalize cohorts that are out of bounds:

$$r_{\mathrm{b}}(X_t) := -\sum_{i=1}^{n} \mathbb{1}_{\{X_{i,t} \notin [\ell_i, u_i]\}}, \tag{3}$$

where $\ell, u \in \mathbb{N}^n_{\geq 0}$ are lower and upper bounds for all n cohorts. Next, we define a component for achieving the target span of control. It is similar to the objective for target headcounts in Eq. (9):

$$r_{\mathrm{soc}}(X_t) := \exp\left(\frac{-\alpha_{\mathrm{soc}}\left(\mathrm{soc}(X_t) - G_{\mathrm{soc}}\right)^2}{G^2_{\mathrm{soc}}}\right), \tag{4}$$

where $G_{\mathrm{soc}} > 0$ is a target average span of control, $\alpha_{\mathrm{soc}} > 0$ a precision parameter, and $\mathrm{soc}(X_t)$ a function that returns the average span of control for X_t:

$$\mathrm{soc}(X_t) := \frac{X_{(n/2+1,t)} + \cdots + X_{(n,t)}}{X_{(1,t)} + \cdots + X_{(n/2,t)}}. \tag{5}$$

The third and final component can be formalized based on a function $\mathrm{sal}(X_t)$ that returns the estimated total salary cost for a workforce X_t. This final component has the lowest priority. Therefore, we only assign a positive value based on salary if the span of control component is sufficient, as expressed by a lower bound $\ell_{\mathrm{soc}} \in [0, 1]$:

$$r_{\mathrm{sal}}(X_t) := \begin{cases} r'_{\mathrm{sal}}(X_t), & \text{if } r_{\mathrm{soc}}(X_t) > \ell_{\mathrm{soc}}, \\ 0, & \text{otherwise,} \end{cases} \tag{6}$$

for a salary normalized to $[0, 1]$ based on the cohort bounds ℓ, u:

$$r'_{\mathrm{sal}}(X_t) := \mathrm{clip}\left(\frac{\mathrm{sal}(X_t) + \mathrm{sal}(\ell)}{\mathrm{sal}(\ell) - \mathrm{sal}(u)}, 0, 1\right). \tag{7}$$

[1] The average number of direct reports of managers in the organization.

[2] A metric to express responsibilities and expectations of a role in the organization, usually associated with compensation in some way.

The *strategic* objective is composed of the sub goals in Eqs. (3)–(6). We combine the components to reflect all sub goals states earlier:

$$r_{\mathrm{s}}(X_t) := r_{\mathrm{b}}(X_t) + r_{\mathrm{soc}}(X_t) + r_{\mathrm{sal}}(X_t). \tag{8}$$

The simulation-optimization approach proposed in this work targets the direct optimization of objectives that reflect an organization's *strategic* workforce goals and that may be non-linear and composed of arbitrary workforce metrics.

Operational Workforce Goals. Another type of workforce goal is to meet a particular known demand for employees in each cohort. This type of goal is relatively easy to optimize for but hard to specify in practice. For this goal, a reward can be assigned based on a distance between the current workforce X_t and the known target composition $X^* = (X_1^*, \ldots, X_n^*)$ for all n cohorts. To ensure that the cohorts contribute uniformly to this reward, headcounts need to be scaled to $[0, 1]$. Now, the following rewards an observed headcount $X_{i,t}$ for a single cohort i based on its target X_i^*:

$$r_{\mathrm{c}}(X_{i,t}) := \begin{cases} \exp\left(\frac{-\alpha(X_{i,t}-X_i^*)^2}{(X_i^*)^2}\right), & \text{if } X_i^* > 0, \\ \exp\left(-\alpha X_{i,t}^2\right), & \text{if } X_i^* = 0, \end{cases} \tag{9}$$

where the so-called *precision* parameter $\alpha > 0$ specifies how strictly to penalize sub-optimal headcounts. A simple operational reward averages over all n cohorts:

$$r_{\mathrm{o}}(X_t) := \frac{1}{n} \sum_{i=1}^{n} r_{\mathrm{c}}(X_{i,t}), \tag{10}$$

These *operational* workforce goals are generally easy to optimize for using established optimization techniques since they can be cast as linear optimization problems. However, defining the required headcounts for all cohorts to meet high-level workforce goals is very hard in practice.

3 Simulation-Optimization with Deep Reinforcement Learning

We propose a simulation-optimization loop for solving SWP problems. Figure 1 contains a visualization of this loop. First, the user specifies the strategic workforce goals of the organization as a reward function to maximize. This function may be any arbitrary, e.g., a non-linear function defined over a cohort representation of the workforce. Next, a policy is learned by a DRL agent by interacting with a simulator. This simulator can be any suitable black-box simulator that outputs a cohort representation of the workforce and can take into account the decisions made by the agent. By using DRL for optimization, the strategic goals are optimized for directly, and, hence, the resulting policy informs the user in taking the right workforce decisions for their strategic workforce goals. If historical data of the workforce is available, then this simulator can be learned from data as described in Sect. 3.2.

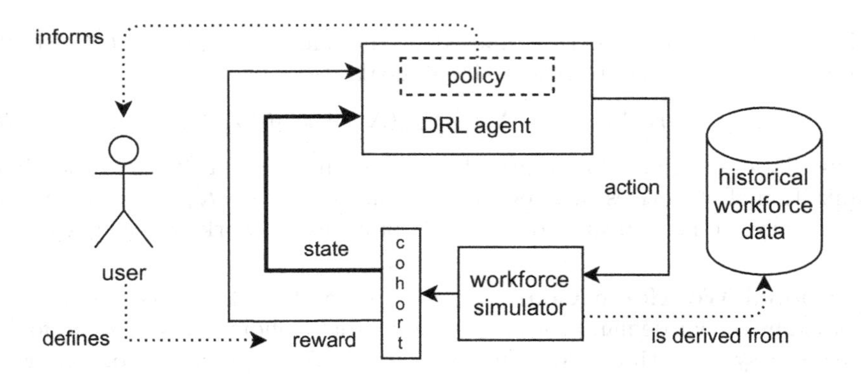

Fig. 1. Overview of the simulation-optimization approach. A user specifies the organization's strategic workforce goal. A black-box workforce simulator is then used to find a policy that directly optimizes for the goal with DRL. This policy helps the user making informed workforce decisions.

3.1 Deep Reinforcement Learning for Workforce Planning

Formally, we cast the SWP problem as a Markov decision process (MDP) $\langle \mathcal{S}, \mathcal{A}, \mathcal{P}, r, \gamma \rangle$, where $\mathcal{S}$ is a state space, $\mathcal{A}$ an action space, $\mathcal{P} : \mathcal{S} \times \mathcal{A} \times \mathcal{S} \rightarrow [0,1]$ a transition function, $r : \mathcal{S} \times \mathcal{A} \rightarrow \mathbb{R}$ a reward function, and $\gamma \in (0,1]$ a discount factor to balance immediate and future rewards. The decisions of the agent are defined by its policy $\pi_\theta : \mathcal{S} \times \mathcal{A} \rightarrow [0,1]$, which depends on a parameter vector θ which can, e.g., be a neural network. The goal of the agent is to maximize the expected discounted return $J(\theta) := \mathbb{E}_{\pi_\theta}\left[\sum_{t=0}^{T-1} \gamma^t r_{t+1}\right]$, which can be done by tuning parameters θ with an algorithm that alternates simulating experience in the environment and optimizing the policy. Here $r_{t+1} = r(s_t, a_t)$ and $\mathbb{E}_{\pi_\theta}$ indicates that $s_{t+1} \sim \mathcal{P}(\cdot, a_t, s_t)$ and $a_t \sim \pi_\theta(\cdot|s_t)$.

In the proposed framework, the state space of the MDP is equal to the state space of the Markov chain over headcounts, i.e., $\mathcal{S} = \{s \in \mathbb{N}^n_{\geq 0} | s \leq (X_{\max})^n\}$. The optimization algorithm uses a neural network to evaluate the value of each state. To help the convergence of the network and significantly reduce training time, the inputs to the network are normalized. Hence, we implement the state space of the MDP as a continuous space $\hat{\mathcal{S}} = [0,1]^n$, where states are defined as $s_t = \left(\frac{X_{1,t}}{X_1^{max}}, \ldots, \frac{X_{n,t}}{X_n^{max}}\right)$ for training. The action space is given by the controls over the workforce as described in Sect. 2, for example, a multi-discrete set of numbers of employees that enter or leave the organization for each cohort. For the purposes of optimization, the dynamic model $\mathcal{P}$ is assumed to be unknown so that any suitable simulator can be used. The reward function is defined by an end user based on the organization's strategic workforce goals. It can be composed of arbitrary and non-linear workforce metrics of interest to the organization, see Sect. 2.2 for details and examples.

In the optimization step a policy is updated to optimize the given objective. This update is performed by approximate gradient ascent on θ, i.e., iteratively update $\theta_{k+1} = \theta_k + \eta\widehat{\nabla_\theta J(\theta)}$. The gradient $\nabla_\theta J(\theta)$ is estimated by $\hat{\mathbb{E}}_t\left[\nabla_\theta \log \pi_\theta(a_t|s_t)\hat{A}_t\right]$, where $\hat{\mathbb{E}}_t$ denotes an empirical estimate over a batch of samples collected over time and $\hat{A}_t$ is an estimator of the advantage function. While our approach is generic to various optimization algorithms, we propose to use Proximal Policy Optimization (PPO) as it has shown to be suitable in high-dimensional settings with non-linear rewards [19].

3.2 Simulating the Workforce

This section details how the dynamics of a cohort model from Sect. 2.1 can be estimated from data. Estimation is necessary for two reasons. Firstly, the dynamics may simply not be available to the organization. Secondly, it may be problematic to fully elaborate the dynamics up-front due to the complexity of the problem. Specifically, the size of the state space of the cohort model Markov chain grows exponentially in the number of cohorts. As a result, it becomes infeasible to analytically define it fully for reasonably large organizations.[3] Hence, we estimate the dynamics from data with simplifications that apply to the cohort model.

In many cases, Eq. (1) can be simplified by assuming limited control of the workforce by management. For example, if we only model management-controlled hires and leavers, N_t becomes equal to the zero matrix and $A_t := H_t - L_t$ for the combined movement of hires and leavers by the organization. The part of the transition function that is out of management control is now given by $X_{t+1} = X_t - O_t + \mathbb{1}M_t - \mathbb{1}M_t'$. Note that the diagonal entries of M_t can be chosen arbitrary, since $(M_t - M_t')_{i,i} = 0$ for all $i = 1, \ldots, n$. By realizing that the numbers of employees that remain in cohort i is equal to the headcount of cohort i minus the number of employees that move to any other cohort or organically leave the organization, we may set $M_{i,i,t} = X_{i,t} - \sum_{j=1,j\neq i}^{n} M_{i,j,t}' - O_{i,t}$, or $X_t = \mathbb{1}M_t' + O_t$. It follows that we can then simply write $X_{t+1} = \mathbb{1}M_t$ for the stochastic dynamics of the cohort model in general.

We observe that all employees within a certain cohort are indistinguishable for the purposes of SWP. Hence, approximation of the dynamics of Eq. (1) at cohort level is sufficient for the purposes of this work. We, therefore, model the movement of employees between cohorts based on the attributes that describe the cohorts. We define the transition probability matrix $P(t) \in [0,1]^{n\times n}$ by letting $p_{i,j}(t)$ be the probability that an employee moves from cohort i at time t to cohort j at time $t+1$. We additionally assume time-homogeneous transition probabilities, i.e., $P(t) \equiv P$. Under these assumptions, the rows of the random matrix M_t follow a multinomial distribution, i.e., for $i = 1, \ldots, n$, $(M_{i,1,t}, \ldots, M_{i,n,t}) \sim \text{Mult}(X_{i,t}, P_i)$, where P_i denotes the i-th row of P.

The transition probability matrix P can be estimated from data that takes record of the cohorts of individual employees over a time period $t = 1, \ldots, T$.

[3] For a model with $n = 30$ cohorts and $X_{\max} = 100$ maximum employees per cohort, the number of transitions in the Markov chain is $|\mathcal{S} \times \mathcal{S}| = \prod_{i=1}^{n}(S_{\max}+1)^2 \approx 10^{120}$.

Let $m_{i,j,t}$ denote the number of employees that are in cohort i at time $t-1$ and in cohort j at time t. Then the maximum likelihood estimator of $p_{i,j}$ is

$$\hat{p}_{i,j} = \frac{\sum_{t=1}^{T} m_{i,j,t}}{\sum_{t=1}^{T} X_{i,t}}. \tag{11}$$

For any time step t and action A_t, the dynamics of the workforce over time can now be simulated by sampling the movement matrix M_t from the multinomial distribution described above and computing

$$X_{t+1} = \mathbb{1}^n M_t + A_t. \tag{12}$$

4 Experimental Setup

This section details the experimental setup, which was designed to answer the following research questions:

1. How does the proposed simulation-optimization approach perform,
   (a) on an operational workforce objective?
   (b) on a strategic workforce objective?
   (c) for a varying employee mobility?
2. Are firing constraints best implemented with a masked policy or an updated objective (penalty for illegal fires)?

We compare the results on a baseline based on linear programming (LP) proposed recently [6]. We evaluate these approaches in two cases. The first is a synthetic organization, and the second is a real-life use case from an international bank to validate the results in practice. We first describe the overall setup, then the baseline, detail the organizations, and include implementation details.[4]

To investigate research question 1a and 1b, we train a reinforcement learning agent for both the operational and strategic tasks in Eq. (9) and Eq. (8). We evaluate both the trained agent and the heuristic baseline described in Sect. 4.1 and compare the performance based on the average reward metric, in the manner as described in Sect. 4.2.

4.1 Baseline

We devise a baseline based on linear programming to compare the performance of the proposed simulation-optimization approach. This baseline was proposed in [6] and makes a number of additional assumptions that allow for efficient solving of the SWP problem. We describe this baseline in detail in this section.

Due to the size of the state space of the Markov chain that describes the workforce dynamics, this stochastic model cannot be used directly with a linear solver. Therefore, we consider a deterministic approximation of Eq. (12), by

[4] Code and data for hypothetical use case available at https://github.com/ysmit933/swp-with-drl-release. Real-life use case data will be made available upon request.

replacing the random variables involved with their expectation. This operation, known as mean-field approximation, is justified for large-scale organizations as a result of the functional law of large numbers; see, e.g., [6]. For Eq. (12) we obtain

$$X_{t+1} \approx \mathbb{E}\left[\mathbb{1}M_t + A_t\right] = X_t P_{\cdot,i} + A_t, \tag{13}$$

where $P_{\cdot,i}$ denotes the i-th column of the transition probability matrix P. Additionally, we optimize for one time step at a time instead of the whole trajectory $t = 0, \ldots, T$. This is reasonable when the rewards do not depend on time and are given at each time step. In that case, there are no situations where it is required to sacrifice short-term gains for long-term profit.

Consider the target level reward defined in Eq. (9) and assume for simplicity that $X_i^* > 0$ for all $i = 1, \ldots, n$. Under the aforementioned assumptions, this version of the SWP problem is given by: find

$$A_t^* = \arg\max_{A_t} \frac{1}{n} \sum_{i=1}^{n} \exp\left(\frac{-\alpha(X_{i,t+1} - X_i^*)^2}{(X_i^*)^2}\right), \tag{14}$$

such that $X_{i,t+1} = \sum_{j=1}^{n} p_{ji} X_{j,t} + A_{i,t}$ for $t = 1, \ldots, T$. Substituting the latter expression in the former, and by noting that each term of the sum is maximized when the term in the exponential is equal to zero, we see that $A_{i,t}^* = X_i^* - \sum_{j=1}^{n} p_{ji} X_{j,t}$. The optimal continuous actions are then mapped to the discrete set of possible hiring options $\mathcal{A} = \mathcal{A}_1 \times \cdots \times \mathcal{A}_n$. Hence, the decision rule becomes

$$A_{i,t}^* = \Pi_{\mathcal{A}_i}\left(X_i^* - \sum_{j=1}^{n} p_{ji} X_{j,t}\right), \tag{15}$$

where $\Pi_{\mathcal{A}_i}(a) := \arg\min_{a' \in \mathcal{A}_i} |a - a'|$.

To develop a heuristic for the combined reward function, we make the same assumptions as for the target level heuristic. We then consider the state that yields the highest immediate reward, given by $X^* = \arg\max_{x \in \mathcal{S}} r(x)$. We use this as a target level to aim for by applying the target level heuristic Eq. (15).

4.2 Training Setup

The reinforcement learning agent is trained for a maximum number of training steps T^{max} specified by the user. At the start of each episode, a random starting state in the neighborhood of X_0 is generated to ensure sufficient exploration of all relevant parts of the state space. This is done by uniformly sampling a state from the interval $[(1-\beta)X_i(0)/X_i^{max}, (1+\beta)X_i, 0/X_i^{max}]$, where $\beta \in [0, 1]$ determines the random spread across the state space. The episode ends after T time steps, at which point the environment resets to a new random starting state. After each T^{eval} number of time steps, the agent is evaluated on an evaluation environment, which is identical to the training environment except for a deterministic start at X_0 and the best performing agent is stored.

When the training process has terminated, we test the trained model on the evaluation environment with a fixed starting state (as a default for 1,000 episodes) and collect several metrics to assess the quality of the model. In particular, during an episode of T time steps, we collect the average reward $\frac{1}{T}\sum_{t=1}^{T} r(X_t)$ and the number of constraint violations $\sum_{t=1}^{T}\sum_{i=1}^{n} \mathbb{1}_{\{A_i(t) \text{ is illegal}\}}$.

4.3 Hypothetical Organization

For the hypothetical organization, we consider a model with four cohorts, labeled by M1, M2, C1, and C2 (two cohorts of managers and two cohorts of contributors). We suppose the probability transition matrix is given by

$$P = \begin{pmatrix} 0.98 & 0 & 0 & 0 \\ 0.01 & 0.93 & 0 & 0 \\ 0 & 0.04 & 0.92 & 0.005 \\ 0 & 0.01 & 0.01 & 0.96 \end{pmatrix},$$

and we let $X_0 = (20, 50, 100, 300)$ be the starting state. The hiring options are set to $\mathcal{A}_1 = \{-2, -1, 0, 1, 2\}$, $\mathcal{A}_2 = \{-5, -1, 0, 1, 5\}$, $\mathcal{A}_3 = \{-10, -2, 0, 2, 10\}$, and $\mathcal{A}_4 = \{-25, -5, 0, 5, 25\}$. The maximum cohort sizes are $X^{max} = 2X_0$, the random starting state percentage is $\beta = 0.25$, the time horizon is $T = 60$, and the salary costs are set to $C^{sal} = (10000, 6000, 4000, 2000)$. The target level objective is $X^* = X_0$ (and remain at the same levels as the starting state), with a default precision of $\alpha = 10$. The combined reward parameters are given by $\ell = 0.75X_0$, $u = 1.25X_0$, $G_{soc} = 7$, $\ell_{soc} = 0.9$.

On top of the transitions introduced before, we vary the employee mobility in order to answer research question 1c. In order to answer this question, we evaluate the approach on transition matrices

$$P_\ell = \begin{pmatrix} 1-\ell & 0 & 0 & 0 \\ \ell/2 & 1-\ell & 0 & 0 \\ 0 & \ell/2 & 1-\ell & 0 \\ 0 & 0 & \ell/2 & 1-\ell \end{pmatrix},$$

for mobility rates $\ell \in \{0, 0.01, \ldots, 0.1\}$. For each of these environments, and for both the operational and strategic tasks, a reinforcement learning agent is trained and evaluated. Next, its performance, based on the average reward obtained, is compared to the heuristic baseline.

4.4 Real-Life Use Case

To investigate the performance of our solution method on a real-life use case, we use the following model based on actual headcounts in one particular department of the Bank. The 14,105 employees of this segment of the organization are divided into cohorts based on manager status (manager or contributor) and based on five distinct job levels, resulting in a cohort model consisting of $n = 10$ cohorts. We

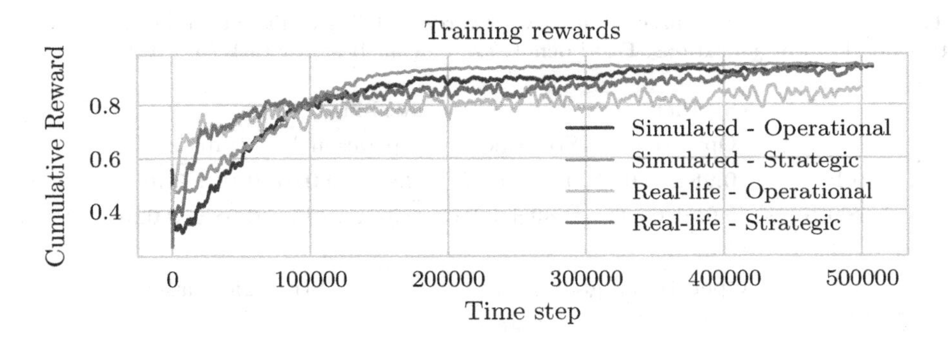

Fig. 2. Normalized cumulative training rewards.

label the cohorts as Manager-1, ..., Manager-5, Contributor-1, ..., Contributor-5. The transition probabilities between these cohorts are estimated based on monthly employee data for a period of 48 months. For both tasks, starting state X_0 is set to the workforce at the beginning of the period and target state X^* to the workforce at the end of the period for the operational goal.

For the strategic goal, we use cohort bounds $\ell_i = 0.75X_i(0)$ and $u_i = 1.25X_i(0)$, and the goal for span of control is $G_{\text{soc}} = 7$, with $\ell_{\text{soc}} = 0.9$. Costs associated with salary and management initiated hires and leavers were set in collaboration with an expert in the organization. The hiring options were chosen based on the cohort sizes and include the option to hire or fire zero, a few, many, or a moderate number of employees. The maximum number of employees that could be hired or fired was roughly ten percent of the starting cohort size. For example, the hiring options for cohort Manager-1 were given by the set $\mathcal{A}_1 = \{-25, -5, -1, 0, 1, 5, 25\}$.

To investigate research question 2, we implement three methods to constrain the choices for management-initiated leavers. The first method is a masked policy, for which the illegal actions are removed from the action space by setting the corresponding action probabilities to zero. For the second method, the agent receives a large negative reward for selecting an illegal action. Finally, we constrain the agent to hires only, i.e. in which all leaving employees do so organically. We then train reinforcement learning agents for both the operational and strategic tasks and compare the performance of the unconstrained agent, the masked agent, the penalty-receiving agent, the no-fire agent, and the baseline heuristic.

5 Results

In this section, we look at all results associated with research questions 1a-2 presented in the previous section. We first look at the convergence of the proposed approach in Fig. 2 and find that the proposed SO approach converges quickly. Next, we compare the resulting policies with an LP baseline on a test set. Table 1 shows that the proposed approach performs close to the optimum

Table 1. Average normalized cumulative rewards and 95% confidence interval for both tasks on both organizations. **Bold** denotes significant best per task ($p = 0.99$).

	Synthetic		Real-life	
	Operational	Strategic	Operational	Strategic
LP	$\mathbf{0.98 \pm 0.030}$	0.41 ± 0.374	$\mathbf{0.99 \pm 0.010}$	0.12 ± 0.106
SO (ours)	0.94 ± 0.033	$\mathbf{0.83 \pm 0.213}$	0.92 ± 0.015	$\mathbf{0.98 \pm 0.026}$

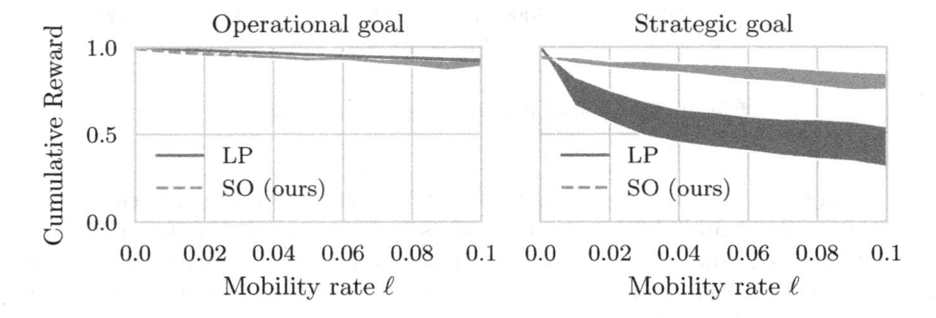

Fig. 3. Normalized cumulative rewards for varying mobility rates.

of the LP baseline on the operational objective and significantly outperforms the baseline on the strategic objective. We move on to research question 1c by looking at the effect of increasing the employee mobility in Fig. 3. It shows that the proposed SO approach is robust against a wide range of mobility levels and that its benefits increase with increasing workforce mobility. The proposed approach shows to be more robust to the stochastic nature of SWP for this nonlinear optimization objective than the LP baseline.

Finally, we compare our approach in a setting with constraints on the organization's control of leavers in Table 2. Here, we find that we can effectively take the organization's constraints into account using either masking, with a negative reward (penalty) or by only including hires in the action space. Out of these, the 'only hires' variant yields the best results with respect to reward and constraint adherence, with rewards close to its unconstrained counterparts without any constraint violations.

6 Discussion

In this work, we have presented a simulation-optimization approach to strategic workforce planning. The approach optimizes workforce decisions with DRL by interacting with a simulator. Any suitable simulator can be used because the optimization step does not depend on its internals. We propose to use a Markov chain simulator *learned* from historical data. By doing so, the full loop only requires a data set of historical workforce compositions and the organization's

Table 2. Average normalized cumulative rewards and constraint violations (% of total decisions), with 95% confidence intervals. **Bold** denotes significant best ($p = 0.99$).

	Operational		Strategic	
	Reward	# Violations (%)	Reward	# Violations (%)
LP	$\mathbf{0.99 \pm 0.010}$	16.83 ± 3.05	0.12 ± 0.106	13.84 ± 2.96
Unconstrained	0.92 ± 0.015	6.94 ± 1.56	$\mathbf{0.98 \pm 0.026}$	21.72 ± 4.39
Masked	0.74 ± 0.030	$\mathbf{0.00 \pm 0.00}$	0.75 ± 0.135	$\mathbf{0.00 \pm 0.00}$
Penalty	0.87 ± 0.046	0.34 ± 0.07	0.74 ± 0.060	$\mathbf{0.00 \pm 0.00}$
Only hires	0.79 ± 0.041	$\mathbf{0.00 \pm 0.00}$	0.94 ± 0.061	$\mathbf{0.00 \pm 0.00}$

objective as inputs. These objectives may be composed of arbitrary workforce metrics of interest that may be non-linear in the workforce composition. The approach optimizes these objectives *directly*, so that the resulting policy can easily be used to ensure a high impact of the SWP efforts.

We have evaluated the proposed approach on a synthetic and a real-world organization and found that it converges quickly. More so, we compared the quality of the obtained policy to a baseline from the literature. In this comparison, we first targeted an objective composed of workforce metrics. Such objectives are easy to define and accurately reflect the organization's strategic goals. We found that our approach significantly outperforms the baseline on this *strategic* objective and that the difference grows as mobility of the workforce increases. We secondly targeted an *operational* goal, in which the optimal workforce composition is known up-front. Such goals are easy to optimize for with established optimization approaches but hard to define in practice. Our approach performed close to the baseline in this setting. We additionally showed how the approach can take into account realistic constraints by limiting the ability of the organization to control leavers in the organization and found that removing the ability to do so has a very limited impact on overall performance.

We have shown that the proposed simulation-optimization approach is suitable for SWP. Additionally, it opens up various avenues for future work. Firstly, the approach is capable of optimizing for strategic objectives composed of arbitrary workforce metrics. It would be interesting to extend the approach with multi-objective reinforcement learning in order to compute a set of Pareto optimal policies [17]. This will increase the organization's understanding of the trade-offs involved and allow them to fine-tune their strategy. Secondly, the approach currently finds a policy that is optimal on average. While this is suitable for many use-cases, there may be some organizations that prefer a probabilistic guarantee on the minimum number of employees to, e.g., meet service level agreements. Here, risk-sensitive DRL can be employed instead of regular DRL [8]. Additionally, organizational constraints can be formalized and used within approaches that guarantee safety of the resulting policy [12]. We believe that, with the proposed approach, these challenging and interesting research directions that will further increase the impact of SWP have become feasible in practice.

References

1. April, J., Better, M., Glover, F.W., Kelly, J.P., Kochenberger, G.A.: Ensuring workforce readiness with optforce (2013). Unpublished manuscript retrieved from opttek.com
2. Banyai, T., Landschutzer, C., Banyai, A.: Markov-chain simulation-based analysis of human resource structure: how staff deployment and staffing affect sustainable human resource strategy. Sustainability **10**(10), 3692 (2018)
3. Bhulai, S., Koole, G., Pot, A.: Simple methods for shift scheduling in multiskill call centers. Manuf. Serv. Oper. Manage. **10**(3), 411–420 (2008)
4. Burke, E.K., De Causmaecker, P., Berghe, G.V., Van Landeghem, H.: The state of the art of nurse rostering. J. Sched. **7**(6), 441–499 (2004)
5. Cotten, A.: Seven steps of effective workforce planning. IBM Center for the Business of Government (2007)
6. Davis, M., Lu, Y., Sharma, M., Squillante, M., Zhang, B.: Stochastic optimization models for workforce planning, operations, and risk management. Serv. Sci. **10**(1), 40–57 (2018)
7. De Feyter, T., Guerry, M., et al.: Optimizing cost-effectiveness in a stochastic Markov manpower planning system under control by recruitment. Ann. Oper. Res. **253**(1), 117–131 (2017)
8. Fei, Y., Yang, Z., Chen, Y., Wang, Z., Xie, Q.: Risk-sensitive reinforcement learning: Near-optimal risk-sample tradeoff in regret. Adv. Neural. Inf. Process. Syst. **33**, 22384–22395 (2020)
9. Gaimon, C., Thompson, G.: A distributed parameter cohort personnel planning model that uses cross-sectional data. Manage. Sci. **30**(6), 750–764 (1984)
10. Grinold, R., Stanford, R.: Optimal control of a graded manpower system. Manage. Sci. **20**(8), 1201–1216 (1974)
11. Heger, J., Voss, T.: Dynamically changing sequencing rules with reinforcement learning in a job shop system with stochastic influences. In: 2020 Winter Simulation Conference (WSC), pp. 1608–1618 (2020)
12. den Hengst, F., François-Lavet, V., Hoogendoorn, M., van Harmelen, F.: Planning for potential: efficient safe reinforcement learning. Mach. Learn. **111**, 1–20 (2022)
13. Jaillet, P., Loke, G.G., Sim, M.: Strategic workforce planning under uncertainty. Oper. Res. **70**, 1042–1065 (2021)
14. Jnitova, V., Elsawah, S., Ryan, M.: Review of simulation models in military workforce planning and management context. J. Defense Model. Simul. **14**(4), 447–463 (2017)
15. Kant, J.D., Ballot, G., Goudet, O.: WorkSim: an agent-based model of labor markets. J. Artif. Soc. Soc. Simul. **23**(4), 4 (2020)
16. Rao, P.P.: A dynamic programming approach to determine optimal manpower recruitment policies. J. Oper. Res. Soc. **41**(10), 983–988 (1990)
17. Roijers, D.M., Vamplew, P., Whiteson, S., Dazeley, R.: A survey of multi-objective sequential decision-making. J. Artif. Intell. Res. **48**, 67–113 (2013)
18. Romer, P.: Human capital and growth: theory and evidence (1989)
19. Schulman, J., Wolski, F., Dhariwal, P., Radford, A., Klimov, O.: Proximal policy optimization algorithms. arXiv preprint arXiv:1707.06347 (2017)
20. Sing, C., Love, P., Tam, C.: Stock-flow model for forecasting labor supply. J. Constr. Eng. Manag. **138**(6), 707–715 (2012)

Enhanced Association Rules and Python

Petr Máša and Jan Rauch(✉)

Department of Information and Knowledge Engineering,
Prague University of Economics and Business, Prague, Czechia
masa@petrmasa.com, rauch@vse.cz

Abstract. Association rules mining and apriori algorithm are very known tools of data mining since the 1990s. However, enhanced association rules were introduced more than 20 years earlier as a tool of mechanizing hypothesis formation – an approach to exploratory data analysis. Various, both practical and theoretical results, were achieved in relation to enhanced association rules that provide much more general rules than apriori. A short overview of these results is presented. A new implementation of an analytical procedure dealing with enhanced association rules is introduced as well as novel algorithm to find association rules not by setting quantifiers but by setting expected number of association rules which corresponds with typical data scientists' needs.

Keywords: Mechanizing hypothesis formation · GUHA method · Association rules · CleverMiner project

1 Introduction

Association rules were introduced in [2] (1993) with the goal of better understand a purchasing behaviour of customers in supermarkets. The idea of association rules was later generalized to data matrices. Rows of a data matrix correspond to observed objects. Columns correspond to attributes of the objects. Attributes have a finite number of possible values, i.e. categories. The association rule is understood as an expression $Ant \rightarrow Suc$ where both Ant and Suc are conjunctions of attribute-value pairs $A(a)$. The algorithm apriori introduced in [2] is widely used to produce association rules. The package *arules* [3] is a very known implementation of apriori available in the R language.

However, the concept of association rules was introduced and studied since 1960s as an inherent part of the GUHA method research [4,6]. GUHA is an abbreviation of *General Unary Hypotheses Automaton*. *General* refers both to a wide field of applications and to generality of results, *Unary* corresponds to the form of analysed data. *Hypotheses* points to hypothesis formation and *Automaton* refers to the use of computers. The method is realized by GUHA procedures. Input of such procedure consists of a definition of a set of relevant patterns and of an analysed data. Output consists of a set of all patterns true in the input data.

G. Nicosia et al. (Eds.): LOD 2022, LNCS 13811, pp. 123–138, 2023.
https://doi.org/10.1007/978-3-031-25891-6_10

Association rules are understood as relations $\varphi \approx \psi$ between general Boolean attributes φ, ψ derived from columns of an analysed data matrix. A GUHA procedure ASSOC [5] was developed to mine for association rules $\varphi \approx \psi$. Its implementation uses an algorithm based on strings of bits [12] which is different from the apriori algorithm. Early applications of the ASSOC procedure are described in [8,9,18]. Since advent of data mining, the GUHA method is developed as a method of data mining [7].

The probably most-used implementation of the ASSOC procedure is the 4ft-Miner procedure implemented in the LISp-Miner system [17,20]. A comparison of the apriori approach and the GUHA approach is available in [17]. GUHA procedures dealing with histograms [15], action rules [11] and additional patterns are also implemented in the LISp-Miner system [7]. Various results [13,14,16,19] relevant to data mining and the GUHA method were achieved.

Development of the CleverMiner system [1,10] has been started recently. Its goal is to implement the GUHA procedures in the Python language which is very popular in the Machine Learning and Data Science community. A new simplified version of the 4ft-Miner procedure has been implemented as a part of the CleverMiner system. The first goal is to introduce this implementation in details and to show that it deals with useful association rules, more general than the association rules the apriori algorithm [2] deals with. The main goal is to present an iterative algorithm based on the 4ft-Miner. Its goal is to ensure that the number of output association rules is in a given range. Such iterative process can eliminate problems with usually very large output of the apriori algorithm. If we refer the 4ft-Miner procedure in this paper, we mean CleverMiner implementation of this procedure unless implicitly stated otherwise.

The structure of the paper is as follows. Association rules and the general features of the 4ft-Miner procedure are introduced in Sect. 2 together with the data set *Adult* used in the examples. Two applications of the 4ft-Miner implementation in CleverMiner are presented in Sect. 3 together with comparison of 4ft-Miner and apriori. An iterative process the goal of which is to identify input parameters of the 4ft-Miner procedure resulting in the number of output association rules in a given range is introduced in Sect. 4. Conclusions and plans for further work are available in Sect. 5.

2 Association Rules and 4ft-Miner Procedure

The 4ft-Miner procedure deals with data matrices and Boolean attributes defined in Sect. 2.1. We use data matrix *Adult* described in Sect. 2.2. Association rules the 4ft-Miner deals with are introduced in Sect. 2.3. Parameters defining a set of relevant association rules to be verified are described in Sect. 2.4.

2.1 Data Matrix and Boolean Attributes

All the implemented GUHA procedures deal with data matrices in a form shown in Fig. 1. Rows of the data matrix correspond to observed objects. Columns of

$\mathcal{M}$	attributes A_1	A_2	...	A_K	Boolean attributes $A_1(1)$	$A_2(4,5)$	$A_1(1) \wedge A_2(4,5)$	$A_1(1) \vee A_2(4,5)$	$\neg A_K(6)$
o_1	1	4	...	2	1	1	1	1	1
o_2	1	7	...	6	1	0	0	1	0
o_3	2	5	...	3	0	1	0	1	1
$\vdots$	$\vdots$	$\vdots$	$\ddots$	$\vdots$	$\vdots$	$\vdots$	$\vdots$	$\vdots$	$\vdots$
o_n	4	1	...	6	0	0	0	0	0

Fig. 1. Data matrix $\mathcal{M}$ and examples of Boolean attributes

the data matrix correspond to attributes describing particular objects. Each attribute has a finite number of possible values called *categories*. Data matrix $\mathcal{M}$ shown in Fig. 1 has K columns – attributes $A_1, \ldots A_K$

Association rules concern Boolean attributes derived from columns – attributes of a given data matrix. Basic Boolean attributes are created first. A *basic Boolean attribute* is an expression $A(\alpha)$ where $\alpha \subset \{a_1, \ldots a_t\}$ and $\{a_1, \ldots a_t\}$ is a set of all categories of the attribute A. The set α is a *coefficient* of the basic Boolean attribute $A(\alpha)$. A *basic Boolean attribute* $A(\alpha)$ *is true in a row* o of $\mathcal{M}$ if $A[o] \in \alpha$. If $A[o] \notin \alpha$, then $A(\alpha)$ is *false in a row* o. Here $A[o]$ denotes a value of the attribute A in the row o. Basic Boolean attribute is called a *literal.*

Each basic Boolean attribute is a Boolean attribute. If φ and ψ are Boolean attributes, then $\varphi \wedge \psi$, $\varphi \vee \psi$ and $\neg\varphi$ are also Boolean attributes. Their values are defined in a usual way. Expressions $A_1(1)$ and $A_2(4,5)$ in Fig. 1 are examples of basic Boolean attributes. Expressions $A_1(1) \wedge A_2(4,5)$, $A_1(1) \vee A_2(4,5)$, and $\neg A_K(6)$ are examples of Boolean attributes.

2.2 Data Set *Adult*

The *Adult* data matrix resulted from data set available from the UCI machine learning repository(https://archive.ics.uci.edu/ml/datasets/Adult). The data matrix has 48 842 rows corresponding to persons and thirteen columns, i.e. attributes. The attributes are introduced in Table 1 together with their categories. Categories of attributes Age, Hours_per_week, Capital_gain, and Capital_loss were defined as intervals on values of original columns. Four attributes have missing values.

2.3 GUHA Association Rules

The 4ft-Miner procedure deals with association rules $\varphi \approx \psi$ and conditional association rules $\varphi \approx \psi/\chi$. Here φ, ψ and χ are Boolean attributes – conjunctions or disjunctions of literals $A(\alpha)$ where α is a subset of categories of the attribute A. φ is called *antecedent*, ψ is *succedent* (*consequent*), and χ is a condition. The symbol $\approx$ is a *4ft-quantifier.*

Table 1. Attributes of the *Adult* data matrix

Attribute	Categories
Sex	*Female, Male*
Age	$\langle 10;20\rangle, \langle 20;30\rangle, \langle 30;40\rangle, \langle 40;50\rangle, \langle 50;60\rangle, \langle 60;70\rangle, \langle 70;90\rangle$
Education	*Preschool, 1st-4th, 5th-6th, 7th-8th, 9th, 10th, 11th, 12th, HS-grad, Some-college, Assoc-voc, Assoc-acdm, Bachelors, Masters, Prof-school, Doctorate*; there are 1 691 missings
Marital_status	*Divorced, Married-AF-spouse, Married-civ-spouse, Married-spouse-absent, Never-married, Separated, Widowed*
Relationship	*Husband, Not-in-family, Other-relative, Own-child, Unmarried, Wife*
Native_country	*United-States, Mexico, Philippines, Germany, Puerto-Rico, Canada, El-Salvador, India*, ...; total 41 countries; there are 857 missings
Race	*Amer-Indian-Eskimo, Asian-Pac-Islander, Black, White, Other*
Workclass	*Federal-gov, Local-gov, Without-pay, Never-worked, Private, Self-emp-inc, Self-emp-not-inc, State-gov*; there are 2 799 missings
Occupation	*Adm-clerical, Craft-repair, Exec-managerial, Machine-op-inspct, Other-service*, ...; total 14 occupations; there are 2 809 missings
Hours_per_week	$\langle 0;10\rangle, \langle 10;20\rangle, \langle 20;30\rangle, \langle 30;40\rangle, \ldots, \langle 60;70\rangle, \langle 70;100\rangle$
Capital_gain	0, $\langle 0;2000\rangle$, $\langle 2000;3000\rangle$, $\langle 3000;4000\rangle$, $\langle 4000;5000\rangle$, $\langle 5000;7000\rangle$, $\langle 7000;10000\rangle$, $\langle 10000;20000\rangle$, $\langle 20000;99000\rangle$, $\langle 99000;100000\rangle$
Capital_loss	0, $\langle 0;1500\rangle$, $\langle 1500;1600\rangle$, ..., $\langle 2400;2500\rangle$, $\langle 2500;4400\rangle$
Target	$\leq 50K$, $> 50K$; (if income exceeds \$50K/yr based on census data)

Each 4ft-quantifier $\approx$ is related to a condition concerning 4ft-tables. A *4ft-table* $4ft(\varphi,\psi,\ \mathcal{M})$ of Boolean attributes φ and ψ in a data matrix $\mathcal{M}$ is a quadruple $4ft(\varphi, \psi, \mathcal{M}) = \langle a, b, c, d\rangle$ where a is the number of rows of $\mathcal{M}$ satisfying both φ, and ψ, etc., see Fig. 2.

$\mathcal{M}$	A_1	A_2	...	A_K	φ	ψ
o_1	1	4	...	2	0	1
$\vdots$	$\vdots$	$\vdots$	$\ddots$	$\vdots$	$\vdots$	$\vdots$
o_n	4	1	...	6	1	0

Data matrix $\mathcal{M}$

$\mathcal{M}$	ψ	$\neg\psi$
φ	a	b
$\neg\varphi$	c	d

$4ft(\varphi,\psi,\ \mathcal{M})$

Fig. 2. Data matrix $\mathcal{M}$ and *4ft-table* $4ft(\varphi,\psi,\ \mathcal{M})$ of φ and ψ in $\mathcal{M}$

An *association rule* $\varphi \approx \psi$ *is true in a data matrix* $\mathcal{M}$ if a condition related to the 4ft-quantifier $\approx$ is satisfied for a 4ft-table $4ft(\varphi,\psi,\ \mathcal{M})$. A *conditional association rule* $\varphi \approx \psi/\chi$ *is true in a data matrix* $\mathcal{M}$ if the association rule

$\varphi \approx \psi$ is true in a data sub-matrix $\mathcal{M}/\chi$. Data sub-matrix $\mathcal{M}/\chi$ consists of all rows of a data matrix $\mathcal{M}$ satisfying χ. There are dozens of 4ft-quantifiers defined in [6,13] including 4ft-quantifiers corresponding to statistical hypothesis tests.

2.4 Definition of a Set of Relevant Rules

Input of the 4ft-Miner procedure consists of a data matrix $\mathcal{M}$ to be analysed and of a definition of a set of relevant association rules. We do not use conditional association rules in our examples. Thus, a definition of a set of relevant association rules is given by (i) a list of basic 4ft-quantifiers (ii) a definition of a set ANT of relevant antecedents (iii) a definition of a set $SUCC$ of relevant succedents. A set of relevant association rules to be generated and verified is defined as a set of all rules $\varphi \approx \psi$ satisfying

- a condition corresponding to a conjunction of conditions given by listed basic 4ft-quantifiers
- $\varphi \in ANTE$, $\psi \in SUCC$ and φ, ψ have no common attributes.

There are several basic 4ft-quantifiers implemented in the 4ft-Miner procedure embedded into the CleverMiner system, see Table 2.

Table 2. Basic 4ft-quantifiers

Name	Parameter	Definition	Name	Parameter	Definition
`Base`	B	$B \geq a$	`RelBase`	s	$\frac{a}{a+b+c+d} \geq s$
`conf`	c	$\frac{a}{a+b} \geq c$	`aad`	q	$\frac{a}{a+b} \geq (1+q)\frac{a+c}{a+b+c+d}$
`bad`	q	$\frac{a}{a+b} \leq (1-q)\frac{a}{a+b+c+d}$	–	–	–

Note that the 4ft-quantifier aad with parameter q corresponds condition that $lift = \frac{a(a+b+c+d)}{(a+b)(a+c)} \geq (1+q)$. The list of basic 4ft-quantifiers is written in a form: `quantifiers=` $\{$`'Name`$_1$`':Parameter`$_1$`,...,'Name`$_u$`':Parameter`$_u\}$.

Both the definition of a set $ANTE$ of relevant antecedents and the definition of a set $SUCC$ of relevant succedents are in a form of a *definition of a set of relevant cedents*. The core of a definition of a set of relevant cedents is a list of definitions of sets of relevant literals. Each definition of a set of relevant literals is in one of the following forms:
`{'name':'`*Att*`','type':'`*Tp*`','minlen':` L_{min}`,'maxlen':` L_{max}`}` or `{'name':'`*Att*`','type':'one','value':` *Category*`'}`. Here *Att* is a name of attribute, *Tp* is a *type of coefficients*, L_{min} and L_{max} are natural numbers satisfying $1 \leq L_{min} \leq L_{max}$, and *Category* is a category of the attribute *Att*.

Recall that a literal is an expression $A(\alpha)$ where A is an attribute and α is a coefficient of $A(\alpha)$ – a subset of a set of categories of A. The number of categories in α is called a *length of coefficient* α. The numbers L_{min} and L_{max} define a *minimal length of coefficient* and a *maximal length of coefficient*.

There are five types of coefficients: `subset`, `one`, `seq`, `lcut`, and `rcut`. We use attributes Race and Education introduced in Sect. 2.2 to present examples of coefficients of particular types. Each coefficient α of a Boolean attribute $A(\alpha)$ is of type `subset`. Thus, Race(*White*) is an example of literal with a coefficient of type `subset` of length 1. A basic Boolean attribute Race(*Black, White*) is an example of literal with a coefficient of type `subset` of length 2. Type `one` is called *one category.* It is specified by a concrete particular category. The category *White* specifies a literal *Race*(*White*).

The remaining types of coefficients are suitable for ordinal attributes. An attribute is *ordinal* if there is a meaningful ranking of its categories. The attribute Education is ordinal. His 16 categories have a following natural ranking: *Preschool*, *1st-4th*, ..., *Prof-school*, *Doctorate*. Let us further assume that an ordinal attribute A has t categories ranked as follows: $a_1, a_2, \ldots, a_{t-1}, a_t$.

The type `seq` is called *sequence.* A coefficient α of a literal $A(\alpha)$ is of type *sequence* if it holds $A(\alpha) = A(a_u, \ldots, a_v)$ where $1 \leq u \leq v \leq t$. Note that $v-u+1$ is a length of the coefficient of $A(a_u, a_{u+1}, \ldots, a_{v-1}, a_v)$. A coefficient of a literal Education(*1st-4th*, *5th-6th*, *7th-8th*) is of the type sequence. A coefficient of a literal Education(*Bachelor*, *Doctorate*) is not of the type sequence.

The type `rcut` is called *right cut.* It corresponds to categories "*cut from the right*". A coefficient α of a literal $A(\alpha)$ is of type *right cut* if it holds $A(\alpha) = A(a_{t-v+1}, a_{t-v+2}, \ldots, a_{t-1}, a_t)$ where $1 \leq t < v$. Note that v is a length of the coefficient of $A(a_{t-v+1}, a_{t-v+2}, \ldots, a_{t-1}, a_t)$. A literal education(*Masters,Prof-school,Doctorate*) is an examples of a literal with coefficients of type *right cut.*

The type `lcut` is called *left cut.* It is analogous to `rcut`. It corresponds to categories "*cut from the left*".

Let L denote a definition of a set of relevant literals. Then $\mathcal{L}(L)$ denotes a set of literals defined by L. A definition of a set of relevant cedents is an expression `'attributes':[`$L_1, \ldots, L_u$`],'minlen':`C_{min}`,'maxlen':`C_{max}`,'type':'`TC`'`. Here C_{min} and C_{max} are positive integers satisfying $C_{min} \leq C_{max}$. TC denotes a *type of cedent.* It holds either TC= `con` or TC= `dis`.

If TC = `con`, then the definition of a set of relevant cedents defines a set of relevant cedents as a set of conjunctions $A_{i_1}(\alpha_{i_1}) \wedge \cdots \wedge A_{i_1}(\alpha_{i_u})$ such that $i_1 < \cdots < i_u$, $A_{i_1}(\alpha_{i_j}) \in \mathcal{L}(L_{i_j})$ for $j = 1, \ldots, u$ and $C_{min} \leq u \leq C_{max}$. If TC = `dis`, then a set of relevant cedents is defined as a set of disjunctions $A_{i_1}(\alpha_{i_1}) \vee \cdots \vee A_{i_1}(\alpha_{i_u})$ such that $i_1 < \cdots < i_u$, $A_{i_1}(\alpha_{i_j}) \in \mathcal{L}(L_{i_j})$ for $j = 1, \ldots, u$ and $C_{min} \leq u \leq C_{max}$. An example of a definition of set of relevant cedents is shown in Table 3.

3 CleverMiner Implementation of 4ft-Miner

General features of the 4ft-Miner procedure are introduced in Sect. 3.1. Two related analytical questions are then solved by this procedure. The first question concerns searching for segments of persons with high relative frequency of rich persons, see Sect. 3.2. The second question concerns searching for segments of persons with high relative frequency of really rich persons, see Sect. 3.3. Literals

Table 3. Example of a definition of set of relevant antecedents

Cedent: `ante`		`minlen`: 1		`maxlen`: 5		`type`: `con`	
`name`	`type`	`minlen`	`maxlen`	`name`	`type`	`minlen`	`maxlen`
Capital_Gain	seq	1	3	Age	seq	1	3
Capital_Loss	seq	1	3	Education	seq	1	4
Hours_per_week	seq	1	3	Sex	subset	1	1
Occupation	subset	1	1	Country	subset	1	1
Marital_Status	subset	1	1	Race	subset	1	1
Relationship	subset	1	1	Workclass	subset	1	1

of type sequence and right cuts are used to solve both tasks. A comparison of the CleverMiner and apriori is presented in Sect. 3.4. It is shown that it is practically not possible to solve these tasks by the apriori.

3.1 General Features

CleverMiner is an implementation of several GUHA procedures in Python. As input, it requires

df - input pandas dataframe
cedents - definition of sets of Boolean attributes from which patterns to be mined are generated
quantifiers - list of basic 4ft-quantifiers.

CleverMiner procedure prepares dataset, processes it into efficient internal binary form and verifies all[1] relevant patterns. As a result, dictionary with following items is returned:

taskinfo – summary information - task type, cedents, quantifiers, etc.
summary_statistics – key statistics about patterns mining - number of verified patterns, number of patterns in result, timing
result – resulting patterns
datalabels – list of labels for categorical variables.

3.2 Segments with High Relative Frequency of Rich Persons

First, we needed to read data. It is straightforward in Python, we used data directly downloaded from https://archive.ics.uci.edu/ml/machine-learning-databases/adult/adult.data and https://archive.ics.uci.edu/ml/machine-learni ng-databases/adult/adult.test. We combined both datasets and set categories. For attribute Education, we have used ordering by Education_Num attribute

[1] In fact, not all rules are verified when optimization is in place but result is the same as if all rules would be verified.

contained in data. For continuous attributes Age, Hours_per_week, Capital_gain, and Capital_loss we used binning with category boundaries according to Table 1.

We are interested in segments of persons with high relative frequency of rich persons. A person is considered as rich if his/here income exceeds $50K/yr i.e. if the literal Target($> 50K$) is true. Thus, we search for true association rules $A \Rightarrow_{c,B} Target(> 50K)$, where A is a suitable Boolean attribute defining a segment of persons. Only 23.9 per cent of persons satisfy Target($> 50K$). We use a definition of a set of relevant cedents introduced in Table 3. The 4ft-quantfier $\Rightarrow_{c,B}$ is an abbreviation of a couple of basic 4ft-quantifiers $\{$`'conf'`$: c,$ `'Base'`$: B\}$. It corresponds to a condition $\frac{a}{a+b} \geq c \wedge a \geq B$.

The number of output true rules strongly depends on values of parameters c and B. When using rules mining, typically too low or too many rules are returned. If the number of rules is too low, we rarely get interesting finding. If the number of rules is too high, there is very hard to cherry-pick the interesting one due to time needed by analyst. In Sect. 4, an iteration process to get the number of output rules in a desired range 30–50 is described. We use values of parameters `conf` and `Base` resulting from this iteration process. This means that we use values `req_conf` = 0.7921875 and `req_base` = 1922 (reason for these values will be shown later). Input of the 4ft-Miner procedure with these parameters is shown in Fig. 3. However, only several definitions of sets of relevant literals are shown. All used definitions are introduced in Table 3. The run of the 4ft-Miner resulted in 47 rules. The rule with the highest value of confidence (i.e. relative frequency of rich peoples) is shown in Fig. 4.

```
clm = cleverminer(df=df, proc='4ftMiner',
quantifiers={'pim': req_conf, 'Base': req_base},
ante={
  'attributes': [
    {'name': 'Education', 'type': 'seq', 'minlen': 1, 'maxlen': 4},
    ........
    {'name': 'Workclass', 'type': 'subset', 'minlen': 1, 'maxlen': 1},
  ], 'minlen': 1, 'maxlen': 5, 'type': 'con'},
succ={
  'attributes': [
    {'name': 'Target', 'type': 'one', 'value': '>50K'}
  ], 'minlen': 1, 'maxlen': 1, 'type': 'con'}
)
print(clm.result)
```

Fig. 3. Mining for segments with high relative frequency of rich persons

```
"rule_id": 17,
"cedents": {
    "cond": "---",
    "ante": "Hours_per_week_b((30,40> (40,50> (50,60>) &
        Martial_Status(Married-civ-spouse) &
        Age_b((40,50> (50,60>) &
        Education(Bachelors Masters Prof-school Doctorate)
        & Race(White)",
    "succ": "Target(>50K)"
},
"params": {
    "base": 2159,
    "rel_base": 0.04420375905982556,
    "conf": 0.8058977230309817,
    "aad": 2.3679863599109443,
    "bad": -2.3679863599109443,
    "fourfold": [2159, 520, 9528, 36635] },
```

Fig. 4. The output rule with the highest value of confidence

This rule can be written as $A_1 \Rightarrow_{0.8058977230309817,2159} \mathsf{T}arget(>50K)$, where $A_1 = \mathsf{Hours\_per\_week}(\langle 31;40\rangle, \langle 41;50\rangle, \langle 51;60\rangle) \wedge \mathsf{Marital\_status}(\textit{Married-civ-spouse}) \wedge \mathsf{Education}(\textit{Bachelors, Masters, Prof-school, Doctorate}) \wedge \mathsf{Race}(\textit{White})$. It holds $4ft(A_1, \mathsf{T}arget(>50K), Adult) = \langle 2159, 520, 9528, 36635\rangle$.

In entire result set (list of rules), there are 47 output rules with an interesting structure:

- 35 rules are in the form $\mathsf{Education}(\textit{Bachelors, Masters, Prof-school, Doctorate}) \wedge \alpha \Rightarrow_{c,B} \mathsf{T}arget(>50K)$
- 11 rules out of 47 output rules are in the form $\mathsf{Education}(\textit{Bachelors, Masters, Prof-school}) \wedge \alpha \Rightarrow_{c,B} \mathsf{T}arget(>50K)$
- the last rule is $\mathsf{Marital\_status}(\textit{Married-civ-spouse}) \wedge \mathsf{Education}(\textit{Masters, Prof-school, Doctorate}) \Rightarrow_{c,B} \mathsf{T}arget(>50K)$

Here α is in the form $\mathsf{Hours\_per\_week}(\kappa) \wedge \mathsf{Age}(\lambda) \wedge \omega$ where κ and λ are coefficients – sequences of length at least two and ω is one of the literals $\mathsf{Relationship}(u)$, $\mathsf{Native\_country}(u)$, $\mathsf{Marital\_status}(u)$, $\mathsf{Sex}(u)$, $\mathsf{Race}(u)$ or a conjunction of two from these literals. In addition, u denotes a category of a corresponding attribute.

Let us emphasize, that each of output rules contains a literal of type sequence created from attribute Education of length at least three. In addition, 46 rules contain two additional literals of type sequence created from attributes Hours_per_week and Age of length at least two.

3.3 Segments with High Relative Frequency of Really Rich Persons

We are interested in segments of persons with high relative frequency of really rich persons. A person is considered as really rich if both his/here income exceeds \$50K/yr and his/here capital gain is positive. Note that only with 9.3% of persons have a positive capital gain. We search for true association rules $A \Rightarrow_{0.2771,209} \mathsf{T}arget(>50K) \wedge \mathsf{C}apital\_gain(>0)$, where A is a suitable Boolean attribute defining a segment of persons.

We use a definition of a set of relevant antecedents introduced in Table 3 without the attribute Capital_gain. The set of relevant succedents is defined as a set of all conjunctions $\mathsf{T}arget(>50K) \wedge \mathsf{C}apital\_gain(\alpha)$ where α is a set of all right cuts of length 1–9. Recall that the attribute Capital_gain has 9 categories corresponding to positive values, see Table 1. The 4ft-quantfier $\Rightarrow_{0.2771,209}$ is a result of an iterative with a goal to get 30–50 true association rules, see Sect. 4 and Table 6. The run of the 4ft-Miner with these parameters resulted in 40 rules. The rule with the highest value of confidence (i.e. relative frequency of really rich peoples) can be written as
$A_1 \Rightarrow_{0.290,221} \mathsf{T}arget(>50K) \wedge \mathsf{C}apital\_gain(\langle 0;2000\rangle,\ldots,(99000;100000\rangle)$ where $A_1 =$ Capital_loss(0) $\wedge$ Age($(40;50\rangle,(50;60\rangle,(60;70\rangle$)) $\wedge$ Education(*Prof-school, Doctorate*). Categories of the attribute Capital_loss and their frequencies are available in Fig. 5. The category 0 is not shown in this figure, its frequency is 44 807 i.e. 92%.

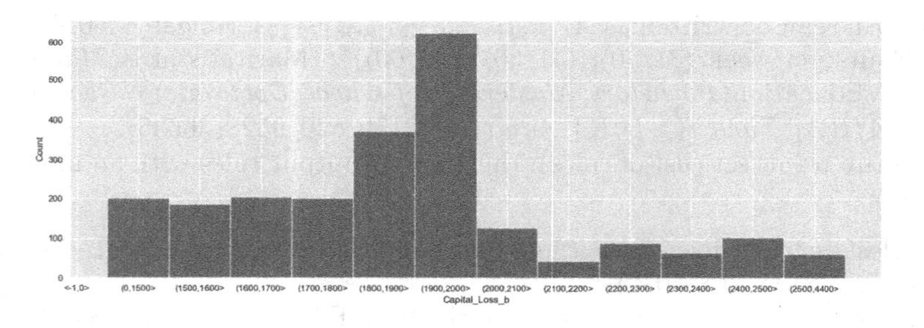

Fig. 5. Distribution of categories for attribute Capital_loss

There are 40 output rules with the following structure:

- Antecedents of 31 rules contain the literal Education(*Prof-school, Doctorate*).
- Antecedents of 24 rules contain a literal Capital_loss(κ) where κ is a coefficient – sequence of length at least two.
- Antecedents of 16 rules contain a literal Age(λ) where λ is a coefficient – sequence of length at least two.
- 3 rules are in a form
  Capital_loss(0) $\wedge$ Education(*Prof-school*)
  $$\Rightarrow_{c,B} \mathsf{T}arget(>50K) \wedge \text{Capital\_gain}(\tau)$$
  where τ are coefficients – right cuts of length at least seven. Antecedents of all remaining 37 rules contain at least one literal with a coefficient – sequence of length at least two.
- Succedents of all 40 rules are in a form Target($> 50K$) $\wedge$ Capital_gain(τ) where τ are coefficients – right cuts of length as follows:

- 10 rules have a coefficient with length 9 i.e. Capital_gain(> 0)
- 10 rules have a coefficient with length 8 i.e. Capital_gain(> 2000)
- 10 rules have a coefficient with length 7 i.e. Capital_gain(> 3000)
- 6 rules have a coefficient with length 6 i.e. Capital_gain(> 4000)
- 3 rules have a coefficient with length 5 i.e. Capital_gain(> 5000)
- 1 rule has a coefficient with length 4 i.e. Capital_gain(> 4000).

Thus, we can again conclude that none of the output rules can be produced by the apriori algorithm nor by *arules* package.

3.4 Comparing CleverMiner and Apriori

The task solved in Sects. 3.2 and 3.3 cannot practically be solved by the apriori algorithm. The core of the problem is the fact that the apriori algorithm deals with literals $A(a)$ only where A is an attribute and a is one of categories. The literals $A(\alpha)$ where α is a general subset of a set of all categories are not allowed.

To solve the mentioned tasks by apriori algorithm, we need to prepare additional attributes with suitable defined categories - subsets of original categories and use them in suitable defined tasks for apriori. The attribute Education has 16 categories, see Table 1. We use coefficients - sequences of length 1–4 for this attribute in our tasks. This way, 15 + 14 + 13 = 42 literals Education(α) with coefficients α - sequences of length 2–4 are automatically generated and used in tasks. This is similar for attributes Age, Hours_per_week, Capital_gain, and Capital_loss.

It is practically not possible to define dozens of necessary derived attributes with suitable categories and to combine them into suitable apriori tasks as well as *arules* package. For more details see [17].

Let summarize advantages of CleverMiner over the standard arules package, apriori-like algorithms and LISp-Miner as well. We selected several features and summarized them in Table 6.

Feature	apriori	arules	LISp-Miner	CleverMiner
Analysis of frequent itemsets	Y	Y	Y	Y
Analysis of rules (conjunction of simple literals)		Y	Y	Y
Succedent can have more literals			Y	Y
Literal can have multiple attribute values			Y	Y
Support for disjunction			Y	Y
Support for different OS	Y	Y		Y
Results processable by standard tools				Y

Fig. 6. Summarization of advantages of CleverMiner

4 Iterative Process with 4ft-Miner

We expect that new algorithm will help to cover part where human-processable number of rules is at input so business expert can focus on finding knowledge or adjusting business patterns and to automate tasks like adjusting base and confidence to find reasonable number of rules. We propose an algorithm for 4ft-Miner dealing with association rules $\varphi \Rightarrow_{p,Base} \psi$. This algorithm can be extended to conditional association rules with additional quantifiers as well as for additional CleverMiner procedures.

4.1 Iterative Algorithm

We have proposed new iterative algorithm to find the desired number of rules. Idea is to use multiple runs of CleverMiner procedure, look at the number of rules in result and adjust parameters accordingly. The algorithm is described for the quantifier $\Rightarrow_{c,B}$ introduced above.

The change is that inputs are not minimal confidence p and support $Base$ but expected number of rules, namely range $MinR$ to $MaxR$ where this number of rules will belong to. Quantifiers parameters c and B and will change appropriatelly to get the desired number of rules. We will have two additional parameters c_{mult} and B_{mult} that drives how much will parameters c and B change to get a bigger or lower number of rules, but in general, any function that increases or decreases the number of rules can be used. As this algorithm is iterative and there may be situations that c and B for a desired number of rules does not exist (or exists but the algorithm will not find it), I_{max} is the maximal number of iterations. Variable i is the number of a current iteration. R and R_{pr} are the numbers of output rules for the current and previous runs of the 4ft-Miner respectively. c and c_{pr} are values of parameter `conf` for the current and previous runs of the 4ft-Miner respectively. The same is true for B, B_{pr} and `Base`.

The algorithm starts by reading $MinR$ and $MaxR$, and optionally with I_{max} and initial values c and B of parameters `conf` and `Base`, c_{mult} and B_{mult} (these optional parameters are predefined with default values). Then it runs according to the following principles:

- The 4ft-Miner procedure is run with parameters `conf` $= c$ and `Base` $= B$.
- If the number of output rules is in the desired interval, then a report is written and stop.
- If the maximal number of iterations is reached, then a report is written and stop.
- Otherwise
  - In first phase, find two consecutive steps where $min(R_{pr}, R) < R_{min}$ and $max(R_{pr}, R) > R_{max}$ (i.e. we will do following until this condition is met or stopping criteria above are met)
    * values of c_{pr} and B_{pr} are set to c and B respectively
    * values of `conf` and `Base` are modified as described in Table 4

  * value of R_{pr} is set to R, the iteration number is increased and the the 4ft-Miner is run with new parameters.
- When first phase is finished, set the following and switch to second phase
  * set $R_{lower} = min(R_{pr}, R), R_{upper} = max(R_{pr}, R), c_{lower} = max(c_{pr}, c), c_{upper} = min(c_{pr}, c), s_{lower} = max(s_{pr}, s), s_{upper} = min(s_{pr}, s)$
- In second phase, we will halving the interval
  * values of R_{lower}, c_{lower}, B_{lower}, R_{upper}, c_{upper}, B_{upper}, `conf` and `Base` are modified as described in Table 4, the iteration number is increased and the 4ft-Miner is run with new parameters.

Table 4. Modification of parameters `conf` and `Base`

#	Phase	Condition	Value of $R_{lower,upper}$	Value c of conf	Value B of Base
1	1	$R < MinR$	–	$c = c * (1 - c_{mult})$	$B = B/B_{mult}$
2		$R > MaxR$	–	$c = c + (1 - c) * c_{mult}$	$B = B * B_{mult}$
3	2	$R < MinR$	$R_{lower} = R$	$c_{lower} = c$	$B_{lower} = B$
4				$c = (c_{lower} + c_{upper})/2$	$B = (B_{lower} + B_{upper})/2$
5		$R > MaxR$	$R_{upper} = R$	$c_{upper} = c$	$B_{upper} = B$
4				$c = (c_{lower} + c_{upper})/2$	$B = (B_{lower} + B_{upper})/2$

We have used these iterations for tasks described in Sects. 3.2 and 3.3 and asked for 30–50 rules. Iteration progress is shown in Tables 5 and 6.

Table 5. Task 1 – modification of parameters `conf` and `Base`

Step	conf	Base	Rules	Phase
1	0.7	1 000	15 591	1
2	0.8	2 000	11	$1 \rightarrow 2$
3	0.75	1 500	1 643	2
4	0.775	1 750	349	2
5	0.788	1 875	88	2
6	0.794	1 938	26	2
7	0.791	1 907	54	2
8	0.792	1 922	47	2

Table 6. Task 2 – modification of parameters `conf` (rounded to 3 digits) and `Base`

Step	`conf`	`Base`	Rules	Phase
1	0.7	1000	0	1
2	0.467	500	0	1
3	0.311	250	0	1
4	0.207	125	42 770	1→ 2
5	0.259	188	1 137	2
6	0.285	219	10	2
7	0.272	204	161	2
8	0.279	211	23	2
9	0.275	208	69	2
10	0.277	209	40	2

5 Conlusions and Further Work

An approach to association rules related to the GUHA method was introduced in Sect. 2. The Python implementation of the 4ft-Miner procedure dealing with the enhanced association rules was described in Sect. 3.1. Two examples of applications of the 4ft-Miner procedure were presented in Sects. 3.2 and 3.3. It was shown in Sect. 3.4 that it is practically impossible to get answers to these analytical questions by the apriori algorithm. The iterative process based on the 4ft-Miner procedure resulting in the number of output association rules in a given range was presented in Sect. 4.

First direction of our further work concerns enhancement of the iteration process to additional 4ft-quantifiers introduced in Table 2. We also plan to investigate additional strategies of modification of quantifiers of parameters. It is possible to define ranges for possible new values of particular quantifiers.

The second direction of further work is inspired by the results presented in Sects. 3.2 and 3.3. The results can be informally formulated as "*if education increases, then capital_gain increases too*". This assertion can be informally written as "Education ↑↑ Capital_gain" It is natural to consider it as an item of domain knowledge. Most rules resulting from the applications described in Sects. 3.2 and 3.3 can be seen as consequences of this item of knowledge. Thus, it is natural to try to filter out all consequences of item of domain knowledge Education ↑↑ Capital_gain. Our goal is to modify the iteration process described in Sect. 4 such that the consequences of items of domain knowledge will be considered in the way described in [14].

References

1. Cleverminer system. https://www.cleverminer.org/

2. Agrawal, R., Imielinski, T., Swami, A.N.: Mining association rules between sets of items in large databases. In: Proceedings of the 1993 ACM SIGMOD International Conference on Management of Data, Washington, DC, USA, 26–28 May 1993, pp. 207–216 (1993). https://doi.org/10.1145/170035.170072
3. Hahsler, M., Chelluboina, S., Hornik, K., Buchta, C.: The arules R-package ecosystem: analyzing interesting patterns from large transaction data sets. J. Mach. Learn. Res. **12**, 2021–2025 (2011). http://dl.acm.org/citation.cfm?id=2021064
4. Hájek, P., Havel, I., Chytil, M.: The GUHA method of automatic hypotheses determination. Computing **1**(4), 293–308 (1966). https://doi.org/10.1007/BF02345483
5. Hájek, P., Havránek, T.: The GUHA method – its aims and techniques (twenty-four questions and answers). Int. J. Man Mach. Stud. **10**(1), 3–22 (1978). https://doi.org/10.1016/S0020-7373(78)80031-5
6. Hájek, P., Havránek, T.: Mechanising Hypothesis Formation - Mathematical Foundations for a General Theory. Springer, Heidelberg (1978). https://www.springer.com/gp/book/9783540087380
7. Hájek, P., Holeňa, M., Rauch, J.: The GUHA method and its meaning for data mining. J. Comput. Syst. Sci. **76**(1), 34–48 (2010). https://doi.org/10.1016/j.jcss.2009.05.004
8. Havránek, T., Chyba, M., Pokorný, D.: Processing sociological data by the GUHA method - an example. Int. J. Man Mach. Stud. **9**(4), 439–447 (1977). https://doi.org/10.1016/S0020-7373(77)80012-6
9. Ivánek, J.: Some examples of transforming ordinal data to an input for GUHA-procedures. Int. J. Man Mach. Stud. **15**(3), 309–318 (1981). https://doi.org/10.1016/S0020-7373(81)80014-4
10. Máša, P., Rauch, J.: Mining association rules between sets of items in large databases. In: Proceedings of the 12th Workshop on Uncertainty Processing, pp. 147–158. MatfyzPress (2022). http://wupes.utia.cas.cz/2022/Proceedings.pdf
11. Powell, L., Gelich, A., Ras, Z.W.: The construction of action rules to raise artwork prices. In: Helic, D., Leitner, G., Stettinger, M., Felfernig, A., Raś, Z.W. (eds.) ISMIS 2020. LNCS (LNAI), vol. 12117, pp. 11–20. Springer, Cham (2020). https://doi.org/10.1007/978-3-030-59491-6_2
12. Rauch, J.: Some remarks on computer realizations of GUHA procedures. Int. J. Man Mach. Stud. **10**(1), 23–28 (1978). https://doi.org/10.1016/S0020-7373(78)80032-7
13. Rauch, J.: Observational Calculi and Association Rules. Studies in Computational Intelligence, vol. 469. Springer, Heidelberg (2013). https://doi.org/10.1007/978-3-642-11737-4
14. Rauch, J.: Expert deduction rules in data mining with association rules: a case study. Knowl. Inf. Syst. **59**(1), 167–195 (2019)
15. Rauch, J., Šimunek, M.: Data mining with histograms and domain knowledge - case studies and considerations. Fundam. Inform. **166**(4), 349–378 (2019). https://doi.org/10.3233/FI-2019-1805
16. Rauch, J., Šimůnek, M.: Learning association rules from data through domain knowledge and automation. In: Bikakis, A., Fodor, P., Roman, D. (eds.) RuleML 2014. LNCS, vol. 8620, pp. 266–280. Springer, Cham (2014). https://doi.org/10.1007/978-3-319-09870-8_20
17. Rauch, J., Šimunek, M.: Apriori and GUHA - comparing two approaches to data mining with association rules. Intell. Data Anal. **21**(4), 981–1013 (2017). https://doi.org/10.3233/IDA-160069

18. Renc, Z., Kubát, K., Kouřim, J.: An application of the GUHA method in medicine. Int. J. Man Mach. Stud. **10**(1), 29–35 (1978). https://doi.org/10.1016/S0020-7373(78)80033-9
19. Turunen, E., Dolos, K.: Revealing drivers natural behavior - a GUHA data mining approach. Mathematics **9**(15), 1818 (2021). https://doi.org/10.3390/math9151818
20. Šimůnek, M.: Academic KDD project LISp-miner. In: Abraham, A., Franke, K., Köppen, M. (eds.) Intelligent Systems Design and Applications. ASC, vol. 23, pp. 263–272. Springer, Heidelberg (2003). https://doi.org/10.1007/978-3-540-44999-7_25

Generating Vascular Networks: A Reinforcement Learning Approach

João Braz Simões[1(✉)], Rui Travasso[2], Ernesto Costa[1], and Tiago Baptista[1]

[1] Centre for Informatics and Systems of the University of Coimbra, Coimbra, Portugal
jbsimoes@student.dei.uc.pt, {ernesto,baptista}@dei.uc.pt

[2] CFisUC, Department of Physics, University of Coimbra, Coimbra, Portugal
ruit@uc.pt

Abstract. Vascular networks are essential for providing nutrients and oxygen to the cells. Cancer progression and type-2 diabetes are examples of pathological conditions which rely on vascular growth. Among various mechanisms, angiogenesis - the process by which new blood vessels grow from existing ones - plays a crucial role. Over the past decades, several mathematical and computational models have been proposed to study different aspects of this biological phenomenon. Most of those models have addressed vascular growth in a bottom-up fashion, for instance by considering biological and physical mechanisms that are involved in the interactions between cells. Instead, we propose a top-down approach, based on reinforcement learning, to generate vascular networks that fully irrigate a given tissue whilst minimising the complexity of the network. In it, vascular networks are represented as graphs composed of multiple nodes and edges. A description of the irrigation provided by the vessel network is utilised to evaluate the quality of the generated vascular networks. In contrast to other models found in the literature, we are not interested in the underlying physical and biochemical mechanisms that drive the vascular network growth, but rather to find the optimal vascular networks that properly irrigate tissue cells. The experimental results show how promising is the proposed approach in the generation of efficient vascular networks. We also study how different reward formulations affect the capacity of the model to generate better solutions.

Keywords: Reinforcement learning · Vascular growth · Angiogenesis · Blood flow · Network optimisation

1 Introduction

Blood vessels are one of the main components of the circulatory system [14]. Their function is to supply oxygen and nutrients throughout the body [14]. Blood vessels are arranged into multi-connected networks, commonly called vascular networks. Blood vessels are formed by endothelial cells (EC), which are the ones responsible for facilitating exchanges between the bloodstream and the surrounding tissue of cells [14].

G. Nicosia et al. (Eds.): LOD 2022, LNCS 13811, pp. 139–153, 2023.
https://doi.org/10.1007/978-3-031-25891-6_11

Vascular network expansion occurs primarily via sprouting angiogenesis, a process by which new blood vessels are formed and expanded from existing ones. This is a key process in physiological processes and pathological scenarios such as tumour growth, diabetic retinopathy and type-2 diabetes. It occurs when cells are experiencing a lack of nutrients and oxygen, releasing pro-angiogenic factors (e.g. VEGF). The VEGF secreted by hypoxic cells forms a well-defined gradient in the tissue. As consequence, the ECs of the surrounding vessels are activated by this pro-angiogenic cocktail, starting the formation of a new sprout. The sprout is led by a tip cell that migrates in the direction of increasing VEGF concentration. The cells behind the tip cell, namely stalk cells, proliferate and start building a new functional blood vessel.

Over the last decades, the process of vascular growth has been studied and modelled by many researchers [6,7,15]. Such models have been crucial to better understand the functional and structural properties of vascular networks, serving as a complement to biological studies. Several different types of models have been utilised to study this complex system. Partial differential equations (PDE), ordinary differential equations (ODE), agent-based and phase-field models are some of the most widely adopted approaches.

Whereas most of the models for vascular growth follow a bottom-up approach, our proposal resorts to a top-down approach. We are not concerned about the interactions between neighbouring cells, but rather about the functional properties of the resulting vascular networks. In fact, structural properties have shown to be crucial in the vascular remodelling process [5,8,9]. In it, vessel network stabilisation requires remodelling and hierarchisation, where blood flow is one of the key actors promoting this process. Previous studies have shown that defects of vascular stabilisation are associated with severe pathologies, thus encouraging the emergence of computational models to predict and simulate efficient vascular networks [8,9].

By following a top-down approach, we propose a Reinforcement Learning (RL) model to generate vascular networks. This biological phenomenon is formulated as a graph generation process, where an RL agent is trained to generate vascular networks by establishing links between pre-existing nodes. We, therefore, consider a set of nodes scattered throughout a 2D environment, and a measure of quality for the generated vascular networks. We then evaluate the capacity of the resulting networks to maintain the minimum oxygen concentration in every tissue cell. At the same time, we seek to build vascular networks with a minimum number of blood vessels. The reward function thus mirrors this requirement and guides the agent to find the optimal vascular network. Since it is a major component of the RL model, our experiments are mainly focused on the influence of different reward functions over the agent's policy.

The results of our experiments show that the proposed model is promising to generate vascular networks with a reduced number of blood vessels while maintaining the minimum acceptable value of oxygen in every cell. From our experiments, we observed an improvement in the resulting networks after mitigating excessive branching.

2 Existing Models for Sprouting Angiogenesis

Several computational models have been proposed to generate vascular networks through the process of sprouting angiogenesis. Whereas most of them are based on mathematical models, others have resorted to agent-based architectures to provide new insights into this biological process. Moreira et al. [6] and Travasso et al. [15] presented studies of vessel growth in tissue of cells, proposing phase-field models to define the interactions between ECs and the extra-cellular matrix. Olsen et al. [7] introduced a multi-scale agent-based model by considering that EC and tissue cells are agents that interact in a simulation environment. Szabó et al. [13] explicitly represented the shape of ECs using a 2D Cellular-Potts model, a very common approach to deal with movements of close-packed cells.

Whereas the previously mentioned models explicitly incorporate the interactions between ECs and tissue cells, we propose to adopt a top-down approach. Instead of defining the properties of the cells and their interactions, we pretend to search for the optimal vascular networks that better irrigate a tissue of cells. For this purpose, we propose a reinforcement learning algorithm that learns how to generate optimal vascular networks.

3 Reinforcement Learning

Reinforcement Learning (RL) is a subfield of Machine Learning that aims to solve problems that involve sequential learning and decision-making under uncertainty [12]. As shown in Fig. 1, RL comprises two main entities - agent and environment - that communicate and interact between them via observations, actions and rewards.

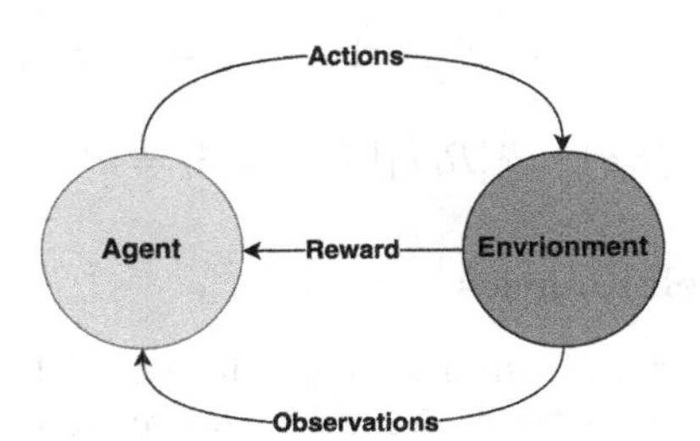

Fig. 1. Reinforcement Learning entities and interactions between them

An agent is an entity that interacts with the environment by making observations, executing actions and then being rewarded. Such a reward is usually a scalar value, positive or negative, that instructs the agent on how good a certain action was. The reward is a local property, reflecting the success of a very specific action and not the overall performance of the agent.

A core aspect of RL is policy. A policy is a set of rules that guide the agent's actions. As the main objective of the agent is to maximise the expected return, different policies can lead to different amounts of return.

In terms of taxonomy, RL models are usually classified as either [4,12]: model-free or model-based, value-based or policy-based and on-policy or off-policy.

For this work, we adopt Q-Learning - a model-free, value-based and off-policy algorithm. As a model-free method, the agent learns the consequences of its actions, carrying out an action multiple times and adjusting its policy. The algorithm is value-based, since it calculates the value of each action and performs the action with the best value, not providing a probability for every action. As an off-policy method, it evaluates and improves a policy different from that used to generate the data.

In overview, an RL agent learns how to act greedily towards a specific goal determined by the reward function. When in a given state, the agent decides which action to take from a joint probability distribution $P(s, a, s')$ of moving to state s' after executing action a in state s. The goal of the agent is to maximise the expected accumulated reward for all interactions with the environment. Such a set of interactions is usually called a trajectory.

Markov Decision Processes (MDPs) formally describe a reinforcement learning environment [12], assuming that the agent fully observes the environment and therefore every state s completely describes the environment. Formally, MDP comprises:

- A finite state space S
- An action set A
- A state transition probability model:

$$\mathcal{P}_{ss'} = P[S_{t+1} = s' | S_t = s, A_t = a] \tag{1}$$

  where s is the current state, s' is the successor state and a is the chosen action.
- A reward function, $r(s, a) = \mathbb{E}[R_{t+1} | S_t = s, A_t = a]$

3.1 Q-Learning Preliminaries

One of the most popular and simplest algorithms is Q-Learning, a model-free, value-based and off-policy method. Some advantages of Q-Learning are its simplicity and support for empirical historical data while providing good results in many application fields.

The Q-Learning algorithm was initially proposed as a tabular method. The maximum expected value for executing action a when in state s, $Q(s, a)$, is stored in a tabular structure. In overview, Q-Learning encompasses the following steps:

1. An empty table maps states to the value of actions
2. Through a simulation process, the agent obtains experience (s, a, r, s') where s is the current state, a is the action, r is the assigned reward and s' is the next start after executing action a.

3. $Q(s,a)$ value is updated as follows: $Q(s,a) = r + \gamma \max_{a' \in A} Q(s',a')$

The tabular approach, as previously described, is suitable for simple environments where the number of states is relatively low. However, the same is not true for complex environments where the number of all possible states is larger or even uncountable. Indeed, it struggles in situations where the tabular approach is unfeasible and impractical.

To tackle this issue, non-linear representations have been utilised as a substitute for tabular approaches. The field of Deep Q-Learning has shown promising results when using neural networks (NN) to approximate the state-value function $Q(s,a)$. In it, the problem of learning the agent's policy is converted into a regression problem, where a neural network estimates the maximum expected reward of every feasible action.

Training Q-Network. By adopting a neural network to learn the agent's policy, the training phase involves updating the parameters against a loss function derived from Bellman's equation:

$$Q(s,a) = r(s,a) + \gamma * max_{a' \in A} Q(s',a') \tag{2}$$

where A stands for the action set and γ is a discount factor. The Bellman equation states that when acting greedily we are not only concerned about the immediate reward, but also the long-term value.

The Q-Network is then trained to better predict the expected long-term reward by using the immediate reward ($r(s,a)$) and the discounted maximum long-term (when acting greedily in the next state) as the ground truth. In this sense, the loss function depends on whether the episode is over (or not) in the next step. The mean squared error (MSE) is used to approximate the $Q(s,a)$ function.

For every intermediate step in the episode, the loss function is computed as:

$$\mathcal{L} = (Q(s,a) - (r + \gamma * max_{a' \in A} Q(s',a')))^2 \tag{3}$$

At the end of the episode, the loss function is computed as:

$$\mathcal{L} = (Q(s,a) - r)^2 \tag{4}$$

where $Q(s,a)$ is the value of the state-action pair, r is the immediate reward, and γ is a discount factor. The network is then trained through mini-batch gradient descent, where each mini-batch is sampled from a replay buffer.

An analysis of Q-Learning applications to other domains suggests that some techniques can be adopted from the beginning. Regarding the exploration, the agent is trained to explore the environment by using the epsilon-greedy method. This method is based on the probability of choosing a random action. The agent switches between greedy and random actions with a probability of ϵ. This probability is decreased over time, which means that the agent will choose random actions more often at the very beginning of the training stage.

The correlation between steps is crucial. As mini-batch gradient descent is used to update the neural network parameters, the most common technique to reduce this correlation is the inclusion of a replay buffer and bootstrapping. The first one allows accumulating experience in a large buffer and sample training data from it. The second one introduces a target network that is periodically updated with the parameters of the main network and used to compute $Q(s', a')$ value in the Bellman equation.

4 Generating Vascular Networks

Unlike the models outlined in Sect. 2, we propose a novel approach to generating vascular networks. We free the model from any explicit specification of the interactions between cells or other local phenomena underlying sprouting angiogenesis. The biological phenomenon is then formulated as a Markovian Decision Process (MDP), resorting to a Q-Learning approach to train an agent capable of generating vascular networks that fully irrigate a given tissue whilst minimising the number of vessels.

4.1 Model Architecture

We formulate this problem as a graph generation task, thus applying a Q-Learning algorithm to generate vascular networks. In this sense, the environment is composed of a set of nodes arbitrarily placed in a two-dimensional space. Each node can then be connected to other nodes by an undirected link. Every change in the environment is performed by the agent, which can establish new connections between nodes. We focus on the functional and structural properties of the resulting vascular network. Blood pressure, blood flow and irrigation are considered for evaluating the generated solutions. We are thus interested in generating vascular networks capable of irrigating the surrounding tissue while resorting to a minimal amount of blood vessels.

Inspired by previous applications of reinforcement learning for graph generation [1], the decision model is split into two independent models: one for selecting the start node and another to select the end node. In fact, despite being constant, the number of nodes can be very large. Arguably, it would be more difficult to train a model capable of selecting two pairs of nodes at the same time. In such a scenario, the action space would grow exponentially with the number of nodes, containing all $N \times N$ pairs of nodes, where N is the number of nodes in the environment. To avoid this potential issue, we formulate the MDP as a two-stage sequential process, considering that the agent initially selects a start node and then an end node. A new link is established between the two previously selected nodes.

The formulation of the MDP process to generate vascular networks is as follows:

- **State space:** At time t, the state S_t contains the current graph G_t and an indicator of which type of node is going to be chosen (start node or end node).

- **Action space:** The action space A_t involves adding a new edge to the current graph G_t. This is a two-step process in which the agent first selects a start node and then an end node:
  - If the agent is selecting a start node, the available actions are the nodes in the graph G_t.
  - If the agent is selecting an end node, the available actions are constrained by the neighbourhood of the selected start node, as specified by an auxiliary neighbourhood map.
- **Transition function:** $T^a_{ss'} = P[S_{t+1} = s'|S_t = s, A_t = a]$
- **Reward function:** The reward function depends on whether the agent selected a start or an end node.
  - A **start node** was selected:

$$R_t = \begin{cases} -1, & D_{start\_node} \geq 3 \\ 0, & \text{otherwise} \end{cases} \tag{5}$$

    where D_{start_node} is the degree of the selected start node, as discussed in Sect. 5.
  - An **end node** was selected:

$$R_t = -1 + (I_t - I_{t-1}) \tag{6}$$

    where I_t is the irrigation score at time t, I_{t-1} is the irrigation score at the previous time step that an edge was added. That is to say, we capture the irrigation improvement after adding an edge. Note that the reward can also be negative, meaning that the irrigation got worse after adding a new edge.

The architecture of the start node selection and the end node selection are similar. Each one comprises two modules: graph encoding and q-value prediction. The graph encoding module is responsible for transforming a graph structure into a fixed-length representation. This representation captures the existence of edges from each node to its neighbourhood. All possible neighbours are pre-defined in an auxiliary neighbourhood map, remaining unchanged until the end of the simulation. The Q-Value prediction model, as the name suggests, is responsible for determining the value of each possible action in each state. It accepts a fixed-length representation of the graph, as generated by the encoding module, and outputs the maximum expected reward for each available action.

Figure 2 shows an overview of the model. Both Q-Networks are composed of two linear layers and a rectified linear activation function is applied to the output of the first layer. The process of generating a new edge between nodes 1 and 3 is described in Fig. 2. Initially, the current graph and the corresponding neighbourhood map are processed by the encoder, outputting a fixed-length representation. For each node, a vector with the same length as the neighbourhood size is initialised. All existing neighbours are marked as 1, whereas the others remain 0.

Apart from the neighbourhood encoding, we keep track of which node was selected in the start node selection step. For this purpose, a one-hot encoding with size N, where N is the number of nodes in the entire graph, is utilised to represent which node was selected in the previous step. Note that this information is irrelevant when selecting a start node, but only necessary when selecting the end node. For simplicity, we are not excluding this information from the encoder used in the start node selection stage.

Concerning the output layers, both models output the same action set, which corresponds to all nodes in the graph. For instance, a graph composed of N nodes results in an action set of size N. In other words, the Q-Network is estimating the long-term reward when selecting each node in the graph.

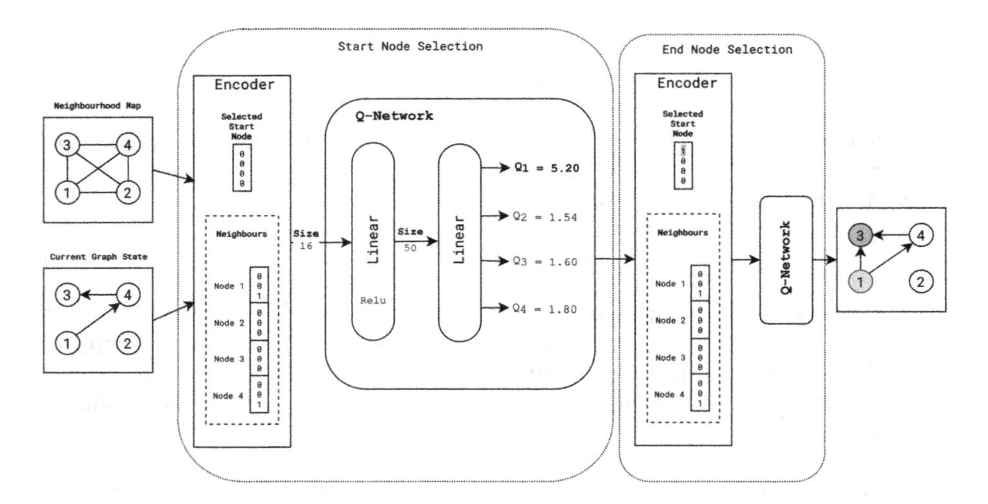

Fig. 2. Overview of the model. It is composed of two independent Q-Networks, both following the same architecture. The first one is responsible for selecting the start node, while the second one is responsible for selecting the end node. A graph encoder is used to transform the graph structure into a fixed-length representation.

Vascular networks are represented as graph-like structures, considering that nodes are places where the blood vessels pass through and branch off. Edges are then formed between adjacent nodes, which can either represent segments of a given vessel, or branches of the network.

4.2 Evaluating Vascular Networks

The evaluation of vascular networks depends on local tissue irrigation. The environment is discretised into a grid of equal-sized cells, each of which is assigned a concentration of oxygen. The tissue is considered fully irrigated when the concentration of oxygen in every cell is above a critical value.

For this purpose, we consider a description of the irrigation provided by the vessel network which is capable of identifying the main vessels responsible for irrigating the tissue. The physiological oxygen diffusion length is considered as well. This description comprises the following two steps: blood flow calculation and oxygen diffusion.

1. **Blood flow calculation:** We consider that the vessels are cylindrical and we take the approximation of laminar flow. In this way, the flow through a vessel follows Poiseuille's law, and it is proportional to the pressure difference between the vessel inlet and outlet. In this context, the vessel flow resistance is proportional to the vessel length and inversely proportional to the fourth power of the vessel radius [3,6]. Here, all vessels between two neighbouring points are considered equal. The blood flow in each vessel is calculated as a function of the constant inlet and outlet pressures for the whole network using Kirchhoff's laws at every bifurcation [3].
2. **Oxygen diffusion:** *In vivo*, oxygen is carried by erythrocytes in the blood and is partly dissolved in blood plasma. The rate of oxygen transport from each vessel to the tissue depends in general on the level of oxygen in the blood (typically higher in vessels with higher pressure), on the tissue oxygen concentration and it increases with the blood flow through the vessel [10,11]. The oxygen then diffuses and is consumed by the tissue cells:

$$\nabla^2 c_O - \frac{c_O}{L^2} + s = 0 \; , \tag{7}$$

where c_O is the oxygen concentration, L is the oxygen diffusion length (we take the approximation of low oxygen concentration where the consumption term is linear in c_O [3,10]) and s is a function of position describing the local oxygen transfer rate from the vascular network to the tissue. Instead of calculating s iteratively [3,10,11], we use the following description for the oxygen transfer rate to the tissue:

$$s(x,y) = \begin{cases} QP_m \; , & (x,y) \in \text{vessel} \\ 0 \; , & \text{otherwise} \end{cases} \; , \tag{8}$$

where Q is the flow and P_m is the average pressure in the vessel. Using (8), vessels with a larger flow and with arterial blood (which is correlated to higher vessel pressure) transfer more oxygen to the tissue.

5 Experiments

A core component of any reinforcement learning approach is the reward design. Our preliminary experiments have been focused on evaluating different approaches to assigning rewards.

Whereas the irrigation goal is an episode stop criterion, the network efficiency is the major focus of the reward design. Observations on real data show that vessel network stabilisation requires remodelling and hierarchisation, where blood flow is one of the key factors promoting network stabilisation [2]. Consequently, severe pathologies have been associated with defective vessel stabilisation, encouraging further research on this topic. In this sense, we are interested in the formation of new blood vessels that minimise the size of the network, while maintaining the minimum oxygen levels in the entire tissue.

Experimental Setup. The experiments are conducted in a simulation environment composed of 100 nodes equally distributed over a 10×10 environment. In every experiment, the episode starts with two fully functional networks with independent inlet and outlet nodes. Both networks are irrigating distinct parts of the environment, as shown in Fig. 3. The parameters of the Q-learning model are specified in Table 1. We evaluate the performance of the agent at each 5000 training steps, where a single episode is executed using the Q-Network trained so far. Each episode stops as soon as every cell of the tissue has reached a minimum oxygen concentration of 2.0.

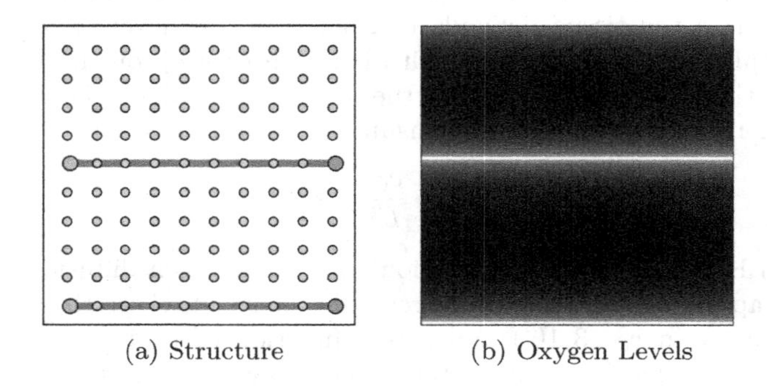

(a) Structure (b) Oxygen Levels

Fig. 3. The images above show the initial state of the simulation environment

Table 1. The parameters of the Q-Learning model used in the experiments.

Target-net sync rate	Epsilon start prob.	Epsilon end prob.	Epsilon decay last step	Replay buffer size	Discount factor	Batch size	Warm-up steps
10^4	1.0	0.1	10^6	10^6	0.99	32	10^5

Episodic Reward. We initially considered an episodic reward, where the relative number of edges added to the network is calculated at the end of each episode. All the intermediate steps were given a zero reward.

As shown in Fig. 4, no interesting results were achieved from this approach. The experience samples used to train the agent are mostly from intermediate steps whose reward is zero. This, in turn, led to the agent being unable to learn how to minimise the network size. Intermediate steps are indistinguishable, and final rewards are not sufficient to motivate the agent to minimise the network size.

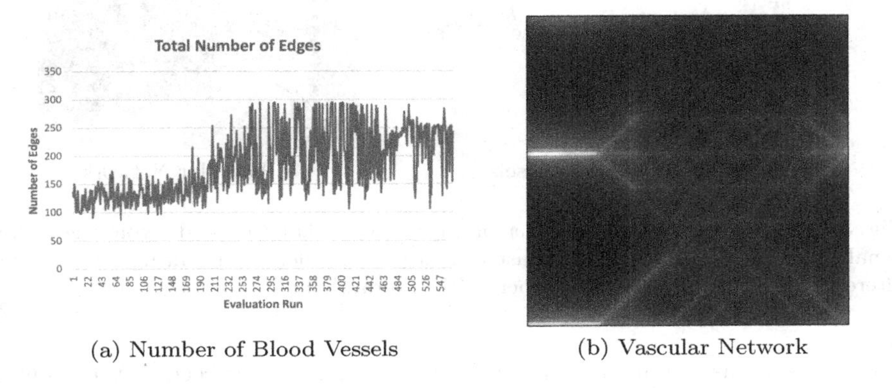

(a) Number of Blood Vessels (b) Vascular Network

Fig. 4. The results from the episodic reward approach. The agent is not able to generate vascular networks with a reduced number of vessels.

Penalising Edge Additions. Our second formulation of the reward was to explicitly penalise every edge addition. In this sense, a reward of -1 is assigned after selecting an end node. This approach is inspired by previous applications of reinforcement learning, where the agent is constantly penalised for performing actions that do not lead to the end of the episode (e.g., maze solving).

Figure 5 shows the number of edges in the vascular network over the evaluation episodes. In the early stages of the training, the number of edges is high and unstable, indicating that the agent is mostly acting randomly. As the training proceeds, it becomes evident that the agent starts learning a strategy to irrigate the environment by using a minimum number of edges.

Nonetheless, we observed that some unnecessary branches are generated. Apart from the negative reward for adding an edge, neither the irrigation function nor the reward function penalises unnecessary branches. The irrigation function presented in Sect. 4.2 is an over-simplified version that considers constant inlet and outlet pressures. Whereas an irrigation function that assumes a constant inlet and outlet flow would penalise redundant branches, the proposed method is not capturing such properties.

Mitigating Redundant Branches. Our initial attempt was to mitigate redundant branches by constraining the maximum number of edges per node. In practice, a maximum degree of 3 is imposed for every node in the network. This constraint is directly applied to the action set when selecting the next start

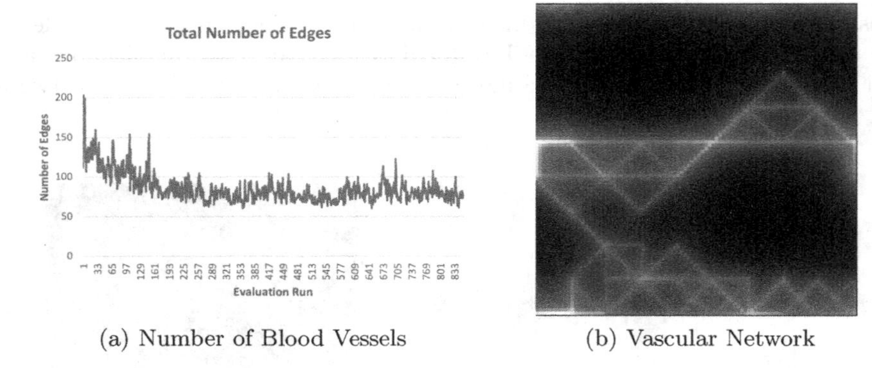

(a) Number of Blood Vessels (b) Vascular Network

Fig. 5. The evolution of the number of edges shows that the model converges after penalising edge additions. In the last evaluation episode, the resulting network has interesting structural properties, albeit with several redundant vessels.

node. As soon as the maximum degree of a given node is exceeded, the agent is not allowed to add more edges to such a node.

The number of edges in the network over the evaluation episodes is shown in Fig. 6. Most of the redundant edges and branches have been mitigated.

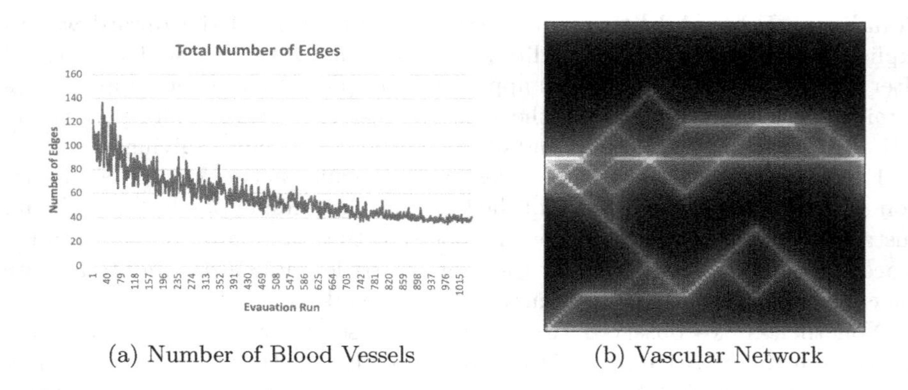

(a) Number of Blood Vessels (b) Vascular Network

Fig. 6. Results after constraining the maximum number of edges per node to 3.

Moving Redundant Branches to the Reward Function. So far, the reward function is always zero every time a start node is selected. After achieving better results by constraining the maximum number of edges, we made a further improvement by moving the redundant branches to the reward function. Instead of forbidding nodes with degrees greater than 3, the agent learns to avoid redundant branches by receiving a negative reward for excessive branching.

In this sense, we assigned a fixed penalty of -1 to every start node selection whose corresponding degree is greater than 3. The obtained performance in the evaluation episodes is shown in Fig. 7.

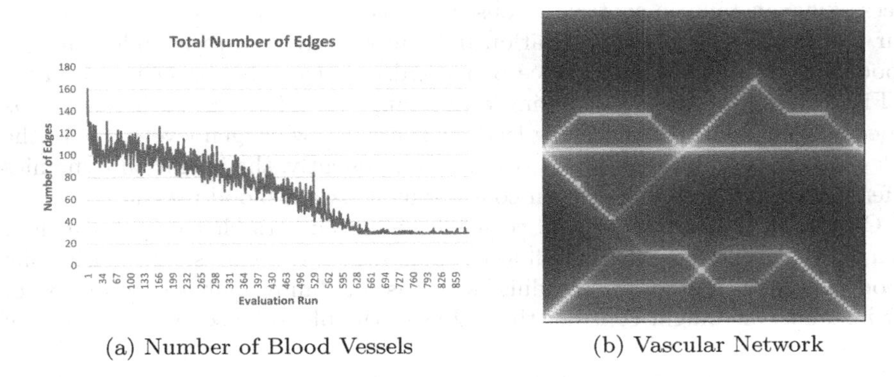

(a) Number of Blood Vessels (b) Vascular Network

Fig. 7. The inclusion of redundant branching penalty in the reward function led to a new minimum number of vessels in the evaluation episodes. The final vascular network (b) contains a reduced number of blood vessels while irrigating the whole tissue.

When compared to the results obtained in Sect. 5, we observe that penalising excessive branching promotes the model to converge faster while generating vascular networks with fewer redundant branches in a more consistent manner.

6 Conclusion and Future Work

In this work, we addressed the problem of generating a vascular network given the goal of irrigating a group of cells by utilising a minimum number of blood vessels. For the first time, a reinforcement learning approach was applied as means of modelling this biological phenomenon. We formulated a reinforcement learning approach to generate vascular networks represented as graphs, considering that an agent is tasked with establishing connections between already existing nodes. As proof of concept, we studied the capacity of this approach to generate vascular networks in a controlled simulation environment. Our experiments were focused on how different reward functions affect the performance of the agent. The obtained results are promising, as they show the ability of the agent to generate functional networks containing a reduced number of edges. An inclusion of irrigation improvement in the reward function appears to have a positive impact on the performance of the model, thus promoting better convergence properties and reducing the number of edges.

Several improvements to the proposed approach have been identified. The proof of concept presented in this work can be extended in terms of biological properties, reinforcement learning methodology and experimental design.

6.1 Biological Properties

As discussed in Sect. 4.2, the proposed method to calculate the blood flow, and corresponding oxygen levels, is assuming constant blood pressure in both the inlet and the outlet nodes. Consequently, the blood flow at these nodes is not necessarily constant, in contrast to observations made in real vascular networks. Our method is a good approximation but can be extended to include constant blood flow. Oddly, the abundance of redundant edges and branches observed in Fig. 5 decreases after considering a constant blood flow. In such cases, it is expected that excessive branching leads to zones of low oxygen levels, due to the lack of blood flow in many blood vessels. Consequently, the irrigation dynamics differ from what is observed when considering constant blood pressure.

On the other hand, real blood vessels have indeed variable calibres, which in turn plays a crucial role in blood flow dynamics. So far, we are solely considering blood vessels with a constant radius, which is a definite point to be improved. For instance, one might consider the radius of the blood vessel as a property of the graph edges.

6.2 Reinforcement Learning Methodology

The proposed method to encode the state of the environment consists in converting the graph into a fixed-length representation, although not independent of the number of nodes and the neighbourhood of each node. In other terms, the Q-Network input varies as the number of nodes and the neighbourhood of each node change. In the context of Graph Neural Networks (GNN), several methods have been proposed to encode graphs into fixed-length and permutation-invariant representations. Such methods allow us to learn embedding representations of graphs that capture structural properties, thus allowing us to generalize across multiple graphs with a variable number of nodes and neighbourhoods.

6.3 Experimental Design

Our preliminary experiments were conducted in an environment with 100 nodes. The next step is to extend the problem to larger environments or even higher density of nodes. On the other hand, we are considering a single initial configuration. It would be interesting to train the agent on a wider range of initial configurations and evaluate the performance for each of them.

Finally, the focus of our experiments was to provide a proof of concept for the proposed method. However, it is expected that this work will be extended to include a comparison with other models. This is an important step in the development of the proposed approach since it allows a better understanding of its main advantages.

Funding. This work is funded by the FCT - Foundation for Science and Technology, I.P./MCTES through the Ph.D. grant SFRH/BD/122607/2016 and national funds (PIDDAC), within the scope of CISUC R&D Unit - UIDB/00326/2020 or project code UIDP/00326/2020.

References

1. Dai, H., et al.: Adversarial attack on graph structured data. In: International Conference on Machine Learning, pp. 1115–1124. PMLR (2018)
2. Farnsworth, R.H., Lackmann, M., Achen, M.G., Stacker, S.A.: Vascular remodeling in cancer. Oncogene **33**(27), 3496–3505 (2014)
3. Gandica, Y., Schwarz, T., Oliveira, O., Travasso, R.D.: Hypoxia in vascular networks: a complex system approach to unravel the diabetic paradox. PLoS ONE **9**(11), e113165 (2014)
4. Lapan, M.: Deep Reinforcement Learning Hands-On. Packt Publishing (2020)
5. McGrath, J.C., et al.: New aspects of vascular remodelling: the involvement of all vascular cell types. Exp. Physiol. **90**(4), 469–475 (2005)
6. Moreira-Soares, M., Coimbra, R., Rebelo, L., Carvalho, J., DM Travasso, R.: Angiogenic factors produced by hypoxic cells are a leading driver of anastomoses in sprouting angiogenesis-a computational study. Sci. Rep. **8**(1), 1–12 (2018)
7. Olsen, M.M., Siegelmann, H.T.: Multiscale agent-based model of tumor angiogenesis. Procedia Comput. Sci. **18**, 1016–1025 (2013)
8. Owen, M.R., Alarcón, T., Maini, P.K., Byrne, H.M.: Angiogenesis and vascular remodelling in normal and cancerous tissues. J. Math. Biol. **58**(4), 689–721 (2009)
9. Renna, N.F., De Las Heras, N., Miatello, R.M.: Pathophysiology of vascular remodeling in hypertension. Int. J. Hypertens. **2013** (2013)
10. Secomb, T.W., Alberding, J.P., Hsu, R., Dewhirst, M.W., Pries, A.R.: Angiogenesis: an adaptive dynamic biological patterning problem. PLoS Comput. Biol. **9**(3), e1002983 (2013)
11. Secomb, T.W., Hsu, R., Park, E.Y., Dewhirst, M.W.: Green's function methods for analysis of oxygen delivery to tissue by microvascular networks. Ann. Biomed. Eng. **32**(11), 1519–1529 (2004)
12. Sutton, R.S., Barto, A.G.: Reinforcement Learning: An Introduction. MIT Press, Cambridge (2018)
13. Szabó, A., Czirók, A.: The role of cell-cell adhesion in the formation of multicellular sprouts. Math. Model. Nat. Phenom. **5**(1), 106–122 (2010)
14. Thomas, K.: Angiogenesis. In: Bradshaw, R.A., Stahl, P.D. (eds.) Encyclopedia of Cell Biology, pp. 102–116. Academic Press, Waltham (2016). https://doi.org/10.1016/B978-0-12-394447-4.40019-2. https://www.sciencedirect.com/science/article/pii/B9780123944474400192
15. Travasso, R.D., Corvera Poiré, E., Castro, M., Rodrguez-Manzaneque, J.C., Hernández-Machado, A.: Tumor angiogenesis and vascular patterning: a mathematical model. PLoS ONE **6**(5), e19989 (2011)

Expressive and Intuitive Models for Automated Context Representation Learning in Credit-Card Fraud Detection

Kanishka Ghosh Dastidar[1(✉)], Wissam Siblini[2], and Michael Granitzer[1]

[1] Faculty of Computer Science and Mathematics, University of Passau, Passau, Germany
{kanishka.ghoshdastidar,michael.granitzer}@uni-passau.de
[2] Development and Innovation, Worldine, Lyon, France

Abstract. The fraud detection literature unanimously shows that the use of a cardholder's transaction history as context improves the classification of the current transaction. Context representation is usually performed through either of two approaches. The first, manual feature engineering, is expensive, restricted, and hard to maintain as it relies on human expertise. The second, automatic context representation, removes the human dependency by learning new features directly on the fraud data with an end-to-end neural network. The LSTM and the more recent Neural Feature Aggregate Generator (NAG) are examples of such an approach. The architecture of the NAG is inspired by manual feature aggregates and addresses several of their limitations, primarily because it is automatic. However, it still has several drawbacks that we aim to address in this paper. In particular, we propose to extend the NAG in the following two main manners: (1) By expanding its expressiveness to model a larger panel of functions and constraints. This includes the possibility to model time constraints and additional aggregation functions. (2) By better aligning its architecture with the domain expert intuition on feature aggregates. We evaluate the different extensions of the NAG through a series of experiments on a real-world credit-card dataset consisting of over 60 million transactions. The extensions show comparable performance to the NAG on the fraud-detection task, while providing additional benefits in terms of model size and interpretability.

Keywords: Representation learning · Fraud detection · Deep learning

1 Introduction

E-commerce credit card fraud is characterized by the unauthorized use of card details to make purchases online. Fraud detection aims to combat this by analyzing the stream of incoming transactions and raising alerts or creating a ranked list of the most suspicious transactions. Following this, human investigators verify the raised alerts and either follow up with the customer or block the card.

G. Nicosia et al. (Eds.): LOD 2022, LNCS 13811, pp. 154–168, 2023.
https://doi.org/10.1007/978-3-031-25891-6_12

Within fraud detection, we distinguish between the two primary data-driven methods: transaction-based and context-based approaches. A transaction-based approach uses data samples comprising only the attributes of the transaction being processed. On the other hand, a context-based approach is one where each data sample consists not only of the transaction to be processed but also additional features based on the context in which the transaction took place. An example of such context would be the account activity preceding a particular transaction. This can be termed as the 'user context' in which the current transaction took place. Directly using these past transactions as inputs to a machine learning model would lead to a very high dimensional feature space. Therefore, several works have proposed ways to define a compact representation for the past behaviour of the card-holder. One commonly used approach in academia and industry is to represent this past information through manually engineered feature aggregates [6, 23]. These aggregates are defined and updated based on input from domain experts, which is an expensive and time-consuming process. Additionally, they suffer from other limitations outlined in [12]. An alternative is to directly learn these representations on the fraud task itself using a gradient-descent based approach. This would entail an end-to-end learner, removing the need for manual feature engineering. However, despite its limitations and the apparent benefits of automatic approaches, manual feature aggregates continue to be broadly used in real-world fraud detection systems. This can be attributed to the fact that commonly used automatic approaches (a recurrent network, for example) learn representations that are hard to decipher and explain. Therefore, there is a trade-off between the cost of human expertise and the loss of explainability. One approach, in particular, the Neural Aggregate Generator (NAG) proposed by us in [12] and extensively studied in [11], is a novel method that addresses this by specifically being designed based on the structure of manual aggregates making it more intuitive for domain experts.

Our main contribution in this paper is based on extending the initial proposal of the NAG. Here, we aim to ask several questions. Firstly, can the architectural design of the NAG be refined to be more analogous to manual aggregates? Secondly, what are the different types of aggregation functions the NAG can be extended to model? Thirdly, can such extensions improve the predictive performance on the fraud task?

2 Background

The bulk of the research in credit card fraud detection focuses on the specific peculiarities of fraud detection data. As described in the extensive reviews by Lucas and Jurgovsky [16], and Rymann-Tub et al. [2], this includes the imbalanced nature of the data, the shift in the data distribution across time, and the sequential properties of the data. The class imbalance problem is addressed in different ways i.e. through different sampling methods [13] or through the use of models that are robust to imbalance [7]. Due to the evolving nature of buying behaviour (co-variate shift) and fraud strategies (concept drift), the i.i.d assumption doesn't necessarily hold for credit card data. This has shown to affect the performance of classifiers in predicting fraud. There have been different works that detect and/or address this shift [8,18].

Several classifiers have been used for the task of fraud detection. Awoyemi et al. [4] compare the performance of k-nearest neighbor, naive bayes, and logistic regression classifiers on a European credit-card dataset. Randhawa et al. [21] use a hybrid approach consisting of different algorithms such as a feed-forward network, random forest, linear regression, etc., in conjunction with AdaBoost. Sahin and Duman [21] compare the performance of support vector machines (SVM) and decision trees, with decision trees detecting more frauds than the SVM. Random forest classifiers have been used in several works [14,24]. However, as shown by Jurgovsky et al., the performance of the random forest at transaction level is significantly worse than using a sequence learner on sequences of transactions.

Deep learning approaches have also been increasingly proposed in recent years. Pumsirirat and Yan [20] use an unsupervised approach using an auto-encoder and restricted Boltzmann machine. Fiore et al. [9] use a generative adversarial network to mimic the minority class samples in order to address the class imbalance problem. Kim et al. [15] show that a feed-forward network outperforms a hybrid ensemble approach within a champion-challenger setting.

As outlined in Sect. 1, several recent works focus on a context-based approach, where the context is defined as the past transactions of a card-holder within a particular window of time. In the following section, we introduce the important definitions and limitations of the two primary forms of context representation.

3 Context Representation for Fraud Detection

In this section, we will analyze the structure of manual feature aggregates and the architecture of the Neural Aggregate Generator. Based on this analysis, we will derive the key aspects in which the NAG can be extended.

3.1 Feature Engineering for Fraud Detection

The broad template for feature aggregations for credit card fraud detection was proposed in the work of Whitrow et al. [23]. Here, different aspects of a transaction that are relevant to fraud, such as the value of the transaction or whether the transaction was verified by magnetic stripe, are identified through exploratory data analysis. Based on this, the authors aggregate these aspects across the recent transactions resulting in an account-level summary. This method assumes that the order of the transactions is not vital for fraud detection. Bahnsen et al. [6] extend this method by aggregating information only over past transactions that share the same value of a particular categorical attribute as the current transaction. This work also introduces the use of combinations of categorical features for the aggregates. An example of such an aggregate would be 'the sum of all transactions that took place at the *same* merchant and the *same* country as the current transaction'. Similar types of aggregate features have found broad usage as seen in a number of studies [1,3,14].

For this work, we borrow the definition of aggregates based on the proposal given by Bahnsen et al. [6]. As an example, we take the aggregate that we mentioned in the previous paragraph. Such an aggregate is composed of three key aspects. The first is the choice of categorical feature constraints. Domain experts identify these attributes through data analysis as relevant to fraud prediction. They include the country of the terminal, the terminal ID, or the local currency of the transaction. Another aspect is the length of the aggregation period. Only those transactions that fulfill the constraints contribute to the computed aggregate. The next step is the aggregation of a particular continuous feature of the selected transactions. This introduces the aspect of the choice of aggregation function itself. Therefore, the selection of the aggregate features depends on the combination criteria for the aforementioned aspects. The choice of combination of constraints, time-periods and aggregate functions gives us different perspectives of customer spending behaviour. For the formal definition of feature aggregates, we would point the reader to the previously mentioned study by Bahnsen et al. [6].

3.2 Neural Aggregate Generator

Recent works propose different automatic approaches to represent the user context of a particular transaction. Jurgovsky et al. [14] use an LSTM to model the user behaviour across a sequence of transactions. In the study, the LSTM is compared to transaction-based models which use feature aggregates and shows comparable performance on a Worldline e-commerce dataset. Forough et al. [10] also adopt a sequential approach to classification. However, in their method, they incorporate the outputs for each sample in the sequence to predict the current transaction. Lucas et al. [17] use hidden markov models (HMMs) to model different types of sequences (sequences of transactions from genuine/fraudulent terminals and card-holders). They use the likelihood of a card-holder's sequence of transactions to be generated by a given HMM as an additional feature. Cheng et al. [5] model both the account activity and spatial context using a spatio-temporal attention-based neural network. Unlike the aforementioned approaches, the architecture of the NAG is based on the structure of manual feature aggregates. The description of the NAG closely follows our original proposal in [12]. The context-representation module of the NAG has three key components:

Computation of Transaction Similarities: Similar to manual aggregates, the NAG performs a selection of previous transactions based on some categorical constraint features. Unlike manual feature aggregates, where a hard-matching takes place, this is realized through a soft similarity between the feature value of the current transaction and each of the previous transactions in the sequence. In order to do this, the categorical features must be projected into a continuous,

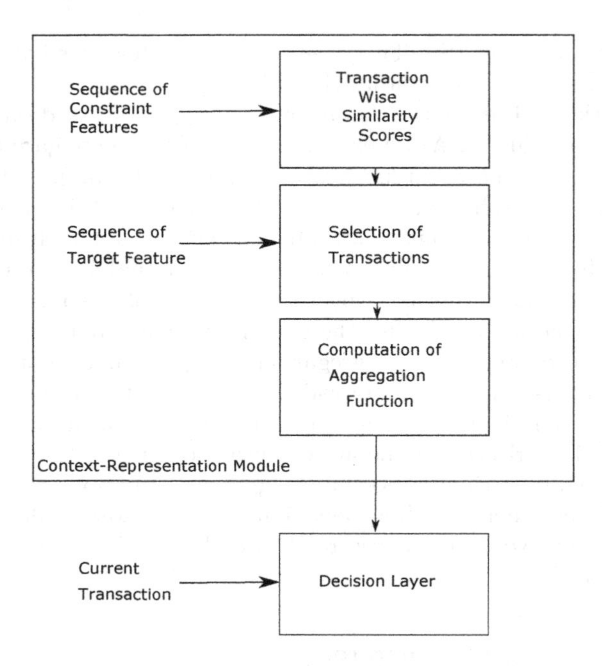

Fig. 1. The architecture of the Neural Aggregate Generator. The Context-Representation module consists of 3 parts: The computation of transaction similarities, the selection of transactions, and the computation of the aggregation function. Finally, the computed aggregates and the current transactions are fed as inputs to the decision layer.

real-valued space (one option here is to use pre-trained feature embeddings as shown by Russac et al. [22]). The inner product between the entity embeddings is used as a similarity measure. Given a sequence of transactions from a card-holder $(\mathbf{x}_i)_{1:n}$, an encoded sequence is obtained $(\mathbf{y}_i^f)_{1:n}, \forall \mathbf{y}_i^f \in \mathbb{R}^{d_f}$ for each categorical feature f with embedding dimension d_f. The index of the categorical feature is dropped for notational convenience. The similarity scores are therefore given as the inner product of the n-th feature vector and those of each of the previous transactions of the sequence.

$$\mathbf{s} = \mathbf{y}_n^\top \cdot [\mathbf{y}_1, \mathbf{y}_2, \ldots, \mathbf{y}_n] \tag{1}$$

In a similar manner, the similarity scores for each categorical feature are calculated. Therefore for each categorical feature $f \in \{1, \ldots, F\}$, as given by 1, we have a set of similarity score vectors $\mathbf{s}^f \in \mathbb{R}^n$. Let $S = [\mathbf{s}^1, \ldots, \mathbf{s}^F] \in \mathbb{R}^{n \times F}$ denote a similarity matrix, where each row i contains the similarity scores of transaction i. In the NAG, the feature similarity scores are combined into a single scalar similarity score for the transaction. This is realized through a weighted linear combination, where the weights are parameters that are learned on the fraud task. The weights modulate the contribution of different features to the overall transaction similarity score:

$$\mathbf{m} = \sigma(S \cdot \mathbf{w} + b_w) \tag{2}$$

Here, $\mathbf{w} \in \mathbb{R}^F$ and b_w are model parameters. w, is shared across all transactions of the sequence. $\mathbf{m}$ is a vector of transaction similarity scores, corresponding to a soft selection of previous transactions.

Selection of Transactions: To modulate the contribution of each of the previous transactions to the aggregate, the chosen continuous feature (target feature) of the transactions is scaled by the corresponding similarity score. Let $\mathbf{z} = (z_1, \ldots, z_n) \in \mathbb{R}^n$ be the values of the continuous feature. A soft selection of components of z is performed using the transaction similarity scores:

$$\tilde{\mathbf{z}} = (m_1 \cdot z_1, \ldots, m_n \cdot z_n) \tag{3}$$

Computation of the Aggregation Function: Finally, an aggregation function must be applied to the scaled targets. In the NAG, choice of aggregation function is left up to the optimization algorithm through a parameterization of the target feature vector from (3):

$$a_n = \sigma(\tilde{\mathbf{z}}^\top \cdot \mathbf{v} + b_v) \tag{4}$$

Here, $\mathbf{v}$ and b_v are model parameters. a_n is the computed aggregate. The index n indicates that the computed aggregate is associated to the n-th or last transaction of the sequence. Multiples of parameters $\mathbf{w}$ and $\mathbf{v}$ can be used to create several different aggregates.

4 Extensions

While the NAG offers a good level of automation by representing the context in such a way that leads to even better performance than feature engineering, the NAG still has certain limitations: the process with which it generates the aggregates has several differences from the one used by experts, which makes it less interpretable for them. Moreover, it is unable to model time dependencies and certain aggregation functions. In this section, we outline more precisely those limitations and propose extensions to address them.

4.1 Shared Functions Across Constraints

In the NAG, for each constraint (parameters $\mathbf{w}$ in Eq. (2)), a corresponding aggregation function(parameters $\mathbf{v}$ in (4)) is computed. However, in practice, manual feature aggregates are constructed in a manner such that aggregation functions are shared across different constraints. For example, in the set of expert aggregates that we use for our experiments, the mean of amounts is used for multiple aggregates comprising of different constraint features. Therefore, we encode this property into the architecture of the NAG. We formalize this extension as follows:

In Eq. (2) we have a parameter vector $\mathbf{w}$ which modulates the contributions of each constraint feature to the overall feature similarity score. We also have a corresponding vector of scaled amounts $\mathbf{v}$ in Eq. (4). Consider we want to compute aggregates with c constraints. Instead of having c couples $(\mathbf{w}_i, \mathbf{v}_i)$ of (constraint, function) like in the original NAG, in our extension we propose to have a set of c constraints $\mathbf{w}_1, \mathbf{w}_2, ..., \mathbf{w}_c$ and a set of g aggregate functions $\mathbf{v}_1, \mathbf{v}_2, ..., \mathbf{v}_g$.

After the operations given by Eqs. (2) and (3), we obtain a matrix $\tilde{Z} = [\tilde{\mathbf{z}}_1, \tilde{\mathbf{z}}_2, ..., \tilde{\mathbf{z}}_c] \in \mathbb{R}^{n \times c}$ of scaled amounts. For each constraint, we use each $\mathbf{v}$ parameter, $\mathbf{v}_i$, shared across the different constraints, akin to a 1-d convolution with multiple filters.

$$A = \sigma(\tilde{\mathbf{Z}}^\top \cdot V + b_V) \tag{5}$$

Therefore we obtain a matrix A of aggregates, where $A \in \mathbb{R}^{c \times g}$ with a number of aggregates equal to the product between the number of constraints c and the number of aggregation functions g. This process of factorizing the functions across the aggregates allows for more aggregates with less parameters.

4.2 Selection of Constraints

From an expert perspective, the selection of constraint features serves as the first step of the aggregation creation process. In the NAG, this selection step is not explicitly defined. Instead, the contribution of each feature is modulated during the computation of the overall transaction similarity. As an alternative approach, we propose an extension to the NAG with a constraint selection gate. For each of the constraint features, we have a set s of current-to-past feature similarities as given in Eq. (1). For example for the country feature we have a vector of similarities $\mathbf{s}^{Country} = [s_1, s_2, ..., s_{10}]$.

Instead of combining all the features similarities into a global transaction similarity before computing a global selection gate as in Eq. (2), we propose to compute a feature selection gate for each constraint feature:

$$\mathbf{m}^f = \sigma(\mathbf{s}^f \cdot u + b_u) \tag{6}$$

Here $\mathbf{m}^f \in \mathbb{R}^F$ is the vector of activations from our feature selection gate. u and b_u are model parameters. σ is the sigmoid activation.

We then sum up the individual feature selection scores into an overall transaction selection score $\mathbf{m}$. The following steps (computation of scaled amounts and aggregation) are identical to the original description of the NAG. The interest of such an extension is that it improves interpretability by bringing the model closer to the natural process of designing aggregates: constraints are not conditions on combinations of features but rather combinations of conditions on single features.

Additionally, we also provide scope for interpretability through an inspection of the parameters (u,b_u) of the feature selection gate. We consider the feature selection gate $\mathbf{m}_f$ as a type of soft selection condition. In particular $\mathbf{m}_f$ will tend to 1 (selection condition is true) if $s_i^f \cdot u >> -b_u$ and 0 if $s_i^f \cdot u << -b_u$. Here, s_i^f is the similarity of the feature f between transaction i and the current transaction.

4.3 Extending Modelling Capabilities of the NAG

The final three extensions that we propose aim at improving the modeling power of the NAG by increasing the expressiveness of the aggregation part of the model and by integrating a time encoding. More precisely:

1. The original NAG computes a weighted sum of the scaled amounts (see Eq. (4)). Therefore we postulate that the modelling capabilities of the NAG remain limited to simple sums of the amounts of the selected transactions. To allow it modelling higher-order moments on the amounts of selected transactions, we increase the depth of the network at the aggregation computation part. The hypothesis here is that the depth of the network will enable the approximation of more complex functions on the target feature.
2. Experts sometimes design aggregates which compute the number of selected transactions based on a set of given constraints. An example would be "The count of transactions that took place in the same country". To extend the NAG with this ability, we add the possibility to drop the use of the target feature (i.e. the amount) and compute the aggregation function directly over the transaction similarity scores (i.e. replacing $\tilde{\mathbf{z}}^\top$ with $\mathbf{m}$ in (4)). This enables the NAG to model the context of the transaction in terms of how anomalous it is in relation to the usual behaviour of the card-holder. Therefore the aggregation computation step for this extension can be defined as follows:
$$a_n = \sigma(\mathbf{m}^\top \cdot \mathbf{v} + b_v) \tag{7}$$
Here the $\mathbf{m}$ is the vector of transaction similarities computed in (2). The rest of the notations are the same as in (4).
3. In the original proposal of the NAG, there was no explicit description regarding the manner in which the time information of the samples is exploited. Based on expert input, we know that most fraudulent transactions occur in quick succession. This is due to the fact that fraudsters want to maximize their gain from a compromised card. Therefore it becomes important to model these bursts of transaction in the NAG. In previous works in the domain, such as the one by Jurgovsky et al. [14] this was done through a simple feature engineering step where the time between two transactions in the sequence was added as an additional feature:
$$tdelta_i = x_i^{(time)} - x_{i-1}^{(time)}. \tag{8}$$

However, since the NAG computes the context through the perspective of the current-to past similarities in the sequence, we instead use the time interval between the current transaction and each of the previous transactions. Therefore, for a sequence of length n we build a set T, where $T = \{x_n^{(time)} - x_k^{(time)}\}_{k<n}$, and feed it as an additional input to the decision layer.

5 Experiments

In this section, we describe the experimental protocol for our study. We closely follow the outline given by Ghosh Dastidar et al. [12] in order to preserve comparability.

5.1 Dataset

In our experiments, we use a real-world dataset from Worldline. It consists of over 60 million e-commerce credit card transactions, primarily from European countries. Each transaction is composed of transaction, card, and merchant attributes. A detailed description of the raw attributes can be found in Table 1 of [11]. Apart from this, we create some additional aggregate attributes. These are based on the structure described in Sect. 3.1. The choice of constraints and functions are defined based on the input of domain experts at *anonymized*. The constraints include the merchant code, the terminal and the country while the aggregation functions we use include the sum, mean, count, etc. Each transaction additionally has an associated ground-truth label of genuine or fraudulent. We group the transactions into card-holder accounts. Each account is associated to a unique card-holder ID and contains one or more transactions.

5.2 Pre-processing

As the NAG and its extensions compute soft similarities between the values of the categorical feature constraints, we must encode these features in a continuous space. In order to do this, we use pre-trained embeddings of the categorical feature values using Word2Vec [22]. For the continuous features such as the amount or the feature aggregates we standardize the values to a mean of 0 and unit variance.

5.3 Dataset Splits and Alignment

We train on 2 months of data, followed by a 7-day validation period, and then test on 1 month (April, May, June and July 2017) of data. This 7-day gap between train and test data is usually needed by the human verifiers to label the transactions. Another aspect we must consider is the alignment of transactions that are predicted on across both context-based and transaction-based approaches. This is due to the limitation that context-based approaches can only predict on transactions that have $n - 1$ preceding transactions, where n is the chosen sequence length. Therefore, in order to ensure that the experiments are comparable across approaches, we also restrict the transaction-based approaches to only predict on the transactions with at least $n - 1$ preceding transactions.

5.4 Model Evaluation

We evaluate our approaches based on the three aspects. The first is in terms of the expressiveness of the model. We define a multi-output regression experiment where the target features are a set of diverse manual feature aggregates. We evaluate a selected set of our extensions based on their ability to model these aggregates. Here, we report the result as the Mean Absolute Error between the true and predicted values. The second evaluation is based on the predictive accuracy of the models. The third evaluation is for certain extensions, where we evaluate the performance at very small parameter budgets. For the second and third evaluation, we use the Area under the Precision-Recall curve (AUCPR) as it is suitable for tasks with imbalanced data [19] and has been used in several other works on fraud detection [12,14,17].

6 Results and Discussion

In this section, we present the results of the series of experiments that we perform to evaluate the various extensions of the NAG. For the sake of readability, we denote the NAG extensions as follows: We denote the extension of shared functions across constraints as $\text{NAG}_{\text{shared}}$, the extension with selection of constraints as $\text{NAG}_{\text{select}}$, the extension with a deeper decision layer as NAG_{deep}, and the extension which computes aggregations directly over transaction similarities as $\text{NAG}_{\text{count}}$. Additionally, whenever we use a model with the current-to-past time-deltas as part of the feature set we denote it with an additional t in the subscript (For e.g., $\text{NAG}_{\text{deep,t}}$).

6.1 Approximation of Manual Aggregates

The first step in our evaluation of the different extensions is to determine if the extensions can model the manual feature aggregates. In this experiment, we only use the context-representation module of the NAG (see Fig. 1) and its extensions for this task. We use the mean absolute error between the predicted value and the true value of the aggregate(s) as our cost function. As one of the limitations of the NAG is that it requires sequences of a fixed length (as opposed to aggregates, where one can define different time-windows as a constraint), we fix the number of previous transactions that we derive the manual aggregates from. Similarly, we fix the sequence length for the input to the NAG and its extensions to the same number. We use 8 different expert aggregates (normalized to a mean of zero and variance of one) as our target features. Our aggregates consist of 2 constraint features (country and merchant code) and 4 aggregate functions (sum, count, mean and variance). Each result is reported as the average Mean Absolute Error across 5 runs. From Table 1, we see that the NAG_{deep} comes closest to approximating the different manual aggregates. Here we postulate that the added complexity of the model allows it to be more expressive. We see that the performance of the $\text{NAG}_{\text{select}}$ extension is similar to the NAG. This is encouraging as it shows that even with a more intuitive architecture from an expert perspective, there is no trade-off in expressiveness. The $\text{NAG}_{\text{shared}}$ shows significantly worse performance than each of the other approaches.

Table 1. Results of the regression task. We use 8 different expert aggregates (normalized to a mean of zero and variance of one) as our target features. Each result is reported as the average Mean Absolute Error across 5 runs.

	Mean Absolute Error
NAG	0.490 ± 0.002
NAG_{deep}	0.174 ± 0.042
$\text{NAG}_{\text{shared}}$	0.675 ± 0.012
$\text{NAG}_{\text{select}}$	0.488 ± 0.002

6.2 Fraud Classification Task

The second step is to evaluate the different models directly on the fraud classification task. We use two feature sets. The base feature set and the feature set using aggregates (denoted by 'Aggs') For this task, we compare the extensions of the NAG to both the NAG and several other context-based (LSTM) and transaction-based classifiers (random forest, feed-forward neural network). As our experimental protocol (including the train and test splits) is identical to that of Ghosh et al. [11], we also compare our results to those presented in that study. We present the results in Table 2. As we see in the results from [11], both forms of context representation (i.e. aggregates or automatic approaches) improve the fraud detection performance. As the AUCPR is sensitive to the fraud ratio of the test data, the results across different months are starkly different (each of the months have different densities of fraud). We see that, like the NAG, each of the extensions significantly improves the fraud detection performance over the transaction-based approaches, even when aggregate features are used. Among the extensions, the $\text{NAG}_{\text{count}}$ has the lowest performance across all months. This would indicate that the amounts spent on past transactions are indeed an important attribute needed to model the user's activity in the fraud context. The NAG_{deep} is of particular interest, as in our regression experiment we showed that it is better at modeling a wide variety of aggregates than the original NAG. This improvement in expressiveness however does not translate to an improvement in the fraud detection accuracy over the NAG. This confirms the view of domain experts on the choice of aggregation functions. In the

Table 2. Results on the Fraud Classification Task: We compare the extensions of the NAG to both transaction-based classifiers and other automatic context representation approaches (including the NAG). All reported results are the mean AUCPR across 5 runs. The score of the model with the best performance for a particular month is marked in bold. The results of models marked with an * are taken from [11].

		April	May	June	July
Transaction-Based	FFN*	0.230 ± 0.001	0.168 ± 0.003	0.126 ± 0.001	0.126 ± 0.003
	FFN (Aggs)*	0.300 ± 0.030	0.205 ± 0.024	0.157 ± 0.006	0.179 ± 0.009
	RF*	0.251 ± 0.001	0.157 ± 0.002	0.121 ± 0.001	0.128 ± 0.005
	RF (Aggs)*	0.308 ± 0.002	0.221 ± 0.004	0.138 ± 0.001	0.178 ± 0.005
Automatic	LSTM*	0.326 ± 0.007	0.225 ± 0.003	0.154 ± 0.003	0.182 ± 0.020
	NAG*	**0.338 ± 0.008**	0.246 ± 0.007	0.201 ± 0.006	0.246 ± 0.020
Extensions	NAG_{shared}	0.333 ± 0.009	0.239 ± 0.005	0.198 ± 0.009	0.227 ± 0.020
	NAG_{deep}	0.333 + 0.010	0.248 ± 0.007	0.201 ± 0.012	0.229 ± 0.010
	NAG_{select}	0.311 ± 0.007	0.247 ± 0.007	0.201 ± 0.015	0.223 ± 0.022
	NAG_{count}	0.315 ± 0.001	0.231 ± 0.007	0.183 ± 0.005	0.205 ± 0.008
	NAG_{t}	0.337 ± 0.005	**0.262 ± 0.006**	0.211 ± 0.006	0.244 ± 0.008
	$NAG_{deep,t}$	0.326 ± 0.003	0.258 ± 0.009	0.212 ± 0.005	0.246 ± 0.004
	$NAG_{select,t}$	0.331 ± 0.008	0.255 ± 0.006	**0.217 ± 0.003**	**0.250 ± 0.010**

set of expert aggregates used by our industrial partner, there are no aggregates where the aggregation function is a higher-order moment. Two of our extensions, NAG_{shared} and NAG_{select} have similar or slightly worse performance than the NAG. This can be explained by the fact that while both these extensions are closer to the expert intuition of feature aggregates, they were not specifically designed to improve predictive performance. Finally, we discuss the impact of adding the time-delta between the current and each of the past transactions. We see that in almost every case, the inclusion of the time-delta feature leads to an increase in predictive performance. This aligns with the knowledge that many frauds occur in bursts. As proposals of a time-delta feature in previous works were computed between consecutive transactions, these results confirm the importance of our alternative for any model that exploits current-to-past relationships. Therefore, across our experiments, we see no clear front-runner in terms of performance among the extensions. With each extension, we see a trade-off between either expressiveness and intuitiveness. Therefore, instead of replacing the NAG with one of the proposed extensions, we would propose use-case based solutions. For example, for the NAG_{shared}, there is an advantage in terms of model size (see Sect. 6.3). Additionally, we believe that as NAG_{shared} and NAG_{select} are further aligned with the expert intuition than the NAG, it opens up the possibility for future works on interpretability.

6.3 Parameter Budget Study

As mentioned in Sect. 4.1, for the NAG_{shared}, factorizing the functions across the aggregates allows for more aggregates with less parameters. To evaluate this, we designed a parameter budget study to compare the NAG_{shared} to the NAG_{select} and the NAG at very small model sizes. Since the architectures of the models differ, we can only compare them at approximately the same number of parameters (budgets). Here we count the parameters for the context representation module only. As we see in Table 3 the NAG_{shared} outperforms the NAG and NAG_{select} at very small parameter budgets. In spite of the fact that its performance worsens relative to the other two approaches at much larger model sizes, these results indicate the suitability of the NAG_{shared} in settings where resources are constrained.

Table 3. Parameter Budget Study: A comparison of the NAG, NAG_{shared} and NAG_{select} at very small parameter budgets. The month of July 2017 has been used for testing. The results are given in AUCPR.

Parameter budget	NAG	NAG_{shared}	NAG_{select}
150	0.194	0.225	0.206
200	0.202	0.217	0.198
250	0.212	0.223	0.206
300	0.213	0.225	0.210

7 Conclusion

In this work, we extend the state-of-the-art in automatic context-representation for credit-card fraud detection, the Neural Aggregate Generator (NAG), in several ways. These extensions are based on our analysis of the limitations of the NAG in terms of expressiveness, and the alignment of the architecture with expert intuition. We evaluate our extensions on a real-world credit-card dataset with over 60 million transactions. In this evaluation, we show that one of the extensions (NAG_{deep}) improves the original NAG in approximating a variety of manual feature aggregates. We also show that the extensions NAG_{select} and NAG_{share} show similar or only slightly worse performance compare to the NAG while having additional benefits (less parameters and scope for interpretability). We also show that adding a current-to-past time interval to our feature set improves the performance for the NAG and each of its extensions. We believe that this work lays the groundwork for defining a search space of possible architectures for a neural architecture search for credit-card fraud detection.

References

1. APATE: A novel approach for automated credit card transaction fraud detection using network-based extensions. Decis. Support Syst. **75**, 38–48 (2015). https://doi.org/10.1016/j.dss.2015.04.013. https://www.sciencedirect.com/science/article/pii/S0167923615000846
2. How artificial intelligence and machine learning research impacts payment card fraud detection: a survey and industry benchmark. Eng. Appl. Artif. Intell. **76**, 130–157 (2018). https://doi.org/10.1016/j.engappai.2018.07.008. https://www.sciencedirect.com/science/article/pii/S0952197618301520

3. Alazizi, A., Habrard, A., Jacquenet, F., He-Guelton, L., Oblé, F., Siblini, W.: Anomaly detection, consider your dataset first an illustration on fraud detection. In: 2019 IEEE 31st International Conference on Tools with Artificial Intelligence (ICTAI), pp. 1351–1355 (2019). https://doi.org/10.1109/ICTAI.2019.00188
4. Awoyemi, J.O., Adetunmbi, A.O., Oluwadare, S.A.: Credit card fraud detection using machine learning techniques: a comparative analysis. In: 2017 International Conference on Computing Networking and Informatics (ICCNI), pp. 1–9 (2017). https://doi.org/10.1109/ICCNI.2017.8123782
5. Cheng, D., Xiang, S., Shang, C., Zhang, Y., Yang, F., Zhang, L.: Spatio-temporal attention-based neural network for credit card fraud detection. In: Proceedings of the AAAI Conference on Artificial Intelligence, vol. 34, no. 01, pp. 362–369 (2020)
6. Correa Bahnsen, A., Aouada, D., Stojanovic, A., Ottersten, B.: Feature engineering strategies for credit card fraud detection. Expert Syst. Appl. **51**, 134–142 (2016)
7. Dal Pozzolo, A.: Adaptive machine learning for credit card fraud detection (2015)
8. Dal Pozzolo, A., Boracchi, G., Caelen, O., Alippi, C., Bontempi, G.: Credit card fraud detection and concept-drift adaptation with delayed supervised information. In: 2015 International Joint Conference on Neural Networks (IJCNN), pp. 1–8 (2015). https://doi.org/10.1109/IJCNN.2015.7280527
9. Fiore, U., De Santis, A., Perla, F., Zanetti, P., Palmieri, F.: Using generative adversarial networks for improving classification effectiveness in credit card fraud detection. Inf. Sci. **479**, 448–455 (2019)
10. Forough, J., Momtazi, S.: Sequential credit card fraud detection: a joint deep neural network and probabilistic graphical model approach. Expert Syst. e12795. https://doi.org/10.1111/exsy.12795. https://onlinelibrary.wiley.com/doi/abs/10.1111/exsy.12795
11. Ghosh Dastidar, K., Jurgovsky, J., Siblini, W., Granitzer, M.: NAG: neural feature aggregation framework for credit card fraud detection. Knowl. Inf. Syst. **64**(3), 831–858 (2022)
12. Ghosh Dastidar, K., Jurgovsky, J., Siblini, W., He-Guelton, L., Granitzer, M.: NAG: neural feature aggregation framework for credit card fraud detection. In: 2020 IEEE International Conference on Data Mining (ICDM), pp. 92–101 (2020). https://doi.org/10.1109/ICDM50108.2020.00018
13. Hordri, N.F., Yuhaniz, S.S., Azmi, N.F.M., Shamsuddin, S.M.: Handling class imbalance in credit card fraud using resampling methods. Int. J. Adv. Comput. Sci. Appl. **9**(11) (2018). https://doi.org/10.14569/IJACSA.2018.091155
14. Jurgovsky, J., et al.: Sequence classification for credit-card fraud detection. Expert Syst. Appl. **100**, 234–245 (2018)
15. Kim, E., et al.: Champion-challenger analysis for credit card fraud detection: hybrid ensemble and deep learning. Expert Syst. Appl. **128**, 214–224 (2019). https://doi.org/10.1016/j.eswa.2019.03.042. https://www.sciencedirect.com/science/article/pii/S0957417419302167
16. Lucas, Y., Jurgovsky, J.: Credit card fraud detection using machine learning: a survey (2020)
17. Lucas, Y., et al.: Multiple perspectives hmm-based feature engineering for credit card fraud detection. In: Proceedings of the 34th ACM/SIGAPP Symposium on Applied Computing, SAC 2019, pp. 1359–1361. Association for Computing Machinery, New York (2019). https://doi.org/10.1145/3297280.3297586
18. Lucas, Y., et al.: Dataset shift quantification for credit card fraud detection. In: 2019 IEEE Second International Conference on Artificial Intelligence and Knowledge Engineering (AIKE), pp. 97–100 (2019). https://doi.org/10.1109/AIKE.2019.00024

19. Ozenne, B., Subtil, F., Maucort-Boulch, D.: The precision-recall curve overcame the optimism of the receiver operating characteristic curve in rare diseases. J. Clin. Epidemiol. **68**(8), 855–859 (2015). https://pubmed.ncbi.nlm.nih.gov/25881487/
20. Pumsirirat, A., Yan, L.: Credit card fraud detection using deep learning based on auto-encoder and restricted Boltzmann machine. Int. J. Adv. Comput. Sci. Appl. **9**(1), 18–25 (2018)
21. Randhawa, K., Loo, C.K., Seera, M., Lim, C.P., Nandi, A.K.: Credit card fraud detection using adaboost and majority voting. IEEE Access **6**, 14277–14284 (2018). https://doi.org/10.1109/ACCESS.2018.2806420
22. Russac, Y., Caelen, O., He-Guelton, L.: Embeddings of categorical variables for sequential data in fraud context. In: Hassanien, A.E., Tolba, M.F., Elhoseny, M., Mostafa, M. (eds.) AMLTA 2018. AISC, vol. 723, pp. 542–552. Springer, Cham (2018). https://doi.org/10.1007/978-3-319-74690-6_53
23. Whitrow, C., Hand, D.J., Juszczak, P., Weston, D., Adams, N.M.: Transaction aggregation as a strategy for credit card fraud detection. Data Min. Knowl. Discov. **18**(1), 30–55 (2009)
24. Xuan, S., Liu, G., Li, Z., Zheng, L., Wang, S., Jiang, C.: Random forest for credit card fraud detection. In: 2018 IEEE 15th International Conference on Networking, Sensing and Control (ICNSC), pp. 1–6 (2018). https://doi.org/10.1109/ICNSC.2018.8361343

Digital Twins: Modelling Languages Comparison

Abdul Wahid([✉]), Jiafeng Zhu, Stefano Mauceri, Lei Li, and Minghua Liu

National University of Ireland, Galway, Ireland
abdul.wahid@insight-centre.org,
{jiafeng.zhu1,lilei029,harbor.liuminghua}@huawei.com,
stefano.mauceri1@huawei-partners.com

Abstract. A digital twin (DT) is a virtual representation of a physical entity that can be used to improve and automate decision-making. For instance, a DT can be used for real-time performance monitoring and optimisation. In this work, we investigate DTs from the point of view of the telecommunication industry (TI). First, we provide an introduction to the main concepts, tools, and applications related to DTs. Then, we clarify the benefits of DTs for the TI. Also, we identify the main challenges hindering widespread adoption of DTs within the TI. Finally, we argue that modelling languages are the key to overcome some of the main challenges. This provides a good starting point for discussion and further research.

Keywords: Digital twin · Cyber-physical systems · System modelling · Decision support systems · Smart maintenance

1 Introduction

DTs have applications in a variety of fields such as healthcare [10,46], real-time remote monitoring and control [1], transportation and energy [23,24], manufacturing [22], aerospace industries [9], and in general are relevant to any cyber-physical system [3]. A DT is the digital representation of a physical entity. Such representation is meant to capture the behaviour and the characteristics of the related physical entity [45]. The main purpose of DTs is to make physical objects and their associated data sources accessible to software and users on ad-hoc digital platforms [2].

DTs have been widely adopted in a variety of industries for modelling complex and dynamic systems [38]. The core advantage of a DT is that it allows optimisation and monitoring of a complex system without having to interact with it. In this way, any decision can be tested in the digital world before to be deployed in the physical world. Besides the advantages for decision making systems (e.g. operations and maintenance systems), DTs could boost innovation while keeping costs under control. In the TI, DTs used to build virtual representations of customers, products, services, and resources, are gaining attention as tools to improve decision making and accelerate a data-centric innovation [4].

G. Nicosia et al. (Eds.): LOD 2022, LNCS 13811, pp. 169–178, 2023.
https://doi.org/10.1007/978-3-031-25891-6_13

The present work focuses on the modelling languages that are used to implement DTs, as such a survey is currently missing in the literature [55]. We consider that it is important to investigate the modelling languages used to implement DTs in order to facilitate the integration and cooperation of DTs that have been independently developed with varying standards, and protocols. To this end, we provide an overview of main modelling languages.

The rest of the paper is organized as follows. Section 2 presents the DT concept. Section 3 discusses the main challenges to widespread DTs adoption. Section 4 compares different modelling languages. Finally, Sect. 5 presents the conclusions of this work.

2 Digital Twin Background

This section presents a brief overview of DT concepts and applications.

2.1 Digital Twin Concept

The concept of digital twin was first introduced by Grieves [15] in 2003. However, the roots of digital twin can be traced back to aviation and space industry [16].

In time, a number of publications have focused on the DT technology, and as a result a variety of DT definitions have been proposed. Table 1 summarises various DT definitions and their implementation domains. A DT is defined as a digital representation of a physical entity [17]. It receives input from its real-world counterpart through a series of sensors. The DT uses this data to simulate the physical entity and to provide insights about its performance, possible problems that may arise due to faulty or invalid behaviours, monitoring and analysis of different attributes, and prediction of future events.

Real-time monitoring, control, and data collection are transmitted between the physical and the digital systems through the DT, allowing for monitoring of all changes and acquisition of all relevant data for business decisions [48]. The goal of the DT is to use multiple technologies (e.g., modelling, simulation, blockchain, edge and cloud computing, artificial intelligence) to jointly optimise the cost and performance of the overall process [47].

2.2 Digital Twin in Industry

Virtual representation of machines or systems are revolutionizing several industries where there is widespread agreement on the need of digitalisation [38]. Many industries are now adopting DTs to detect problems and boost productivity [38]. A digital twin can be used in a variety of ways: below we give a brief overview of some of the industries that have adopted the DTs. Table 2 summarises the values and benefits enabled by DTs.

Healthcare: DTs can detect faults with equipment, and support diagnostic procedures. The large amount of data generated can be used to identify trends and anomalies, it can be combined with the patient's medical history resulting in a faster diagnosis [49].

Manufacturing: The planning of manufacturing processes, gathering appropriate process information, and subsequent development of said process consume the most time in the manufacturing industry [22]. Linking of production systems with their digital counterparts allows automated monitoring and optimisation with limited human intervention [50].

Smart Cities: DTs are used to create a virtual model of the city and to allow visualisation of all city resources as well as the interaction of people and vehicles. DTs can be used to monitor infrastructures, utilities, and support future development planning [51].

Next Generation Networks: Testing and optimisation of telecommunication networks before deployment is a difficult task [48]. DTs can support network deployment working as simulation tools [52].

Aviation: Airline fleets require a lot of maintenance work. These operations are costly and need meticulous planning to avoid significant downtime for planes that could otherwise be employed [9]. DTs can be used to improve planes performance and reduce costs [53].

3 DTs Adoption Challenges

Widespread adoption of DTs faces a number of challenges as summarised below.

- **Infrastructure**: The DT cannot be implemented, maintained, and operated without a reliable IT infrastructure [55].
- **Data quality**: The data consumed by the DT should be of high quality e.g. with minimum amount of noise and missing values. Also, the data should capture any relevant aspect of the related physical entity. If the data is inconsistent with the behaviour of the physical entity or with the required decision making task it is likely that the DT will under-perform [38].
- **Data connections**: DTs need real-time two-way connections to ingest data from the related physical objects and to transmit back the results of the data analysis. This can be challenging due to the volume of data, data analysis time, and limited computational and networking resources.
- **Modelling**: Two main challenges about DT modelling concern (a) standardisation, and (b) domain knowledge. (a) Standardisation is needed to enable cooperation of independently developed DTs part of the same complex system. (b) Domain knowledge is fundamental to support data-driven decision making in the context of a specific application.
- **Other tools**: Ad-hoc tools may be needed to allow visualisation of the DT functioning and to allow human-machine interaction.

Table 1. Different definition of digital twins.

Definition	References	Fields
DT is a novel way of managing IoT devices and IoT systems-to-systems across their entire life-cycle	[6]	Industrial IoT
DT is a technology that connects the worlds	[7]	Healthcare
DT is the digital representation of data and information that connects the virtual and real-world entities together	[15]	General
A complete physical and functional description of a component, system, or product	[8]	Simulation
DT is a complete re-engineering of structural life prediction and management of a physical system	[9]	Aerospace
A secure and safe environment in which you can evaluate the impact of proposed changes on the system performance	[10]	Healthcare
The DT links the physical system with its virtual equivalent. It helps to define and understand the system behavior and determines the test to evaluate the progress made throughout the product life cycle	[15]	General
DT's have the capacity to properly replicate events on different scales of space and time. They are based on expert knowledge and real-time data collected from all deployed systems of their type	[18]	Simulation
DT is the virtual replica of a physical asset that can be used for real-time prediction, optimisation, and improved decision-making using data and simulations	[19]	Prediction and Diagnosis
DT is a way to build an effective communication between physical and digital world using large amounts of data generated by the physical world	[20]	Smart Manufacturing
A DT is a self-contained cyber-physical system that attempts to impersonate a human workforce by means of dynamic adapted values of a database representing the sample features, preferences, skill set, and working schedules	[21]	Cyber-physical Systems
A DT is an ever-evolving digital profile of the historical and current behavior of a physical system responsible for optimizing the business	[11]	IoT
A DT is a probabilistic multi-physics, multi-scale simulation of a complex product that uses the best available physical models, and sensors	[17]	General

Table 2. Benefits of digital twin technology.

What digital twin achieves	Outcomes
Reduced Complexity	– Real-time monitoring and continuous data collection – Identification, diagnosis, and prediction of anomalies – Fault management and self-healing – Human-machine interaction
Minimized Operations	– Improving product design and engineering – Improving equipment performance – Reducing operations and process variability – Reducing operations and maintenance workforce
Forecasting and Optimisation	– Improving forecasting to eliminate over/under generation – Increasing production and reducing defects – What-if risk analysis and cost analysis – Risk mitigation
Innovation	– Knowledge accumulation – Simulation – Cost-effective testing and research
Visualisation and Analysis	– Visualisation and analysis of the health of remote assets (spacecrafts, offshore wind turbines, power stations, and manufacturer-owned machines operating in customer plants)

4 Digital Twin Modelling Languages

Modelling is fundamental to allow the DT to behave consistently with its related physical entity. We summarise the main characteristics of the most common modelling languages in Table 3.

The DT model should be created in a way that makes it simple for it to adapt to various conditions. Large-scale sensor networks and real-time data are used with a well-designed DT model to enhance the consistency of the DT with its related physical entity.

A physics-based modelling approach has largely guided the engineering community so far. This method entails observing a physical occurrence of interest, gaining a partial knowledge of it, converting that understanding into mathematical equations, and then solving them. However, a number of new modelling methods have been proposed in the literature. Tao et al. [38] introduced a 5-D modelling approach describing the physical, virtual, data, connection, and service modelling. A four layer model, proposed by Datong et al. [56], included a modelling calculation and data assurance layer, a DT functioning layer, and an immersive experience layer. A DT modelling architecture with the following layers: device, user interface, web service, query, and data repository was proposed by Schroeder et al. [57]. The modelling approach introduced by Fei et al. [38] focuses on data considering physical, virtual, service, and historical data and it is used to drive physical objects, and services.

The Systems Modelling Language (SysML) [27] is a graphical modelling language for complex systems analysis, specification, design, verification, and validation. It is used to model the interrelations, structures, and behavior of a system based on messages, states, and functions.

The Computational Fluid Dynamic (CFD) model and the Finite Element Model (FEM) [9] are two physics-based theories/models that have been developed to disclose the mapping relationship between the input and the behavior. Tao et al. [38] suggested that behavioral models describe the mechanical correspondence of production equipment under the circumstances of a given numerical control program and disturbance, such as human interference.

Modelica is an object-oriented programming language [58]. Modelica is primarily concerned with system modelling. A given model can be automatically converted into a mathematical model and then into an executable program [29].

Table 3. Summary of different modelling languages and their main characteristics.

Modelling languages	Features
Computer Aided Designs (CAD) [30]	– Define the geometry and design of a product – Provides digital hardware description simulating the system behavior – Provides modelling tools such as PTC Creco, CATIA, or Autodesk
Yang Modelling Language [31,32]	– Allows users to write the specification for what the interface between a client and server should be on a particular topic – Focuses on the data that a client manipulates and observes using standardized operations – Incorporates a level of extensibility and flexibility not present in other modelling languages – Human readable – Data modularity through modules and sub modules – A data modelling language used to model configuration, state data, and administrative actions manipulated by the NETCONF protocol
SysML [30]	– A graphical modelling language that enables engineers to model the system architecture and facilitates the application of MBSE – Supports analysis, design, verification, and validation of a broad range of systems and systems-of-systems. These systems may include hardware, software, information, processes, personnel, and facilities – Modelling at multiple levels of the system – It is a subset of the Unified modelling Language (UML) extended with functionality for systems engineering that is realized as a graphical modelling language – Enables engineers to model the system architecture and behavior facilitating the application of MBSE – Features four diagram types: (1) structure: Block Definition Diagrams (BDDs), Internal Block Diagrams (IBDs), and Package Diagrams; (2) behavior: Sequence Diagrams, State Machines, and Activity Diagrams; (3) parametric (constraints): parametric diagrams to specify physical properties and dependencies between parameters; and (4) requirements: diagrams that provide a graphical notation to capture requirements of a system

(*continued*)

Table 1. (*continued*)

Modelling languages	Features
Unified Modelling Language (UML) [33]	– It is a general purpose modelling language – Allows object-oriented analysis and design – Used to visualise the workflow of the system
Digital Twin Definition-Language (DTDL) [34]	– It creates a "neural network" data structure – It enables DTs to leverage higher-order analytics, machine learning, and complex computational tasks – It is a platform language for B2B integration – It is made up of a set of meta-model classes which are used to define the behavior of all DTs
Extensible Markup Language (XML) [35]	– It is a markup language that defines a set of rules for encoding documents in a format that is both human-readable and machine-readable – XML is able to work with different data types
Topology and Orchestration Specification for Cloud Applications (TOSCA) [36]	– Defines portable deployment and automated management of services on variety of platforms – Uses model-based descriptions of services, platforms, and data components along with their configuration, requirements, capabilities, relationships, and operational policies – It is a declarative language

5 Conclusions

This work provides an overview of recent research and technological advances in the field of DTs. We summarise the most widely accepted definitions of DT as well as some of the main applications. We argue that modelling languages are the most important aspect of DTs development. This is because modelling languages are crucial to allow a DT to evolve in line with its related physical entity and to cooperate with other DTs in the context of a complex system. We provide an overview of the most widely adopted modelling languages.

DTs show the potential to deeply improve decision making while mitigating some of the challenges related to automation. Future research is needed to prototype increasingly complex use cases that will involve DTs and artificial intelligence to practically demonstrate theoretical advantages.

Acknowledgement. This research is funded by Huawei Technologies Ireland.

References

1. Laaki, H., Miche, Y., Tammi, K.: Prototyping a digital twin for real time remote control over mobile networks: application of remote surgery. IEEE Access **7**, 20325–20336 (2019)
2. Datta, S.P.A.: Emergence of digital twins. arXiv preprint arXiv:1610.06467 (2016)
3. Alam, K.M., El Saddik, A.: C2PS: a digital twin architecture reference model for the cloud-based cyber-physical systems. IEEE Access **5**, 2050–2062 (2017)
4. Enhancing Innovation in Telecom with Digital Twins (2022). https://www.tcs.com/content/dam/tcs/pdf/Industries/communication-media-and-technology/Abstract/digital-twins-drive-innovation-telecom.pdf

5. Top, G.: Strategic technology trends for 2019. David Cearley, Brian Burke (10)
6. Canedo, A.: Industrial IoT lifecycle via digital twins. In: Proceedings of the Eleventh IEEE/ACM/IFIP International Conference on Hardware/Software Codesign and System Synthesis, p. 1 (2016)
7. Michelfeit, F.: Exploring the possibilities offered by digital twins in medical technology. Siemens Healthcare GmbH-Communc. RSNA (2018)
8. Boschert, S., Rosen, R.: Digital twin—the simulation aspect. In: Hehenberger, P., Bradley, D. (eds.) Mechatronic Futures, pp. 59–74. Springer, Cham (2016). https://doi.org/10.1007/978-3-319-32156-1_5
9. Tuegel, E.J., Ingraffea, A.R., Eason, T.G., Michael Spottswood, S.: Reengineering aircraft structural life prediction using a digital twin. Int. J. Aerosp. Eng. **2011** (2011)
10. Science Service - Dr. Hempel GmbH. Healthcare solution testing for future—Digital Twins in healthcare. Dr. Hempel Digital Health Network (2017). https://www.dr-hempel-network.com/digital-health-technolgy/digital-twins-in-healthcare/
11. Parrott, A., Warshaw, L.: Industry 4.0 and the digital twin. In: Deloitte University Press, pp. 1–17 (2017)
12. GE Launches the Next Evolution of Wind Energy Making Renewables More Efficient, Economic: the Digital Wind Farm—GE News (2015). https://www.Ge.Com/News/Press-Releases/Ge-Launches-next-Evolution-Wind-Energy-Making-Renewables-More-Efficient-Economic
13. GE Digital: Digital Twin for the Digital Power Plant - ge.com (n.d.). https://www.ge.com/digital/sites/default/files/download_assets/Digital-Twin-for-the-digital-power-plant-.pdf. Accessed 18 May 2022
14. Qi, Q., et al.: Enabling technologies and tools for digital twin. J. Manuf. Syst. **58**, 3–21 (2021)
15. Grieves, M.: Digital twin: manufacturing excellence through virtual factory replication. White Pap. **1**, 1–7 (2014)
16. Singh, M., et al.: Digital twin: origin to future. Appl. Syst. Innov. **4**(2), 36 (2021)
17. Ríos, J., et al.: Product avatar as digital counterpart of a physical individual product: literature review and implications in an aircraft. Transdisciplinary Lifecycle Anal. Syst. 657–666 (2015)
18. Gabor, T., Belzner, L., Kiermeier, M., Beck, M.T., Neitz, A.: A simulation-based architecture for smart cyber-physical systems. In: 2016 IEEE International Conference on Autonomic Computing (ICAC), pp. 374–379. IEEE (2016)
19. Liu, Z., Meyendorf, N., Mrad, N.: The role of data fusion in predictive maintenance using digital twin. In: AIP Conference Proceedings, vol. 1949, no. 1, p. 020023. AIP Publishing LLC (2018)
20. Guo, J., Zhao, N., Sun, L., Zhang, S.: Modular based flexible digital twin for factory design. J. Ambient. Intell. Humaniz. Comput. **10**(3), 1189–1200 (2019)
21. Graessler, I., Pöhler, A.: Integration of a digital twin as human representation in a scheduling procedure of a cyber-physical production system. In: 2017 IEEE International Conference on Industrial Engineering and Engineering Management (IEEM), pp. 289–293. IEEE (2017)
22. Rüßmann, M., et al.: Industry 4.0: the future of productivity and growth in manufacturing industries. Boston Consult. Group **9**(1), 54–89 (2015)
23. Besselink, B., et al.: Cyber-physical control of road freight transport. Proc. IEEE **104**(5), 1128–1141 (2016)
24. Lu, H., Guo, L., Azimi, M., Huang, K.: Oil and Gas 4.0 era: a systematic review and outlook. Comput. Ind. **111**, 68–90 (2019)

25. Moshood, T.D., Nawanir, G., Sorooshian, S., Okfalisa, O.: Digital twins driven supply chain visibility within logistics: a new paradigm for future logistics. Appl. Syst. Innov. **4**(2), 29 (2021)
26. Rasheed, A., San, O., Kvamsdal, T.: Digital twin: values, challenges and enablers from a modelling perspective. IEEE Access **8**, 21980–22012 (2020)
27. Makarov, V.V., Frolov, Y.B., Parshina, I.S., Ushakova, M.V.: The design concept of digital twin. In: 2019 Twelfth International Conference Management of Large-Scale System Development (MLSD), pp. 1–4. IEEE (2019)
28. Hughes, T.J.R., Cottrell, J.A., Bazilevs, Y.: Isogeometric analysis: CAD, finite elements, NURBS, exact geometry and mesh refinement. Comput. Methods Appl. Mech. Eng. **194**(39–41), 4135–4195 (2005)
29. Vöth, S., Vasilyeva, M.: Potential of Modelica for the creation of digital twins. In: Advances in Raw Material Industries for Sustainable Development Goals, pp. 386–389. CRC Press (2020)
30. Dalibor, M., Jansen, N., Rumpe, B., Wachtmeister, L., Wortmann, A.: Model-driven systems engineering for virtual product design. In: 2019 ACM/IEEE 22nd International Conference on Model Driven Engineering Languages and Systems Companion (MODELS-C), pp. 431–436. IEEE (2019)
31. Bjorklund, M.: YANG-a data modelling language for the network configuration protocol (NETCONF) (2010)
32. Xu, H., Xiao, D.: Data modelling for NETCONF-based network management: XML schema or YANG. In: 2008 11th IEEE International Conference on Communication Technology, pp. 561–564. IEEE (2008)
33. Azangoo, M., Taherkordi, A., Blech, J.O.: Digital twins for manufacturing using UML and behavioral specifications. In: 2020 25th IEEE International Conference on Emerging Technologies and Factory Automation (ETFA), vol. 1, pp. 1035–1038. IEEE (2020)
34. B. DTDL models - Azure Digital Twins. Microsoft Docs (2022). https://docs.microsoft.com/en-us/azure/digital-twins/concepts-models
35. Schroeder, G.N., Steinmetz, C., Pereira, C.E., Espindola, D.B.: Digital twin data modelling with automationml and a communication methodology for data exchange. IFAC-PapersOnLine **49**(30), 12–17 (2016)
36. Lipton, P., Palma, D., Rutkowski, M., Tamburri, D.A.: Tosca solves big problems in the cloud and beyond!. IEEE Cloud Comput. (2018)
37. Schneider, G.F., Wicaksono, H., Ovtcharova, J.: Virtual engineering of cyber-physical automation systems: the case of control logic. Adv. Eng. Inform. **39**, 127–143 (2019)
38. Tao, F., Zhang, H., Liu, A., Nee, A.Y.C.: Digital twin in industry: state-of-the-art. IEEE Trans. Ind. Inform. **15**(4), 2405–2415 (2018)
39. Wu, Y., Zhang, K., Zhang, Y.: Digital twin networks: a survey. IEEE Internet Things J. **8**(18), 13789–13804 (2021)
40. Barricelli, B.R., Casiraghi, E., Fogli, D.: A survey on digital twin: definitions, characteristics, applications, and design implications. IEEE Access **7**, 167653–167671 (2019)
41. The value of Digital Twin Technology. The value of digital twin technology (n.d.). https://www.siemens-healthineers.com/en-us/services/value-partnerships/asset-center/white-papers-articles/value-of-digital-twin-technology. Accessed 16 June 2022
42. Glatt, M., Sinnwell, C., Yi, L., Donohoe, S., Ravani, B., Aurich, J.C.: modelling and implementation of a digital twin of material flows based on physics simulation. J. Manuf. Syst. **58**, 231–245 (2021)

43. Costa, L.D.F., et al.: Analyzing and modelling real-world phenomena with complex networks: a survey of applications. Adv. Phys. **60**(3), 329–412 (2011)
44. Wen, J., Gabrys, B., Musial, K.: Towards digital twin oriented modelling of complex networked systems and their dynamics: a comprehensive survey. arXiv preprint arXiv:2202.09363 (2022)
45. Malakuti, S., et al.: Digital twins for industrial applications. Definition, Business Values, Design Aspects, Standards and Use Cases. Version 1, pp. 1–19 (2020)
46. Liu, Y., et al.: A novel cloud-based framework for the elderly healthcare services using digital twin. IEEE Access **7**, 49088–49101 (2019)
47. Khan, L.U., Han, Z., Saad, W., Hossain, E., Guizani, M., Hong, C.S.: Digital twin of wireless systems: overview, taxonomy, challenges, and opportunities. arXiv preprint arXiv:2202.02559 (2022)
48. Mashaly, M.: Connecting the twins: a review on digital twin technology & its networking requirements. Procedia Comput. Sci. **184**, 299–305 (2021)
49. Dabrowski, K.: What is a Digital Twin? Benefits & Examples—PGS Software. PGS Software—Application Development, Outsourcing Offshore Software Development Company, Outsourcing.NET, Java, Nearshoring (2021). https://www.pgs-soft.com/blog/digital-twin-explained-the-next-thing-after-iot
50. Uhlemann, T.H.-J., Schock, C., Lehmann, C., Freiberger, S., Steinhilper, R.: The digital twin: demonstrating the potential of real time data acquisition in production systems. Procedia Manuf. **9**, 113–120 (2017)
51. Kent, L., Snider, C., Hicks, B.: Early stage digital-physical twinning to engage citizens with city planning and design. In: 2019 IEEE Conference on Virtual Reality and 3D User Interfaces (VR), pp. 1014–1015. IEEE (2019)
52. Nguyen, H.X., Trestian, R., To, D., Tatipamula, M.: Digital twin for 5G and beyond. IEEE Commun. Mag. **59**(2), 10–15 (2021)
53. Domone, J.: Digital twin for life predictions in civil aerospace. Technical report, Atkins, Epsom, UK (2018)
54. Anbalagan, A., Shivakrishna, B., Srikanth, K.S.: A digital twin study for immediate design/redesign of impellers and blades: Part 1: CAD modelling and tool path simulation. Mater. Today Proc. **46**, 8209–8217 (2021)
55. Fuller, A., Fan, Z., Day, C., Barlow, C.: Digital twin: enabling technologies, challenges and open research. IEEE Access **8**, 108952–108971 (2020)
56. Liu, D., Guo, K., Wang, B., Peng, Y.: Summary and perspective survey on digital twin technology. Chin. J. Sci. Instrum. **39**(11), 1–10 (2018)
57. Schroeder, G., et al.: Visualising the digital twin using web services and augmented reality. In: 2016 IEEE 14th International Conference on Industrial Informatics (INDIN), pp. 522–527. IEEE (2016)
58. Fritzson, P., Engelson, V.: Modelica—a unified object-oriented language for system modeling and simulation. In: Jul, E. (ed.) ECOOP 1998. LNCS, vol. 1445, pp. 67–90. Springer, Heidelberg (1998). https://doi.org/10.1007/BFb0054087
59. Trauer, J., Pfingstl, S., Finsterer, M., Zimmermann, M.: Improving production efficiency with a digital twin based on anomaly detection. Sustainability **13**(18), 10155 (2021)

Algorithms that Get Old: The Case of Generative Deep Neural Networks

Gabriel Turinici(✉)

CEREMADE, Université Paris Dauphine - PSL, CNRS, Paris, France
gabriel.turinici@dauphine.fr
https://turinici.com

Abstract. Generative deep neural networks used in machine learning, like the Variational Auto-Encoders (VAE), and Generative Adversarial Networks (GANs) produce new objects each time when asked to do so with the constraint that the new objects remain similar to some list of examples given as input. However, this behavior is unlike that of human artists that change their style as time goes by and seldom return to the style of the initial creations.

We investigate a situation where VAEs are used to sample from a probability measure described by some empirical dataset.

Based on recent works on Radon-Sobolev statistical distances, we propose a numerical paradigm, to be used in conjunction with a generative algorithm, that satisfies the two following requirements: the objects created do not repeat and evolve to fill the entire target probability distribution.

Keywords: Variational auto-encoder · Generative adversarial network · Statistical distance · Vector quantization · Deep neural network · Measure compression

1 Motivation

Consider a distribution μ and $\mu_e = \frac{1}{M}\sum_{\ell=1}^{M}\delta_{x_\ell}$ an empirical measure sampling this distribution given by a collection of objects x_m, $m = 1,..., M$ where $x_m \in \mathbb{R}^N$ are independent and follow the law μ; in some sense to be defined latter (cf. discussion on statistical distances) μ_e is close to μ; we focus on generative deep neural network architectures that, given μ_e can produce samples from the distribution μ. One such neural network class are the Variational Auto Encoders (cf. [8,15] for an introduction) that, after some training, output two functions (that in practice are implemented as neural networks): the encoder function $E : x \in \mathbb{R}^N \mapsto z \in \mathbb{R}^L$ and the decoder function $D : z \in \mathbb{R}^L \mapsto y \in \mathbb{R}^N$; the decoder function has the property that the image of a multi-dimensional Gaussian on the latent space $\mathbb{R}^L$ through E is close to μ_e thus to μ. Some recent proposals to construct such a VAE are presented in [16], which will also be our inspiration for the statistical distance used in this work (see also [14]). The quality of a VAE is given

G. Nicosia et al. (Eds.): LOD 2022, LNCS 13811, pp. 179–187, 2023.
https://doi.org/10.1007/978-3-031-25891-6_14

1. by the proximity of the $D \circ E$ to the identity operator (at least on the support of the target measure μ);
2. and the small distance between target distribution μ and $D(\mathcal{N}(O, I_L))$ (here $\mathcal{N}$ is the L-dimensional standard Gaussian);

However, although in general the VAEs (same thing applies to the Generative Adversarial Networks - GANs) obtain very good quality results by the previous criteria, the sampling performed at the exploitation phase is, because of the construction, done in an independent way: each time a new $y \sim \mu$ is required, a $z \simeq \mathcal{N}_L$ is sampled and $D(z)$ computed. But such a procedure is at odds with what we observe in real life: the painters do not paint the same landscape again (but still paint pictures), the musical composers' productions vary in style over the years, etc., in general some evolution is witnessed with time. Such a phenomena is probably due to **taking into account the objects previously created**. Our goal is to be able to mimic such an evolution and propose a generative algorithm that

1. is able to create new objects from some target distribution μ (that for VAEs and GANs is the latent distribution);
2. is able to "recall" having created previous objects; **This second point will therefore synthetically induce an "artificial age" for an AI because the process is irreversible.**

A non-aging generative algorithm, when asked to produce, e.g. one new result will likely produce the same object (or similar) over and over again: think of the situation of a standard $1D$-Gaussian: most likely the origin will be drawn over and over again, one has to wait a long time to obtain let's say, a value at 3 standard deviations from the mean. The main goal of this paper is to speed up this waiting time. The advantages of such a process is to allow some "maturation" for the results i.e. to be able to create new results, not the same ones again; this comes at the price of a irreversibility and additional computation cost.

1.1 Relation to Previous Literature

Technically, our proposal has some similarities with different areas in computational statistics: first one can invoke the "vector quantization" procedures (see [7,9,10] and references therein) that, given a distribution, find a set of objects that represent it as a sum of Dirac masses. However, there the technical solution (Voronoi diagrams for instance) is naturally oriented to use for probability measures (or more generally finite positive measures) which is not our situation (our effort involves signed measures); in the same vein see also [3] in the context of machine learning algorithms. On the other hand some efforts have been made to generalize the quantile idea to multi-dimensional distributions; in a one-dimensional situation our procedure and these techniques give similar results but they diverge as soon as the dimension is increased, see [5,6,12].

On another hand and from a completely different perspective, the notion of "age" of a task in a queue is used in scheduling to ensure execution of low priority processes, see [13].

1.2 Technical Goal of the Paper

Continuing the works above, and given the discussion on the generative algorithms, we need a procedure that can incrementally find a good representation of a target measure μ as a number of K Dirac masses (K is given and fixed) centered at some x_k, $k = 1, ..., K$ while taking into account a set of points $Y = (y_j)_{j=1}^{K_p}$ already available. The points Y are called **historical points**. To put it otherwise we want to find the multi-point $X = (x_k)_{k=1}^{K} \in \mathbb{R}^{N \times K}$ $(k = 1, ..., K)$ that minimizes the distance from the total empirical distribution $\frac{\sum_{k=1}^{K_p} \delta_{y_k} + \sum_{l=1}^{K} \delta_{x_k}}{K_p + K}$ to the target measure μ (here the points y_k are not submitted to optimization); this can be written as minimizing the distance $d(\delta_X, \eta)^2$ from the distribution $\delta_X = \frac{1}{K} \sum_{l=1}^{K} \delta_{x_k}$ to the signed measure

$$\eta = \frac{(K_p + K)\mu - K_p \delta_Y}{K_P + K}, \text{ where } \delta_Y = \frac{1}{K_p} \sum_{k}^{K_p} \delta_{y_k}. \tag{1}$$

We present in Sect. 2 our choice of distance d and a theoretical result ensuring that, under appropriate hypotheses, the minimum with respect to X exits. The algorithm to find such a minimum is presented in Sect. 3 together with some numerical results. Final remarks are the object of Sect. 4.

2 Theoretical Results

In order to present the theoretical framework we need to define the distance d between signed measures ζ and η. Note that in fact ζ is a probability measure and the total mass of both is set to 1.

We will take a kernel-based metric given as follows: choose $h(\cdot)$ a conditionally negative definite kernel (see [11] for the precise definition and an introduction), taken here to be $\sqrt{a^2 + |x|^2} - a$ for some given constant $a \geq 0$), see [4,16] for some use cases in machine learning; for any η_1, η_2 signed measures such that $\int (1 + |X|)\eta_i(dX) < \infty$ $(i = 1, 2)$ we define:

$$d(\eta_1, \eta_2) = \sqrt{\int\int -h(|X - Y|)(\eta_1 - \eta_2)(dX)(\eta_1 - \eta_2)(dY)}. \tag{2}$$

The fact that the quantity inside the square root is positive is a consequence of the fact that h is a conditionally negative definite kernel.

Note that in particular, if both η_1 and η_2 are sums of (signed) Dirac masses such that $\eta_1 - \eta_2 = \sum_{k=1}^{K} p_k \delta_{z_k}$ (with $K < \infty$) then Eq. (2) can be written (see [16]):

$$d(\eta_1, \eta_2)^2 = - \sum_{k,\ell=1}^{K} p_k p_\ell h(|z_k - z_\ell|). \tag{3}$$

Once the distance is defined, a legitimate question is whether, given a target signed measure η one can indeed find a uniform sum of Dirac masses ζ that minimizes $d(\zeta, \eta)^2$. This question is settled in the following

Proposition 1. *Suppose K is a fixed positive integer. Let η be a signed measure such that $\int(1+|X|)\eta(dX) < \infty$. For any vector $Z = (z_j)_{j=1}^J \in \mathbb{R}^{N\times J}$ denote*

$$\delta_Z := \frac{1}{J}\sum_{j=1}^J \delta_{z_j},\ f(Z) := d\left(\delta_Z,\eta\right)^2. \tag{4}$$

Then the minimization problem:

$$\inf_{X=(x_k)_{k=1}^K \in \mathbb{R}^{N\times K}} f(X) \tag{5}$$

admits at least one solution.

Remark 1. The previous result only states the existence of a solution, the uniqueness is not necessarily true as one can observe by taking e.g. a rotation invariant measure: any solid rotation of a minimum will still be a minimum.

Proof. Let us denote

$$m_\eta := \inf_{X=(x_k)_{k=1}^K \in \mathbb{R}^{N\times K}} f(X). \tag{6}$$

Take a point X such that $f(X) \le m_\eta + 1$ (the existence of X is guaranteed by the definition of m_η). Then (denoting by 0 the null vector in $\mathbb{R}^{N\times K}$):

$$m_\eta + 1 \ge f(X) = d\left(\delta_X,\eta\right)^2 \ge \frac{d\left(\delta_X,\delta_0\right)^2 - 2d\left(\delta_0,\eta\right)^2}{2}, \tag{7}$$

which implies

$$d\left(\delta_X,\delta_0\right)^2 \le 2(m_\eta+1) + 2d\left(\delta_0,\eta\right)^2. \tag{8}$$

But, using Eq. (3), we obtain:

$$d\left(\delta_X,\delta_0\right)^2 = \frac{2}{K}\sum_{k=1}^K h(|x_k|) - \frac{1}{K^2}\sum_{k,k'=1}^K h(|x_k - x_{k'}|) \tag{9}$$

$$\ge \frac{2}{K}\sum_{k=1}^K h(|x_k|) - \frac{1}{K^2}\sum_{k,k'=1,k\neq k'}^K [a + h(|x_k|) + h(|x_{k'}|)] \tag{10}$$

$$\ge \frac{2}{K^2}\sum_{k=1}^K h(|x_k|) - a, \tag{11}$$

where for the passage from (9) to (11) we used the inequality $h(|x-y|) \le h(|x|) + h(|y|) + a$ true for any $x, y \in \mathbb{R}^N$. We obtain that, when $f(X) \le m_\eta + 1$ there exists a constant $C_0 = K^2\left(a/2 + m_\eta + 1 + d\left(\delta_0,\eta\right)^2\right)$ such that $\sum_{k=1}^K h(|x_k|) \le C_0$; therefore any minimizing sequence $(X_n)_{n\ge1}$ (that is, any sequence such that $\lim_{n\to\infty} f(X_n) = m$) is bounded. This sequence will have a sub-sequence $(X_{n_k})_{k\ge1}$ that is convergent to some $X^\star$; but since the distance is continuous, we obtain that $f(X^\star) = m$ which means that $\delta_{X^\star}$ is a solution of the minimization problem (5). □

3 Algorithm and Numerical Results

3.1 Algorithm Formulation

Consider now a target distribution μ and a set of previously constructed points Y (of cardinal K_p); these **historical points** are explicitly known; we will propose an algorithm that, given a number K of points to be constructed, will find a multi-point $X \in \mathbb{R}^{N \times K}$ such that the overall measure $\delta_{X \cup Y}$ minimizes the distance to the target measure μ.

Algorithm A1. History aware (signed measure) compression algorithm : HAW-C

1: **procedure** HAW-C
2: • set batch size B, parameter $a = 10^{-6}$,
3: • load the historical points y_k, $k = 1, ..., K_p$
4: • initialize points x_k, $k = 1, ..., K$ sampled at random from μ, denote $X = (x_k)_{k=1}^{K}$ (considered as vector in $\mathbb{R}^{N \times K}$)
5: **while** (max iteration not reached) **do**
6: • sample $z_1, ..., z_B \sim \mu$ (i.i.d).
7: • compute the global loss [1] using formula (3):
$$L(X) := d\left(\frac{1}{K}\sum_{l=1}^{K} \delta_{x_k}, \frac{K_p+1}{B}\sum_{b=1}^{B} \delta_{z_b} - \sum_{j=1}^{K_p} \delta_{y_j}\right)^2;$$
8: • backpropagate the loss $L(X)$ in order to minimize $L(X)$ and update X.
9: **end while**
10: **end procedure**

3.2 Numerical Results

A Python code implementing the algorithm in both history unaware and history aware compression modes can be consulted at [2].

History Unaware Compression of a 2D Gaussian Mix Distribution
We first test the algorithm without any historical points i.e., $K_p = 0$. When the target measure is positive, the HAW-C algorithm A1 allows to compress any given (probability) measure, as illustrated in Fig. 1 for the situation of a uniform mixture of 16 lattice-centered $2D$ normal variables. Good results are obtained: without any previous knowledge, the algorithm can unveil the mixing structure and allow a coherent compression.

History Aware Multi-dimensional Gaussian Compression and Application to Generative Algorithms
We move now to a test where incremental compression is performed: we consider a $2D$ Gaussian centered at origin. First we compress it with a single point u_1; then we use $K_p = 1$ and $y_1 = u_1$ as history and compress the signed measure :

[1] The global loss is the distance from $\delta_X = \frac{1}{K}\sum_{l=1}^{K} \delta_{x_k}$ to the signed measure η.

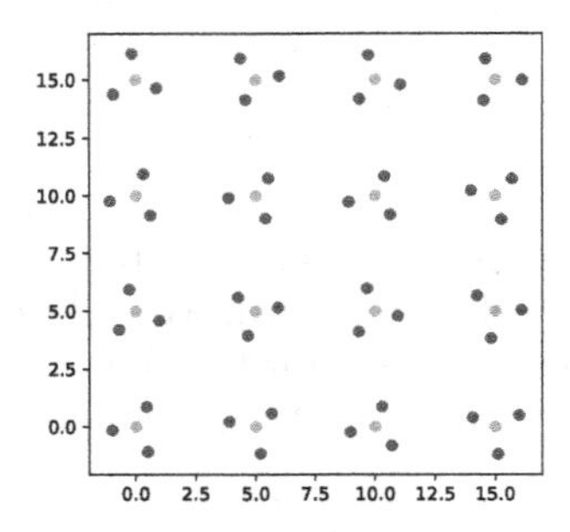

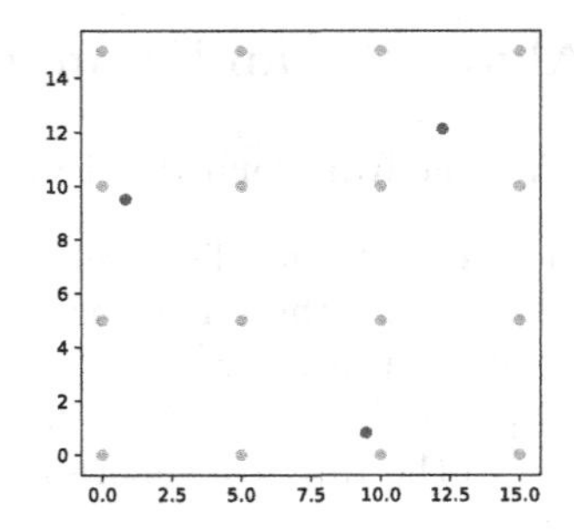

Fig. 1. Test without any historical points, $K_p = 0$. An example of compression for an uniform Gaussian mixture of 16 Gaussians centered on points of a 4×4 grid (red points are the centers of the Gaussians, blue points are the compressed points). We used K points to summarize the distribution: $K = 48$ (**left image**) or $K = 3$ (**right image**). Good quality results are obtained as the algorithm "understands" the mixing structure: for instance for $K = 48$ the algorithm allocates precisely 3 points per Gaussian mixture term. (Color figure online)

initial Gaussian minus the first obtained point u_1, as detailed in Eq. (1), with another supplementary point u_2; then consider $K_p = 2$ and $y_i = u_i$ $(i = 1, 2)$ and compress the Gaussian measure minus the sum of Dirac masses in u_i with another point u_3; the procedure is then continued recursively, each step being an application of the algorithm A1. The results are presented in Fig. 2.

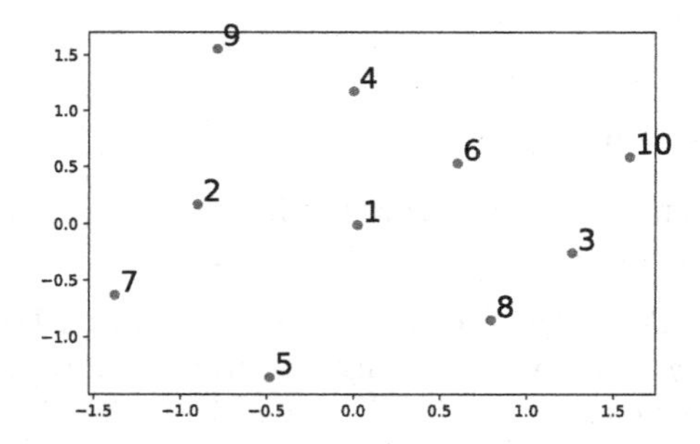

Fig. 2. An example of recursive compression of a $2D$ standard Gaussian (see text for details). **Left image:** the result of the compression after 10 iterations. Each point u_i is labeled by its corresponding index i when it was found.

In order to test these results on a practical case, we used the CVAE procedure from the Tensorflow documentation [1] with default parameters (except that we used 100 epochs instead of 10 because the results with 10 epochs are very fuzzy). The code was executed once in order to construct the encoder/ decoder networks and then the sampling was done in the latent space using either a multi-dimensional sampling of 10 objects or an incremental sampling; once the sampling is done the data is propagated through the decoder network and the resulting images are presented in Fig. 3. We note that the history aware sampling retains a good diversity with respect to the uniform sampling and avoids some repetitions:

- the figures 1 and 2 that are repeated in the propagated random samples and only appear once in the incremental sampling;
- the figure 9 appears only very unclearly in the left sample and more clearly in the right sample
- the incremental sampling avoids the symbol in row 2 column 2 (left image) which is not a figure.

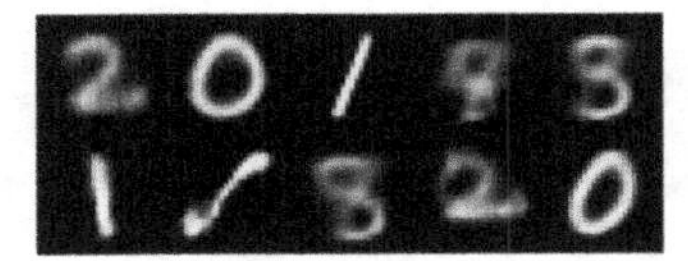

Fig. 3. Images from [1] obtained by taking either a random sampling of 10 points from a 2D Gaussian (left image) or the sampling obtained in Fig. 2 (right image). The decoder is the one obtained by a run of the code from [1]. The right image appears more faithful of the database. Improved quality results are available in Fig. 4.

Note that this out-of-the-box C-VAE is not good enough to make figures too precise which explains large numbers of fuzzy images - resembling to a 8, 6 or 9- present in both results). To improve the result, we re-run the CVAE for 20 epochs but increased, as recommenced in the documentation, all 'filters' numbers to 512 (instead of 32 or 64 in the initial setup). We obtain the results in Fig. 4: the quality of the VAE is indeed increased and the same conclusions hold for the comparison of the i.i.d. sampling with the incremental sampling.

4 Final Remarks

We explored in this work the construction of objects from generative algorithms (like VAEs and GANs); more specifically the construction was incremental in the sense that each new sampling from the latent space considers the previous samples, called the **history** and tries to both respect the desired target distribution of the latent space but also stays away from the points already sampled. We described some theoretical properties of the procedure and tested it both on general data and on a C-VAE benchmark.

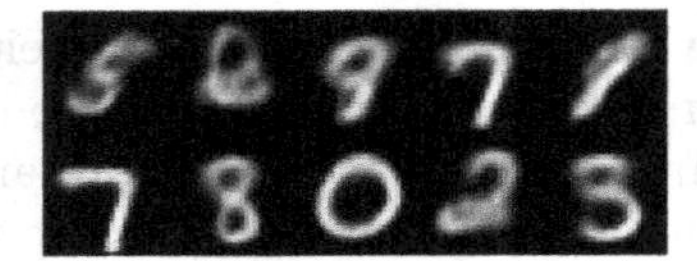

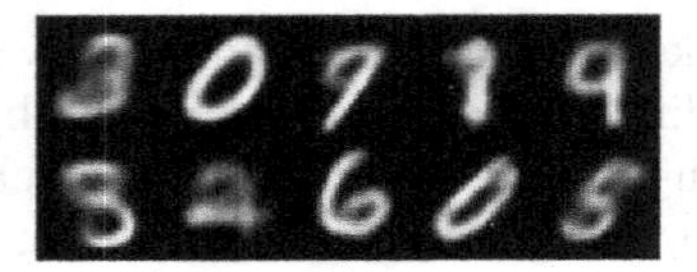

Fig. 4. Results of the same procedure as in Fig. 3 but obtained for an improved network (512 filters everywhere and 20 epochs): **left image:** random sampling; **right image:** decoding of the incremental sampling from Fig. 2. The left figure has several repetitions (for instance figure 7) that are absent from the right figure but more importantly, some figures abondant in the database and not present in the left figure appear in the right one, like the figures 1 and 6.

More experiments are required to characterize fully the applicability domain of the proposed procedure, but the present results provide encouraging arguments to do so.

References

1. cVAE, Tensorflow Documentation. https://www.tensorflow.org/tutorials/generative/cvae. Accessed 30 Jan 2022
2. Python implementation for measure compression algorithm described in arxiv:2202.03008 [stat.ML] by Turinici, G. https://doi.org/10.5281/zenodo.6671523
3. Chazal, F., Levrard, C., Royer, M.: Optimal quantization of the mean measure and applications to statistical learning (2021)
4. Deshpande, I., et al.: Max-sliced Wasserstein distance and its use for GANs. In: The IEEE Conference on Computer Vision and Pattern Recognition (CVPR) (2019)
5. Fraiman, R., Pateiro-Lopez, B.: Quantiles for finite and infinite dimensional data. J. Multivar. Anal. **108**, 1–14 (2012). https://www.sciencedirect.com/science/article/pii/S0047259X12000176
6. Glazer, A., Lindenbaum, M., Markovitch, S.: q-OCSVM: a q-quantile estimator for high-dimensional distributions. In: Burges, C.J.C., Bottou, L., Welling, M., Ghahramani, Z., Weinberger, K.Q. (eds.) Advances in Neural Information Processing Systems. vol. 26. Curran Associates, Inc., New York (2013), https://proceedings.neurips.cc/paper/2013/file/819f46e52c25763a55cc642422644317-Paper.pdf
7. Graf, S., Luschgy, H.: Foundations of Quantization for Probability Distributions. Springer, Berlin (2007)
8. Kingma, D.P., Max, W.: An Introduction to Variational Autoencoders. Now Publishers Inc, Netherlands (2019)
9. Kohonen, T.: Learning vector quantization. In: Kohonen, T. (ed.) Self-Organizing Maps. Springer Series in Information Sciences, vol. 30, pp. 175–189. Springer, Berlin, Heidelberg (1995). https://doi.org/10.1007/978-3-642-97610-0_6
10. Gray, R.: Vector quantization. IEEE ASSP Mag. **1**(2), 4–29 (1984). https://doi.org/10.1109/MASSP.1984.1162229
11. Sejdinovic, D., Sriperumbudur, B., Gretton, A., Fukumizu, K.: Equivalence of distance-based and RKHS-based statistics in hypothesis testing. Ann. Stat. **41**(5), 2263–2291 (2013). https://www.jstor.org/stable/23566550

12. Serfling, R.: Quantile functions for multivariate analysis: approaches and applications. Stat. Neerl. **56**(2), 214–232 (2002)
13. Silberschatz, A., Gagne, G., Galvin, P.B.: Operating System Concepts, 10th edn. John Wiley & Sons Inc, Hoboken (2018)
14. Székely, G.J., Rizzo, M.L.: Energy statistics: a class of statistics based on distances. J. Stat. Plann. Infer. **143**(8), 1249–1272 (2013)
15. Tabor, J., Knop, S., Spurek, P., Podolak, I.T., Mazur, M., Jastrzkebski, S.: Cramer-Wold AutoEncoder. CoRR abs/1805.09235 (2018). https://arxiv.org/abs/1805.09235
16. Turinici, G.: Radon-Sobolev variational auto-encoders. Neural Netw. **141**, 294–305 (2021). https://doi.org/10.1016/j.neunet.2021.04.018, https://www.sciencedirect.com/science/article/pii/S0893608021001556

A Two-Country Study of Default Risk Prediction Using Bayesian Machine-Learning

Fabio Incerti[1(✉)], Falco J. Bargagli-Stoffi[2], and Massimo Riccaboni[1]

[1] IMT School for Advanced Studies of Lucca, Lucca, Italy
{fabio.incerti,massimo.riccaboni}@imtlucca.it
[2] Harvard University, Cambridge, MA, USA
fbargaglistoffi@hsph.harvard.edu

Abstract. This paper tackles the prediction problem of firm default based on financial accounts and other firm features. We propose to exploit a novel machine-learning algorithm, the Bayesian Additive Regression Tree with Missingness Incorporated in Attributes (BART-MIA), which has been recently shown to outperform many traditional algorithms in analogous prediction tasks. We address the issue from an international perspective to assess its performance in both Netherlands and Italy over recent years. Despite the structural differences in the financial accounts in the two countries, we find the BART-MIA can take advantage of country-level missingness patterns and outperform state-of-the-art econometric and machine-learning models.

Keywords: Supervised machine-learning · Bayesian statistical learning · Firm default risk · Bankruptcy · Italy · Netherlands

1 Introduction

Predicting and anticipating a company's failure is crucial. The call for the development of methodologies able to spot non-viable firms and anticipate financial distress conditions that can lead otherwise viable firms to bankruptcy has preponderantly risen in the last few decades. In this spirit, the recent Directive (EU) 2019/1023 has introduced a framework for pre-insolvency restructuring by promoting a rescue-oriented culture of turnaround management to avoid unnecessary bankruptcies. On the other hand, recent artificial intelligence (AI) developments have led to increasingly advanced methodologies for specific predictive tasks. A canonical example is forecasting a company's failure using its past financial statements. This challenge has been addressed in numerous works belonging to a relevant research line to study the so-called zombie firms – i.e. non-viable firms dodging failure despite their high financial distress. According to [6], this stream of research has followed two main directions:

1. The reasons and the effects of the banks' behaviour leading to credit misallocation by supporting non-viable firms [1,8,17];
2. The existence of zombie firms and their effects on the economy and growth [2,10,18,19].

G. Nicosia et al. (Eds.): LOD 2022, LNCS 13811, pp. 188–192, 2023.
https://doi.org/10.1007/978-3-031-25891-6_15

As the literature emphasises, the zombie firms theory is still developing [18]. The evidence from Europe is scarce, limiting comparability and discussion among EU countries. Thus, despite growing literature and emerging studies from European countries, spotting non-viable firms within a common AI framework remains an underresearched topic with great real-world significance and potential.

More recently, a machine-learning (ML) approach, the so-called BART-MIA, has been introduced to estimate the probability of default and spot non-viable firms in Italy [5]. The work tackles the issue as a prediction problem by detecting non-viable firms as those persisting at a high risk of default from which the chances of shifting to more sustainable conditions are minimal. The ML approach proposed is convenient for several reasons. It reports great performances compared to several popular ML algorithms [4], and it is robust to non-random patterns of missing values to predict the outcome. This robustness is crucial when dealing with non-random missingness in the financial accounts, often ascribable to the firm management's missed release of crucial financial information.

This paper proposes to test whether the BART-MIA approach can work across countries with structural differences in financial statement standards. In light of the increasing harmonisation of the insolvency regimes across the EU member states [11,16], we propose to feature Italy and the Netherlands in our analysis. Indeed, despite a convergence towards a common approach in the insolvency regimes, the cross-country comparability of accounting data is still challenging because of the different data coverage and financial reporting rules [3].

2 The Model

We propose to exploit an ensemble method, the so-called Bayesian Additive Regression Tree with Missingness Incorporated in Attributes (BART-MIA) [14]. The algorithm is based on the aggregation of different independent trees and incorporates information about patterns of missing values. This attribute makes it particularly suitable for missing financial statement entries where missingness is informative for the prediction and should be captured as a relevant feature. BART-MIA has three regularizing priors designed to prevent overfitting [9,13]: (i) the prior on the tree structure describing the probability a node is split at depth d is $\alpha(1+d)^{-\beta}$, where $\alpha \in (0,1)$, $\beta \in [0,\infty)$, and according to the recommendation of [9,13], the values of the hyper-parameters are set $\alpha = 0.95$ and $\beta = 2$; (ii) the prior on the leaf parameters given the tree structure has normal density $\mathcal{N}(0,\sigma_\mu^2)$, where σ_μ is empirically computed such that the density yields with 95% of probability that $Pr[Y=1|x]$ lies in the observed range of the dependent variable; (iii) the prior on the error variance is $\sigma \sim$Inv-$\chi^2 = \nu\lambda/\chi^2_\nu$ where the recommended values for the hyperparameters are $\nu = 3$ and λ is tuned from data in such a way there is 90% of prior likelihood that the model improves upon the root mean squared error (RMSE) from an OLS regression of Y on x.

We consider the problem of inferring on an unknown function $f(\cdot)$ predicting firm i failure at time t based on previous year predictors

$$Y_{i,t} = f\left(\mathbf{x}_{i,t-1}\right) + \epsilon_{i,t} \tag{1}$$

where $Y_{i,t}$ is a dummy variable assuming 1 if firm i failed in year t and 0 otherwise, $\mathbf{x}_{i,t}$ is a P-dimensional vector of firm-level predictors and $\epsilon_{i,t} \sim \mathcal{N}(0, \sigma^2)$. We approximate $f(\mathbf{x}_{i,t-1}) = Pr[Y_{i,t} = 1 \mid \mathbf{x}_{i,t-1}]$ with the BART-MIA model:

$$f(\mathbf{x}_{i,t-1}) \approx \mathcal{T}_1^{\mathcal{M}}(\mathbf{x}_{i,t-1}) + \mathcal{T}_2^{\mathcal{M}}(\mathbf{x}_{i,t-1}) + \ldots + \mathcal{T}_q^{\mathcal{M}}(\mathbf{x}_{i,t-1}). \tag{2}$$

where $\mathcal{T}_i^{\mathcal{M}}$ with $i \in \{1, ..., q\}$ is a distinct decision tree and $\mathcal{M}$ represents the whole set of leaf parameters.

3 Empirical Analysis

We analyze a Bureau Van Dijk sample of 368,901 Italian firms and 466,284 Dutch firms. We compare the model performances in the Netherlands with a standard econometric model, i.e. the Logit, and three popular ML algorithms, i.e. the Conditional Inference Tree (CTree) [12], the Random Forest [7], and the Super Learner [15]. However, many financial statement entries exhibit huge missingness compared to Italy, and traditional algorithms work with no missing values[1]. Hence, we perform the exercise with only a few predictors for which the number of missing is not so critical as to wipe out the sample dimension. For the sake of comparison, since we use only companies with non-missing values for standard models, the BART-MIA is trained on a random sample of the same dimension. However, this time we include the missing values. We fine-tune priors and related hyperparameters according to the recommendation of [9]. Please note that BART-MIA could be trained on a much larger sample as the model can handle them. To test the performance of the models, we use hold-one-out cross-validation by assigning 90% of the sample to the training set and 10% to the test set.

Tables 1 and 2 report the performance in each country for 2018 and 2019. We predict firm default with past year predictors each year. We use the area under the Receiver Operating Characteristic (ROC) curve (AUC) as a threshold-free performance metric and the Balanced Accuracy (BACC) as a threshold-based classification performance measure that we set equal to the median of the estimated default risk distribution. The BART-MIA reached high scores in both years and both countries. The lead from traditional models is more marked in the Netherlands than in Italy, where higher performances are achieved with the help of a broader set of features[2]. This result witnesses the BART-MIA's potential to use a much larger number of features and, consequently, an even more significant number of observations.

[1] The CTree [12] can manage missing values but is not designed for high missingness settings. Indeed, rather than incorporate the missingness patterns in the prediction model, it uses surrogate splits to compensate for the missingness in a feature. For this reason, we do not exploit this option.

[2] Since Italy does not exhibit extreme missingness, we can use the whole set of available features. We add the following variables: turnover, financial revenues, cash flow, EBITDA, depreciation, added value, cost of employees, net income and interest paid. For 2018 the BART-MIA scores 0.9059 for the AUC and 0.7508 for the BACC. Likewise, for 2019, the AUC is 0.9550, and the BACC is 0.7507.

Table 1. Performance across models in Italy. The predictors involved are: employees, total assets, liquidity ratio, current liabilities, non-current liabilities, fixed assets, current assets and shareholders funds.

Year	2018		2019	
	AUC	BACC	AUC	BACC
Logit	0.5000	0.5000	0.6435	0.6054
CTree	0.5605	0.5605	0.5332	0.5332
Random forest	0.7448	0.7462	0.7721	0.7682
Super learner	0.8393	0.8333	0.8322	0.6844
BART-MIA	0.8596	0.7208	0.8850	0.7166
Training obs.	251,011		263,785	
Test obs.	27,891		29,310	

Table 2. Performance across models in the Netherlands. The predictors involved are: employees, total assets, liquidity ratio, current liabilities, non-current liabilities, fixed assets, current assets and shareholders funds.

Year	2018		2019	
Netherlands	AUC	BACC	AUC	BACC
Logit	0.5211	0.5000	0.4238	0.4404
CTree	0.5622	0.5717	0.5119	0.5116
Random forest	0.4909	0.4909	0.6790	0.6669
Super learner	0.4777	0.4777	0.6763	0.6398
BART-MIA	0.9163	0.7506	0.8259	0.7075
Training obs.	155,374		152,273	
Test obs.	17,264		16,920	

4 Conclusions

The heterogeneity across European countries regarding accounting standards, disclosure requirements and the availability of such data poses a challenge to prediction tasks based on firm-level data involving different countries. The development and research of AI methodologies, flexible and adaptable to structural differences between countries, remains a relevant topic in the context of ever-greater European economic integration. With this exercise, we show how the BART-MIA is a robust cross-border method compared to traditional ML models to estimate the risk of bankruptcy at the firm level. Future research steps should test the predictive capabilities in other European countries and provide adequate measures to manage the trade-off between cross-country adaptability and predictive performance.

References

1. Ahearne, A.G., Shinada, N.: Zombie firms and economic stagnation in Japan. IEEP **2**(4), 363–381 (2005). https://doi.org/10.1007/s10368-005-0041-1
2. Asanuma, D.: An examination on the zombie theory: an agent-based-approach. Int. Bus. Manag. **9**(5), 719–725 (2015)
3. Bajgar, M., Berlingieri, G., Calligaris, S., Criscuolo, C., Timmis, J.: Coverage and representativeness of Orbis data (2020)
4. Bargagli-Stoffi, F.J., Niederreiter, J., Riccaboni, M.: Supervised learning for the prediction of firm dynamics. In: Consoli, S., Reforgiato Recupero, D., Saisana, M. (eds.) Data Science for Economics and Finance, pp. 19–41. Springer, Cham (2021). https://doi.org/10.1007/978-3-030-66891-4_2
5. Bargagli-Stoffi, F.J., Riccaboni, M., Rungi, A.: Machine learning for zombie hunting. Firms' Failures and Financial Constraints (2020)
6. Blažková, I., Dvouletỳ, O.: Zombies: who are they and how do firms become zombies? J. Small Bus. Manage. **60**(1), 119–145 (2022)
7. Breiman, L.: Random forests. Mach. Learn. **45**(1), 5–32 (2001). https://doi.org/10.1023/a:1010933404324
8. Caballero, R.J., Hoshi, T., Kashyap, A.K.: Zombie lending and depressed restructuring in Japan. Am. Econ. Rev. **98**(5), 1943–77 (2008)
9. Chipman, H.A., George, E.I., McCulloch, R.E.: BART: Bayesian additive regression trees. Ann. Appl. Stat. **4**(1), 266–298 (2010)
10. De Veirman, E., Levin, A.T.: When did firms become more different? Time-varying firm-specific volatility in Japan. J. Jpn. Int. Econ. **26**(4), 578–601 (2012)
11. Ghio, E., Boon, G.J., Ehmke, D., Gant, J., Langkjaer, L., Vaccari, E.: Harmonising insolvency law in the EU: new thoughts on old ideas in the wake of the COVID-19 pandemic. Int. Insolv. Rev. **30**(3), 427–459 (2021)
12. Hothorn, T., Hornik, K., Zeileis, A.: Unbiased recursive partitioning: a conditional inference framework. J. Comput. Graph. Stat. **15**(3), 651–674 (2006)
13. Kapelner, A., Bleich, J.: bartMachine: machine learning with Bayesian additive regression trees. arXiv preprint arXiv:1312.2171 (2013)
14. Kapelner, A., Bleich, J.: Prediction with missing data via Bayesian additive regression trees. Can. J. Stat. **43**(2), 224–239 (2015)
15. Van der Laan, M.J., Polley, E.C., Hubbard, A.E.: Super learner. Stat. Appl. Genet. Mol. Biol. **6**(1) (2007)
16. McGowan, M.A., Andrews, D.: Design of insolvency regimes across countries (2018)
17. Peek, J., Rosengren, E.S.: Unnatural selection: perverse incentives and the misallocation of credit in Japan. Am. Econ. Rev. **95**(4), 1144–1166 (2005)
18. Urionabarrenetxea, S., Garcia-Merino, J.D., San-Jose, L., Retolaza, J.L.: Living with zombie companies: do we know where the threat lies? Eur. Manag. J. **36**(3), 408–420 (2018)
19. Urionabarrenetxea, S., San-Jose, L., Retolaza, J.L.: Negative equity companies in Europe: theory and evidence. Bus. Theor. Pract. **17**(4), 307–316 (2016)

Enforcing Hard State-Dependent Action Bounds on Deep Reinforcement Learning Policies

Bram De Cooman[1(✉)], Johan Suykens[1], and Andreas Ortseifen[2]

[1] KU Leuven, ESAT - STADIUS, Leuven, Belgium
{bram.decooman,johan.suykens}@esat.kuleuven.be
[2] Ford Research and Innovation Center, Aachen, Germany
aortseif@ford.com

Abstract. Imposing hard constraints on deep reinforcement learning policies trained with model-free methods is a challenging task. In this paper we specifically focus on constraining the policy's actions, by imposing state-dependent action bounds. Such bounds allow the designer to incorporate prior domain knowledge into the model-free learning framework and can be used to improve the stability or safety of the learned policies. The approach is applied to two benchmark environments and a more complicated autonomous driving problem. When correctly applied, the state-dependent action bounds can provide strong safety guarantees, as well as improve the convergence speed.

Keywords: Reinforcement learning · Hard constraints · Action bounds · Safe control · Autonomous driving

1 Introduction

An advantage of model-free reinforcement learning (RL) is that no model knowledge is required and policies can be trained purely through interactions with an unknown environment. This, however, comes with a cost, as the lack of a model makes these methods less sample efficient than their model-based counterparts. Moreover, it is harder to constrain policies to certain (safe) regions in state-action space without the guidance of a model. In this paper we propose a novel way to embed prior domain knowledge into the model-free learning scheme using state-dependent bounds on the agent's actions. When correctly applied, these bounds can improve sample efficiency as well as provide strong safety guarantees, while keeping the benefits of the model-free learning setting.

1.1 Motivation

The potential benefits of state-dependent action bounds can be illustrated by a virtual driver example in which the agent needs to learn a highway-driving

Supplementary Information The online version contains supplementary material available at https://doi.org/10.1007/978-3-031-25891-6_16.

G. Nicosia et al. (Eds.): LOD 2022, LNCS 13811, pp. 193–218, 2023.
https://doi.org/10.1007/978-3-031-25891-6_16

policy — this setup is also analyzed in more depth in Sect. 4. Obviously, if the agent starts with zero domain knowledge, many collisions will occur early on in the training process. Incorporating domain knowledge by providing common rules of traffic to the agent helps preventing obvious mistakes (Fig. 1a). As the resulting state-dependent action bounds are effectively hard constraints, it might improve the trust of the general public in the learned policies. Beside the safety aspect, these bounds also speed up training, by constraining the agent to only explore the relevant, safe regions of state-action space. Finally, such bounds can also be used to ensure compliance with traffic regulations required for vehicle homologation in certain markets, such as keeping a minimum safety distance during automated lane changes (Fig. 1b).

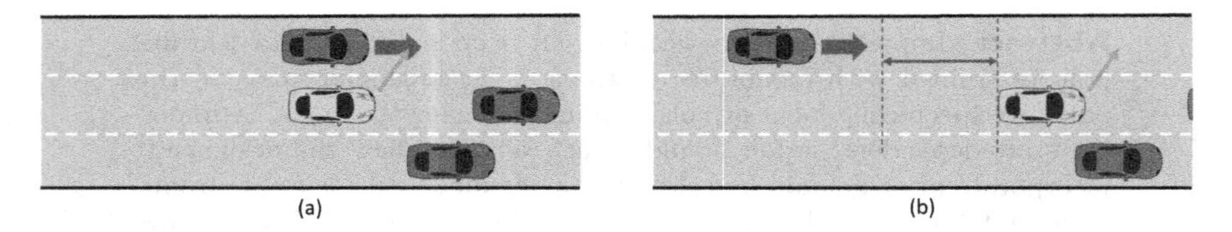

Fig. 1. Schematic overview of two situations on a highway. On the left a situation where an unconstrained virtual driver (yellow) might decide to move towards the occupied left lane. State-dependent action bounds can prevent the virtual driver from executing such hazardous manoeuvres. On the right a minimum safety distance for an automated lane change is visualized, which can be ensured by state-dependent action bounds. (Color figure online)

1.2 Main Contributions and Related Work

Safe reinforcement learning is an active field of research for which an extensive overview is given by Garcia et al. [13]. Some approaches consider the setting in which safety must be learned through environmental interactions, which means safety constraints may be violated during training [7,25]. Other methods try to incorporate the safety constraints *throughout* the learning phase, ensuring exploration itself is also safe [3,19,27]. In this paper we follow the latter approach and enforce the state-dependent bounds both during training and deployment.

Closely related to the safe RL domain are methods solving Constrained Markov Decision Processes (CMDP), which beside optimizing for performance (long-term accumulated reward) also take an arbitrary amount of policy constraints into account [1,7,25]. Our approach could be considered a special case under this CMDP framework, as we specifically focus on state-dependent action bounds, i.e. upper and lower bound constraints on actions only. One benefit of restricting us to this specific subclass of problems is that we can guarantee hard constraint satisfaction, while the methods solving the more general (and difficult) CMDP problem typically only have near-constraint satisfaction.

For discrete action spaces a commonly used approach is to prune the available action set, such that unsafe actions are excluded. The Deep Constrained Q-learning method [16] can be used to learn such policies. The pruning can either be done in a preprocessing step, by providing the agent with a safe set of actions to choose from; or in a postprocessing step, after the agent has already made a decision [2]. In the latter case, the agent assigns a list of priorities to each discrete action from which the *safe* action with highest priority is chosen in the pruning step [15,21]. While our approach is only applicable to continuous action spaces, it could be classified as *preemptive shielding*: the agent is only allowed to select safe actions that lie within the predefined range of state-dependent bounds.

Postposed shielding approaches for continuous action spaces are provided by Dalal et al. [8] and Cheng et al. [6]. In both setups the control pipeline is extended with a "safety controller", which maps potentially unsafe actions selected by the agent to the closest safe actions. This requires an optimization problem to be solved at each timestep, which may not be feasible if computing resources are limited. To alleviate this, Dalal et al. [8] suggest solving the optimization problem analytically, under the extra assumption that only a single constraint is active at the same time.

Chandak et al. [5] study the setting where the action space is stochastic. Such a setting is more generic than ours, as the range of available actions (defined through the state-dependent bounds) is considered to be deterministic and defined before learning starts. Finally, it should be mentioned that many environments make use of *static* action bounds, for example to model actuator saturation. Most standard RL methods are able to deal with that, for instance through using bounded activation functions in the last layer of the actor network. However, such bounds are constant for all states. Hence, to the best of our knowledge, our work is the first to apply *state-dependent* action bounds.

Enforcing such hard, state-dependent bounds on policy actions is the central contribution of this work. These bounds not only prevent exploration and operation in irrelevant parts of the state-action space, but also allow the designer to embed prior domain knowledge in the model-free learning setting; ultimately leading to policies with an improved performance, sample efficiency and safety.

1.3 Organization

This paper is organized as follows. The necessary reinforcement learning essentials are first described in Sect. 2, followed by an in-depth introduction of state-dependent action bounds in Sect. 3. Our approach is then evaluated on an autonomous driving problem in the experiments of Sect. 4. Extra background information and additional experiments can be found in the appendices.

2 Reinforcement Learning

Formally, the Reinforcement Learning (RL) problem can be described by a Markov Decision Process (MDP), written down as the tuple $(\mathcal{S}, \mathcal{A}, \iota, \tau, r, \gamma)$ with $\mathcal{S} \subset \mathbb{R}^S$ denoting the state space, $\mathcal{A} \subset \mathbb{R}^A$ the action space, $\iota(\boldsymbol{s}_0) : \mathcal{S} \to [0; 1]$ the initial-state distribution, $\tau(\boldsymbol{s}_{t+1}|\boldsymbol{s}_t, \boldsymbol{a}_t) : \mathcal{S} \times \mathcal{S} \times \mathcal{A} \to [0; 1]$ the state transition model, $r(\boldsymbol{s}_t, \boldsymbol{a}_t, \boldsymbol{s}_{t+1}) : \mathcal{S} \times \mathcal{A} \times \mathcal{S} \to \mathbb{R}$ the reward model and $\gamma \in [0; 1)$ the discount factor. The state transition model $\tau(\boldsymbol{s}_{t+1}|\boldsymbol{s}_t, \boldsymbol{a}_t)$ describes the probability of going to state s_{t+1} in the next time step, when selecting action a_t in state s_t at the current time step. The reward model $r(s_t, a_t, s_{t+1})$ provides the agent with a scalar reward for transitioning towards state s_{t+1} when selecting action a_t in state s_t. It is this reward signal that provides the agent with feedback, which can be used to improve its policy $\pi(a|s) : \mathcal{A} \times \mathcal{S} \rightarrow [0; 1]$. This policy describes the probability of taking action a when the agent perceives state s. For deterministic policies we will also use the notation $a = \pi(s)$.

The goal of the agent is to maximize the future discounted return, given by

$$R_t = \sum_{k=0}^{\infty} \gamma^k r(\mathbf{s}_{t+k}, \mathbf{a}_{t+k}, \mathbf{s}_{t+k+1}),$$

for all possible initial states and subsequent visited state-action pairs by following its policy π. The optimal policy π^* is thus found as

$$\pi^* = \arg\max_{\pi} \mathbb{E}_{\pi,\tau,\iota}[R_0], \tag{1}$$

where the expectation $\mathbb{E}_{\pi,\tau,\iota}$ is taken over the probability distribution of actions $a_t \sim \pi(\cdot|s_t)$, induced by the policy, and over the probability distribution of states $s_0 \sim \iota(\cdot)$ and $s_{t+1} \sim \tau(\cdot|s_t, a_t)$, induced by the environment.

Instead of working directly with the objective (1), it is often useful to consider the *state-value* function $V_\pi(s) = \mathbb{E}_{\pi,\tau}[R_t \| s_t = s]$ and *action-value* function $Q_\pi(s, a) = \mathbb{E}_{\pi,\tau}[R_t | s_t = s, a_t = a]$. Both of these value functions can also be defined recursively through a relationship known as the Bellman equation

$$V_\pi(\mathbf{s}) = \mathbb{E}_{\pi,\tau}[r(s_t, a_t, s_{t+1}) + \gamma V_\pi(s_{t+1}) | s_t = \mathbf{s}],$$
$$Q_\pi(\mathbf{s}, \mathbf{a}) = \mathbb{E}_{\pi,\tau}[r(s_t, a_t, s_{t+1}) + \gamma Q_\pi(s_{t+1}, a_{t+1}) | s_t = \mathbf{s}, a_t = \mathbf{a}].$$

More background information regarding the fundamentals of reinforcement learning can be found in the book by Sutton & Barto [24].

There are many different RL algorithms to find an estimate of the optimal policy or value function for the previously described MDP (see [28] for a taxonomy of different approaches). Enforcing state-dependent action bounds on the policies, requires the usage of RL methods with explicit actor networks (policy-based or actor-critic), supporting both continuous state and action spaces. In this work, we specifically focus on two model-free, off-policy, actor-critic methods, which are briefly introduced in the following two subsections. Being actor-critic methods, both an actor network $\mu(s; \theta_\mu)$, modelling the policy, and a critic network $Q(s, a; \theta_Q)$, estimating the optimal state-action value function, are jointly trained. Extra target networks (denoted by primes on the weight vectors θ_i') are also introduced to improve the stability of the learning process and are updated using Polyak averaging. As the environment dynamics ι, τ remain unknown to the agent in the model-free setting, the agent is required to explore the state-action space during training. The *behavioural policy* $\beta(a|s)$ is used for this exploration, collecting experience tuples (s_t, a_t, r_t, s_{t+1}) which are stored in a replay buffer $\mathcal{B}$. The actor and critic networks are then trained using batches of experience tuples from this replay buffer.

2.1 Deterministic Methods

Two popular deterministic, actor-critic methods are 'Deep Deterministic Policy Gradient' (DDPG) [20] and 'Twin Delayed DDPG' (TD3) [12]. In these methods, the actor network represents the deterministic policy $\boldsymbol{a} = \mu(\boldsymbol{s}; \boldsymbol{\theta}_\mu)$. To ensure sufficient exploration of the state-action space, an external source of stochasticity is necessary during training, as a deterministic policy would always take the same actions in the same states. Hence, the behavioural policy $\beta(\boldsymbol{s}) = \mu(\boldsymbol{s}; \boldsymbol{\theta}_\mu) + \boldsymbol{\epsilon}$ with exploration noise $\boldsymbol{\epsilon} \sim \mathcal{N}(\mathbf{0}, \boldsymbol{\sigma}_{expl} I)$ is used to collect experience during training.

The critic network is updated by minimizing a squared temporal difference error

$$L_Q(\boldsymbol{\theta}_Q) = \mathbb{E}_{(\boldsymbol{s}_t, \boldsymbol{a}_t, r_t, \boldsymbol{s}_{t+1}) \sim \mathcal{B}} \left[\left(Q(\boldsymbol{s}_t, \boldsymbol{a}_t; \boldsymbol{\theta}_Q) - y_t \right)^2 \right],$$
$$y_t = r_t + \gamma Q(\boldsymbol{s}_{t+1}, \mu(\boldsymbol{s}_{t+1}; \boldsymbol{\theta}'_\mu); \boldsymbol{\theta}'_Q).$$

The actor network is updated by minimizing $L_\mu(\boldsymbol{\theta}_\mu) = -\mathbb{E}_{\boldsymbol{s}_t \sim \mathcal{B}} [Q(\boldsymbol{s}_t, \mu(\boldsymbol{s}_t; \boldsymbol{\theta}_\mu); \boldsymbol{\theta}_Q)]$, resulting in an approximation of the deterministic policy gradient [23]

$$\nabla_{\boldsymbol{\theta}_\mu} L_\mu \approx \mathbb{E}_{\boldsymbol{s}_t \sim \mathcal{B}} \left[\nabla_{\boldsymbol{\theta}_\mu} \mu(\mathbf{s}; \boldsymbol{\theta}_\mu)|_{\mathbf{s}=\boldsymbol{s}_t} \nabla_{\mathbf{a}} Q(\boldsymbol{s}, \boldsymbol{a}; \boldsymbol{\theta}_Q)|_{\mathbf{s}=\boldsymbol{s}_t, \mathbf{a}=\mu(\boldsymbol{s}_t; \boldsymbol{\theta}_\mu)} \right].$$

2.2 Stochastic Methods

The 'Soft Actor-Critic' (SAC) [14] method is an example of a stochastic, actor-critic algorithm. In this method, the actor network provides the parameters for the probability distribution used to model the stochastic policy $\pi(\boldsymbol{a}_t | \boldsymbol{s}_t; \boldsymbol{\theta}_\mu)$. Typically a Gaussian distribution is used, taking $\boldsymbol{a}_t \sim \mathcal{N}(\boldsymbol{\mu}_t, \mathrm{diag}[\boldsymbol{\sigma}_t])$ with $[\boldsymbol{\mu}_t; \boldsymbol{\sigma}_t] = \mu(\boldsymbol{s}_t; \boldsymbol{\theta}_\mu)$. To ensure a sufficient exploration of the state-action space, this method does not only optimize for long-term reward accumulation, but also tries to maximize the policy's entropy $\pi^* = \arg\max_\pi \mathbb{E}_{\pi, \mathcal{T}, \iota} [R_0 + \alpha H_0]$, with $H_t = \sum_{k=0}^{\infty} \gamma^k \mathcal{H}[\pi(\cdot | \mathbf{s}_{t+k})]$ the future discounted entropy. The critic network estimates the optimal soft Q-function, which takes this extra entropy term into account. The resulting critic loss can then be written down as the soft Bellman residual

$$L_Q(\boldsymbol{\theta}_Q) = \mathbb{E}_{(\boldsymbol{s}_t, \boldsymbol{a}_t, r_t, \boldsymbol{s}_{t+1}) \sim \mathcal{B}} \left[\left(Q(\boldsymbol{s}_t, \boldsymbol{a}_t; \boldsymbol{\theta}_Q) - y_t \right)^2 \right],$$
$$y_t = r_t + \gamma \mathbb{E}_{\boldsymbol{a}_{t+1} \sim \pi} \left[Q(\boldsymbol{s}_{t+1}, \boldsymbol{a}_{t+1}; \boldsymbol{\theta}'_Q) - \alpha \log[\pi(\boldsymbol{a}_{t+1} | \boldsymbol{s}_{t+1}; \boldsymbol{\theta}_\mu)] \right].$$

To improve the policy, the Kullback-Leibler divergence between the policy and the exponential of the soft Q-function is minimized, leading to the actor loss

$$L_\mu(\boldsymbol{\theta}_\mu) = \mathbb{E}_{\boldsymbol{s}_t \sim \mathcal{B}} \left\{ \mathbb{E}_{\boldsymbol{a}_t \sim \pi} \left[\alpha \log[\pi(\boldsymbol{a}_t | \boldsymbol{s}_t; \boldsymbol{\theta}_\mu)] - Q(\boldsymbol{s}_t, \boldsymbol{a}_t; \boldsymbol{\theta}_Q) \right] \right\}. \quad (2)$$

Finally, the temperature parameter α can be tuned either manually or automatically adapted throughout training, see Haarnoja et al. [14] for further details.

3 Proposed Methodology

Two modifications to the usual reinforcement learning (RL) control pipeline are necessary to enforce state-dependent bounds on policy actions, as shown in Fig. 2. First of all, the behavioural policy β and learned policy π are constrained such that their sampled *normalized actions* $\tilde{\boldsymbol{a}}$ always lie within a predetermined, *fixed* interval. In a second step, these normalized actions are rescaled using a chosen mapping σ and the *state-dependent* action bounds $a_{\mathrm{L}}(\boldsymbol{s})$ and $a_{\mathrm{U}}(\boldsymbol{s})$ — encoding the prior domain knowledge to be embedded. The resulting rescaled actions $\boldsymbol{a}$ are then executed in the environment, leading to a next state in the following timestep.

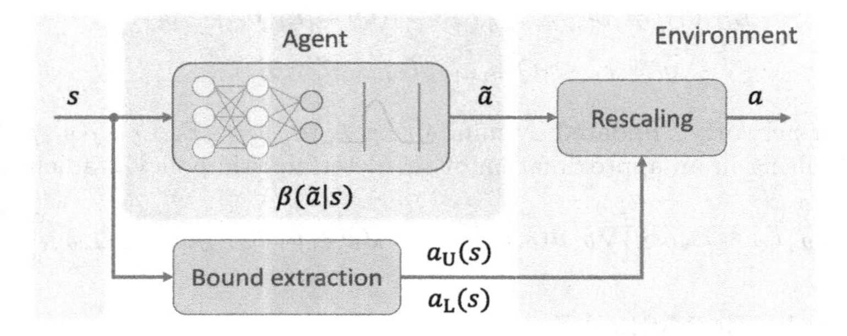

Fig. 2. Schematic overview of the reinforcement learning control pipeline with enforced state-dependent action bounds.

The chosen reinforcement learning algorithm is applied using the normalized actions $\tilde{\boldsymbol{a}}$ instead of the rescaled actions $\boldsymbol{a}$, i.e. $\tilde{\boldsymbol{a}}$ is stored in the replay buffer $\mathcal{B}$ and used for the calculation of the actor (and critic) losses. As a result, from the agent's perspective, the rescaling operation is part of the environment. In case the mapping σ is differentiable with respect to the normalized action inputs, it would be possible to work directly with the rescaled actions, making the rescaling operation part of the agent instead. To simplify the comparison of different (possibly non-differentiable) rescaling functions, we do not investigate such a setup in this paper and leave this for future work.

3.1 Normalized Actions

The specific implementation of policies with bounded normalized actions depends on the chosen RL method used for policy optimization. In essence, it consists of the following two steps. First, squash the outputs of the actor network representing the mean actions $\boldsymbol{\mu}$ to the fixed interval $[\tilde{a}_{\mathrm{L}}; \tilde{a}_{\mathrm{U}}]$ using a bounded activation function, such as tanh, in the last layer of the network. Secondly, to make sure no actions outside this fixed interval are sampled, use a bounded probability distribution, such as the truncated Gaussian distribution $\mathcal{N}_{\tilde{a}_{\mathrm{L}}}^{\tilde{a}_{\mathrm{U}}}(\boldsymbol{\mu}, \sigma I)$ with finite support $[\tilde{a}_{\mathrm{L}}; \tilde{a}_{\mathrm{U}}]$ [4].

Below, two implementations are presented for a state-of-the-art deterministic and stochastic actor-critic method respectively. Note that while we stick to the tanh activation function with bounds $\tilde{a}_{\mathrm{L}} = -1$, $\tilde{a}_{\mathrm{U}} = 1$ in this paper, any other bounded and differentiable activation function can be used instead.

TD3 (Deterministic). In deterministic actor-critic methods, such as TD3, the output of the actor network corresponds to the deterministic action to be taken for a given state $\boldsymbol{a}_t = \mu(\boldsymbol{s}_t; \boldsymbol{\theta}_\mu)$. To properly explore the state-action space during training, an external source of stochasticity is used, typically a spherical Gaussian with mean $\boldsymbol{a}_t$ and a decaying variance σ: $\boldsymbol{a}_t^\epsilon \sim \mathcal{N}(\mu(\boldsymbol{s}_t; \boldsymbol{\theta}_\mu), \sigma I)$.

Bounding the mean action outputs on the actor network, thus corresponds to using tanh activation in the last layer of the network, resulting in normalized action outputs $\tilde{\boldsymbol{a}}_t$. To ensure satisfaction of the bounds during exploration, the truncated Gaussian distribution is used.

$$\begin{aligned} \pi &: \tilde{\boldsymbol{a}}_t = \mu(\boldsymbol{s}_t; \boldsymbol{\theta}_\mu) \\ \beta &: \tilde{\boldsymbol{a}}_t^\epsilon \sim \mathcal{N}_{-1}^{1}(\mu(\boldsymbol{s}_t; \boldsymbol{\theta}_\mu), \sigma I) \end{aligned}$$

SAC (Stochastic). In stochastic actor-critic methods, such as SAC, the output of the actor network corresponds to the parameters of the distribution from which stochastic actions can be sampled. Using a Gaussian distribution for example, we have $[\boldsymbol{\mu}_t; \boldsymbol{\sigma}_t] = \mu(\boldsymbol{s}_t; \boldsymbol{\theta}_\mu)$ and $\boldsymbol{a}_t \sim \mathcal{N}(\boldsymbol{\mu}_t, \mathrm{diag}[\boldsymbol{\sigma}_t])$.

In this case it suffices to use a tanh activation in the last layer of the μ-head of the actor network, leading to normalized mean action outputs $\tilde{\boldsymbol{\mu}}_t$. Once again, the truncated Gaussian distribution is used to ensure satisfaction of the bounds for all sampled actions. During evaluation, the mode of the learned stochastic policy β is used to retrieve a deterministic policy π.

$$\begin{aligned} \pi &: \tilde{\boldsymbol{a}}_t = \tilde{\boldsymbol{\mu}}_t \\ \beta &: \tilde{\boldsymbol{a}}_t^\epsilon \sim \mathcal{N}_{-1}^{1}(\tilde{\boldsymbol{\mu}}_t, \mathrm{diag}[\boldsymbol{\sigma}_t]) \end{aligned} \qquad \begin{bmatrix} \tilde{\boldsymbol{\mu}}_t \\ \boldsymbol{\sigma}_t \end{bmatrix} = \mu(\boldsymbol{s}_t; \boldsymbol{\theta}_\mu)$$

The SAC method requires the calculation of differentiable log probabilities for the sampled actions (2). For the standard Gaussian distribution, such gradients can be calculated straightforwardly using the *reparametrization trick* [18]. Such a reparametrization trick can not be applied, however, when using the truncated Gaussian distribution. Fortunately, differentiable log probabilities can still be calculated using implicit reparametrization gradients [11].

Note that in Appendix C of the soft-actor critic paper [14] an alternative way of bounding the actor network's outputs is described. The "tanh trick" uses samples from the regular Gaussian distribution which are afterwards passed through the tanh function, squashing each sample to the bounded interval $[-1; 1]$. While this is also a viable approach, we work with the truncated Gaussian distribution in this paper for consistency with the deterministic methods.

3.2 State-Dependent Rescaling

The sampled normalized actions are afterwards rescaled to the interval $[a_{\rm L}(\boldsymbol{s}); a_{\rm U}(\boldsymbol{s})]$ determined by the predefined, state-dependent action bounds $a_{\rm L}$ and $a_{\rm U}$. Different rescaling functions $\sigma(\tilde{a}; \boldsymbol{s}) : [\tilde{a}_{\rm L}; \tilde{a}_{\rm U}] \times \mathcal{S} \rightarrow [a_{\rm L}(\boldsymbol{s}); a_{\rm U}(\boldsymbol{s})]$ can be used for that purpose. We specifically focus on rescaling functions that are monotonically increasing bijections, to guarantee the structure of the underlying Markov Decision Process (MDP) is not modified (see Subsect. 3.3). In this subsection we introduce three such mappings. For simplicity, we assume scalar (normalized) actions, but this can be easily generalized to vectors by applying the σ functions elementwise.

The most straightforward way to rescale a variable is using the linear function

$$\sigma_{\rm lin}(\tilde{a}; \boldsymbol{s}) = a_{\rm L}(\boldsymbol{s}) + \frac{a_{\rm U}(\boldsymbol{s}) - a_{\rm L}(\boldsymbol{s})}{\tilde{a}_{\rm U} - \tilde{a}_{\rm L}}(\tilde{a} - \tilde{a}_{\rm L}).$$

The benefit of this mapping is its generality, as it is able to handle any range of bounds $[a_{\rm L}(\boldsymbol{s}); a_{\rm U}(\boldsymbol{s})]$. A drawback is that there is no fixed *anchor point* within the bounds, more precisely there is no $\tilde{a}_0 \in [\tilde{a}_{\rm L}; \tilde{a}_{\rm U}]$ that is mapped to a fixed a_0. For some control problems such a fixed anchor point can be useful or provide extra (stability) guarantees.[1] The following piecewise linear rescaling function allows to specify such an anchor point

$$\sigma_{\rm pwl}(\tilde{a}; \boldsymbol{s}) = \begin{cases} a_0 + \frac{a_{\rm L}(\boldsymbol{s}) - a_0}{\tilde{a}_{\rm L} - \tilde{a}_0}(\tilde{a} - \tilde{a}_0) & \tilde{a} < \tilde{a}_0 \\ a_0 + \frac{a_{\rm U}(\boldsymbol{s}) - a_0}{\tilde{a}_{\rm U} - \tilde{a}_0}(\tilde{a} - \tilde{a}_0) & \tilde{a} \geq \tilde{a}_0 \end{cases}.$$

Remark that a_0 is fixed and hence $a_{\rm L}(\boldsymbol{s}) \leq a_0 \leq a_{\rm U}(\boldsymbol{s})$ should be satisfied for all $\boldsymbol{s}$, making this mapping slightly more restrictive than the previous one. A drawback of this rescaling function is that it is not differentiable at $\tilde{a} = \tilde{a}_0$. Hence, a smoother variant is given by the hyperbolic interpolation function

$$\begin{aligned} \sigma_{\rm hyp}(\tilde{a}; \boldsymbol{s}) &= a_0 + \frac{p(\boldsymbol{s})(\tilde{a} - \tilde{a}_0)}{q(\boldsymbol{s})(\tilde{a} - \tilde{a}_0) + r(\boldsymbol{s})}, \\ p(\boldsymbol{s}) &= (a_{\rm L}(\boldsymbol{s}) - a_0)(a_{\rm U}(\boldsymbol{s}) - a_0)(\tilde{a}_{\rm U} - \tilde{a}_{\rm L}), \\ q(\boldsymbol{s}) &= (a_{\rm L}(\boldsymbol{s}) - a_0)(\tilde{a}_{\rm U} - \tilde{a}_0) - (a_{\rm U}(\boldsymbol{s}) - a_0)(\tilde{a}_{\rm L} - \tilde{a}_0), \\ r(\boldsymbol{s}) &= (\tilde{a}_{\rm L} - \tilde{a}_0)(\tilde{a}_{\rm U} - \tilde{a}_0)(a_{\rm U}(\boldsymbol{s}) - a_{\rm L}(\boldsymbol{s})). \end{aligned}$$

Fig. 3 shows the different rescaling functions for $\tilde{a}_{\rm L} = -1$, $\tilde{a}_0 = 0$, $\tilde{a}_{\rm U} = 1$, $a_{\rm L} = -2$, $a_0 = 0$ and $a_{\rm U} = 5$. The inverse mapping $\sigma^{-1}(\cdot; \boldsymbol{s})$, from actions a to normalized actions $\tilde{a}$, can be easily calculated for any of the presented rescaling functions by swapping the normalized bounds and anchor $\tilde{a}_{\rm L}$, $\tilde{a}_0$, $\tilde{a}_{\rm U}$ with the rescaled bounds and anchor $a_{\rm L}$, a_0, $a_{\rm U}$. A proof for this property is provided in Appendix A.

[1] For example, if actions in a local neighbourhood of a_0 stabilize the system around an equilibrium point $\boldsymbol{s}^*$ it might be desirable to map $\tilde{a}_0 = \mathbb{E}[\pi(\cdot|\boldsymbol{s}^*)]$ to a_0.

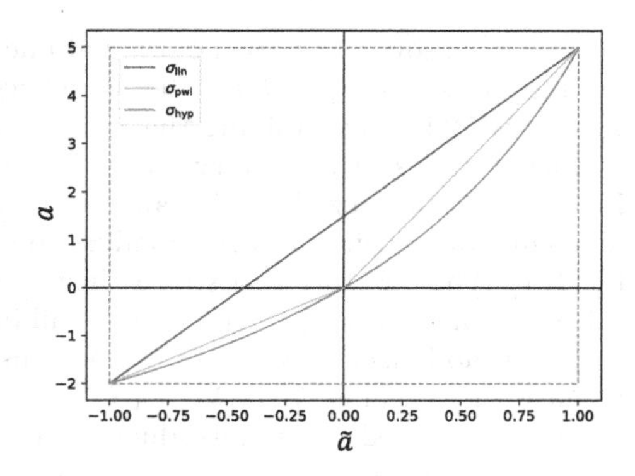

Fig. 3. Comparison of the different rescaling functions used throughout this paper. In this plot the bounds were set to $\tilde{a}_{\mathrm{L}} = -1$, $\tilde{a}_{\mathrm{O}} = 0$, $\tilde{a}_{\mathrm{U}} = 1$, $a_{\mathrm{L}} = -2$, $a_{\mathrm{O}} = 0$ and $a_{\mathrm{U}} = 5$.

3.3 Preservation of MDP

The environment dynamics and underlying MDP are fully preserved when using the state-dependent action bounds procedure outlined above. Indeed, as the rescaling operation is a bijection, every original action is mapped to a unique normalized action (and vice versa). By constraining the policy to only support normalized actions within a fixed interval $[\tilde{a}_{\mathrm{L}}; \tilde{a}_{\mathrm{U}}]$, actions outside the range $[a_{\mathrm{L}}(\boldsymbol{s}); a_{\mathrm{U}}(\boldsymbol{s})]$ can however no longer be selected, leading to a pruned MDP from the agent's perspective.

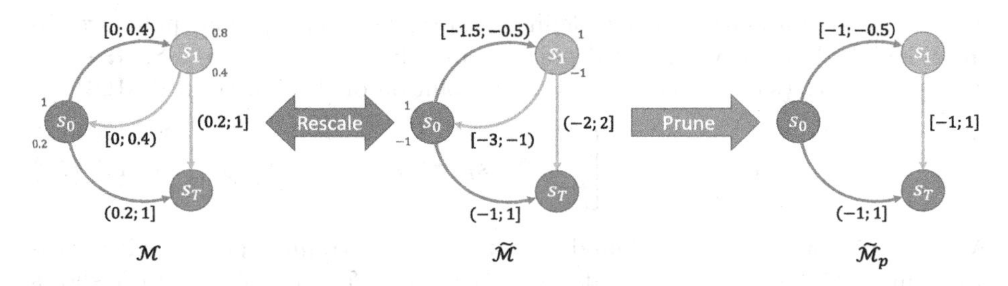

Fig. 4. Effect of the rescaling operation and imposed bounds on the actor network on the MDP of an example environment. The used rescaling function is σ_{lin} with bounds $a_{\mathrm{L}}(\boldsymbol{s}_0) = 0.2, a_{\mathrm{U}}(\boldsymbol{s}_0) = 1, a_{\mathrm{L}}(\boldsymbol{s}_1) = 0.4, a_{\mathrm{U}}(\boldsymbol{s}_1) = 0.8$.

To give some more intuition, let us consider the effect of state-dependent action bounds on a simple MDP as shown in Fig. 4. The original MDP $\mathcal{M}$ consists of three states $\boldsymbol{s}_0$, $\boldsymbol{s}_1$ and $\boldsymbol{s}_T$. In every non-terminal state, the agent can select a continuous action a in the range $[0; 1]$. Each arrow between two states denotes a transition potentially triggered by a certain range of actions. If an action can lead to multiple transitions, each such transition is equally likely.

Using the σ_{lin} rescaling function with bounds $a_{\mathrm{L}}(\boldsymbol{s}_0) = 0.2$, $a_{\mathrm{U}}(\boldsymbol{s}_0) = 1$, $a_{\mathrm{L}}(\boldsymbol{s}_1) = 0.4$, $a_{\mathrm{U}}(\boldsymbol{s}_1) = 0.8$, a rescaled MDP $\tilde{\mathcal{M}}$ can be constructed with normalized actions instead of the original actions. Because σ is a bijection, this normalized MDP is completely equivalent to the original one. Advantageously, the action bounds become state-*independent*, i.e. the same fixed bounds for each state, in this normalized MDP. By constraining the search for optimal policies to the set of policies with outputs in the interval $[-1;1]$, not every transition is still applicable. Hence, the normalized MDP can be further pruned to the final MDP $\tilde{\mathcal{M}}_p$ by removing all transitions corresponding to normalized actions outside the interval $[-1;1]$. This last pruning step is however not enforced on the environment itself but rather on the policies. In fact, all original transitions are still there, but some can no longer be selected by the constrained policies.

More formally we have the original MDP $\mathcal{M} = (\mathcal{S}, \mathcal{A}, \iota, \tau, r, \gamma)$, from which the rescaled MDP $\tilde{\mathcal{M}}$ is constructed — by introducing the rescaling function σ in the environment — as $\left(\mathcal{S}, \tilde{\mathcal{A}}, \iota, \tilde{\tau}, \tilde{r}, \gamma\right)$ with $\tilde{\mathcal{A}} = \{\tilde{\boldsymbol{a}} | \exists \boldsymbol{s} \in \mathcal{S}, \exists \boldsymbol{a} \in \mathcal{A} : \sigma(\tilde{\boldsymbol{a}}; \boldsymbol{s}) = \boldsymbol{a}\}$, $\tilde{\tau}(\boldsymbol{s}_{t+1}|\boldsymbol{s}_t, \tilde{\boldsymbol{a}}_t) : \mathcal{S} \times \mathcal{S} \times \tilde{\mathcal{A}} \to [0;1] = \tau(\boldsymbol{s}_{t+1}|\boldsymbol{s}_t, \sigma(\tilde{\boldsymbol{a}}_t; \boldsymbol{s}_t))$ and $\tilde{r}(\boldsymbol{s}_t, \tilde{\boldsymbol{a}}_t, \boldsymbol{s}_{t+1}) : \mathcal{S} \times \tilde{\mathcal{A}} \times \mathcal{S} \to \mathbb{R} = r(\boldsymbol{s}_t, \sigma(\tilde{\boldsymbol{a}}_t; \boldsymbol{s}_t), \boldsymbol{s}_{t+1})$. The usage of bounded probability distributions and bounded activation functions in the actor network constrains the set of normalized policies as $\tilde{\Pi}_C = \left\{\tilde{\pi} \forall \boldsymbol{s} \in \mathcal{S}, \forall \tilde{\boldsymbol{a}} \in \tilde{\mathcal{A}} : \tilde{\boldsymbol{a}}_{\text{L}} \leq \tilde{\boldsymbol{a}} \leq \tilde{\boldsymbol{a}}_{\text{U}} \vee \tilde{\pi}(\tilde{\boldsymbol{a}}|\boldsymbol{s}) = 0\right\}$. Hence, applying the chosen RL method to the experience tuples $(\boldsymbol{s}_t, \tilde{\boldsymbol{a}}_t, \tilde{r}_t, \boldsymbol{s}_{t+1})$ boils down to finding an optimal normalized policy as the solution of the constrained MDP

$$\tilde{\pi}^* = \underset{\tilde{\pi} \in \tilde{\Pi}_C}{\arg\max} \, \mathbb{E}_{\tilde{\pi}, \tilde{\tau}, \iota} \left[\sum_{k=0}^{\infty} \gamma^k \tilde{r}(\boldsymbol{s}_k, \tilde{\boldsymbol{a}}_k, \boldsymbol{s}_{k+1}) \right]. \qquad (\tilde{\mathcal{M}} + \tilde{\Pi}_C)$$

This optimal *normalized* policy defines a corresponding optimal policy π^* in the original MDP $\mathcal{M}$ with added policy constraints $\Pi_C = \{\pi | \forall \boldsymbol{s} \in \mathcal{S}, \forall \boldsymbol{a} \in \mathcal{A} : \boldsymbol{a}_{\text{L}}(\boldsymbol{s}) \leq \boldsymbol{a} \leq \boldsymbol{a}_{\text{U}}(\boldsymbol{s}) \vee \pi(\boldsymbol{a}|\boldsymbol{s}) = 0\}$, i.e. the solution of the constrained MDP

$$\pi^* = \underset{\pi \in \Pi_C}{\arg\max} \, \mathbb{E}_{\pi, \tau, \iota} \left[\sum_{k=0}^{\infty} \gamma^k r(\boldsymbol{s}_k, \boldsymbol{a}_k, \boldsymbol{s}_{k+1}) \right]. \qquad (\mathcal{M} + \Pi_C)$$

As a result, we effectively found an optimal constrained policy, with variable support (based on the state-dependent action bounds). Figure 5 shows such transformed policies, after application of different rescaling functions. The transformed policies can be easily calculated from their normalized analogues using the 'change of variable' formula from probability theory (requiring the inverse rescaling function σ^{-1} to be differentiable) $\pi(a|\boldsymbol{s}) = \tilde{\pi}(\sigma^{-1}(a; \boldsymbol{s})|\boldsymbol{s}) \left|\partial_a \sigma^{-1}(a; \boldsymbol{s})\right|$. Samples from this rescaled policy can be readily calculated using the procedure outlined in the previous two subsections, i.e. by transforming samples from the normalized policy using the chosen rescaling function σ.

As a final remark, let us briefly consider the case of a differentiable rescaling function, which allows to directly store the rescaled actions **a** in the replay buffer rather than the normalized actions. In such a situation, both the bounding and rescaling operations can be enforced on the policy (agent), leaving the environment untouched, effectively solving $(\mathcal{M} + \Pi_C)$ directly.

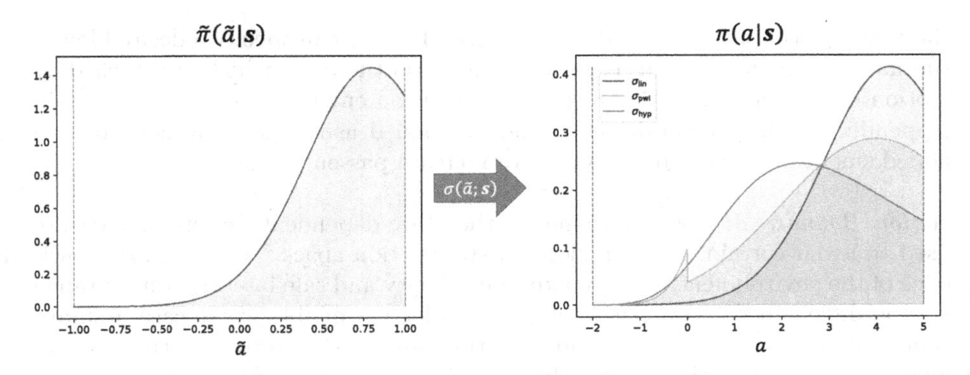

Fig. 5. Effect of the rescaling operation on the probability density function. On the left the normalized policy $\tilde{\pi}$ and on the right the rescaled policy π for different rescaling functions. The used bounds are $a_{\mathrm{L}}(\boldsymbol{s}) = -2, a_{\mathrm{U}}(\boldsymbol{s}) = 5$ with anchor point $a_0 = 0$ for $\tilde{a}_0 = 0$.

3.4 Implementation

A reference implementation of our approach in pytorch is available here[2]. In this implementation we specifically use the SAC method for learning, while the tanh function and truncated Gaussian distribution are used to constrain the policies. All of the discussed rescaling functions are available, allowing to easily reproduce the experiments of Appendix B.

4 Experimental Results

The effectiveness of the state-dependent action bounds is demonstrated in a virtual autonomous driving environment. A patent application for this specific application of state-dependent action bounds has been filed [10]. Additional experiments on two commonly used openAI gym environments can be found in Appendix B.

Every experiment is repeated ten times, using ten different seeds for initialization. Throughout training, E independent evaluation episodes are executed to get an estimate of the learned policy's average performance ($E = 5$ for the autonomous driving environment and $E = 10$ for the gym environments). These $10E$ datapoints are used to construct the plots (using exponential average smoothing with $\alpha = 0.1$) and report the performances.

4.1 Highway Driving

The considered simulated environment is a three lane highway in which a virtual driver (the agent) has to efficiently navigate traffic by issuing high-level steering commands based on the current traffic situation in its neighbourhood. Through

[2] https://github.com/dcbr/sdab.

the continuous actions v_{ref} and d_{ref} the virtual driver can specify a desired longitudinal velocity and lateral reference position on the road, which are tracked by motion controllers. More details of the simulation environment can be found in Appendix C. The virtual driver's policy is trained using the TD3 method with added smoothness regularization [9], to increase passenger comfort.

Action Bounds. In this environment, the state-dependent action bounds are used to avoid certain unsafe regions in state-action space.[3] Using prior knowledge of the environment, a safe maximum velocity and safe lateral position range can be determined, preventing collisions under reasonable worst case assumptions. Enforcing such state-dependent action bounds thus provides strong safety guarantees, allowing the agent to focus on its core task of efficiently navigating traffic.

Results. Two constrained policies using the state-dependent action bounds with rescaling functions σ_{lin} and σ_{pwl} are trained.[4] Both constrained policies are compared with an unconstrained policy, which has to learn all aspects of driving on its own, using only the negative penalties in the reward signal obtained when the virtual driver chooses unreasonable or dangerous actions. Figure 6 summarizes the results of this comparison.

The plots show the average obtained reward and average episode length of each evaluation episode. By default, an episode lasts for 5000 timesteps, but the episode ends immediately when the virtual driver crashes. From these results, we can conclude both constrained policies are never involved in a crash throughout the whole training process, even at the very start. On the other hand, without guiding action bounds, the unconstrained policies cause many crashes early on and throughout the training process. Moreover, these crashes never truly vanish, even after extensive training. As a result, the best unconstrained policies never achieve the same level of performance as the constrained policies. Additionally, there is a significant difference in convergence speed. The constrained policies can focus on the 'core task' of efficiently navigating traffic, leading to a fast convergence after roughly $2 \cdot 10^5$ training steps. The unconstrained policies, on the other hand, have to simultaneously learn common traffic rules and how to avoid crashes, leading to a much slower convergence after roughly $8 \cdot 10^5$ training steps.

These results clearly highlight the benefits of state-dependent action bounds. First of all, it allows to effectively enforce *hard* constraints (bounds) on sampled policy actions, which are always satisfied, both during training and evaluation. Secondly, as the relevant domain knowledge can be enforced directly on the policies, it can keep reward functions simple, focusing on the core learning tasks. This not only considerably speeds up training, by only focusing on the relevant parts of state-action space, but can also result in better performing policies, as compared to unconstrained learning.

[3] Similar to the avoidance objective for the pendulum environment in Appendix B.
[4] These are the best performing rescaling functions in the experiments of Appendix B.

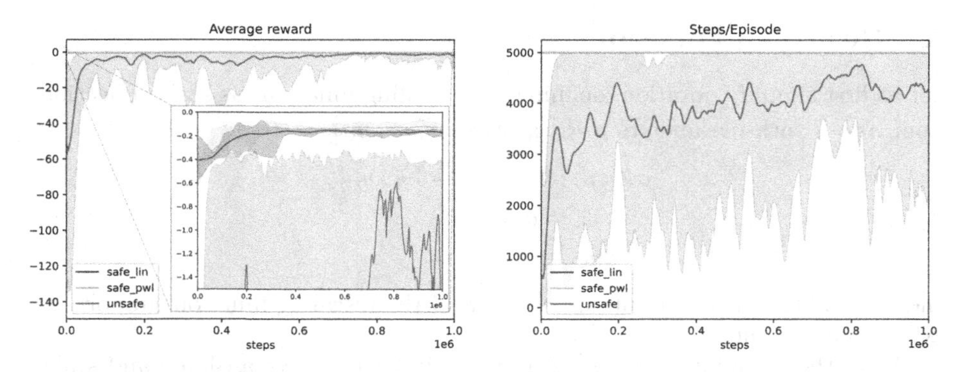

Fig. 6. Average obtained reward (left) and episode length (right) during evaluation as a function of training steps for two constrained policies with state-dependent safety bounds (blue and orange) and one unconstrained policy (green). (Color figure online)

5 Conclusion

In this paper we proposed a novel way to enforce certain hard constraints on policies and prevent an agent from exploring and operating in irrelevant parts of the state-action space. More specifically, state-dependent action bounds allow the designer to restrict the range of available actions in certain states to a predefined (safe) interval. It is thus a tool to incorporate prior domain knowledge for a specific set of states, while keeping the model-free learning scheme's benefits. The proposed two-step procedure is both flexible and generic, making it easily combinable with existing policy-based and actor-critic reinforcement learning algorithms, as well as applicable to various domains. Experiments showed the effectiveness of the bounds, resulting in both faster learning and better performing policies.

A limitation of the current approach is the usage of a single interval of allowed actions for each state. In some situations there might however be multiple permissible action intervals. For example, an obstacle in front can be avoided by going either left or right. An extension to support multiple, disjoint action intervals might therefore be an interesting avenue for future research. Further extensions to the type of constraints that can be enforced is another potential path forward, for instance by allowing to bound the derivatives of policy actions.

Acknowledgements. The presented results were primarily obtained under Ford Alliance Project KUL0076, funded by Ford. This research was partially supported by TAILOR, a project funded by EU Horizon 2020 research and innovation programme under GA No 952215. This research received funding from the Flemish Government (AI Research Program). Johan Suykens is affiliated to Leuven. AI - KU Leuven institute for AI, B-3000, Leuven, Belgium. The resources and services used in this work were provided by the VSC (Flemish Supercomputer Center), funded by the Research Foundation - Flanders (FWO) and the Flemish Government.

A Rescaling Functions

Let us first simplify notation, defining the rescaling functions as scalar bijections from x to y with parameters $\rho = \left[x_{\mathrm{L}}\ x_0\ x_{\mathrm{U}}\ y_{\mathrm{L}}\ y_0\ y_{\mathrm{U}}\right]^\top$,

$$\begin{aligned}\sigma(x;\rho) &: [x_{\mathrm{L}}; x_{\mathrm{U}}] \to [y_{\mathrm{L}}; y_{\mathrm{U}}],\\ y_{\mathrm{L}} &= \sigma(x_{\mathrm{U}};\rho),\\ y_{\mathrm{U}} &= \sigma(y_{\mathrm{U}};\rho),\end{aligned}$$

and the additional constraint $y_0 = \sigma(x_0;\rho)$ if the rescaling function supports an extra anchor point.

All of the rescaling functions introduced in Sect. 3 are translated and scaled *involutions*. An involution is a bijection whose inverse is equal to the function itself, i.e. $f(x) = f^{-1}(x)$ for all x in a certain set.

Lemma 1. *Denoting by* $m(x;\tau_{\mathrm{y}},\sigma_{\mathrm{y}},\tau_{\mathrm{x}},\sigma_{\mathrm{x}}) = \tau_{\mathrm{y}} + \sigma_{\mathrm{y}} f(\sigma_{\mathrm{x}} x + \tau_{\mathrm{x}})$ *the translated and scaled variant of an involution* f*, we can write its inverse as*

$$m^{-1}(y;\tau_{\mathrm{y}},\sigma_{\mathrm{y}},\tau_{\mathrm{x}},\sigma_{\mathrm{x}}) = m\left(y;\frac{-\tau_{\mathrm{x}}}{\sigma_{\mathrm{x}}},\frac{1}{\sigma_{\mathrm{x}}},\frac{-\tau_{\mathrm{y}}}{\sigma_{\mathrm{y}}},\frac{1}{\sigma_{\mathrm{y}}}\right).$$

Proof. The inverse of m is given by

$$\begin{aligned}m^{-1}(y;\tau_{\mathrm{y}},\sigma_{\mathrm{y}},\tau_{\mathrm{x}},\sigma_{\mathrm{x}}) &= \frac{1}{\sigma_{\mathrm{x}}} f^{-1}\left(\frac{y-\tau_{\mathrm{y}}}{\sigma_{\mathrm{y}}}\right) - \frac{\tau_{\mathrm{x}}}{\sigma_{\mathrm{x}}}\\ &= \frac{-\tau_{\mathrm{x}}}{\sigma_{\mathrm{x}}} + \frac{1}{\sigma_{\mathrm{x}}} f\left(\frac{y}{\sigma_{\mathrm{y}}} + \frac{-\tau_{\mathrm{y}}}{\sigma_{\mathrm{y}}}\right)\\ &= m\left(y;\frac{-\tau_{\mathrm{x}}}{\sigma_{\mathrm{x}}},\frac{1}{\sigma_{\mathrm{x}}},\frac{-\tau_{\mathrm{y}}}{\sigma_{\mathrm{y}}},\frac{1}{\sigma_{\mathrm{y}}}\right),\end{aligned}$$

where we used the property $f(x) = f^{-1}(x)$ of involutions.

Denoting by $P_{x\leftrightarrow y}$ the permutation matrix swapping x and y entries in ρ, i.e. $\tilde{\rho} = P_{x\leftrightarrow y}\rho = \left[y_{\mathrm{L}}\ y_0\ y_{\mathrm{U}}\ x_{\mathrm{L}}\ x_0\ x_{\mathrm{U}}\right]^\top$, we can prove the following Lemma.

Lemma 2. *Each of the considered rescaling functions* σ_{lin}*,* σ_{pwl} *and* σ_{hyp} *satisfy following relationship* $\sigma^{-1}(y;\rho) = \sigma(y;P_{x\leftrightarrow y}\rho)$*, allowing to easily calculate the inverse rescaling by swapping* x *and* y *entries in the parameter vector.*

Proof. The above relationship $\sigma^{-1}(y;\rho) = \sigma(y;\tilde{\rho})$ can be proven for each rescaling function by rewriting it as a scaled and translated involution

$$\sigma(x;\rho) = \tau_{\mathrm{y}}(\rho) + \sigma_{\mathrm{y}}(\rho) f(\sigma_{\mathrm{x}}(\rho)x + \tau_{\mathrm{x}}(\rho)),$$

satisfying

$$\tau_{\mathrm{y}}(\tilde{\rho}) = -\frac{\tau_{\mathrm{x}}(\rho)}{\sigma_{\mathrm{x}}(\rho)},\quad \sigma_{\mathrm{y}}(\tilde{\rho}) = \frac{1}{\sigma_{\mathrm{x}}(\rho)},\quad \tau_{\mathrm{x}}(\tilde{\rho}) = -\frac{\tau_{\mathrm{y}}(\rho)}{\sigma_{\mathrm{y}}(\rho)},\quad \sigma_{\mathrm{x}}(\tilde{\rho}) = \frac{1}{\sigma_{\mathrm{y}}(\rho)},\tag{3}$$

and applying Lemma 1.

Both linear rescaling functions $\sigma_{\text{lin}}, \sigma_{\text{pwl}}$ can be rewritten as scaled and translated variants of the involution $f(x) = x$ using following parameters

$$\begin{aligned}
\sigma_{\text{lin}}: \quad & \tau_{\text{y}}(\rho) = y_{\text{L}}, \quad \sigma_{\text{y}}(\rho) = y_{\text{U}} - y_{\text{L}}, \quad \tau_{\text{x}}(\rho) = \frac{-x_{\text{L}}}{x_{\text{U}} - x_{\text{L}}}, \quad \sigma_{\text{x}}(\rho) = \frac{1}{x_{\text{U}} - x_{\text{L}}} \\
\sigma_{\text{pwl,L}}: \quad & \tau_{\text{y}}(\rho) = y_0, \quad \sigma_{\text{y}}(\rho) = y_{\text{L}} - y_0, \quad \tau_{\text{x}}(\rho) = \frac{-x_0}{x_{\text{L}} - x_0}, \quad \sigma_{\text{x}}(\rho) = \frac{1}{x_{\text{L}} - x_0} \\
\sigma_{\text{pwl,U}}: \quad & \tau_{\text{y}}(\rho) = y_0, \quad \sigma_{\text{y}}(\rho) = y_{\text{U}} - y_0, \quad \tau_{\text{x}}(\rho) = \frac{-x_0}{x_{\text{U}} - x_0}, \quad \sigma_{\text{x}}(\rho) = \frac{1}{x_{\text{U}} - x_0}
\end{aligned}$$

for which the conditions (3) can be readily checked. Note that the inverse of the piecewise function σ_{pwl} is given by the inverse of its constituents in each of their non-overlapping domains (both σ_{pwl} and its constituents are bijections).

Similarly, the hyperbolic rescaling function σ_{hyp} can be rewritten as a transformed version of the involution $f(x) = \frac{1}{x}$ using following scales and translations

$$\begin{aligned}
\tau_{\text{y}}(\rho) &= \frac{(y_{\text{U}} - y_0)(x_{\text{L}} - x_0)y_{\text{L}} - (y_{\text{L}} - y_0)(x_{\text{U}} - x_0)y_{\text{U}}}{(y_{\text{U}} - y_0)(x_{\text{L}} - x_0) - (y_{\text{U}} - y_0)(x_{\text{U}} - x_0)}, \\
\sigma_{\text{y}}(\rho) &= \frac{(y_{\text{U}} - y_0)(y_{\text{U}} - y_0)(y_{\text{U}} - y_{\text{L}})}{(y_{\text{U}} - y_0)(x_{\text{L}} - x_0) - (y_{\text{L}} - y_0)(x_{\text{U}} - x_0)}, \\
\tau_{\text{x}}(\rho) &= \frac{(y_{\text{U}} - y_0)(x_{\text{L}} - x_0)x_{\text{U}} - (y_{\text{L}} - y_0)(x_{\text{U}} - x_0)x_{\text{L}}}{(x_{\text{U}} - x_0)(x_{\text{L}} - x_0)(x_{\text{U}} - x_{\text{L}})}, \\
\sigma_{\text{x}}(\rho) &= -\frac{(y_{\text{U}} - y_0)(x_{\text{L}} - x_0) - (y_{\text{L}} - y_0)(x_{\text{U}} - x_0)}{(x_{\text{U}} - x_0)(x_{\text{L}} - x_0)(x_{\text{U}} - x_{\text{L}})},
\end{aligned}$$

for which the conditions (3) can be verified.

B State-Dependent Bounds in Gym Environments

To provide some intuition, we compare different implementations of the state-dependent action bounds on two commonly used openAI gym environments[5]: `Pendulum-v0` and `LunarLanderContinous-v2`. For these experiments, we used a customized version of the latest Stable-Baselines implementation[6].

B.1 Inverted Pendulum

The objective in this `Pendulum-v0` environment is to swing up and stabilize an inverted pendulum around the upward position. The observable state by the agent $\boldsymbol{s} = \begin{bmatrix}\cos(\theta) \; \sin(\theta) \; \dot{\theta}\end{bmatrix}^\top$ contains information about the pendulum's orientation angle θ and angular velocity $\dot{\theta}$. The scalar action a denotes the torque to apply on the pendulum and is bounded by the interval $[-2; 2]$ (larger or smaller values are clipped by the environment). We consider two different applications of state-dependent action bounds on this environment.

[5] https://gym.openai.com.
[6] https://github.com/DLR-RM/stable-baselines3.

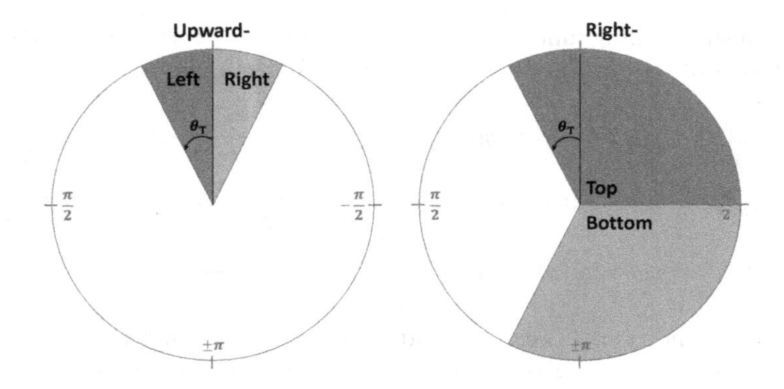

Fig. 7. Visualization of the state space (orientation angle component) partitioning imposed by the Upward and Right predicates in the `Pendulum-v0` environment.

Aiding Stabilization. The first set of state-dependent action bounds aims to aid in the stabilization of the pendulum around the upward position. As prior domain knowledge, we know that at the upward equilibrium point, the velocity of the pendulum should be low and small deviations to the left or right should be corrected by a negative or positive torque respectively. This extra domain knowledge is encoded in the chosen state-dependent bounds of this setup. Using the resulting bounds, the agent still has to figure out how to swing up the pendulum, but once the pendulum is close to the upward position, tighter action bounds aid in the stabilization by slowing down the pendulum and preventing it from falling over.

$$a_{\mathrm{L}}(\boldsymbol{s}) = \begin{cases} 2-\epsilon & \text{if } \mathrm{Upward}(\boldsymbol{s}) \wedge \mathrm{FastNeg}(\boldsymbol{s}) & (S.1) \\ -(2-\epsilon)\frac{\sin(\theta)}{\theta_{\mathrm{T}}} & \text{if } \mathrm{UpwardRight}(\boldsymbol{s}) \wedge \mathrm{Slow}(\boldsymbol{s}) & (S.2) \\ -2 & \text{else} \end{cases}$$

$$a_{\mathrm{U}}(\boldsymbol{s}) = \begin{cases} -2+\epsilon & \text{if } \mathrm{Upward}(\boldsymbol{s}) \wedge \mathrm{FastPos}(\boldsymbol{s}) & (S.1) \\ (-2+\epsilon)\frac{\sin(\theta)}{\theta_{\mathrm{T}}} & \text{if } \mathrm{UpwardLeft}(\boldsymbol{s}) \wedge \mathrm{Slow}(\boldsymbol{s}) & (S.2) \\ 2 & \text{else} \end{cases}$$

with predicates

$$\begin{aligned} \mathrm{UpwardLeft}(\boldsymbol{s}) &= \cos(\theta) > 0 \wedge 0 \leq \sin(\theta) < \theta_{\mathrm{T}}, \\ \mathrm{UpwardRight}(\boldsymbol{s}) &= \cos(\theta) > 0 \wedge -\theta_{\mathrm{T}} < \sin(\theta) \leq 0, \\ \mathrm{Upward}(\boldsymbol{s}) &= \mathrm{UpwardLeft}(\boldsymbol{s}) \vee \mathrm{UpwardRight}(\boldsymbol{s}), \\ \mathrm{FastNeg}(\boldsymbol{s}) &= \dot{\theta} < -\dot{\theta}_{\mathrm{T}}, \\ \mathrm{FastPos}(\boldsymbol{s}) &= \dot{\theta} > \dot{\theta}_{\mathrm{T}}, \\ \mathrm{Slow}(\boldsymbol{s}) &= \neg\mathrm{FastNeg}(\boldsymbol{s}) \wedge \neg\mathrm{FastPos}(\boldsymbol{s}). \end{aligned}$$

The *slowdown* conditions and bounds *(S.1)* ensure the pendulum is slowed down when it is approaching the upward position, while the *stabilization* conditions and bounds *(S.2)* attempt to prevent the pendulum from falling over when it is close to the upward equilibrium position. See the left side of Fig. 7 for a visualization of the state space partitioning originating from the Upward predicates. In the conducted experiments we chose $\epsilon = 0.1$, $\theta_{\mathrm{T}} = 0.3$ and $\dot{\theta}_{\mathrm{T}} = 0.3$.

Avoiding One Side. For this environment we also analyze a second set of state-dependent action bounds to show their usage in avoiding certain (e.g. unsafe) regions in state space. More specifically, in this setup the chosen bounds push the pendulum away from the right side[7], leaving only the left side available for swinging up the pendulum.

$$a_{\mathrm{L}}(\boldsymbol{s}) = \max \begin{cases} (2-\epsilon)\left(2\frac{\min\{-\dot{\theta},\dot{\theta}_{\mathrm{M}}\}}{\dot{\theta}_{\mathrm{M}}} - 1\right) & \text{if } \mathrm{RightTop}(\boldsymbol{s}) \wedge \mathrm{FastNeg}(\boldsymbol{s}) \quad (A.1) \\ (2-\epsilon)\frac{\max\{\sin(\theta),-\theta_{\mathrm{T}}\}}{\theta_{\mathrm{T}}} & \text{if } \mathrm{RightTop}(\boldsymbol{s}) \quad (A.2) \\ -2 & \text{else} \end{cases}$$

$$a_{\mathrm{U}}(\boldsymbol{s}) = \min \begin{cases} (-2+\epsilon)\left(2\frac{\min\{\dot{\theta},\dot{\theta}_{\mathrm{M}}\}}{\dot{\theta}_{\mathrm{M}}} - 1\right) & \text{if } \mathrm{RightBottom}(\boldsymbol{s}) \wedge \mathrm{FastPos}(\boldsymbol{s}) \quad (A.1) \\ (-2+\epsilon)\frac{\max\{\sin(\theta),-\theta_{\mathrm{T}}\}}{\theta_{\mathrm{T}}} & \text{if } \mathrm{RightBottom}(\boldsymbol{s}) \quad (A.2) \\ 2 & \text{else} \end{cases}$$

with extra predicates

$$\begin{aligned} \mathrm{RightTop}(\boldsymbol{s}) &= \sin(\theta) < \theta_{\mathrm{T}} \wedge \cos(\theta) > 0, \\ \mathrm{RightBottom}(\boldsymbol{s}) &= \sin(\theta) < \theta_{\mathrm{T}} \wedge \cos(\theta) \leq 0, \\ \mathrm{Right}(\boldsymbol{s}) &= \mathrm{RightTop}(\boldsymbol{s}) \vee \mathrm{RightBottom}(\boldsymbol{s}), \end{aligned}$$

where we used the special notation min{ and max{ to denote taking the minimum/maximum of any (possibly multiple) active cases. The *slowdown* conditions and bounds *(A.1)* ensure the pendulum is slowed down when it is approaching the right side with high velocity, while the *avoidance* conditions and bounds *(A.2)* force the pendulum away from the right side. See the right side of Fig. 7 for a visualization of the state space partitioning caused by the Right predicates. In the conducted experiments we chose $\epsilon = 0.1$, $\theta_{\mathrm{T}} = 0.3$, $\dot{\theta}_{\mathrm{T}} = 0$ and $\dot{\theta}_{\mathrm{M}} = 6$.

Results. Figure 8 shows the average evaluation return under both setups. Different SAC policies with state-dependent action bounds (using different rescaling functions) are compared against an unconstrained SAC policy. In this relatively simple environment there are no significant differences between the different rescaling functions, as all policies solve the environment in less than 50000 training steps.

[7] Leaving only some slack for the swing up movements, as otherwise, with too strict bounds, the environment is unsolvable.

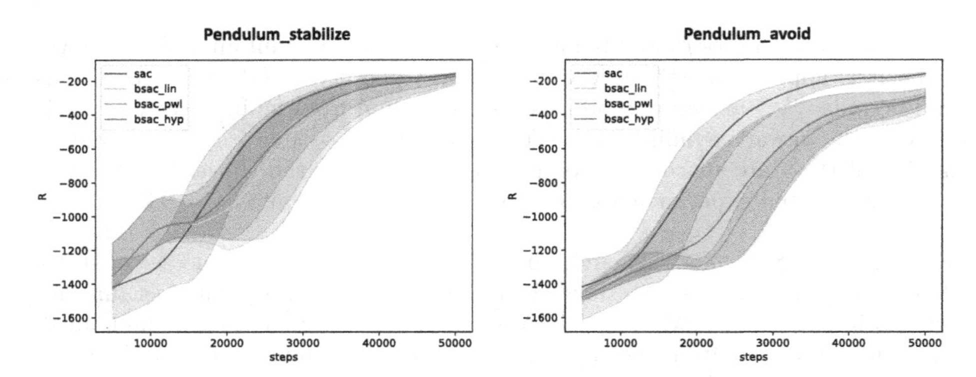

Fig. 8. Average evaluation return on the `Pendulum-v0` environment using the stabilization (left) and avoidance (right) state-dependent bounds for various rescaling functions.

While the stabilizing bounds provide an initial benefit to the policies (giving higher rewards in the initial training phase), this does not lead to faster training convergence (left plot). The constrained policies require extra training time to figure out the effect of the state-dependent action bounds on the chosen normalized actions. As the unconstrained SAC policy converges already very fast in this environment, there is little benefit in aiding the stabilization process through state-dependent action bounds.

On the right plot of Fig. 8 we can also notice the slightly lower performance of the constrained policies after convergence using the avoidance setup. Such a drop is to be expected as we are effectively solving a harder problem. For certain initial states, it would be faster to swing up along the right side, but this is made impossible by the action bounds. Hence, the constrained policies take a longer time to reach the upward position for certain initial states, which results in a slightly lower return. The most important question under this setup is however if the bounds actually succeeded in avoiding the right side as much as possible. Figure 9 shows that this is indeed the case. The unconstrained policies roughly utilize the left and right side equally, as the left state histogram is almost symmetric about the vertical radial axis. On the other hand, the right state histogram shows that the constrained policies only use the right side for swing up (high velocities in the lower right quadrant) and to move away from initial states[8] (low velocities in the upper right quadrant).

B.2 Lunar Lander

The agent's objective in the `LunarLanderContinuous-v2` environment is to safely land a lunar module within a marked landing zone on the moon's surface in a simplified 2D setting. The observable state of the agent $\mathbf{s}$ consists, among other signals, of the lander's downward velocity v, its orientation angle θ and

[8] Initial states are not enforced to lie on the left side.

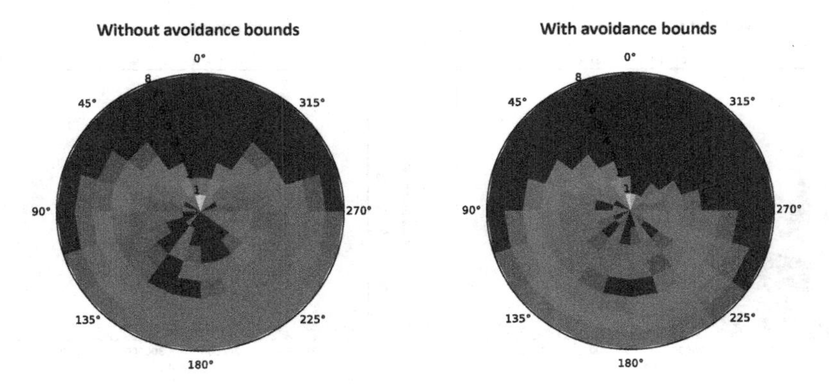

Fig. 9. Polar histogram of the pendulum's state space using states visited by all evaluation episodes at the end of training. The angular axis represents the pendulum's orientation angle θ, while the radial axis represents the pendulum's angular velocity $\dot{\theta}$. The left plot shows the results for an unconstrained SAC policy, while the right plot shows the results for a constrained SAC policy, using σ_{pwl} as the rescaling function.

angular velocity $\dot{\theta}$. The action vector $\mathbf{a}$ consists of two components bounded to the interval $[-1; 1]$ (imposed by the environment). The first component handles the main engine (at the module bottom), while the second component controls the two side engines (at the left and right). The relative power $P \in [0; 1]$ of each engine is determined as follows

$$P_{\text{main}} = \begin{cases} 0 & -1 \leq a_1 \leq 0 \\ \frac{1+a_1}{2} & 0 < a_1 \leq 1 \end{cases},$$

$$P_{\text{left}} = \begin{cases} 0 & -1 \leq a_2 \leq 0.5 \\ a_2 & 0.5 < a_2 \leq 1 \end{cases}, \qquad P_{\text{right}} = \begin{cases} -a_2 & -1 \leq a_2 < -0.5 \\ 0 & -0.5 \leq a_2 \leq 1 \end{cases}.$$

Stabilizing Bounds. The left plot in Fig. 10 shows the fraction of evaluation episodes ending either in a crash or astray (too far from the landing zone) for an unconstrained SAC policy. As can be seen, at the start of training, most evaluation episodes end without a correct landing on the lunar module's legs. To help the agent in correctly landing on its legs, stabilizing state-dependent bounds are enforced on its actions. Similar to the pendulum environment these bounds ensure the lunar module does not tilt too much to the left or right by activating the appropriate side engines. Additionally, the bounds ensure the main engine is activated when the downward velocity exceeds a certain threshold value.

The bounds on the first action component ensure the main engine is turned on whenever the downward velocity is too large, thereby slowing down the descent.

$$a_{\text{L},1}(\boldsymbol{s}) = \begin{cases} \epsilon & \text{if } v > v_{\text{T}} \\ -1 & \text{else} \end{cases} \qquad\qquad a_{\text{U},1}(\boldsymbol{s}) = 1$$

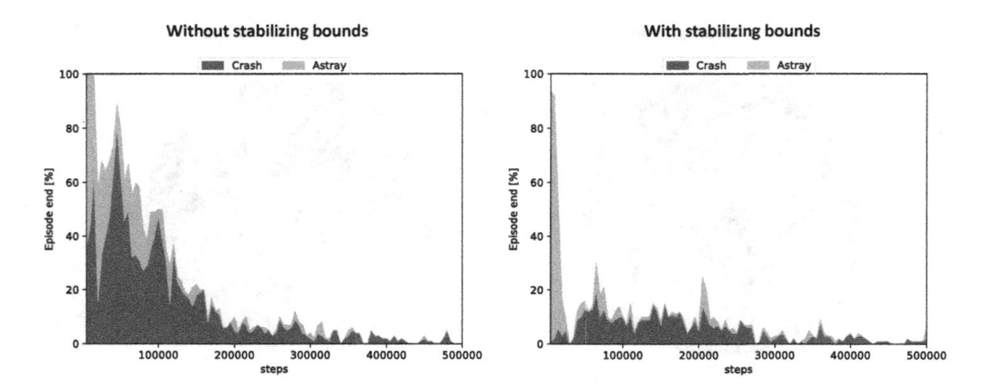

Fig. 10. Fraction of evaluation episodes ending in a crash (excessive downward velocity on touchdown or not landed on legs) or astray (too far away from the landing zone) for an unconstrained SAC policy (left) and a SAC policy with stabilizing action bounds, using σ_{lin} as the rescaling function (right).

The bounds on the second action component ensure the side engines are turned on when the lander tilts too much to the left or right without adjusting for it, thereby stabilizing the landing.

$$a_{\text{L},2}(\boldsymbol{s}) = \begin{cases} 0.5 + \epsilon & \text{if TiltLeft}(\boldsymbol{s}) \wedge \neg\text{SpinRight}(\boldsymbol{s}) \\ -1 & \text{else} \end{cases}$$

$$a_{\text{U},2}(\boldsymbol{s}) = \begin{cases} -0.5 - \epsilon & \text{if TiltRight}(\boldsymbol{s}) \wedge \neg\text{SpinLeft}(\boldsymbol{s}) \\ -1 & \text{else} \end{cases}$$

using predicates

$$\text{TiltLeft}(\boldsymbol{s}) = \theta > \theta_{\text{T}}, \qquad \text{TiltRight}(\boldsymbol{s}) = \theta < -\theta_{\text{T}},$$
$$\text{SpinLeft}(\boldsymbol{s}) = \dot{\theta} > \dot{\theta}_{\text{T}}, \qquad \text{SpinRight}(\boldsymbol{s}) = \dot{\theta} < -\dot{\theta}_{\text{T}}.$$

In the conducted experiments we used $\epsilon = 0.01$, $v_{\text{T}} = 0.2$, $\theta_{\text{T}} = 0.2$ and $\dot{\theta}_{\text{T}} = 0.12$. With these bounds in place, the resulting constrained policies have significantly fewer evaluation episodes ending in a crash or astray (right plot in Fig. 10). The remaining job for the agent is now to steer the lunar module towards the landing zone and prevent any remaining crashes[9].

Results. Fig. 11 compares the average evaluation returns of different constrained SAC policies with an unconstrained SAC policy. Similar to the results on the pendulum environment, the constrained policies have a head start. The imposed state-dependent bounds prevent most of the crashes, giving higher returns early

[9] The chosen bounds do not guarantee a crash-free experience, but could be further tightened if needed.

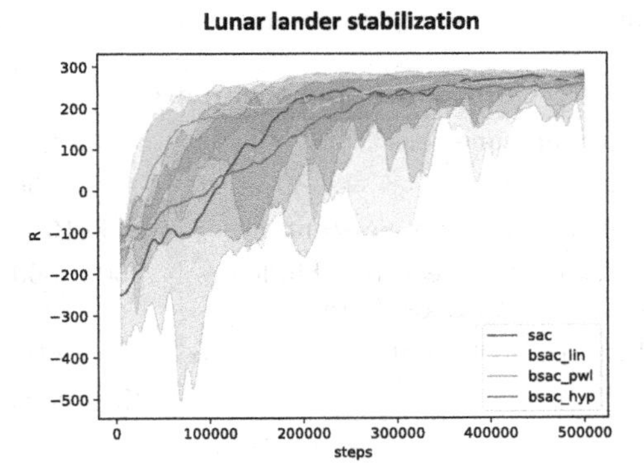

Fig. 11. Average evaluation return on the `LunarLanderContinuous-v2` environment using the stabilizing state-dependent bounds for various rescaling functions.

on in the training process. Contrary to the pendulum results, the extra stabilization benefit of most constrained policies *does* lead to a faster overall convergence in this case. The state-dependent bounds allow the agent to focus on the most relevant parts of the state-action space during training and exploration, instead of loosing many timesteps on the irrelevant parts that lead to crashes. Only the hyperbolic rescaling function σ_{hyp} cannot turn its initial head start in a faster overall convergence.

B.3 SAC Hyperparameters

The tables below show the used hyperparameters for the SAC algorithm on the two used OpenAI gym environments. Most of these values correspond to the tuned hyperparameters of the Stable-Baselines3 repository[10], the differences are highlighted in bold.

Table 1. Overview of the used hyperparameters for each environment. The shown hyperparameters are: maximum timesteps per episode k_M, total training timesteps $k_M \cdot T_M$, discount factor γ, replay buffer size $|\mathcal{B}|$, learning rate (for both actor and critic networks) η.

Environment	k_M	$k_M T_M$	γ	$\|\mathcal{B}\|$	η
LunarLanderContinuous-v2	1000	$5 \cdot 10^5$	0.99	$1 \cdot 10^6$	$\mathbf{7 \cdot 10^{-4}}$
Pendulum-v0	200	$\mathbf{5 \cdot 10^4}$	**0.98**	$\mathbf{2 \cdot 10^5}$	$1 \cdot 10^{-3}$

[10] https://github.com/DLR-RM/rl-baselines3-zoo/blob/master/hyperparams/sac.yml.

Table 2. Overview of the used hyperparameters, common across both environments.

Common hyperparameters		
Warmup timesteps		10000
Batch size	B	256
Polyak averaging constant	τ	$5 \cdot 10^{-3}$
Network architecture – hidden dimensions (actor + critic)		400×300
Entropy coefficient	α	Automatic adjustment

C Autonomous Highway Driving Environment

The results shown in Sect. 4 for the highway driving environment are obtained using a proprietary highway simulator. In this section the most relevant components of this simulator are briefly discussed.

C.1 Roads

All experiments were conducted on a three lane highway, shaped as a closed-loop circuit, with both straight and curved segments. The maximum speed limit was set to 30 m/s in all lanes, although some vehicles were instructed to slightly deviate from this limit, to get more varying situations on the road.

C.2 Vehicles

Every vehicle in the simulator follows the kinematic bicycle model (KBM) [22] to update its state based on the selected inputs

$$\begin{bmatrix} \dot{x} \\ \dot{y} \\ \dot{\psi} \\ \dot{v} \end{bmatrix} = \begin{bmatrix} v\cos(\psi+\beta) \\ v\sin(\psi+\beta) \\ \frac{v}{l_r}\sin\beta \\ \frac{a}{\cos\beta} \end{bmatrix} \qquad \beta = \arctan\left(\frac{l_r}{l_f+l_r}\tan\delta\right).$$

The vehicle's local state vector consists of an absolute x and y position, a heading angle ψ and velocity v. The vehicle can be controlled through its inputs, consisting of a steering angle δ and a longitudinal acceleration a. To make the control task of the virtual driver (agent) easier, extra vehicle motion controllers are used to stabilize the vehicle on the road, allowing the agent to select high-level steering actions $\boldsymbol{a}$, consisting of a desired longitudinal velocity v_{ref} and desired lateral position d_{ref}, to solve the driving task. To take correct high level steering decisions, the virtual driver needs some extra information about other traffic participants in its neighbourhood. This information is gathered in the agent's observation vector $\boldsymbol{s}$, containing local information such as the vehicle's offset w.r.t. different lane centers and its velocity components; and relative information such as relative distances and velocities w.r.t. neighbouring traffic.

C.3 Policies

Every vehicle is controlled by a policy, mapping observations $\boldsymbol{s}$ to suitable high-level actions $\boldsymbol{a}$. The policy of the autonomous vehicle is learned using any of the described reinforcement learning methods in this paper. The policies of the other vehicles in the simulation environment are fixed beforehand. A mixture of vehicles equipped with a custom rule-based policy and a policy implementing the 'Intelligent Driver Model' (IDM) [26] and 'Minimizing Overall Braking Induced by Lane change' (MOBIL) [17] is used. Both policies try to mimic rudimentary human driving behaviour, although being fully deterministic.

C.4 Reward

The used reward signal is calculated as a weighted sum of different penalties

$$r = \frac{w_{\mathrm{F}} r_{\mathrm{F}} + w_{\mathrm{V}} r_{\mathrm{V}} + w_{\mathrm{C}} r_{\mathrm{C}} + w_{\mathrm{R}} r_{\mathrm{R}}}{w_{\mathrm{F}} + w_{\mathrm{V}} + w_{\mathrm{C}} + w_{\mathrm{R}}}.$$

The first 'frontal' component r_{F} gives a penalty whenever the following distance to the leading vehicle is smaller than a predefined threshold. The 'velocity' component r_{V} gives a penalty whenever the virtual driver is not travelling at or near the maximum allowed speed. The third 'center' component r_{C} gives a penalty whenever the vehicle is not travelling central in the lane. To force the virtual driver to keep right whenever possible, the 'right' penalty r_{R} is given whenever there is a free lane to the right available. Finally, an extra penalty is given when the virtual driver collides with other vehicles or the highway boundaries.

C.5 Action Bounds

In this environment, the state-dependent action bounds are used to avoid potentially unsafe regions in state-action space. To determine the bounds, following braking criterion is used

$$\min\left(\Delta x, \Delta x + \frac{v_{\mathrm{L}}^2}{2b} - \frac{v_{\mathrm{F}}^2}{2b}\right) > \Delta x_{\mathrm{SAFE}}, \tag{4}$$

where Δx is the following distance between 2 vehicles, v_{L} and v_{F} are the velocities of the leading and following vehicle respectively, b is the maximum deceleration of both vehicles and Δx_{SAFE} is a minimum safe distance to keep between both vehicles. For simplicity we assume following vehicles can react instantly (no delay) to changing behaviour of the leading vehicle and the maximum deceleration b is assumed to be the same for both leading and following vehicles. Complying with the braking criterion then ensures the following vehicle is always able to avoid a crash with the leading vehicle, even in case of an emergency brake (braking with maximum deceleration b until standstill).

The upper bound for the longitudinal reference velocity can be derived from (4) as

$$v_{\mathrm{U}} = \sqrt{v_{\mathrm{L}}^2 + 2b(\Delta x - \Delta x_{\mathrm{SAFE}})}.$$

This expression is evaluated for all visible[11] leading vehicles having a lateral overlap with the virtual driver (Fig. 12.a). The lowest encountered value for v_{U} is then taken as the final velocity upper bound.

For the bounds on the lateral reference position, let us first consider the set of safe lateral intervals $\mathcal{D}_{\text{SAFE}} = \{[d_{\text{l}}; d_{\text{r}}] \subset [d_{\text{left}}; d_{\text{right}}]\, \forall d \in [d_{\text{l}}; d_{\text{r}}] : (4) \text{ holds}\}$ where d_{left} and d_{right} denote the lateral distance to the left and right road boundary respectively. The braking criterion is evaluated for all visible (See footnote 11) vehicles having some overlap with lateral position d, with the virtual driver as leading or following vehicle depending on the other vehicle's longitudinal position with respect to the virtual driver. The bounds for the lateral reference position, are then constructed such that $[d_{\text{L}}; d_{\text{U}}] \in \mathcal{D}_{\text{SAFE}}$ is the nearest (with respect to the virtual driver's current lateral position) and largest safe interval (Fig. 12b).

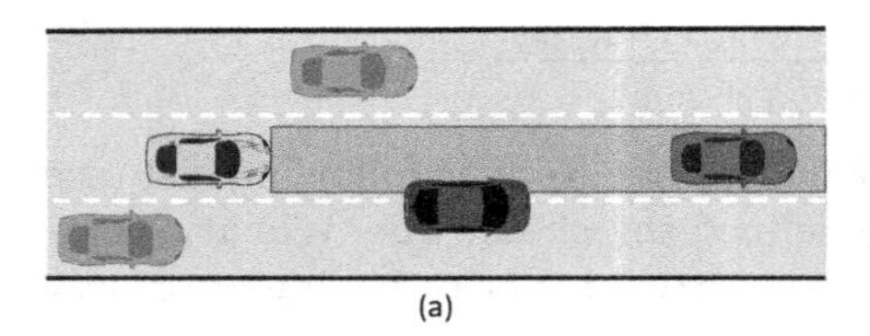

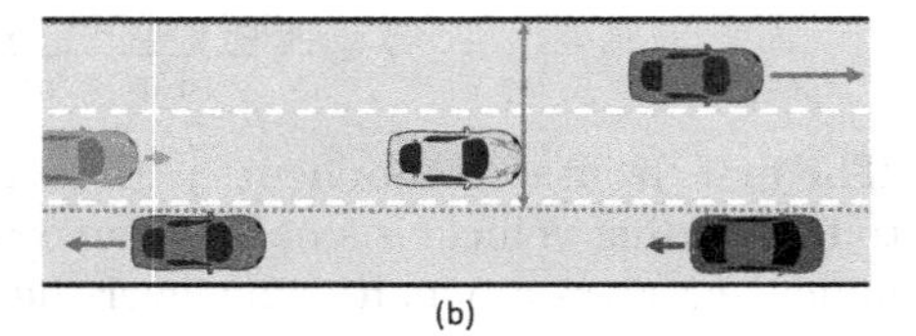

(a) (b)

Fig. 12. Illustration of safe action bound determination on highways using the braking criterion (4). The yellow car is driven by the virtual driver; the gray cars are not considered for determining the safe action range; and the red cars are the limiting entities, responsible for the lowest velocity upper bound or tightest lateral action range. The arrows indicate the relative velocity of each car with respect to the virtual driver's velocity. The left side (a) illustrates the determination of the longitudinal velocity upper bound, by evaluating the braking criterion for all leading vehicles with lateral overlap (green area). The right side (b) illustrates the determination of the safe lateral interval, by evaluating the braking criterion for each visible car. The green lines indicate the resulting safe action range: allowing the virtual driver to move towards the left lane — as the only vehicle there is quickly moving away from the virtual driver — but preventing movement towards the right lane — as this lane is occupied by the slower red vehicle. (Color figure online)

References

1. Achiam, J., Held, D., Tamar, A., Abbeel, P.: Constrained policy optimization. In: 34th International Conference on Machine Learning, ICML 2017, vol. 1, pp. 30–47 (2017)
2. Alshiekh, M., Bloem, R., Ehlers, R., Könighofer, B., Niekum, S., Topcu, U.: Safe reinforcement learning via shielding. In: 32nd AAAI Conference on Artificial Intelligence, AAAI 2018, pp. 2669–2678 (2018)
3. Berkenkamp, F., Turchetta, M., Schoellig, A.P., Krause, A.: Safe model-based reinforcement learning with stability guarantees. In: Advances in Neural Information Processing Systems, vol. 30, pp. 909–919 (2017)

[11] Visible means part of the state vector in this context.

4. Burkardt, J.: The truncated normal distribution, department of scientific computing. Fla. State Univ. 1–35 (2014)
5. Chandak, Y., Theocharous, G., Metevier, B., Thomas, P.S.: Reinforcement learning when all actions are not always available. In: AAAI 2020–34th AAAI Conference on Artificial Intelligence, pp. 3381–3388 (2020). https://doi.org/10.1609/aaai.v34i04.5740
6. Cheng, R., Orosz, G., Murray, R.M., Burdick, J.W.: End-to-end safe reinforcement learning through barrier functions for safety-critical continuous control tasks. In: 33rd AAAI Conference on Artificial Intelligence, AAAI 2019, pp. 3387–3395 (2019)
7. Chow, Y., Nachum, O., Faust, A., Duenez-Guzman, E., Ghavamzadeh, M.: Lyapunov-based safe policy optimization for continuous control (2019)
8. Dalal, G., Dvijotham, K., Vecerik, M., Hester, T., Paduraru, C., Tassa, Y.: Safe exploration in continuous action spaces (2018)
9. De Cooman, B., Suykens, J., Ortseifen, A.: Improving temporal smoothness of deterministic reinforcement learning policies with continuous actions. In: 33rd Benelux Conference on Artificial Intelligence, BNAIC 2021, pp. 217–240 (2021)
10. De Cooman, B., Suykens, J., Ortseifen, A., Subramanya, N.: Method for autonomous driving of a vehicle, a data processing circuit, a computer program, and a computer-readable medium, E.U. Patent Application EP22151063.9. 11 Jan 2022
11. Figurnov, M., Mohamed, S., Mnih, A.: Implicit reparameterization gradients. In: Advances in Neural Information Processing Systems, vol. 31, pp. 441–452 (2018)
12. Fujimoto, S., Van Hoof, H., Meger, D.: addressing function approximation error in actor-critic methods. In: 35th International Conference on Machine Learning, ICML 2018, vol. 4, pp. 2587–2601 (2018)
13. García, J., Fernández, F.: A comprehensive survey on safe reinforcement learning (2015)
14. Haarnoja, T., et al.: Soft actor-critic algorithms and applications (2018)
15. Hoel, C.J., Driggs-Campbell, K., Wolff, K., Laine, L., Kochenderfer, M.J.: Combining planning and deep reinforcement learning in tactical decision making for autonomous driving. Technical report (2019)
16. Kalweit, G., Huegle, M., Werling, M., Boedecker, J.: Deep constrained Q-learning (2020)
17. Kesting, A., Treiber, M., Helbing, D.: General lane-changing model MOBIL for car-following models. Transp. Res. Rec. **1999**, 86–94 (2007). https://doi.org/10.3141/1999-10
18. Kingma, D.P., Salimans, T., Welling, M.: Variational dropout and the local reparameterization trick. In: Advances in Neural Information Processing Systems, vol. 28, pp. 2575–2583 (2015)
19. Koller, T., Berkenkamp, F., Turchetta, M., Krause, A.: Learning-based model predictive control for safe exploration. In: Proceedings of the IEEE Conference on Decision and Control, pp. 6059–6066 (2019). https://doi.org/10.1109/CDC.2018.8619572
20. Lillicrap, T.P.: Continuous control with deep reinforcement learning. In: 4th International Conference on Learning Representations, ICLR 2016 - Conference Track Proceedings (2016)
21. Mirchevska, B., Pek, C., Werling, M., Althoff, M., Boedecker, J.: High-level decision making for safe and reasonable autonomous lane changing using reinforcement learning. In: IEEE Conference on Intelligent Transportation Systems, Proceedings, ITSC, pp. 2156–2162 (2018). https://doi.org/10.1109/ITSC.2018.8569448

22. Rajamani, R.: Vehicle Dynamics and Control, 2nd edn. Springer, Cham (2012). https://doi.org/10.1007/978-1-4614-1433-9
23. Silver, D., Lever, G., Heess, N., Degris, T., Wierstra, D., Riedmiller, M.: Deterministic policy gradient algorithms. In: 31st International Conference on Machine Learning, ICML 2014. vol. 1, pp. 605–619 (2014)
24. Sutton, R.S., Barto, A.G.: Reinforcement Learning: An Introduction, A Bradford book, vol. 258, 1st edn. MIT Press, Cambridge (1998)
25. Tessler, C., Mankowitz, D.J., Mannor, S.: Reward constrained policy optimization. In: 7th International Conference on Learning Representations, ICLR 2019 (2019)
26. Treiber, M., Hennecke, A., Helbing, D.: Congested traffic states in empirical observations and microscopic simulations. Phys. Rev. E Stat. Phys. Plasmas Fluids Relat Interdiscip. Topics **62**(2), 1805–1824 (2000). https://doi.org/10.1103/PhysRevE.62.1805
27. Wachi, A., Sui, Y.: Safe reinforcement learning in constrained markov decision processes. In: 37th International Conference on Machine Learning, ICML 2020, vol. Part F16814, pp. 9739–9748 (2020)
28. Zhang, H., Yu, T.: Taxonomy of reinforcement learning algorithms. In: Dong, H., Ding, Z., Zhang, S. (eds.) Deep Reinforcement Learning, pp. 125–133. Springer, Singapore (2020). https://doi.org/10.1007/978-981-15-4095-0_3

A Bee Colony Optimization Approach for the Electric Vehicle Routing Problem with Drones

Nikolaos A. Kyriakakis, Themistoklis Stamadianos, Magdalene Marinaki, Nikolaos Matsatsinis, and Yannis Marinakis(✉)

School of Production Engineering and Management, Technical University of Crete, University Campus, Akrotiri 73100, Greece
{nkyriakakis,tstamadianos}@isc.tuc.gr, magda@dssl.tuc.gr, {nikos,marinakis}@ergasya.tuc.gr

Abstract. In recent years, studies of vehicle routing problems utilizing novel means of transportation, such as electric ground vehicles and unmanned aerial vehicles, also known as drones, have been increasing in popularity. This paper presents a Bee Colony Optimization (BCO) approach for the recently introduced Electric Vehicle Routing Problem with Drones (EVRPD). The EVRPD utilizes electric vans as mother-ships, from which drones are deployed in order to perform the last-mile delivery to customers. The objective of the EVRPD is to minimize the total energy consumption of the operation and it considers packages of various weights. The Bee Colony Optimization algorithm is one of the few swarm intelligence algorithms designed particularly for solving combinatorial optimization problems. In this approach the exploratory properties of the BCO swarm are combined with the strong exploitative capabilities of the Variable Neighborhood Descent, used as a local search procedure. The results are compared with the other algorithmic approaches in the literature.

Keywords: Bee colony optimization · Energy minimizing · Electric vehicles · Drones

1 Introduction

The Electric Vehicle Routing Problem with Drones (EVRPD) was recently introduced by [14] and it combines the EVs with unmanned aerial vehicles, commonly known as drones, with the objective of minimizing energy consumption. The EVRPD considers travel range limitations, quantity capacities and maximum payload limits for both vehicle types. The packages to be delivered belong to different weight classes, and thus, each package affects the energy consumption rate differently. By extension, each package class has a different impact on the vehicles' maximum travel distance and total energy consumption.

G. Nicosia et al. (Eds.): LOD 2022, LNCS 13811, pp. 219–233, 2023.
https://doi.org/10.1007/978-3-031-25891-6_17

Drones share many of their environmental benefits with their ground counterparts, but have additional limitations, as the weight of batteries today restricts their flight time significantly. Furthermore, the maximum payload they can transport is limited. Nevertheless, drones are very efficient for carrying lightweight packages and are not affected by traffic congestion, nor do they contribute to it. Thus, they are perfect to operate in an urban environment, not only offering greener transportation, but also faster deliveries.

To compensate for the drones' drawbacks and to overcome their limitations, the EVRPD, combines EVs and drones, leveraging on the strengths of each vehicle type. The EVs are utilized to carry drones, as mother-ships, to pre-designated launch and retrieval locations, in order to perform the last-mile deliveries to customers. With this joint operation scheme, the effective range of the drones can be significantly increased, while retaining the benefits they offer. The combined utilization of these novel vehicle types offers advantages beyond the capabilities of traditional means of transportation.

In this paper, a new formulation for the EVRPD is presented, which considers payload weight and quantity attributes separately, and the EVRPD is solved using a Bee Colony Optimization (BCO) algorithm. The first algorithm of the BCO family of algorithms was the Bee System, proposed by [18]. The BCO metaheuristic algorithm was later proposed by [38]. Unlike most swarm intelligence algorithms, which were designed to optimize continuous optimization functions, the BCO family of algorithms utilizes swarm behavior to solve combinatorial optimization problems in particular.

The BCO algorithm developed for solving the EVRPD is inspired by the approach of [42], originally proposed for solving the Traveling Salesman Problem. Unlike the original BCO metaheuristic algorithm, the bees in this version are memory-less and their waggle dance to notify the rest of the swarm is performed only after a full solution has been constructed. To complement the BCO strategy, a strong local search scheme has been utilized based on the well known Variable Neighborhood Descent algorithm [24]. The implemented BCO algorithm for solving the EVRPD is tested on 21 benchmark instances found in the literature.

The rest of the paper is structured as follows. Section 2 presents the recent related literature on studies utilizing EVs and drones. In Sect. 3, the EVRPD is defined and its mathematical formulation is presented. In Sect. 4 the developed BCO approach for solving the EVRPD is described. Section 5 presents and discusses the experimental results. Lastly, Sect. 6 presents the conclusions of the research.

2 Related Literature

2.1 Electric Ground Vehicle Routing Problems

The first variant of the EVRP is proposed in [37], featuring both recharging stations and time windows. Since then, many papers around the subject of EVRP have been presented, the most recent of which are discussed below.

Ant Colony algorithms have been very popular on many variants of VRP. [44] minimized the energy consumption of the EVRP with an Ant Colony Optimization (ACO) metaheuristic, while [23] used an ACO algorithm enriched with local search algorithms in a unique problem, allowing the driver to choose if the battery gets swapped or recharged. Energy consumption, a very important parameter of the EVRP, has interested many researchers such as [3], who also considered the effect that the charging stations would have to the solution, given the need to recharge. [2] used a machine learning algorithm to calculate the State-of-Charge of the vehicles.

Many extensions have been made on the EVRP to try and represent a more realistic view of the world. Uncertain charging station waiting time and potential consequences have been the subject of research in [11], solving the EVRP with time windows. Vehicle-2-Grid technology has seen great developments lately, and was examined in the research of [17], where electricity prices are not fixed. The concept of a distribution center that would also serve as a charging station with renewable energy production on site, was presented in [29], and evaluated through a case study. Other variants have been proposed by [36], showcasing mid-haul logistics with EVs and by [1] who employed mixed vehicle fleets in city logistics. Further analysis of the presented literature may be found in the review of [43], who examined many of the parameters EVRPs are concerned with.

2.2 Routing Problems with Drones

In the context of supply chains, the VRP seeks to determine the best customer visiting order for a vehicle fleet while minimizing one or more objectives. The first variant of VRP employing drones was presented in [27]. They proposed a Traveling Salesman Problem with a Flying Sidekick (FSTSP). The Traveling Salesman Problem (TSP) is a case of VRP with only one vehicle. The multiple FSTSP variant was presented in [13] and also studied in [28] and [32]. [9] took into account both the payload and mandatory detours, [31] used drones to resupply their vehicles, and [20] allowed for multiple drone visits per trip.

VRP with drones (VRPD) was first introduced in [41]. [8,35], and [30] route trucks and drones separately. [34] emphasized the correlation between drone range and their potential financial benefits. [5] employed only one drone per truck, while [25] employed multiple drones per truck. In [10] and [33] only drones make deliveries. Two echelon approaches have been proposed in [12] and [16], using the trucks as mobile depots. [39] considered the weather in drone deliveries. [45] considered the height the drones fly at, when making deliveries, which can have an impact on battery life. [21] and [7] presented humanitarian drone applications. A significant number of reviews have been presented on the subject of drone usage in VRP. The most recent are the following: [4,6,22,26,40] and [15].

3 Mathematical Formulation of the EVRPD

The mathematical formulation of EVRPD aims to minimize the total energy consumption of both types of vehicles combined. The model is similar to two-echelon problems, since the ground vehicles transport the drones to their launch sites (satellite locations), however, they wait for them to return before continuing. Unlike the original EVRPD formulation, this model does not follow the abstracted description of payload, and considers payload quantity and payload weight separately. The sets of nodes, the parameters and the variables of the model are presented in Table 1.

The mathematical formulation is the following:

$$\min f = \sum_{(i,j)\in A_1} \sum_{k\in K^{EV}} (d_{ij} \times (1 + f^1_{ijk}) \times x_{ijk}) + \sum_{(i,j)\in A_2} \sum_{k\in K^D} \sum_{s\in V_S} (d_{ij} \times (1 + f^2_{ijsk}) \times z_{ijsk}) \tag{1}$$

Subject to:

$$\sum_{j\in(V'_D\cup V_S)} x_{ijk} = \sum_{j\in(V'_D\cup V_S)} x_{jik}, \forall i \in (V_S \cup V_D), k \in K^{EV} \tag{2}$$

$$\sum_{j\in(V_C\cup V'_S)} z_{ijsk} = \sum_{j\in(V_C\cup V'_S)} z_{jisk}, \forall i \in V_C, s \in V_S, k \in K^D \tag{3}$$

$$\sum_{k\in K^D} \sum_{s\in V_S} \sum_{j\in(V_C\cup V'_S)} z_{ijsk} = 1, \forall i \in V_C \tag{4}$$

$$\sum_{i\in(V_S\cup V_D)} x_{isk} \leq 1, \forall s \in V_S, k \in K^{EV} \tag{5}$$

$$\sum_{j\in(V_S\cup V'_D)} x_{v_d jk} = 1, \forall k \in K^{EV} \tag{6}$$

$$\sum_{i\in(V_D\cup V_S)} x_{iv'_d k} = 1, \forall k \in K^{EV} \tag{7}$$

$$w_{ik} = \sum_{j\in(V'_D\cup V_S)} f^1_{jik} - \sum_{j\in(V'_D\cup V_S)} f^1_{ijk}, \forall i \in V_S, k \in K^{EV} \tag{8}$$

$$0 \leq f^1_{ijk} \leq W^{EV} \times x_{ijk}, \forall (i,j) \in A_1, k \in K^{EV} \tag{9}$$

$$\sum_{i\in(V_C\cup V_S)} \sum_{j\in V_C} z_{ijsk} \leq Q^D, \forall s \in V_S, k \in K^D \tag{10}$$

$$\sum_{i\in(V_D\cup V_S)} \sum_{j\in V_S} x_{ijk} \leq Q^{EV}, \forall k \in K^{EV} \tag{11}$$

$$p_i = \sum_{j \in (V_C \cup V_S')} f^2_{jisk} - \sum_{j \in (V_C \cup V_S')} f^2_{ijsk}, \forall i \in V_C, s \in V_S, k \in K^D \tag{12}$$

$$0 \leq f^2_{ijsk} \leq W^D \times z_{ijsk}, \forall (i,j) \in A_2, s \in V_S, k \in K^D \tag{13}$$

$$\sum_{k \in K^D} \sum_{(i,j) \in A_2} p_i \times z_{ijsk} \times TD_{lks+} = w_{sl}, \forall s \in V_S, l \in K^{EV} \tag{14}$$

$$\sum_{i \in V_S} \sum_{j \in V_C} \sum_{k \in K^D} z_{ijsk} \times TD_{lks+} \leq k_d, \forall s \in V_S, l \in K^{EV} \tag{15}$$

$$\sum_{i \in (V_D \cup V_S)} \sum_{j \in (V_C \cup V_S')} (1 + f^1_{ijk}) \times d_{ij} \times x_{ijk} \leq E^{EV}, \forall k \in K^{EV} \tag{16}$$

$$\sum_{(i,j) \in A_2} \sum_{s \in V_S} (1 + f^2_{ijsk}) \times d_{ij} \times z_{ijsk} \leq E^D, \forall k \in K^D \tag{17}$$

$$\sum_{i \in (V_S \cup V_S')} \sum_{j \in (V_S \cup V_S')} z_{ijsk} = 0, \forall s \in V_S, \forall k \in K^D \tag{18}$$

$$TD_{ijs+} = TD_{ijs-}, \forall i \in K^{EV}, j \in K^D, s \in V_S \tag{19}$$

$$\sum_{i \in K^{EV}} TD_{ijs+} = 1, \forall j \in K^D, s \in V_S \tag{20}$$

$$TD_{ijs+} = TD_{ijs'+}, \forall i \in K^{EV}, j \in K^D, s \in V_S, s' \in \{V_S | s' \neq s\} \tag{21}$$

$$TD_{ijs+}, TD_{ijs-} \in \{0,1\}, \forall i \in K^{EV}, j \in K^D, s \in (V_S \cup V_{S'}) \tag{22}$$

$$x_{ijk} \in \{0,1\}, \forall (i,j) \in A_1, k \in K^{EV} \tag{23}$$

$$z_{ijsk} \in \{0,1\}, \forall (i,j) \in A_2, s \in V_S, k \in K^D \tag{24}$$

Constraints (2) and (3) force each node to have the same number of inbound and outbound arcs, for trucks and drones respectively. Constraints (4) ensure each customer node is visited exactly once, while constraints (5) ensure each satellite location is visited at most once by each truck. Constraints (6) and (7) make sure each EV's route starts from the depot and returns at the depot after completion. The payload weight transported by each EV to each satellite location is calculated by constraints (8), while (9) force the EV's payload to be less than its maximum payload weight capacity, if the arc is traversed, otherwise it sets it to 0. Constraints (10) limit the payload quantity of each drone at each satellite location. Likewise, constraints (11) limit the payload quantity of carried by each EV. Constraints (12) and (13) limit the payload weight carried by the drones. The payload consistency between trucks and drones is ensured by constraints (14). Constraints (15) limit the number of available drones per EV. Energy consumption limitations are enforced by constraints (16) and (17), for the EVs and drones respectively. To prevent connections between satellite locations and their dummies, constraints (18) are used. Each drone is paired to an EV uniquely by constraints (19), (20) and (21). Decision variables are restricted using constraints (22) to (24).

Table 1. Sets, parameters, and decision variables of the EVRPD formulation.

Symbol	Description
Node Sets & Characteristics	
V_D, V'_D	Depot set, $V_D = \{v_D\}$ and its dummy $V'_D = \{v'_D\}$
V_S, V'_S	Satellites set, $V_S = \{v_{S1}, v_{S2}, ..., v_{Sn_s}\}$ and its dummy $V'_S = \{v'_{S1}, v'_{S2}, ..., v'_{Sn_s}\}$
V_C	Customers, $V_C = \{v_{C1}, v_{C2}, ..., v_{Cn_c}\}$
A_1	First echelon arcs set, $A_1 = \{(i,j) \mid i \in V_D \cup V_S, j \in V'_D \cup V_S, i \neq j\}$
A_2	Second echelon arcs set, $A_2 = \{(i,j) \mid i \in V_C \cup V_S, j \in V_C \cup V'_S, i \neq j\}$
n_c	Number of customers
n_s	Number of satellites
EV and drone Characteristics	
K^{EV}	Electric Vehicles (EVs) set
K^D	Drones sets
k_{EV}	Number of EVs
k_d	Number of drones
Q^{EV}	Maximum payload quantity capacity of EVs
Q^D	Maximum payload quantity capacity of drones
W^{EV}	Maximum payload weight capacity of EVs
W^D	Maximum payload weight capacity of drones
E^{EV}	Maximum energy capacity of EVs
E^D	Maximum energy capacity of drones
Node Characteristics	
d_{ij}	Distance from node i to node j
p_i	payload weight demand of node i
Variables	
x_{ijk}	Describes whether or not EV k traverses the arc(i,j)
z_{ijsk}	Describes whether or not drone k traverses the arc(i,j), beginning from satellite s
w_{ik}	Payload weight delivered to satellite i by vehicle k
f^1_{ijk}	Stores the payload weight in EV k that arrives from i to j
f^2_{ijsk}	Stores the payload weight in drone k that arrives from i to j, beginning from satellite s
TD_{ijs+}	$\{= 1\}$ when EV i arrives at satellite s transporting drone j
TD_{ijs-}	$\{= 1\}$ when EV i departs from satellite s transporting drone j

4 Bee Colony Optimization Approach

The BCO is inspired by the foraging behaviour of bee colonies in nature and simulates their strategy in order to effectively explore the solution space. Bees communicate with each other, relaying information on the quality of the food source by performing the bee waggle dance in a characteristic figure eight-pattern. Depending on the angle of the dance relative to the sun's position and the length of waggle dance, bees can inform each other about the direction and the distance of the food source. The BCO uses this mechanism in order to exchange information on solutions and to search around promising areas in order to escape the local minima.

The BCO approach implemented for the EVRPD is based on the BCO algorithm of [42] for solving the Traveling Salesman Problem. The outline of the implemented BCO algorithm is presented in Algorithm 1. For the initialization of the bee colony, a greedy randomized solution construction process is utilized. Afterwards, at each iteration, each bee in the colony has three tasks: it observes a dance of another bee, it constructs a solution based on the observation, and it performs a dance depending to the solution quality obtained. The solution obtained by the best bee of each iteration is further improved by a local search procedure.

Algorithm 1: BCO outline

```
Data: Instance, BCO_Iters, β, numBees, λ, N = {N_1, N_2, ..N_k}, LS_Iters
Initialize Bees;
for it ← 0 to BCO_Iters do
    foreach bee in Swarm do
        S_pref ← ObserveDance(Swarm);
        S_bee ← ConstructBeeSolution(S_pref, β, λ);
        PerformWaggleDance(Swarm, S_bee);
        S_itBest ← UpdateItBest(S_bee);
    S_itBest ← LocalSearch(S_itBest, LS_Iters, N);// N the set of operators
    S_gBest ← UpdateGlobalBest(S_bee);
return S_gBest;
```

4.1 Solution Construction

Each bee represents a complete solution of the EVRPD. In the initialization phase, the bee solutions are constructed using a simple randomized greedy heuristic. For the rest of the BCO algorithm the following solution generation procedure is applied: Bees start constructing their tour from the depot adding nodes (satellites or customers, depending on route type) to the route until a further addition violates one of the constraints. The bees then return to the starting node (Depot or EV) and begin a second tour with the same drone or a new one, depending on the constraints in the case of a drone route, or with a new EV in the case of EV route. This process is repeated until all customers have been serviced.

During the construction process, the choice of the next node to be added in the route is based on two parameters. The implemented BCO utilizes parameter λ to control randomness, and parameter β to control the greediness by adjusting the importance of the heuristic information $1/d_{ij}$, where d_{ij} is the distance between nodes i and j.

The transition probability from node i to node j for the proposed BCO algorithm is presented in Eq. (25):

$$p_{ij} = \begin{cases} \dfrac{[\rho_{ij}][\frac{1}{d_{ij}}]^{\beta}}{\sum_{l \in L_i}[\rho_{il}][\frac{1}{d_{il}}]^{\beta}}, & \texttt{if } j \in L_i \\ 0, & \texttt{otherwise} \end{cases} \tag{25}$$

where L_i is the set of customers not visited yet, in the case of drone routes or satellites for EV routes and ρ_{ij} is the fitness of arc (i, j) defined by Eq. (26):

$$\rho_{ij} = \begin{cases} \lambda, & \texttt{if } j \in F_i \\ \dfrac{1 - \lambda \cdot |L_i \cap F_i|}{|L_i| - |L_i \cap F_i|}, & \texttt{otherwise} \end{cases} \tag{26}$$

where F_i is the preferred node to transition to, based on the observed dance and $\lambda \in (0, 1)$.

4.2 Waggle Dance Mechanism

After the solution construction process is completed, the bees may perform the waggle dance in order to inform the other bees on the quality of the solution it found. The bee is allowed to dance, only if the new solution found improves on its previous. The other bees observing the dance will choose to follow or not the dancing bee depending on the quality of the solution it has found. The bees with solutions of lower cost have a higher chance of being chosen. For the choosing process, the score rules determining the probability of a bee being followed found in [42] have been used and are presented in Table 2. The score SC of each bee, based on which the choice probabilities are calculated, is given by the inverse of its solution cost, $SC_{bee} = 1/cost(S_{bee})$. The average score of the population is given by $\bar{SC}_{colony} = \dfrac{\sum_{bee \in bees} SC_{bee}}{numBees}$

Table 2. Probability of bees being followed

Score	$P_{followed}$
$SC_{bee} < 0.5\bar{SC}_{colony}$	0.60
$0.5\bar{SC}_{colony} \leq SC_{bee} < 0.65\bar{SC}_{colony}$	0.20
$0.65\bar{SC}_{colony} \leq SC_{bee} < 0.85\bar{SC}_{colony}$	0.02
$0.85\bar{SC}_{colony} \leq SC_{bee}$	0.00

4.3 Local Search

To complement the optimization process, a strong local search procedure is utilized based on the Variable Neighborhood Descent (VND) algorithm. It is a deterministic variant of the variable neighborhood search metaheuristic framework originally proposed by [24]. This implementation uses the Pipe VND (P-VND) variant, where the search continues in the same neighborhood until no further improvement can be made, before it proceeds to the next neighborhood. The process terminates when no further improvement can be made by the last neighborhood. The neighborhood structures used in the BCO implementation are the following:

- **Intra-EV-Intra-Drone 1-1 Intra-route Swap:** Two customers of a single drone route, swap positions.
- **Intra-EV-Intra-Drone 1-1 Inter-route Exchange:** Two customers, from different routes of a single drone, exchange their positions.
- **Intra-EV-Intra-Drone 1-0 Inter-route Relocation:** A customer is removed from a drone route and is inserted in a different route of the same drone.
- **Intra-EV-Inter-Drone 1-1 Inter-route Exchange:** Two customers, from different drones routes, exchange their positions.
- **Intra-EV-Inter-Drone 1-0 Inter-route Relocation:** A customer is removed from a drone route and is inserted in a route of another drone.
- **Inter-EV-Inter-Drone 1-1 Inter-route Exchange:** Same as the intra-route variant, but for drones belonging to different EVs.
- **Inter-EV-Inter-Drone 1-0 Inter-route Relocation:** Same as the intra-route variant, but for drones belonging to different EVs.
- **EV-route 2-Opt Intra-route:** The order of the customers in a range of positions in the EV route are reversed.

5 Experimental Results

The algorithm was coded in C++ and compiled with GCC 11. The experiments were conducted using a 2014 Intel®Core i7-4770 CPU (3.40 GHz) with 7.7 GB RAM on the Fedora Workstation 35 OS. Each instance was solved independently 10 times. The parameters tested are displayed in Table 3.

Table 3. Parameter description and settings

Parameter	Description	Values tested
BCO_{Iters}	Number of iterations	10000
$numBees$	Bee colony population	10
λ	Controls greediness during solution construction	$\{0.7, 0.8, 0.9\}$
β	Controls the importance of heuristic information	$\{1.0, 2.0, 3.0\}$
LS_{Iters}	Number of the local search iterations	50

Figure 1 presents the average gap between the best solutions found and the best known value (BKV) for each instance. Although more than one combinations of parameter values are able to provide quality results, the implemented BCO algorithm benefits from higher values for parameter β. The algorithm is observed to be less sensitive to change in the values of parameter λ, with value 0.7 being the best overall choice.

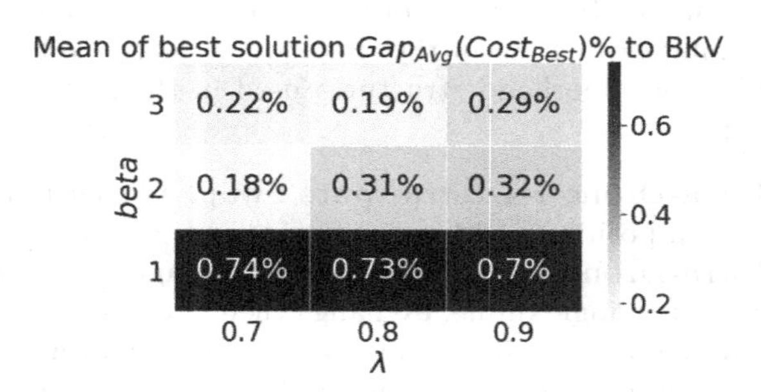

Fig. 1. Average gap between the best solutions and the BKV, for different parameter value combinations.

Table 4 presents the results obtained by the implemented BCO algorithm on the EVRPD instances previously solved by ACO algorithms in [14]. Column 1 denotes the instance name. Column 2 denotes the best known solution values (BKV) from the literature while Columns 3 and 4 present the best and the average solution values obtained, respectively. Finally, column 5 indicates the average computational time required by the BCO in seconds.

The BCO was able to obtain three new best known solutions and match one BKV from the literature, all of them in the small instances of 22 nodes. For the medium and larger instances the performance of the BCO is not as competitive. As instances grow larger, the gap between the obtained results and the BKVs also becomes bigger. The competitive, or even superior in some instances, performance of the BCO compared to the ACO implementations in the literature may be attributed to its memory-less nature. Unlike the ACO algorithms in which premature pheromone convergence on a local optima can hinder the performance of the algorithms, the BCO promotes the search around a newly found improved solution. Nevertheless, the same advantage can become a drawback on larger instances, since the algorithm does not intensify the search in any particular promising area of the solution space.

Table 5 summarizes the average results of the five total algorithms for solving the EVRPD. Overall, the MMAS was able to obtain the best average results both on the best and average solution cost. The BCO is second to last in the average best cost found and last in terms of average solution values. For the computational time required, the BCO is the third best, less than a second behind the second best HMMAS.

Table 4. BCO results on EVRPD instances (BKS in bold).

Instance		BCO		
	BKV	$Cost_{best}$	$Cost_{avg}$	$Time_{avg}$
EVRPD-n22-k4-s10-14	1144.28	**1139.10**	1146.98	24.32
EVRPD-n22-k4-s11-12	**1403.94**	1406.06	1412.19	23.87
EVRPD-n22-k4-s12-16	**1240.95**	1241.12	1246.34	24.58
EVRPD-n22-k4-s6-17	1610.70	**1604.59**	1621.69	22.90
EVRPD-n22-k4-s8-14	**1191.20**	**1191.20**	1199.60	23.72
EVRPD-n22-k4-s9-19	1874.20	**1873.95**	1877.78	22.45
EVRPD-n33-k4-s1-9	**3599.16**	3601.61	3608.19	40.85
EVRPD-n33-k4-s14-22	**4033.19**	4035.32	4038.20	41.36
EVRPD-n33-k4-s2-13	**3428.85**	3430.99	3439.99	38.58
EVRPD-n33-k4-s3-17	**3307.26**	3312.22	3331.69	40.64
EVRPD-n33-k4-s4-5	**3795.61**	3797.89	3802.25	40.21
EVRPD-n33-k4-s7-25	**3819.62**	3820.01	3827.62	39.34
EVRPD-n51-k5-s11-19	**3061.89**	3084.43	3141.16	60.94
EVRPD-n51-k5-s11-19-27-47	**1916.57**	1922.41	1935.76	60.55
EVRPD-n51-k5-s2-17	**2891.04**	2926.24	2968.56	58.93
EVRPD-n51-k5-s2-4-17-46	**2895.94**	2937.20	2982.40	58.75
EVRPD-n51-k5-s27-47	**1917.50**	1920.97	1935.56	59.66
EVRPD-n51-k5-s32-37	**4918.59**	4923.96	4955.67	62.51
EVRPD-n51-k5-s4-46	**4170.25**	4181.82	4315.67	58.81
EVRPD-n51-k5-s6-12	**2540.91**	2571.65	2620.27	59.34
EVRPD-n51-k5-s6-12-32-37	**2543.73**	2572.56	2634.08	57.60

Table 5. Results summary of algorithmic approaches in the literature.

Algorithm	Avg. $Cost_{best}$	Avg. $Cost_{avg}$	Avg. $Time_{avg}$
ACS	2736.84	2749.76	167.99
HACS	2738.77	2752.68	73.72
MMAS	2730.26	2745.36	41.75
HMMAS	2734.24	2752.78	43.39
BCO	2737.87	2763.88	43.81

Table 6. Wilcoxon signed-rank test of BCO and other approaches on EVRPD instances

Other Algorithm	ACS	ACSVND	MMAS	MMASVND
# instances	21	21	21	21
W-value	62.5	76.0	30.0	81
Significance level α_s	0.05	0.05	0.05	0.05
p-value	0.1125	0.2789	0.0051	0.2428
H_0 rejected	No	No	Yes	No

Table 6 displays the results obtained by comparing the BCO with the ACO algorithms from the literature using the Wilcoxon signed-rank test. Row 1 indicates the number of instances included. Row 2 displays the W-value. The third row declares the significance level α_s tested and Row 4 displays the corresponding p-value. The last row indicates whether the null hypothesis H_0 can be rejected with the risk to reject the null hypothesis when H_0 is true, being less than 5%.

The null hypothesis H_0 assumes that the true mean of the other algorithm is equal to the mean of BCO and H_1 assumes the true mean of the two algorithms differ. The statistical test results suggest that the null hypothesis H_0 can be rejected with a risk less than 5% for the comparison between the MMAS and BCO algorithms. Therefore, BCO's performance is statistically inferior to the MMAS algorithm and statistically indifferent for the other ACO variants.

6 Conclusions

The use of alternative vehicles in VRP has gained a lot of interest recently. This research presents a VRP using both electric vehicles and drones for joined operation. While, these types of vehicles differ significantly in size and abilities, they have some common traits such as zero local greenhouse gas emissions and low noise pollution. Battery capacity and maximum payload capacity are also shared limitations, however, they are a greater impediment for drones, given their small size. Nonetheless, the combination of these vehicles offers some unique opportunities for energy savings, since these vehicles complement each other greatly. Drones are perfect for transporting light packages over small distances, while electric trucks can carry heavier payloads but are bound by traffic and the road network. In the EVRPD, electric trucks are used as mother-ships. They transport the drones and the packages to designated launch and retrieval sites. From there, drones carry the items to their destination and return to the truck. This operation scheme is a great fit for city logistics. It shortens the truck trips while offering faster deliveries at the same time.

The mathematical model of the EVRPD was presented, along with the detailed description of its variables. The objective is to minimize the total energy consumption of the operation. The most significant controllable factor of the energy consumption is the payload weight of the transported packages, therefore, the objective function considers them in the calculations. A Bee Colony Optimization algorithm was implemented and compared against four ACO algorithms found in the literature. Its performance was competitive in the small instances tested, and it was able to obtain three new best known solutions. In larger instances its performance was lackluster compared to the top performing algorithms. Overall, the BCO was observed to offer good exploratory capabilities in the small instances due to its memory-less nature, but was lacking the intensification properties that memory structures offer, such as pheromone in the ACO.

The future research possibilities for the EVRPD are many. For example, the EVRPD can be further expanded and account for more parameters that may have an effect on deliveries, such as weather conditions that prohibit drone flights. Other possible expansions may include real-time battery depletion updates from the drones, which would affect the planned route and necessitate a dynamic approach to routing.

Acknowledgments

The research work was supported by the Hellenic Foundation for Research and Innovation (HFRI) under the HFRI PhD Fellowship grant (Fellowship Number: 6334.)

References

1. Al-dalain, R., Celebi, D.: Planning a mixed fleet of electric and conventional vehicles for urban freight with routing and replacement considerations. Sustain. Urban Areas **73**, 103105 (2021)
2. Basso, R., Kulcsár, B., Sanchez-Diaz, I.: Electric vehicle routing problem with machine learning for energy prediction. Transp. Res. Part B Methodol. **145**, 24–55 (2021)
3. Chakraborty, N., Mondal, A., Mondal, S.: Intelligent charge scheduling and eco-routing mechanism for electric vehicles: a multi-objective heuristic approach. Sustain. Urban Areas **69**, 102820 (2021)
4. Cheikhrouhou, O., Khoufi, I.: A comprehensive survey on the multiple traveling salesman problem: applications, approaches and taxonomy. Comput. Sci. Rev. **40**, 100369 (2021)
5. Chiang, W.-C., Li, Y., Shang, J., Urban, T.L.: Impact of drone delivery on sustainability and cost: realizing the UAV potential through vehicle routing optimization. Appl. Energy **242**, 1164–1175 (2019)
6. Chung, S.H., Sah, B., Lee, J.: Optimization for drone and drone-truck combined operations: a review of the state of the art and future directions. Comput. Oper. Res. **123**, 105004 (2020)
7. Ghelichi, Z., Gentili, M., Mirchandani, P.B.: Logistics for a fleet of drones for medical item delivery: a case study for Louisville, KY. Comput. Oper. Res. **135**, 105443 (2021)
8. Hu, M., et al.: On the joint design of routing and scheduling for vehicle-assisted multi-UAV inspection. Futur. Gener. Comput. Syst. **94**, 214–223 (2019)
9. Jeong, H.Y., Song, B.D., Lee, S.: Truck-drone hybrid delivery routing: payload-energy dependency and no-fly zones. Int. J. Prod. Econ. **214**, 220–233 (2019)
10. Karak, A., Abdelghany, K.: The hybrid vehicle-drone routing problem for pick-up and delivery services. Transp. Res. Part C Emerg. Technol. **102**, 427–449 (2019)
11. Keskin, M., Çatay, B., Laporte, G.: A simulation-based heuristic for the electric vehicle routing problem with time windows and stochastic waiting times at recharging stations. Comput. Oper. Res. **125**, 105060 (2021)
12. Kitjacharoenchai, P., Min, B.-C., Lee, S.: Two echelon vehicle routing problem with drones in last mile delivery. Int. J. Prod. Econ. **225**, 107598 (2020)
13. Kitjacharoenchai, P., Ventresca, M., Moshref-Javadi, M., Lee, S., Tanchoco, J.M., Brunese, P.A.: Multiple traveling salesman problem with drones: mathematical model and heuristic approach. Comput. Ind. Eng. **129**, 14–30 (2019)

14. Kyriakakis, N.A., Stamadianos, T., Marinaki, M., Marinakis, Y.: The electric vehicle routing problem with drones: An energy minimization approach for aerial deliveries. Clean. Logist. Supply Chain **4**, 100041 (2022)
15. Li, H., Chen, J., Wang, F., Bai, M.: Ground-vehicle and unmanned-aerial-vehicle routing problems from two-echelon scheme perspective: a review. Eur. J. Oper. Res. **294**, 1078–1095 (2021)
16. Li, H., Wang, H., Chen, J., Bai, M.: Two-echelon vehicle routing problem with time windows and mobile satellites. Transp. Res. Part B Methodol. **138**, 179–201 (2020)
17. Lin, B., Ghaddar, B., Nathwani, J.: Electric vehicle routing with charging/discharging under time-variant electricity prices. Transp. Res. Part C Emerg. Technol. **130**, 103285 (2021)
18. Lučić, P., Teodorović, D.: Bee system: modeling combinatorial optimization transportation engineering problems by swarm intelligence. In: Preprints of the TRISTAN IV Triennial Symposium on Transportation Analysis, pp. 441–445. Sao Miguel, Azores Islands, Portugal (2001)
19. Lučić, P., Teodorović, D.: Computing with bees: attacking complex transportation engineering problems. Int. J. Artif. Intell. Tools **12**(3), 375–394 (2003)
20. Luo, Z., Poon, M., Zhang, Z., Liu, Z., Lim, A.: The multi-visit traveling salesman problem with multi-drones. Transp. Res. Part C Emerg. Technol. **128**, 103172 (2021)
21. Macias, J.E., Angeloudis, P., Ochieng, W.: Optimal hub selection for rapid medical deliveries using unmanned aerial vehicles. Transp. Res. Part C Emerg. Technol. **110**, 56–80 (2020)
22. Macrina, G., Pugliese, L.D.P., Guerriero, F., Laporte, G.: Drone-aided routing: a literature review. Transp. Res. Part C Emerg. Technol. **120**, 102762 (2020)
23. Mao, H., Shi, J., Zhou, Y., Zhang, G.: The electric vehicle routing problem with time windows and multiple recharging options. IEEE Access **8**, 114864–114875 (2020)
24. Mladenović, N., Hansen, P.: Variable neighborhood search. Comput. Oper. Res. **24**(11), 1097–1100 (1997)
25. Moshref-Javadi, M., Hemmati, A., Winkenbach, M.: A truck and drones model for last-mile delivery: a mathematical model and heuristic approach. Appl. Math. Model. **80**, 290–318 (2020)
26. Moshref-Javadi, M., Winkenbach, M.: Applications and research avenues for drone-based models in logistics: a classification and review. Expert Syst. Appl. **177**, 114854 (2021)
27. Murray, C.C., Chu, A.G.: The flying sidekick traveling salesman problem: optimization of drone-assisted parcel delivery. Transp. Res. Part C Emerg. Technol. **54**, 86–109 (2015)
28. Murray, C.C., Raj, R.: The multiple flying sidekicks traveling salesman problem: parcel delivery with multiple drones. Transp. Res. Part C Emerg. Technol. **110**, 368–398 (2020)
29. Napoli, G., Micari, S., Dispenza, G., Andaloro, L., Antonucci, V., Polimeni, A.: Freight distribution with electric vehicles: a case study in Sicily. RES, infrastructures and vehicle routing. Transp. Eng. **3**, 100047 (2021)
30. Nguyen, M.A., Dang, G.T.-H., Hà, M.H., Pham, M.-T.: The min-cost parallel drone scheduling vehicle routing problem. Eur. J. Oper. Res. **299**, 910–930 (2022)
31. Pina-Pardo, J.C., Silva, D.F., Smith, A.E.: The traveling salesman problem with release dates and drone resupply. Comput. Oper. Res. **129**, 105170 (2021)

32. Raj, R., Murray, C.: The multiple flying sidekicks traveling salesman problem with variable drone speeds. Transp. Res. Part C Emerg. Technol. **120**, 102813 (2020)
33. Rossello, N.B., Garone, E.: Carrier-vehicle system for delivery in city environments. IFAC-PapersOnLine **53**(2), 15253–15258 (2020)
34. Sacramento, D., Pisinger, D., Ropke, S.: An adaptive large neighborhood search metaheuristic for the vehicle routing problem with drones. Transp. Res. Part C Emerg. Technol. **102**, 289–315 (2019)
35. Schermer, D., Moeini, M., Wendt, O.: A matheuristic for the vehicle routing problem with drones and its variants. Transp. Res. Part C Emerg. Technol. **106**, 166–204 (2019)
36. Schiffer, M., Klein, P.S., Laporte, G., Walther, G.: Integrated planning for electric commercial vehicle fleets: a case study for retail mid-haul logistics networks. Eur. J. Oper. Res. **291**(3), 944–960 (2021)
37. Schneider, M., Stenger, A., Goeke, D.: The electric vehicle routing problem with time windows and recharging stations. Transp. Sci. **48**(4), 500–520 (2014)
38. Teodorovic, D., Dell'Orco, M.: Bee colony optimization-a cooperative learning approach to complex transportation problems. Adv. OR AI Methods Transp. **51**, 60 (2005)
39. Thibbotuwawa, A., Bocewicz, G., Nielsen, P., Zbigniew, B.: Planning deliveries with UAV routing under weather forecast and energy consumption constraints. IFAC-PapersOnLine **52**(13), 820–825 (2019)
40. Vidal, T., Laporte, G., Matl, P.: A concise guide to existing and emerging vehicle routing problem variants. Eur. J. Oper. Res. **286**(2), 401–416 (2020)
41. Wang, Z., Sheu, J.-B.: Vehicle routing problem with drones. Transp. Res. Part B Methodol. **122**, 350–364 (2019)
42. Wong, L.P., Low, M.Y.H., Chong, C.S.: A bee colony optimization algorithm for traveling salesman problem. In: 2008 Second Asia International Conference on Modelling & Simulation (AMS), pp. 818–823. IEEE (2008)
43. Xiao, Y., Zhang, Y., Kaku, I., Kang, R., Pan, X.: Electric vehicle routing problem: a systematic review and a new comprehensive model with nonlinear energy recharging and consumption. Renew. Sustain. Energy Rev. **151**, 111567 (2021)
44. Zhang, S., Gajpal, Y., Appadoo, S., Abdulkader, M.: Electric vehicle routing problem with recharging stations for minimizing energy consumption. Int. J. Prod. Econ. **203**, 404–413 (2018)
45. Zhen, L., Li, M., Laporte, G., Wang, W.: A vehicle routing problem arising in unmanned aerial monitoring. Comput. Oper. Res. **105**, 1–11 (2019)

Best Practices in Flux Sampling of Constrained-Based Models

Bruno G. Galuzzi[1,2], Luca Milazzo[3], and Chiara Damiani[1,2(✉)]

[1] Department of Biotechnology and Biosciences, University of Milano-Bicocca, 20125 Milan, Italy
chiara.damiani@unimib.it
[2] SYSBIO Centre of Systems Biology/ISBE.IT, Milan, Italy
[3] Department of Informatics, Systems and Communication, University of Milano-Bicocca, 20125 Milan, Italy

Abstract. Random sampling of the feasible region defined by knowledge-based and data-driven constraints is being increasingly employed for the analysis of metabolic networks. The aim is to identify a set of reactions that are used at a significantly different extent between two conditions of biological interest, such as physiological and pathological conditions. A reference constraint-based model incorporating knowledge-based constraints on reaction stoichiometry and a reasonable mass balance constraint is thus deferentially constrained for the two conditions according to different types of -omics data, such as transcriptomics and/or proteomics. The hypothesis that two samples randomly obtained from the two models come from the same distribution is then rejected/confirmed according to standard statistical tests. However, the impact of under-sampling on false discoveries has not been investigated so far. To this aim, we evaluated the presence of false discoveries by comparing samples obtained from the very same feasible region, for which the null hypothesis must be confirmed. We compared different sampling algorithms and sampling parameters. Our results indicate that established sampling convergence tests are not sufficient to prevent high false discovery rates. We propose some best practices to reduce the false discovery rate. We advocate the usage of the CHRR algorithm, a large value of the thinning parameter, and a threshold on the fold-change between the averages of the sampled flux values.

Keywords: Metabolic network · Flux sampling · Constrained-based modelling

1 Introduction

Metabolic networks represent powerful instruments to study metabolism and cell physiology under different conditions. Such models are generally studied with Constrained-Based Reconstruction and Analysis (COBRA) methods. The starting point of COBRA modelling is the information embedded in the metabolic

G. Nicosia et al. (Eds.): LOD 2022, LNCS 13811, pp. 234–248, 2023.
https://doi.org/10.1007/978-3-031-25891-6_18

network, which can be represented with a stoichiometric matrix $m \times r$, where m is the number of metabolites and r is the number of reactions. The entries in each column are the stoichiometric coefficients of the metabolites participating in a reaction. The flux through all of the reactions in a network is represented by the vector $\boldsymbol{v}$. The set of vectors for which $S\boldsymbol{v} = 0$, i.e. the null space of the stoichiometric matrix, mathematically represents the mass balance for each intracellular metabolite, and expresses all flux distributions that can be achieved by a given metabolic network. Additional constraints, such as thermodynamics or capacity constraints, can also be incorporated. All these constraints lead to a space of feasible flux distributions, each flux distribution representing a feasible state.

Assuming that cell behavior is optimal with respect to an objective, optimization methods, such as Flux Balance Analysis (FBA) [13] can be used to calculate an optimal flux distribution that allows to achieve a specific objective. However, defining one or multiple objective functions intrinsically introduces an observer bias to the main "goal" of the cell, in the context of the analysis.

Nowadays, Flux Sampling (FS) is increasingly being used [1,10,11,16,18] to explore the feasible flux solutions in metabolic networks without the need of setting an objective function. In a nutshell, FS generates a sequence of feasible solutions, called chain, that satisfy the network constraints, with the aim of covering the entire solution space. With this procedure, one can obtain information on the range of feasible flux solutions, as well as on the statistical properties of the network, using e.g. the marginal flux distributions. When *-omics* data such as proteomics or transcriptomics data, coming e.g. from different biological samples, are integrated in a metabolic network, one can use FS to generate a dataset of feasible fluxes for each different network [5,15]. Then, one can use statistical analysis to perform cluster analysis or to obtain a list of marker fluxes for each network. Another common operation consists in using a statistical test, such as Mann-Whitney (MW) or Kolmogorov-Smirnov tests, to test if a marginal flux distribution of a dataset significantly differs from that of another dataset. Thus, if the associated p-value results below a specific threshold, the null hypothesis (i.e. the first distribution is not significantly different from the second distribution) is rejected [3,5].

There are several Monte Carlo sampling methods that are currently available in the literature for studying metabolic networks, including the Hit-and-Run algorithm (HR) [2], the Artificial Centering Hit-and-Run (ACHR) [8], an OPTimized General Parallel sampler (OPTGP) [12], and Coordinate Hit-and-Run with Rounding (CHRR) [9]. All these algorithms can suffer from convergence problems, i.e. the number of samples is insufficient or the sampling does not explore the entire solution space (e.g., in case of irregular shape of the solution space). In this case, if this effect is not properly taken into account, one can make false discoveries. For example, one could reject the null hypothesis of a statistical test not because the marginal flux distribution of two different networks are different, but because the metabolic networks are under-sampled.

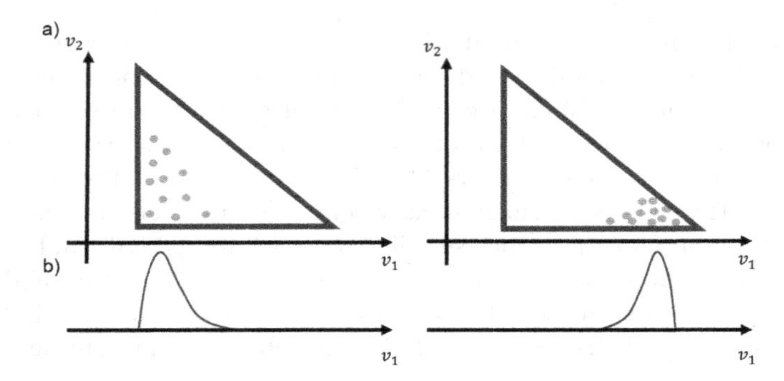

Fig. 1. a) Example of a toy network and two different samples that do not explore the feasible region entirely. b) Corresponding marginal flux distributions for v_1.

Figure 1 illustrates the key concepts of this problem. If we consider two different samples of the same network in which we do not explore the solution space entirely, we could conclude that the flux marginal distribution of v_1 for the first sample is different from that of the second one, even if the metabolic network is the same.

Some works have evaluated the performance of sampling algorithms in terms of computational time and convergence [7,12]. However, to the best of our knowledge, the impact of under-sampling on the results of statistical analyses has not been investigated yet. To this aim, we propose to study the possible effects of improper flux sampling, by comparing different samples obtained from the very same constraint-based metabolic model. More in detail, we want to investigate if different executions of the same sampling technique produce similar statistical properties, or, on the contrary, if such properties vary, thus causing a possibly high rate of false discoveries. We report in this work the results of this approach obtained with a small metabolic network, using different sample size and sampling strategies.

2 Material and Methods

2.1 Metabolic Network Model

In this study, we used the small metabolic network model ENGRO1 [4]. The network includes the catabolic pathways of glucose and glutamine and the anabolic reactions necessary for the main building blocks of biomass, i.e. amino acids and fatty acids. It consists of 84 reactions (62 irreversible and 22 reversible reactions) and 67 metabolites, all belonging to a single, lumped intracellular compartment that includes cytosolic, mitochondrial and membrane reactions. The obtained model is structurally free from thermodynamically infeasible loops and has no blocked reactions. The medium is made by six elements and includes: oxygen, glucose, glutamine, arginine, metionine, and tetrahydrofolate. We set the flux boundaries for these nutrients as in [4], for the internal reactions to $[0, 1000]$ for irreversible reaction, and to $[-1000, 1000]$ for reversible reactions, respectively.

2.2 Sampling Algorithms

HR. The first approach proposed to sample the feasible region of metabolic network is based on the HR, which is a Markov Chain Monte Carlo method. The standard HR algorithm collects samples from a given N-dimensional convex set $P \subset \mathbb{R}^N$ by choosing an arbitrary starting point $\boldsymbol{v}_0 \in P$. Setting $i = 0$, where i is the iteration number, the algorithm repeats iteratively the following three steps:

1. choosing an arbitrary direction $\boldsymbol{\theta}^i$ uniformly distributed on the boundary of then unit sphere in $\mathbb{R}^N$.
2. finding the minimum $\lambda_{\min}$ and maximum $\lambda_{\max}$ values such that $\boldsymbol{v}^i + \lambda_{\min}\boldsymbol{\theta}^i \in P$ and $\boldsymbol{v}^i + \lambda_{\max}\boldsymbol{\theta}^i \in P$.
3. generating a new sample $\boldsymbol{v}^{i+1} = \boldsymbol{v}^i + \lambda^i\boldsymbol{\theta}^i$ such that $\boldsymbol{v}^{i+1} \in P$ and $\lambda^i \in [\lambda_{\min}, \lambda_{\max}]$.

Even if convergence to the target distribution is guaranteed for this algorithm, usually the simple HR algorithm is not widely used because of the *slow-mixing* effect, that is HR creates very small steps and consequently new samples close to the previous ones in case of narrow regions, i.e. polytopes which results very narrow in some directions.

ACHR. The ACHR algorithm was the first algorithm proposed to cope the *slow-mixing* effect. The main idea of ACHR consists of using optimal direction choices in HR to allow for larger steps when the current point of the chain is situated in a narrow region. In a nutshell, after $i \geq N$ iterations of the standard HR algorithm, ACHR approximates the center of the feasible region by computing the average $\overline{\boldsymbol{v}}^{j-1} = i^{-1}\sum_{j=0}^{i-1} \boldsymbol{v}^j$ of all the samples $\boldsymbol{v}^0, \cdots, \boldsymbol{v}^{i-1}$ generated so far for each coordinate. Then, ACHR chooses a sample $\boldsymbol{v}^i$ from all the previous samples, and finds a new direction θ^a as the normalized difference between the selected sample and the current approximated center $\overline{\boldsymbol{v}}^{j-1}$, as follows:

$$\theta^a = \frac{\boldsymbol{v}^i - \overline{\boldsymbol{v}}^{j-1}}{||\boldsymbol{v}^i - \overline{\boldsymbol{v}}^{j-1}||}. \tag{1}$$

Thus, it generates a new sample as $v^{i+1} = v^i + \lambda^i\theta^a$, where λ^i is chosen as in the standard HR. Finally, to reduce the possible auto-correlation of the chain, it is possible to select a sample at each k iterates, where k is called the thinning parameter.

Even if ACHR is able to obtain chains characterized by larger steps along the elongated direction, the direction θ^a is dependent on all previous iterations and directions, then ACHR is not a Markovian algorithm. Therefore it is not guaranteed that the sequence of samples converges toward the target distribution. Moreover, for large metabolic networks, the computational time of ACHR becomes much higher compared to the HR algorithm.

OPTGP. The OPTGP algorithm was introduced to create a faster and parallel version of the ACHR algorithm. To do this, first the flux through each reaction is maximized and minimized to generate the $2N$ points. From these points, this algorithm generates multiple p chains in parallel, where p is the number of processes to be used in parallel. The sampling strategy of each chain is similar to that of ACHR, and we can select a sample at each k iterates also in this case. The use of the parallel computing allows to generate shorter chains of length Tk/p, where T is the number of desired samples and k is the thinning parameter.

CHRR. The CHRR is the most recent sampling strategy and consists of two steps: rounding and sampling. In the rounding phase, a maximum volume inscribed ellipsoid is built to match closely the polytope P. Then, the polytope is rounded through transforming the inscribed ellipsoid to a unit ball. In the sampling phase, a variant of HR algorithm known as Coordinate Hit-and-Run (CHR) is used to sample from the rounded polytope. In the CHR algorithm the direction θ^a is selected randomly among the coordinate directions. After running the CHR algorithm, the sampled points are transformed back to the original space through an inverse transformation. Since the CHRR uses the CHR algorithm for sampling, its convergence to the target distribution is guaranteed in contrast to ACHR and OPTGP based algorithms. Also in this case, it is possible to select a sample at each k iterates.

2.3 Convergence Diagnostics

For this study, we made use of two diagnostic tests, namely the Gekewe and the Raftery-Lewis diagnostics. The first one is used to see if the chain generated by a sampling algorithm does not convergence in distribution. The second one is used to measure the possible auto-correlation between the samples of the chain.

Geweke Diagnostic. This diagnostic proposed a single-chain convergence diagnostic which compares the average value of the first and last segments of a chain. If we denote B_1 the first 10% of the samples, and B_2 the last 50%, then the Geweke diagnostic computes the following quantity:

$$Z = \frac{\mu_1 - \mu_2}{\sqrt{\overline{\sigma}_1^2 + \overline{\sigma}_2^2}}, \tag{2}$$

where μ_1 and μ_2 are the average of the two sub-chains, and $\overline{\sigma}_1$ and $\overline{\sigma}_2$ are the associated standard deviations. The idea of this test is that when the sample size increases, then $\mu_1 \approx \mu_2$, and thus Z will follow a standard normal distribution, with $|Z| \approx 0$. It is common to assume convergence if $|Z| \leq 1.28$.

Raftery-Lewis Diagnostic. This diagnostic provided a dependence factor I such that:

$$I = \frac{M_{warm} + N_{\max}}{N_{\min}}, \tag{3}$$

where M_{warm} represents the number of iterations in the warm up phase, $N_{\max}$ is the required number of further iterations to estimate the quantile q to within a precision of $\pm e$ with probability p, and $N_{\min}$ is the minimum number of iterations that should be run as a pilot chain assuming independent samples. This measure represents a measure of dependency between consecutive samples (auto-correlation). It is common to assume auto-correlation if $I > 5$.

2.4 Statistical Analysis

We made use of several statistics to compare two executions of a sampling algorithm. More in detail, we used the Pearson Correlation and the Kullback-Leibler divergence to measure the similarity between two samples, and the MW test to compare the marginal flux distributions coming from two different samples.

Pearson Correlation. The Pearson Correlation between two samples i and j is given as:

$$r_{i,j} = \frac{\sum_{l=1}^{N} (\overline{\boldsymbol{v}}_{l,i} - \tilde{\boldsymbol{v}}_i)(\overline{\boldsymbol{v}}_{l,j} - \tilde{\boldsymbol{v}}_j)}{\sqrt{\sum_{l=1}^{N} (\overline{\boldsymbol{v}}_{l,i} - \tilde{\boldsymbol{v}}_i)^2 (\overline{\boldsymbol{v}}_{l,j} - \tilde{\boldsymbol{v}}_j)^2}}, \tag{4}$$

where $\overline{\boldsymbol{v}}_{l,i}$ is the sample average for the $l-$flux for the sample i, and $\tilde{\boldsymbol{v}}_i$ is the across-flux average.

Kullback-Leibler Divergence. The Kullback-Leibler divergence (KLD) represents a measure of dissimilarity between two distributions. Given two samples i and j, if we consider the two marginal distribution $p_i(v_k)$ and $p_j(v_k)$ of a specific flux k, then the KL divergence is defined as

$$KLD\left(p_i | p_j\right) = \int_{-\infty}^{\infty} \ln\left(\frac{p_i(v_k)}{p_j(v_k)}\right) p_i(v_k) dv_k. \tag{5}$$

The KLD is zero only if p_i and p_j are identical functions. We can classify the difference between the two distributions as low if KLD < 0.05, moderate if $0,05 \leq KLD \leq 0.5$ and high if $KLD > 0.5$, as in [7].

Statistical Tests. We used the MW test to compare the possible statistical differences between the marginal distributions of two samples. The null hypothesis is that the two distributions come from the same distribution. We rejected the hypothesis if the associated p-value results less than 0.01. We adjusted such a value applying the Benjamini-Hochberg Procedure.

However, even if the two distributions result different (i.e. adjusted $p < 0.01$), the effective difference between their averages could be so small to not represent a significant change. To take into account this fact, we considered also the fold-change between the two distributions i and j, defined as:

$$FD_{i,j}(l) = \left| \frac{\overline{\boldsymbol{v}}_{l,i} - \overline{\boldsymbol{v}}_{l,j}}{\overline{\boldsymbol{v}}_{l,j}} \right|. \tag{6}$$

where $\overline{v}_{l,i}$ is the sample average for the $l-$flux and sample i, and $\overline{v}_{k,j}$ is the sample average for the $l-$flux and sample j. This value can be used to filter out all statistical tests for which the Fold Change value results less than a certain threshold.

3 Experimental Setting

We applied flux sampling of the ENGRO1 metabolic network. For the sampling, we varied:

- the sampling algorithm (ACHR, OPTGP, CHRR)
- the thinning parameter k $(1, 10, 100)$
- the sample size n between 1000 and 30000 with step 1000.

For each of these configurations, we performed 20 different executions to simulate a possible statistical analysis among different samples. Thus, we computed the following quantities:

- the Pearson Correlation between any of the possible couples (190 in total) which can be extracted from the 20 samples, as in Eq. 4.
- for each reaction, the MW test between all possible couples (190 in total), obtained from the 20 samples. The total number of tests results 190×84 for each possible configuration. We introduced the False Discovery Rate (FDR) as the fraction of MW tests which reject the null hypothesis. Such quantity should be virtually 0 since we consider the same metabolic network.
- for each sample, the rejection-rate for the Geweke and Raftery-Lewis diagnostics obtained for the 84 marginal chains. This quantity is computed as the fraction of non passed diagnostic tests.

For ACHR and OPTGP, we used the implementation provided by the COBRApy library [6]. For CHRR, we used the implementation available in the COBRA toolbox [17]. The diagnostic tests were computed using the corresponding functions for the CODA package [14]. The statistical tests were computed using the corresponding functions of the Scipy library [19]. All the computations were run on an Intel(R) Xeon @3.5 GHz (32 GB RAM).

4 Results

4.1 CHRR Mitigates the Risk of False-Discoveries

To assess the coherence between the different executions of each sampling algorithm, we first computed the average over the 190 Pearson Correlation coefficients obtained across all reactions. Table 1 reports the average values obtained for the different n and k under study. It can be noticed that the Pearson Correlation results over $0,99$ already when the $n = 1000$, exception made for the cases ACHR, with $k = 1$ and $k = 10$, and CHRR with $k = 1$.

Table 1. Average over the Pearson Correlation coefficients across all reactions, obtained for different values of n and k.

Algorithm	Thinning	$n = 1000$	$n = 5000$	$n = 10000$	$n = 30000$
ACHR	1	0.928	0.966	0.976	0.987
ACHR	10	0.979	0.993	0.996	0.998
ACHR	100	0.997	0.999	1	0.999
OPTGP	1	0.998	0.998	0.997	0.998
OPTGP	10	0.996	0.998	0.999	1
OPTGP	100	0.998	1	1	1
CHRR	1	0.968	0.990	0.995	0.999
CHRR	10	0.996	0.999	0.999	1
CHRR	100	1	1	1	1

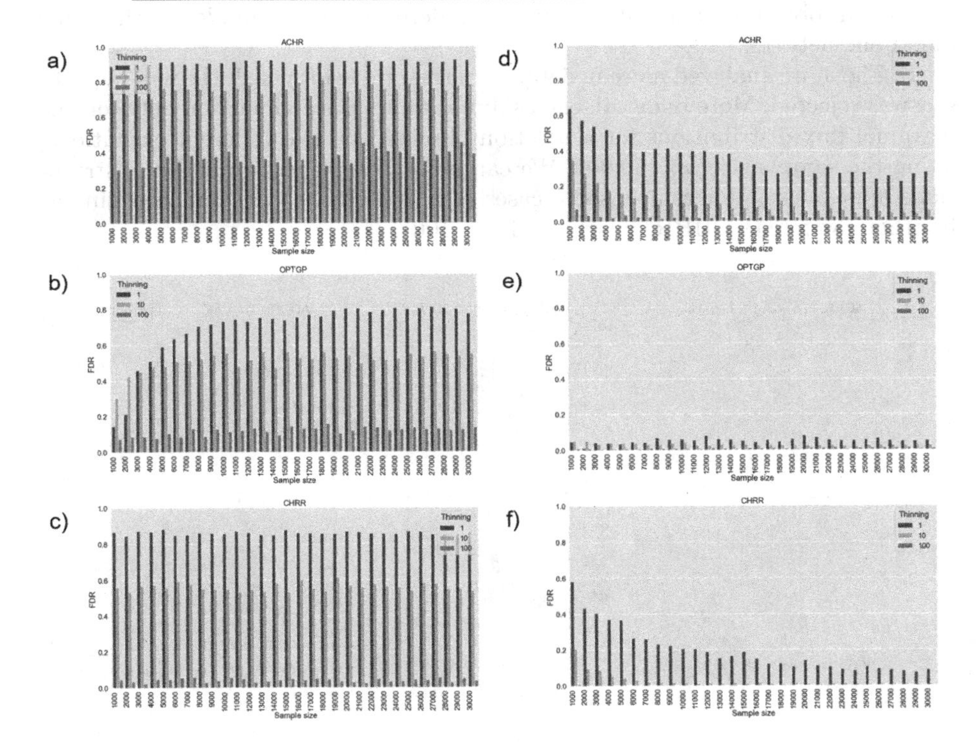

Fig. 2. On the left, bar-plots representing the FDR using ACHR (a), OPTGP (b), and CHRR (c), as a function of n and using different colors for the different k values. On the right, the same bar-plots representing the FDR using ACHR (d), OPTGP (e), and CHRR (f), with the application of the fold-change filter.

Despite the strong correlation between the different samples obtained before, different samples could create different marginal flux distributions with the risk of a non-negligible FDR. In Fig. 2, we reported the bar-plots representing the FDR using ACHR (a), OPTGP (b) and CHRR (c), as a function of the sample size n and using different colors for value of k.

Surprisingly, high values of FDR could be observed and there is no significant trend as a function of the sample size n, but rather as a function of the thinning value k. In particular, there is always a non-negligible value of FDR for all the algorithms and, with $k = 1$, the FDR even reaches a value between 0.8 and 0.9. However, when k increases, the FDR decreases considerably and important differences between the sampling strategies become more evident. The CHRR ($k = 100$) shows the best performances having the lowest FDR value, followed by OPTGP ($k = 100$) and ACHR ($k = 100$). From these results, we can assume that an improper sampling can cause the presence of false statistical discoveries, i.e. we can obtain statistically significant differences from samples of the same metabolic network.

In Fig. 3, we analyzed more in detail some reactions for which the null hypothesis was rejected. More in detail, each subplot shows the comparison between the marginal flux distributions for a reaction obtained by two different executions, fixing the sample size $n = 30.000$. We can note that the marginal flux distributions appear very different is some cases (e.g. ACHR, $k = 1$), but very similar in other cases (e.g. CHRR, $k = 10$ and $k = 100$).

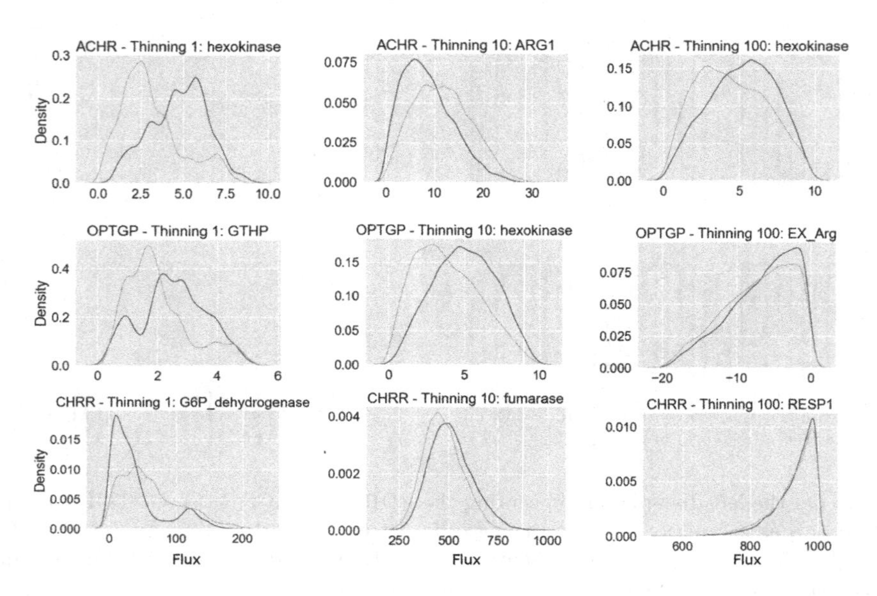

Fig. 3. Marginal flux distributions of a specific reaction of the ENGRO1 model, obtained from two different executions of the same sampling procedure.

4.2 Filtering on Fold-Change Reduces the Risk of False Discoveries

To reduce the FDR, we introduced a filter on the statistical tests based on the fold-change. Hence, we rejected the null hypothesis only if the adjusted p-value is less than 0.01 and the fold-change of the averages is upper than 0.2. In Fig. 2d–f, we reported the bar-plots representing the FDR but with fold-change applied. An important reduction of FDR can be appreciated. In particular, the FDR for CHRR, with $k = 100$, becomes virtually 0. Moreover, a significant decreasing trend can be observed for ACHR and CHRR for thinning 1 as a function of n. From these results, we concluded that the FDR can be reduced by applying the fold-change filter and using a thinning value of 100. Moreover, CHRR is confirmed better than ACHR and OPTGP in exploring the feasible region.

4.3 Standard Diagnostic Analysis Does Not Prevent False-Discoveries

To investigate the possible causes of these false discoveries, we investigated the level of the convergence and auto-correlation of the various samples. In Fig. 4, we reported the bar-plots representing the rejection-rate for the Geweke and Rafery-Lewis diagnostic[1], using ACHR, OPTGP, and CHRR, as a function of n and k. Surprisingly, both the level of convergence and the auto-correlation seem to be strongly affected by k but not by n. Concerning the Raftery-Lewis diagnostic, all the algorithms assure a negligible rejection-rate when $k = 100$ is used. The Geweke rejection-rate decrease as a function of k, and CHRR results to be the algorithm which assures the lower level of Geweke rejection-rate, followed by ACHR and OPTGP. However, even if we use CHRR with $k = 100$ and $n = 30000$, the Geweke rejection-rate remains upper than 0.15. Moreover, the rejection-rates are very different as a function of the algorithm and the configuration parameter n and k. For example, for the OPTGP algorithm with $k = 1$, the Geweke rejection-rate for $n = 15000$ results even higher than $n = 1000$.

We investigated if the FDR results related to the Geweke rejection-rate of marginal chains generated by a sampling strategy. Thus, for any configuration of the sampling strategy (i.e. type of algorithm and thinning value), we computed the correlations between the FDR across all the reactions and samples sizes, and the average value of Geweke rejection-rate of the corresponding marginal chains. We reported the values of the correlations in Table 2. We did not consider CHRR thinning 100 because of the very low FDR value. A moderate level of correlation between FDR and the Geweke rejection-rate is observed, especially for the ACHR algorithm. In this case, a way to mitigate these false discoveries could be to not consider all tests for which one of marginal flux distribution does not reach the convergence. However, such a correlation appear quite low and so, this strategy could be not sufficient to remove the FDR completely.

---

[1] Note that it was not possible to compute the Raftery-Lewis tests when the n results less than 3.746.

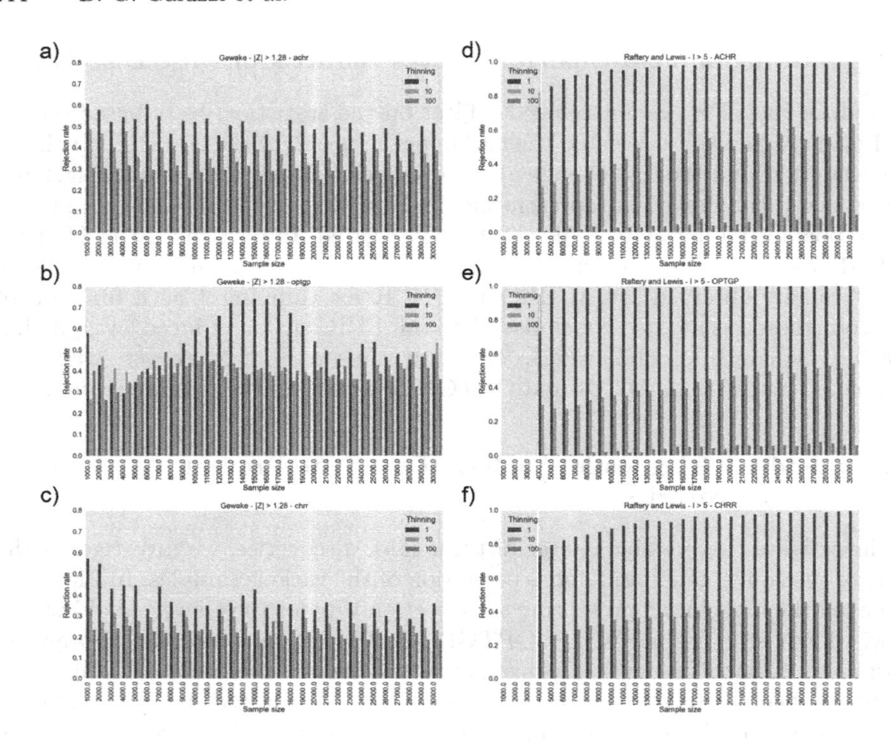

Fig. 4. On the left, bar-plots representing the rejection-rate for the Geweke diagnostics using ACHR (a), OPTGP (b), and CHRR (c), as a function of the n and using different colors for different k values. On the right, bar-plots representing the rejection-rate for Raftery-Lewis diagnostics using ACHR (d), OPTGP (e), and CHRR (f).

Table 2. Pearson Correlation between FDR and Geweke rejection-rate, using different algorithms and thinning values.

Algorithm	Thinning	FDR vs Geweke
ACHR	1	0.39
ACHR	10	0.42
ACHR	100	0.32
OPTGP	1	0.09
OPTGP	10	0.25
OPTGP	100	0.31
CHRR	1	0.14
CHRR	10	0.05

4.4 Sample Size Increases Similarity but Does Not Decrease FDR

The results reported so far suggest that a large value of the thinning parameter k is much more important than a large sample size n to mitigate the incidence of false discoveries. Remarkably, it seems that increasing the sample size does not reduce the FDR at all. To further investigate this interesting result, we analyzed how k and n affect the similarity between the different executions of the algorithms. To this aim we performed dimensionality reduction analysis and analyzed the cumulative distribution of the Kullback-Leibler divergence (KLD-CDF), focusing on the CHRR algorithm.

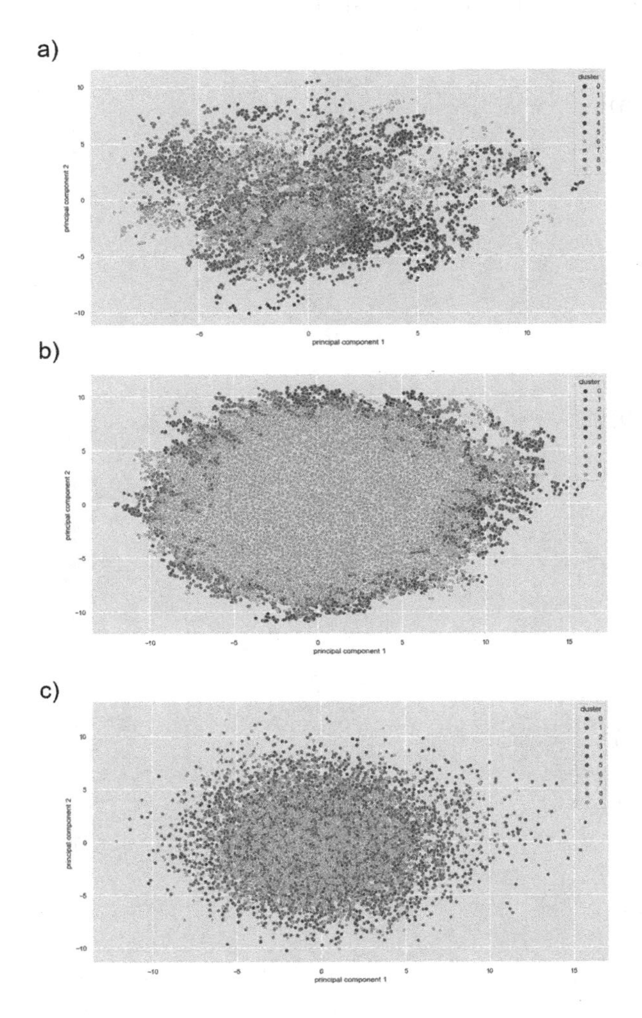

Fig. 5. PCA representation of the dataset formed by the first 10 executions of CHRR using: $k = 1$ and $n = 1.000$ (a), $k = 1$ and $n = 30000$ (b), $k = 100$ and $n = 1000$ (c).

Figure 5 reports the Principal Component Analysis (PCA) representation of the dataset formed by the first 10 executions, for different value of k and/or n. We can note that for a small value of k and n (Fig. 5a), the points sampled within the same execution tend to cluster together, regardless of the sample size. When increasing either the thinning value or the sample size (Fig. 5b and c) this cluster effect vanishes, indicating that different executions similarly explore the solution space.

Similar conclusions can be derived from the KLD-CDF analysis. Figure 6 reports the KLD-CDF for different values of k and/or n. It can be observed that, the KLD values tend to be lower for lower values of n or k. However, the effect is more accentuated for the thinning value k.

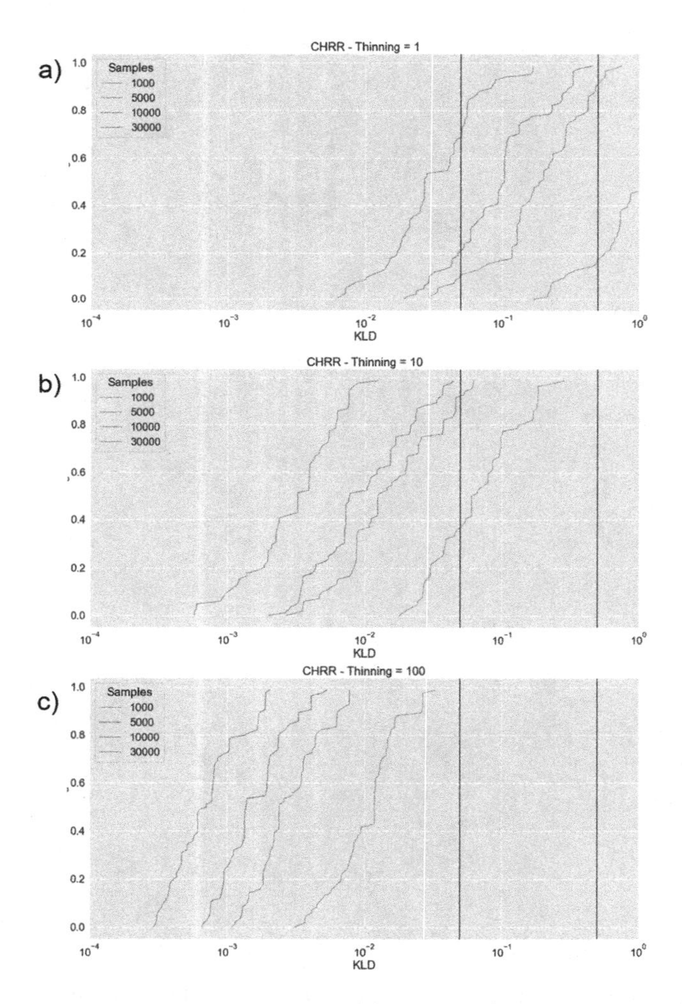

Fig. 6. KLD-CDF for CHRR using $k = 1$ (a), $k = 10$ (b) and $k = 100$ (c). In each subplot, different colors identify different sample sizes.

All in all these results indicate that although increasing the sample size does increase the similarity of the samples obtained with different executions of the same sampling algorithm, this fact is not sufficient to reduce the incidence of false discoveries.

5 Conclusions

Flux sampling represents a powerful tool to integrate omics data into metabolic networks with the aim of characterizing the behaviour of different biological conditions. However, the statistical results coming from a flux sampling must be handled with care. Indeed, we have showed that, without the necessary cautions, one might observe statistically significant differences even if when analyzing samples coming from the very same feasible region. The experimental results obtained varying the sampling algorithm, the sample size, and the thinning parameter, led us to the following conclusions: CHRR represents the most promising sampling algorithm because of its better ability to explore the solution space; increase the thinning value is very important to reduce the auto-correlation of the chains and the false discovery rate; the use of a filter based on the fold-change could help to reduce the false discovery rate. Moreover, we also showed that performing standard diagnostic analyses does not exclude the risk of high FDR values. Hence, we recommend to execute different executions of the same sampling in order to identify the thinning value that keeps the FDR below an acceptable level.

As a further work, we will repeat our investigation on genome-wide metabolic networks, including tens of thousands of reactions. The problem of false discoveries could indeed amplify when one considers more complex feasible regions. We will also investigate corner-based sampling strategies, based on random objective functions, which we expect to mitigate the problem.

References

1. Almaas, E., Kovacs, B., Vicsek, T., Oltvai, Z., Barabási, A.: Global organization of metabolic fluxes in the bacterium Escherichia coli. Nature **427**, 839–843 (2004)
2. Bélisle, C., Romeijn, H., Smith, R.: Hit-and-run algorithms for generating multivariate distributions. Math. Oper. Res. **18**, 255–266 (1993)
3. Bordel, S., Agren, R., Nielsen, J.: Sampling the solution space in genome-scale metabolic networks reveals transcriptional regulation in key enzymes. PLoS Comput. Biol. **6**, e1000859 (2010)
4. Damiani, C., et al.: A metabolic core model elucidates how enhanced utilization of glucose and glutamine, with enhanced glutamine-dependent lactate production, promotes cancer cell growth: the WarburQ effect. PLoS Comput. Biol. **13**, e1005758 (2017)
5. Di Filippo, M., et al.: INTEGRATE: model-based multi-omics data integration to characterize multi-level metabolic regulation. PLoS Comput. Biol. **18**, e1009337 (2022)
6. Ebrahim, A., Lerman, J., Palsson, B., Hyduke, D.: COBRApy: constraints-based reconstruction and analysis for python. BMC Syst. Biol. **7**, 1–6 (2013)

7. Fallahi, S., Skaug, H., Alendal, G.: A comparison of monte Carlo sampling methods for metabolic network models. PLoS ONE **15**, e0235393 (2020)
8. Kaufman, D., Smith, R.: Direction choice for accelerated convergence in hit-and-run sampling. Oper. Res. **46**, 84–95 (1998)
9. Haraldsdóttir, H., Cousins, B., Thiele, I., Fleming, R., Vempala, S.: CHRR: coordinate hit-and-run with rounding for uniform sampling of constraint-based models. Bioinformatics **33**, 1741–1743 (2017)
10. Herrmann, H., Dyson, B., Miller, M., Schwartz, J., Johnson, G.: Metabolic flux from the chloroplast provides signals controlling photosynthetic acclimation to cold in Arabidopsis thaliana. Plant Cell Environ. **44**, 171–185 (2021)
11. Herrmann, H., Dyson, B., Vass, L., Johnson, G., Schwartz, J.: Flux sampling is a powerful tool to study metabolism under changing environmental conditions. NPJ Syst. Bio. Appl. **5**, 1–8 (2019)
12. Megchelenbrink, W., Huynen, M., Marchiori, E.: optGpSampler: an improved tool for uniformly sampling the solution-space of genome-scale metabolic networks. PLoS ONE **9**, e86587 (2014)
13. Orth, J., Thiele, I., Palsson, B.: What is flux balance analysis? Nat. Biotechnol. **28**, 245–248 (2010)
14. Plummer, M., Best, N., Cowles, K., Vines, K.: CODA: convergence diagnosis and output analysis for MCMC. R News **6**, 7–11 (2006)
15. Režen, T., Martins, A., Mraz, M., Zimic, N., Rozman, D., Moškon, M.: Integration of omics data to generate and analyse COVID-19 specific genome-scale metabolic models. Comput. Biol. Med. **145**, 105428 (2022)
16. Schellenberger, J., Palsson, B.: Use of randomized sampling for analysis of metabolic networks. J. Biol. Chem. **284**, 5457–5461 (2009)
17. Schellenberger, J., et al.: Quantitative prediction of cellular metabolism with constraint-based models: the COBRA Toolbox v2.0. Nat. Protoc. **6**, 1290–1307 (2011)
18. Scott, W., Smid, E., Block, D., Notebaart, R.: Metabolic flux sampling predicts strain-dependent differences related to aroma production among commercial wine yeasts. Microb. Cell Fact. **20**, 1–15 (2021)
19. Virtanen, P., et al.: SciPy 1.0: fundamental algorithms for scientific computing in Python. Nat Methods **17**, 261–272 (2020)

EFI: A Toolbox for Feature Importance Fusion and Interpretation in Python

Aayush Kumar[1], Jimiama M. Mase[1](✉), Divish Rengasamy[1], Benjamin Rothwell[1], Mercedes Torres Torres[2], David A. Winkler[3], and Grazziela P. Figueredo[1]

[1] The University of Nottingham, Nottingham, UK
psxjmma@nottingham.ac.uk
[2] B-Hive Innovations, Lincoln, UK
[3] Monash Institute for Pharmaceutical Sciences, Parkville, Australia

Abstract. This paper presents an open-source Python toolbox called Ensemble Feature Importance (EFI) to provide machine learning (ML) researchers, domain experts, and decision makers with robust and accurate feature importance quantification and more reliable mechanistic interpretation of feature importance for prediction problems using fuzzy sets. The toolkit was developed to address uncertainties in feature importance quantification and lack of trustworthy feature importance interpretation due to the diverse availability of machine learning algorithms, feature importance calculation methods, and dataset dependencies. EFI merges results from multiple machine learning models with different feature importance calculation approaches using data bootstrapping and decision fusion techniques, such as mean, majority voting and fuzzy logic. The main attributes of the EFI toolbox are: (i) automatic optimisation of ML algorithms, (ii) automatic computation of a set of feature importance coefficients from optimised ML algorithms and feature importance calculation techniques, (iii) automatic aggregation of importance coefficients using multiple decision fusion techniques, and (iv) fuzzy membership functions that show the importance of each feature to the prediction task. The key modules and functions of the toolbox are described, and a simple example of their application is presented using the popular Iris dataset.

Keywords: Feature importance · Fuzzy logic · Decision fusion · Interpretability · Machine learning interpretation · Responsible AI

1 Introduction

Machine Learning (ML) systems are providing very useful autonomous and intelligent solutions in diverse science and technology domains. In areas where the decisions from the ML systems ultimately affect human lives (e.g., healthcare,

---

A. Kumar and J. M. Mase—Equally contributed to the work.

G. Nicosia et al. (Eds.): LOD 2022, LNCS 13811, pp. 249–264, 2023.
https://doi.org/10.1007/978-3-031-25891-6_19

transport, law and security), it is important to understand how and why the decisions are made for system verification [1], regulatory compliance [11], elucidation of ethical concerns, trustworthiness [6], and system diagnostics [13]. A potential solution to the problem of understanding system decisions is model interpretability that aims to understand decisions of complex ML architectures not readily interpretable by design (e.g. neural networks, support vector machines and ensembles of decision trees). A popular approach used in model interpretation is feature importance (FI) analysis, which estimates the contribution of each data feature to the model's output [13]. Several comprehensive open-source libraries already exist for ML automation [5,7,16] and explainability [2,3,9,10]. However, the available libraries do not cover ensemble feature importance and fuzzy logic interpretability.

The wide availability of ML algorithms and diversity of FI techniques complicates the selection of ML and FI approaches and the reliability of interpretations. Different ML models may generate different FI values due to variations in their learning algorithms. Similarly, different FI techniques may produce different importances for the same ML algorithms, and different data samples may produce different importances for the same ML algorithms and FI techniques. Additionally, for models in which the response surface is significantly nonlinear (curved in multidimensional space), FI is a local rather than global property.

To address these uncertainties in the selection and interpretation of models, ensemble feature importance (EFI) methods have been proposed that combine the results from multiple ML models coupled with different feature importance quantifiers to produce more robust, accurate, and interpretable estimates of FI [4,8,12,13,15,18]. The latest ensemble methods apply multiple model-agnostic feature importance methods to multiple trained ML algorithms, and aggregate the resulting FI coefficients using crisp [13] or/and fuzzy [12] decision fusion strategies.

The fuzzy ensemble feature importance (FEFI) method extends the crisp ensemble methods by modelling the variance in feature importance generated by the different ML methods, FI techniques, and data spaces using fuzzy sets. FEFI has shown to provide more reliable FI values for high-dimensional datasets with non-linear feature relationships, and noise injected by system dynamics [12]. In addition, FEFI provides better explanations of FI using linguistic terms: 'low', 'moderate' and 'high' importance.

To further enhance the utility of fuzzy approaches to FI, here we present an Ensemble Feature Importance (EFI) toolbox, developed in the Python programming environment (available online[1]) that implements the crisp and fuzzy ensemble feature importance strategies. It includes automatic data preprocessing, model training and optimisation, ensemble feature importance calculation, and feature importance interpretation. The following sections present the main modules and functions of EFI toolbox as an approach to assist ML researchers, domain experts, and decision makers to obtain reliable interpretations of the importance of features. It is also a diagnostic tool for identifying data subsets

[1] https://github.com/jimmafeni/EFI-Toolbox.

with extreme cases of feature importance or with significant variation of feature importance.

2 EFI Toolbox

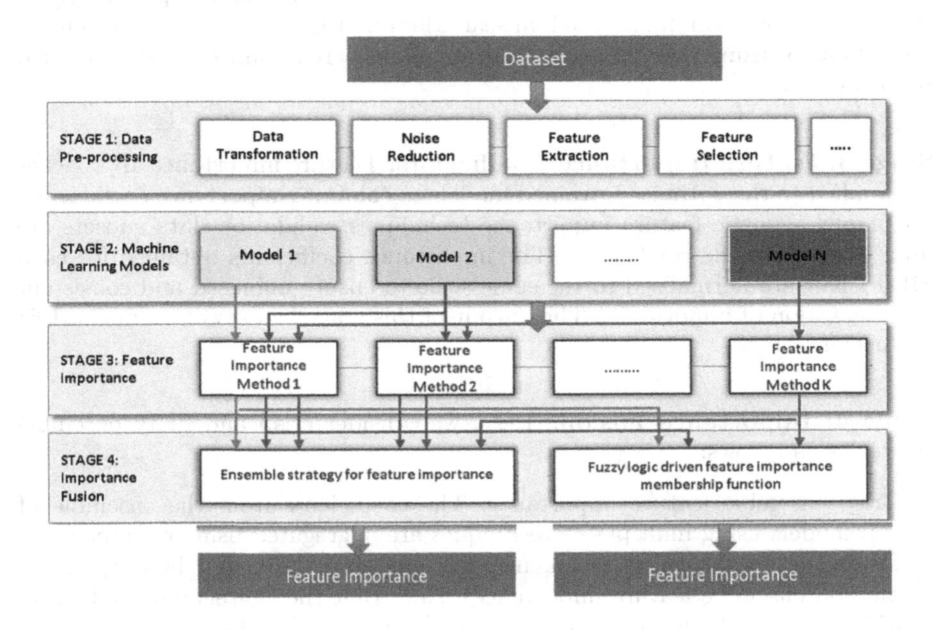

Fig. 1. The four stages of EFI framework.

2.1 An Overview of EFI Framework

Currently, the EFI toolbox is implemented for classification problems based on the EFI framework proposed for crisp [13] and fuzzy ensemble strategies [12]. However, the toolbox can be easily adapted for regression problems. The EFI framework generates a set of FI coefficients using an ensemble of ML models coupled with multiple FI techniques and data bootstrapping. The FI coefficients are aggregated using crisp decision fusion strategies to produce robust and accurate FIs or/and using fuzzy ensemble (FEFI) for a more robust interpretation of FIs. A flowchart of the EFI framework is shown in Fig. 1 and the four stages of the process are described in the following subsections.

Stage 1: Data Preprocessing. Data are preprocessed to manage missing values, outliers, irrelevant features, and highly correlated features. Next, they are normalised to ensure that all features are of the same magnitude. The preprocessed data are partitioned into k equally sized data samples for model optimisation i.e., k-fold cross validation.

Stage 2: Machine Learning Model Optimisation. Here, an ensemble of optimised ML models is generated using cross-validation and hyperparameter tuning (e.g., grid search and random search) on multiple ML algorithms. This is important to ensure that the most accurate models are obtained for the specified problem.

Stage 3: Feature Importance Coefficients. Feature importance approaches are applied to the optimised, trained models to compute importance coefficients. For model-agnostic feature importance techniques, validation data subsets are used to produce the coefficients. The importance coefficients obtained for each ML-FI pair are normalised to the same scale to ensure unbiased and consistent representation of importance. The output of this stage is a set of normalised FI coefficients.

Stage 4: Importance Fusion. Here, we consider crisp and fuzzy ensemble strategies as follows:

- Crisp ensemble feature importance: The coefficients from the ensemble of ML models using multiple FI techniques are aggregated using crisp decision fusion methods such as mean, median, mode etc. (denoted here as multi-method ensemble feature importance). Note that the aggregated coefficients of individual ML algorithms can also be obtained by filtering coefficients from a particular ML algorithm (denoted here as model-specific ensemble feature importance).
- Fuzzy ensemble feature importance: FEFI generates membership functions (MFs) from the set of coefficients, labelling the importance of features as 'low', 'moderate' and 'high'. For each ML approach, FEFI assigns the labels low, moderate and high importance to the feature set. Each feature is assigned low, moderate and high importance relative to the other features in the data for each ML approach and for all ML approaches combined.

2.2 Toolbox Modules

A modular programming approach is employed in developing the toolbox for easy documentation, debugging, testing, and extensibility as shown in Fig. 2, with a *Main()* module for executing the various stages of EFI framework. The main modules of EFI toolbox are: -

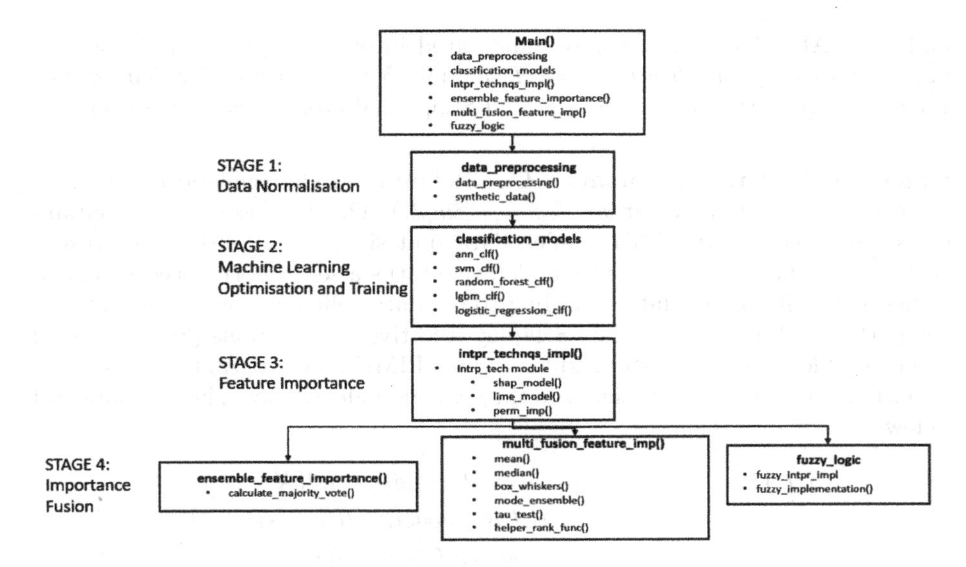

Fig. 2. Layout of EFI toolbox.

Data Preprocessing. The essential task of automatic preprocessing of the data (i.e. normalise and encode target labels) occurs in the data_preprocessing module. This module can also generate synthetic data, using synthetic_data(), for testing the toolbox. The following is an example of the application of this module, where a feature set X is preprocessed and target labels Y are encoded:

$$x, y = data\_preprocessing(X, Y, loc\_specified = 'No')$$

A user can also pass the data location and target column name as X and Y parameters and set *loc_specified* to 'Yes'.

Model Training and Optimisation. The *classification_models* module automatically optimises ML algorithms using grid search hyperparameter tuning and cross validation. The best performing hyperparameters are used to train and return the model. A user can choose which ML algorithm to optimise and train by passing as parameters the dataset and number of folds, as shown in the command below:

$$ann = ann\_clf(X\_train, Y\_train, X\_test, Y\_test, k)$$

The command returns an optimised neural network by performing k-fold cross-validation on the training set (X_train, Y_train). The optimised model is evaluated using (X_test, Y_test) and its performance (classification accuracy and ROC-AUC curve) is displayed and saved in output files. There exist similar functions in *classification_models* module for optimising Gradient Boost, Logistic Regressor, Random Forests and Support Vector Machine classifiers.

Calculate Feature Importance Coefficients. Model-agnostic FI methods are run using the function *intpr_technqs_impl*(). This function takes as parameters a dataset, a trained ML model, data split size for computing importance coefficients, and the name of the model. It returns a set of normalised FI coefficients. It partitions the data according to the data split size parameter and uses the partitioned data to calculate SHapley Additive exPlanations (SHAP), Local Interpretable Model-agnostic Explanations (LIME), and Permutation Importance (PI) importance coefficients on the trained model as shown in the command below:

$$SHAP,\ LIME,\ PI = intpr\_technqs\_impl(X,\\ Y, LR\_model, datasplitsize = 0.2,\\ 'LogisticRegression')$$

The above command returns the SHAP, LIME, and PI importance coefficients of the features in dataset *(X, Y)* when 80% of the dataset is used to train and optimise a logistic regression model and the remaining 20% of the dataset is used to calculate FI coefficients.

The module uses functions within *intrp_tech* module (i.e., shap_model(), lime_model(), perm_imp()) to calculate the importance coefficients for SHAP, LIME, and PI respectively. For example,

$$PI = perm\_imp(X, Y, ANN\_model\\ model\_name)$$

returns PI importance coefficients of the features in dataset *(X, Y)* on the trained ML model (i.e. ANN_model). These functions are implemented using Python functions for model-agnostic feature importance quantification:

- *permutation_importance():* This function is found in sklearn.inspection module. It calculates the importance of features in a given dataset by measuring the increase in the prediction error of the trained model after permuting the values of features [13].
- *KernelExplainer():* KernelExplainer function is found in the shap library for calculating the SHAP values for a set of samples. SHAP determines the average marginal contribution of features to a model's prediction by considering all possible combinations of the features [13].
- *LimeTabularExplainer(.):* This function is found in lime library for calculating the LIME values of features of randomly selected instances in the dataset. LIME determines the average contribution of features of randomly selected instances in the dataset by generating a new dataset consisting of perturbed samples with corresponding predictions from the original model, and training

interpretable models on the new dataset [14]. The perturbed samples are weighted according to their proximity from the randomly selected instances.

Model-Specific Ensemble Feature Importance. After calculating the feature importance coefficients, the toolbox provides a function to combine the feature importance coefficients for individual ML algorithms using a majority vote ensemble method i.e. *ensemble_feature_importance*(). This model-specific ensemble method is useful to obtain the importance of features for a specific ML algorithm with higher classification accuracy compared to the other ML algorithms. For example, the command

$$LR\_fi = ensemble\_feature\_importance(SHAP,\\ LIME, PI, model\_name, num\_features)$$

combines the importance coefficients obtained by applying *SHAP*, *LIME*, and *PI* techniques on the ML algorithm specified using 'model_name' parameter. For example, if the *model_name* is equal to 'Logistic Regression', the command will return the aggregated importance coefficients for the Logistic Regression classifier. The value of the parameter num_features ranges from 0 to 1 and it determines the number of features to consider in the ranking process of majority vote ensemble [13].

Multi-method Ensemble Feature Importance. The EFI toolbox provides a function, *multi_fusion_feature_imp*(), for combining the importance coefficients from multiple ML algorithms using multiple FI techniques. This is suitable for problems where the ML algorithms do not provide required predictive performance and there exist significant variations in the prediction of instances and the quantification of FIs. With such problems, a multi-method ensemble usually achieves more accurate and robust interpretations of FIs. For example, this command:

$$fi\_df = multi\_fusion\_feature\_imp(SHAP,\\ LIME, PI, X, models\_selected)$$

combines the importance coefficients obtained by applying *SHAP*, *LIME*, and *PI* techniques on the ML algorithms specified using the *models_selected* parameter. Therefore, if '*models_selected*' is equal to ['Logistic Regression','Random Forests'], the feature will combine the importance coefficients obtained from both models.

The following decision fusion methods are employed within this module: Mode, Median, Mean, Box-Whiskers, Tau Test, Majority Vote, RATE-Kendall Tau, RATE-Spearman Rho. For more information regarding these methods, please refer to [13]. The methods are implemented using the following functions within *multi_fusion_feature_imp*():

- *majority_vote_func*(): Ranks features based on their coefficients produced by each ML model and FI technique. Calculate the mean of the coefficients in the most common ranks for each feature.
- *mean*(): Averages importance coefficients obtained by FI techniques for each feature.
- *median*(): Sorts importance coefficients for each feature and returns the middle coefficients as the final coefficient of features.
- *box_whiskers*(): Averages coefficients that lie between the lower and upper whiskers of their distribution. The whiskers are obtained by adding the interquartile range to the third quartile (upper whisker) or subtracting the interquartile range from the first quartile (lower whisker) to eliminate potential outliers in the data.
- *mode_ensemble*(): Uses a probability density function to get an estimate of the mode of the coefficients for each feature.
- *tau_test*(): Uses Thompson Tau test to detect and remove anomalies, and uses the mean to determine the final coefficient of each feature.
- *helper_rank_func*(): Uses Kendall and Spearman statistical tests to eliminate outliers and employs a majority vote ensemble to determine the final coefficient of each feature.

Fuzzy Ensemble Feature Importance. The toolbox provides a module called *fusion_logic* for combining the importance coefficients using fuzzy logic. This prevents loss of information and provides a more understandable representation of importance as linguistic terms i.e. 'low', 'moderate' and 'high' importance. The main function in this module, $fuzzy\_implementation()$, takes as parameters a dataset of FIs obtained using cross-validation and a list of model names to generate their membership functions [17]. For example, the command:

$$fuzzy\_implementation(FIs, model\_names)$$

combines the importance coefficients (FIs) obtained by applying feature importance techniques on the ML algorithms to produce membership functions determining:

1. for each ML approach listed in 'model_names', the range of low, moderate and high importance in the feature set after training;
2. for each feature, its low, moderate, and high importance relative to the other features in the data for each ML approach;
3. for each feature, its low, moderate, high importance after training and after combining with importances from multiple ML approaches.

3 Case Study

In this section, we will use the popular Iris dataset to illustrate the application of our toolbox for feature importance fusion and interpretation.

3.1 The Iris Dataset

The Iris dataset is an open-source flower classification dataset that consists of three types of flowers i.e. Setosa, Versicolour, and Virginica. The dataset is made up of 50 samples from each of the three types of iris flowers and for each sample, four features are reported: sepal length, sepal width, petal length and petal width. Figure 3 presents a code snippet to load the dataset in Python.

```
#Load iris data
import pandas as pd
from sklearn.datasets import load_iris

data = load_iris()
X = pd.DataFrame(data.data, columns=data.feature_names)
y = pd.DataFrame(data.target, columns=['Iris_type'])
```

Fig. 3. Code snippet to load Iris data

3.2 Data Pre-processing

We normalise the features in the iris dataset and transform all target labels to normalised integer class representations. Figure 4 presents a code snippet to pre-process the loaded iris dataset *(X,y)* using our toolbox:

```
import data_preprocessing as dp
x, y = dp.data_preprocessing(X, y,loc_specified='No' )
```

Fig. 4. Code snippet to pre-process Iris data using EFI

3.3 Model Optimisation and Training

We optimise and train random forest, support vector machine, and neural network models to classify iris flowers as shown in Fig. 5.

The optimised random forest classifier (model_rf) has an average accuracy of 90% and an average f1-score of 88.4% with hyperparameters: {'bootstrap': False, 'criterion of split': 'entropy', 'minimum samples at leaf node': 5, 'number of trees': 300}. The optimised support vector machine classifier (model_svm) produces an average accuracy of 90% and an average f1-score of 89.5% with hyperparameters: {'regularisation coefficient': 1, 'kernel': 'rbf', 'kernel coefficient': 0.1,}. The optimised neural network model (model_nn) produces an average accuracy of 91% and an average f1-score of 83.1% with hyperparameters: {'learning rate': 0.001, 'batch size': 2, 'epochs': 25}.

```
import classification_models as cm

#test data
data_size_for_testing = 0.4
# K-Fold Cross validation for model
cv = 5
# split into train test sets as per the configuration
x_train, x_test, y_train, y_test = train_test_split(x, y,
                 test_size=data_size_for_testing, random_state=42,
                 shuffle=True,stratify=y)

#Optimise and train Random Forest, support Vector machine
# and neural network classifiers
model_rf = cm.random_forest_clf(x, y, x_train, x_test, y_train,y_test, cv)
model_svm = cm.svm_clf(x, y, x_train, x_test, y_train,y_test, cv)
model_nn = cm.ann_clf(x, y, x_train, x_test, y_train,y_test, cv)
```

Fig. 5. EFI code to optimise machine learning models

3.4 Feature Importance Coefficients

We calculate the contribution (importance) of each feature (sepal length, sepal width, petal length and petal width) in classifying the three types of iris flowers by applying the FI techniques on the models trained above. Figure 6 illustrates the calculation of FI coefficients of random forest, support vector machines and artificial neural networks.

```
#Extract names of all ML models implemented in the toolbox
import user_xi as usxi
import interpretability_methods as im

#create list of models that have been trained
model_selected = ['Random Forest classifier',
                  'Support vector machines',
                  'Artificial Neural Network']

SHAP_RESULTS = pd.DataFrame(index=x.columns.values, columns=usxi.models_to_eval)
LIME_RESULTS = pd.DataFrame(index=x.columns.values, columns=usxi.models_to_eval)
PI_RESULTS = pd.DataFrame(index=x.columns.values, columns=usxi.models_to_eval)

for model_name in model_selected:
    if model_name == "Random Forest classifier":
        SHAP_RESULTS[model_name], LIME_RESULTS[model_name], PI_RESULTS[
            model_name] = im.intpr_technqs_impl(x, y,
                                model_rf,data_size_for_testing, model_name)
    elif model_name == 'Support vector machines':
        SHAP_RESULTS[model_name], LIME_RESULTS[model_name], PI_RESULTS[
            model_name] = = im.intpr_technqs_impl(x, y,
                                model_svm,data_size_for_testing, model_name)
    elif model_name == "Artificial Neural Network":
        SHAP_RESULTS[model_name], LIME_RESULTS[model_name], PI_RESULTS[
            model_name] = im.intpr_technqs_impl(x, y,
                                model_nn,data_size_for_testing, model_name)
```

Fig. 6. EFI code to calculate feature importance using SHAP, LIME and PI for multiple machine learning models

The resulting normalised importance coefficients from each FI technique are shown in Tables 1, 2, and 3. The bold coefficients represent the top contributing feature to a model's classification. We observe disagreements in the most important features across ML models and FI techniques. In addition, trusting the importance results of a particular ML algorithm is difficult as all three ML algorithms show similar classification performance, presented in Sect. 3.3 above. Hence, the motivation for EFI.

Table 1. Normalised permutation importance coefficients for Iris features (best contribution in bold).

	Neural network	Random forest	Support vectors
Sepal length (cm)	0.07	0.02	0.00
Sepal width (cm)	0.00	0.00	0.08
Petal length (cm)	**1.00**	0.45	0.91
Petal width (cm)	0.52	**1.00**	**1.00**

Table 2. Normalised shap coefficients for Iris features (best contribution in bold).

	Neural Network	Random Forest	Support Vectors
Sepal length (cm)	0.20	0.12	0.00
Sepal width (cm)	0.00	0.00	0.01
Petal length (cm)	**1.00**	0.84	**1.00**
Petal width (cm)	0.37	**1.00**	0.90

Table 3. Normalised lime coefficients for Iris features (best contribution in bold).

	Neural network	Random forest	Support vectors
Sepal length (cm)	**1.00**	0.11	0.58
Sepal width (cm)	0.56	0.36	0.72
Petal length (cm)	0.46	**1.00**	0.00
Petal width (cm)	0.00	0.00	**1.00**

3.5 Model Specific Ensemble Feature Importance

Let's assume we trust the classification performance of the neural network model in predicting iris flowers over the performance of random forest and support vector machine models, we can combine the coefficients obtained using the different FI techniques as shown in Fig. 7.

```
#dataframe for all coefficients
ENSEMBLE_ML_MODEL = pd.DataFrame(index=x.columns.values,
                                 columns=['SHAP', 'LIME', 'PI'])

model_name = 'Artificial Neural Network'
ENSEMBLE_ML_MODEL['PI'] = PI_RESULTS[model_name]
ENSEMBLE_ML_MODEL['LIME'] = LIME_RESULTS[model_name]
ENSEMBLE_ML_MODEL['SHAP'] = SHAP_RESULTS[model_name]
im.ensemble_feature_importance(ENSEMBLE_ML_MODEL[['SHAP']],
                ENSEMBLE_ML_MODEL[['LIME']],
                ENSEMBLE_ML_MODEL[['PI']],
                model_name,top_feature_majority_voting=2)
```

Fig. 7. EFI ensemble of feature importance coefficients for a specific machine learning model

The coefficients obtained from the neural network model are combined using the majority vote ensemble method that ranks features in each FI technique and computes the average of the coefficients in the most common ranks of each feature. This produces the final importance coefficients in Table 4. We can observe that 'petal length' is the most importance feature in classifying iris flowers using the neural network model, while 'sepal width' is the least important.

Table 4. Final importance coefficients for Iris features computing using majority vote ensemble method (best contribution in bold).

	Artificial neural network
Sepal length (cm)	0.13
Sepal width (cm)	0.00
Petal length (cm)	**1.00**
Petal width (cm)	0.45

Table 5. Final importance coefficients for Iris features computing using crisp ensemble methods (best contribution in bold).

	Box-Whiskers	Majority vote	Mean	Median	Mode	RATE-Kendall Tau	RATE-Spearman Rho	Tau test
Petal length (cm)	**0.74**	**0.74**	**0.74**	**0.91**	**0.74**	**1.00**	**1.00**	**0.83**
Petal width (cm)	0.64	0.72	0.64	0.90	0.64	0.37	0.37	0.64
Sepal length (cm)	0.07	0.10	0.23	0.11	0.23	0.20	0.20	0.05
Sepal width (cm)	0.19	0.18	0.19	0.01	0.19	0.00	0.00	0.00

3.6 Multi-method Ensemble Feature Importance

To exploit the advantages of ensemble learning to obtain more robust and accurate importance coefficients for the features in iris dataset, EFI toolbox's multi-ensemble FI method can be used as shown in Fig. 8:

```
import multi_fusion as mf

model_selected = ['Random Forest classifier',
                  'Support vector machines',
                  'Artificial Neural Network']

ensembl_fi = mf.multi_fusion_feature_imp(SHAP_RESULTS,
                                         LIME_RESULTS, PI_RESULTS,
                                                    x, model_selected)
```

Fig. 8. EFI ensemble of feature importance coefficients of multiple machine learning models

The above code produces the aggregated coefficients in Table 5 using crisp decision fusion methods. We can observe that 'Petal Length' is the most importance feature for all methods. However, the coefficients simply specify the most important feature or importance rank of features. They do not provide an interpretation of how important the features are.

3.7 Fuzzy Ensemble Feature Importance

An understanding of the importance of each feature to the classification of iris flowers is provided by EFI toolbox's fuzzy ensemble FI method. This method is also important in calculating the importance coefficients for high dimensional datasets, non-linear relationships, and the presence of noise because it explores the data space more thoroughly and uses fuzzy logic to capture uncertainties in coefficients.

The fuzzy ensemble FI method consist of 2 steps:

- Compute FI coefficients for different partitions of the data.
- Generate membership functions from the coefficients to explain the range of importance for coefficients in each ML algorithm, the importance of each feature relative to the other features in the dataset and compute the aggregated coefficients of features.

The fuzzy ensemble method provides interpretations for the importance of iris flower features as well as their levels of uncertainty as shown in Fig. 9. We observe that both 'petal width' and 'petal length' have high likelihood of 'moderate' and 'high' importance. However, 'petal width' shows less uncertainty in importance compared to 'petal length' i.e., importance coefficients range from 0.50 to 0.85 for 'petal width' and 0.15 to 0.85 for 'petal length'. 'sepal width' and 'sepal length' have a high likelihood of 'low' and 'moderate' importances, with importance coefficients ranging from 0.1 to 0.55 and 0.1 to 0.65 respectively. With these membership functions, users can diagnose which data subsets or time steps produce the extreme cases of feature importance.

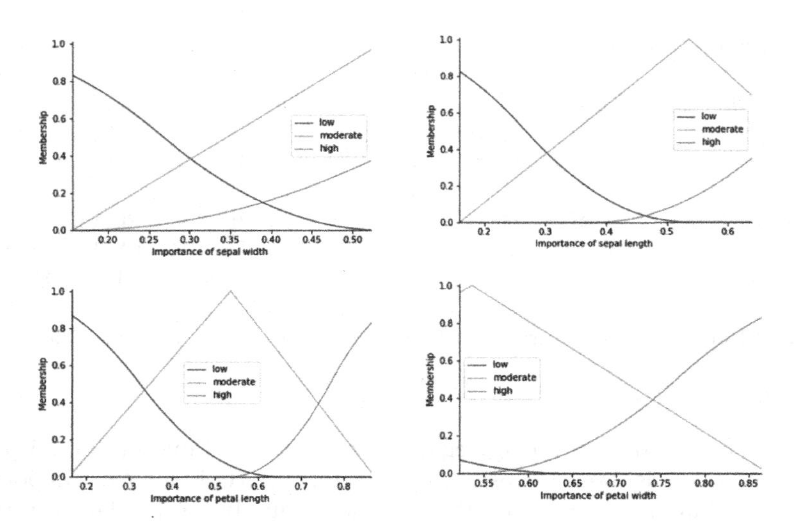

Fig. 9. Membership functions of features generated from Iris flower dataset showing the interpretations of feature importance and their levels of uncertainty

It is important to note that the code snippets used to demonstrate the application of EFI toolbox on the Iris dataset are extracted from different modules in the EFI toolbox because the implementation of EFI follows a modular software engineering approach. In addition, the toolbox can be easily extended to include other ML algorithms, model specific FI techniques and regression problems. Lastly, the toolbox outputs detailed reports of model optimisation and evaluation, and multiple graphs and plots to provide a clear understanding of computation and fusion of FIs by the different FI techniques and ensemble strategies.

4 Conclusion

This paper has presented a novel, extensible open-source toolbox in Python programming language called EFI (Ensemble Feature Importance) that aggregates feature importance coefficients from multiple models coupled and different feature importance techniques using bootstrapping and decision fusion methods. We have described the major modules and functions of the toolbox and provided a step-by-step example using the popular Iris dataset. The outputs of our toolbox (i.e. feature importance coefficient tables, plots, graphs and membership functions) substantially improve the understanding and interpretation of the importance of features to prediction tasks. Due to the complexity and multitude of steps and methods to: (i) optimise multiple ML algorithms, (ii) calculate and visualise the importance of features using various feature importance techniques, (iii) aggregate the importance coefficients using multiple ensemble methods, and (iv) create fuzzy logic systems for capturing uncertainties and interpreting importance, it is important to have tools to make these tasks as simple and robust as possible. As an open-source toolbox, we plan to extend the toolbox to deal with regression tasks in the future and encourage other researchers to contribute to its growth, such as by improving the structure of the toolbox, implementing additional state-of-the-art ML algorithms and implementing model specific feature importance techniques.

References

1. Arrieta, A.B., et al.: Explainable artificial intelligence (XAI): concepts, taxonomies, opportunities and challenges toward responsible AI. Inf. Fusion **58**, 82–115 (2020)
2. Arya, V., et al.: One explanation does not fit all: a toolkit and taxonomy of AI explainability techniques (2019). https://arxiv.org/abs/1909.03012
3. Baniecki, H., Kretowicz, W., Piatyszek, P., Wisniewski, J., Biecek, P.: Dalex: responsible machine learning with interactive explainability and fairness in Python. J. Mach. Learn. Res. **22**(214), 1–7 (2021). http://jmlr.org/papers/v22/20-1473.html
4. Bobek, S., Bałaga, P., Nalepa, G.J.: Towards model-agnostic ensemble explanations. In: Paszynski, M., Kranzlmüller, D., Krzhizhanovskaya, V.V., Dongarra, J.J., Sloot, P.M.A. (eds.) ICCS 2021. LNCS, vol. 12745, pp. 39–51. Springer, Cham (2021). https://doi.org/10.1007/978-3-030-77970-2_4
5. Feurer, M., Eggensperger, K., Falkner, S., Lindauer, M., Hutter, F.: Auto-sklearn 2.0: Hands-free automl via meta-learning. arXiv:2007.04074 (2020)
6. Gille, F., Jobin, A., Ienca, M.: What we talk about when we talk about trust: theory of trust for AI in healthcare. Intell.-Based Med. 1–2, 100001 (2020). https://doi.org/10.1016/j.ibmed.2020.100001. https://www.sciencedirect.com/science/article/pii/S2666521220300016
7. Google: Auto ml tables. https://cloud.google.com/automl-tables/docs Accessed June 2022
8. Huynh-Thu, V.A., Geurts, P.: Optimizing model-agnostic random subspace ensembles. arXiv preprint arXiv:2109.03099 (2021)
9. Klaise, J., Looveren, A.V., Vacanti, G., Coca, A.: Alibi explain: algorithms for explaining machine learning models. J. Mach. Learn. Res. **22**(181), 1–7 (2021). http://jmlr.org/papers/v22/21-0017.html
10. Nori, H., Jenkins, S., Koch, P., Caruana, R.: Interpretml: A unified framework for machine learning interpretability. arXiv preprint arXiv:1909.09223 (2019)
11. Reddy, S., Allan, S., Coghlan, S., Cooper, P.: A governance model for the application of AI in health care. J. Am. Med. Inf. Assoc.: JAMIA **27**(3), 491–497 (2019)
12. Rengasamy, D., Mase, J.M., Torres, M.T., Rothwell, B., Winkler, D.A., Figueredo, G.P.: Mechanistic interpretation of machine learning inference: a fuzzy feature importance fusion approach. arXiv preprint arXiv:2110.11713 (2021)
13. Rengasamy, D., Rothwell, B.C., Figueredo, G.P.: Towards a more reliable interpretation of machine learning outputs for safety-critical systems using feature importance fusion. Appl. Sci. **11**(24), 11854 (2021)
14. Ribeiro, M.T., Singh, S., Guestrin, C.: Model-agnostic interpretability of machine learning. arXiv preprint arXiv:1606.05386 (2016)
15. Ruyssinck, J., Huynh-Thu, V.A., Geurts, P., Dhaene, T., Demeester, P., Saeys, Y.: NIMEFI: gene regulatory network inference using multiple ensemble feature importance algorithms. PLoS ONE **9**(3), e92709 (2014)

16. Wang, Y., et al.: Espresso: a fast end-to-end neural speech recognition toolkit. In: 2019 IEEE Automatic Speech Recognition and Understanding Workshop (ASRU) (2019)
17. Zadeh, L.A.: Fuzzy logic and approximate reasoning. Synthese **30**(3), 407–428 (1975)
18. Zhai, B., Chen, J.: Development of a stacked ensemble model for forecasting and analyzing daily average PM2. 5 concentrations in Beijing, China. Sci. Total Environ. **635**, 644–658 (2018)

Hierarchical Decentralized Deep Reinforcement Learning Architecture for a Simulated Four-Legged Agent

Wadhah Zai El Amri(✉), Luca Hermes, and Malte Schilling

Machine Learning Group, Bielefeld University, Bielefeld, Germany
{wzaielamri,lhermes,mschilli}@techfak.uni-bielefeld.de

Abstract. Legged locomotion is widespread in nature and has inspired the design of current robots. The controller of these legged robots is often realized as one centralized instance. However, in nature, control of movement happens in a hierarchical and decentralized fashion. Introducing these biological design principles into robotic control systems has motivated this work. We tackle the question whether decentralized and hierarchical control is beneficial for legged robots and present a novel decentral, hierarchical architecture to control a simulated legged agent. Three different tasks varying in complexity are designed to benchmark five architectures (centralized, decentralized, hierarchical and two different combinations of hierarchical decentralized architectures). The results demonstrate that decentralizing the different levels of the hierarchical architectures facilitates learning of the agent, ensures more energy efficient movements as well as robustness towards new unseen environments. Furthermore, this comparison sheds light on the importance of modularity in hierarchical architectures to solve complex goal-directed tasks. We provide an open-source code implementation of our architecture (https://github.com/wzaielamri/hddrl).

Keywords: Deep reinforcement learning · Motor control · Decentralization · Hierarchical architecture

1 Introduction

Legged locomotion has been widely used in current mobile robots [6,16] as it provides a high degree of mobility and is adequate for various types of terrains. This makes legged robots useful for different application scenarios, from assisting humans in their daily lives to rescue missions in dangerous situations. Researchers have developed several ideas and algorithms to design and program legged robots. One important inspiration for the field comes from biology as walking animals show high stability, agility, and adaptability. Insects provide one example: They can climb and move in nature while relying on comparably simple control structures. At the same time, they only have limited knowledge of their surroundings. Still, adaptive behavior emerges and enables these animals

G. Nicosia et al. (Eds.): LOD 2022, LNCS 13811, pp. 265–280, 2023.
https://doi.org/10.1007/978-3-031-25891-6_20

to move successfully and efficiently. Such capabilities are desirable for today's robots. A drawback is that the transfer of biological principles usually requires detailed knowledge and explicit setup or programming of control systems. This is difficult to scale towards real world settings covering a wide range of scenarios.

Another approach to robotic control is based on machine learning. While common reinforcement learning approaches have been successfully applied in robot locomotion, these systems often lack generalizability towards multiple tasks [3]. In particular, it has shown to be non-trivial to design reward functions for each task. Ideally, it would be sufficient to simply reward completion of the full task, e.g. reaching a goal. But it is challenging for an algorithm to learn goal-directed behavior from such sparse feedback. Learning from sparse rewards requires temporal abstraction that contributes to solve multiple tasks and adapt to different environments [8]. Animals are successful at such tasks as they rely on a hierarchical organization into modules in their control systems [11]. By splitting the locomotion system into different hierarchical levels (vertical levels), it becomes possible to coordinate, acquire and use knowledge gained in previous tasks. Reinforcement learning algorithms have adopted this notion of hierarchical organization [11] in the form of hierarchical networks and applying transfer learning where primitive behaviors are stored and can be reused in new tasks.

Recently, the idea of multiple decentralized modules (horizontal modularization) running concurrently to ensure high-speed control and fast learning has been introduced to Reinforcement Learning [15]. Such a solution can be found in nature, e.g., in insect locomotion it is assumed that each leg is controlled by a single module that adapts to local disturbances based only on local sensory feedback [12]. Both these types of modularization—into a hierarchical and decentralized organization—have been used separately in state-of-the-art legged mobile robots. In this article, we combine these concepts in a hierarchical decentralized architecture and consider how this helps learning behavior as well as how this positively influences behavior, in particular generalization towards novel tasks and more efficient behavior. The paper is structured as follows: General concepts and ideas are discussed in related work to address the aforementioned challenges. Section 3 introduces the methodological foundations of this work and addresses the experimental setup designed to deal with the research questions. Section 4 shows the main results of the conducted experiments and the various investigated details, such as the emerged new learning paradigms out of this novel concept. Section 5 discusses further aspects and concludes this paper.

2 Related Work

As we aim to take inspiration from animal control, we focus on two key aspects: decentralization and hierarchical organization. Merel et al. [11] provided an analysis of the mammalian hierarchical architecture and gave several design ideas for hierarchical motor control. Quite a number of scientists have already taken advantage of such an hierarchical architecture for legged locomotion. For example, Heess et al. [4] introduced a novel hierarchical deep reinforcement learning

(DRL) architecture. The controller consists of two policies, a high-level and a low-level policy. The low-level controller (LLC) is updated at a high execution frequency while the high-level controller (HLC) operates on a low frequency. The HLC has access to the entire observation space including task-related information such as the goal position. However, the LLC has access only to local inputs corresponding to the agent's state and latent information originating from the HLC. The learned low-level motor behaviors can be transferred to various other, more complex tasks—such as navigation—while getting a sparse reward [14]. Other approaches, e.g. [9], use a hybrid-learning model. In their approach [9], the HLC is model-based, and it selects the appropriate policy to use. Whereas the LLC is trained with an off-policy learning-based method. In [1] a simulated robot learned to walk through environments that differed in terrain or had obstacles. LLCs were trained separately for a specific environment, and a HLC was trained afterwards to select the suitable LLC for the current environment.

Decentralization has been identified as another key principle in biological motor control [12]. It has already been applied in DRL in [13,15]. They proposed a way to take advantage of a decentralized architecture in learning of legged locomotion. DRL was used to train four independent policies for a four-legged agent, one for each leg. They further regulated the information flow for the individual (leg) policies, e.g., instead of directing all available input information to the controller, the information was split, and each of the four agents received only local information relevant to them. The results of the paper clearly demonstrated that the decentralized architecture provides an enhancement of the speed of learning by reaching high performance values compared to centralized architectures. In addition, the learning process is more robust to new unpredictable environment sets, such as changes in the flatness of the terrain. A different approach to decentralization was proposed by Huang et al. [5]. They trained shared modular policies, each responsible for an actuator of the agent. With the help of the messaging between the reusable policies, a communication between the different modules emerges, and this communication ensures a stable movement, which is robust to various changes in the physiology of the agent. But this approach took a long period of time and requires high computational resources to obtain a trained shared policy. Both approaches used a simple 1-D forward movement as a task where the agent received a reward for the speed of movement in the x-direction with other penalties such as contact or control punishments. However, it is important to note that the decentralized architectures in these works are not tested on more complex tasks. In this work, we introduce novel architectures that attempt to combine both approaches and use DRL to train the different policies when applied to more complex tasks.

3 Methods

The aim of this study is to compare how decentralization and hierarchical organization of control structures influence adaptive behavior and learning of behavior. In particular, different architectures for walking of a four-legged simulated agent are evaluated with respect to how fast stable behavior emerges, how robust and efficient learned behavior is, and, finally, how well an architecture supports transfer learning, i.e. being applied in a different setting. We experiment using central and decentral walking policies in hierarchical and non-hierarchical setups. We start by introducing these principles and then describe the specific architectures.

Current reinforcement learning (RL) approaches are usually based on central policies $\pi_{central}$ as the most straight forward approach. Such a central policy (Eq. 1) consists of a single module that receives the full observation vector $\mathbf{s}_t$ at time step t and outputs control signals $\mathbf{a}_t$ for each degree of freedom (DOF). In contrast, a decentral control system consists of separate sub-controllers $\pi^l_{\text{decentral}}$. Each sub-controller only outputs the control signal for the DOFs for its assigned leg l and usually receives only local observations $\mathbf{s}^l_t$, for example, only from the respective leg considering control of a walking agent. As mentioned, decentralization can be found throughout biological control as, for example, it is required to compensate for sensory delays. In RL, usually decentralization is abstracted away and not considered because it is only assumed as an implementation detail. In contrast, our study aims to analyze the contribution of decentralization as we assume that decentralization itself has positive effects. In the case of the four-legged agent used here (see Sect. 3.1), for a decentralized approach the four legs require four sub-controllers that are trained concurrently. Note, that it is possible to choose other patterns of decentralization. Instead of per-leg controllers, per-joint controllers would be an option, too, which is not explored in this work.

$$\pi_{\text{central}}(\mathbf{a}_t|\mathbf{s}_t), \qquad \pi^l_{\text{decentral}}(\mathbf{a}^l_t|\mathbf{s}^l_t) \tag{1}$$

Decentralization introduces a form of modularization into multiple modules on the same level of abstraction that act concurrently (one controller for each leg, but these are equal in structure). A different form of modularization is introduced by an hierarchical structure. In a hierarchical structure higher levels usually drive lower level modules. This has the advantage that lower level modules can be learned as reusable building blocks that can be applied by higher levels in multiple tasks. Such hierarchical approaches facilitate transfer learning. A non-hierarchical setup—as the standard RL approach—consists of a single level of control, as observations are mapped onto actions by a policy neural network. In a hierarchical setup the controller is distributed onto two, or more, levels of a hierarchy. In our case, we use two levels of hierarchy and define our principles of hierarchy following Heess et al. [4]: Low-level controller (LLC) output actions at every time step t and high-level controller (HLC) output a latent vector $\mathbf{m}_\tau$ every τ time steps (Eq. 2). This introduces a form of temporal abstraction. The LLC only gets local features $\mathbf{s}_t$—like the joint angles—as an input, and is in addition conditioned on the latent vector $\mathbf{m}_\tau$ from the HLC. This allows the HLC to modulate the LLC. The HLC only gets global features $\bar{\mathbf{s}}_\tau$, such as the angle between robot orientation and target position.

$$\mathbf{m}_\tau = \pi_{HLC}(\bar{\mathbf{s}}_\tau), \quad \mathbf{a}_t = \pi_{LLC}(\mathbf{m}_\tau \parallel \mathbf{s}_t) \tag{2}$$

We experiment with five different architectures: On the one hand, employing known architectures as centralized, decentralized, and an hierarchical approach.

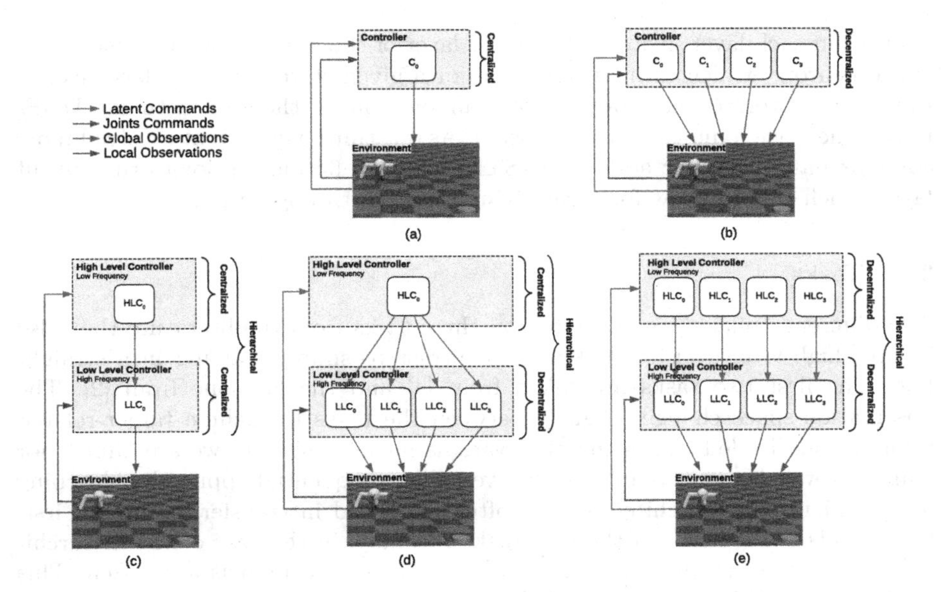

Fig. 1. The different architectures. (a): Centralized architecture with one controller (C_0, termed Cen). (b): Decentralized architecture with four different controllers (C_0, C_1, C_2, C_3, termed Dec). (c): Hierarchical architecture with 1 HLC and 1 LLC (each level consists of a single, centralized controller and there is no decentralization on each level of abstraction; therefore, this controller is abbreviated as C_C). (d): Hierarchical decentralized architecture with 1 HLC and 4 LLCs (abbr. as C_D as the higher level consists of a single, central module and the lower level is decentralized). (e): Hierarchical decentralized architecture with 4 HLCs and 4 LLCs (abbr. as D_D).

On the other hand, as the main novel contribution of this article, we test two combinations of hierarchical and decentralized architectures (Fig. 1 visualizes all five explored architectures). The central architecture (shown in Fig. 1 (a)) serves as the central baseline. The decentralized architecture (shown in Fig. 1 (b)) serves as the decentral baseline, replicating the architecture in [15]. To investigate if decentralization benefits hierarchical controllers, we experiment with combinations of central and decentral policies on two different levels of hierarchy, as shown in Fig. 1 (c)–(e).

3.1 Experimental Setup

The experiments focus on learning locomotion tasks for a four-legged simulated agents. We used the MuJoCo physics simulator [19] (v. 1.50.1.68) to run the experiments and adopt the modified Ant-v3 model from [15]. This version of Ant was made heavier by a factor of ten to replicate the weight of a real robot. We used the RLlib framework (v. 1.0.1) [10] and applied PPO [17] in an A2C setup [7] to optimize our policies. Actor and critic shared a similar architecture. Both were realized as *tanh*-activated two hidden-layer neural networks with 64 neurons in each layer. Note that the last layer of the critic network consists of a single neuron—which estimates a value for a given state—and the last layer of the actor network consists of twice as many neurons as there are actuated DOFs to parameterize a multivariate Gaussian. As our third experiment involved more complex navigation, we added an LSTM cell with 64 units prior to the output layer which should allow for memorizing the path through a maze.

3.2 Tasks

The different policies are evaluated on three tasks from the literature that also involve high-level decision-making in contrast to simply walking in a straight line. The first two tasks are taken from [4] and the last one from [2]. The observation-space changes slightly between the tasks to include target-related information. Table 1 shows the observations for all tasks. As we are aiming for transfer towards more complex tasks, we follow a sequential approach of learning for the hierarchical architectures as often employed in transfer learning. First, the models are trained on the first task. Secondly, in the case of the hierarchical models, the parameters of the LLCs of the trained models are frozen. This means, the LLC is kept fixed and the HLC is trained from scratch for task two and three, respectively, for the hierarchical architectures. This ensures that the LLC learns a universal policy for general leg control. In the non-hierarchical approaches (centralized architecture and decentralized architecture, Fig. 1 (a) and (b)), as there is no distinction between high and low-level, these models are trained directly on tasks two and three and there is no freezing of parameters.

Table 1. Observation spaces of the three tasks divided into global and local information; d denotes the dimensionality of the corresponding vector.

Group	Observation	d	First Task	Second Task	Third Task
Global	Angle between target and robot's "north pole"	1	x	x	x
	Robot's position: x, y, z	3		x	x
	Torso's orientation: quaternion	4		x	x
	Robot's velocity: x, y, z	3		x	x
	Robot's angular velocity: $\omega_x, \omega_y, \omega_z$	3		x	x
Local	Joints angles	8	x	x	x
	Joints angular velocities	8	x	x	x
	Passive forces exerted on each joint	8	x	x	x
	Last executed joint actions	8	x	x	x
Total dims		46	33	46	46

Task 1: In the first task, the agent should reach a specific target location **g**. The initial location of the target and the agent are randomly chosen, such that the distance between robot and target ranges from 0.5 m to 5 m. This task is considered solved if the agent reaches the target within $t_{\max} = 300$ simulation

steps. The target is reached if the robot is within in a 0.25 m radius around the target location. For the two non-hierarchical models as well as the LLC of the hierarchical models, the following equation is used to calculate the continuously provided reward at timestep t

$$r_1(t) = \frac{v_g(t)}{N} - 0.05 \times \|\mathbf{a}(t)\|^2 + r_T(t)$$

$$r_T(t) = \begin{cases} 0.2 \times (t_{\max} - t), & \text{if target reached} \\ 0, & \text{otherwise} \end{cases},$$

where v_g denotes the velocity towards the target, N denotes the number of subcontrollers, i.e. $N = 1$ for the central policy and $N = 4$ for the decentral case, $\mathbf{a}$ denotes the action vector and $\mathbf{p}$ denotes the robot position. The HLCs in hierarchical policies receive the following reward:

$$\bar{r}_1(t) = \cos\Theta(t) \times \|\mathbf{p}(t)\|^2 + 0.1 \cdot r_T(t)$$

where Θ denotes the angle between the orientation of the robot and the direction from robot to target.

Task 2: The second task is again to reach a set goal location, but this time only a sparse reward is given. The policy only collects a reward at the end of an episode on reaching the goal and not after every simulation set. We apply the following equation to calculate the reward at timestep t

$$\bar{r}_2(t) = \begin{cases} \frac{1}{N} - \frac{t}{N \cdot t_{\max}}, & \text{if target reached} \\ 0, & \text{otherwise} \end{cases}.$$

Task 3: In the third task, the robot is put in a 15 $\times$ 15 m maze, as shown in Fig. 2. The robot always spawns in the green area and has to reach the red area within 1600 simulator steps. The initial orientation of the robot is randomized. This task is formulated in a sparse reward setting, but there are a couple of subgoals provided that should help guide the agent towards the goal. The agent gets as a reward

$$\bar{r}_3(t) = \begin{cases} \frac{r_{\text{subgoal}}}{N}, & \text{upon first visit at subgoal} \\ 0, & \text{otherwise} \end{cases},$$

where r_{subgoal} is 1, 2, or 3 for the blue, yellow, and red subgoal, respectively.

4 Results

Three different tasks with increasing complexity have been studied. In each task, the five architectures introduced in Sect. 3 were evaluated by running ten trials.

Fig. 2. Maze environment of task three. The green square is the starting position, the blue, yellow, and red square represent the subgoals. (Color figure online)

During each trial, training ran for 40M simulation time steps. A single episode was terminated as soon as the goal condition was reached. We report the return accumulated over running an episode, the ratio of successful episodes and also power consumption of the agent as quantitative metrics. Throughout the result section, results will use abbreviations to represent task and architectures (first, providing the task number, followed by an underscore and the type of architecture; flat architectures (centralized, decentralized) are abbreviated as Cen and Dec, respectively, while for the hierarchical architectures it will be provided individually for the higher and lower level, if the policies are decentralized (D) or central (C) on that level, see Fig. 1 for all architectures; furthermore, we will use color coding throughout the results to distinguish architectures). The power consumption P is calculated over time as the product of the torque acting on each joint along with its actual velocity, i.e. $P(t) = \sum_{\forall i} a_i \omega_i$, where a_i is the action and ω_i the angular velocity of DOF i. We report the return-specific results for hierarchical and non-hierarchical policies in separate plots as these differ in the reward functions used. Furthermore, we provide in addition the ratio of successful episodes as a fair measure for comparison.

In addition to providing training statistics, all agents were afterwards tested for 100 episodes. This test was repeated with three different randomly generated terrains. The terrains were generated using the algorithm for terrain generation from [18]. It generates random height fields consisting of various superimposed sinusoidal shapes. The *smoothness* parameter determines how uneven the terrain should be. This parameter ranges from 1.0 (flat terrain) to 0.0 (very uneven terrain). Three parameter values were tested in our experiments: flat terrain ($smoothness$ = 1.0), slightly uneven terrain ($smoothness$ = 0.8), and bumpy terrain ($smoothness$ = 0.6).

4.1 First Task: Navigate to a Specific Goal

In the first task, the agent's goal was to reach a random target location and he was provided all the time with a continuous reward guiding him to the target location. Figure 3 (a), (b) show the development during training as the mean over all trials for the flat architectures and the mean return for the hierarchical architectures. The shaded areas show the standard deviation. It can be observed

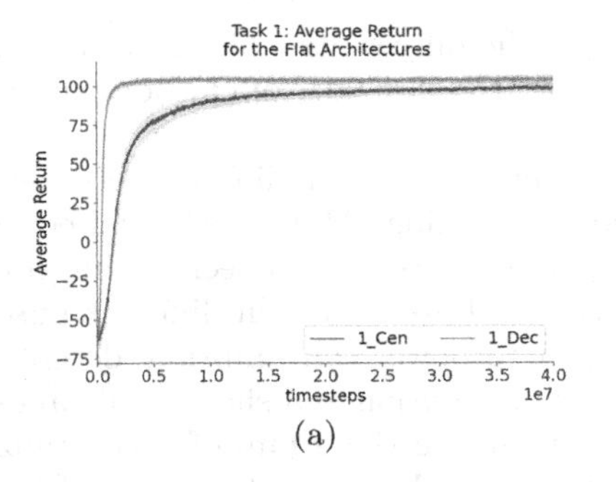

(a)

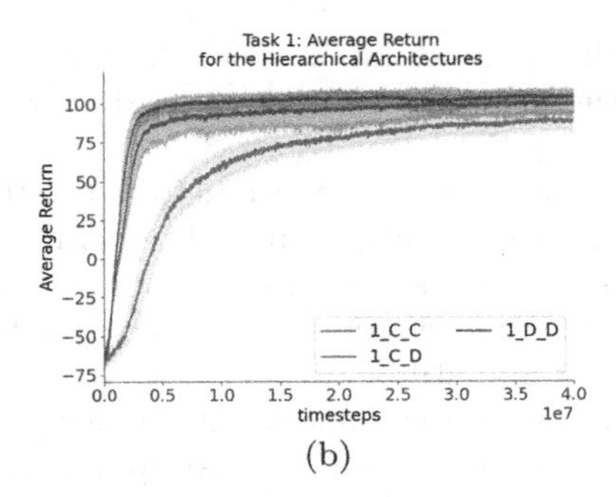

(b)

Fig. 3. Average return of ten trials for the different architectures during learning. (a) The average return of all ten trials for the flat architectures. The shaded area represents the limits of the standard deviation and the plot represents the mean value. (b) The average return of all ten trials for the hierarchical architectures. The shaded area represents the limits of the standard deviation and the plot represents the mean value.

that the decentral architecture improved faster than the central architecture. This replicates the results of Schilling et al. [15], even for the more complicated task of walking towards a random target position. Figure 3 (b) shows that the fully centralized architecture (1_C_C) trained the slowest, whereas the fully decentralized architecture (1_D_D) was learning the fastest. This indicates that the result of [15] also appear to apply for hierarchical cases.

Figure 4 shows that agents with a flat decentralized architecture (1_Dec) learned the task the fastest and had a high success rate. The hierarchical architecture with decentralized HLCs and LLCs (1_D_D) learned slightly slower. A possible explanation may be the hypothesis of Schilling et al. [15] that a larger observation space requires longer training time. In this particular case, the observations space for the hierarchical policies was larger due to the introduction of

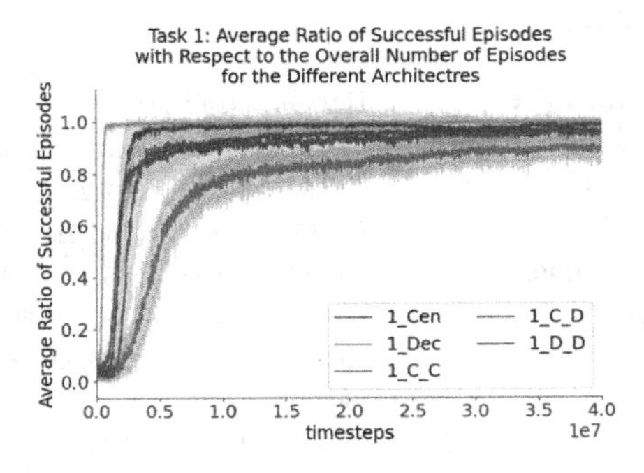

Fig. 4. Visualization of the ratio of successful episodes with respect to the overall number of episodes.

the latent modulation vector from the HLC. The fully central hierarchical policy learned the task the slowest and only finished roughly 90% of episodes successfully.

Table 2 shows the results from the 100 evaluation runs on different terrains for the different architectures after completion of training. All trained architectures achieved high success ratios even on hard, bumpy terrains not seen during training. In other words, the studied architectures all are robust in different unseen terrain and still solve the task. With respect to power consumption, decentralized architectures had a much lower power consumption and show an advantage compared to the centralized approaches. But inside the group of decentralized approaches, introducing furthermore an hierarchical organization showed to be further beneficial. For the flat decentralized architecture (1_Dec) the power consumption was higher than that of both hierarchical architectures with a decentralized lower level (1_C_D and 1_D_D). This reflects how energy efficient the agent is when combining a hierarchical with a decentralized architecture.

4.2 Second Task: Seek Target

In the second experiment, the agent was aiming for reaching a target, but received only a sparse reward when this was reached. In this case, all flat (non hierarchical) architectures were trained from scratch (indicated in the figures and result tables by the suffix _SC). In contrast, hierarchical architectures used LLC with frozen parameters that were pretrained on task 1. Only the randomly initialized HLC was trained (as done in transfer learning, indicated in the results

Table 2. Task 1: Results of the trained controllers after training. Given are mean values and standard deviation (in parenthesis) over 100 episodes for each of the ten trained seeds. Best values are highlighted (as the reward functions differ for the flat and hierarchical architectures, there is one highlight for each of these types (flat and hierarchical architectures). Therefore, these return values can not be compared directly, but we report this value for completeness. $P_{suc.}$ denotes the power consumption in successfully finished episodes.

Architecture (HLC + LLC)	Flat terrain (1.0)			Uneven terrain (0.8)			Bumpy terrain (0.6)		
	Return*	Ratio	$P_{suc.}$	Return*	Ratio	$P_{suc.}$	Return*	Ratio	$P_{suc.}$
Centralized	99.08 (22.48)	0.971	879.47 (364.65)	97.16 (23.57)	0.965	867.22 (356.51)	94.55 (25.36)	0.932	938.32 (407.12)
Decentralized	104.47 (19.45)	**0.993**	578.29 (250.50)	103.12 (19.94)	0.988	585.96 (250.43)	99.41 (21.86)	**0.963**	625.93 (275.78)
Cent. + Cent.	90.01 (31.08)	0.895	1240.55 (577.86)	89.15 (30.79)	0.911	1276.92 (623.84)	83.56 (33.91)	0.836	1355.97 (668.56)
Cent. + Decent.	99.15 (25.68)	0.943	**468.69** (203.27)	97.34 (26.59)	0.949	**501.49** (229.88)	88.76 (28.85)	0.826	**543.43** (249.64)
Decent. + Decent.	103.72 (20.81)	0.985	501.96 (201.18)	104.69 (19.64)	**0.993**	528.25 (209.52)	96.71 (25.35)	0.9	580.76 (240.11)

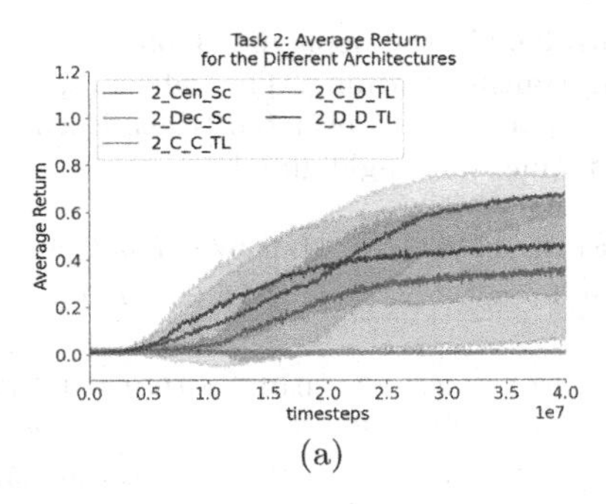

(a)

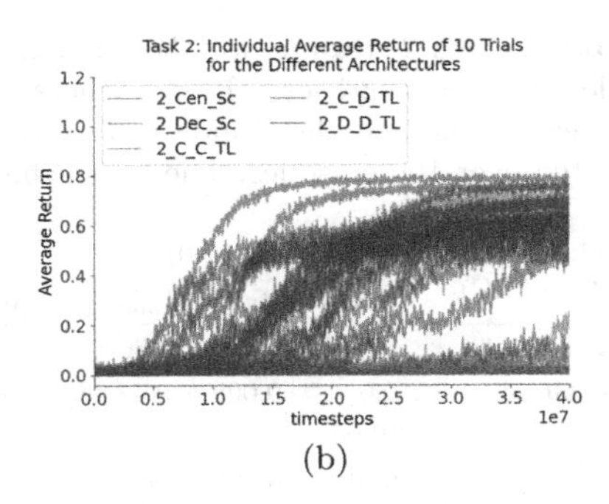

(b)

Fig. 5. Average return of ten trials for the different architectures during learning in task 2. (a) The average return of all ten trials. The shaded area represents the limits of the standard deviation and the plot represents the mean value. (b) Visualizing return for all ten individual seeds for the different architectures in task 2 during training.

by the suffix _TL). This assumes that during training in task 1 on random goals the agent learns to walk in the environment and that this ability is retained now in task 2, but now the agent has to search for the new target position while still being able to reuse already acquired locomotion skills in the lower level. Figure 5 shows that policies trained from scratch did not learn the task properly as should be expected (see as well Fig. 6). When only using a sparse reward, the task became too difficult. However, using a pretrained LLC in the case of the three hierarchical architectures showed much higher returns. The 2_C_D_TL architecture outperformed the other hierarchical architectures, while the hierarchical, both level centralized architecture (2_C_C_TL) trained the slowest.

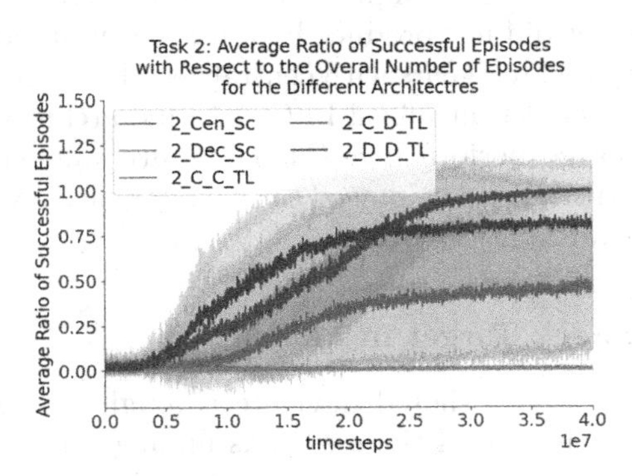

Fig. 6. Visualization of the ratio of successful episodes with respect to the overall number of episodes for the different architectures in task 2 during training.

Table 3 again provides the results from 100 evaluation runs on different terrains after training was finished. This illustrates the importance of transfer learning in such sparse tasks. All the flat architectures that were trained from scratch

Table 3. Task 2: Results for the trained controllers in the deployment phase for controllers trained from scratch and others using transfer learning. The values represent the mean value and the standard deviation (in parenthesis) over ten trials. Each trial is tested over 100 episodes. The best achieved values are highlighted.

Architecture (HLC + LLC)	Flat terrain (1.0)			Uneven terrain (0.8)			Bumpy terrain (0.6)		
	Return	Ratio	$P_{suc.}$	Return	Ratio	$P_{suc.}$	Return	Ratio	$P_{suc.}$
From scratch									
Centralized	0.0 (0.06)	0.005	4433.46 (4825.27)	0.01 (0.09)	0.01	4678.03 (4183.50)	0.01 (0.09)	0.009	3477.72 (1726.77)
Decentralized	0.01	0.008	7995.00	0.01	0.006	3927.36	0.01	0.009	3067.37
Transfer learning (pre-trained on random targets in task 1)									
Cent. + Cent.	0.34 (0.39)	0.449	2991.25 (1963.93)	0.36 (0.39)	0.475	3081.00 (2017.79)	0.27 (0.37)	0.354	3049.25 (1949.27)
Cent. + Decent.	**0.67** (0.16)	**0.996**	**1448.00** (700.28)	**0.64** (0.19)	**0.985**	**1512.89** (736.02)	**0.45** (0.32)	**0.754**	**1698.74** (892.17)
Decent. + Decent.	0.46 (0.28)	0.821	1924.18 (861.37)	0.4 (0.29)	0.765	2091.21 (930.43)	0.26 (0.29)	0.543	2236.82 (1014.62)

failed the task even after 40 million training steps. However, much improved results and positive return values were observed for the hierarchical architectures that used pre-trained weights from previous tasks. Interestingly, the hierarchical architecture with a centralized HLC and decentralized LLCs (2_C_D_TL) achieved the high success ratios. In addition, the values from Table 3 for this architecture demonstrate robustness towards uneven terrain.

Decentralized low-level hierarchical architectures (2_C_D_TL and 2_D_D_TL) again showed lower power consumption $P_{suc.}$ in successful episodes. In these architectures, the agent did not produce high torques in its joints and was more energy efficient compared to agents with centralized HLCs and LLCs (2_C_C_TL. Overall, introducing the decentralized LLCs in a hierarchical architecture using transfer learning boosted performance (higher return values), increased robustness (higher success ratios) and enhanced the energy efficiency of the agent (lower power consumption values).

4.3 Third Task: Seek Target in Maze

In the third task, the agent should navigate through a maze in order to collect a final reward at the end of the maze. As finding a way through the whole maze was too difficult for a random, exploration based search, two subgoals were introduced guiding the agent towards the goal. Figure 7 and 8 show results during training. As the task requires memorizing a path through the maze, the neural network architecture was extended by an additional layer of LSTM units (see Sect. 3.1). In this task, flat architectures were trained either from scratch (3_Cen_Sc, 3_Dec_Sc) or—following the sequential protocol as used in transfer

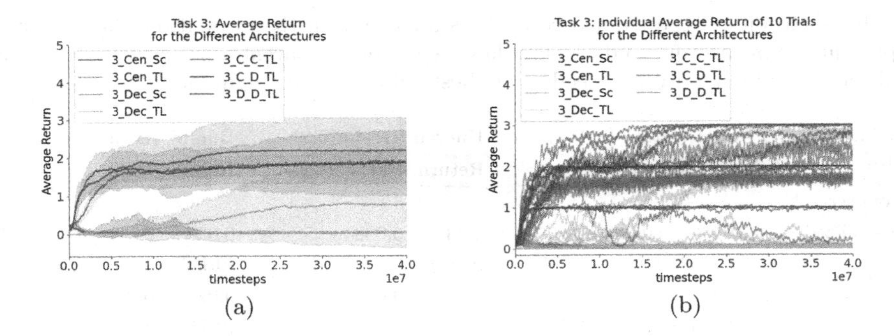

Fig. 7. Average return of ten trials for the different architectures. (a): The average return of all ten trials. The shaded area represents the limits of the standard deviation and the plot represents the mean value. (b): Visualization of the average return of each of the ten seeds trained for every architecture.

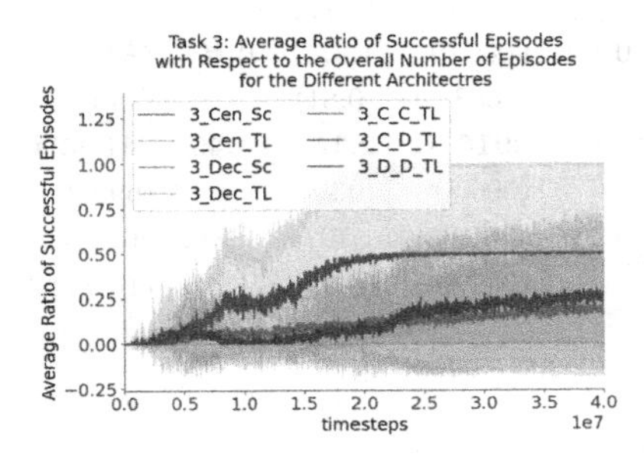

Fig. 8. Visualization of the ratio of successful episodes with respect to the overall number of episodes for the different architectures.

learning approaches—were pretrained for task 1 and shown are results for subsequential training in the maze environment (3_Cen_TL, 3_Dec_TL). All these non-hierarchical approaches showed little to no success. In contrast, hierarchical architectures again showed transfer learning capabilities and were able to collect a considerable return. In particular, agents with a centralized HLC and decentralized LLCs (3_C_D_TL) learned faster and reached higher return values as they were able to solve the maze at a considerable rate (Fig. 8).

Table 4 shows the results from the 100 evaluation trials for different terrains. The results again support the hypothesis that using decentralized and hierarchical architectures is also advantageous for such complex tasks that require forms of memory. The different hierarchical architectures maintain a level of success for generalization to uneven terrain (smoothness 0.8). But this drops considerably when turning towards bumpy terrain (smoothness equal to 0.6). The power consumption values $P_{suc.}$ show again that agents with decentralized hierarchical architectures had lower power consumption which supports the observations already encountered in the previous tasks.

Table 4. Results of the trained controllers for task 3 in the evaluation phase. Values represent the mean value and the standard deviation (in parenthesis) over 100 episodes and for ten different training seeds. The best values are highlighted.

Architecture (HLC + LLC)	Flat terrain (1.0)			Uneven terrain (0.8)			Bumpy terrain (0.6)		
	Return	Ratio	$P_{suc.}$	Return	Ratio	$P_{suc.}$	Return	Ratio	$P_{suc.}$
From scratch									
Centralized	0.03 (0.19)	0.000	– (–)	0.03 (0.18)	0.000	– (–)	0.02 (0.15)	0.000	– (–)
Decentralized	0.03 (0.17)	0.0	– (–)	0.02 (0.13)	0.0	– (–)	0.02 (0.15)	0.0	– (–)
Transfer learning (Random pre-trained trial trained in task 1)									
Centralized	0.76 (1.18)	0.17	3297.79 (1609.73)	0.76 (1.18)	0.17	3538.87 (896.04)	0.45 (0.94)	0.078	4047.24 (1156.85)
Decentralized	0.03 (0.17)	0.0	– (–)	0.03 (0.17)	0.0	– (–)	0.03 (0.16)	0.0	– (–)
Cent. + Cent.	1.87 (0.81)	0.186	6000.01 (1808.74)	1.87 (0.81)	0.18	5847.41 (1467.75)	**1.27** (1.01)	0.099	6181.09 (1399.8)
Cent. + Decent.	**2.19** (0.88)	**0.5**	**2916.61** (590.8)	**2.15** (0.93)	**0.49**	**3169.24** (699.05)	**1.27** (1.15)	**0.227**	**3879.17** (978.29)
Decent. + Decent.	1.86 (0.9)	0.257	4367.02 (962.67)	1.79 (0.92)	0.232	4698.57 (1078.28)	1.01 (0.98)	0.086	4911.36 (942.24)

5 Conclusion

This study investigated the combination of decentralized and hierarchical control architectures for learning of locomotion which was compared to state-of-the-art architectures as the standard centralized architecture or a simple hierarchical approach. Overall, it showed that adding concurrent decentralized modules across different vertical levels of a control hierarchy facilitates the learning process and ensures more robustness as well as more energy efficient behavior. Furthermore, the advantage of hierarchical architectures for transfer learning was maintained. The results demonstrate that the horizontal modularity and vertical temporal abstraction can be used together to improve modern architectures and solve more goal-directed tasks that require transfer learning. This facilitates faster learning, avoids catastrophic forgetting, and allows to reuse learned skills. This study provided a starting point for combination of such modular organizations of control approaches. Future research should consider fine-tuning the networks' architectures, use more layers, or change the degree of decentralization. Another point worth investigating is the robustness of these methods against specific leg failures after training, and whether the trained HLCs can adapt to such disturbances. This can be further explored to test the limits of such architectures. It should be noted that this work refers to a simulated robot. Therefore, it would be also interesting to test these architectures on real robots.

References

1. Azayev, T., Zimmerman, K.: Blind hexapod locomotion in complex terrain with gait adaptation using deep reinforcement learning and classification. J. Intell. Rob. Syst. **99**, 659–671 (2020)
2. Duan, Y., Chen, X., Houthooft, R., Schulman, J., Abbeel, P.: Benchmarking deep reinforcement learning for continuous control. In: Proc. of the 33rd International Conference on Machine Learning. ICML 2016, vol. 48, pp. 1329–1338. JMLR (2016)
3. Frans, K., Ho, J., Chen, X., Abbeel, P., Schulman, J.: Meta Learning Shared Hierarchies. In: International Conference on Learning Representations (2018). https://openreview.net/forum?id=SyX0IeWAW
4. Heess, N., Wayne, G., Tassa, Y., Lillicrap, T., Riedmiller, M., Silver, D.: Learning and transfer of modulated locomotor controllers. arXiv preprint arXiv:1610.05182 (2016)
5. Huang, W., Mordatch, I., Pathak, D.: One policy to control them all: shared modular policies for agent-agnostic control. In: III, H.D., Singh, A. (eds.) Proceedings of the 37th International Conference on Machine Learning, vol. 119, pp. 4455–4464 (2020)
6. Hutter, M., Gehring, C., Jud, D., Lauber, A., Bellicoso, C.D.: ANYmal - a highly mobile and dynamic quadrupedal robot. In: 2016 IEEE/RSJ International Conference on Intelligent Robots and Systems (IROS), pp. 38–44 (2016)
7. Konda, V., Tsitsiklis, J.: Actor-critic algorithms. In: Solla, S., Leen, T., Müller, K. (eds.) Advances in Neural Information Processing Systems, vol. 12. MIT Press (1999)
8. Kulkarni, T.D., Narasimhan, K.R., Saeedi, A., Tenenbaum, J.B.: Hierarchical deep reinforcement learning: integrating temporal abstraction and intrinsic motivation. In: Proceedings of the 30th International Conference on Neural Information Processing Systems, pp. 3682–3690. Curran Associates Inc., Red Hook, NY, USA (2016)
9. Li, T., Lambert, N., Calandra, R., Meier, F., Rai, A.: Learning generalizable locomotion skills with hierarchical reinforcement learning. In: 2020 IEEE International Conference on Robotics and Automation (ICRA), pp. 413–419 (2020)
10. Liang, E., et al.: Ray RLlib: A composable and scalable reinforcement learning library. CoRR abs/1712.09381 (2017), http://arxiv.org/abs/1712.09381
11. Merel, J., Botvinick, M., Wayne, G.: Hierarchical motor control in mammals and machines. Nat. Commun. **10**(1), 1–12 (2019)
12. Schilling, M., Hoinville, T., Schmitz, J., Cruse, H.: Walknet, a bio-inspired controller for hexapod walking. Biol. Cybern. **107**(4), 397–419 (2013)
13. Schilling, M., Konen, K., Ohl, F.W., Korthals, T.: Decentralized deep reinforcement learning for a distributed and adaptive locomotion controller of a hexapod robot. In: Proceedings of IROS, pp. 5335–5342 (2020)
14. Schilling, M., Melnik, A.: An approach to hierarchical deep reinforcement learning for a decentralized walking control architecture. In: Samsonovich, A.V. (ed.) BICA 2018. AISC, vol. 848, pp. 272–282. Springer, Cham (2019). https://doi.org/10.1007/978-3-319-99316-4_36
15. Schilling, M., Melnik, A., Ohl, F.W., Ritter, H.J., Hammer, B.: Decentralized control and local information for robust and adaptive decentralized deep reinforcement learning. Neural Netw. **144**, 699–725 (2021)
16. Schneider, A., Paskarbeit, J., Schilling, M., Schmitz, J.: HECTOR, a bio-inspired and compliant hexapod robot. In: Biomimetic and Biohybrid Systems, vol. 8608, pp. 427–429 (2014)

17. Schulman, J., Wolski, F., Dhariwal, P., Radford, A., Klimov, O.: Proximal policy optimization algorithms. arXiv preprint (2017)
18. Tassa, Y., et al.: Deepmind control suite. CoRR abs/1801.00690 (2018)
19. Todorov, E., Erez, T., Tassa, Y.: MuJoCo: a physics engine for model-based control. In: 2012 IEEE/RSJ International Conference on Intelligent Robots and Systems, pp. 5026–5033 (2012)

Robust PCA for Anomaly Detection and Data Imputation in Seasonal Time Series

Hông-Lan Botterman[1], Julien Roussel[1], Thomas Morzadec[1], Ali Jabbari[1], and Nicolas Brunel[1,2](✉)

[1] Quantmetry, 52 rue d'Anjou, 75008 Paris, France
nicolas.brunel@ensiie.fr
[2] LaMME, ENSIIE, Université Paris Saclay, 1 square de la Résistance, 91025 Evry Cedex, France

Abstract. We propose a robust principal component analysis (RPCA) framework to recover low-rank and sparse matrices from temporal observations. We develop an online version of the batch temporal algorithm in order to process larger datasets or streaming data. We empirically compare the proposed approaches with different RPCA frameworks and show their effectiveness in practical situations.

Keywords: Robust principal component analysis · Anomaly detection · Data imputation · Regularisation · Adaptive estimation

1 Introduction

The ability to generate and collect data has increased considerably with the digitisation of our society. Unfortunately, this data is often corrupted or not collected in its entirety because of *e.g.* faulty data collection systems, data privacy and so on. This is an issue for instance when tackling prediction tasks with a supervised learning algorithm.

In this work, we are interested in anomaly detection and the subsequent imputation in time series. Our goal is to retrieve the smooth signal (plus possibly some white noise) from corrupted temporal data, as a preprocessing step for any downstream learning algorithm. Techniques to achieve these goals are extremely diverse and the interested reader can refer *e.g.* to [1–4] for a review or further details. Here, we reshape a noisy and corrupted time series into an appropriate matrix and we assume the smooth signal can be recovered since it corresponds to a low-rank matrix, while the anomalies are sparse.

Principal component analysis (PCA) is one of the most widely used techniques for dimension reduction in statistical data analysis. Given a collection of samples, PCA computes a linear projection of the samples to a low-dimensional subspace that minimises the ℓ_2-projection error [5,6]. This method is proved effective in many real world applications, *e.g.* [7–9]. However, its fragility in the

G. Nicosia et al. (Eds.): LOD 2022, LNCS 13811, pp. 281–295, 2023.
https://doi.org/10.1007/978-3-031-25891-6_21

face of missing or outlier data, common in everyday applications such as image processing or web data analysis, often puts its validity in jeopardy. In presence of a mixture of normal and abnormal measures, it is also difficult to select the dimension of the projection subspace representing the data.

To address this problem, a plethora of methods have been proposed for recovering low-rank and sparse matrices (aka robust principal component analysis (RPCA) [10]) with incomplete or grossly corrupted observations, such as principal component pursuit (PCP) [11] or outlier pursuit [12]. In principle, these methods aim at minimising an optimisation problem involving both trace norm and ℓ_1-norm, convex relaxations for the rank function and the ℓ_0-norm respectively [11]. Such relaxations bypass the fact that the basic problem is NP-hard. RPCA is proving useful and effective in many applications such as video surveillance [13], image and video processing [14], speech recognition [15] or latent segmentation indexing [16] just to name a few.

We are interested in anomaly detection in univariate time series and signal reconstruction: time series are first sized in matrices of appropriate dimensions (see Sect. 4 for an example) in order to apply RPCA to detect anomalies and impute their values. Although many phenomena are modelled by time series and RPCA is a widely used tool, there is, to our knowledge, very little work specifically concerned with applying RPCA to time series. In [17], RPCA is specifically used to detect anomalies in time series of users' query records, but these anomalies are the result of a particular treatment of the sparse component. Hence, the problem to minimise resumes to the basic RPCA problem. In [18], the authors introduce a compression matrix acting on the sparse one to reveal traffic anomalies. They show that it is possible to exactly recover the low rank and sparse components by solving a non-smooth convex optimisation problem. The work in [19] presents an estimator dedicated to recover sufficiently low-dimensional nominal traffic and sparse enough anomalies when the routing matrix is column-incoherent [11], and an adequate amount of flow counts are randomly sampled. In [20], time is directly taken into account in the objective function: a constraint is added to maintain consistency between the rows of the low-rank matrix. The method is applied to subway passenger flow and is effective for detecting anomalies. However, to effectively recover anomalies, a filtering of the sparse matrix is performed, and the choice of the threshold value is debatable. Furthermore, their method is not directly applicable when there is missing or unobserved data. In this paper, we aim to improve this RPCA approach, augmented by a consistency constraint between rows/columns, by taking into account missing data and noise, thus avoiding the debate of thresholding.

It is worth mentioning that RPCA methods are parametric. These parameters drastically influence the results of the method, *i.e.* the capacity of the algorithm to extract a smooth signal and to detect anomalies. For some formulations, there exist theoretical values that, in principle, allow the *exact* recovery of the desired matrices [21]. However, in practice and for real-world data, these values do not always work and it is often difficult to determine them without very precise knowledge of the data. We here seek to estimate all parameters from the examined data. More precisely, we propose to determine the parameters by cross-validation and consequently, we propose an end-to-end procedure to compute the RPCA.

Batch algorithms suffer from some disadvantages: *i*) they require a large amount of memory since all observed data must be stored in memory in order to be processed; *ii*) they also become too slow to process large data; *iii*) when data arrives continuously, they cannot simply adapt to this influx of information, but must restart from scratch and *iv)* they do not adapt to data drift. Henceforth, an online version of robust PCA method is highly necessary to process incremental data set, where the tracked subspace can be updated whenever new data is received.

Several RPCA methods have been developed but we can mainly group them into three categories/versions: Grassmannian Robust Adaptive Subspace Tracking Algorithm (GRASTA, [22]), Recursive Projected Compress Sensing (ReProCS, [23,24]) and Online Robust PCA via Stochastic Optimization (RPCA-STOC, [25]). GRASTA is built on Grassmannian Rank-One Update Subspace Estimation (GROUSE, [26]) and performs incremental gradient descent on the Grassmannian. ReProCS is mainly designed for solving problems for video sequence: separating a slowly changing background from moving foreground objects on-the-fly. A limitation of this approach is that it requires knowledge of the structure of the correlation model on the sparse part. Both GRASTA and Proc-ReProCS can only handle slowly changing subspaces. RPCA-STOC is based on stochastic optimisation for a close reformulation of the Robust PCA based on Principal Component Pursuit. In particular, RPCA-STOC splits the nuclear norm of the low-rank matrix as the sum of Frobenius norm of two matrices.

Lastly, note that RPCA-STOC works only with stable subspace, which could be a restriction for real applications, *e.g.* data drift. In practice, at time t, RPCA-STOC updates the basis of subspace by minimising an empirical loss which involves all previously observed samples with equal weights. In this work, we propose an online formulation RPCA with temporal regularisations by pursuing the idea of RPCA-STOC. We adapt the algorithm to cope with unstable subspaces by considering an online moving window, *i.e.* by considering an empirical loss based only on the most recent samples.

The rest of the paper is organised as follows. In Sect. 2 we recall some notions and present the different formulations of RPCA and deep dive into the algorithms in Sect. 3. In Sect. 4, we compare the different methods on synthetic and real-world data and we finally conclude in Sect. 5.

2 Background and Problem Formulation

Let $\mathbf{y} = \{y_i\}_{1 \leq i \leq I}$ be a univariate time series, where indices correspond to uniform time steps. Let T_0 be the main seasonality (or period) of this time series. We assume that we observe $n \in \mathbb{N}^+$ periods in $\mathbf{y}$, such that $I = T_0 \times n$; otherwise we add missing values at the end of the signal to fit in. We consider the case where the time series $\mathbf{y}$ is a corrupted version of a true signal: the corruptions can be multiple such as missing data, presence of anomalies, and additive observation noise. We propose to filter and reconstruct the whole signal by using the similarity of n different periods of length T_0. For this reason, we convert $\mathbf{y}$ into a matrix $\mathbf{D} \in \mathbb{R}^{T_0 \times n}$ such that $\mathbf{D} = [\mathbf{d}_1|\ldots|\mathbf{d}_n]$ with $\mathbf{d}_t = [y_{(t-1)T_0+1}, \cdots, y_{(t-1)T_0+T_0}]^\top$, $t = 1, \cdots, n$. We consider that the periodicity of $\mathbf{y}$ and the repetition of patterns across time means that the different periods $\mathbf{d}_t, t = 1, \cdots, n$ lie in a vector space of small dimension r. In favorable situations, this subspace can be estimated with a PCA of the matrix $\mathbf{D}$ but the assumed corruption of the data requires to use a robust version. In addition, a noticeable change with respect to standard PCA or RPCA is that the observations $\mathbf{d}_t, \mathbf{d}_{t+T_k}$ are not independent: we show how we can take advantage of the similarity between $\mathbf{d}_t$ and $\mathbf{d}_{t+T_k}$ for improving the global reconstruction of the signal.

In the following, $\|\mathbf{M}\|_* = \sum_{\sigma \in \mathrm{Sp}(\mathbf{M})} \sigma$ is the nuclear norm of $\mathbf{M}$; $\|\mathbf{M}\|_1 = \sum_{ij} |M_{ij}|$ is the ℓ_1-norm of $\mathbf{M}$ seen as a vector and $\|\mathbf{M}\|_F^2 = \sum_{ij} M_{ij}^2$. The Ω set is the set of observed data and $\mathcal{P}_\Omega(\mathbf{M})$ is the projection of $\mathbf{M}$ on Ω, *i.e.* $\mathcal{P}_\Omega(\mathbf{M})_{ij} = M_{ij}$ if $(i, j) \in \Omega$, 0 otherwise.

2.1 Robust Principal Component Analysis

Suppose we observe some data $\mathbf{y}$ or equivalently some entries of an incomplete, noisy and corrupted matrix $\mathbf{D}$. We seek to compute the following decomposition $\mathbf{D} = \mathbf{X} + \mathbf{A} + \mathbf{E}$, where $\mathbf{X}$ is a low-rank matrix whose rank is upper bounded by $r \leq \min\{T_0, n\}$, $\mathbf{A}$ is the sparse matrix of anomalies and $\mathbf{E}$ represents a Gaussian random fluctuation. In addition, we assume that for whatever reason, we only observe a subset of entries $\Omega \subseteq \{1, \cdots, T_0\} \times \{1, \cdots, n\}$. The RPCA formulation is as follows

$$\min_{\substack{\mathbf{X}, \tilde{\mathbf{A}} \\ \text{s.t. } \mathcal{P}_\Omega(\mathbf{D}) = \mathcal{P}_\Omega(\mathbf{X}+\tilde{\mathbf{A}})}} \|\mathbf{X}\|_* + \lambda_2 \|\tilde{\mathbf{A}}\|_1, \tag{1}$$

where $\lambda_2 > 0$ is the inverse sensitivity parameter. Here we threshold out the *significant* anomalies $\mathbf{A}$ from the noise $\mathbf{E}$, based on the anomaly score $s_{ij} = |\tilde{A}_{ij}|/\sigma_{X_{i:}}$: given a threshold $\alpha > 0$, $\tilde{\mathbf{A}} = \mathbf{A} + \mathbf{E}$ with $A_{ij} = \tilde{A}_{ij} \mathbb{1}_{s_{ij} > \alpha}$ and $E_{ij} = \tilde{A}_{ij} \mathbb{1}_{s_{ij} \leq \alpha}$. Here $\sigma_{X_{i:}}$ denotes the standard deviation of the row i. This sparsifying post-processing on the matrix $\tilde{\mathbf{A}}$ is quite common, see for instance [20].

This separation only makes sense if the low-rank part $\mathbf{X}$ is not sparse. This is related to the notion of *incoherence* for matrix completion problem introduced in [21] which imposes conditions on the singular vectors of the low-rank matrix. There are also issues when $\tilde{\mathbf{A}}$ is low-rank, in particular when data is missing not at random. Under these conditions, the main result of [11] states it is possible to *exactly* recover the low-rank and sparse components.

2.2 Robust Principal Component Analysis with Temporal Regularisations

In the case of independent observations, we only have to take into account the correlations between variables, whereas time series exhibits a dependency structure. When the time series has a seasonality of periodicity T_0, we expect the correctly sized matrix to have low rank and adjacent columns to be similar (see Sect. 4). However, to the best of our knowledge, not much specific temporal or time series oriented RPCA has been developed yet [18–20].

Here we use a set of K terms penalising the differences between columns spaced by given times T_k: $\|\mathbf{XH_k}\|_F^2$, where the $\{\mathbf{H_k}\}_{1\leq k\leq K}$ are Toeplitz matrices with 1 on the diagonal, -1 on the super-T_k diagonal and 0 otherwise. Stated otherwise, T_k is the number of columns between two supposedly similar ones. A typical example is to use $\mathbf{H_1}$ for controlling the similarity between $\mathbf{d}_t$ and $\mathbf{d}_{t+1}$, *i.e.* $T_k = 1$. Another example: if each column of a matrix represents a day, we can put $T_k = 7$ since we expect each day of the week to be "similar", *i.e.* similarity between $\mathbf{d}_t$ and $\mathbf{d}_{t+7}$. In addition and in order to explicitly take noise into account, we relax the constraint $\mathbf{D} = \mathbf{X} + \tilde{\mathbf{A}}$ to eventually get the following problem

$$\min_{\mathbf{X},\mathbf{A}} \frac{1}{2}\|\mathcal{P}_\Omega(\mathbf{D}-\mathbf{X}-\mathbf{A})\|_F^2 + \lambda_1\|\mathbf{X}\|_* + \lambda_2\|\mathbf{A}\|_1 + \sum_{k=1}^{K}\eta_k\|\mathbf{XH_k}\|_F^2 \tag{2}$$

where λ_1, λ_2 and $\{\eta_k\}_k$ are real parameters.

Note that Eq. (2) can be related to RPCA on graph (see *e.g.* [27]), where nodes have at most one outgoing link (our graph is a chain). Indeed, each $\mathbf{H_k}$ can be associated with an adjacency matrix $\hat{\mathbf{H}}_\mathbf{k}$ defined as

$$\hat{\mathbf{H}}_\mathbf{k} := \mathbf{I_n} - \left[\begin{array}{c|c} \mathbf{I}_{\mathbf{T_k}} & \\ & \mathbf{H_k} \\ & \\ \mathbf{0} & \end{array}\right],$$

with $\mathbf{I_a}$ the identity matrix of dimension a. By defining the Laplacian associated with this graph $\mathcal{L}_k = \mathbf{Dg} - \hat{\mathbf{H}}_\mathbf{k}$ ($\mathbf{Dg}$ is the degree matrix), we find a similar formulation to [27] of the regularisation, *i.e.* $\|\mathbf{XH_k}\|_F^2 = \mathrm{Tr}(\mathbf{X}^\top\mathcal{L}_k\mathbf{X})$. However, contrary to their problem, we force the matrix $\mathbf{X}$ to be low-rank. Hence $\mathbf{H_k}$ matrices select an appropriate "neighbourhood" for each observation and measure similarity, and the choice of $\mathbf{H_k}$ offers an avenue for modelling complex dependencies.

One can solve this problem (Eq. (2)) by computing its associated augmented Lagrangian and using the Alternating Direction Method of Multipliers [28].

2.3 Online Robust Principal Component Analysis with Temporal Regularisations

We now turn to the online formulation of the problem (2). For this, recall the nuclear norm of a matrix $\mathbf{X} \in \mathbb{R}^{T_0\times n}$ whose rank is upper bounded by r can be expressed as

$$\|\mathbf{X}\|_* = \inf_{\substack{\mathbf{L}\in\mathbb{R}^{T_0\times r}\\ \mathbf{Q}\in\mathbb{R}^{n\times r}}} \left\{\frac{1}{2}(\|\mathbf{L}\|_F^2 + \|\mathbf{Q}\|_F^2) : \mathbf{X} = \mathbf{LQ}^\top\right\}. \tag{3}$$

$\mathbf{L}$ can be interpreted as the basis for low-rank subspace while $\mathbf{Q}$ represents the coefficients of the observations with respect to that basis. Plugging this expression into Eq. (2), the problem becomes

$$\min_{\substack{\mathbf{X},\mathbf{A},\mathbf{L},\mathbf{Q} \\ \text{s.t. } \mathbf{X}=\mathbf{L}\mathbf{Q}^\top}} \frac{1}{2}\|\mathcal{P}_\Omega(\mathbf{D}-\mathbf{X}-\mathbf{A})\|_F^2 + \frac{\lambda_1}{2}\left(\|\mathbf{L}\|_F^2 + \|\mathbf{Q}\|_F^2\right) + \lambda_2\|\mathbf{A}\|_1 + \sum_{k=1}^{K}\eta_k\|\mathbf{X}\mathbf{H}_\mathbf{k}\|_F^2. \tag{4}$$

Given a finite set of samples $\mathbf{D} = [\mathbf{d}_1, ..., \mathbf{d}_n] \in \mathbb{R}^{T_0\times n}$, solving problem (2) amounts to minimising the following empirical cost function

$$f_n(\mathbf{L}) := \frac{1}{n}\sum_{i=1}^{n} l(\mathbf{d}_i, \mathbf{L}) + \frac{\lambda_1}{2n}\|\mathbf{L}\|_F^2 \tag{5}$$

where the loss function for each sample is defined as

$$\begin{aligned} l(\mathbf{d}_i, \mathbf{L}) &:= \min_{\mathbf{q},\mathbf{a}} h(\mathbf{d}_i, \mathbf{L}, \mathbf{q}, \mathbf{a}) \\ &:= \min_{\mathbf{q},\mathbf{a}} \frac{1}{2}\|\mathcal{P}_\Omega(\mathbf{d}_i - \mathbf{L}\mathbf{q} - \mathbf{a})\|_2^2 + \frac{\lambda_1}{2}\|\mathbf{q}\|_2^2 + \lambda_2\|\mathbf{a}\|_1 \\ &\quad + \sum_{k=1}^{K}\eta_k\|\mathbf{L}\mathbf{q} - \mathbf{L}\mathbf{q}_{-T_k}\|_2^2, \end{aligned} \tag{6}$$

with $\mathbf{q}_{-T_k}$ the T_k-th last computed vector $\mathbf{q}$.

In stochastic optimisation, one is usually interested in minimising the expected cost over all the samples, *i.e.* $f(\mathbf{L}) := \mathbb{E}_\mathbf{d}[l(\mathbf{d}, \mathbf{L})] = \lim_{n\to\infty} f_n(\mathbf{L})$ where the expectation is taken with respect to the distribution of the samples $\mathbf{d}$.

To solve this problem, we borrow ideas from [25] that consist in optimising the coefficients $\mathbf{q}$, anomalies $\mathbf{a}$ and basis $\mathbf{L}$ in an alternative manner, *i.e.* solving one vector while keeping the others fixed. In the t-th time instance $(0 \le t \le n)$, $\mathbf{q}_t$ and $\mathbf{a}_t$ are solutions of

$$\begin{aligned} \{\mathbf{q}_t, \mathbf{a}_t\} &= \operatorname*{argmin}_{\mathbf{q},\mathbf{a}} \frac{1}{2}\|\mathbf{d}_t - \mathbf{L}_{t-1}\mathbf{q} - \mathbf{a}\|_2^2 + \frac{\lambda_1}{2}\|\mathbf{q}\|_2^2 + \lambda_2\|\mathbf{a}\|_1 \\ &\quad + \sum_{k=1}^{K}\eta_k\|\mathbf{L}_{t-1}\mathbf{q} - \mathbf{L}_{t-1}\mathbf{q}_{-T_k}\|_2^2. \end{aligned} \tag{7}$$

We then obtain the estimation of the basis $\mathbf{L}_t$ through minimising the cumulative loss with respect to the previously estimated coefficients $\{\mathbf{q}_i\}_{i=1}^t$ and anomalies $\{\mathbf{a}_i\}_{i=1}^t$. To do so, we define a surrogate function (and an upper bound) of the empirical cost function $f_t(\mathbf{L})$ as

$$\begin{aligned} g_t(\mathbf{L}) &:= \frac{1}{t}\sum_{i=1}^{t} h(\mathbf{d}_i, \mathbf{L}, \mathbf{q}_i, \mathbf{a}_i) + \frac{\lambda_1}{2t}\|\mathbf{L}\|_F^2 \\ &\geq \frac{1}{t}\sum_{i=1}^{t} \min_{\mathbf{q},\mathbf{a}} h(\mathbf{d}_i, \mathbf{L}, \mathbf{q}, \mathbf{a}) + \frac{\lambda_1}{2t}\|\mathbf{L}\|_F^2 = f_t(\mathbf{L}). \end{aligned} \tag{8}$$

Minimising this function yields the estimation of $\mathbf{L}_t$; the solution is given by

$$\mathbf{L}_t = \left[\sum_{i=1}^{t}(\mathbf{d}_i - \mathbf{a}_i)\mathbf{q}_i^\top\right]\left[\sum_{i=1}^{t}\mathbf{q}_i\mathbf{q}_i^\top + 2\sum_{k=1}^{K}\eta_k(\mathbf{q}_t - \mathbf{q}_{t-T_k})(\mathbf{q}_t - \mathbf{q}_{t-T_k})^\top + \lambda_1\mathbf{I}\right]^{-1} \tag{9}$$

In practice $\mathbf{L}_t$ can be quickly updated by block-coordinate descent with warm restarts [29].

One limitation of this method is that it assumes a stable subspace. As it can be seen in applications, for instance background substraction in video surveillance or anomaly detection in financial datasets, the stability of the signal subspace is a restrictive assumption that needs to be relaxed. Our Eq. (8) makes this assumption because we seek to minimise the function considering all the past (from 1 to t), each sample having the same importance. This is obviously undesirable if the underlying subspace changes over time. To avoid such a problem, we combine common idea of time series–namely moving windows–with stochastic RPCA with temporal regularisations. Concretely, we update the basis $\mathbf{L}_t$ from the last n_w (*i.e.* most recent) samples. This translates into a slight modification of Eq. (8) as follow

$$\begin{aligned} g_t^w(\mathbf{L}) := \frac{1}{n_w}\sum_{i=t-n_w}^{t} &\left(\frac{1}{2}\|\mathcal{P}_\Omega(\mathbf{d}_i - \mathbf{L}\mathbf{q}_i - \mathbf{a}_i)\|_2^2 + \frac{\lambda_1}{2}\|\mathbf{q}_i\|_2^2\right. \\ &\left. + \lambda_2\|\mathbf{a}_i\|_1 + \sum_{k=1}^{K}\eta_k\|\mathbf{L}\mathbf{q}_i - \mathbf{L}\mathbf{q}_{i-T_k}\|_2^2\right) + \frac{\lambda_1}{2n_w}\|\mathbf{L}\|_F^2. \end{aligned} \tag{10}$$

This formulation has the ability to *quickly* adapt to changes in the underlying subspace.

3 Method

We present the algorithms for the batch and online RPCA with temporal regularisations. To simplify notation, the period T_0 (time) is m in the algorithms.

3.1 Batch Temporal Algorithm

The associated augmented Lagrangian of Eq. (2) is

$$\begin{aligned} \mathcal{L}_\mu(\mathbf{X}, \mathbf{L}, \mathbf{Q}, \mathbf{A}, \{\mathbf{R_k}\}_{k=1}^{K}, \mathbf{Y}) &= \frac{1}{2}\|\mathcal{P}_\Omega(\mathbf{X} + \mathbf{A} - \mathbf{D})\|_F^2 + \frac{\lambda_1}{2}\|\mathbf{L}\|_F^2 + \frac{\lambda_1}{2}\|\mathbf{Q}\|_F^2 \\ &+ \lambda_2\|\mathbf{A}\|_1 + \sum_{k=1}^{K}\eta_k\|\mathbf{R_k}\|_F^2 + \langle\mathbf{Y}, \mathbf{X} - \mathbf{L}\mathbf{Q}^\top\rangle + \frac{\mu}{2}\|\mathbf{X} - \mathbf{L}\mathbf{Q}^\top\|_F^2 \end{aligned} \tag{11}$$

The Alternating Direction Method of Multipliers (ADMM, [28]) has proven to be efficient in solving this type of problems [11,20]. More precisely, by using the fact that $2\langle \mathbf{a}, \mathbf{b}\rangle + \|\mathbf{b}\|_F^2 = \|\mathbf{a}+\mathbf{b}\|_F^2 - \|\mathbf{a}\|_F^2$ and by zeroing the derivatives, it is straightforward to obtain the closed-form solutions. Moreover, recall that $\mathcal{S}_\alpha(\mathbf{Z}) := \text{sgn}(\mathbf{Z})\max(|\mathbf{Z}| - \alpha, 0)$ is the soft thresholding operator at level α, applied element-wise and is the solution of $\nabla_\mathbf{B}(\|\mathbf{Z}-\mathbf{B})\|_F^2 + \alpha\|\mathbf{B}\|_1) = 0$. All steps are summarised in Algorithm 1. Since $\min(m, n) > r$, time complexity at each iteration is $\mathcal{O}(m^2 n)$ while space complexity equals $\mathcal{O}(mn)$.

Algorithm 1: Batch Temporal Robust Principal Component Analysis

Data: observations $\mathbf{D} \in \mathbb{R}^{m\times n}$, params $\lambda_1, \lambda_2, \{\eta_k\}_{1\leq k\leq K}$
Result: low-rank matrix $\mathbf{X} \in \mathbb{R}^{m\times n}$, sparse matrix $\mathbf{A} \in \mathbb{R}^{m\times n}$
initialisation: $\mathbf{X}^{(0)} = \mathbf{A}^{(0)} = \mathbf{Y}^{(0)} = \mathbf{1} \in \mathbb{R}^{m\times n}$, $\mathbf{L}^{(0)} = \mathbf{1} \in \mathbb{R}^{m\times r}$, $\mathbf{Q}^{(0)} = \mathbf{1} \in \mathbb{R}^{n\times r}$, $\mu =$ 1e-6, $\mu_{max} =$ 1e10, $\rho = 1.1$, $\epsilon =$ 1e-8, max_iter = 1e6, $i = 0$;
while $e > \epsilon$ *and* $i <$ *max_iter* **do**

$$\mathbf{X}^{(i+1)} \leftarrow \left(\mathbf{D} - \mathbf{A}^{(i)} + \mu\mathbf{L}^{(i)}\mathbf{Q}^{(i)\top} - \mathbf{Y}\right)\left((1+\mu)\mathbf{I} + 2\textstyle\sum_k \eta_k \mathbf{H_k}\mathbf{H_k}^\top\right)^{-1};$$
$$\mathbf{A}^{(i+1)} \leftarrow \mathcal{S}_{\lambda_2/2}(\mathbf{D} - \mathbf{X}^{(i+1)});$$
$$\mathbf{L}^{(i+1)} \leftarrow \left(\mu\mathbf{X}^{(i+1)} + \mathbf{Y}^{(i)}\right)\mathbf{Q}^{(i)}\left(\lambda_1\mathbf{I} + \mu{\mathbf{Q}^{(i)}}^\top\mathbf{Q}^{(i)}\right)^{-1};$$
$$\mathbf{Q}^{(i+1)} \leftarrow \left(\mu\mathbf{X}^{(i+1)} + \mathbf{Y}^{(i)}\right)^\top\mathbf{L}^{(i+1)}\left(\lambda_1\mathbf{I} + \mu\mathbf{L}^{(i+1)\top}\mathbf{L}^{(i+1)}\right)^{-1};$$
$$\mathbf{Y}^{(i+1)} \leftarrow \mathbf{Y}^{(i)} + \mu(\mathbf{X}^{(i+1)} - \mathbf{L}^{(i+1)}{\mathbf{Q}^{(i+1)}}^\top);$$
$$e = \max\left\{\|\mathbf{M}^{(t+1)} - \mathbf{M}^{(t)}\|_\infty, \mathbf{M} \in \{\mathbf{X}, \mathbf{A}, \mathbf{L}, \mathbf{Q}\}\right\};$$
$$\mu \leftarrow \min(\rho\mu, \mu_{max});$$
$$i \leftarrow i + 1$$

end

3.2 Online Temporal Algorithm via Stochastic Optimisation

We here present the algorithm for solving the online RPCA with temporal regularisations (Algorithm 2). First, we compute the batch algorithm on the n_{burnin} first columns of $\mathbf{D}$, *i.e.* the columns we already observed. Then, we alternatively update the coefficients, the anomalies and the basis with off-the-shelf convex optimisation solver (Algorithm 3) and block-coordinate descent with warm restarts (Algorithm 4) respectively. Here, time complexity at each iteration of the online part is $\mathcal{O}(mr^2)$ (and for the batch part on the n_{burnin} samples: $\mathcal{O}(m^2 n_{burnin})$) while space complexity equals $\mathcal{O}(mr)$ (and $\mathcal{O}(mn_{burnin})$ for the batch part). When processing *big data*–especially when $n \gg m > r$–there is a computational complexity advantage to using the online algorithm since we will have $\frac{r^2}{mB} < 1$ with B the number of iterations of the temporal batch algorithm.

Algorithm 2: Stochastic optimisation

Data: observations $\mathbf{D} = [\mathbf{d}_1, \mathbf{d}_2, ..., \mathbf{d}_n] \in \mathbb{R}^{m\times n}$ (which are revealed sequentially), params $\lambda_1, \lambda_2, \{\eta_k\}_{1\leq k\leq K}$, $\mathbf{D}_b = [\mathbf{d}_{-n_{burnin}}, ..., \mathbf{d}_{-1}] \in \mathbb{R}^{m\times n_{burnin}}$

Result: low-rank matrix $\mathbf{X} \in \mathbb{R}^{m\times n}$, sparse matrix $\mathbf{A} \in \mathbb{R}^{m\times n}$

initialisation: $\mathbf{X}^{(0)} = \mathbf{A}^{(0)} = \mathbf{Y_0}^{(0)} = \mathbf{1} \in \mathbb{R}^{m\times T_0}$, $\mathbf{L}^{(0)} = \mathbf{1} \in \mathbb{R}^{m\times r}$, $\mathbf{Q}^{(0)} = \mathbf{1} \in \mathbb{R}^{T_0\times r}$, μ = 1e-6, μ_{max} = 1e10, $\rho = 1.1$, ϵ = 1e-8, max_iter = 1e6, $k = 0$, n_{burnin}, n_w ;

Compute RPCA on burn in samples via Algorithm 1:

  $\mathbf{X}_b, \mathbf{A}_b = \text{batch\_RPCA}(\mathbf{D}_b)$;

Compute SVD decomposition of $\mathbf{X}_b$:

  $\mathbf{X}_b = \mathbf{U}\mathbf{\Sigma}\mathbf{V}^\top$;

Initialise $\mathbf{L}$, $\mathbf{B}$ and $\mathbf{C}$:

  $\mathbf{L}_0 \leftarrow \mathbf{U}\mathbf{\Sigma}^{1/2}$

  $\mathbf{B}_0 \leftarrow \sum_{i=1}^{n_{burnin}} \mathbf{q}_i\mathbf{q}_i^\top + 2\sum_{k=1,T_k>i}^{K} \eta_k(\mathbf{q}_i - \mathbf{q}_{i-T_k})(\mathbf{q}_t - \mathbf{q}_{i-T_k})^\top$

  $\mathbf{C}_0 \leftarrow \sum_{i=1}^{n_{burnin}} (\mathbf{q}_i - \mathbf{a}_i)\mathbf{q}_i^\top$;

for $t = 1$ *to* n **do**

  Reveal sample $\mathbf{d}_t$;

  Project this new sample via Algorithm 3:

  $\{\mathbf{q}_t, \mathbf{a}_t\} = \text{argmin}_{\mathbf{q},\mathbf{a}} \frac{1}{2}\|\mathbf{d}_t - \mathbf{L}_{t-1}\mathbf{q} - \mathbf{a}\|_2^2 + \frac{\lambda_1}{2}\|\mathbf{q}\|_2^2 + \lambda_2\|\mathbf{a}\|_1 + \sum_{k=1}^{K} \eta_k\|\mathbf{L}_{t-1}\mathbf{q} - \mathbf{L}_{t-1}\mathbf{q}_{-T_k}\|_2^2$;

  Update $\mathbf{B}$ and $\mathbf{C}$:

    $\mathbf{B}_t \leftarrow \mathbf{B}_{t-1} + \mathbf{q}_t\mathbf{q}_t^\top + 2\sum_{k=1}^{K} \eta_k(\mathbf{q}_t - \mathbf{q}_{t-T_k})(\mathbf{q}_t - \mathbf{q}_{t-T_k})^\top$ - $\mathbf{q}_{t-n_w}\mathbf{q}_{t-n_w}^\top$

    $\mathbf{C}_t \leftarrow \mathbf{C}_{t-1} + (\mathbf{d}_t - \mathbf{a}_t)\mathbf{q}_t^\top$ - $(\mathbf{d}_{t-n_w} - \mathbf{a}_{t-n_w})\mathbf{q}_{t-n_w}^\top$;

  Compute $\mathbf{L}_t$ with $\mathbf{L}_{t-1}$ as warm restart using Algorithm 4:

    $\mathbf{L}_t = \text{argmin}_{\mathbf{L}} \text{Tr}(\mathbf{L}^\top(\mathbf{B}_t + \lambda_1\mathbf{I})\mathbf{L}) - \text{Tr}(\mathbf{L}^\top\mathbf{C}_t)$

Return $\mathbf{X} = \mathbf{L}_n\mathbf{Q}^\top$ (low-rank matrix, $\mathbf{Q}$ is stacked $\mathbf{q}$), $\mathbf{A}$ (sparse matrix, $\mathbf{A}$ is stacked $\mathbf{a}$)

3.3 Hyperparameters Search

There are numerous hyperparameters $\boldsymbol{\lambda} := [\lambda_1, \lambda_2, \{\eta_k\}]$ to be chosen and it is often unclear how to determine them. Without prior knowledge of the data, a cross-validation approach can be used. In practice, we select J random subsets $\Omega_j \subset \Omega$ that we designate as missing entries. The cardinality of each subset is $|\Omega_j| = c|\Omega|$, where $c \in]0, 1[$. To obtain the optimal hyperparameters $\boldsymbol{\lambda}^*$, we use Bayesian Optimisation with Gaussian processes [30] where the complex and expensive function to minimise is the average reconstruction error over the J subsets Ω_j: $\frac{1}{J}\sum_{j=1}^{J} \|\mathcal{P}_{\Omega_j}(D_{ij} - X(\boldsymbol{\lambda})_{ij})\|_1$. We simply require that there are no values already missing from the Ω_j in order to evaluate the reconstruction.

Algorithm 3: Sample projection

Data: Input basis $\mathbf{L} = [\mathbf{l}_1, \mathbf{l}_2, ..., \mathbf{l}_r] \in \mathbb{R}^{m \times r}$, $\mathbf{d} \in \mathbb{R}^m$,
params $\lambda_1, \lambda_2, \{\eta\}_{1 \leq k \leq K}$, tol
Result: projected samples $\mathbf{q} \in \mathbb{R}^r$ and $\mathbf{a} \in \mathbb{R}^m$
$\mathbf{a} \leftarrow 0$;
while *not converged* **do**
  Update the coefficient $\mathbf{q}$:
    $\mathbf{q} \leftarrow (\mathbf{L}^\top \mathbf{L} + \lambda_1 \mathbf{I} + 2\mathbf{L}^\top \mathbf{L} \sum_{k=1}^{K} \lambda_k)^{-1} \mathbf{L}^\top (\mathbf{d} - \mathbf{a} + 2\mathbf{L} \sum_{k=1}^{K} \lambda_k \mathbf{q}_{-T_k})$;
  Update the sparse component $\mathbf{a}$:
    $\mathbf{a} \leftarrow \mathcal{S}_{\lambda_2}(\mathbf{d} - \mathbf{L}\mathbf{q})$;
  Converged if $\max \left\{ \frac{\|\mathbf{q}_{k+1} - \mathbf{q}_k\|}{\|\mathbf{d}\|}, \frac{\|\mathbf{a}_{k+1} - \mathbf{a}_k\|}{\|\mathbf{d}\|} \right\} < tol$
Return $\mathbf{q}, \mathbf{a}$

Algorithm 4: Fast basis update

Data: $\mathbf{L} = [\mathbf{l}_1, \mathbf{l}_2, ..., \mathbf{l}_r] \in \mathbb{R}^{m \times r}$, $\mathbf{B} = [\mathbf{b}_1, \mathbf{b}_2, ..., \mathbf{b}_r] \in \mathbb{R}^{r \times r}$,
  $\mathbf{C} = [\mathbf{c}_1, \mathbf{c}_2, ..., \mathbf{c}_r] \in \mathbb{R}^{m \times r}$
Result: Update basis $\mathbf{L}$
$\tilde{\mathbf{B}} \leftarrow \mathbf{B} + \lambda_1 \mathbf{I}$;
for $j = 1$ *to* r **do**
  $\mathbf{l}_j \leftarrow \frac{1}{\tilde{b}_{jj}} (\mathbf{c}_j - \mathbf{L}\tilde{\mathbf{b}}_\mathbf{j}) + \mathbf{l}_j$
Return $\mathbf{L}$

Another parameter of prime importance is the estimation of the rank r. We have to provide the rank r of $\mathbf{D}$ to initialise the matrices $\mathbf{L}$ and $\mathbf{Q}$. The rank r has to be as small as possible in order to minimise the matrix sparsity and the low-rank error while keeping *sufficient* information. A common practice is to look at the plot of the cumulative sum of the eigenvalues, *e.g.* when one reaches $x\%$ of the nuclear norm (often at least 90%).

4 Experiments

We apply our framework to time series. First, we make up synthetic time series from sine functions. Second, we consider empirical data with a complex structure representing train ticketing validations.

4.1 Synthetic Data

As a first example of time series, we generate J sine functions of N points over $[0, 2\pi N']$ with different frequencies f_j and amplitude a_j, which we corrupt. More precisely, we delete data to create missing data (missing at random), we add anomalies that can reach twice the maximum amplitude of the initial function and finally, we add Gaussian noise. Time series are then added up; the period of the resulting time series equals $P = N/N' \times \text{LCM}\{1/f_j\}_{j \in J}$ with LCM the

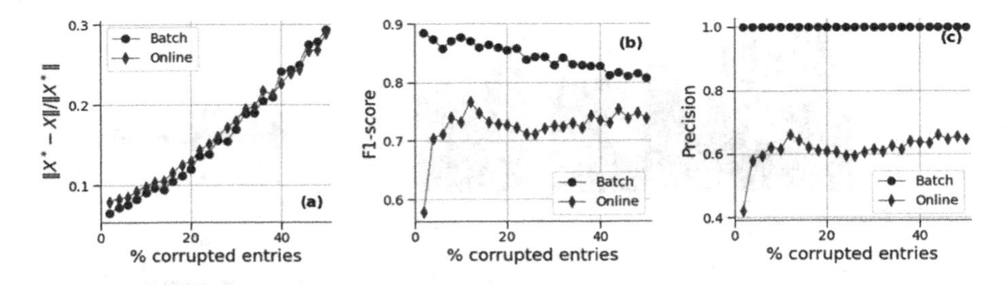

Fig. 1. Comparison of two RPCA formulations: batch or online with additional temporal regularisations. (a) reconstruction errors; (b) F1-scores for the anomaly detection and (c) precision scores for the anomaly detection–with respect to the percentage of corrupted data.

Lowest Commun Multiple. We decide to reshape this time series into matrix of dimension $m \times n$, *i.e.* $T_0 = m$ and $T_1 = P/m$. In this work, we set $N = 10000, N' = 10, m = 100, f_1 = 1, f_2 = 3, f_3 = 1/2, a_1 = a_2 = 1$ and $a_3 = 2$. Hence the resulting matrix is of dimension 100×100, $T_0 = 100$ and $T_1 = 20$. This procedure (corruption generation) is repeated 20 times and results reported are the average of these simulations. We make use of this simple case to observe the ability of the proposed frameworks to correctly disentangle the low-rank and sparse parts of the corrupted data.

We start by evaluating the ability of the algorithms to reconstruct the underlying regular signal $\mathbf{X}$ by imputing the anomalies and missing data. The reconstruction errors are quantified by $\|\mathbf{X}^* - \mathbf{X}\|_F^2/\|\mathbf{X}^*\|_F^2$. The batch and online versions offer similar reconstruction qualities regardless of the data corruption rate (Fig. 1(a)). We also evaluate the performances in terms of anomaly detection. To do so, F1-scores and precision scores are computed. For these metrics, there is a clear superiority of the batch version over the online one (Fig. 1(b) and (c)).

4.2 Real Data: Ticketing Validations Time Series

The methods are evaluated on Automatic Ticket Counting data which is the hourly data of transport ticket validations in stations of Ile-de-France[1]. The data cover the period January 2015–June 2018 at an hourly granularity. They show strong weekly and, to a lesser extent, daily seasonality. The data also shows non stationary behaviours, such as local decreases in the number of validation (Fig. 2(a)–(c)). In order to apply the RPCA algorithms, the time series is transformed into a matrix. Without any prior knowledge of the data, we can compute the time series partial autocorrelation function (PACF, Fig. 2(d)). One sees peaks corresponding to one-day and one-week correlations. For the RPCA algorithm to provide good results, a tip is to take T_0 as the lag giving the largest

[1] The data is available at https://data.iledefrance-mobilites.fr/explore/dataset/histo-validations/information/.

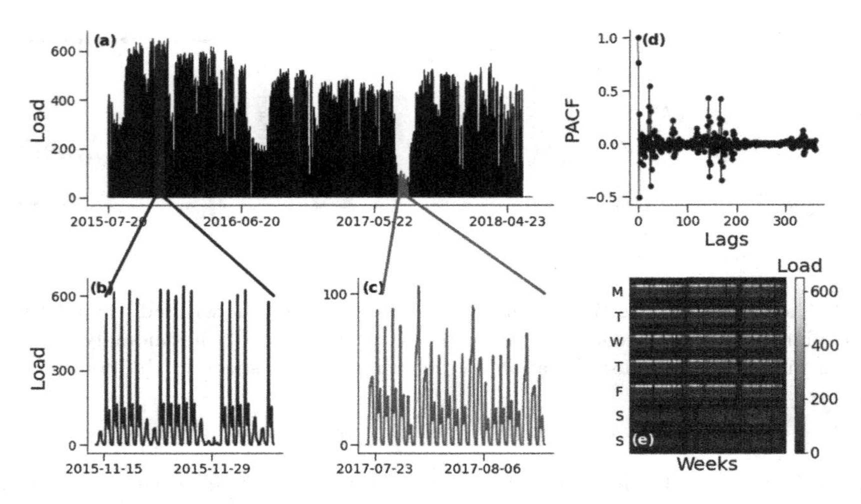

Fig. 2. RPCA on a validation profile. (a) Entire time series of ticketing validations at the Athis-Mons station. (b) and (c) Focus on a subpart of the time series. (d) The time series displays a strong weekly seasonality, and to a lesser extent, a daily seasonality. (e) Matrix **D** associated to the time series in (a). Since the strong weekly seasonality, each column represents a week (dotted red lines separate the days Monday, Tuesday,...), and we can see that the weeks have similar patterns (high values in the morning of the working days). (Color figure online)

PACF value while having a low rank matrix **D**. Here, the resulting matrix has $T_0 = 24 \times 7 = 168$ rows, *i.e.* each week represents a column and we stack the successive 153 weeks vertically (see Fig. 2(e)). We eventually get a matrix of dimension 168×153.

In order to quantify the performance of the proposed methods, we randomly corrupt some data and again inspect the reconstruction errors, F1-score and precision score as a function of the percentage of corrupted data. Temporal regularisations are computed with $T_1 = 1$, *i.e.* similarity between consecutive weeks. Results are reported in Fig. 3. For this dataset, online versions offer better results than batch ones. Online RPCA are used with a moving window of size 50 (almost one year). A possible explanation for this behaviour may lie in the fact that the underlying subspace changes over time and thus the rank of matrix **L**–which includes all observed samples–will increase over time. Retaining all past information may be more harmful than restricting oneself to the n_{win} most recent ones. Indeed, as observed in Fig. 2(e), it is possible that the subspace changes slightly at a yearly rate, giving the advantage to sliding window methods.

We next explore the impact of RPCA filtering on modal decompositions. In particular, we add ten percent of corruptions to the observations **D** and obtain a corrupted matrix $\tilde{\mathbf{D}}$. We apply RPCA algorithms, *e.g.* batch (b), temporal batch (tb), temporal online (to), and temporal online with window (tow) to **D** and $\tilde{\mathbf{D}}$ to get several matrices **X**, *i.e.* $\mathbf{X}_b(\mathbf{D}), \mathbf{X}_o(\mathbf{D}), ..., \mathbf{X}_b(\tilde{\mathbf{D}}), ...$ We then compute the PCA modes by the SVD decomposition on all these matrices with

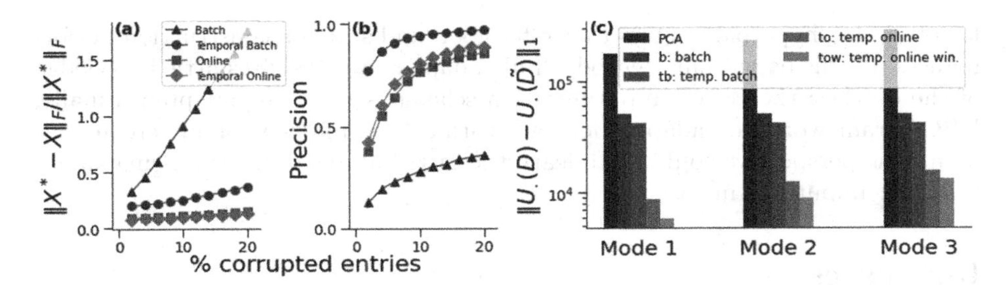

Fig. 3. Comparison of four RPCA formulations on a validation profile: batch or online and with or without additional temporal regularisations ($T_1 = 1$, *i.e.* similarity between consecutive weeks) (a) reconstruction errors and (b) precision scores for the anomaly detection; with respect to the percentage of corrupted data. The online versions offer the best reconstructions whilst the temporal ones are better at avoiding false positives. (c) RPCA filtering on modal decompositions: focus on the first modes computed on the resulting low-rank matrices.

$\mathbf{M} = \mathbf{U}\mathbf{\Sigma}\mathbf{V}^\top$ the SVD decomposition of $\mathbf{M}$. We thus obtained several matrices $\mathbf{U}_{\mathrm{PCA}}(\mathbf{D}), \mathbf{U}_b(\mathbf{D})$, ... Finally, we compute the difference between the first modes for each couple $(\mathbf{U}_.(\mathbf{D}), \mathbf{U}_.(\tilde{\mathbf{D}}))$. As illustrated in Fig. 3(c)), the differences are smaller when applying an RPCA algorithm before calculating the decomposition; this show the interest of using RPCA to effectively recover coherent structures, when directly applying PCA fails. The smallest difference is for the temporal online with window method.

5 Discussion and Conclusion

This work is motivated by real world data whose quality issues hinder their usage by machine learning algorithms. By representing a noisy and corrupted time series by an appropriate matrix, we have assume that the smooth signal can be recovered since it corresponds to a low-rank matrix, potentially perturbed by additive white noise, and some (sparse) anomalies. The temporal aspect of the data under scrutiny has allowed us to introduce a set of Toeplitz matrices $\mathbf{H}_\mathbf{k}$ to improve some general RPCA frameworks. We have also proposed an online formulation that can be solved by a stochastic optimisation algorithm, offering a better computational complexity. The performed simulations have demonstrated the effectiveness of the proposed methods. Besides, we have proposed an end-to-end pipeline with a cross-validation strategy for the hyperparameters tuning.

This work enlightens a new trade-off in the data reconstruction process. One would like to set the parameters that best lead to the recovery of the *theoretical* low-dimensional subspace. In that case, we might have interest in selecting small dimension r. But at the same time, we want to have an exact detection of anomalies in the observed data betting also sparse. The anomaly detection requires a good reconstruction, hence these two sparsity assumptions probably compete. We emphasise the important computational effort in the batch version required for obtaining hyperparameters that offer good and stable reconstruction performances. In this paper, our method is fully adaptive and requires little knowledge of the dataset: the Bayesian optimisation scheme is then well-adapted. Finally, RPCA frameworks are adapted for data with entries missing completely at random. One perspective could be to learn the distributions of these missing values to better impute them.

References

1. Blázquez-García, A., Conde, A., Mori, U., Lozano, J.A.: A review on outlier/anomaly detection in time series data. ACM Comput. Surv. (CSUR) **54**(3), 1–33 (2021)
2. Shaukat, K., et al.: A review of time-series anomaly detection techniques: a step to future perspectives. In: Arai, K. (ed.) FICC 2021. AISC, vol. 1363, pp. 865–877. Springer, Cham (2021). https://doi.org/10.1007/978-3-030-73100-7_60
3. Pratama, I., Permanasari, A.E., Ardiyanto, I., Indrayani, R.: A review of missing values handling methods on time-series data. In: 2016 International Conference on Information Technology Systems and Innovation (ICITSI), pp. 1–6 (2016)
4. Honaker, J., King, G.: What to do about missing values in time-series cross-section data. Am. J. Polit. Sci. **54**(2), 561–581 (2010)
5. Eckart, C., Young, G.: The approximation of one matrix by another of lower rank. Psychometrika **1**(3), 211–218 (1936)
6. Jolliffe, I.T., Cadima, J.: Principal component analysis: a review and recent developments. Philos. Trans. R. Soc. A Math. Phys. Eng. Sci. **374**(2065), 20150202 (2016)
7. Partridge, M., Calvo, R.A.: Fast dimensionality reduction and simple PCA. Intell. Data Anal. **2**(3), 203–214 (1998)
8. Agarwal, M., Agrawal, H., Jain, N., Kumar, M.: Face recognition using principle component analysis, eigenface and neural network. In: 2010 International Conference on Signal Acquisition and Processing, pp. 310–314. IEEE (2010)
9. Parinya, S.: Principal Component Analysis: Multidisciplinary Applications. BoD-Books on Demand (2012)
10. Wright, J., Ganesh, A., Rao, S., Peng, Y., Ma, Y.: Robust principal component analysis: exact recovery of corrupted low-rank matrices via convex optimization. In: NIPS, vol. 58, pp. 289–298 (2009)
11. Candès, E.J., Li, X., Ma, Y., Wright, J.: Robust principal component analysis? J. ACM (JACM) **58**(3), 1–37 (2011)
12. Xu, H., Caramanis, C., Sanghavi, S.: Robust PCA via outlier pursuit. arXiv preprint arXiv:1010.4237 (2010)
13. Luan, X., Fang, B., Liu, L., Yang, W., Qian, J.: Extracting sparse error of robust PCA for face recognition in the presence of varying illumination and occlusion. Pattern Recognit. **47**(2), 495–508 (2014)
14. Bouwmans, T., Javed, S., Zhang, H., Lin, Z., Otazo, R.: On the applications of robust PCA in image and video processing. Proc. IEEE **106**(8), 1427–1457 (2018)
15. Gavrilescu, M.: Noise robust automatic speech recognition system by integrating robust principal component analysis (RPCA) and exemplar-based sparse representation. In: 2015 7th International Conference on Electronics, Computers and Artificial Intelligence (ECAI), pp. S-29. IEEE (2015)

16. Deerwester, S., Dumais, S.T., Furnas, G.W., Landauer, T.K., Harshman, R.: Indexing by latent semantic analysis. J. Am. Soc. Inf. Sci. **41**(6), 391–407 (1990)
17. Jin, Y., Qiu, C., Sun, L., Peng, X., Zhou, J.: Anomaly detection in time series via robust PCA. In: 2017 2nd IEEE International Conference on Intelligent Transportation Engineering (ICITE), pp. 352–355. IEEE (2017)
18. Mardani, M., Mateos, G., Giannakis, G.B.: Recovery of low-rank plus compressed sparse matrices with application to unveiling traffic anomalies. IEEE Trans. Inf. Theory **59**(8), 5186–5205 (2013)
19. Mardani, M., Giannakis, G.B.: Robust network traffic estimation via sparsity and low rank. In: 2013 IEEE International Conference on Acoustics, Speech and Signal Processing, pp. 4529–4533. IEEE (2013)
20. Wang, X., Zhang, Y., Liu, H., Wang, Y., Wang, L., Yin, B.: An improved robust principal component analysis model for anomalies detection of subway passenger flow. J. Adv. Transp. 2018 (2018)
21. Candes, E., Recht, B.: Exact matrix completion via convex optimization. Found. Comput. Math. **9**(6), 717–772 (2009)
22. He, J., Balzano, L., Lui, J.: Online robust subspace tracking from partial information. arXiv preprint arXiv:1109.3827 (2011)
23. Guo, H., Qiu, C., Vaswani, N.: An online algorithm for separating sparse and low-dimensional signal sequences from their sum. IEEE Trans. Signal Process. **62**(16), 4284–4297 (2014)
24. Qiu, C., Vaswani, N., Lois, B., Hogben, L.: Recursive robust PCA or recursive sparse recovery in large but structured noise. IEEE Trans. Inf. Theory **60**(8), 5007–5039 (2014)
25. Feng, J., Xu, H., Yan, S.: Online robust PCA via stochastic optimization. In: Advances in Neural Information Processing Systems, vol. 26 (2013)
26. Balzano, L., Nowak, R., Recht, B.: Online identification and tracking of subspaces from highly incomplete information. In: 2010 48th Annual Allerton Conference on Communication, Control, and Computing (Allerton), pp. 704–711. IEEE (2010)
27. Shahid, N., Kalofolias, V., Bresson, X., Bronstein, M., Vandergheynst, P.: Robust principal component analysis on graphs. In: Proceedings of the IEEE International Conference on Computer Vision, pp. 2812–2820 (2015)
28. Boyd, S., Parikh, N., Chu, E.: Distributed Optimization and Statistical Learning Via the Alternating Direction Method of Multipliers. Now Publishers Inc, Netherlands (2011)
29. Loshchilov, I., Hutter, F.: SGDR: Stochastic gradient descent with warm restarts. arXiv preprint arXiv:1608.03983 (2016)
30. Pedregosa, F., et al.: Scikit-learn: machine learning in Python. J. Mach. Learn. Res. **12**, 2825–2830 (2011)

Monte Carlo Optimization of Liver Machine Perfusion Temperature Policies

Angelo Lucia[1(✉)] and Korkut Uygun[2]

[1] Department of Chemical Engineering, University of Rhode Island, Kingston, RI 02881, USA
alucia@uri.edu

[2] Center for Engineering in Medicine, Massachusetts General Hospital, Boston, MA 02114, USA

Abstract. In this work, a constrained multi-objective function formulation of liver machine perfusion (MP) based on widely accepted viability criteria and network metabolic efficiency is described. A novel Monte Carlo method is used to improve machine perfusion (MP) performance by finding optimal temperature policies for hypothermic machine perfusion (HMP), mid-thermic machine perfusion (MMP), and subnormothermic machine perfusion (SNMP). It is shown that the multi-objective function formulation can exhibit multiple maxima, that greedy optimization can get stuck at a local optimum, and that Monte Carlo optimization finds the best temperature policy in each case.

Keywords: Multi-objective optimization · Monte Carlo · Machine perfusion

1 Introduction

Machine perfusion (MP), in which an organ is placed in chamber at *fixed* temperature and constantly supplied with nutrients (called a perfusate) containing dissolved oxygen through recirculation of the perfusate, is rapidly gaining acceptance as a better alternative to static cold storage (SCS) for the preservation of organs (i.e., kidney, liver, heart) for transplantation. Machine perfusion protocols are often classified based on the temperature of the perfusion chamber [1]. The most common protocols are Hypothermic (HMP, 0–12 °C), Mid-thermic (MMP, 13–20 °C), Subnormothermic (SNMP, 21–34 °C), and Normothermic (NMP, 35–37 °C) MP. There are also various metrics used for establishing organ viability for transplantation. However, those based on the energy state of the organ (i.e., ATP content or energy charge) seem to be the most widely accepted metrics for predicting transplantation success [2, 3]. In this paper, the terms protocol, profile, and policy, have the same meaning.

While conventional temperature protocols for MP use a fixed temperature, recent clinical studies [4, 5] in kidney MP have shown that gradual rewarming to body temperature (37 °C) improves the energy state of kidneys compared to conventional HMP and MMP. Recent optimization results using greedy and Monte Carlo optimization [7] and a metabolic model of the liver [6] also yield *discrete* policies that systematically raise the temperature of the perfusion chamber to body temperature and show improvements

G. Nicosia et al. (Eds.): LOD 2022, LNCS 13811, pp. 296–303, 2023.
https://doi.org/10.1007/978-3-031-25891-6_22

in the energy state of liver cells compared to conventional SNMP and NMP. See Fig. 1. We note that the temperature policy for Monte Carlo optimization shown in Fig. 1 has the same shape as the results for gradual rewarming for kidney shown in Fig. 3A in [5].

The constrained optimization problem studied in [7] was based only on the liver viability criteria in [8] and formulated as follows:

$$\max_{T(t)} R\colon pH > 7.3;\ [lactate] < 2.3\text{mM};\ T_1 \leq T_2 \leq \cdots \leq T_N \tag{1}$$

where $T(t) = \{T_1, T_2, \ldots, T_N\}$ is a discrete temperature policy, N is the number of discrete temperature adjustments, and the reward, R, is defined by

$$R = w_1|Glc| + w_2ATP + w_3Mev + w_4EC \tag{2}$$

Glc in Eq. 2 denotes glucose consumption, ATP is net ATP synthesis, Mev denotes mevalonate production, which was used as a measure of bile synthesis, EC is energy charge, and w_1 through w_4 are weights.

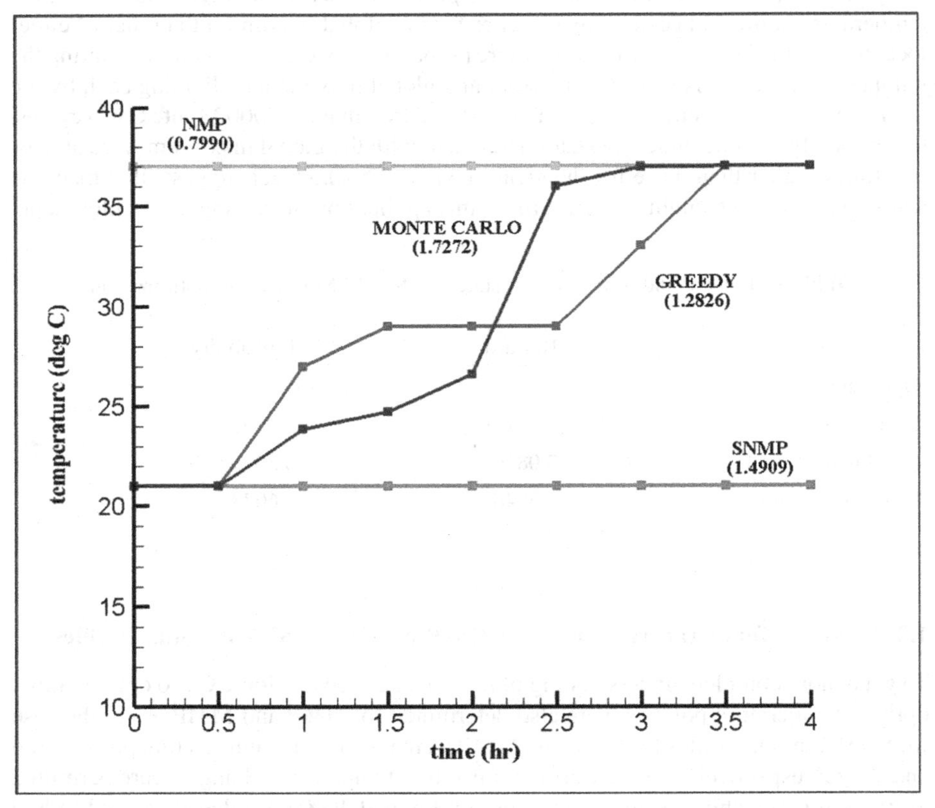

Fig. 1. Liver machine perfusion temperature policies for SNMP, NMP, GREEDY, and MONTE CARLO. Numbers in parenthesis represent amounts of ATP generated in pmol/cell.

1.1 SNMP Monte Carlo Temperature Policy Optimization and Statistics

The SNMP temperature policy optimizations in [7] were initialized to a flat temperature protocol of 21 °C. Random temperature changes for each Monte Carlo cycle (or episode) were permitted every 30 min over the course of 4 h and 3000 Monte Carlo cycles (or episodes) were used for each optimization run, which proved sufficient to find the optimal temperature policy shown in Fig. 1.

One of the more interesting aspects of Monte Carlo temperature optimization from an SNMP starting profile from the study conducted in [7] was the observation that there were three distinct levels of reward (i.e., three maxima in the reward function) – two local maxima at ~2.87 and 3.08 and a global maximum at ~3.55. Moreover, the existence of multiple optimal solutions did not appear to be an artifact of the modeling since both greedy and Monte Carlo optimization gradually warmed the perfusion chamber to body temperature and improved SNMP performance. In addition, we observed that Monte Carlo optimization bounced between the two local and global maxima quite regularly, indicating that the barriers between the three optima are small. However, as the number of cycles (i.e., episodes) was increased, the *probability* associated with states (i.e., the temperature policies) became much higher for the global maximum than that of either local maximum, as shown in Table 1, where probabilities were computed by counting the number of production cycles for the local and global maxima and dividing each by the total number of production cycles. Table 1 also shows that for 3000 Monte Carlo cycles, the probability of encountering states associated with the global maximum were at least two times higher than those for the local maxima. This last fact suggests that machine learning approaches might prove useful in this application in training and deployment.

Table 1. Reward and probability of states for SNMP Monte Carlo optimization

	Reward	Probability
Optimum		
Local maximum 1	2.8716	0.0862
Local maximum 2	3.0816	0.3029
Global maximum	3.5548	0.6078

1.2 Optimal Temperature Policies for HMP, MMP and SNMP Initial Profiles

To get a more complete understanding of the performance of Monte Carlo optimization, optimal temperature policies were also determined for HMP and MMP. As in the case for SNMP, initial profiles for HMP and MMP were taken as flat temperature profiles of 6 and 16 °C, respectively, Monte Carlo optimization temperature changes were permitted every 30 min, machine perfusion was run for 4 h, and 3000 episodes were used to find the optimal temperature policy.

Figure 2 shows a comparison of all three cases where it is interesting to note the following: (1) the optimal temperature policy is different in each case, (2) the colder the initial temperature profile is, the more rapid is the rise during the first 2.5 h of operation, (3) all global optimal solutions have an energy charge at the low end of the normal range [0.6, 0.95], and (4) it is somewhat surprising that mid-thermic machine perfusion synthesizes the largest amount of net ATP.

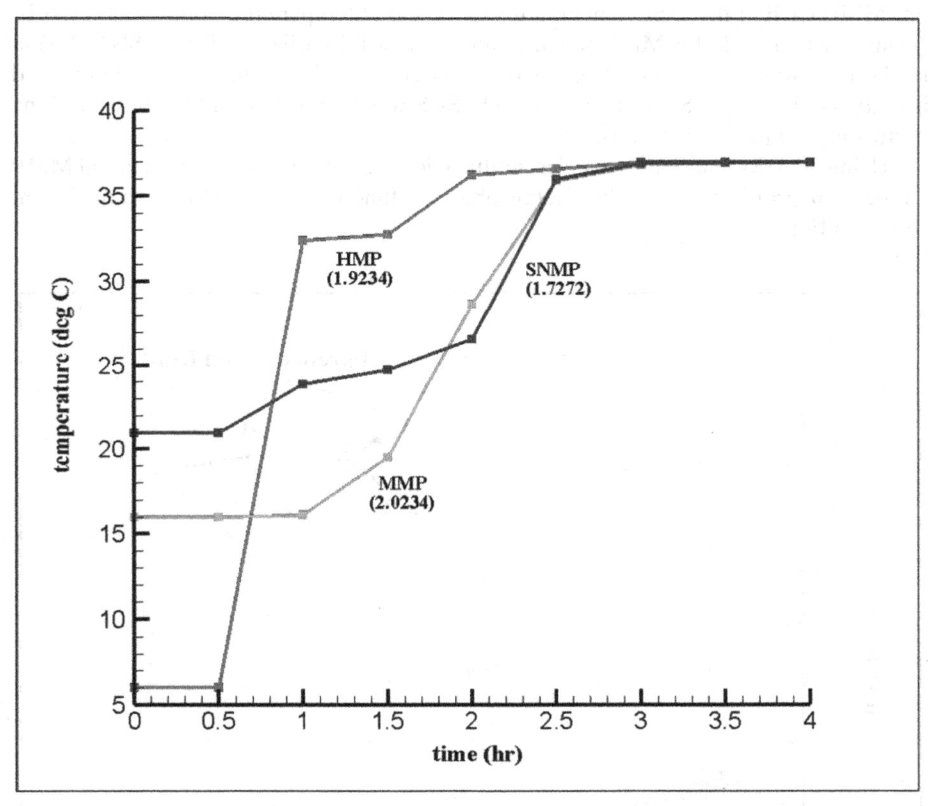

Fig. 2. Liver machine perfusion globally optimal temperature policies from initial profiles for HMP, MMP, and SNMP. Numbers in parenthesis represent amounts of ATP generated in pmol/cell.

2 Multi-objective Optimization of MP Temperature Profiles

The constrained optimization problem described in Eqs. 1 and 2 does not consider network metabolic efficiency, E. Here we include network metabolic efficiency by adding the following objective to the problem formulation

$$\max E = \frac{net\,ATP}{|Glc|} \tag{3}$$

Equations 1, 2 and 3 strike a balance between pure reward and network metabolic efficiency and give rise to a Pareto optimal front. To illustrate this, we chose to apply multi-objective optimization to MMP since it provided the 'best' optimal solution in Fig. 2 as measured by net ATP produced.

2.1 Multi-objective Optimization of MMP

For MMP, each of the eight temperatures in the initial temperature policy was set to 16 °C and each episode for MMP was the same as that for earlier studies of SNMP. That is, the liver was flushed with University of Wisconsin (UW) solution and then placed in static cold storage (SCS) at 4 °C for 6 h. SCS was then followed by 4 h of machine perfusion. See [6, 7] for details.

Using 3000 Monte Carlo episodes, multi-objective optimization with an initial MMP temperature profile produced the discrete objective function sets and Pareto optimal front shown in Fig. 3.

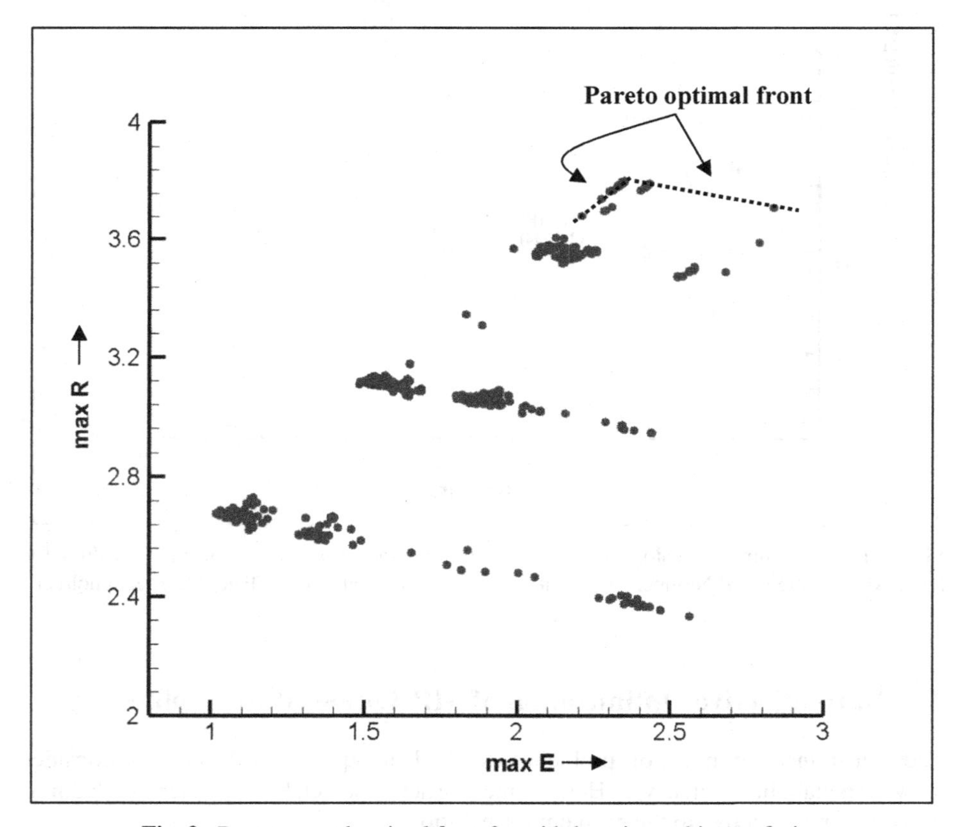

Fig. 3. Pareto set and optimal front for mid-thermic machine perfusion

Note that the Pareto optimal set (dotted lines) is rather flat in R and shows more variation with respect to network metabolic efficiency, E. In addition, there is virtually no discernable difference in the best temperature policies along the Pareto optimal front. On the other hand, the most significant differences lie in the metabolic behavior of glycolysis and malonyl-CoA synthesis, the latter of which is a driver of fatty acid synthesis and β-oxidation. Comparable results were obtained for the Pareto sets and optimal fronts computed from initial flat HMP and SNMP temperature profiles.

3 Discussion of Results

In this section, the machine perfusion optimization results are discussed along with the performance of the Monte Carlo algorithm.

3.1 Machine Perfusion Optimization

The results shown in Figs. 1, 2 and 3 show that Monte Carlo temperature policy optimization is an effective tool for improving the performance of liver machine perfusion. Using a constrained multi-objective function derived from (1) a reward function based on clinical liver viability criteria [8] and (2) network metabolic efficiency measured in terms of net ATP synthesized per amount of glucose consumed, temperature policy optimization improved net ATP synthesis, energy charge, and bile production while clearing lactate and maintaining pH within the physiological range for healthy liver cells from the following static cold storage results: lactate concentration was 4.11 mM, the energy charge of the cell was 0.4395, and the pH was 8.93. In addition, the results in Figs. 1 and 2 are consistent with recent clinical results in gradual warming [4, 5].

3.2 Algorithm Performance

The constrained optimization formulation given by Eqs. 1–3 is novel in that it is a multi-objective function that accounts for gradual warming, constraints on the temperature policy, viability constraints, and network metabolic efficiency.

In this work, Monte Carlo acceptance ratios ranged from 0.1 to 0.37, which is within the usual range of 'good' acceptance ratios for Monte Carlo methods. In all cases, the algorithm found the global maximum in under 3000 Monte Carlo cycles. Computer times required for the temperature policy optimizations with 3000 Monte Carlo cycles averaged 1335 CPU sec (0.37 CPU hrs.). However, it is important to note that episodes tend to jump back and forth between the global and local maxima suggesting that the barriers between all maxima (i.e., valleys) were shallow and presented little resistance to movement on the reward surface.

Another point of interest is the fact that the global maximum in reward found from an initial SNMP temperature profile was *different from* the global maximum in reward found from either the initial HMP or MMP temperature profile and this was a bit confusing at first. Two additional considerations are important here. First, when the number of Monte Carlo cycles for the optimization starting with an SNMP temperature profile was increased to 5000, the algorithm still did not find a global maximum with a reward value of ~3.8. Second, using a different initial temperature profile of 28 °C, which is midway in the region classified as SNMP, the resulting global maximum reward changed only slightly from 3.5548 to 3.5619. Thus, it appears that using initial SNMP temperature profiles may preclude the existence of the global maximum that occurs when initial HMP and MMP temperature profiles are used.

4 Conclusions

A multi-objective function consisting of (1) a reward function based on liver viability criteria [8] and (2) a measure of network metabolic efficiency along with Monte Carlo optimization were used to determine globally optimal temperature policies for a mathematical model of liver metabolism during machine perfusion. Initial temperature policies corresponded to traditional hypothermic, mid-thermic, and subnormothermic temperature profiles. In each case, the global optimum temperature policy was determined within a reasonable number of Monte Carlo cycles (episodes). See Fig. 2. However, optimization results varied depending on the initial temperature profile.

No single initial temperature profile gave the best results for all terms in the reward function. The best overall results were obtained when the Monte Carlo optimizations were initiated from a MMP temperature profile. In addition, the probability of finding the global maximum reward of ~3.8 was much higher in this case than when the optimizations were started from an initial HMP temperature profile. It also came as a surprise that starting from an initial SNMP temperature profile yielded results that were not as good as those for MMP.

A Pareto set and optimal front were determined for mid-thermic machine perfusion. Results here showed that the objective function set is discrete, that the Pareto optimal front is flat in terms of the reward function and that all solutions along the front result in essentially the same optimal temperature policy. On the other hand, the network metabolic efficiency varied considerably along the Pareto optimal front, suggesting that other factors in addition to those in the multi-objective formulation may play a role in determining liver metabolism during machine perfusion.

Acknowledgement. This material is partially based upon work supported by the National Science Foundation under Grant No. EEC 1941543. Support from the US National Institutes of Health (grants R01DK096075 and R01DK114506) and the Shriners Hospitals for Children is gratefully acknowledged.

References

1. Petrenko, A., et al.: Organ preservation into the 2020s: the era of dynamic intervention. Tranfus. Med. Hemother. **46**, 151–172 (2019)
2. Berendsen, T.A., et al.: A simplified subnormothermic machine perfusion system restores ischemically damaged liver grafts in rat model of orthotopic liver transplantation. Transplant. Res. **1**, 6 (2012)
3. Bruinsma, B.G., Berendsen, T.A., Izamis, M.-L., Yarmush, M.L., Uygun, K.: Determination and extension of the limits of static cold storage using subnormothermic machine perfusion. Int. J. Artif. Organs **36**(11), 775–780 (2013)
4. Gallinat, A., Lu, J., von Horn, C., et al.: Transplantation of cold stored porcine kidneys after controlled oxygenated rewarming. Artif. Organs **42**, 647–654 (2018)
5. Mahboub, P., Aburawi, M., Karimian, N., et al.: The efficacy of HBOC-201 in ex situ gradual rewarming kidney perfusion in a rat model. Artif. Organs **44**, 81–90 (2020)
6. Lucia, A., Ferrarese, E., Uygun, K.: Modeling energy depletion in rat livers using Nash equilibrium metabolic pathway analysis. Sci. Rep. **12**, 3496 (2022)

7. Lucia, A., Uygun. K.: Optimal temperature protocols for liver machine perfusion using a Monte Carlo method. In: FOSBE 2022 (2022)
8. Laing, R.W., Mergental, H., Yap, C., et al.: Viability testing and transplantation of marginal livers (VITTAL) using normothermic machine perfusion: study protocol for an open-label, non-randomised, prospective, single-arm trial. BMJ Open **7**, e017733 (2017)

A Comparison of SVM Against Pre-trained Language Models (PLMs) for Text Classification Tasks

Yasmen Wahba[1(✉)], Nazim Madhavji[1], and John Steinbacher[2]

[1] Western University, London, ON, Canada
{ywahba2,nmadhavji}@uwo.ca
[2] IBM Canada, Toronto, ON, Canada
jstein@ca.ibm.com

Abstract. The emergence of pre-trained language models (PLMs) has shown great success in many Natural Language Processing (NLP) tasks including text classification. Due to the minimal to no feature engineering required when using these models, PLMs are becoming the de facto choice for any NLP task. However, for domain-specific corpora (e.g., financial, legal, and industrial), fine-tuning a pre-trained model for a specific task has shown to provide a performance improvement. In this paper, we compare the performance of four different PLMs on three public domain-free datasets and a real-world dataset containing domain-specific words, against a simple SVM linear classifier with TFIDF vectorized text. The experimental results on the four datasets show that using PLMs, even fine-tuned, do not provide significant gain over the linear SVM classifier. Hence, we recommend that for text classification tasks, traditional SVM along with careful feature engineering can provide a cheaper and superior performance than PLMs.

Keywords: Text classification · Pre-trained language models · Machine learning · Domain-specific datasets · Natural language processing

1 Introduction

Text classification is the task of classifying text (e.g., tweets, news, and customer reviews) into different categories (i.e., tags). It is a challenging task especially when the text is 'technical'. We define 'technical' text in terms of the vocabulary used to describe a given document, e.g., classifying health records, human genomics, IT discussion forums, etc. These kinds of documents require special pre-processing since the basic NLP pre-processing steps may remove critical words necessary for correct classification, resulting in a performance drop of the deployed system [1].

Recently, pre-trained language models (PLMs) such as BERT [2] and ELMO [3] have shown promising results in several NLP tasks, including spam filtering, sentiment analysis, and question answering. In comparison to traditional models, PLMs require less feature engineering and minimal effort in data cleaning. Thus becoming the consensus for many NLP tasks [4].

G. Nicosia et al. (Eds.): LOD 2022, LNCS 13811, pp. 304–313, 2023.
https://doi.org/10.1007/978-3-031-25891-6_23

With an enormous number of trainable parameters, these PLMs can encode a substantial amount of linguistic knowledge that is beneficial to contextual representations [4]. For example, word polysemy (i.e., the coexistence of multiple meanings for a word or a phrase –e.g., 'bank' could mean 'river bank' or 'financial bank') in a domain-free text.

In contrast, in a domain-specific text that contains technical jargon, a word has a more precise meaning (i.e., monosemy) [5]. For example, the word 'run' in an IT text would generally only mean 'execute' and not 'rush'. Thus, it appears that domain-specific text classification will likely not benefit from the rich linguistic knowledge encoded in PLMs.

Despite the widespread use of PLMs in a broad range of downstream tasks, their performance is still being evaluated by researchers for their drawbacks [6]. For example: (i) the large gap between the pre-training objectives (e.g., predict target words) and the downstream objectives (e.g., classification) limits the ability to fully utilize the knowledge encoded in PLMs [7], (ii) the high computational cost and the large set of trainable parameters make these models impractical for training from scratch, (iii) dealing with rare words is a challenge for PLMs [8], and (iv) the performance of PLMs may not be generalizable [9].

Thus, this paper evaluates the performance of different pre-trained language models (PLMs) against a linear Support Vector Machine (SVM) classifier. The motivation for this comparative study is rooted in the fact that: (i) while PLMs are being used in text classification tasks [10, 11], they are more computationally expensive than the simpler SVMs, and (ii) PLMs have been used predominantly on public or domain-free datasets and it is not clear how they fare against simpler SVMs on domain-specific datasets.

The findings of our study suggest that the problem of classifying domain-specific or generic text can be addressed efficiently using old traditional classifiers such as SVM and a vectorization technique such as TFIDF bag-of-words that do not involve the complexity found in neural network models such as PLMs. To the best of our knowledge, no such comparative analysis has so far been described in the scientific literature.

The rest of the paper is organized as follows. Section 2 describes related work. Section 3 describes the empirical study. Section 4 presents the research results. Section 5 concludes the paper.

2 Related Work

In this section, we give an overview of the existing literature on the applications of PLMs and some of the drawbacks reported.

Pre-trained language models (PLMs) are deep neural networks trained on unlabeled large-scale corpora. The motivation behind these models is to capture rich linguistic knowledge that could be further transferred to target tasks with limited training samples (i.e., fine-tuning). BERT [2], XLM [12], RoBERTa [13], and XLNet [14] are examples of PLMs that have achieved significant improvements on a large number of NLP tasks (e.g., question answering, sentiment analysis, text generation).

Nevertheless, the performance of these models on domain-specific tasks was questioned [15] as these models are trained on general domain corpora such as Wikipedia, news websites, and books. Hence, fine-tuning or fully re-training PLMs for downstream tasks has become a consensus. Beltagi et al. [16] released SciBERT that is fully retrained on scientific text (i.e., papers). Lee et al. [17] released BioBERT for biological text. Similarly, Clinical BERT [18, 19] was released for clinical text, and FinBERT [20] for the financial domain.

Other researchers applied PLMs by fine-tuning the final layers to the downstream task. For example, Elwany et al. [21] report valuable improvements on legal corpora after fine-tuning. Lu [22] fine-tuned RoBERTa for Commonsense Reasoning and Tang et al. [23] fine-tuned BERT for multi-label sentiment analysis in code-switching text. Finally, Yuan et al. [24] fine-tuned BERT and ERNIE [25] for the detection of Alzheimer's Disease.

However, Gururangan et al. [15] show that simple fine-tuning of PLMs is not always sufficient for domain-specific applications. Their work suggests that the second phase of pre-training can provide significant gains in task performance. Similarly, Kao et al. [26] suggest that duplicating some layers in BERT prior to fine-tuning can lead to better performance on downstream tasks.

Another body of research focuses on understanding the weaknesses of PLMs by either applying them to more challenging datasets or by investigating their underlying mechanisms. For example, McCoy et al. [9] report the failure of BERT when evaluated on the HANS dataset. Their work suggests that evaluation sets should be drawn from a different distribution than the train set. Also, Schick et al. [8] introduce WNLaMPro (WordNet Language ModelProbing) dataset to assess the ability of PLMs to understand rare words. Lastly, Olga et al. [27] show redundancy in the information encoded by different heads in BERT, and manually disabling attention in certain heads will lead to performance improvement.

This paper adds to the growing literature on evaluating PLMs. In particular, our investigative question is: How does a linear classifier such as SVM compare against the state-of-the-art PLMs on both general and technical domains?

3 Empirical Study

In this section, we describe the empirical study that we conducted. In particular, we describe the infrastructure used, the datasets, and the different PLMs used. Finally, we describe the SVM algorithm used, and the pre-processing steps done prior to applying SVM. The experimental algorithms are written in Python 3.8.3. The testing machine is Windows 10 with an Intel Core i7 CPU 2.71 GHz and 32 GB of RAM.

3.1 Text Classification Datasets

Our experiments were evaluated on four datasets:

1. BBC News [28]: a public dataset originating from BBC News. It consists of 2,225 documents, categorized into 5 groups, namely: business, entertainment, politics, sport, and tech.
2. 20NewsGroup [29]: a public dataset consisting of 18,846 documents, categorized into 20 groups.
3. Consumer Complaints [30]: a public benchmark dataset published by the Consumer Financial Protection Bureau; it is a collection of complaints about consumer financial products and services. It consists of 570,279 documents categorized into 15 classes.
4. IT Support tickets: a private dataset obtained from a large industrial partner. It is composed of real customer issues related to a cloud-based system. It consists of 194,488 documents categorized into 12 classes.

Table 1 summarizes the properties of the four datasets.

Table 1. Dataset properties

Dataset	# of classes	# of instances	# of features (n-gram = 1)	# of features (n-gram = 3)
BBC News	5	2,225	26,781	811,112
20NewsGroup	20	18,846	83,667	2,011,358
Consumer Complaints	15	570,279	53,429	6,112,905
IT Support tickets	12	194,488	16,011	3,185,796

The IT Support tickets dataset will be referred to hereon as the 'domain-specific' dataset. This dataset suffers from a severe imbalance as seen in Fig. 1. However, we prefer to avoid the drawbacks of sampling techniques [31, 32] and keep the distribution as is.

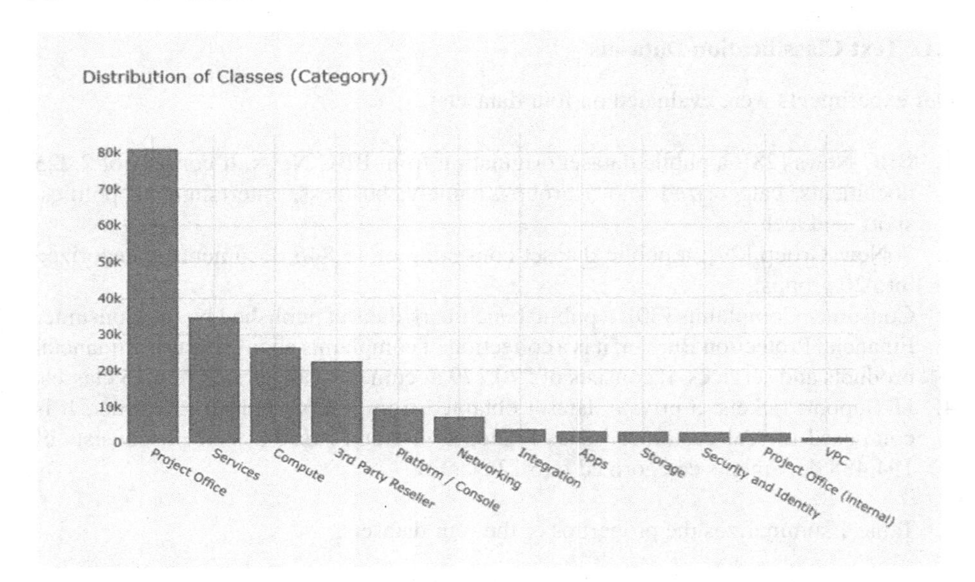

Fig. 1. Class distribution of the domain-specific dataset showing imbalance

Another problem with this dataset is the presence of a large number of technical words (i.e., jargon) related to the Cloud terminologies (e.g., Bluemix, Kubernetes, Iaas, Vmware, etc.). These words are not found in the PLMs vocabulary and hence, they get broken down into subwords using a subword tokenization algorithm. For instance, BERT uses a WordPiece tokenizer [33] which handles non-technical words quite well. However, we notice that it fails to tokenize technical words and domain-specific abbreviations in our domain-specific dataset. For example:

"Kubernetes" ⇒ ['ku', '##ber', '##net', '##es'].

"configuration" ⇒ "config" ⇒ ['con', '##fi', '##g'].

3.2 Pre-trained Language Models (PLMs)

The following PLMs were considered for this study:

1. BERT [2]: A widely used pre-training language model that is based on a bidirectional deep Transformer as the main structure. BERT achieved state-of-the-art results on 11 different NLP tasks including question answering and named entity recognition (NER).
2. DistilBERT [34]: A lighter, smaller, and faster version of BERT. By reducing the size of the BERT model by 40%, while keeping 97% of its language understanding capability, it's considered 60% faster than BERT.
3. RoBERTa [13]: One of the successful variants of BERT that achieved impressive results on many NLP tasks. By changing the MASK pattern, discarding the NSP task, and using a larger batch size and longer training sentences.

4. XLM [12]: Designed specifically for cross-lingual classification tasks by leveraging bilingual sentence pairs. XLM uses a known pre-processing technique (BPE) and a dual-language training mechanism.

For this study, we fine-tuned all the PLMs to the domain-specific dataset and the three generic datasets. In all our experiments, we use the following hyperparameters for fine-tuning: maximum sequence length of 256, adam learning rate (lr) of $1e^{-5}$, batch size of 16, and a train-test split ratio of 80:20.

Support Vector Machines (SVM). A Support Vector Machine is a popular supervised margin classifier, reported as one of the best algorithms for text classification [35, 36]. We chose the LinearSVC algorithm in the *Scikit-learn* library [37], which implements a one-versus-all (OVA) multi-class strategy. This algorithm is suitable for high-dimensional datasets and is characterized by a low running time [38].

Unlike PLMs, traditional machine learning models require pre-processing data cleaning steps. In our study, we used the following pre-processing steps on the four datasets: (i) removing missing data; (ii) removing numbers and special characters; (iii) lower casing; (iv) tokenization; (v) lemmatization; and (vi) word vectorization using TFIDF[1].

It is important to note that when applying the TFIDF vectorizer, we tried different N-grams. An 'N-gram' is simply a sequence of N words that predicts the occurrence of a word based on the occurrence of its (N – 1) previous words. The default setting is Unigrams. In our study, we used trigrams which means that we included feature vectors consisting of all unigrams, bigrams, and trigrams.

4 Results

In this section, we discuss the results of applying four different fine-tuned PLMs (i.e., BERT, DistilBERT, RoBERTa, XLM) and a linear SVM classifier on the four datasets described in Sect. 3.1.

Table 2 shows the F1-scores obtained when applying the four PLMs and a linear SVM classifier on the four datasets. When evaluating PLMs, we used 3 epochs because we observed that when the number of epochs exceeds 3, the training loss decreases with each epoch and the validation loss increases. This translates to overfitting. Thus, all our experiments are run for 3 epochs only.

For the domain-specific dataset, it is clear how the linear SVM achieves a comparable performance (0.79) as any of the fine-tuned PLMs. Similarly, for the BBC dataset, SVM surprisingly achieves the same F1-score (0.98) as RoBERTa on the third epoch. However, we expected that PLMs would significantly outperform SVM on general domain datasets.

For the 20NewsGroup, SVM outperformed all PLMs with an F1-score of 0.93. This accuracy score was a result of considering the meta-data (i.e., headers, footers, and

[1] TFIDF stands for Term Frequency-Inverse Document Frequency, which is a combination of two metrics:1. Term frequency (*tf*): a measure of how frequently a term *t*, appears in a document *d*.2. Inverse document frequency(idf): a measure of how important a term is. It is computed by dividing the total number of documents in our corpus by the document frequency for each term and then applying logarithmic scaling on the result.

quotes) as part of the text that is fed to the classifier. However, when we ignored the meta-data, there was a performance drop of 15%.

The last dataset is the Consumer Complaints which is the largest dataset (570,279 instances) as described in Table 1. The accuracy of the linear SVM (0.82) was very close to the highest accuracy of 0.85 obtained by BERT and RoBERTa. While 0.82 is very competitive, we believe there is room for improvement if feature selection techniques were considered as this dataset is characterized by a large feature set.

The accuracy scores of PLMs are generally higher on generic datasets that do not contain domain-specific or rare words. Also, we notice a small gap between the accuracy scores of all PLMs in the third epoch for all datasets.

In summary, the key points are:

- Linear SVM proved to be comparable to PLMs for text classification tasks.
- PLMs accuracy scores are generally higher on generic datasets.
- The importance of feature engineering for text classification is highlighted by including meta-data.

Table 2. Comparison of four PLMs against SVM Linear classifier in terms of accuracy (f1-score)

Dataset	Model	Epoch 1	Epoch 2	Epoch 3
		Accuracy (F1-score)		
IT Support Tickets	*BERT*	0.78	0.79	0.79
	DistilBERT	0.77	0.78	0.79
	XLM	0.77	0.79	0.79
	RoBERTa	0.77	0.78	0.79
	LinearSVM (n-gram = 3)	0.79		
BBC	*BERT*	0.97	0.97	0.97
	DistilBERT	0.97	0.97	0.97
	XLM	0.88	0.96	0.97
	RoBERTa	0.97	0.97	0.98
	LinearSVM (n-gram = 3)	0.98		
20NewsGroup	*BERT*	0.85	0.91	0.92
	DistilBERT	0.82	0.90	0.90
	XLM	0.89	0.91	0.92
	RoBERTa	0.84	0.87	0.90
	LinearSVM	0.93		

(continued)

Table 2. (*continued*)

Dataset	Model	Epoch 1	Epoch 2	Epoch 3
		Accuracy (F1-score)		
Consumer complaints	*BERT*	0.83	0.84	0.85
	DistilBERT	0.82	0.84	0.84
	XLM	0.80	0.82	0.83
	RoBERTa	0.83	0.84	0.85
	LinearSVM	0.82		

5 Conclusions

The study described in this paper compares the performance of several fine-tuned PLMs (see Sect. 3.2) against that of a linear SVM classifier for the task of text classification. The datasets used in the study are: a domain-specific dataset of real-world support tickets from a large organization as well as three generic datasets (see Table 1).

To our surprise, we found that a pre-trained language model does not provide significant gains over the linear SVM classifier. We expected PLMs to outperform SVM on the generic datasets, however, our study indicates comparable performance for both models (see Table 2). Also, our study indicates that SVM outperforms PLMs on one of the generic datasets (i.e., 20NewsGroup).

Our finding goes against the trend of using PLMs on any NLP task. Thus, for text classification, we recommend prudence when deciding on the type of algorithms to use. Since our study seems to be the first comparative study of PLMs against SVM on generic datasets as well as on a domain-specific dataset, we encourage replication of this study to create a solid body of knowledge for confident decision-making on the choice of algorithms.

References

1. Brundage, M.P., Sexton, T., Hodkiewicz, M., Dima, A., Lukens, S.: Technical language processing: unlocking maintenance knowledge. Manuf. Lett. **27**, 42–46 (2021)
2. Devlin, J., Chang, M.W., Lee, K., Toutanova, K.: BERT: pre-training of deep bidirectional transformers for language understanding. In: Proceedings of the Conference of the North American Chapter of the Association for Computational Linguistics: Human Language Technologies. Minneapolis, pp. 4171–4186 (2019)
3. Peters, M.E., et al.: Deep contextualized word representations. In: Proceedings of the Conference of the North American Chapter of the Association for Computational Linguistics: Human Language Technologies, New Orleans, pp. 2227–2237 (2018)
4. Han, X., et al.: Pre-trained models: past, present and future. AI Open **2**, 225–250 (2021)
5. Aronoff, M., Rees-Miller, J. (eds.): The Handbook of Linguistics. Wiley, Hoboken (2020)
6. Acheampong, F.A., Nunoo-Mensah, H., Chen, W.: Transformer models for text-based emotion detection: a review of BERT-based approaches. Artif. Intell. Rev. **54**(8), 5789–5829 (2021). https://doi.org/10.1007/s10462-021-09958-2

7. Han, X., Zhao, W., Ding, N., Liu, Z., Sun, M.: PTR: prompt tuning with rules for text classification. arXiv preprint arXiv:2105.11259 (2021)
8. Schick, T., Schütze, H.: Rare words: A major problem for contextualized embeddings and how to fix it by attentive mimicking. In: Proceedings of the AAAI Conference on Artificial Intelligence, vol. 34, no. 05, pp. 8766–8774 (2020)
9. McCoy, R.T., Pavlick, E., Linzen, T.: Right for the wrong reasons: diagnosing syntactic heuristics in natural language inference. In: Proceedings of the 57th Annual Meeting of the Association for Computational Linguistics, Florence, Italy (2019)
10. Zhao, Z., Zhang, Z., Hopfgartner, F.: A comparative study of using pre-trained language models for toxic comment classification. In: Companion Proceedings of the Web Conference, pp. 500–507 (2021)
11. Zheng, S., Yang, M.: A new method of improving BERT for text classification. In: Cui, Z., Pan, J., Zhang, S., Xiao, L., Yang, J. (eds.) IScIDE 2019. LNCS, vol. 11936, pp. 442–452. Springer, Cham (2019). https://doi.org/10.1007/978-3-030-36204-1-37
12. Conneau, A., Lample, G.: Cross-lingual language model pretraining. In: Proceedings of the Advances in Neural Information Processing Systems, Vancouver, pp. 7057–7067 (2019)
13. Liu, Y., et al.: RoBERTa: a robustly optimized BERT pretraining approach. arXiv preprint arXiv:1907.11692 (2019)
14. Yang, Z., Dai, Z., Yang, Y., Carbonell, J., Salakhutdinov, R.R., Le, Q.V.: XLNet: generalized autoregressive pretraining for language understanding. In: Proceedings of the Advances in Neural Information Processing Systems, Vancouver, pp. 5754–5764 (2019)
15. Gururangan, S., et al.: Don't stop pretraining: adapt language models to domains and tasks. In: Proceedings of ACL (2020)
16. Beltagy, I., Lo, K., Cohan, A.: SciBERT: a pretrained language model for scientific text. In: Proceedings of the Conference on Empirical Methods in Natural Language Processing, Hong Kong, pp. 3613– 3618 (2019)
17. Lee, J., et al.: BioBERT: a pre-trained biomedical language representation model for biomedical text mining. Bioinformatics **36**, 1234–1240 (2020)
18. Huang, K., Altosaar, J., Ranganath, R.: ClinicalBERT: modeling clinical notes and predicting hospital readmission. arXiv:1904.05342 (2019)
19. Alsentzer, E., et al.: Publicly available clinical BERT embeddings. In: Proceedings of the 2nd Clinical Natural Language Processing Workshop, pp. pp.72–78. Association for Computational Linguistics, Minneapolis, Minnesota, USA (2019)
20. Araci, D.: FinBERT: financial sentiment analysis with pre-trained language models. arXiv preprint. arXiv:1908.10063 (2019)
21. Elwany, E., Moore, D., Oberoi, G.: Bert goes to law school: quantifying the competitive advantage of access to large legal corpora in contract understanding. In: Proceedings of NeurIPS Workshop on Document Intelligence (2019)
22. Lu, D.: Masked reasoner at SemEval-2020 Task 4: fine-tuning RoBERTa for commonsense reasoning. In: Proceedings of the Fourteenth Workshop on Semantic Evaluation, pp. 411–414 (2020)
23. Tang, T., Tang, X., Yuan, T.: Fine-tuning BERT for multi-label sentiment analysis in unbalanced code-switching text. IEEE Access **8**, 193248–193256 (2020)
24. Yuan, J., Bian, Y., Cai, X., Huang, J., Ye, Z., Church, K.: Disfluencies and fine-tuning pre-trained language models for detection of Alzheimer's disease. In: INTER-SPEECH, pp. 2162–2166 (2020)
25. Sun, Y., et al.: ERNIE 2.0: a continual pre-training framework for language understanding. In: Proceedings of the AAAI Conference on Artificial Intelligence, vol. 34, no. 05, pp. 8968–8975 (2020)

26. Kao, W.T., Wu, T.H., Chi, P.H., Hsieh, C.C., Lee, H.Y.: BERT's output layer recognizes all hidden layers? Some Intriguing Phenomena and a simple way to boost BERT. arXiv preprint arXiv:2001.09309 (2020)
27. Kovaleva, O., Romanov, A., Rogers, A., Rumshisky, A.: Revealing the dark secrets of BERT. In: Proceedings of the 2019 Conference on Empirical Methods in Natural Language Processing and the 9th International Joint Conference on Natural Language Processing (EMNLP-IJCNLP), Hong Kong, China (2019)
28. Greene, D., Cunningham, P.: Practical solutions to the problem of diagonal dominance in kernel document clustering. In: Proceedings of the 23rd International Conference on Machine Learning, pp. 377–384 (2006)
29. Newsgroups Data Set Homepage. http://qwone.com/~jason/20Newsgroups/. Accessed March 2022
30. Consumer Complaint Database Homepage. https://www.consumerfinance.gov/data-research/consumer-complaints/.Online. Accessed March 2022
31. Zhou, Z.H., Liu, X.Y.: Training cost-sensitive neural networks with methods addressing the class imbalance problem. IEEE Trans. Knowl. Data Eng. **18**(1), 63–77 (2006)
32. He, H., Ma, Y.: Imbalanced Learning: Foundations, Algorithms, and Applications, 1st edn. Wiley-IEEE Press, New York (2013)
33. Wu, Y., et al.: Google's neural machine translation system: Bridging the gap between human and machine translation. arXiv preprint arXiv:1609.08144 (2016)
34. Sanh, V., Debut, L., Chaumond, J., Wolf, T.: DistilBERT, a distilled version of BERT: smaller, faster, cheaper and lighter. In: Proceedings of the 5th Workshop on Energy Efficient Machine Learning and Cognitive Computing (EMC2) co-located with the Thirty-third Conference on Neural Information Processing Systems (NeurIPS 2019), pp. 1–5 (2019)
35. Joachims, T.: Text categorization with support vector machines: learning with many relevant features. In: Nédellec, C., Rouveirol, C. (eds.) ECML 1998. LNCS, vol. 1398, pp. 137–142. Springer, Heidelberg (1998). https://doi.org/10.1007/BFb0026683
36. Telnoni, P.A., Budiawan, R., Qana'a, M.: Comparison of machine learning classification method on text-based case in Twitter. In: Proceedings of International Conference on ICT for Smart Society: Innovation and Transformation Toward Smart Region, ICISS (2019)
37. 4. Support Vector Machines—scikit-learn 0.23.1 documentation. https://scikit-learn.org/stable/modules/svm.html. Accessed March 2022
38. Chauhan, V.K., Dahiya, K., Sharma, A.: Problem formulations and solvers in linear SVM: a review. Artif. Intell. Rev. **52**(2), 803–855 (2018). https://doi.org/10.1007/s10462-018-9614-6

Normalization in Motif Discovery

Frederique van Leeuwen[1,2](✉), Bas Bosma[3], Arjan van den Born[1,2], and Eric Postma[1,2]

[1] Jheronimus Academy of Data Science, 's-Hertogenbosch, The Netherlands
f.c.a.v.leeuwen@JADS.nl
[2] Tilburg University, Tilburg, The Netherlands
[3] VU Amsterdam, Amsterdam, The Netherlands

Abstract. Motif discovery can be used as a subroutine in many time series data mining tasks such as classification, clustering and anomaly detection. A motif represents two or more highly similar subsequences of a time series. The vast majority of the motif discovery methods implicitly assume that subsequences need to be normalized before determining their similarity. While normalization is widely adopted, it may affect the discovery of motifs. We examine the effect of normalization on motif discovery using 96 real-world time series. To determine if the discovered motifs are meaningful, all time series are assigned labels that indicate the states of the system generating the time series. Our experiments show that in over half of the considered cases, normalization affects motif discovery negatively by returning motifs that are not meaningful. We therefore conclude that the assumption underlying normalization does not always hold for real-world time series and thus should not be uncritically adopted.

Keywords: Time series · Motif discovery · Normalization · Distance metrics

1 Introduction

As the number of connected devices and sensors grows exponentially, so does the volume of streaming data. The purpose of time series data mining is to try to extract as much meaningful knowledge from these data streams as possible. Especially the task of *motif discovery*, i.e. the detection of repeating patterns in time series data, has received much attention in the past decade [13–15,17]. Additionally, motif discovery can be used as subroutine in many time series data mining tasks such as classification [19,21], clustering, anomaly detection and time series labeling [8,11]. As such, motif discovery provides a key ingredient in understanding many real-world challenges.

Motif discovery research so far is mainly concerned with (I) how to rank motifs [6,10], (II) how to increase the efficiency of motif discovery [9,15,23], and (III) how to choose a relevant subsequence length m [5,6,10,13,21,22]. The vast majority of methods implicitly assumes that all subsequences need to be

G. Nicosia et al. (Eds.): LOD 2022, LNCS 13811, pp. 314–325, 2023.
https://doi.org/10.1007/978-3-031-25891-6_24

normalized in order to find *meaningful* motifs [7]. Typically, this normalization is considered to be part of the distance metric used to compare subsequences. As far as we are aware, there is no research dedicated to understanding the actual effect of the normalization step on motif discovery. Our aim is to demonstrate how normalization affects the discovery of meaningful motifs.

Due to the different characteristics of real-world time series, the motifs discovered may depend on the specific type of normalization. To further investigate this, we make use of 96 real-world labeled time series. The labels indicate the states of the system generating the time series. As a consequence, we can examine to what extent normalization impacts the detection of relevant states. This, then, allows us to draw conclusions on the effect of normalization on the detection of *meaningful motifs*. We conclude with recommendations on when to use normalization and when not.

The rest of the paper is organized as follows. In Sect. 2 we define the key terms in motif discovery. Subsequently, in Sect. 3 we review two distance metrics, a commonly-used normalized metric and a non-normalized one. In Sect. 4 we consider a wide variety of time series for which we discuss the effect of using the normalized versus the non-normalized metric on motif discovery. In Sect. 5, we conclude and suggest directions for future research.

2 Time Series, Subsequences, and Motifs

We start by defining some key terminology used in time series motif discovery [3].

Definition 1. (*Time series*). A *time series* $T = \{t_1, \ldots, t_n\}$ is an ordered sequence of n real-valued numbers, often measured at fixed intervals.

Definition 2. (*Subsequence*). Given any time series T of length n, a *subsequence* S of length $m \leq n$ is a contiguous sampling $S = \{t_k, \ldots, t_{k+m-1}\}$, with $0 < k \leq n - m + 1$.

Motifs are defined originally by [16, p. 370] as "previously unknown, frequently occurring patterns in time series". While motif discovery is a commonly used technique, existing algorithms may value motifs differently. Based on the literature so far, two classes of algorithms are considered. One class mainly considers the most similar motifs [12–15], the other the most frequently occurring motifs [16,17,21].[1] Despite the class, a motif can generally be defined as follows.

Definition 3. (*Motif*). Given any time series T of length n, distance metric D, and threshold δ, non-overlapping subsequences of length m are called *motif* iff for all S_i and S_j, with $i \neq j$, $D(S_i, S_j) \leq \delta$.[2]

[1] These classes are often referred to as *pair* motifs and *set* motifs respectively [6,10].

[2] In the practical implementation of motif discovery, an *exclusion zone* of length $m/2$ before and after the location of the subsequence of interest is commonly set [20]. This ensures that so-called *trivial matches* are avoided.

In this paper we focus on all $n - m + 1$ subsequences and their nearest neighbor in time series T. The vector including this information is also known as the *Matrix Profile* [20,23]. We define the *Most-Similar Subsequence* (MSS) as the nearest neighbour under distance metric D.

3 Distance Metrics

The distances between subsequences can be measured by a wide variety of measures [3]. The most commonly used measure within the motif discovery literature is the Euclidean Distance. Keogh and Kasetty [7, p. 361] argued that "without normalization time series similarity has essentially no meaning". Specifically, they support their argument by showing that if the mean is not subtracted from subsequences, the resulting motifs may look dissimilar. Therefore, their interpretation of "normalization" includes subtracting the mean. In a more recent paper [20] the normalization is defined as the z-normalized Euclidean Distance, $\mathbf{ED_z}$, which essentially corresponds to applying the z-transform to both subsequences separately, before computing their Euclidean Distance (**ED**).

Through normalization the effect of offsets and magnitudes on motif discovery declines. In many, but certainly not all situations, as is visualized in [2, p. 126, Fig. 3], this is beneficial for the detection of meaningful motifs. Specifically, whenever the time series has a trend or when the relevant subsequences vary in magnitude, normalization is beneficial. But there may also exist situations in which the average values of subsequences can be relevant for the detection of meaningful motifs. For instance, in a time series representing daily temperature measurements of a patient, subsequences associated with periods of flu might be matched with subsequences of non-flu periods. Hence, flu-specific motifs may go undetected because their main distinguishing feature is the elevation (offset) of the temperature. Also the amplitudes of subsequences may be of importance. For instance, in a geological time series, subsequences associated with earthquakes may be matched with small non-earthquake tremors. The earthquake-specific motifs may not be detected because amplitude (magnitude) is their main characteristic feature.

As these examples suggest, normalization can lead to desired but potentially also to undesired motif-discovery results. The aim of this paper is to demonstrate the effect of normalization on motif discovery by examining 96 real-world time series, 6 of which will be examined in detail. While more distance metrics can be used for motif detection [3], we focus on the most commonly used one: the Euclidean distance. As far as we are aware, there is no previous research dedicated to understanding the actual differences caused by subsequence scaling.

The comparative evaluation of both distance metrics has to take into account the subsequence length m, which is of relevance to motif discovery [5,6,13]. We will examine how the value of m affects both distance metrics. Based on the labeled time-series, we examine the ability to detect meaningful MSSs in relation to the distance metric and value of m.

4 ED Versus $\mathbf{ED_z}$

In this section we analyze the effect of normalization on motif discovery for 96 real-world time series. As the time series are labeled, the states of subsequences are known. This allows us to determine to what extent the detected MSSs capture relevant states in the time series and thus are meaningful or not.

We start with explaining the experiments and focus first on six diverse time series. Subsequently, we analyze to what extent **ED** and $\mathbf{ED_z}$ produce similar results. Based on this and the additional 90 labeled time series, we examine which metric is able to detect the most meaningful MSSs. We conclude with an analysis and recommendations on which metric to use when.

4.1 Real-World Time Series

We selected the following six real-world time series for a detailed illustration as they cover as broad a range of characteristics as possible.

1. **ACP1:** Entomology time series. Data from an Asian Citrus Psyllid (snippet 1).
2. **ACP5:** Entomology time series. Data from an Asian Citrus Psyllid (snippet 5).
3. **EER:** Epilepsy time series representing distinct epilepsy episodes.
4. **ECG:** Electrocardiogram time series. Each series traces the electrical activity recorded during one heartbeat.
5. **SLC:** Part of the StarLight Curves time series.
6. **HCS:** A Hydraulic Control System time series.

The time series $ACP1$ (length $n = 5,000$), $ACP5$ ($n = 10,000$), and EER ($n = 2,700$) are retrieved from [11], who created a set of completely annotated time series to mimic expert feedback for time series labeling.[3] Time series ECG ($n = 9,600$) and SLC ($n = 10,000$) are from the UCR Archive [1], which includes 128 labeled time series commonly used in time series classification research [4,18].[4] Lastly, HCS ($n = 10,000$) is a real-world industrial time series collected and labeled by the authors of this paper.[5] While all n observations of the considered time series are labeled by experts in the field, they may include some error and/or bias. Nevertheless, the labels help to understand how the time series are divided into different states and thus to which state(s) a certain subsequence belongs. Figure 1 depicts the six time series. The vertical dashed lines demarcate transitions in labels, representing real-world state changes. This means that all observations in between two dashed lines have the same label and

[3] The time series are downloaded from the website of the authors, where the names of some time series differ from the original naming convention.

[4] Due to the nature of the ECG and SLC time series, we concatenated the separate sequences into a single time series. Due to the large size of the SLC time series, we took a subset including two different states occurring at least twice.

[5] The selected time series, code and results are available at Github.

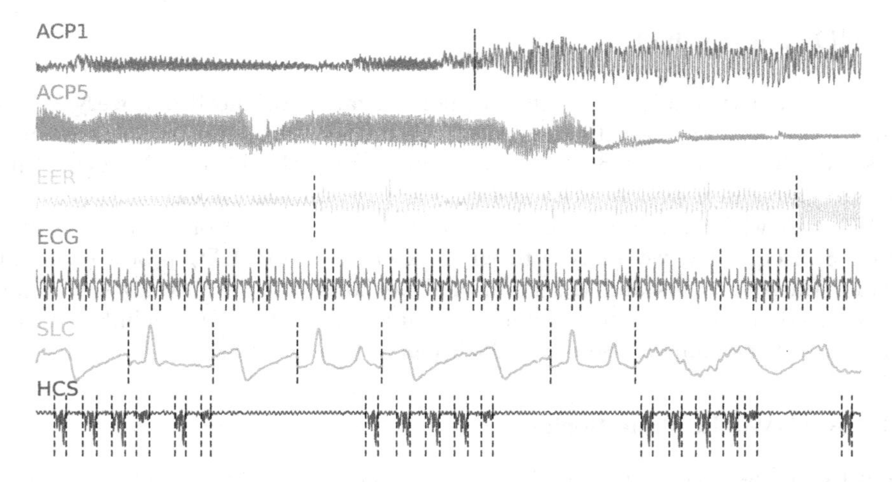

Fig. 1. The six real-world time series. Each row shows one of the time series. The dashed vertical lines demarcate transitions of real-world states. Please note that the time scales are different, due to the different time series lengths.

thus belong to the same state. All considered time series consist of either two ($ACP1$, $ACP5$, ECG, and SLC) or three different states (EER and HCS). If the number of vertical lines is higher than the number of states—as is the case for ECG, SLC, and HCS—it means that the time series include recurring states.

We examine a wide range of different subsequence lengths proportional to the time series length: $m = \{0.005n, 0.0075n, \ldots, 0.12n\}$. Then, for all abovementioned time series and subsequence length combinations, we examine the MSS of **ED** and $\mathbf{ED_z}$ for all $n - m + 1$ subsequences of the time series. In this way we are able to get insights into the differences and similarities between using **ED** and $\mathbf{ED_z}$ for the discovery of MSSs.

4.2 Differences in MSS

The two distance metrics **ED** and $\mathbf{ED_z}$ are likely to result in the detection of different MSSs. To determine the degree of disagreement of both metrics, we start by examining the percentage of different MSSs per time series for subsequence lengths $m \in \{50, 51, \ldots, 500\}$. Figure 2 shows the percentage of *different* results obtained by **ED** and $\mathbf{ED_z}$ ($\%ED \neq ED_z$, y-axis) as a function of subsequence length m (x-axis). Each curve represents one of the time series. For example, for $m = 50$, the percentage of different MSSs for the $ACP1$ time series (red curve) is up to 50%. This means that of the total number of $(n - m + 1) = 4,951$ subsequences, only half of them return the same MSSs for **ED** and $\mathbf{ED_z}$.

Overall, the curves depicted in Fig. 2 reveal that there is substantial disagreement between MSSs discovered with **ED** and $\mathbf{ED_z}$ for all time series. The largest disagreement is found for SLC (light blue curve). For $m = 50$ the disagreement

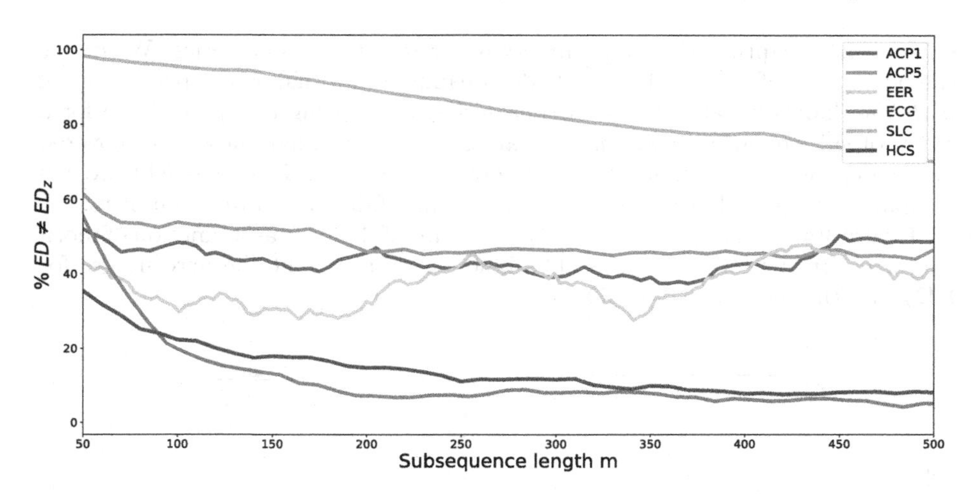

Fig. 2. Percentage of different MSSs returned by **ED** and $\mathbf{ED_z}$. (Color figure online)

is almost 100%. For the other five time series, the disagreement ranges from about 35% to 62% for $m = 50$. For all time series, the percentage disagreement becomes smaller with increasing subsequence length. This is most likely due to the reduction in motif variability for larger subsequences.

The similarity between the curves for HCS (blue) and ECG (green) are not surprising as the characteristics of both time series are similar; repetitive with constant offsets and magnitudes for the considered subsequences. The latter also holds for the curves of $ACP1$, $ACP5$ and EER time series, but these time series are more noisy. The SLC time series shows a repeating pattern, but as can be seen in Fig. 1 it has offset and magnitude variations. This results in a high disagreement between the MSSs detected with **ED** versus those detected with $\mathbf{ED_z}$, especially for small m. With increasing m, the downward trend continues and becomes stable as soon as m reaches the pattern length ($m \approx 1,000$, not shown in the figure).

Based on the large differences in MSSs, we can conclude that using **ED** versus $\mathbf{ED_z}$ makes a difference for motif discovery. But what difference?

4.3 Meaningful MSS

To determine which of the two metrics returns the most meaningful MSSs, we compare for each MSS the labels of its constituent pair of subsequences.[6] As all n observations of the considered time series are labeled, we know to which state(s) all m observations of a subsequence belong. When the labels match for all m observations, we consider the MSS to be meaningful, because both its

[6] It can be the case that several subsequences have minimum distance to the subsequence of interest. In this case only one subsequence is randomly chosen to be directly compared.

subsequences capture the same underlying state of the time series. We define the percentage of meaningful motifs (% Meaningful Motifs) as the percentage of *matching* labels for all MSSs of a time series. Computing this percentage for a range of subsequence lengths m, reveals how the meaningfulness of discovered motifs depends on m. In addition, by doing this for both $\mathbf{ED_z}$ and $\mathbf{ED}$, we can compare how both differ in terms of the meaningfulness as a function of m.

Figure 3(a–f) shows the percentage of meaningful motifs as a function of motif length m for our six time series. The blue curves represent the percentages for $\mathbf{ED_z}$ and the red curves for $\mathbf{ED}$.

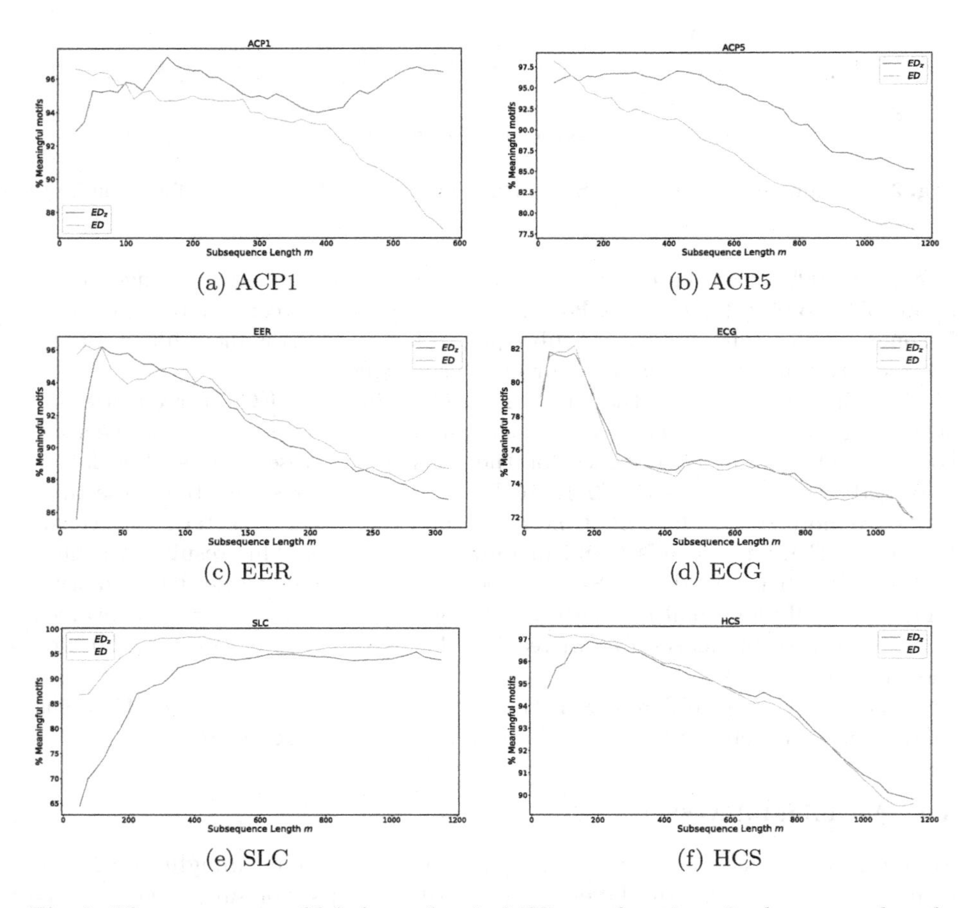

(a) ACP1 (b) ACP5

(c) EER (d) ECG

(e) SLC (f) HCS

Fig. 3. The percentage of label matches in MSS as a function of subsequence length m, for **ED** and $\mathbf{ED_z}$. (a–f) display the results for the six time series. The higher the percentage (y-axis), the more meaningful the MSSs are. (Color figure online)

We observe that for small values of m, **ED** (red curve) generally returns a higher percentage of meaningful MSSs than $\mathbf{ED_z}$ (blue curve). For $ACP1$ and $ACP5$, the choice between **ED** and $\mathbf{ED_z}$ depends on m: **ED** performs better for small m, but $\mathbf{ED_z}$ is preferred as m increases. The same is observed for EER, except that there is a turn in performance again. The results for ECG and HCS are fairly close for both metrics. Only for SLC holds that, despite the subsequence length, one of the two metrics is superior, namely **ED**. From this figure we can conclude that (I) the subsequence length plays an important role in the detection of meaningful MSSs and (II) that normalization ($\mathbf{ED_z}$) is not always superior. Actually, **ED** equally often returned meaningful MSSs as $\mathbf{ED_z}$ in the considered cases.

Given that $\mathbf{ED_z}$ is unquestioningly adopted in the motif discovery research community, it is surprising that **ED** competes with $\mathbf{ED_z}$. To add more evidence to this finding, we tested the metrics on an additional 90 time series retrieved from the UCR Archive [1].[7] For each considered time series we examine five different subsequence lengths proportional to the time series length: $m = \{0.005n, 0.01n, 0.02n, 0.03n, 0.05n\}$. This means that for a time series with length $n = 5,000$, we examine $m = \{25, 50, 100, 150, 250\}$, and for a time series with length $n = 10,000$ we examine $m = \{50, 100, 200, 300, 500\}$, and so on. Then, for all above-mentioned time series and subsequence length combinations, we (again) examine the labels of the constituent pair of subsequences of the MSSs. Based on the average percentage of label mismatches in MSS for **ED** and $\mathbf{ED_z}$ per subsequence length, we count the number of cases for which the metrics return the most meaningful MSS.

In total we considered 450 (90×5) cases. Not normalizing subsequences (**ED**) is most beneficial in 57.8% of the cases. For 31.1% of the cases it holds that $\mathbf{ED_z}$ returned the most meaningful MSSs. And in 11.1% of the cases the metrics are equally effective. Evidently, the assumption that all subsequences need to be normalized in order to find meaningful motifs, does not always hold for real-world time series. Offsets and magnitudes proved to be crucial for meaningful motif discovery in over half of the considered cases.

4.4 Which Distance Metric to Use When?

To better understand which metric to use when, the highest percentage of MSSs per distance metric per time series are visualized in Fig. 4. As the distance metrics depend on the chosen subsequence length m, we only plotted the best result of one of the five considered subsequence lengths mentioned above ($m = \{0.005n, 0.01n, 0.02n, 0.03n, 0.05n\}$). Hence, Fig. 4 includes 96 dots representing the highest obtained percentage with $\mathbf{ED_z}$ on the y-axis and **ED** on the x-axis. The diagonal dotted line represents the case $\mathbf{ED_z} = \mathbf{ED}$ and only the names of the time series for which hold that the one distance metric returned

[7] Due to the demanding running time of the calculations, the maximum length of the time series is set to $n = 20,000$. We only consider time series without missing values and for which hold that the (sub)set consists of two or more states.

5% more meaningful MSSs than the other are listed. The higher density of dots below the diagonal visualizes the conclusion made in the previous subsection about $\mathbf{ED_z}$ versus **ED**. Namely, that not normalizing subsequences (**ED**) is most beneficial in the majority of the cases.

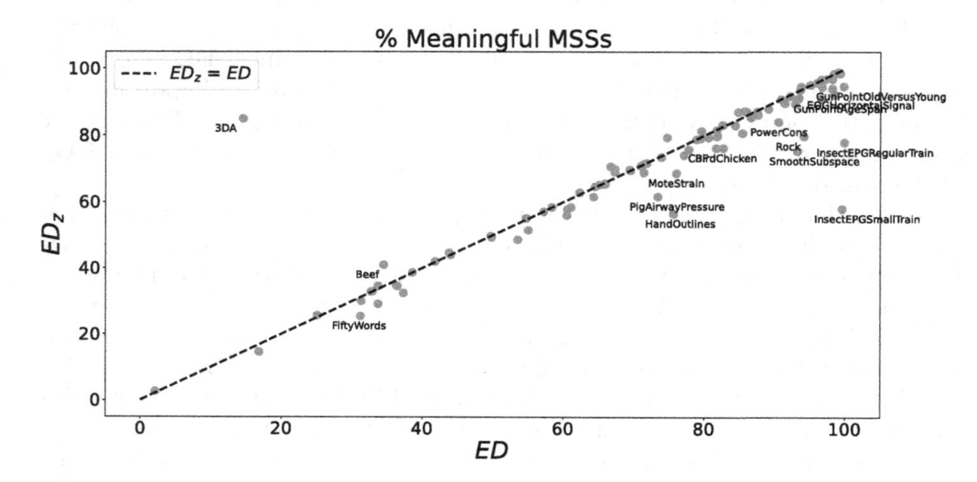

Fig. 4. Percentage of meaningful MSSs returned by **ED** and $\mathbf{ED_z}$. The 96 dots represent the best obtained results—regardless of the subsequence length m—for all considered time series. The names of the time series for which hold that the one distance metric returned 5% more meaningful MSSs than the other are listed.

For almost half (46%) of the considered time series, the difference in results between the two metrics is smaller than 1%. The characteristics of these time series are similar in that the time series has stationary mean and stationary variance. Or in other words, the average offset and magnitude of all subsequences of length m remains stable (e.g. $OSULeaf$, $coffee$, and $OliveOil$ in Fig. 5). As soon as offset and (to a lesser extent) magnitude changes exist, the dots deviate further from the diagonal line. The upper and lower time series in Fig. 5 represent the two extremes on both sides of the diagonal. The $3DA$ time series is an example for which hold that normalization is required: the offset is not representative for the different states. On the contrary, the time series in blue ($InsectEPGSmallTrain$) is divided into three states, easily recognizable by the different offsets. In such situations, normalization is undesired.

To detect changes in the offset, one can consider the standard deviation of all subsequences of length $m = 0.01n$ of min-max scaled time series (i.e. $TS_{scaled} = \frac{TS - TS_{min}}{TS_{max} - TS_{min}}$). As long as the time series is constant, the standard deviation remains low. If the offset changes, the standard deviation will rise. By taking the ratio of maximum standard deviation to the average of all subsequences, we obtain a value SD. This value has a correlation of 0.78 with the absolute difference in results of both metrics. Hence, if the offset and magnitude are stable for the entire time series—and thus the value of SD is low (often < 2)—,

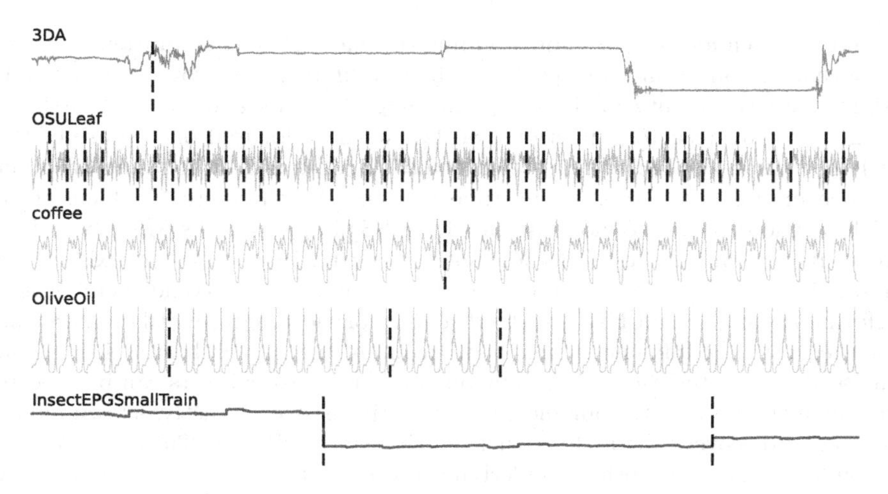

Fig. 5. Examples of time series where $\mathbf{ED_z} > \mathbf{ED}$ (red), $\mathbf{ED_z} \approx \mathbf{ED}$ (orange) and $\mathbf{ED_z} < \mathbf{ED}$ (blue). The gray dashed vertical lines demarcate transitions of real-world states. Please note that the time scales are different, due to the different time series lengths. (Color figure online)

one can chose for either $\mathbf{ED_z}$ or **ED**. But if the time series has non-stationary mean and variance, one should consider whether the offset and magnitude are representative for a certain class. If this is true, **ED** should be used. Otherwise, $\mathbf{ED_z}$ is recommended.

As mentioned before, the subsequence length m plays an important role in the detection of meaningful MSSs. For 63 out of 96 considered time series, both metrics returned the best results for the same subsequence length. For the remaining time series, $\mathbf{ED_z}$ performs better with higher values for m than **ED**. The latter observation is related to the actual pattern length in a time series. If the subsequence length m is smaller than the actual pattern length, the offset and magnitude matter and thus not normalizing is more favorable. In absolute terms, **ED** is more robust to different subsequence lengths than $\mathbf{ED_z}$. While the subsequence length should always be carefully chosen, this is especially important when $\mathbf{ED_z}$ is used.

5 Conclusion and Future Directions

Most motif discovery methods implicitly assume that subsequences need to be normalized in order to find meaningful motifs. Experiments with 96 real-world labeled time series showed that this assumption does not hold for the majority of the cases. Plain **ED** returns in 68.9% of the considered cases equally or more meaningful MSSs than $\mathbf{ED_z}$. This implies that subsequence offsets and magnitudes are often relevant for motif discovery. We therefore conclude that the assumption underlying normalization does not always hold for real-world time series and thus should not be uncritically adopted.

Our research focused on understanding the effect of the assumption underlying normalization on motif discovery. While preliminary, it provided important insights into the ability of detecting meaningful motifs and serves as a basis for many future research directions. So far, we only focused on the use of the Euclidean distance. This examination could be continued by focusing on other distance metrics such as dynamic time warping. Furthermore, we focused on which metric returns the most meaningful MSSs in general. As MSSs are often part of motifs, they are essential for motif discovery. However, there exist MSSs for which $D(S_i, S_j) \leq \delta$ does not hold. In future work one could focus on the performance of **ED** and $\mathbf{ED_z}$ on the discovery of meaningful motif clusters and how the value of δ influences this. Another interesting research direction is how time series characteristics such as entropy and or statistical tests can be used to determine the best metric for meaningful motif discovery. Lastly, as commonly used motif discovery algorithms such as the Matrix Profile [20] and SAX [9] are built on the assumptions underlying normalization, we will examine how normalization affects these commonly used methods.

References

1. Dau, H.A., et al.: The UCR time series archive. IEEE/CAA J. Autom. Sin. **6**(6), 1293–1305 (2019). https://doi.org/10.1109/jas.2019.1911747
2. Dau, H.A., Keogh, E.: Matrix profile V: a generic technique to incorporate domain knowledge into motif discovery. In: Proceedings of the 23rd ACM SIGKDD International Conference on Knowledge Discovery and Data Mining, pp. 125–134 (2017)
3. Esling, P., Agon, C.: Time-series data mining. ACM Comput. Surv. **45**(1), 12 (2012)
4. Ismail Fawaz, H., Forestier, G., Weber, J., Idoumghar, L., Muller, P.-A.: Deep learning for time series classification: a review. Data Min. Knowl. Disc. **33**(4), 917–963 (2019). https://doi.org/10.1007/s10618-019-00619-1
5. Gao, Y., Lin, J.: HIME: discovering variable-length motifs in large-scale time series. Knowl. Inf. Syst. **61**(1), 513–542 (2018). https://doi.org/10.1007/s10115-018-1279-6
6. Gao, Y., Lin, J., Rangwala, H.: Iterative grammar-based framework for discovering variable-length time series motifs. In: 2017 IEEE International Conference on Data Mining (ICDM), pp. 111–116. IEEE (2017)
7. Keogh, E., Kasetty, S.: On the need for time series data mining benchmarks: a survey and empirical demonstration. Data Min. Knowl. Disc. **7**(4), 349–371 (2003)
8. van Leeuwen, F., Bosma, B., van den Born, A., Postma, E.: RTL: a robust time series labeling algorithm. In: Abreu, P.H., Rodrigues, P.P., Fernández, A., Gama, J. (eds.) IDA 2021. LNCS, vol. 12695, pp. 414–425. Springer, Cham (2021). https://doi.org/10.1007/978-3-030-74251-5_33
9. Lin, J., Keogh, E., Lonardi, S., Chiu, B.: A symbolic representation of time series, with implications for streaming algorithms. In: Proceedings of the 8th ACM SIGMOD Workshop on Research Issues in Data Mining and Knowledge Discovery, pp. 2–11 (2003)
10. Linardi, M., Zhu, Y., Palpanas, T., Keogh, E.: Matrix profile X: VALMOD-scalable discovery of variable-length motifs in data series. In: Proceedings of the 2018 International Conference on Management of Data, pp. 1053–1066 (2018)

11. Madrid, F., Singh, S., Chesnais, Q., Mauck, K., Keogh, E.: Matrix profile XVI: efficient and effective labeling of massive time series archives. In: 2019 IEEE International Conference on Data Science and Advanced Analytics (DSAA), pp. 463–472 (2019). https://doi.org/10.1109/DSAA.2019.00061
12. Mohammad, Y., Nishida, T.: Exact discovery of length-range motifs. In: Nguyen, N.T., Attachoo, B., Trawiński, B., Somboonviwat, K. (eds.) ACIIDS 2014. LNCS (LNAI), vol. 8398, pp. 23–32. Springer, Cham (2014). https://doi.org/10.1007/978-3-319-05458-2_3
13. Mueen, A., Chavoshi, N.: Enumeration of time series motifs of all lengths. Knowl. Inf. Syst. **45**(1), 105–132 (2015)
14. Mueen, A., Keogh, E.: Online discovery and maintenance of time series motifs. In: Proceedings of the 16th ACM SIGKDD International Conference on Knowledge Discovery and Data Mining, pp. 1089–1098 (2010)
15. Mueen, A., Keogh, E., Zhu, Q., Cash, S., Westover, B.: Exact discovery of time series motifs. In: Proceedings of the 2009 SIAM International Conference on Data Mining, pp. 473–484. SIAM (2009)
16. Patel, P., Keogh, E., Lin, J., Lonardi, S.: Mining motifs in massive time series databases. In: 2002 IEEE International Conference on Data Mining, 2002. Proceedings, pp. 370–377. IEEE (2002)
17. Senin, P., et al.: GrammarViz 2.0: a tool for grammar-based pattern discovery in time series. In: Calders, T., Esposito, F., Hüllermeier, E., Meo, R. (eds.) ECML PKDD 2014. LNCS (LNAI), vol. 8726, pp. 468–472. Springer, Heidelberg (2014). https://doi.org/10.1007/978-3-662-44845-8_37
18. Shifaz, A., Pelletier, C., Petitjean, F., Webb, G.I.: TS-CHIEF: a scalable and accurate forest algorithm for time series classification. Data Min. Knowl. Disc. **34**(3), 742–775 (2020)
19. Wang, X., et al.: RPM: representative pattern mining for efficient time series classification. In: EDBT, pp. 185–196 (2016)
20. Yeh, C.C.M., et al.: Matrix profile I: all pairs similarity joins for time series: a unifying view that includes motifs, discords and shapelets. In: 2016 IEEE 16th International Conference on Data Mining (ICDM), pp. 1317–1322. IEEE (2016)
21. Yin, M.S., Tangsripairoj, S., Pupacdi, B.: Variable length motif-based time series classification. In: Boonkrong, S., Unger, H., Meesad, P. (eds.) Recent Advances in Information and Communication Technology. AISC, vol. 265, pp. 73–82. Springer, Cham (2014). https://doi.org/10.1007/978-3-319-06538-0_8
22. Yingchareonthawornchai, S., Sivaraks, H., Rakthanmanon, T., Ratanamahatana, C.A.: Efficient proper length time series motif discovery. In: 2013 IEEE 13th International Conference on Data Mining, pp. 1265–1270. IEEE (2013)
23. Zhu, Y., Yeh, C.C.M., Zimmerman, Z., Kamgar, K., Keogh, E.: Matrix profile XI: SCRIMP++: time series motif discovery at interactive speeds. In: 2018 IEEE International Conference on Data Mining (ICDM), pp. 837–846. IEEE (2018)

Tree-Based Credit Card Fraud Detection Using Isolation Forest, Spectral Residual, and Knowledge Graph

Phat Loi Tang[1,2], Thuy-Dung Le Pham[1,2], and Tien Ba Dinh[1,2](✉)

[1] Faculty of Information Technology, University of Science, Ho Chi Minh City, Vietnam
{tlphat18,pltdung18}@apcs.fitus.edu.vn
[2] Vietnam National University, Ho Chi Minh City, Vietnam
dbtien@fit.hcmus.edu.vn

Abstract. Credit card frauds have constantly been a significant threat to the worldwide economy since the advent of non-cash payments, leading to high demand for fraud detection research. Almost existing studies on supervised credit card fraud classifiers cannot leverage domain characteristics such as fraud rings and sudden changes in payment traffic. They also often fail to detect unseen fraud patterns. In this paper, we introduce a set of features, targeting both traditional and modern fraud patterns, generated from Isolation Forest, Spectral Residual, and Knowledge Graph to boost the performance of tree-based models. We evaluate different tree-based models on the Kaggle public dataset and demonstrate the enhancement of our feature scores in the enterprise payment context. The proposed feature set increases the AUC-ROC and AUC-PR of XGBoost by up to 4.5% and 47.89%, respectively, on the enterprise dataset with only 0.0028% of fraud samples. Our results suggest a method for further improvement on a wide range of fraud detection problems based on domain analysis and feature extractors, especially Knowledge Graph.

Keywords: Credit card fraud detection · Cash-out · Supervised learning · Unsupervised learning · Knowledge graph · Feature construction

1 Introduction

Since the 1970s, credit cards have witnessed significant growth over the world [11] and specifically in Vietnam during the last decade [17]. Unfortunately, it is followed by a serious rise in the number of credit card frauds (CCF), resulting in a total of US$ 28.6 million loss in 2020 [22]. Therefore, transaction processors nowadays have a high demand for credit card fraud detection (CCFD). Due to

P. L. Tang and T.-D. Le Pham—The first two authors contribute equally to this work.

G. Nicosia et al. (Eds.): LOD 2022, LNCS 13811, pp. 326–340, 2023.
https://doi.org/10.1007/978-3-031-25891-6_25

the diversity and unpredictability of frauds [4], it remains a challenging problem. In Vietnam, besides the familiar types of CCF such as identity theft and counterfeiting, most illegal transactions are either credit card cash-out or refinancing. Section 2 explains in details these sophisticated fraud types in Vietnam.

Many studies and surveys evaluate different methods for CCFD [18]. However, the existing benchmark dataset from Kaggle contains only numerical attributes after PCA transformation, hence cannot demonstrate the capability of models in production. Furthermore, each machine learning approach can only deal with one aspect of the problem. While supervised learning models can extract the fraud patterns in the training dataset, the unsupervised learning approaches rely on isolation and abnormality principles that are more flexible in discovering unseen patterns. Consequently, combining the best of both approaches is a potential way to improve classification performance. Among the choices of supervised models, tree-based is the most suitable approach for structured transaction data with many correlated categorical attributes. We discuss those existing solutions to CCFD in Sect. 2.

In this paper, we evaluate the performance of different tree-based models on the public dataset and propose a feature-enhancing pipeline to improve their predictions. Observing that CCFs are either anomalies or closely related to the other frauds, we introduce extra features generated by Isolation Forest (iForest), Spectral Residual (SR), and Knowledge Graph (KG). Although our work is not the first attempt to extend the feature set [6], our contribution is the set of scores combined from multiple research domains, especially KG, to help the baseline detect fraud behaviors from multiple perspectives. Section 3 clearly describes our proposed method.

We apply the proposed pipeline to the Kaggle public CCF dataset, showing that the performance of Decision Tree, Random Forest, and XGBoost can increase by no less than 0.3% in terms of AUC-PR. Then we apply the method to the enterprise dataset and show that our method outperforms the models given the full attribute information. As we expected, SR boosts the performance of XGBoost by up to 47.89% in terms of AUC-PR. Our results suggest further investigation into more effective extractors from multiple areas of work to increase CCF detecting performance and robustness. Section 4 and 5 illustrates our experiments and result. Finally, Sect. 6 presents our conclusion.

Contributions. In summary, our work makes the following contributions.

- Evaluating different tree-based models (Decision Tree, Random Forest, and XGBoost) on a public dataset.
- Proposing extra features and demonstrating their enhancement, especially given the context of the attributes.
- One of the first applications of Knowledge Graph and Spectral Residual in detecting credit card cash-out and refinancing behaviors (to the best of our knowledge). Our results highlight the potential of the method to be applied to fraud detection systems in Vietnam.

2 Background and Related Work

Credit cardholders in Vietnam can make a purchase using an interest-free advance money amount from the banks within 45 days [24]. However, banks apply a heavy interest rate for withdrawing cash directly from the credit card [1]. Cardholders avoid the extra fee by committing a fictitious payment at retail stores, getting cash from the merchants instead of products. Meanwhile, the stores charge their customers a small service fee [14]. Another similar behavior is that the retail stores help customers pay their credit card debt and then commit a fictitious payment to consolidate the new debt in 45 days. Due to the collusion between fraudsters and merchants in such illegal actions, modern fraud strategies, in general, are sophisticated and often challenging to identify. For this reason, payment firms in Vietnam are careful when selecting partners (mostly large merchants with high reputations). Nevertheless, the companies will soon cooperate with more small and medium-size merchants, where those illegal behaviors are more likely to occur. Our study is a part of the project aiming to implement a fraud detection system in the payment service providers, which can help the enterprises detect and prevent potential loss in their future expansion.

Due to the labeled data shortage and the class imbalance problem, unsupervised approaches are reasonable solutions to detect fraud. Liu et al. [16] propose the iForest, suggesting that the fewer partitions to isolate a certain point from the rest, the more likely it is an anomaly. Auto-Encoder (AE) [3] aims to estimate an identified function that approximates the input using the encoding and decoding phases. When applying AE to fraud detection, a high reconstruction error indicates that the given point is fraudulent. Numerous studies experiment iForest and AE to detect illicit transactions of credit cards [12,18,26]. Overall, despite giving acceptable performance, unsupervised approaches almost fail to utilize existing labels to learn about fraud patterns.

Various researchers use supervised learning methods to fully exploit the available labeled data in CCFD. Random Forest [5] combines the predicting capability of multiple equally-contributed decision trees training on a different random bag of datasets. XGBoost is a tree-ensemble boosting system that optimizes the loss function by adding regularization to handle the sparse data and weighted quantile sketch for tree learning. It is not only a prevalent challenge winning solution on Kaggle and KDDCup 2015 [8] but also the superior boosting algorithms in CCFD [10]. The notable deficiencies of supervised approaches are the heavy reliance on the training patterns and the neglect of domain-specific properties, such as the traditional fraudulent behaviors are often outliers among the usual transactions.

Some researchers propose hybrid approaches that combine the best of both approaches. Li et al. suggest a hybrid framework that mainly handles the class imbalance with the overlap in CCF [15]. The framework uses an unsupervised anomaly detection model to learn the principle of minority samples and a supervised model to distinguish between anomaly transactions and normal ones in the overlapping subset. Meanwhile, Wu et al. carefully select five fundamental statistical feature sets to enhance the supervised baseline performance in detecting credit card cash-out in China. Their analysis motivates us to experiment with more advanced feature sets in detecting the same theme of CCF [25].

All machine learning methods mentioned above share two common drawbacks. They ignore the relative behaviors in the temporal domain since they consider the timestamps as only the numerical attribute. Additionally, they cannot extract the association between multiple transactions. For example, a payment from the same customer and the same store with a fraudulent one is also more likely to be fraudulent behavior.

3 Proposed Method

To address the issues mentioned in Sect. 2, we propose a feature construction framework to enhance the performance of the supervised learning fraud classifier. The extra information that we attach to the feature space includes the fraud scores, anomaly score, and temporal saliency scores of each transaction. We expect that when training on a new set of features, the supervised learning baseline can achieve a better result than the original pipeline, especially on a large, heterogeneous enterprise dataset where merchants and customers are both involved in committing fraudulent behaviors.

3.1 Overview

As illustrated in Fig. 1, our score generator consists of three different modules:

1. We first construct the KG that interprets the relationship between different entities involved in the transactions. Then we calculate the shortest distance from each transaction to the nearest fraud (**fraud distance score**) and the total number of neighbors that the transaction share with fraudulent ones (**fraud similarity score**). These measures reflect the connection between the transaction and the set of illegitimate transactions.
2. Next, we segment the data into different chunks, one for each merchant. We run the iForest algorithm on each segment to calculate the **anomaly score** of each transaction. It tells the oddness of the behavior among those committed in the same merchant.
3. Regarding the temporal dimension, we compose two time series for each chunk, corresponding to the traffic and total amount at each point in time. Then, we apply SR [21] to convert them into saliency maps. From those two maps, respectively, we can extract the **temporal traffic score** and **temporal amount score**.
4. Finally, the fraud scores, anomaly score, and temporal saliency scores extracted in the previous steps are combined with the original feature space to enhance the performance of the supervised learning baseline.

The fraud scores expose the multi-hop relations between different transactions quantitatively, which can help highlight the ring of abnormal behaviors committed by an organization of fraudsters. Moreover, the scores interpreted

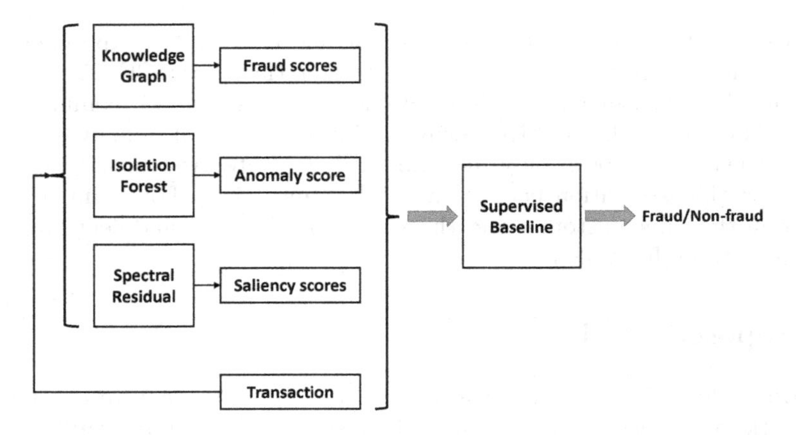

Fig. 1. The proposed feature engineering pipeline with three score generating modules.

from the temporal dimension and merchant granularity can help reduce the overfitting of the baseline on the training patterns. Therefore, our framework can help raise the performance of supervised learning on the dataset consisting of subtle fraudulent acts such as credit card cash-out and refinancing.

3.2 Fraud Scores from the Knowledge Graph

Paying attention to the relationship between multiple transactions, we observe that fraudulent ones are often closely related to each other. Hence, KG [23], which represents the spatial relations between entities, will more or less contribute to the baseline prediction where categorical information of transactions is available.

Figure 2 depicts the ontology and a KG snapshot of the first 20 records in our enterprise dataset. The nodes represent certain entities such as the store, transaction, cardholder, and the like, while the edges demonstrate the relationship between those entities. Consulted by the domain experts on the fraud patterns in Vietnam, we only retain the most relevant information to form the ontology. They include card number, cardholder name, phone number, issuing bank, card type, store, and the meta-information related to store (address, merchant, sector).

We measure the fraud scores according to two following principles:

1. The shorter the path from a transaction to any fraud, the higher chance that it is fraudulent.
2. The more neighbors a transaction shares with the frauds, the higher chance that it is fraudulent.

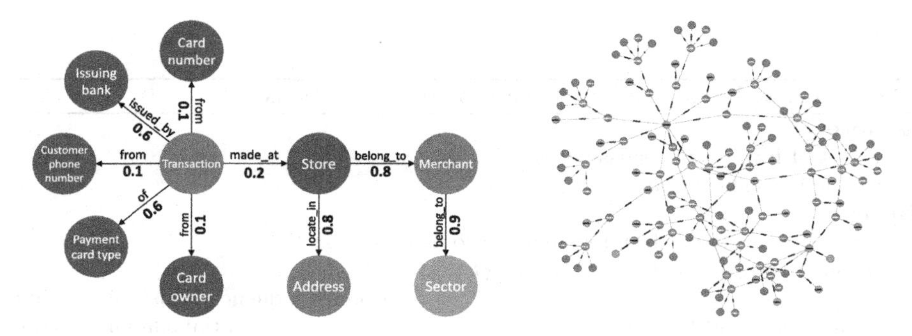

Fig. 2. The ontology and a KG snapshot of the enterprise data. Consulted by the domain experts on the fraud patterns, we propose the set of edge weights that represent the restriction on the impact from the source to the destination node. The lower the weight, the more likely the fraudulent source leads to the fraudulent destination.

Several studies in the KG domain use the resistance distance algorithm [13] to calculate entity similarity following the above principles. However, it requires the construction of the Laplacian matrix and its Moore-Penrose pseudo-inverse. Such expensive calculations are inappropriate for a large-scale dataset with millions of nodes. Therefore, we adopt two simpler algorithms to calculate the fraud distance score and fraud similarity score, each corresponding to a single principle.

For path distance calculation, we assign the edge weights based on the impact of each attribute on the whole fraud ring. The lower the weight, the more likely the fraudulent source leads to the fraudulent destination. Through discussions with financial experts, we propose the set of weights as illustrated in Fig. 2. Specifically, we assign the lowest weight (0.1) to the relations that uniquely identify the fraudsters, namely card information and phone number. Stores might collude with customers to commit credit card cash-out, so we assign a slightly higher weight (0.2) to the transaction-store relations. Other relations to payment type and issuing bank are assigned a moderate weight (0.6). Finally, the least relevant entities such as store address (0.8), merchant (0.8), and sector (0.9) are supposed to be the furthest away from the transactions.

We adopt the Dijkstra algorithm [9] starting from the set of fraud transactions in the training set to calculate the fraud distance score for each transaction. In addition, we count the number of times a transaction can be reached from all fraud transactions after visiting one intermediate attribute node. Algorithm 1 illustrates the details of our second procedure.

3.3 Anomaly Score from the Isolation Forest

We construct the Isolation Tree (iTree) by splitting the data randomly along one dimension at each step. The earlier the point is isolated, the higher chance that it is an anomaly. Figure 3 demonstrates the idea of iTree [16].

Formally, we define the path length $h(x)$ of a point x in an iTree as the number of edges we need to traverse from the root to the leaf node whose value

Algorithm 1. Fraud similarity scores of all transactions

Require:
F: set of fraudulent transactions
G: the knowledge graph

Ensure:
similar: an array of fraud similarity scores (with size $|V|$)

```
 1: function FraudSimilarityScores(F, G)
 2:     q ← F                                   ▷ Initialize a queue consisting of frauds
 3:     similar ← [0, 0, ..., 0]                ▷ Initial similarity scores
 4:     while q is not empty do
 5:         u ← pop an element fromq
 6:         for all v which is neighbor of u in G do
 7:             if v is a transaction node then
 8:                 similar[v] ← similar[v] + 1
 9:             end if
10:             if u is in F then               ▷ Only explore nodes directly connecting to frauds
11:                 Add v to q
12:             end if
13:         end for
14:     end while
15:     return similar
16: end function
```

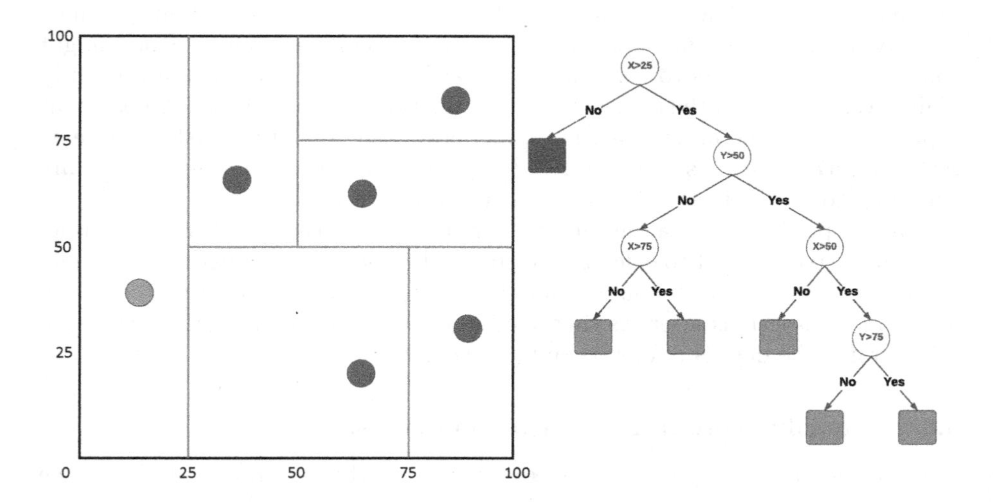

Fig. 3. Visualization of the isolation process in a single iTree.

is x. Then, $E(h(x))$ is defined as the average of $h(x)$ from iForest (a collection of iTrees). Given the dataset of n instances, we calculate the anomaly score of x using the following formula:

$$s(x) = 2^{-\frac{E(h(x))}{2H(n-1)-2(n-1)/n}}, \tag{1}$$

where $H(i)$ is the harmonic number estimated by the sum of the natural logarithmic function and Euler's constant.

$$H(i) = \sum_{i=1}^{n} \frac{1}{i} \approx \ln i + 0.5772156649 \tag{2}$$

The denominator of (1) can be interpreted as the estimated average path length of an iTree constructed in data of size n. Given that the path length is always positive, $s(x)$ only takes a real value from 0 to 1. The lower the average path length of x, the higher value that $s(x)$ receives. In other words, the earlier a point is isolated in the forest, the more likely it is to be an anomaly.

The transaction anomaly scores extracted from different stores using iForest provide a new dimension for our binary classifier. We expect them to help reduce the model variance, especially when training on very few fraud examples.

3.4 Saliency Scores from the Spectral Residual

Given a sequence of real values $x = x_1, x_2, ..., x_n$, we derive the saliency map $S(x)$ by calculating the spectral residual in log amplitude spectrum. Mathematically, the algorithm consists of the following steps [21]:

$$A(f) = Amplitude(F(x)) \tag{3}$$

$$P(f) = Phase(F(x)) \tag{4}$$

$$L(f) = \log A(f) \tag{5}$$

$$AL(f) = h_q(f) * L(f) \tag{6}$$

$$R(f) = L(f) - AL(f) \tag{7}$$

$$S(x) = \|F^{-1}(exp(R(f) + iP(f)))\| \tag{8}$$

F and F^{-1} denote Fourier Transform and Inverse Fourier Transform, respectively. First, we convert the original input into the frequency domain, with the corresponding amplitude spectrum $A(f)$ and phase spectrum $P(f)$. Then, we calculate the log amplitude spectrum $L(f)$ and convolute it with the vector $h_q(f)$ of dimension q defined as follows to get the average log spectrum:

$$h_q(f) = \frac{1}{q}\left[1\ 1\ ...\ 1\right] \tag{9}$$

Subtracting $AL(f)$ from the log amplitude spectrum, we get the residual $R(f)$. Finally, converting the residual back to the spatial domain gives us the saliency map $S(x)$ of the original time series. Figure 4 illustrates an example where we calculate the saliency map of the sine waves with two innovative points. Looking at the conversed map, we observe a significant increase in the saliency score at the corresponding timestamps of the anomalies in the original series.

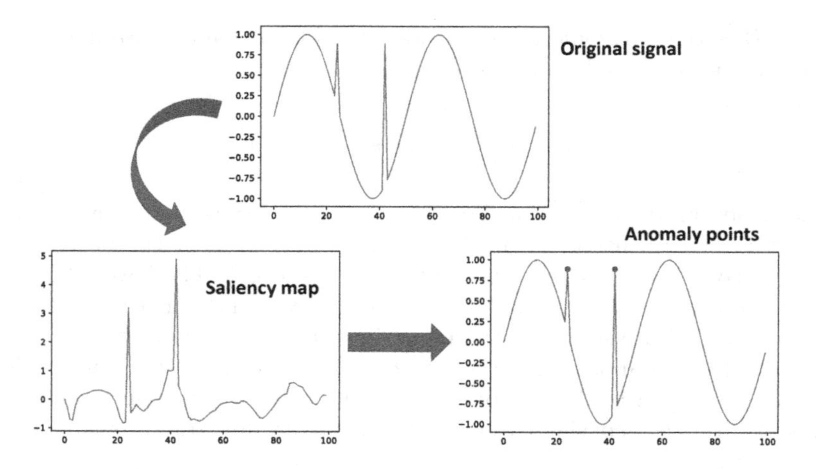

Fig. 4. Saliency map obtained by applying SR on a sine wave.

We extract the two time series for each merchant and aggregate the corresponding scores of two maps to get the temporal traffic score and temporal amount score for each transaction. Since the extra information from the saliency scores allows the classifier to look at the abnormality in the temporal dimension, we expect the supervised learning model can detach the time frames where the traffic or payment amount suddenly rises or drops.

4 Experimental Setup

4.1 Datasets

We use two real-world datasets to evaluate the proposed pipeline, Kaggle Credit Card Fraud[1] (KCCF) and Enterprise Credit Card Fraud (ECCF). Both datasets contain the credit card transactions in a tabular format with high dimensional feature space and are heavily imbalanced with less than 0.2% fraud samples. Table 1 summarizes the statistics of KCCF and ECCF.

KCCF released on Kaggle is a small-scale dataset recorded in September 2013 by European cardholders. It contains only numerical attributes as the results of PCA transformation. For confidentiality issues, there is no background information about the original data features. Experiments on KCCF demonstrate the improvement of our proposed pre-processing pipeline on the baseline

[1] https://www.kaggle.com/datasets/mlg-ulb/creditcardfraud.

Table 1. Statistics of the datasets.

Dataset	Total records	Anomaly records	Number of features
KCCF	284,807	492 (0.1727%)	30
ECCF	2,957,255	82 (0.0028%)	17

performance. In addition, the baseline produces the highest prediction capability and efficiency is applied to the second dataset.

ECCF is an internal large-scale payment transaction dataset collected by a payment service provider in Vietnam in the last quarter of 2021. The annotation of fraud samples refers to the alerts from the acquiring and issuing banks that possess commercial fraud detectors. As we mentioned earlier in Sect. 2, payment companies in Vietnam are very careful when selecting their partners, which results in a low number of fraud samples. Unlike KCCF, ECCF provides a raw representation of high-dimensional transactions with many categorical fields such as the retail stores, working sector, address, and card operation type. As we described earlier in Sect. 2, the fraud behaviors in ECCF are subtle and deceptive to the traditional classification approaches. We expect our enhancement of the feature space, especially with knowledge extracted from the transaction graph, can increase the classification capability of the original model.

4.2 Metrics

Instead of accuracy measure, we use the three most prevalent metrics, the harmonic mean of precision and recall (F1 score), Area Under Receiver Operating Characteristic Curve (AUC-ROC), and Area Under Precision-Recall Curve (AUC-PR), which better highlight the number of false predictions in the imbalanced datasets. Both AUC-ROC and AUC-PR are suitable for measuring the performance without considering the classification threshold. However, when we want to emphasize the positive samples (fraud transactions), AUC-PR is a better choice. It is also often more challenging to optimize the AUC-PR metric [19].

4.3 Performance Evaluation

Scientists often use cross-validation to demonstrate how well the classifier performs on unseen data. However, to generate the temporal saliency scores, we need to look at the records in a continuous range of time. Considering the use of the SR module, the traditional rotation cross-validation is no longer appropriate. To address this issue, we adapt the time-series rolling version of cross-validation. Specifically, we start with a small continuous training set and test set. Then we aggregate the test set to the training set and select the data in the next chunk of time to be the test set in the next split. The overall performance is an average of the individual results.

4.4 Supervised Learning Baseline

We conduct the experiments on the Decision Tree [20], Random Forest [5], and XGBoost [8] group of models. Both Random Forest and XGBoost are ensemble versions of the Decision Tree but with two different approaches. Random Forest uses the bagging technique in which random subsets of the training samples are selected with replacement to train many trees. Meanwhile, XGBoost adopts boosting to learn from the error of previous trees. We do not conduct experiments on neural networks since they often fail to provide the optimal solution with tabular datasets and often require data pre-processing [2]. Furthermore, we also prefer the tree-based models considering their interpretability, which is a crucial aspect of business applications.

Within each group, we benchmark the results of the original baseline and different combinations of generated features. For the KCCF dataset with a small number of heavily anonymized transactions, we evaluate all three baseline groups to analyze the pros and cons of different approaches. Then, we select the best group to apply to the real-world scenario with a large and heterogeneous dataset.

5 Experiments and Evaluation

We conduct all the experiments in Python 3.9 with Intel(R) Core(TM) i7-9700K CPU @ 3.60 GHz, NVIDIA(R) GeForce(R) RTX 2080 Ti GPU 11 GB, 64 GB RAM, and 1 TB disk space available.

5.1 Experiments on KCCF Dataset

As we mentioned earlier, this dataset does not contain any information about the attributes except amount and time. Therefore, we simplify the pre-processing pipeline by omitting the knowledge graph model, treating the whole dataset as coming from the same entity, and applying only SR with iForest on the global granularity. We select 100 as both the sliding window size in SR and the number of base estimators in iForest. In addition, due to the high imbalance of the dataset, we apply the SMOTE oversampling technique [7] to reduce bias in the negative examples.

Table 2 reports the average training time, best F1 score (among the choices of thresholds), AUC-ROC, and AUC-PR of each model after applying the 5-fold rolling cross-validation (as discussed in Subsect. 4.3). We highlight the best performance results of each group. Notably, enhancing the feature space can slightly improve the F1 score, AUC-ROC, or AUC-PR of all tree-based models.

Specifically, by augmenting the appropriate scores, we can increase the AUC-ROCs, AUC-PRs, and F1 scores by no less than 0.1%, 0.3%, and 0.7%, respectively. Notably, in the case of a naive model like Decision Tree, the set of anomaly score and temporal traffic score boosts the AUC-PR measure by 15.8% and the F1 score by 10.2%. It is a significant enhancement since AUC-PR is difficult to increase and is more critical than other metrics when we care about positive data points like fraud detection (as mentioned earlier in Subsect. 4.2).

Table 2. Result comparison on test data in KCCF dataset. DT, RF, and XGB stand for Decision Tree, Random Forest, and XGBoost. Within each baseline group, we also report the enhancement versions using the anomaly score from iForest (iF), temporal traffic score (traffic), temporal amount score (amount), and a combination of all three scores (all).

Model	Best F1 score	AUC-ROC	AUC-PR	Training time (s)
DT	0.4270	0.8318	0.2345	14.3
DT (+iF)	0.4365	**0.8398**	0.2351	23.7
DT (+iF, traffic)	**0.4705**	0.8247	**0.2717**	24.5
DT (+traffic, amount)	0.4401	0.8212	0.2432	24.3
DT (+all)	0.4333	0.8143	0.2390	25.0
RF	0.8286	0.9460	0.7688	112.9
RF (+iF)	0.8247	0.9392	**0.7806**	122.9
RF (+iF, traffic)	**0.8346**	0.9437	0.7765	123.07
RF (+traffic, amount)	0.8299	0.9432	0.7690	121.2
RF (+all)	0.8309	**0.9485**	0.7659	126.1
XGB	0.8116	0.9633	0.7767	17.2
XGB (+iF)	**0.8173**	**0.9674**	**0.7793**	17.7
XGB (+iF, traffic)	0.8149	0.9661	0.7753	18.0
XGB (+traffic, amount)	0.8147	0.9663	0.7774	18.3
XGB (+all)	0.8136	0.9656	0.7723	20.1

Judging from only the results in KCCF, the anomaly score component has a positive contribution to the baseline, while temporal scores alone decrease the model performance. In addition, the combination of all scores also impairs the strength of anomaly score enhancement.

5.2 Experiments on ECCF Dataset

Comparing three results from different tree-based models on KCCF, we observe that Random Forest achieves the best performance in terms of AUC-PR. However, XGBoost is better suited for practical applications due to better efficiency (with lower training time) and has roughly equal AUC results with Random Forest. Therefore, for the enterprise dataset, whose size is ten times larger, we only evaluate the XGBoost baseline with their enhanced versions.

Since we have all background information on the attributes of ECCF, including the meaning of each categorical feature, we include the scores generated from the knowledge graph module and apply the merchant-level granularity to both the iForest and SR modules. We conduct experiments on the configuration of XGBoost that produces the best AUC-PR. The number of iTree estimators and SR window size is selected empirically. Table 3 summarizes the best F1 score, AUC-ROC, AUC-PR, and training time (including the time for constructing the

Table 3. Result comparison on test data in ECCF dataset. XGB stands for XGBoost. We also report the enhancement versions using the anomaly score from iForest (iF), temporal scores from SR (SR), fraud score from knowledge graph (KG), and combinations of them.

Model	Best F1 score	AUC-ROC	AUC-PR	Training time (s)
XGB	0.1117	0.7173	0.0495	1.4
XGB (+iF)	**0.1452**	0.6959	0.0426	138.8
XGB (+KG)	0.1009	**0.7496**	0.0381	627.5
XGB (+SR)	0.1192	0.7315	**0.0732**	29.2
XGB (+iF, SR)	0.1259	0.6656	0.0558	159.6
XGB (+iF, KG)	0.1193	0.7217	0.0549	769.8
XGB (+SR, KG)	0.1117	0.7349	0.0558	659.1
XGB (+iF, SR, KG)	0.1111	0.6585	0.0607	795.5

graph) of different score combinations. We highlight the best measures of those three and the models that outperform the baseline in all metrics. As a side note, the training time of the baseline is much faster due to the adoption of GPU implementation of XGBoost.

AUC-PR and F1 score is heavily affected by the imbalance of the test set. The random classifier can achieve 0.5 on AUC-ROC regardless of fraud proportion, but only up to r on AUC-PR and no more than $2r/(r+1)$ on F1 score, with r being the ratio of positive-negative samples. Therefore, the measures of AUC-PR and F1 scores are much smaller and more challenging to improve on ECCF with only 0.0028% of fraud transactions.

In general, each additional feature improves either the F1 score, AUC-ROC, or AUC-PR of the XGBoost baseline, and they outperform XGBoost by a considerable margin when combined together. Notably, the adoption of SR in the time-series anomaly detection domain and KG in the information retrieval domain can extract the subtle abnormalities, increasing the original AUC-ROC and AUC-PR by 4.5% and 47.89%, respectively. Meanwhile, those combinations include the anomaly score from Isolation Forest degrades such enhancements.

This degradation further consolidates the statement we made earlier in Sect. 2 about the modern distribution of CCF, which is getting similar to genuine behaviors and more sophisticated to detect. Our experiments suggest that feature enhancement can better cope with the concept drift in CCFD than simply retraining the same model on the new data. To the best of our knowledge, our work is one of the first applications of Knowledge Graph on CCFD. Our success motivates further research in this domain to help extract crucial behavioral features of merchants and customers.

6 Conclusion

Concentrating on the class of subtle frauds, we introduce a set of features to improve the classification capability of the original supervised baseline. Our feature construction pipeline includes the anomaly score from iForest, the temporal scores from SR, and the fraud scores from KG. We also evaluate different tree-based baseline and their enhanced versions on the well-known Kaggle public dataset to select the most appropriate baseline for the enterprise transaction dataset in Vietnam. Experiments on both datasets demonstrate the utility of our pipeline by increasing the performance of XGBoost by up to 4.5% and 47.89% in terms of AUC-ROC and AUC-PR, respectively. Our success can motivate further research on the application of Knowledge Graph and Spectral Residual to the field of credit card fraud detection, as well as improve the quality of the existing fraud detection system in Vietnam.

This research is a part of the industry project that builds the fraud detection system for the payment firm. In the future, we aim to expand the feature set, especially the KG scores, to upgrade the credit card detection pipeline. In addition, we are going to integrate the pipeline into the real-world system, expecting to help the enterprise be aware of and respond timely to the fraudulent risks.

Acknowledgement. This research is partially supported by the research funding from the Faculty of Information Technology, University of Science, Ho Chi Minh city, Vietnam. We are also grateful to the payment service provider for their support, including the large-scale ECCF dataset (due to confidentiality issues, we do not expose the name of the company). Researchers who work in the same domain and wish to access the enterprise dataset can contact the corresponding author for more information.

References

1. Agarwal, S., Driscoll, J.C., Gabaix, X., Laibson, D.: Learning in the credit card market. Tech. rep., National Bureau of Economic Research (2008). https://doi.org/10.3386/w13822
2. Arık, S.O., Pfister, T.: TabNet: attentive interpretable tabular learning. In: AAAI, vol. 35, pp. 6679–6687 (2021)
3. Baldi, P.: Autoencoders, unsupervised learning, and deep architectures. In: Guyon, I., Dror, G., Lemaire, V., Taylor, G., Silver, D. (eds.) Proceedings of ICML Workshop on Unsupervised and Transfer Learning. Proceedings of Machine Learning Research, vol. 27, pp. 37–49. PMLR, Bellevue, Washington, USA (2012)
4. Bhatla, T.P., Prabhu, V., Dua, A.: Understanding credit card frauds. Cards Bus. Rev. **1**(6), 1–15 (2003)
5. Breiman, L.: Random forests. Mach. Learn. **45**(1), 5–32 (2001). https://doi.org/10.1023/A:1010933404324
6. Carcillo, F., Le Borgne, Y.A., Caelen, O., Kessaci, Y., Oblé, F., Bontempi, G.: Combining unsupervised and supervised learning in credit card fraud detection. Inf. Sci. **557**, 317–331 (2021)
7. Chawla, N.V., Bowyer, K.W., Hall, L.O., Kegelmeyer, W.P.: Smote: synthetic minority over-sampling technique. J. Artif. Intell. Res. **16**, 321–357 (2002)

8. Chen, T., et al.: Xgboost: extreme gradient boosting. R package version 0.4-2 1(4), 1–4 (2015)
9. Dijkstra, E.W., et al.: A note on two problems in connexion with graphs. Numer. Math. **1**(1), 269–271 (1959)
10. Divakar, K., Chitharanjan, K.: Performance evaluation of credit card fraud transactions using boosting algorithms. Int. J. Electron. Commun. Comput. Eng. **10**(6), 262–270 (2019)
11. Garcia, G.: Credit cards: an interdisciplinary survey. J. Consum. Res. **6**(4), 327–337 (1980). https://doi.org/10.1086/208776
12. John, H., Naaz, S.: Credit card fraud detection using local outlier factor and isolation forest. Int. J. Comput. Sci. Eng. **7**(4), 1060–1064 (2019). https://doi.org/10.26438/ijcse/v7i4.10601064
13. Klein, D.J., Randić, M.: Resistance distance. J. Math. Chem. **12**(1), 81–95 (1993). https://doi.org/10.1007/BF01164627
14. Li, Y., Sun, Y., Contractor, N.: Graph mining assisted semi-supervised learning for fraudulent cash-out detection. In: Proceedings of the 2017 IEEE/ACM International Conference on Advances in Social Networks Analysis and Mining 2017, pp. 546–553 (2017). https://doi.org/10.1145/3110025.3110099
15. Li, Z., Huang, M., Liu, G., Jiang, C.: A hybrid method with dynamic weighted entropy for handling the problem of class imbalance with overlap in credit card fraud detection. Expert Syst. Appl. **175**, 114750 (2021). https://doi.org/10.1016/j.eswa.2021.114750
16. Liu, F.T., Ting, K.M., Zhou, Z.H.: Isolation forest. In: 2008 8th IEEE International Conference on Data Mining, pp. 413–422 (2008). https://doi.org/10.1109/ICDM.2008.17
17. Nguyen, O.D.Y., Cassidy, J.F.: Consumer intention and credit card adoption in Vietnam. Asia Pac. J. Mark. Logist. (2018). https://doi.org/10.1108/APJML-01-2017-0010
18. Niu, X., Wang, L., Yang, X.: A comparison study of credit card fraud detection: supervised versus unsupervised. arXiv preprint arXiv:1904.10604 (2019)
19. Pang, G., Shen, C., van den Hengel, A.: Deep anomaly detection with deviation networks. In: Proceedings of the 25th ACM SIGKDD International Conference on Knowledge Discovery & Data Mining, pp. 353–362 (2019). https://doi.org/10.1145/3292500.3330871
20. Quinlan, J.R.: Learning decision tree classifiers. ACM Comput. Surv. **28**(1), 71–72 (1996). https://doi.org/10.1145/234313.234346
21. Ren, H., et al.: Time-series anomaly detection service at Microsoft. In: Proceedings of the 25th ACM SIGKDD International Conference on Knowledge Discovery & Data Mining, pp. 3009–3017 (2019). https://doi.org/10.1145/3292500.3330680
22. Report, N.: The Nilson report (2021)
23. Singhal, A.: Introducing the knowledge graph: things, not strings (2020)
24. Trinh, N.H., TRan, H.H., Vuong, Q.D.H.: Perceived risk and intention to use credit cards: a case study in Vietnam. J. Asian Finance Econ. Bus. **8**(4), 949–958 (2021). https://doi.org/10.13106/JAFEB.2021.VOL8.NO4.0949
25. Wu, Y., Xu, Y., Li, J.: Feature construction for fraudulent credit card cash-out detection. Decis. Support Syst. **127**, 113155 (2019). https://doi.org/10.1016/j.dss.2019.113155
26. Zamini, M., Montazer, G.: Credit card fraud detection using autoencoder based clustering. In: 2018 9th International Symposium on Telecommunications (IST), pp. 486–491. IEEE (2018). https://doi.org/10.1109/ISTEL.2018.8661129

A Variational Algorithm for Quantum Single Layer Perceptron

Antonio Macaluso[1](✉), Filippo Orazi[2], Matthias Klusch[1], Stefano Lodi[2], and Claudio Sartori[2]

[1] German Research Center for Artificial Intelligence (DFKI), Saarbruecken, Germany
{antonio.macaluso,matthias.klusch}@dfki.de
[2] University of Bologna, Bologna, Italy
filippo.orazi@studio.unibo.it, {stefano.lodi,claudio.sartori}@unibo.it

Abstract. Hybrid quantum-classical computation represents one of the most promising approaches to deliver novel machine learning models capable of overcoming the limitations imposed by the classical computing paradigm. In this work, we propose a novel variational algorithm for quantum Single Layer Perceptron (qSLP) which allows producing a quantum state equivalent to the output of a classical single-layer neural network. In particular, the proposed qSLP generates an exponentially large number of parametrized linear combinations in superposition that can be learnt using quantum-classical optimization. As a consequence, the number of hidden neurons scales exponentially with the number of qubits and, thanks to the universal approximation theorem, our algorithm opens to the possibility of approximating any function on quantum computers. Thus, the proposed approach produces a model with substantial descriptive power and widens the horizon of potential applications using near-term quantum computation, especially those related to quantum machine learning. Finally, we test the qSLP as a classification model against two different quantum models on two different real-world datasets usually adopted for benchmarking classical algorithms.

Keywords: Quantum machine learning · Quantum computing · Machine learning · Neural networks

1 Background and Motivation

Machine learning (ML) can be considered as one of the most disruptive technology of the last decades. However, the ever-increasing size of datasets and Moore's law coming to an end emphasizes that the current computational tools will no longer be sufficient in the near future. An opportunity for ML to overcome these limitations is to leverage quantum computation, where quantum effects allow solving selected problems that are intractable using classical machines. The intersection between ML and quantum computing is known as Quantum

A. Macaluso and F. Orazi—Both authors equally contributed to this research.

G. Nicosia et al. (Eds.): LOD 2022, LNCS 13811, pp. 341–356, 2023.
https://doi.org/10.1007/978-3-031-25891-6_26

Machine Learning (QML) where a quantum algorithm is employed to solve typical ML tasks. In particular, QML algorithms are designed to tackle optimization problems using both classical and quantum resources and they are composed by three main ingredients: *i*) a parametrized quantum circuit $U(x;\Theta)$, *ii*) a quantum output $f(x;\Theta)$ and *iii*) an updating rule for Θ.

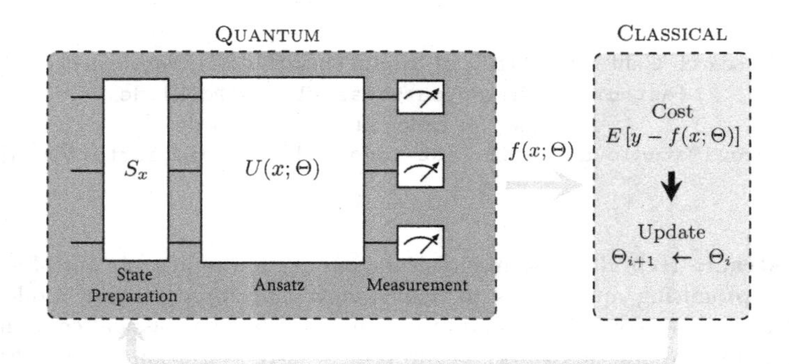

Fig. 1. Scheme of a hybrid quantum-classical algorithm for supervised learning. The quantum variational circuit is depicted in green, while the classical component is represented in blue. (Color figure online)

The hybrid quantum-classical approach (Fig. 1) proceeds as follows. The data, x, are initially encoded as a quantum state through the use of a quantum routine S_x. Then the quantum hardware computes $U(x;\Theta)$ with randomly initialized parameters Θ. After multiple executions of $U(x;\Theta)$, the classical component post-processes the measurements and generates a prediction $f(x;\Theta)$. Finally, the parameters are updated, and the whole cycle is run multiple times in a closed loop between the classical and quantum hardware.

1.1 Neural Network as Universal Approximator

A Single Layer Perceptron (SLP) [1] is a single-layer neural network suitable for classification and regression problems. Given a training point $(\mathbf{x}, y)$, the output of a SLP containing H neurons can be expressed as:

$$f(\mathbf{x};\beta,\theta) = \sigma_{\text{out}}\left(\sum_{j=1}^{H} \beta_j \sigma_{\text{hid}}\left[L(\mathbf{x};\boldsymbol{\theta}_j)\right]\right) = \sum_{j=1}^{H} \beta_j g(\mathbf{x};\theta_j). \tag{1}$$

where σ_{out} is the identify function in case of a continuous target variable y. Each hidden neuron j computes a linear combination, $L(\cdot)$, of the input features $\mathbf{x} \in \mathbb{R}^p$ with coefficients given by the p-dimensional vector $\boldsymbol{\theta}_j$. This operation is performed for all neurons, and the results are individually fed into the inner activation function σ_{hid}. The outputs of the previous operation are then linearly combined with coefficients β_j. Finally, a task-dependent outer activation function, σ_{out}, is applied.

Despite being less utilized than the deep architectures, the SLP model can be very expressive. According to the *universal approximation theorem* [2], a SLP with an arbitrarily large number of hidden neurons and a non-constant, bounded and continuous activation function can approximate any continuous function on a closed and bounded subset of $\mathbb{R}^p$. In spite of this crucial theoretical result, the SLP are rarely adopted in practice due to the unfeasibility of training large amounts of hidden neurons on classical devices. Quantum computers, however, could leverage the superposition of states to scale the number of hidden neurons exponentially with the number of available qubits. Starting from these considerations, cleverly implementing a quantum SLP would therefore enable a real chance to benefit from the universal approximation property.

1.2 Activation Function

The implementation of a proper activation function – in the sense of the Universal Approximation Theorem – is one of the major issues for building a complete quantum neural network. Recently, the idea of quantum splines (QSplines) [3] has been proposed to approximate non-linear activation functions by means of a fault-tolerant quantum algorithm. Although a QSpline provides a method to store the value of a non-linear function in the amplitudes of a quantum state, it uses the HHL [4] as subroutine, a quantum algorithm for matrix inversion which imposes several practical limitations [5].

In this work, we do not discuss how to implement in practice a non-linear activation function. Nevertheless, we provide a framework that permits to train a quantum SLP with an exponentially large number of hidden neurons for a given activation function Σ. Furthermore, our architecture is naturally capable of incorporating new implementations of non-linear activation functions as long as they fit in the learning paradigm.

2 Related Works

Variational circuits can be considered as the composition of multiple layers of connected computational units controlled by trainable parameters. This definition is similar to the characterization of neural networks, but the comparison between the two poses several challenges. In fact, classical neural networks require a non-linear activation function which is a limit for quantum computation since quantum operations are unitary and therefore linear. Furthermore, it is impossible to access the quantum state at intermediate points during computation. Thus, backpropagation [6], which is the standard to work on large-scale models, cannot be used to train quantum circuits.

In practice, several attempts for building a quantum neural network are discussed in the literature. Although a potential quantum advantage of quantum neural networks over classical ones has been recently investigated [7], to the best of our knowledge, there are no trainable quantum algorithms that can effectively implement a complete neural network in a quantum setting. A concrete implementation of a quantum circuit to solve a typical ML task using a near-term processor is illustrated in [8], where the authors introduced a model for binary classification using a modified version of the perceptron updating rule. Another approach is represented by Tensor Networks inspired by quantum many-body physics. Tensor networks [9] are methods to represent quantum states whose entanglement is constrained by local interactions. This approach enables the numerical treatment of systems through layers of abstraction, as for deep neural networks. Some of the most studied tensor networks have been adopted for classification and generative modeling [10,11]. Recently, the idea of building a parametrized quantum circuit to generate a quantum state equivalent to the output of a two-layer neural network has been proposed [12]. However, the generalization to more than two neurons is not provided.

In general, the standard approach for quantum neural networks (QNN) consists of building a quantum circuit on top of existing approaches for variational circuits which solve a supervised learning task [13]. Methods within this approach take care of the supervised task end-to-end, where the input quantum state is fed through a parametrized unitary operator and the final measurement provides the estimate of the target variable of interest.

Alternatively, quantum support vector machines (QSVM) [14] use a quantum circuit to map classical data into a quantum feature space. The resulting features are then fed into a classical linear model which calculates the final output.

2.1 Contribution

In this work, we propose the quantum Single Layer Perceptron (qSLP), a novel quantum algorithm that reproduces a quantum state equivalent to the output of a classical SLP with an arbitrarily large number of hidden neurons. In particular, building on top of the approach described in [12], we design an efficient variational algorithm that performs unitary parametrized transformations of the input in superposition. The results are then passed through an activation function with just one application. The flexible architecture of the qSLP enables to plug in custom implementations of the activation function routine. Thus, thanks to the possibility of learning the parameters for a given task, the proposed algorithm allows training models that can potentially approximate any function if provided of a proper activation function.

However, we do not address the problem of implementing a non-linear activation function. Our goal is to provide a framework that generates multiple linear combinations in superposition entangled with a control register. As a consequence, instead of executing a given activation function for each hidden neuron, a single application is needed to propagate it to all of the quantum states. This allows scaling the number of hidden neurons exponentially with the number of qubits, while increasing linearly the depth of the correspondent quantum circuit and candidating the qSLP to be a concrete alternative for approximating complex and diverse functions.

Finally, we test the qSLP as classification model against two different quantum approaches on two different real-world datasets usually adopted for benchmarking classical ML algorithms.

3 Generalized Quantum Single Layer Perceptron

3.1 Preliminaries

Gates as Linear Operators. As discussed in Sect. 1.1, an SLP applies multiple linear combinations on the same input based on different sets of parameters. In case of qSLP, we can consider the following parametrized quantum gate to map a generic input quantum state any possible other quantum states:

$$U_3(\boldsymbol{\theta}) = U_3(\alpha, \beta, \gamma) = \begin{pmatrix} e^{i\beta} cos(\alpha/2) & e^{i\gamma} sin(\alpha/2) \\ -e^{-i\gamma} sin(\alpha/2) & e^{-i\beta} cos(\alpha/2) \end{pmatrix}. \tag{2}$$

The unitary $U_3(\boldsymbol{\theta})$ represents a bounded linear operation where the set of parameters vector $\boldsymbol{\theta} = \{\alpha, \beta, \gamma\}$ can be learnt using classical optimization procedures. Importantly, the adoption of $U_3(\boldsymbol{\theta})$ as linear operator implies transforming the input data using complex coefficients. Therefore, it describes a more general operation with respect to the classical SLP, that only allows for linear combinations with real-valued coefficient. Nonetheless, one can still parametrize the circuit using Pauli-Y rotation to restrict the computation to the real domain.

Ansatz for Linear Operators in Superposition. The original proposal of the qSLP [12] creates entanglement between the *control* and the *data* register using two CSWAP operations and an additional *temp* register to generate a superposition of two different linear transformations. Although this approach allows obtaining the desired result, it implies the adoption of twice the number of qubits to encode the input data (*temp* register). Furthermore, using the CSWAP gates introduces a linear overhead with respect to the size of the two registers *data* and *temp* [15], in terms of gate complexity. To reduce such complexity, we propose a different approach (illustrated in Fig. 2) which does not require the additional *temp* register and the use of the CSWAP gates.

3.2 Quantum Single Layer Perceptron

Intuitively, a generalized qSLP can be implemented into five steps: *state preparation, entangled linear operators in superposition, application of the activation function, measurement step, post processing optimization.* To this end, two quantum registers are necessary: *control* (d qubits), *data* (n qubits).

Step 1: State Preparation. The *control* register is turned into a non-uniform superposition parameterized by the d-dimensional vector $\hat{\boldsymbol{\beta}}$ by means of d

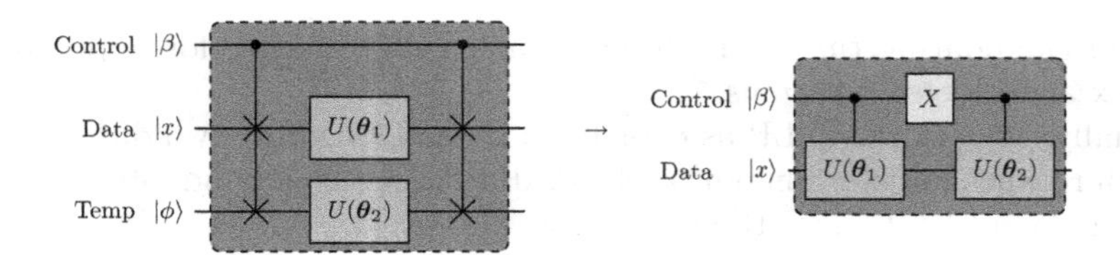

Fig. 2. Alternative ansatz for linear operators in superposition. On the left the quantum gates adopted in the first version of the two-layer qSLP [12]. On the right the proposed schema to generate two different linear operation in superposition each entangled with one of the basis states of the *control* quibit.

Pauli-Y rotation gates. Also, a single p-dimensional training point is encoded into the n-qubit *data* register through the quantum gate S_x:

$$
\begin{aligned}
|\Phi\rangle_{\text{SP}} &= \left(\bigotimes_{i=1}^{d} R_y(\hat{\beta}_i) \otimes S_x\right) |0\rangle_{\text{control}}^{\otimes d} \otimes |0\rangle_{\text{data}}^{\otimes n} \\
&= \bigotimes_{i=1}^{d} (a_i |0\rangle + b_i |1\rangle) \otimes |x\rangle = \bigotimes_{i=1}^{d} |c_i\rangle \otimes |x\rangle, \qquad (3)
\end{aligned}
$$

where $\{a_i, b_i\}$ are real numbers such that $|a_i|^2 + |b_i|^2 = 1$. Importantly, the algorithm is completely independent by the encoding strategy chosen for S_x.

Step 2: Entangled Linear Operators in Superposition. The second step generates a superposition of 2^d linear operations with different parameters, each entangled with a basis state of the *control* register. In particular, once the two registers are initialized, each qubit in the *control* register is entangled with two different random transformations of the *data* register. As a result, 2^d different transformations in superposition of the input are generated. Here we describe the procedure assuming 3 *control* qubits where the ansatz described in Sect. 3.1 is applied only $d = 3$ times in order to obtain $2^d = 8$ different parametrized unitary transformations of the input in superposition.

Step 2.1 ($d = 1$) The first step applies the unitary $U(\boldsymbol{\theta}_{1,1}, \boldsymbol{\theta}_{1,2})$ to the initialized quantum system, which is defined as follows:

$$
U(\boldsymbol{\theta}_{1,1}, \boldsymbol{\theta}_{1,2}) = \left[\mathbb{1}^{\otimes d-1} \otimes C\text{-}U(\boldsymbol{\theta}_{1,1})\right] \left[\mathbb{1}^{\otimes d-1} \otimes X \otimes \mathbb{1}\right] \left[\mathbb{1}^{\otimes d-1} \otimes C\text{-}U(\boldsymbol{\theta}_{1,2})\right]
$$

where $\mathbb{1}$ is the identity matrix and X is the NOT gate (i.e., Pauli-X gate). Thus, the first step ($d = 1$) leads to the following quantum state:

$$
\begin{aligned}
|\Phi\rangle_{(d=1)} &= U(\boldsymbol{\theta}_{1,1}, \boldsymbol{\theta}_{1,2}) |\Phi\rangle_{\text{SP}} \\
&= \bigotimes_{i=1}^{2} |c_i\rangle \otimes (a_1 |1\rangle U(\boldsymbol{\theta}_{1,2}) |x\rangle + b_1 |0\rangle U(\boldsymbol{\theta}_{1,1}) |x\rangle). \qquad (4)
\end{aligned}
$$

The basis states of the first *control* qubit is then entangled with two parametrized unitary transformations of the *data* register.

Step 2.2 ($d = 2$) The second step employs another *control* qubit and other two sets of parameters $\boldsymbol{\theta}_{2,1}$ and $\boldsymbol{\theta}_{2,2}$. Then the same procedure is applied through the unitary operations $U(\boldsymbol{\theta}_{2,1}, \boldsymbol{\theta}_{2,2})$:

$$
\begin{aligned}
|\Phi\rangle_{(d=2)} &= U(\boldsymbol{\theta}_{2,1}, \boldsymbol{\theta}_{2,2})\, |\Phi\rangle_{(d=1)} \\
&= \frac{1}{\sqrt{4}} \Big[b_2 a_1 \,|00\rangle\, U(\boldsymbol{\theta}_{2,1}) U(\boldsymbol{\theta}_{1,1})\, |x\rangle + b_2 b_1 \,|01\rangle\, U(\boldsymbol{\theta}_{2,1}) U(\boldsymbol{\theta}_{1,2})\, |x\rangle \\
&\quad + a_2 a_1 \,|10\rangle\, U(\boldsymbol{\theta}_{2,2}) U(\boldsymbol{\theta}_{1,1})\, |x\rangle + a_2 b_1 \,|11\rangle\, U(\boldsymbol{\theta}_{2,2}) U(\boldsymbol{\theta}_{1,2})\, |x\rangle \Big]
\end{aligned}
\tag{5}
$$

with $U(\boldsymbol{\theta}_{2,1}, \boldsymbol{\theta}_{2,2}) = [\mathbb{1} \otimes C \otimes \mathbb{1} \otimes U(\boldsymbol{\theta}_{2,1})]\big[\mathbb{1} \otimes X \otimes \mathbb{1}^{\otimes n+1}\big][\mathbb{1} \otimes C \otimes \mathbb{1} \otimes U(\boldsymbol{\theta}_{2,1})]$. The position of the C gate indicates the *control* qubit used to perform the controlled operations $C\text{-}U(\boldsymbol{\theta}_{2,1})$ and $C\text{-}U(\boldsymbol{\theta}_{2,2})$. As a consequence of this second step, four parametrized transformations of the input $|x\rangle$ are generated, each results from the product of two $U(\boldsymbol{\theta}_{i,k})$ gates, for $i, k \in \{1, 2\}$. Furthermore, these transformations are entangled with the basis states of the two *control* qubits.

Step 2.3 ($d = 3$) We perform the same procedure for the third *control* qubit, using the unitary $U(\boldsymbol{\theta}_{3,1}, \boldsymbol{\theta}_{3,2})$:

$$
\begin{aligned}
|\Phi\rangle_{(d=3)} &= U(\boldsymbol{\theta}_{3,1}, \boldsymbol{\theta}_{3,2})\, |\Phi\rangle_{(d=2)} \\
&= \frac{1}{\sqrt{8}} \Big[\beta_1 \,|000\rangle\, U(\boldsymbol{\theta}_{3,1}) U(\boldsymbol{\theta}_{2,1}) U(\boldsymbol{\theta}_{1,1})\, |x\rangle + \beta_2 \,|001\rangle\, U(\boldsymbol{\theta}_{3,1}) U(\boldsymbol{\theta}_{2,1}) U(\boldsymbol{\theta}_{1,2})\, |x\rangle \\
&\quad + \beta_3 \,|010\rangle\, U(\boldsymbol{\theta}_{3,1}) U(\boldsymbol{\theta}_{2,2}) U(\boldsymbol{\theta}_{1,1})\, |x\rangle + \beta_4 \,|011\rangle\, U(\boldsymbol{\theta}_{3,1}) U(\boldsymbol{\theta}_{2,2}) U(\boldsymbol{\theta}_{1,2})\, |x\rangle \\
&\quad + \beta_5 \,|100\rangle\, U(\boldsymbol{\theta}_{3,2}) U(\boldsymbol{\theta}_{2,1}) U(\boldsymbol{\theta}_{1,1})\, |x\rangle + \beta_6 \,|101\rangle\, U(\boldsymbol{\theta}_{3,2}) U(\boldsymbol{\theta}_{2,1}) U(\boldsymbol{\theta}_{1,2})\, |x\rangle \\
&\quad + \beta_7 \,|110\rangle\, U(\boldsymbol{\theta}_{3,2}) U(\boldsymbol{\theta}_{2,2}) U(\boldsymbol{\theta}_{1,1})\, |x\rangle + \beta_8 \,|111\rangle\, U(\boldsymbol{\theta}_{3,2}) U(\boldsymbol{\theta}_{2,2}) U(\boldsymbol{\theta}_{1,2})\, |x\rangle \Big] \\
&= \frac{1}{\sqrt{8}} \sum_{j=1}^{8} \beta_j \,|j\rangle\, U(\boldsymbol{\Theta}_j)\, |x\rangle ,
\end{aligned}
\tag{6}
$$

where $U(\boldsymbol{\theta}_{3,1}, \boldsymbol{\theta}_{3,2}) = \big[C \otimes \mathbb{1}^{\otimes 2} \otimes U(\boldsymbol{\theta}_{2,1})\big]\, [X \otimes \mathbb{1} \otimes \mathbb{1}]\, \big[C \otimes \mathbb{1}^{\otimes 2} \otimes U(\boldsymbol{\theta}_{2,1})\big]$.

The final quantum state is a superposition of 8 different parametrized unitary transformations of the input, each entangled with a basis state of the *control* register whose amplitudes depend on a set of parameters $\{\beta_i\}_{i=1,\dots,d}$. We can generalize the quantum state $|\Phi\rangle_{(d=3)}$ for a generic d-qubit *control* register:

$$
\begin{aligned}
|\Phi\rangle_d &= \prod_{i=1}^{d} U(\boldsymbol{\theta}_{i,1}, \boldsymbol{\theta}_{i,2}) \left(\bigotimes_{i=1}^{d} |c_i\rangle \otimes |x\rangle \right) \\
&= \frac{1}{\sqrt{2^d}} \sum_{j=1}^{2^d} \beta_j \,|j\rangle\, U(\boldsymbol{\Theta}_j)\, |x\rangle = \frac{1}{\sqrt{2^d}} \sum_{j=1}^{2^d} \beta_j \,|j\rangle\, |L(x; \boldsymbol{\Theta}_j)\rangle ,
\end{aligned}
\tag{7}
$$

where $U(\boldsymbol{\Theta}_j)$ is the product of d unitaries $U(\boldsymbol{\theta}_{i,k})$ for $i = 1, \dots, d$ and $k = 1, 2$.

To summarize, the underlying idea of this procedure is to initialize the *control* register according to a set of weights and assign each weight β_j to a parametrized unitary function $U(\boldsymbol{\Theta}_j)$. This approach is extremely flexible and allows learning all the parameters β_j and $\boldsymbol{\Theta}_j$ for specific use cases. Furthermore, it allows to generate a superposition of 2^d diverse unitary transformation while increasing linearly the depth correspondent quantum circuit.

Step 3: Activation. A further step consists of applying the Σ gate, representing the quantum version of the classical activation function to the *data* register. Notice that, having all the parametrized unitary operations in superposition allows propagating the application of Σ with a single execution, as follows:

$$\begin{aligned} |\Phi\rangle_{\Sigma} &= (\mathbb{1}^{\otimes d} \otimes \Sigma)\,|\Phi_d\rangle \rightarrow \frac{1}{\sqrt{2^d}} \sum_{j=1}^{2^d} \beta_j\,|j\rangle\,|\sigma[L(x;\theta_j)]\rangle\rangle \\ &= \frac{1}{\sqrt{H}} \sum_{j=1}^{H} \beta_j\,|j\rangle\,|g(x;\boldsymbol{\Theta}_j)\rangle\,, \end{aligned} \tag{8}$$

where $H = 2^d$ and $\mathbb{1}^{\otimes d}$ is the identity matrix. In this way, the result of the algorithm corresponds to the output of the SLP (Eq. (1)) with 2^d hidden neurons that can be accessed by measuring the *data* register only.

Step 4: Measurement. Finally, the expectation measurement on the *data* register is performed:

$$\begin{aligned} \langle M\rangle &= \langle \Phi_\Sigma \,|\, \mathbb{1}^{\otimes d} \otimes M \,|\Phi_\Sigma\rangle \\ &= \sum_{j=1}^{2^d} \beta_j^{'}\,\langle g(x;\boldsymbol{\Theta}_j)\,|\,M|g(x;\boldsymbol{\Theta}_j)\rangle \\ &= \sum_{j=1}^{2^d} \beta_j^{'}\,\langle M_j\rangle = \sum_{j=1}^{2^d} \beta_j^{'} g_j = f(\mathbf{x};\boldsymbol{\beta},\boldsymbol{\Theta}), \end{aligned} \tag{9}$$

where M is a generic measurement operator (e.g., the Pauli σ_z), the function $g(x;\boldsymbol{\Theta}_j) = \sigma[L(x;\theta_j)]$ and $\beta_j^{'} = |\beta_j|^2$ with $\sum_j |\beta_j|^2 = 1$. Although we do not measure the *control* register, the j-th transformation of the input is associated to a specific amplitude β_j of the *control* register. The parameters of the quantum circuit $\{\hat{\beta}_i\}_{i=1,\dots,d}$ and $\{\boldsymbol{\theta}_{i,1},\boldsymbol{\theta}_{i,2}\}_{i=1,\dots,d}$, that indirectly determine the parameters in Eq. (9), can be randomly initialized and hybrid quantum-classical optimization process can be exploited (Sect. 1).

Thus, we extended the proposed approach of the qSLP [12] to an exponentially large number of neurons in the hidden layer. In fact, the entanglement of linear combinations to the basis states of the *control* register implies that the number of different linear combinations is equal to the number of basis states of the *control* register. This translates in a number of hidden neurons H which scales exponentially with the number of qubits of the *control* register, d. This is a consequence of each hidden neuron being represented by a single independent trajectory (Eq. (8)). The exponential scaling alongside the ability to freely learn the parameters, enables the construction of quantum neural network with an arbitrary large number of hidden neurons as the amount of available qubits increases. In other terms, we can build qSLP with an incredible descriptive power that may be really capable of being an universal approximator. The quantum circuit to implement the qSLP ($d = 3$) is depicted in Fig. 3.

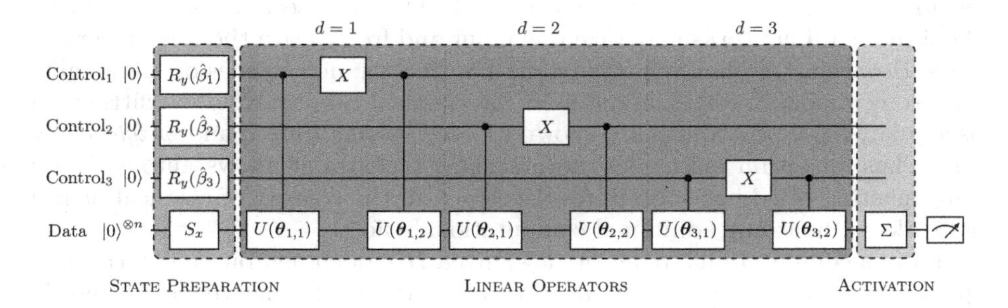

Fig. 3. Quantum circuit for a qSLP with 8 ($d = 3$) hidden neurons.

Step 5: Optimization. The post-processing is task-dependent and performed classically. In particular, given the measurement of the quantum circuit $f(\mathbf{x}; \boldsymbol{\beta}, \boldsymbol{\Theta})$, a cost function is computed and then classical approaches to update the parameters of the quantum circuit $\{\boldsymbol{\beta}, \boldsymbol{\Theta}\}$ are employed. As loss function, in case of supervised problem, the *Sum of Squared Errors* (SSE) between the predictions and the true values y is usually adopted:

$$SSE = Loss(\beta, \boldsymbol{\Theta}; D) = \sum_{i=1}^{N} \left[y_i - f\left(\mathbf{x}_i; \boldsymbol{\beta}, \boldsymbol{\Theta}\right)\right]^2, \tag{10}$$

where N is the total number of observations in the training set.

To summarize, the variational algorithm described above allows reproducing a classical neural network with one hidden layer on a quantum computer. In particular, it includes a variational circuit adopted for encoding the data, performing the linear combinations of input neurons and applying the same activation function to their results with just one execution. A single iteration during the learning process is then completed to measure the output of the network, compute the loss function and update the parameters. The whole process is repeated iteratively until convergence, as for classical neural networks.

3.3 Discussion

The algorithm provided in the previous section allows building a generalized qSLP with an exponentially large number of neurons, increasing the depth of the quantum circuit linearly. Furthermore, the model supports amplitude encoding strategy, which translates into a exponential advantage in terms of space complexity (i.e., number of qubits) when encoding data into the *data* quantum register. This implies a potential polylogarithmic advantage in terms of the number of parameters with respect to its classical counterpart [16].

There are two main differences between the classical and quantum SLP, deriving from the normalization constraint introduced when dealing with the amplitudes of quantum systems. In the qSLP, the input data $\mathbf{x}$ and the weight vector $\boldsymbol{\beta} = \{\beta_j\}_{j=1,\dots,2^d}$ are normalized to 1. This may seem a limitation since classical neural networks may take raw input and freely learn the weight parameters. However, rescaling the inputs and limiting the magnitudes of the weights are two common strategies adopted in the classical case to avoid overfitting. In particular, these procedures are known as *batch normalization* and *weight decay* [17]. Thus, quantum mechanics' normalization constraint allows automatically implementing of ad-hoc procedures developed in the context of classical neural networks without any additional computational effort.

From a computational point of view, given H hidden neurons and L training epochs, the training of a classical SLP scales (at least) linearly in H and L, since the output of each hidden neuron needs to be calculated explicitly to obtain the final output. Also, if H is too large (a necessary condition for an SLP to be a universal approximator [2]), the problem becomes NP-hard [18]. The proposed qSLP, instead, allows scaling linearly with respect to $log_2(H) = d$ thanks to the entanglement between the two quantum registers that generates an exponentially large number of quantum trajectories in superposition. However, the cost of state preparation (gate S_x) needs to be taken into consideration, as well as the cost implementing the quantum activation function Σ. Nonetheless, once the optimal set of parameters of the qSLP is obtained for a specific task, the whole quantum algorithm can be employed as subroutine in other quantum algorithms to reproduce the function for which it is trained.

4 Experiments

To test the performances of the qSLP, we implemented the circuit illustrated in Fig. 3 using IBM Qiskit environment on two different real-world datasets. After training the qSLP on the simulator, we executed the pre-trained algorithm on two real quantum devices. In addition, QNNs [7] and QSVMs [19] are adopted as benchmark to compare the proposed qSLP with state-of-the-art QML models[1].

4.1 Dataset Description

The simulation of a quantum system on a classical device is a challenging task even for systems of moderate size. For this reason, the experiments consider only

[1] All code to generate the data, figures and analyses is available at github.com/filorazi/qSLP-quantum-Single-Layer-Perceptron.

datasets with a relatively small number of observations (100–200) that will be split in training (80%) and test (20%) set. Furthermore, the PCA is performed to reduce the number of features to 2. We consider five different binary classification problems from two real-world datasets (MNIST and Iris) usually adopted for benchmarking classical machine learning algorithms. The Iris dataset consists of 50 examples for three species of Iris flower (*Setosa, Virginica, Versicolor*) described by four different features. The MNIST dataset contains $60k$ images of ten possible handwritten digits, each represented by 28×28 pixels. Since these two datasets exhibit a multiclass target variable, but the current implementations of qSLP, QNN, and QSVM solve a binary classification problem, we consider five different datasets extracted from the original multiclasses problems (Table 1).

Table 1. Datasets description. The columns represent five different binary classification datasets. Each row describes a characteristic of the data: the number of training points (Training set), the number of test points (Test set), the number of features in the original data (Raw features), the number of PCA features, the explained variance of the PCA (% Variance) and the target classes chosen as the binary target variable (Target classes).

	MNIST09	**MNIST38**	**SeVe**	**SeVi**	**VeVi**
Training set	160	160	80	80	80
Test set	40	40	20	20	20
Raw features	784	784	4	4	4
PCA features	2	2	2	2	2
% Variance	31%	20.3%	98%	98.4%	92.2%
Target classes	0 vs 9	3 vs 8	Setosa vs Versicolor	Setosa vs Virginica	Versicolor vs Virginica

4.2 State Preparation for the qSLP

The most efficient encoding strategy adopted in QML is amplitude encoding that associates quantum amplitudes with real vectors at the cost of introducing a normalization constraint to the raw input. Formally, a normalized vector $\mathbf{x} \in \mathbb{R}^p$ can be described by the amplitudes of a quantum state $|x\rangle$ as:

$$|x\rangle = \sum_{k=1}^{p} x_k |k\rangle \longleftrightarrow \mathbf{x} = [x_1, \ldots, x_p]^T . \tag{11}$$

Two different approaches are used for amplitude encoding in the qSLP: *Padded state preparation* and *Single-qubit state preparation.*

Single-qubit State Preparation. This approach encodes a two-dimensional real vector $\mathbf{x}$ into the amplitudes of a qubit by performing two parametric rotations $R_y(x_1), R_z(x_2)$, where the angles $\{x_1, x_2\}$ are the two components of the (normalized) feature vector $\mathbf{x} = (x_1, x_2)$ [20]. In this case, the parametrized quantum gate $U(\boldsymbol{\theta}_{i,k})$ of the qSLP is directly implemented using the gate U_3 (Eq. (2)).

Padded State Preparation. A second strategy for amplitude encoding consists of mapping an input state $|x\rangle$ to the all-zero state $|0\ldots0\rangle$. Once the circuit is obtained, all of the operations are inverted and applied in the reversed order. In this case, a two-dimensional input data $\mathbf{x}$ is first mapped into a four-dimensional vector adding constant values and then normalized. Thus, following the procedure described in [21], a set of angles are defined in such a way to apply a sequence of R_y and CNOT gates to generate a quantum state whose amplitudes are equivalent to the input (four-dimensional) feature vector. The quantum circuit for *padded state preparation* is depicted in Fig. 4. When using this strategy, since the *data* register is made up of 2 qubits, the unitary operator $U(\boldsymbol{\theta}_{i,k})$ of the qSLP is implemented by two U_3 gates (one per qubit) and a CNOT gate.

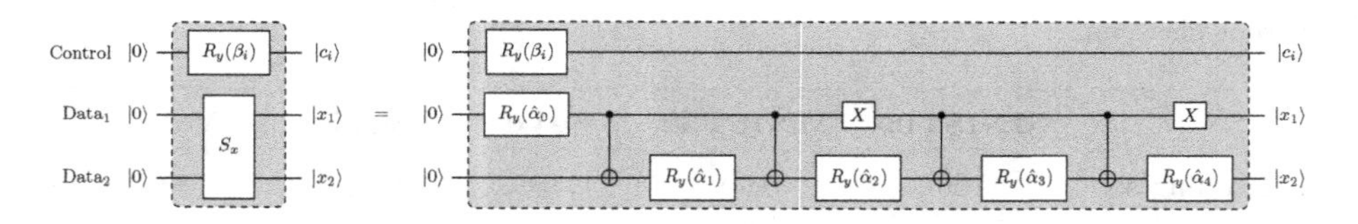

Fig. 4. Quantum circuit for *padded state preparation* in the qSLP. The $\{\hat{\alpha}_i\}_{i=0,\ldots,4}$ parameters are determined following the procedure described in [21].

4.3 Results

We consider nine different algorithms based on the three QML models. A QSVM [19] is trained using a two-layer quantum feature map (*ZZFeature-map*). Two types of QNNs are tested: the first one (QNNC v1) uses single-layer quantum feature map (*ZFeature-map*) as state preparation routine and *Real amplitudes* approach as classification ansatz; the second type (QNNC v2) differs from the first for the use of *ZZFeature-map* as state preparation. Regarding the qSLP, six different configurations are tested based on two possible implementations for amplitude encoding (Sect. 4.2) and three values for the parameter *d*. To the best of our knowledge, there is no quantum routine suitable for near-term quantum computation capable of approximating the behaviour of an activation function. For this reason, as Σ gate (Eq. (8)), we adopt the identity gate[2].

The training is performed on a QASM simulator, a backend that simulates the execution of a quantum algorithm in a fault-tolerant quantum device. The

[2] Importantly, the code already allows the embedding of a gate Σ different from the identity gate, as quantum activation function.

number of measurements for each run of the quantum circuits is fixed to 1024. The results for each model are reported in Table 2.

Table 2. Training and test results of three different models (qSLP, QNN, QSVM) on five different binary classification problems.

		MNIST09		MNIST38		SeVe		SeVi		VeVi	
Model	d	Train	Test	Train	Test	Train	Test	Train	Test	Train	Test
QSVM		.97	**.91**	.83	**.84**	1.0	**1.0**	1.0	**1.0**	.98	**.95**
QNN (v1)		.87	.86	.55	.61	.95	.90	.98	.95	.92	.90
QNN (v2)		.84	.84	.74	.70	1.0	.95	1.0	**1.0**	.85	.80
(padded) qSLP	1	.92	.89	.89	.81	1.0	**1.0**	1.0	**1.0**	.82	.70
	2	.93	**.91**	.87	.82	1.0	**1.0**	1.0	**1.0**	.82	.70
	3	.92	**.91**	.86	**.84**	1.0	**1.0**	1.0	**1.0**	.82	.70
(Single-qubit) qSLP	1	.83	.85	.86	**.84**	1.0	**1.0**	1.0	**1.0**	.80	.70
	2	.83	.85	.87	**.84**	1.0	**1.0**	1.0	**1.0**	.78	.65
	3	.83	.85	.87	.82	1.0	**1.0**	1.0	**1.0**	.80	.65

Assuming a perfect quantum device, the QSVM and the QNN outperform the qSLP considering the VeVi dataset. This means that for specific cases the use of a quantum feature map provides a practical advantage with respect to standard amplitude encoding. However, in case of MNIST09, MNIST38 and SeVe the qSLP outperforms the QNNs, thanks to its ability to aggregate multiple and diverse unitary functions. Thus, the flexible architecture of the ansatz of the qSLP allows achieving a better performance when comparing full quantum models[3]. Nonetheless, the qSLP and QSVM equally perform on four cases (MNIST09, MNIST38, SeVe, SeVi), although the implementation of the qSLP is not complete due to the lack of a proper activation function and a limited number of hidden neurons ($H = 2^3 = 8$).

Furthermore, the trained algorithms have been executed on real IBM quantum devices. Results are reported in Table 3. The deterioration in the performance of the *(Single-qubit) qSLP* is negligible. While the results of the *(Padded) qSLP*, which uses a higher number of qubits and deeper quantum circuits, deteriorate as d increases. Instead, the QSVM seems not to be affected by the use of a real device since the quantum component only generates the new features, while the classification is performed classically. However, being each feature represented by a single qubit (as for QNN), the use of the quantum feature map on real-world datasets seems to be prohibitive even when considering quantum devices with hundreds of (noisy) qubits. Instead, the amplitude encoding strategy adopted in the qSLP represents an optimal strategy to efficiently encode a

[3] The QSVM uses a quantum circuit to translate the classical data into quantum states while the classification is performed classically.

Table 3. Test performances using real devices (*ibmq_lima* and *ibmq_quito*).

Model	d	MNIST09	MNIST38	SeVe	SeVi	ViVe
QSVM	-	**.91**	**.84**	**1.0**	**1.0**	**.95**
QNN (v1)	-	.89	.64	.95	**1.0**	**.95**
QNN (v2)	-	.89	.68	**1.0**	**1.0**	.75
(padded) qSLP	1	.89	.81	**1.0**	.90	.70
	2	.81	.65	**1.0**	.60	.50
	3	.81	.65	**1.0**	.60	.50
(Single-qubit) qSLP	1	.82	.78	**1.0**	**1.0**	.50
	2	.82	.70	**1.0**	**1.0**	.55
	3	.84	.78	**1.0**	.95	.45

large dimensional input data in a small number of qubits. Thus, the *(Single-qubit) qSLP* is the best compromise when using a real device, since it seems to preserve its performance while adopting the amplitude encoding strategy.

Importantly, these results are an indication of how the tested QML models perform and do not represent an exhaustive evaluation of their performance.

5 Conclusion and Outlook

In this work, we proposed a quantum algorithm to generate the quantum version of the Single Layer Perceptron (qSLP). The key idea is to use a single state preparation routine and apply different linear combinations in superposition, each entangled with a control register. This allows propagating a generic activation function to all the basis states with only one execution. As a result, a model trained through our algorithm is potentially able to approximate any desired function as long as enough hidden neurons and a non-linear activation function are available. Furthermore, we provided a practical implementation of our variational algorithm that reproduces a quantum SLP for classification with different possible hidden neurons and an identity function as activation.

In addition, we performed a comparative analysis between our algorithm and two quantum baselines on real-world data and demonstrated that the qSLP outperforms other full quantum approaches (QNN) and matches with the QSVM.

The main challenges to tackle in the near future is the design of a routine that reproduces a non-linear activation function, as well as the adoption of the qSLP for regression tasks. Yet, recently it has been shown that quantum feature maps alongside functions aggregation are able to achieve universal approximation [22]. Thus, another promising future work is the study of the qSLP on top of the quantum feature map, to obtain a universal approximator, without implementing a non-linear quantum activation function.

In conclusion, we are still far from proving that machine learning can benefit from quantum computation in practice. However, thanks to the flexibility of variational algorithms, the hybrid quantum-classical approach may be the ideal setting to make universal approximation possible in quantum computers.

Acknowledgments. This work has been partially funded by the German Ministry for Education and Research (BMB+F) in the project QAI2-QAICO under grant 13N15586.

References

1. Friedman, J., Hastie, T., Tibshirani, R.: The Elements of Statistical Learning, vol. 1. Springer, New York (2001). https://doi.org/10.1007/978-0-387-21606-5
2. Hornik, K., Stinchcombe, M., White, H., et al.: Multilayer feedforward networks are universal approximators. Neural Netw. **2**(5), 359–366 (1989)
3. Macaluso, A., Clissa, L., Lodi, S., Sartori, C.: Quantum splines for non-linear approximations. In: Proceedings of the 17th ACM International Conference on Computing Frontiers, pp. 249–252 (2020)
4. Harrow, A.W., Hassidim, A., Lloyd, S.: Quantum algorithm for linear systems of equations. Phys. Rev. Lett. **103**(15), 150502 (2009)
5. Aaronson, S.: Read the fine print. Nat. Phys. **11**(4), 291 (2015)
6. Rumelhart, D.E., Hinton, G.E., Williams, R.J.: Learning representations by back-propagating errors. Nature **323**(6088), 533–536 (1986)
7. Abbas, A., Sutter, D., Zoufal, C., Lucchi, A., Figalli, A., Woerner, S.: The power of quantum neural networks. Nat. Comput. Sci. **1**(6), 403–409 (2021)
8. Tacchino, F., Macchiavello, C., Gerace, D., Bajoni, D.: An artificial neuron implemented on an actual quantum processor, zak1998quantum. NPJ Quantum Inf. **5**(1), 26 (2019)
9. Grant, E., et al.: Hierarchical quantum classifiers. NPJ Quantum Inf. **4**(1), 1–8 (2018)
10. Huggins, W., Patil, P., Mitchell, B., Whaley, K.B., Stoudenmire, E.M.: Towards quantum machine learning with tensor networks. Quantum Sci. Technol. **4**(2), 024001 (2019)
11. Liu, D., et al.: Machine learning by unitary tensor network of hierarchical tree structure. New J. Phys. **21**(7), 073059 (2019)
12. Macaluso, A., Clissa, L., Lodi, S., Sartori, C.: A variational algorithm for quantum neural networks. In: Krzhizhanovskaya, V.V., et al. (eds.) ICCS 2020. LNCS, vol. 12142, pp. 591–604. Springer, Cham (2020). https://doi.org/10.1007/978-3-030-50433-5_45
13. Benedetti, M., Lloyd, E., Sack, S., Fiorentini, M.: Parameterized quantum circuits as machine learning models. Quantum Sci. Technol. **4**(4), 043001 (2019)
14. Havlicek, V., et al.: Supervised learning with quantum enhanced feature spaces. Nature (2018)
15. Smolin, J.A., DiVincenzo, D.P.: Five two-bit quantum gates are sufficient to implement the quantum Fredkin gate. Phys. Rev. A **53**(4), 2855 (1996)
16. Schuld, M., Bocharov, A., Svore, K.M., Wiebe, N.: Circuit-centric quantum classifiers. arXiv preprint arXiv:1804.00633 (2018)
17. Goodfellow, I., Bengio, Y., Courville, A.: Deep Learning, vol. 1. MIT Press, Cambridge (2016)
18. Judd, J.S.: Neural Network Design and the Complexity of Learning. MIT Press, Cambridge (1990)

19. Havlíček, V., et al.: Supervised learning with quantum-enhanced feature spaces. Nature **567**(7747), 209–212 (2019)
20. Shende, V.V., Prasad, A.K., Markov, I.L., Hayes, J.P.: Synthesis of quantum logic circuits. IEEE Trans. Comput. Aided Des. Integr. Circuits Syst. (2006)
21. Mottonen, M., Vartiainen, J.J., Bergholm, V., Salomaa, M.M.: Transformation of quantum states using uniformly controlled rotations. arXiv preprint quant-ph/0407010 (2004)
22. Goto, T., Tran, Q.H., Nakajima, K.: Universal approximation property of quantum feature map. arXiv preprint arXiv:2009.00298 (2020)

Preconditioned Gradient Method for Data Approximation with Shallow Neural Networks

Nadja Vater(✉) and Alfio Borzì

Institut für Mathematik, Universität Würzburg, Würzburg, Germany
{nadja.vater,alfio.borzi}@mathematik.uni-wuerzburg.de

Abstract. A preconditioned gradient scheme for the regularized minimization problem arising from the approximation of given data by a shallow neural network is presented. The construction of the preconditioner is based on random normal projections and is adjusted to the specific structure of the regularized problem.

The convergence of the preconditioned gradient method is investigated numerically for a synthetic problem with a known local minimizer. The method is also applied to real problems from the Proben1 benchmark set.

Keywords: Nonlinear least squares · Regularization · Gradient descent · Preconditioning · Neural networks

1 Introduction

In machine learning, neural networks provide flexible models to perform a variety of tasks. In particular, for a given set of data points $X = \{(x^\ell, y^\ell) \mid \ell = 1, \dots, m\} \subseteq \mathbb{R}^{n_x} \times \mathbb{R}^{n_y}$ they represent nonlinear models of the desired map $g^* : \mathbb{R}^{n_x} \to \mathbb{R}^{n_y}$ that provides a best fit, in the sense that $g^*(x^\ell) \approx y^\ell$, $\ell = 1, \dots, m$. Specifically, this function is defined by a chosen architecture and a set of weights θ, such that the map sought minimizes a quadratic loss function

$$\sum_{\ell=1}^{m} \|y^\ell - g(\theta, x^\ell)\|^2,$$

where $\|\cdot\|$ denotes the Euclidean norm. Thus, we consider the setting, where the representation ability of the network is measured by the mean squared error as in [13]. Typically, the architecture of the network is fixed while the parameters θ are adjusted to fit the data. Consequently, the resulting minimization task is to solve a nonlinear least squares problem, which is non-convex in general. Therefore, it is convenient to introduce a regularization term that aims at smoothing the optimization landscape. In machine learning, various regularization schemes have been introduced to ensure that the model that solves the minimization task also performs well on related out-of-sample data; see, e.g., [4]. A well-known strategy applied for nonlinear least squares problems as well as in machine learning is

G. Nicosia et al. (Eds.): LOD 2022, LNCS 13811, pp. 357–372, 2023.
https://doi.org/10.1007/978-3-031-25891-6_27

Tikhonov regularization. For this reason, we consider a related approach as in [2] that is suitable for the underdetermined nonlinear least squares problem arising from models with more parameters than given data samples.

In any case, assuming differentiability of the resulting regularized loss functional, a standard approach to determine a minimizer is by applying a gradient method. Unfortunately, this method often suffers from a slow-down of convergence, and several preconditioning techniques are suggested to accelerate this scheme. Usually, a right preconditioned scheme ist obtained by rescaling the gradient by multiplying a suitable matrix from the left. Newton and quasi-Newton methods follow this approach with the scaling matrix being the inverse of the Hessian or an approximation of it. These schemes have provably a faster local convergence rate than gradient descent [1]. However, in cases where the cost of this computation is expensive, such as for large-scale problems, the faster convergence in terms of iterations is mitigated by the increased computational cost, which may even result in a slower convergence with respect to time. Instead, several computationally cheaper methods are suggested. For stochastic gradient descent, which is the variant of gradient methods often used in machine learning, several diagonal preconditioning schemes are available [3,11,14]. Often preconditioners are designed for a specific application such as neural network training [9], image registration [14], or matrix factorization [16].

In this paper, we propose a new preconditioning scheme that is tailored to underdetermined regularized nonlinear least squares problems and based on random normal projections similar to [10] for linear least squares. Following their approach for underdetermined systems we apply the transformation matrix as a left preconditioner, which is possible due to the least squares structure of our problem and is different from the right preconditioning schemes described above. Additionally, the construction of our preconditioner exploits the structure of the gradient specific to the regularized problem to reduce the computational effort. Finally, we include the preconditioning in the gradient scheme to obtain our preconditioned gradient (PGD) method for regularized underdetermined nonlinear least squares problems.

In the next section, we introduce the regularized nonlinear least squares problem for fitting with a shallow neural network, which is considered throughout the paper. In Sect. 3, we review the gradient algorithm and its convergence behavior. Based on this discussion, we suggest our preconditioning scheme. This section is followed by a numerical investigation of the performance of the proposed preconditioned gradient method. Experiments with synthetic as well as real data sets are included. A section of conclusion completes the work.

2 Neural Network Learning Problem

We consider the data approximation problem arising in supervised learning. Let $X = \{(x^\ell, y^\ell) \mid \ell = 1, \ldots, m\} \subseteq \mathbb{R}^{n_x} \times \mathbb{R}^{n_y}$ be a set of given data points. The goal is to find a function $g^* : \mathbb{R}^{n_x} \to \mathbb{R}^{n_y}$ from a function class $\mathcal{Y}$ that approximates these points best, that is $g^*(x^\ell) \approx y^\ell$ for $\ell = 1, \ldots, m$. Our set of functions is defined by a feed-forward neural network with one hidden layer, that is $\mathcal{Y} = \{g(\theta,\cdot) : \mathbb{R}^{n_x} \to \mathbb{R}^{n_y} \mid \theta \in \mathbb{R}^n\}$, where

$$g(\theta, x)_j = \sum_{i=1}^{n_h} \theta_{(n_x+1)n_h+(i-1)n_y+j}\, \sigma \left(\sum_{k=1}^{n_x} \theta_{(i-1)n_x+k}\, x_k + \theta_{n_x n_h+i} \right) \quad (1)$$

for $j = 1, \ldots, n_y$, with activation function $\sigma(x) = \frac{1}{1+\exp(-x)}$, the element-wise logistic sigmoid function. The parameter vector $\theta \in \mathbb{R}^n$ contains all weights and bias parameters of the network. It combines the $n_x n_h$ weights from input to hidden layer, the n_h bias values of the hidden nodes, and the $n_h n_y$ weights from hidden to output layer in this order. We call the mapping $g(\theta,\cdot)$ in (1) a shallow neural network with architecture n_x - n_h - n_y. Although we restrict our discussion to this setting, our scheme is applicable to more general neural network architectures with differentiable activation functions.

We denote the function mapping the parameter vector to the error per sample (x^ℓ, y^ℓ) and element $y_1^\ell, \ldots, y_{n_y}^\ell$ by $F : \mathbb{R}^n \to \mathbb{R}^{mn_y}$, that is let

$$F(\theta)_i = \frac{1}{\sqrt{mn_y}} \left(y_j^\ell - g(\theta, x^\ell)_j \right) \quad \text{with} \quad i = n_y(\ell - 1) + j. \quad (2)$$

With this notation the parameter vector θ of a neural network $g(\theta,\cdot)$ that represents the exact relationship between x^ℓ and y^ℓ solves the nonlinear system of equations $F(\theta) = 0$. We assume that the problem is underdetermined, that is, $n > mn_y$.

A network that fits the given data best can be found by minimizing the nonlinear least squares problem given by $\min_{\theta\in\mathbb{R}^n} \frac{1}{2}\|F(\theta)\|^2$. The objective function $\frac{1}{2}\|F(\theta)\|^2 = \frac{1}{2}\frac{1}{mn_y}\sum_{\ell=1}^{m}\sum_{j=1}^{n_y}(y_j^\ell - g(\theta,x^\ell)_j)^2$ measures the squared error scaled by the number of equations. If there exists a solution $\theta^* \in \mathbb{R}^n$ to the system of nonlinear equations $F(\theta^*) = 0$, then it is also a minimizer of the least squares problem.

Additionally, we include a regularization term in the objective function. For a linear least squares problem it is known that the minimizer θ^* of the regularized problem $\frac{1}{2}\|A\theta - b\|^2 + \frac{\lambda^2}{2}\|\theta\|^2$ is given by the first part of the solution $\begin{pmatrix}\theta^* \\ p^*\end{pmatrix}$ of the problem

$$\min \frac{1}{2} \left\| \begin{pmatrix} \theta \\ p \end{pmatrix} \right\|^2 \quad \text{subject to} \quad A\theta + \lambda p = b; \quad (3)$$

see, e.g., [8,10]. The latter formulation preserves the underdetermined nature of the problem. Applying the gradient method with appropriate step size to minimize $\frac{1}{2}\|A\theta + \lambda p\|_2^2$ with respect to (θ, p) converges to the solution of (3) if started from $(0,0)$; see, e.g., [2]. Consequently, we consider minimizing the regularized function $\frac{1}{2}\|F(\theta) + \lambda p\|^2$ with respect to (θ, p) resulting in the regularized underdetermined nonlinear least squares problem given by

$$\min_{\theta,p} J_\lambda(\theta, p) := \frac{1}{2}\|F_\lambda(\theta, p)\|^2 := \frac{1}{2}\|F(\theta) + \lambda p\|^2. \quad (4)$$

This reformulation is also used in [2], where it is shown that $J_\lambda(\theta, p)$ is invex and satisfies the Polyak-Łojasiewicz inequality. Empirically, this problem formulation allows gradient-based methods to reach a low value of $\frac{1}{2}\|F(\theta)\|^2$ independently from the choice of the regularization factor λ, which is not the case for the standard formulation $\frac{1}{2}\|F(\theta)\|^2 + \frac{\lambda}{2}\|\theta\|^2$.

The regularized objective function J_λ in (4) is twice continuously differentiable and its gradient is given by

$$\nabla J_\lambda(\theta_k, p_k) = F'_\lambda(\theta_k, p_k)^\top \left(F(\theta_k) + \lambda p_k\right), \tag{5}$$

where

$$F'_\lambda(\theta_k, p_k) = \begin{pmatrix} F'(\theta_k) & \lambda I \end{pmatrix} \in \mathbb{R}^{mn_y \times (n+mn_y)} \tag{6}$$

denotes the Jacobian of F_λ at (θ_k, p_k). In the discussion that follows, in order to define a local approximation of the objective function at a point $(\hat{\theta}, \hat{p})$, we consider the following functional

$$\hat{J}_\lambda(\theta, p; \hat{\theta}, \hat{p}) = \frac{1}{2}\left\| F_\lambda(\hat{\theta}, \hat{p}) + F'_\lambda(\hat{\theta}, \hat{p}) \begin{pmatrix} \theta - \hat{\theta} \\ p - \hat{p} \end{pmatrix} \right\|^2. \tag{7}$$

We use this function for the analysis of convergence of the iterative method introduced in the next section.

3 A Preconditioned Gradient Method

Gradient descent (GD) is a well-known iterative optimization method. It can be applied to find a local minimum of the differentiable function $J_\lambda(\theta, p)$. This method generates a sequence of parameters $\{(\theta_k, p_k)\}_{k\in\mathbb{R}}$ by

$$\begin{pmatrix} \theta_{k+1} \\ p_{k+1} \end{pmatrix} = \begin{pmatrix} \theta_k \\ p_k \end{pmatrix} - \eta \nabla J_\lambda(\theta_k, p_k) \tag{8}$$

with arbitrary initial value $(\theta_0, p_0) \in \mathbb{R}^n \times \mathbb{R}^{mn_y}$.

The convergence to a local minimum of $J_\lambda(\theta, p)$ can be shown under the assumption that F_λ is continuously differentiable with Lipschitz-continuous Jacobian F'_λ for a sufficiently small step size η; see, e.g., [7].

The choice of the step size η is crucial for the convergence of the algorithm. Selecting a small η slows down convergence while setting η to a too large value can cause divergence of the method. For the theoretical investigation of an appropriate step size, we consider the local approximation of the problem at a local minimum $(\theta^*, p^*) \in \mathbb{R}^n \times \mathbb{R}^{mn_y}$ as defined in (7), that is $\hat{J}_\lambda(\theta, p; \theta^*, p^*)$. If (θ_k, p_k) is sufficiently close to (θ^*, p^*), then the gradient update is approximated by

$$\begin{pmatrix} \theta_{k+1} \\ p_{k+1} \end{pmatrix} = \begin{pmatrix} \theta_k \\ p_k \end{pmatrix} - \eta F'_\lambda(\theta^*, p^*)^\top \left(F_\lambda(\theta^*, p^*) + F'_\lambda(\theta^*, p^*) \begin{pmatrix} \theta_k - \theta^* \\ p_k - p^* \end{pmatrix} \right). \tag{9}$$

Using $0 = \nabla J_\lambda(\theta^*, p^*) = F'_\lambda(\theta^*, p^*)^\top F_\lambda(\theta^*, p^*)$, we obtain the relation

$$\begin{pmatrix} \theta_{k+1} - \theta^* \\ p_{k+1} - p^* \end{pmatrix} = \left(I - \eta F'_\lambda(\theta^*, p^*)^\top F'_\lambda(\theta^*, p^*)\right) \begin{pmatrix} \theta_k - \theta^* \\ p_k - p^* \end{pmatrix}. \tag{10}$$

Let $F'_\lambda(\theta^*, p^*)^\top F'_\lambda(\theta^*, p^*) = \sum_{i=1}^{mn_y} \lambda_i v_i v_i^\top$ denote the eigenvalue decomposition. Then the components of the error $e_k = \begin{pmatrix} \theta_k - \theta^* \\ p_k - p^* \end{pmatrix}$ in direction of the eigenvectors change approximately as follows

$$v_i^\top \begin{pmatrix} \theta_{k+\ell} - \theta^* \\ p_{k+\ell} - p^* \end{pmatrix} = (1 - \eta\lambda_i)^\ell \, v_i^\top \begin{pmatrix} \theta_k - \theta^* \\ p_k - p^* \end{pmatrix}. \tag{11}$$

Therefore, the error in direction v_i decreases as long as $|1 - \eta\lambda_i| < 1$. Thus, the norm of the error $\|e_k\|$ converges to zero if $\eta < \frac{1}{\max_i |\lambda_i|}$. In this case, a large value of $\frac{\max_i |\lambda_i|}{\min_i |\lambda_i|}$, the square of the condition number of the matrix $F'_\lambda(\theta^*, p^*)$, causes a large variation in the convergence speeds of the error in the direction of different eigenvectors [15].

A common approach to improve the convergence rate of an iterative method is preconditioning. We can obtain a preconditioned method by applying the iterative scheme to a transformation of our nonlinear least squares problem (4), that has the same solutions as (4). The condition number of the transformed problem indicates the convergence speed of the preconditioned method. The effectiveness of this strategy in terms of a reduced number of iterations and a good approximation of the exact solution depends on a well chosen transformation operator. Basically, we can either transform the domain or the codomain of the function $F_\lambda(\theta, p)$ or both. The resulting schemes are called left, right and split preconditioning, respectively. Since we consider an underdetermined setting, we focus on a transformation of the codomain of $F_\lambda(\theta, p)$ as in [10] for underdetermined linear least squares problems, that is, we apply a left preconditioner. Let $M \in \mathbb{R}^{mn_y \times r}$ with $r \leq mn_y$, then a transformation of our nonlinear least squares system is given by

$$\frac{1}{2}\|M^\top F_\lambda(\theta, p)\|^2. \tag{12}$$

If M has full row rank ($r \geq mn_y$), then this function has exactly the same minimizers as $\frac{1}{2}\|F_\lambda(\theta, p)\|^2$. Applying the gradient update from (8) yields

$$\begin{pmatrix} \theta_{k+1} \\ p_{k+1} \end{pmatrix} = \begin{pmatrix} \theta_k \\ p_k \end{pmatrix} - \eta \, F'_\lambda(\theta_k, p_k)^\top M M^\top F_\lambda(\theta_k, p_k). \tag{13}$$

We call the resulting scheme preconditioned gradient (PGD) for the regularized nonlinear least squares problem. A pseudocode of PGD is given in Algorithm 1.

This scheme can also be applied batchwise by computing a transformation matrix for each batch leading to a preconditioned stochastic gradient method. Convergence results for stochastic gradient descent, as given in, e.g., [5], can be transferred to this setting.

Algorithm 1: Preconditioned Gradient Method (PGD) for the regularized problem

Input: Initial guess θ_0; stopping criteria; step size η
regularization parameter λ; update frequency μ_M
Output: θ_K, p_K
Initialize $p_0 = 0$.
for $k = \{0, \dots, K-1\}$ **do**
  if $0 \equiv k \mod \mu_M$ **then**
    Compute a transformation matrix M.
  end
  Compute the update $\begin{pmatrix}\theta_{k+1}\\ p_{k+1}\end{pmatrix} = \begin{pmatrix}\theta_k\\ p_k\end{pmatrix} - \eta\, F'_\lambda(\theta_k, p_k)^\top M M^\top F_\lambda(\theta_k, p_k)$.
end

Notice that, the gradient $\nabla J_\lambda(\theta, p)$ at an iterate (θ_k, p_k) is the same as the gradient $\nabla \hat{J}_\lambda(\theta, p; \theta_k, p_k)$ of the approximation at this point. Therefore, the PGD update in (13) equals the gradient update at (θ_k, p_k) for the transformed linear least squares system

$$\frac{1}{2}\left\| M^\top F_\lambda(\theta_k, p_k) + M^\top F'_\lambda(\theta_k, p_k)\begin{pmatrix}\theta - \theta_k\\ p - p_k\end{pmatrix}\right\|^2. \tag{14}$$

In this case, the transformation M should be adapted to the system matrix $F'_\lambda(\theta_k, p_k)$ to provide a transformed problem with lower condition number and approximately the same minimal-norm minimizer as $J_\lambda(\theta, p; \theta_k, p_k)$. This observation suggests to update the transformation M in every iteration or at least regularly in order to obtain the most appropriate matrix at every iteration. Therefore, we introduce the parameter $\mu_M \in \mathbb{N}$ in Algorithm 1, that determines the update frequency of the transformation matrix.

Next, we investigate the construction of a suitable transformation matrix M. We suggest a method based on a random normal projection as used in the LSRN method introduced in [10] for a linear least squares problem $\min_x \|Ax + b\|$. In order to become familiar with our approach, we start revising the preconditioner construction in the original setting.

Let $A \in \mathbb{R}^{m_A \times n_A}$ have more columns than rows, that is $m_A < n_A$, and assume that it has full row rank. First, a randomized transform $G \in \mathbb{R}^{n_A \times s}$ with $s = \lceil \gamma m_A \rceil$ and oversampling factor $\gamma \geq 1$ is applied to the system matrix A, such that the space spanned by the columns of $\tilde{A} = AG$ represents the column space of A with high probability. The projection matrix G has entries that are independently sampled from the standard normal distribution and is called random normal projection. The transformation can be viewed as a random mixing of the columns of A followed by a random sampling of the mixed columns. Then, the matrix M is computed from the singular value decomposition of AG in the following way. Denote the singular value decomposition of $\tilde{A} = AG$ by $\tilde{A} = \tilde{U}\tilde{\Sigma}\tilde{V}^\top$, where $\tilde{U} \in \mathbb{R}^{m_A \times m_A}$ and $\tilde{V} \in \mathbb{R}^{m_A \times \lceil \gamma m_A \rceil}$ have orthonormal columns and $\tilde{\Sigma} \in \mathbb{R}^{m_A \times m_A}$ is a diagonal matrix. Then, set

$$M = \tilde{U}\tilde{\Sigma}^{-1}, \tag{15}$$

where $\tilde{\Sigma}^{-1}$ is the diagonal matrix obtained by inverting the entries of the diagonal matrix $\tilde{\Sigma}$. The resulting transformed problem $\min_x \|M^\top Ax + M^\top b\|$ has almost surely the same minimal-norm solution as the original problem $\min_x \|Ax + b\|$. At the same time the condition number of $M^\top A$ is independent from the spectrum of A. This property ensures that convergence is achieved in fewer iterations [10].

For an approximately rank-deficient matrix $A = A_r + E$, where A_r denotes the best rank-r approximation to A and E is a small perturbation, we are interested to find the minimal-norm solution of $\min_x \|A_r x + b\|$. An approximation to this minimizer is given by the solution of the transformed system $\|M_r^\top Ax + M_r^\top b\|$, where $M_r = \tilde{U}_r\tilde{\Sigma}_r^{-1}$ is computed from the truncated singular value decomposition $(AG)_r = \tilde{U}_r\tilde{\Sigma}_r\tilde{V}_r^\top$. In this way the computational effort is reduced from $\mathcal{O}(m_A^2 n_A)$ computations for the full decomposition to $\mathcal{O}(m_A n_A r)$ for the truncated decomposition [6].

Now, we are prepared to consider our nonlinear least squares problem. Since the transformation M at iteration k should be suitable for the Jacobian $F'_\lambda(\theta_k, p_k)$, an obvious application of the described construction would be to replace the matrix A by $F'_\lambda(\theta_k, p_k)$ in the above reasoning. Nevertheless, we obtain a simplified scheme by taking into account the structure of the Jacobian and its singular value decomposition. In our case, it holds $F'_\lambda(\theta_k, p_k) = \big(F'(\theta_k)\ \lambda I\big)$, see (6), and we denote the corresponding singular value decomposition by $F'_\lambda(\theta_k, p_k) = U\Sigma V^\top$. We discard the fixed index k in the notation of the decomposition in order to avoid confusion with the notation of the truncated version. We show, that we can obtain U, Σ, and V from the singular value decomposition of $F'(\theta_k)$, denoted by $F'(\theta_k) = \bar{U}\bar{\Sigma}\bar{V}^\top$, and the regularization parameter λ. Then, we use this relation to construct a preconditioner for $F'_\lambda(\theta_k, p_k)$ from the singular value decomposition of a random normal projection of $F'(\theta_k)$, which is denoted by $F'(\theta_k)G = \tilde{U}\tilde{\Sigma}\tilde{V}^\top$.

Let $i \in \{1, \ldots, mn_y\}$. Denote the ith column of $\bar{U}$ and $\bar{V}$ by $\bar{u}_i$ and $\bar{v}_i$, respectively, and the ith diagonal entry of $\bar{\Sigma}$ by $\bar{\sigma}_i = \bar{\Sigma}_{ii}$. Then, it holds

$$F'(\theta_k)^\top \bar{u}_i = \bar{\sigma}_i \bar{v}_i \quad \text{and} \quad F'(\theta_k)\bar{v}_i = \bar{\sigma}_i \bar{u}_i \tag{16}$$

as well as $\|\bar{u}_i\| = 1$ and $\|\bar{v}_i\| = 1$. Moreover, using (16), for the Jacobian $F'_\lambda(\theta_k, p_k)$ of the regularized nonlinear least squares problem we have

$$F'_\lambda(\theta_k, p_k)^\top \bar{u}_i = \begin{pmatrix} F'(\theta_k)^\top \bar{u}_i \\ \lambda \bar{u}_i \end{pmatrix} = \begin{pmatrix} \bar{\sigma}_i \bar{v}_i \\ \lambda \bar{u}_i \end{pmatrix}.$$

Denote the normalized vector on the right hand side by v_i, that is, let $v_i := \frac{1}{\sqrt{\bar{\sigma}_i^2 + \lambda^2}} \begin{pmatrix} \bar{\sigma}_i \bar{v}_i \\ \lambda \bar{u}_i \end{pmatrix}$. Then, it holds

$$F'_\lambda(\theta_k, p_k)^\top \bar{u}_i = \sqrt{\bar{\sigma}_i^2 + \lambda^2}\ v_i$$

Moreover, we have

$$F'_\lambda(\theta_k, p_k)v_i = \frac{1}{\sqrt{\bar{\sigma}_i^2 + \lambda^2}} \left(\bar{\sigma}_i F'(\theta_k)\bar{v}_i + \lambda^2 \bar{u}_i\right) = \sqrt{\bar{\sigma}_i^2 + \lambda^2}\, \bar{u}_i.$$

Therefore, the singular values of $F'_\lambda(\theta_k, p_k)$ are given by

$$\Sigma_{ii} = \sqrt{\bar{\sigma}_i^2 + \lambda^2} = \sqrt{\bar{\Sigma}_{ii}^2 + \lambda^2} \quad \text{for } i = 1, \ldots, mn_y, \tag{17}$$

and the unitary matrices containing the corresponding singular vectors satisfy the relations

$$U = \bar{U} \quad \text{and} \quad V = \begin{pmatrix} \bar{V}\Sigma^{-1}\bar{\Sigma} \\ \lambda\bar{U}\Sigma^{-1} \end{pmatrix}. \tag{18}$$

Next, we investigate the effect of the random normal projection. In [10] it is established that the spectrum of the random normal projection AG of a matrix A approximates the spectrum of A scaled by $\sqrt{s}$. Thus, it holds

$$\bar{\Sigma}_{ii}\sqrt{s} \approx \tilde{\Sigma}_{ii} \quad \text{for } i = 1, \ldots, mn_y, \tag{19}$$

where $\tilde{\Sigma}_{ii}$ is the ith singular value of $F'(\theta_k)G$. By inserting this relation into (17) we obtain an approximation $\hat{\Sigma}_{ii} \approx \Sigma_{ii}$ to the ith singular value of $F'_\lambda(\theta_k, p_k)$ by defining

$$\hat{\Sigma}_{ii} := \sqrt{\tilde{\Sigma}_{ii}^2/s + \lambda^2}. \tag{20}$$

Having established the relation $U = \bar{U}$ between the singular vectors of $F'_\lambda(\theta_k, p_k)$ and $F'(\theta_k)$ we consider the approximation $\tilde{U}$ of $\bar{U}$ from the singular value decomposition of the random normal projected matrix $F'(\theta_k)G$ to define the preconditioner as follows

$$M := \tilde{U}\hat{\Sigma}^{-1}. \tag{21}$$

The main difference compared to the direct construction of the preconditioner from $F'_\lambda(\theta_k, p_k)$ is that a suitable random projection matrix G has far less rows when applied to $F'(\theta_k)$ compared to $F'_\lambda(\theta_k, p_k)$ resulting in a computationally cheaper scheme.

As for the linear least squares problem the computational effort can be further reduced by constructing a preconditioner with rank $r < mn_y$ from a truncated singular value decomposition. For this reason, we define the rank-r preconditioner by

$$M_r := \tilde{U}_r\hat{\Sigma}_r^{-1}, \tag{22}$$

where $\tilde{U}_r \in \mathbb{R}^{mn_y \times r}$ and $\hat{\Sigma}_r^{-1} \in \mathbb{R}^{r \times r}$ are the truncated versions of the matrices $\tilde{U}$ and $\hat{\Sigma}^{-1}$ used in (21). We call the PGD scheme in Algorithm 1 with M replaced by M_r the rank-r preconditioned gradient method (PGD_r).

For a comparison we also consider the rank-r gradient scheme (GD_r) with the Jacobian $F'_\lambda(\theta_k, p_k)$ replaced by its best rank-r approximation. It is defined by

$$\begin{pmatrix} \theta_{k+1} \\ p_{k+1} \end{pmatrix} = \begin{pmatrix} \theta_k \\ p_k \end{pmatrix} - \eta(F'_\lambda(\theta_k, p_k))_r^\top F_\lambda(\theta_k, p_k). \tag{23}$$

4 Results of Experiments

In this section, we present results of experiments that validate the convergence properties of the suggested PGD method. First, we investigate the performance of the preconditioner for the linear least squares problem arising from minimizing the local approximation $\hat{J}_\lambda(\theta, p; \theta^*, p^*)$ at a fixed point (θ^*, p^*). We are interested to verfiy that our preconditioner is suitable for the regularized problem in the sense that the preconditioned iterative method for a regularized linear least squares problem converges in less iterations to approximately the same solution as the iterative method without preconditioning. Since we consider a linear problem, we choose the iterative solver LSQR [12] as in [10] to obtain insights into the quality of the preconditioner. Afterwards, we turn towards the nonlinear problem and investigate the convergence of the preconditioned method compared to the gradient method. Additionally, we examine the effect of a rank-r approximation by comparing PGD_r to GD_r.

In the experiments we fix the update interval of the preconditioner to be $\mu_M = 10$, which we empirically found to be effective. Additionally, we set the number of columns of G to $s = \lceil \frac{mn_y}{n_x+1+n_y} \rceil (n_x + 1 + n_y)$ to ensure a proper representation of the column space of $F'(\theta_k)$ while keeping the number of columns of the projected matrix $F'(\theta_k)G$ relatively small. This number is chosen such that the $n_h(n_x+1+n_y)$ columns of $F'(\theta)$ are represented by a multiple of (n_x+1+n_y) columns to account for the structure of the neural network.

The experiments are performed for nonlinear least squares problems arising from the search for a best fitting neural network to given data as described in Sect. 2. We consider a synthetic data set as well as real classification data sets from the Proben1 benchmark set [13].

We construct the synthetic data set by randomly initializing a network with $n_x = 15$ input nodes, $n_h = 16$ hidden nodes and $n_y = 1$ output node (architecture 15-16-1) with parameters $\theta^* \in \mathbb{R}^n$. We use Xavier initialization to generate the parameters of the network. Thus, we sample the weights to the hidden layer $\theta_1, \dots, \theta_{n_x n_h}$ i.i.d. from a Gaussian with mean zero and variance $\frac{1}{n_x}$, set the bias parameters $\theta_{n_x n_h+1}, \dots, \theta_{n_h(n_x+1)}$ to zero and sample the weights to the output layer $\theta_{n_h(n_x+1)+1}, \dots, \theta_{n_h(n_x+1+n_y)}$ i.i.d. from a Gaussian with mean zero and variance $\frac{1}{n_h}$. The input data x_i^ℓ are drawn i.i.d. from a Gaussian with mean zero and variance $\frac{1}{n_x}$, so that each sample x^ℓ has length 1 in expectation. We consider $m = 100$ samples. The target values y^ℓ are given by the output of the network with parameters θ^*, that is

$$y^\ell := g(\theta^*, x^\ell). \tag{24}$$

A local minimum of the regularized least squares problem $\min_{\theta,p} J_\lambda(\theta, p)$ defined by this data set is given by $(\theta^*, 0) \in \mathbb{R}^n \times \mathbb{R}^{mn_y}$.

4.1 Preconditioning the Regularized Linear Problem

In order to investigate the effect of the adaption of the preconditioner construction to the regularized problem, we consider the local approximation

$\hat{J}_\lambda(\theta, p; \theta^*, 0)$ at the local minimum $(\theta^*, 0)$ of the synthetic problem described above. To obtain an appropriate preconditioner for this linear least squares problem, we replace $F'_\lambda(\theta_k, p_k)$ by $F'_\lambda(\theta^*, 0)$ in the above derivation of M and M_r.

First, we investigate the connection between the singular values of $F'_\lambda(\theta^*, 0)$, $F'(\theta^*)$ and $F'(\theta^*)G$ as well as the approximation $\hat{\Sigma}$ to the singular values of $F'_\lambda(\theta^*, 0)$ defined in (20) empirically. Finally, we compare the convergence of LSQR and preconditioned LSQR.

The display of the singular value distribution of $F'(\theta^*)$ in Fig. 1 reveals two jumps - one between the 1st and the 2nd value and one between the 16th and 17th value. The positions of these jumps can be related to the structure of the network by $n_y = 1$ and $n_y(n_x + 1) = 16$. An explanation of this connection goes beyond the scope of this discussion, so we leave it for future work and concentrate on the approximation quality of $\hat{\Sigma}$ instead. From Fig. 1, we see that the singular values Σ of $F'_\lambda(\theta^*, 0)$ are represented well by the diagonal entries of $\hat{\Sigma}$. Additionally, the largest scaled singular values $\tilde{\Sigma}/\sqrt{s}$ of $F'(\theta^*)G$ provide a good approximation for the largest singular values $\bar{\Sigma}$ of $F'(\theta^*)$. The gap grows for the smaller singular values.

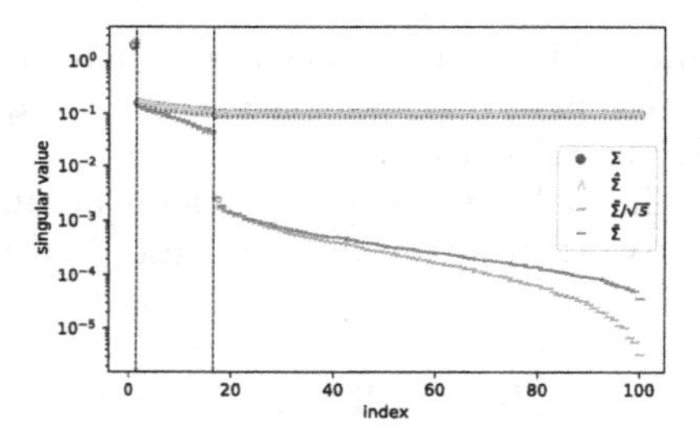

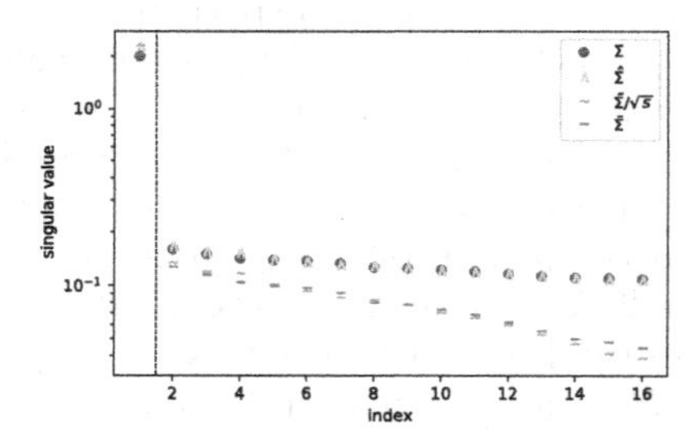

Fig. 1. Singular values Σ of $F'_\lambda(\theta^*, 0)$, $\bar{\Sigma}$ of $F'(\theta^*)$, and $\tilde{\Sigma}$ of $F'(\theta^*)G$ as well as the approximation $\hat{\Sigma} \approx \Sigma$ as defined in (20). Notice that, the $\tilde{\Sigma}$ is scaled by $\sqrt{s}$. The dashed vertical lines after the n_yth and $n_y(n_x + 1)$th value indicate jumps in the distribution of the singular values of $F'(\theta^*)$. The right image shows the largest 16 singular values.

Next, we investigate the quality of the preconditioner for the linear least squares problem arising from minimizing $\hat{J}_\lambda(\theta, p; \theta^*, 0)$. Due to the jump after the 16th singular value of $F'(\theta^*)$, we consider LSQR applied to the rank reduced problem

$$\min_{\theta, p} \frac{1}{2} \left\| F_\lambda(\theta^*, p^*) + (F'_\lambda(\theta^*, p^*))_{16} \begin{pmatrix} \theta \\ p \end{pmatrix} \right\|^2 \tag{25}$$

and its preconditioned variant, that uses the rank-16 preconditioner from (22). Table 1 shows the condition number of the matrices $F'_\lambda(\theta^*, p^*)_{16}$ and $M_{16}^\top F'_\lambda(\theta^*, p^*)$, the number of iterations as well as the relative differences between the norm of the residual and the norm of the values obtained by the two methods. We expect the preconditioned method to need less iterations and $M_{16}^\top F'_\lambda(\theta^*, p^*)$

to have a lower condition number than $F'_\lambda(\theta^*, p^*)_{16}$. At the same time the solutions obtained by LSQR and preconditioned LSQR should be approximately the same, indicated by small relative differences in the table. We observe, that the condition number of $F'_\lambda(\theta^*, p^*)_{16}$ and the number of LSQR iterations grow with decreasing regularization parameter λ. In contrast the preconditioned LSQR and the condition number of $M_{16}^\top F'_\lambda(\theta^*, p^*)$ are less sensitive to the change of λ. Preconditioned LSQR needs less iterations compared to LSQR in all cases. The effect of the preconditioner is mitigated by an increased regularization parameter λ. The approximation quality measured by the relative difference in the norm of the residuals and the norm of the solution vector is worse than the approximation error presented in [10]. This issue may be explained by the problem size, which is much smaller in our setting. Nevertheless, the preconditioning scheme seems to work sufficiently well for the nonlinear least squares system that we discuss next.

Table 1. Condition number of A_r and $M_r^\top A$ and number of LSQR and preconditioned LSQR iterations with $A = F'_\lambda(\theta^*, p^*)$, $b = F_\lambda(\theta^*, p^*)$, $x = (\theta^\top, p^\top)^\top$, $r = 16$ for different values of λ. The rightmost columns show the relative differences in the norm of the residuals and norm of the values x^* obtained by LSQR and x obtained by preconditioned LSQR.

λ	cond(A)		Iterations		$\frac{\lvert\lVert Ax^*+b\rVert - \lVert Ax+b\rVert\rvert}{\lVert Ax^*+b\rVert}$	$\frac{\lvert\lVert x^*\rVert - \lVert x\rVert\rvert}{\lVert x^*\rVert}$
	A_r	$M_r^\top A$	A_r	$M_r^\top A$		
0.000	208.10	11.54	19	11	4.64e–06	2.15e–05
0.001	208.08	11.54	19	11	2.38e–06	1.28e–05
0.010	205.71	11.55	19	11	6.76e–04	7.04e–06
0.100	86.82	8.07	14	8	5.53e–02	1.07e–05
1.000	6.16	4.00	5	4	2.74e–02	1.25e–07

4.2 Performance for the Synthetic Problem

To investigate convergence in a nonlinear setting we use the same synthetic data set as before. We consider the set of shallow neural networks with $n_h = 16$ hidden nodes for the approximation and set the regularization parameter of the resulting objective function $J_\lambda(\theta, p)$ to $\lambda = 0.01$. We initialize the parameters with $\theta_0 = \theta^* + 10^{-5}\lVert\theta^*\rVert\epsilon$, where $\epsilon \in \mathbb{R}^n$ has entries i.i.d. sampled from the standard normal distribution, and $p_0 = 0 \in \mathbb{R}^{mn_y}$. We set the step size to $\eta = 0.1$, which slightly exceeds $\frac{1}{\sigma_1} \approx \frac{1}{1.1}$ (see Fig. 1), but still leads to a descending scheme (see Fig. 2). We stop after $K = 50000$ iterations or when the value of $J_\lambda(\theta_k, p_k)$ falls below 10^{-16}.

Figure 2 shows the value of the objective function J_λ during optimization with gradient descent and the preconditioned variant. The preconditioning scheme reaches $J_\lambda(\theta_k, p_k) < 10^{-16}$ much earlier than the gradient method without preconditioner.

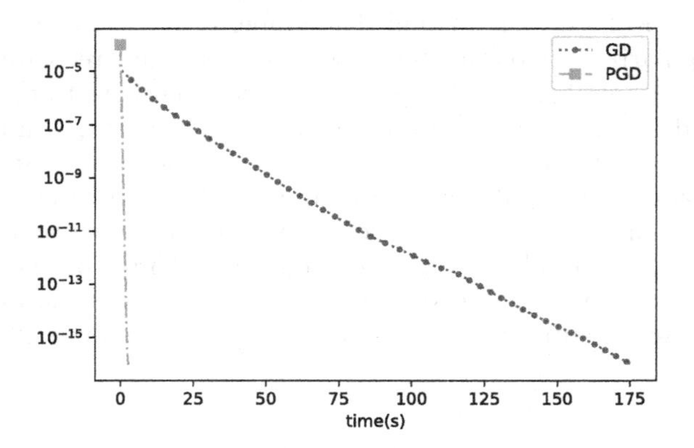

Fig. 2. Value of $J_\lambda(\theta_k, p_k)$ versus time during optimization with GD and PGD. Marker indicate every 500th iteration.

In the same setting, we compare the computationally cheaper method PGD_r with preconditioner M_r as in (22) to GD_r. Based on our knowledge about the jumps in the singular value distribution of $F'(\theta^*)$, which are shown in Fig. 2, we set $r = 16$. In Fig. 3 we can see that the preconditioned variant reaches the same value of $J_\lambda(\theta, p)$ much faster than the gradient scheme GD_{16}.

4.3 Performance on Real Data Sets

Finally, we investigate the convergence of PGD for nonlinear least squares problems arising from real world data sets from the Proben1 benchmark set [13]. We use the training data from the cancer1, card1, glass1 and horse1 data sets in their original setting with the loss function $J_\lambda(\theta, p)$ defined in (4). The number of hidden nodes $n_h \in \{30, 100\}$ is chosen such that the resulting nonlinear system is underdetermined. The dimensions of each problem are given in Table 2.

Table 2. Dimensions of the considered problems from the Proben1 benchmark set.

Data set	n_x	n_y	m	n_h	Number of parameters n	Number of equations mn_y
cancer1	9	2	350	100	1200	699
card1	51	2	345	30	1620	690
glass1	9	6	107	100	1600	642
horse1	58	3	182	30	1860	546

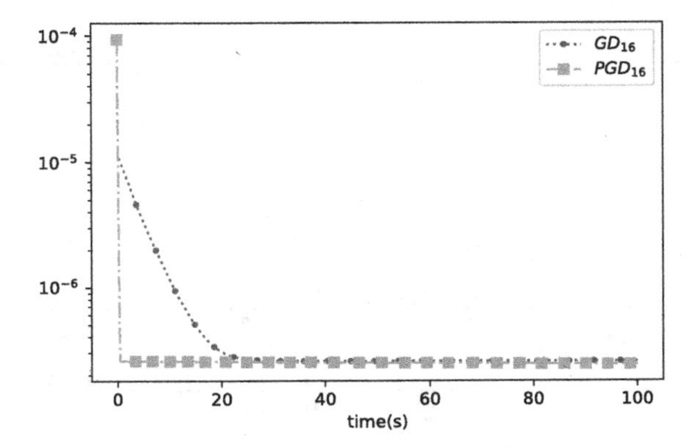

Fig. 3. Value of $J_\lambda(\theta_k, p_k)$ versus time at the beginning of optimization with GD_{16} and PGD_{16}. Marker indicate every 500th iteration. Essentially, no further reduction in later iterations.

In our experiments, we set the regularization parameter to $\lambda = 0.01$ and the maximal number of iterations to $K = 1000$. As for the synthetic problem we base the choice of the step size on the largest singular value of the Jacobian at a randomly initialized point and set it to $\eta = 0.1$. We start from $\theta_0 \in \mathbb{R}^n$ generated with the Xavier initialization as described above and $p_0 = 0 \in \mathbb{R}^{mn_y}$. We compare PGD_r with $r = \lceil 0.1mn_y \rceil$, which approximates only the top 10% singular values of $F'_\lambda(\theta_k, p_k)$, to GD. For all four problems PGD_r reduces the objective function $J_\lambda(\theta_k, p_k)$ to a lower value than GD_r, see Fig. 4. Simultaneously, the computation time for 100 iterations of PGD_r exceeds the time for 100 iterations of GD only slightly. For the glass1 data set we observe that the value of the objective function increases instead of decreases in some iterations. This effect may be caused by an inappropriate step size. We leave the investigation of this phenomenon to the future.

Figure 5 shows the performance of the more computation-intensive PGD method on the problems. This method reaches an approximately zero objective value within 400 iterations for all data sets. Nevertheless, the figure suggests that the time necessary for 100 iterations of PGD is about twice as large as the time for 100 iterations of GD.

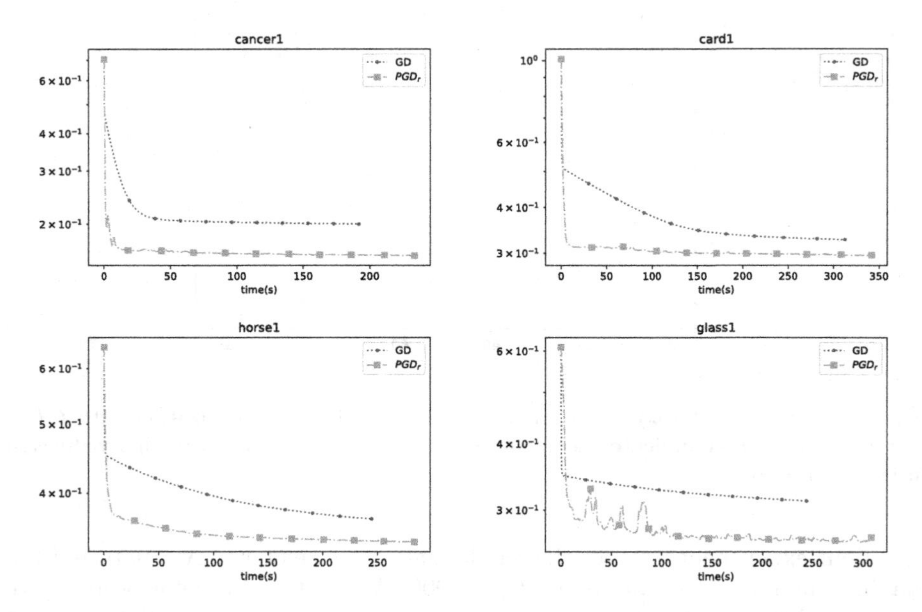

Fig. 4. Value of $J_\lambda(\theta_k, p_k)$ versus time during optimization with GD and PGD_r with $r = \lceil 0.1mn_y \rceil$. Marker indicate every 100th iteration. The top left figure corresponds to the cancer1 data set, the top right figure to the card1 data set, the bottom left figure to the horse1 data set and the bottom right figure to the glass1 data set.

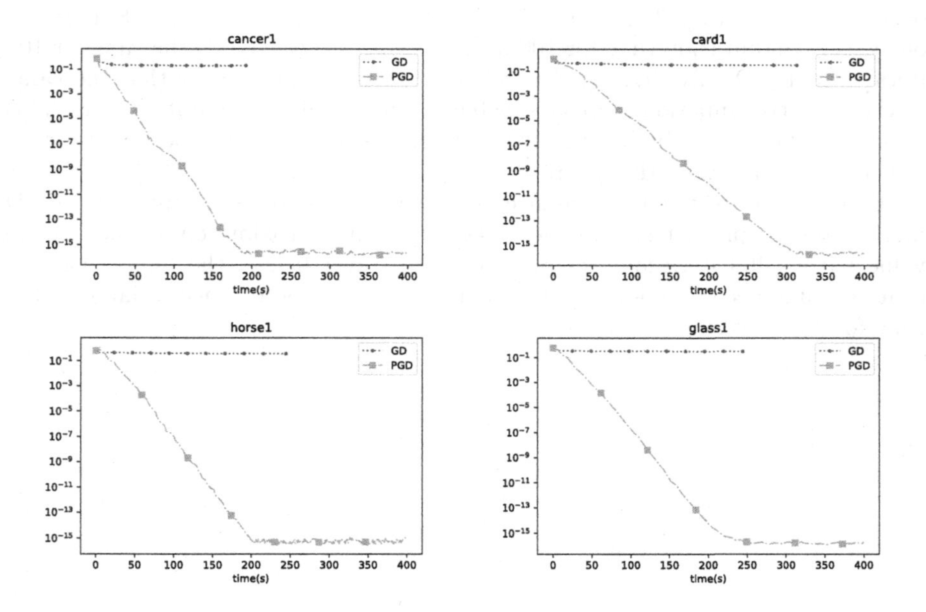

Fig. 5. The same as Fig. 4 with PGD_r replaced with the full scheme PGD.

5 Conclusion

A preconditioning scheme for a gradient method was developed. It was specifically designed for the regularized nonlinear least squares problem arising from approximation with a shallow neural network. The approach was adapted to the underdetermined setting and the structure of the Jacobian of the problem in order to give an effective method. The performance of the scheme was investigated for a synthetic problem. Faster convergence of the PGD method was also validated for classification problems on data sets from the Proben1 benchmark set.

References

1. Broyden, C.G., Dennis, J.E., Jr., Moré, J.J.: On the local and superlinear convergence of quasi-Newton methods. IMA J. Appl. Math. **12**(3), 223–245 (1973)
2. Crane, R., Roosta, F.: Invexifying regularization of non-linear least-squares problems. arXiv preprint arXiv:2111.11027 (2021)
3. Duchi, J., Hazan, E., Singer, Y.: Adaptive subgradient methods for online learning and stochastic optimization. J. Mach. Learn. Res. **12**(7), 2121–2159 (2011)
4. Goodfellow, I., Bengio, Y., Courville, A.: Deep Learning. MIT Press, Cambridge (2016)
5. Gorbunov, E., Hanzely, F., Richtárik, P.: A unified theory of SGD: variance reduction, sampling, quantization and coordinate descent. In: International Conference on Artificial Intelligence and Statistics, pp. 680–690. PMLR (2020)
6. Halko, N., Martinsson, P.G., Tropp, J.A.: Finding structure with randomness: probabilistic algorithms for constructing approximate matrix decompositions. SIAM Rev. **53**(2), 217–288 (2011)
7. Hanke-Bourgeois, M.: Grundlagen der numerischen Mathematik und des wissenschaftlichen Rechnens, 3rd edn. Vieweg + Teubner, Wiesbaden (2009). https://doi.org/10.1007/978-3-8351-9020-7
8. Herman, G.T., Lent, A., Hurwitz, H.: A storage-efficient algorithm for finding the regularized solution of a large, inconsistent system of equations. IMA J. Appl. Math. **25**(4), 361–366 (1980)
9. Lange, S., Helfrich, K., Ye, Q.: Batch normalization preconditioning for neural network training. arXiv preprint arXiv:2108.01110 (2021)
10. Meng, X., Saunders, M.A., Mahoney, M.W.: LSRN: a parallel iterative solver for strongly over-or underdetermined systems. SIAM J. Sci. Comput. **36**(2), C95–C118 (2014)
11. Onose, A., Mossavat, S.I., Smilde, H.J.H.: A preconditioned accelerated stochastic gradient descent algorithm. In: 28th European Symposium on Artificial Neural Networks, Computational Intelligence and Machine Learning (2020)
12. Paige, C.C., Saunders, M.A.: Algorithm 583: LSQR: sparse linear equations and least squares problems. ACM Trans. Math. Softw. **8**(2), 195–209 (1982)
13. Prechelt, L.: Proben1: a set of neural network benchmark problems and benchmarking rules (1994)
14. Qiao, Y., Lelieveldt, B.P., Staring, M.: An efficient preconditioner for stochastic gradient descent optimization of image registration. IEEE Trans. Med. Imaging **38**(10), 2314–2325 (2019)

15. Vater, N., Borzì, A.: Training artificial neural networks with gradient and coarse-level correction schemes. In: Nicosia, G., et al. (eds.) International Conference on Machine Learning, Optimization, and Data Science, pp. 473–487. Springer, Cham (2021). https://doi.org/10.1007/978-3-030-95467-3_34
16. Zhang, J., Fattahi, S., Zhang, R.: Preconditioned gradient descent for over-parameterized nonconvex matrix factorization. Adv. Neural Inf. Process. Syst. **34** (2021)

Grid-Based Contraction Clustering in a Peer-to-Peer Network

Antonio Mariani[1], Italo Epicoco[1,2], Massimo Cafaro[1,2](✉), and Marco Pulimeno[1]

[1] University of Salento, Lecce, Italy
antonio.mariani@studenti.unisalento.it,
{italo.epicoco,massimo.cafaro,marco.pulimeno}@unisalento.it
[2] Euro-Mediterranean Centre on Climate Change, Foundation, Lecce, Italy

Abstract. Clustering is one of the main data mining techniques used to analyze and group data, but often applications have to deal with a very large amount of spatially distributed data for which most of the clustering algorithms available so far are impractical. In this paper we present P2PRASTER, a distributed algorithm relying on a gossip–based protocol for clustering that exploits the RASTER algorithm and has been designed to handle big data in a decentralized manner. The experiments carried out show that P2PRASTER returns perfect results under both optimal and non-optimal conditions, and also provides excellent scalability.

Keywords: Clustering · RASTER · Distributed computing · Peer to peer · Gossip protocol · Big data

1 Introduction

The term Clustering or Cluster Analysis encompasses several data analysis techniques whose aim is to discover the structure inherent on the data, identifying and grouping data similar to each other and distinguishing them from different ones, thus forming the so-called clusters. These techniques fall into the unsupervised category of machine learning which includes algorithms that draw inferences from datasets without relying on labeled input data.

Clustering is used to extract information for various purposes, such as market research, pattern recognition, grouping of customers in marketing, computer vision, social networks analysis and outliers detection. When dealing with big data, some clustering algorithms are impractical due to their computational complexity which makes those algorithms unusable in case of huge datasets.

For example, K-Means [5,6] has complexity equal to $O(n^{dk+1})$, where n denotes the number of objects to be clustered, d is the number of dimensions of the objects and k the desired number of clusters, too high even for a modest amount of data. Similarly, DBSCAN [7] with a complexity of $O(n \log n)$ can have problems in dealing with very large amount of data; moreover, it also requires $O(n^2)$ space, so that handling huge data is anyway difficult [8].

G. Nicosia et al. (Eds.): LOD 2022, LNCS 13811, pp. 373–387, 2023.
https://doi.org/10.1007/978-3-031-25891-6_28

RASTER is a recent clustering algorithm with $O(n)$ time complexity, linear with regard to the number of input data and a space requirement which is even smaller than n, because in its original version the algorithm does not keep the information of the whole dataset but only retains the so-called "most significant tiles". RASTER is a geometric, grid-based clustering algorithm on which the distributed P2P algorithm presented in this work is based. A grid-based clustering algorithm works by summarizing the whole dataset through a geometric grid representation; the clusters are then obtained by merging the grid cells (called tiles in RASTER). In practice, it is the geometric space surrounding the dataset points that is clustered instead of the points. The geometric spatial summary of the dataset, built covering the data points with a grid, is then weighted by assigning to each grid cell a weight strictly related to the cell's density, i.e., the number of points falling within the cell. Finally, adjacent dense cells are merged together to form clusters. In practice this kind of approach is very similar to the one used by density-based algorithms, but instead of being applied to the dataset points it is applied to the grid cells, taking into account both the local cell densities and the neighbouring relationships among the cells. Therefore, it is a very efficient approach, that allows dealing with massive datasets.

Besides looking for algorithms that can work in linear time, the need for parallel and distributed algorithms is rising. Indeed, it is increasingly common to deal with large amounts of heterogeneous data stored on many independent geographically spread computers; another common situation is provided by deployments of Wireless Sensor Networks (WSNs). In many applications the distributed nodes can not send their information to a centralized node for subsequent processing. For instance, limited bandwidth or security reasons prevent this approach. Moreover, in the case of WSNs the transmission of data is severely affected by the distance, so that a given node may only communicate with its neighbours. In general, transmitting a huge amount of data from a node to a centralized node may not even be possible in some applications (e.g., in WSNs the available transmission power limits the effective communication distance, so that only neighbours node within a certain radius can be reached).

A common requirement is that, ideally, the quality of the clustering done in the distributed setting should match the quality obtained in the centralized setting, in which all of the data are stored in a single site. In this paper we design two possible distributed variants of the RASTER algorithm; in particular, the second variant may be thought as an extension of the first. Since we have selected RASTER for its specific properties (discussed in detail in Sect. 3), in this paper we show that our distributed versions of the algorithm retain the same properties and provide identical or very similar results with regard to accuracy to the ones obtained by the sequential version, matching the quality requirement. The authors of the sequential RASTER algorithm already discussed in their work the accuracy and effectiveness of their approach comparing RASTER against the state of the art algorithms; the interested readers can refers to [8] for further analysis and comparison.

The manuscript is organized as follows: in Sect. 2 we discuss related work, recalling the state-of-the-art clustering algorithms; in Sect. 3 we present the sequential version of RASTER, whilst in Sect. 4 we describe the design of our distributed algorithm; in Sect. 5 we discuss the experimental results related to the distributed and sequential versions, evaluating the quality of the clusters obtained in the distributed approach with regard to the ones obtained by the sequential version. We draw our conclusions and final remarks in Sect. 6.

2 Related Work

The clustering algorithms can be classified in the following main categories, though they are not exhaustive: representative-based clustering, hierarchical, density-based, graph-based, spectral, grid-based. An algorithm can fall into more than one category, for example RASTER is both a grid and a density-based algorithm. K-MEANS and EXPECTATION-MAXIMIZATION are representative-based, whilst DBSCAN is density-based. In this Section we recall and discuss some of the most common and used clustering algorithms. We shall also briefly discuss two algorithms that use a spatial data approach similar to RASTER.

K-MEANS is one of the most popular clustering algorithm [5,6]. Its goal is to group the input data into k different clusters, where all of the items belonging to the same cluster are similar to each other but different from the items belonging to other clusters. It is a representative-based algorithm and uses a centroid-based approach to create the clusters. The final result of the algorithm strongly depends on the value of the parameter k that must be established in advance and on the initial choice of centroids.

MEAN-SHIFT [4] is a density-based algorithm. It is also known as *mode-seeking* algorithm. Indeed, it is a procedure for locating the maxima - the modes - of a density function given discrete samples from that function [1]. Unlike K-MEANS, this algorithm does not require prior knowledge of the number of clusters, which is instead automatically inferred by the algorithm. It is a centroid-based algorithm, it assigns the data points to the clusters by shifting the data points towards high-density regions. MEAN-SHIFT is based on the concept of Kernel Density Estimation (KDE), which is a method to estimate the underlying distribution of a set of data.

DBSCAN (Density-Based Spatial Clustering of Applications with Noise) [7] is probably one of the best and most widely used clustering algorithm. It is a density-based algorithm: it groups all of the points whose distance is less than a fixed threshold and validates the group as a cluster only if its number of points is greater than a density threshold. DBSCAN does not need to know in advance the number of clusters we expect to find and does not assign to any cluster the points present in low density areas, which are considered noise and reported as outliers.

The EXPECTATION-MAXIMIZATION (EM) clustering algorithm [3] belongs to the category of the *soft clustering* algorithms. Unlike *hard clustering* algorithms that assign a point exactly to one cluster, in soft clustering, a point x_i is assigned to a cluster C_h with a probability p_{ih}, where $i \in \{1, 2, \cdots, n\}$, $h \in \{1, 2, \cdots, k\}$ and k denotes the number of clusters.

BIRCH (Balanced Iterative Reducing and Clustering using Hierarchies) [10], is the first clustering algorithm proposed in the database area to effectively handle noise (data points that are not part of the underlying pattern). BIRCH is a hierarchical clustering algorithm and it is designed to handle very large datasets. Its working principle is slightly different from probabilistic or centroid-based algorithms: the purpose of the algorithm is to use the memory as efficiently as possible and to ideally cluster a dataset with a single scan pass, possibly using a few additional passes to improve the quality of the obtained clustering.

STING (STatistical INformation Grid approach) [9] is one of the first grid-based spatial data mining cluster algorithms. The workspace is partitioned through a hierarchical structure composed of rectangular cells. The first level of this structure is a single, large rectangle. The characteristic that distinguishes this algorithm is that it is query independent, i.e., the algorithm computes for each level some parameters and then uses them to respond efficiently to the queries that are forwarded to it.

3 Sequential Algorithm

RASTER [8] is a geometric, grid-based clustering algorithm designed to manage big data, since its complexity scales linearly with regard to the input size, therefore its computational complexity is $O(n)$ where n denotes the number of dataset points. The algorithm consists of two main steps: **Projection** (or **Contraction**) and **Clustering**. The algorithm uses four parameters to cluster the dataset: the precision *prec* determines the size of the grid that shall be used in the projection step (a.k.a. the resolution of the projecting grid), the threshold th is used for determining whether a tile is considered significant or not, min_size indicates the minimum number of tiles required to form a cluster and δ represents the maximum distance between a tile that belongs to a cluster and a new tile that could be included in the same tile. A cluster is formed by m agglomerated tiles, with the constraint $m \geq min\_size$, Finally, it is worth recalling here that the algorithm does not need to know in advance the number of clusters to create. The procedures of the RASTER algorithm are shown in pseudocode respectively as Algorithm 1 and 2.

The algorithm starts with the projection phase. In general a value p is projected onto the tile t using a precision *prec* as follows: $t = \left\lfloor \frac{p}{prec} \right\rfloor$.

The choice of the precision parameter *prec* is fundamental for the algorithm. If this value is too small we shall obtain small tiles and therefore there is the risk of producing a significant number of tiles too far apart from each other resulting in an overestimation of clusters, whilst if we choose a value that is too big we get consequently very large tiles with the risk of joining two or more clusters together. The authors of RASTER do not indicate a specific method to define the correct value of *prec*. The parameter must be empirically estimated knowing the distribution of the input dataset. In the case of two or more dimensions,

Algorithm 1: RASTER - Projection

```
   input : Dataset D, prec, th
   output: set of Tiles T
 1 T ← ∅ // set of Tiles
 2 foreach p ∈ D do
 3 |  project p to corresponding tile t
 4 |  t.weight ← t.weight + 1
 5 |  T ← T ∪ {t} // insert t in the set of tiles
 6 end
 7 foreach t ∈ T do
 8 |  if t.weight < th then
 9 |  |  remove t from T
10 |  end
11 end
```

Algorithm 2: RASTER - Clustering

```
   input : Dataset T, distance δ, min_size
   output: Clusters set C
 1 C ← ∅
 2 while T ≠ ∅ do
 3 |  arbitrary select tile t from T
 4 |  c ← c ∪ {t} // initialize a potential cluster
 5 |  remove t from T
 6 |  foreach t ∈ c do
 7 |  |  while ∃ t_n ∈ T : distance(t, t_n) < δ do
 8 |  |  |  c ← c ∪ {t_n}
 9 |  |  |  remove t_n from T
10 |  |  end
11 |  end
12 |  if c.size ≥ min_size then
13 |  |  C ← C ∪ c
14 |  end
15 end
```

the projection is applied to all of the dimensions. After the projection phase, the algorithm prunes all of the tiles with an insignificant weight (i.e. the tiles with a number of projected points less than the threshold th are discarded).

Only the tiles with a significant number of points are considered for the clustering phase. A cluster is built starting from a tile randomly chosen from those that do not belong to any other cluster. The cluster is then enlarged including the neighbouring tiles of the tiles that are part of the cluster. Two tiles are considered neighbours if their distance is less than or equal to δ, where the definition of *distance* can be chosen from the norma-l family. In our implementation we used norma-1 distance (a.k.a. Manhattan distance). Finally, a further phase of pruning is applied where the clusters that include a number of tiles less than min_size are discarded and the points projected in the discarded tiles are considered as noise (i.e. unclustered). It is important to highlight that for each cluster only the list of its tiles is retained: the original version of the algorithm does not retain/store all of the input data. This is to avoid using too much memory, in case we are dealing with extremely large datasets.

4 Distributed Algorithm

We assume that the distributed nodes may not communicate with a centralized node (for instance owing to limited bandwidth or due to security policies); therefore, this rules out any centralized protocol. Moreover, in many different situations (e.g. WSNs) each node may only communicate with neighbouring nodes. Therefore, in this Section we introduce two versions of a distributed clustering algorithm based on RASTER by means of a gossip–based protocol. The two versions differ with regard to the way the clustering phase of RASTER is handled in the distributed environment. The second variant may be also thought as an extension of the first one.

The distributed context is such that each peer independently handles a data stream and, at query time, a global clustering must be provided. The mergeability property of the RASTER algorithm has been discussed by the authors in [8]. This property ensure that two RASTER data structures which represent the set of projected tiles coming from two distinct data streams can be merged together providing a resulting data structure equivalent to the one that would have been obtained if the input data had been processed together in a single stream. Due to this property, the sequential algorithm can be used in a distributed (but also in a parallel) environment without any modification.

A gossip–based protocol [2] is a synchronous distributed algorithm consisting of periodic rounds. In each of the rounds, a process, which is known in this context as a peer (or agent) randomly selects one or more of its neighbours, exchanges its local state with them and finally updates its local state. The information is disseminated through the network by using one of the following possible communication methods: (i) *push*, (ii) *pull* or (iii) *push–pull*. The main difference between push and pull is that in the former a peer randomly selects the peers to whom it wants to send its local state, whilst in the latter it randomly selects the peers from whom to receive the local state. Finally, in the hybrid push–pull communication style, a peer randomly selects the peers to send to and from whom to receive the local state. In this synchronous distributed model it is assumed that updating the local state of a peer is done in constant time, i.e., with $O(1)$ worst-case time complexity; moreover, the duration of a round is such that each peer can complete a push–pull communication within the round. Finally, the performance of a gossip–based protocol is measured as the number of rounds required to achieve convergence.

4.1 The P2PRASTER Algorithm

In this Section a distributed algorithm for clustering the global dataset through a gossip–based approach is described. The distributed algorithm aims at achieving the same accuracy that can be achieved if the whole dataset is processed by the sequential RASTER algorithm. We denote by N_i the number of points assigned to peer i. The global dataset consists of the union of the datasets owned by each peer; its size is $N = \sum_{i=1}^{p} N_i$, where p is the number of peers.

We recall that RASTER consists of two main steps, the projection (or contraction) and the clustering. In the P2PRASTER version only the projection step is executed in the distributed environment, whilst the clustering phase is executed locally after reaching the consensus over the projection of the global dataset. The pseudocode of the algorithm is reported as Algorithm 3. Each peer i starts by computing locally the projection of its data points onto the tiles resulting in the set of tiles $\mathcal{T}_i$. We assume that each peer uses the same value for the parameter *prec*, which represents the tile size. During the gossip phase, the peers follow an averaging protocol to reach the consensus, exchanging their states by means of a *push–pull* communication. The state of a peer is represented by the set of tiles $\mathcal{T}_i$ and the estimated value $\tilde{q}_i$ which is used to derive the number of peers involved in the P2P network. Therefore, for two peers i and j communicating in round r their state at round $r+1$ is computed according to the averaging protocol as follow:

$$\begin{aligned} \mathcal{T}_i(r+1) &= \mathcal{T}_j(r+1) = \mathcal{T}_i(r) \oplus \mathcal{T}_j(r), \\ q_i(r+1) &= q_j(r+1) = \frac{q_i(r) + q_j(r)}{2}, \end{aligned} \tag{1}$$

where the operator $\oplus$ is the merging operator between the sets of tiles and shall be described later. During each round a peer selects fo (the fan-out) neighbours uniformly at random and sends to each of them its local state in a message of type *push*. Upon receiving a message, the peer extracts the message's type, sender and state. A message is processed according to its type as follows. A *push* message is handled in two steps. In the first one, the peer updates its local state by using the state received; this is done by invoking the merging procedure described as Algorithm 4. In the second one, the peer sends back to the sender, in a message of type *pull*, its updated local state. A pull message is handled by a peer setting its local state equal to the state received.

The merging procedure is applied to two different sets of tiles $\mathcal{T}_i$ and $\mathcal{T}_j$ producing the final merged set $\mathcal{T}_{ij}$ which is the union of the sets and each tile τ_{ij} has a weight computed as the average between the corresponding tiles from the sets $\mathcal{T}_i$ and $\mathcal{T}_j$. If one of the sets does not include the tile τ_{ij}, the corresponding weight is considered equal to 0. In the MERGE procedure the state variable $\tilde{q}$ is updated by the average between the values $\tilde{q}_i$ and $\tilde{q}_j$ of the corresponding peers i and j.

According to the averaging protocol, after the execution of a certain number of rounds, all of the peers converge to a common estimate both for $\tilde{q}$ and $\mathcal{T}$ that is equal to the average value among the initial values belonging to the peers.

The state variable $\tilde{q}_i$ is initialized at the beginning with only one peer setting it to 1 whilst all of the others peers set it to 0, hence the final value converges to

$$\begin{aligned} \tilde{q} &= \frac{1}{p}, \\ \tau.weight &= \frac{1}{p} \sum_{i=1}^{p} \tau_i.weight \; \forall \; \tau \in \mathcal{T}. \end{aligned} \tag{2}$$

The value of $\tilde{q}$ is then used to compute the final values of the tiles weight. At the end of the gossip phase, each peer has computed its own estimate of the number of peers and has obtained the complete projection that includes all of the tiles; the algorithm then discards the tiles which are not significant (i.e. it prunes those tiles whose weight is less than the threshold th) and finally it locally executes the clustering phase. We assume that the distributed gossip algorithm has reached convergence, and this implies that each peer has estimated the correct cardinality of each tile and can execute a local clustering.

The Clustering procedure in Algorithm 3 is equal to the sequential version already discussed in Sect. 3. We also propose a different procedure to cluster the tiles. In the distributed version just discussed each peer clusters all of the tiles obtained from the gossip phase. Therefore, if the peers do not reach the consensus, then the information among the peers can be different and each peer can find different clusters. We have therefore designed another distributed version of the algorithm to distribute the load due to clustering among the various peers. Then, the peers communicate with each other to merge the information related to the clusters. In the remainder of the paper we refer to the first distributed version as P2PRASTER_v1 and to the second as P2PRASTER_v2. Algorithm 5 is related to the pseudocode of the clustering procedure for the P2PRASTER_v2 variant. The parameter msq, passed as input to the procedure, denotes the number of squares to be assigned to each peer in a checkerboard partition of the tiles domain.

It is worth noting here how each peer after removing the **non-significant** tiles, calls the procedure `GetPeerProjection`, which removes all of the tiles that do not need to be processed by the peer l and retains only a small portion of tiles denoted by $tiles'$. It is therefore necessary to establish which tiles a peer must process.

The algorithm requires that each peer partitions its tiles domain using a checkerboard partition (obviously this partition must not cut any tile). A peer shall only keep squares whose identifier is $id = k \cdot l$, with $k = \{1, \cdots, msq\}$.

We assume that the square at the bottom left has $id = 0$. The id of squares increases by moving horizontally along the axis denoting one dimension, and at the end of the dimension's domain we go up to the next square until we reach the square at the top right. Using its identifier l, each peer knows which squares it must process, and having access to the entire domain of the tiles it can get the coordinates of its squares. A linear scan of the tiles is enough for a peer to keep only the tiles falling inside its domain.

Algorithm 3: P2PRASTER

```
1  Procedure GOSSIP
2      for r = 0 to R do
3          neighbors ← select fo random neighbors
4          foreach i ∈ neighbors do
5              SEND(push, i, state_{r,l})
6          end
7      end
8  end procedure
9  Procedure ON_RECEIVE(msg)
10     state ← msg.state
11     if msg.type == push then
12         state_{r+1,l} ← MERGE(state, state_{r,l})
13         SEND(pull, msg.sender, state_{r+1,l})
14     end
15     if msg.type == pull then
16         state_{r+1,l} ← state
17     end
18 end procedure
19 Procedure ON_CONSENSUS
20     p̃_{r,l} ← 1 / q̃_{r,l}
21     foreach t ∈ T_{r,l} do
22         t.weight ← t.weight · p̃_{r,l}
23     end
24     T_{r,l} ← ProjectionThreshold(T_{r,l}, th)
25     C_{r,l} ← Clustering(T_{r,l}, min_size)
26 end procedure
```

The algorithm does not definitively discard the tiles out of the squares to be examined, because when a peer clusters its tiles contained in *tiles′*, it can find some clusters that may lie between two squares belonging to two different peers. This cluster shall then be found in part by a peer and in part by another peer. When the algorithm is parallel this problem has an immediate solution: it is enough to set the communication between the peers that have adjacent squares to reconstruct the original cluster (see P-RASTER in [8]). For a distributed algorithm this is impractical, because two peers can communicate with each other only if they are neighbours. To overcome this problem, each peer keeps the "discarded tiles" so it can complete clusters straddling two squares belonging to two different peers. This involves a small trade-off as some clusters shall be "found" by two or more peers. This is why the `ClusteringTiles'` procedure take as input both *tiles* and *tiles′*.

The trade-off arises owing to the need to subsequently identify and manage "duplicate" clusters. To solve this problem, each peer computes the centroid of

Algorithm 4: MERGE

```
    Input: (𝒯_i, q̃_i),( 𝒯_j, q̃_j)
1  q̃_j ← (q̃_i + q̃_j)/2
2  foreach t_i ∈ 𝒯_i do
3  |  t_j ← GetTile(𝒯_j, t_i)
4  |  if ∃ t_j then
5  |  |  t_j.weigth ← t_j.weight + t_i.weight
6  |  else
7  |  |  𝒯_j ← 𝒯_j ∪ {t_i}
8  |  end
9  end
10 /* compute average value for all of the tiles */
11 foreach t ∈ 𝒯_j do
12 |  t.weight ← t.weight/2
13 end
14 return ( 𝒯_j, q̃_j)
```

Algorithm 5: CLUSTERING'

```
1  Procedure CLUSTERING'(msq)
2  |  (tiles, q̃_{r,l}) ← state_{r,l}
3  |  p̃_{r,l} ← 1/q̃_{r,l}
4  |  foreach tile t ∈ tiles do
5  |  |  t.v ← t.v · p̃_{r,l} // compute tile value
6  |  end
7  |  tiles ← ProjectionThreshold(tiles, th)
8  |  /* each peer keeps only a few tiles */
9  |  tiles' ← GetPeerProjection(tiles, l, msq)
10 |  clusters ← ClusteringTiles'(tiles', tiles, ms)
11 |  centroids ← GetCentroids(clusters)
12 end procedure
```

its cluster with a weighted averaging of the tiles. The centroid is also projected onto the tile that contains it. Now the peers begin a new phase of communication according to the gossip protocol. This procedure involves communication between the peers in order to merge the clusters and remove the duplicates. The state of a peer l in round r is given by the clusters and their centroids: $state_{r,l} = (clusters_{r,l}, centroids_{r,l})$. We can merge two sets of clusters from two different peers and detect if there are duplicates. The decision to consider equal, similar or different two clusters is made using a procedure that takes as input the centroids of two clusters $a = (x_1, y_1)$ and $b = (x_2, y_2)$, and returns the Manhattan distance L_1. There are three cases:

1. $L_1(a, b) > \delta \; \forall \; b \in centroids_j$ or viceversa: a cluster is only present in one of the two peers and is therefore maintained;
2. $0 < L_1(a, b) \leq \delta$: in this case the two clusters are similar but not the same, and it is necessary to invoke a procedure that performs the merge of the two clusters; in turn this requires recomputing the new centroid;
3. $L_1(a, b) = 0$: the two clusters can be considered equal, in this case only one cluster is maintained.

At the end of this procedure, all of the peers shall have reached the consensus on the clusters with high probability. It is worth noting here that the check to establish if two clusters are different, equal or similar is carried out considering only the centroids, which are computed preliminarily by each peer and eventually updated only in case of similar clusters. This implies that there is no need to examine all of the clusters every time, saving precious time.

With regard to the previous version, at the end of the algorithm we expect that the information in each peer shall be almost uniform, despite at the cost of greater computational complexity.

5 Experimental Results

In this Section, we compare and discuss the results obtained by the distributed algorithm and the original sequential version. In order to assess the algorithm's performance, we implemented a P2P simulator which is able to run the algorithm fully supporting our distributed gossip–based protocol. The source code is freely available[1] as open source, for reproducibility of the results and experimentation. We shall assume as ground truth the clustering obtained by the sequential algorithm using the same parameters for both versions. The assessment metrics used for validation are the following ones: Precision; Recall; F1-Measure and Conditional Entropy.

Precision measures the fraction of relevant instances among the retrieved instances, *recall* is the fraction of the total amount of relevant instances that were actually retrieved, the F_1 metric (or F_1 score), is defined as the harmonic mean between the precision and recall. In practice this metric tries to balance the two values obtained by combining them together. Finally, the *Conditional Entropy* quantifies the amount of information needed to describe the outcome, a high entropy value indicates that a cluster is split into several partitions. For a perfect clustering this value must be 0, whilst in the worst case it is equal to $\log k$.

In carrying out these tests, two synthetic datasets, respectively consisting of 5 and 8 million points, were used. Both datasets were first clustered using the sequential algorithm RASTER. Then, we executed several runs of the distributed algorithm on both datasets to understand how the results of the distributed algorithm change with regard to the sequential algorithm when

[1] https://github.com/cafaro/P2Praster.

some parameters vary. Specifically, tests were performed for both versions of P2PRASTER and using for each run the Barabási-Albert random graph model.

In each of the following plots we report the run of the sequential and distributed version varying the following parameters: number of peers and number of rounds executed. For each run, one of these parameters is fixed, allowing the other to vary.

Figures 1 shows how the results vary when the number of peers in the network increases from 500 to 2000 whilst the fan-out is fixed to 1 and the number of rounds to 10. We can see that none of the four metrics are affected. The trend of P2PRASTER_v2, however, shows that the peers that do not reach convergence are a small minority. We can therefore deduce that the algorithm suffers very little from the variation of the number of peers.

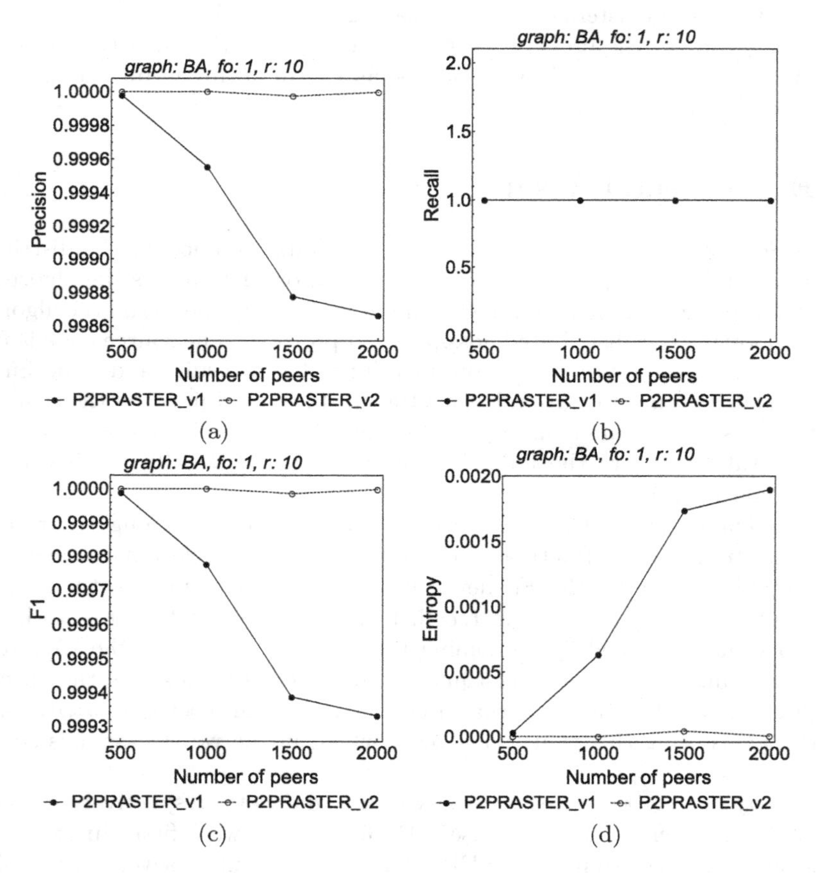

Fig. 1. Results obtained when the number of peers increases in a Barabási-Albert graph

Figure 2 depicts the case in which the number of rounds is allowed to increase from 10 to 22 whilst the fan-out is fixed to 1 and the number of peers to 1000. Also in this case we obtain results in line with the characteristics of the two algorithms. The plots show the presence of an error in both algorithms if we execute a low number of rounds, but the error vanishes for all of the metrics as the number of rounds increases.

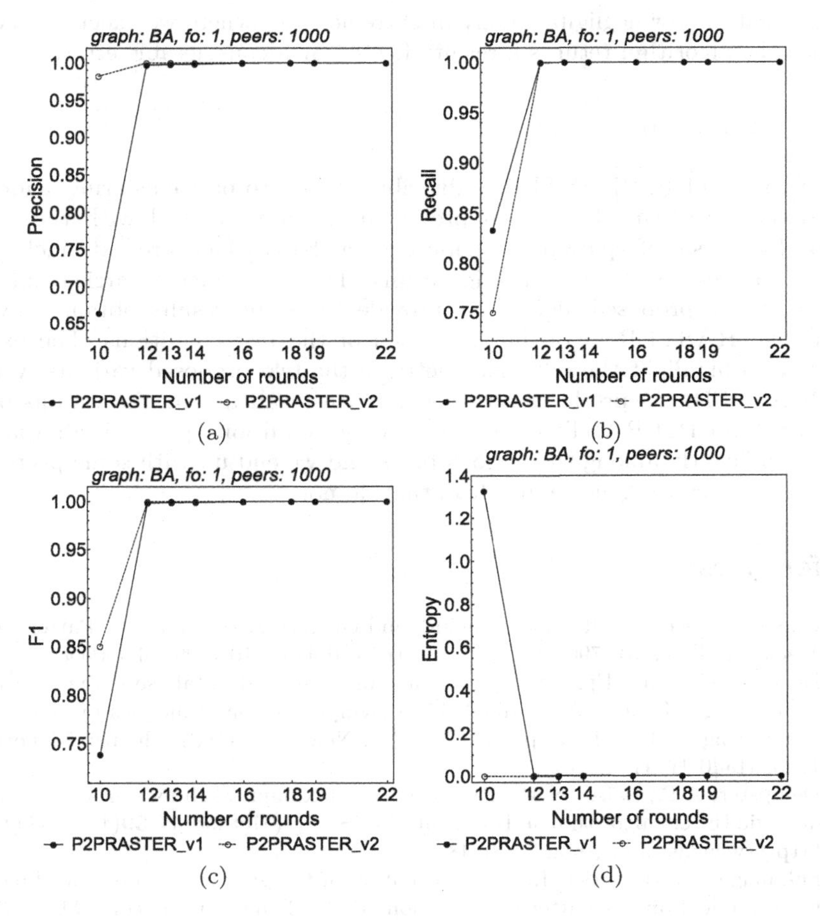

Fig. 2. Results obtained when the number of rounds increases in a Barabási-Albert graph

We compared the algorithm by using also the bigger dataset. Also in this case we obtained the same behaviour as previously described. We do not report here the plots owing to the space constraint of the manuscript. For both datasets we have obtained results almost similar and in accordance with what we expected given the characteristics of the algorithm. These tests have highlighted how the results of the distributed version are practically identical to those of the sequential version, not only if the algorithm is allowed to achieve convergence, but also with a smaller number of rounds. The algorithm needs to perform 22 rounds of communication in the Barabási-Albert graph to obtain an error smaller than the convergence threshold which has been set to 0.01. However, the algorithm converges perfectly in both versions with much less rounds, already using 16 rounds in most tests; therefore we can state that the algorithm has an error equal to 0 or negligible. Only in extreme cases where we execute very few rounds the algorithm returns a result affected by a considerable error.

6 Conclusion

Two variants of P2PRASTER, a distributed peer-to-peer clustering algorithm based on RASTER, have been presented in order to deal with the decentralized analysis of spatially distributed big data, which are extremely difficult to process by traditional algorithms. The experiments carried out confirm that the proposed algorithms provide the same results obtained by the sequential RASTER algorithm under all of the test conditions. The experiments also highlight the difference between the two proposed variants: version P2PRASTER_v1 produces a rather uniform result among the various peers, whilst variant P2PRASTER_v2, in extreme conditions, gives results that can vary significantly among the various peers and we end up with some peers that have reached convergence and others that do not.

References

1. Cheng, Y.: Mean shift, mode seeking, and clustering. IEEE Trans. Pattern Anal. Mach. Intell. **17**(8), 790–799 (1995). https://doi.org/10.1109/34.400568
2. Demers, A., et al.: Epidemic algorithms for replicated database maintenance. In: Proceedings of the Sixth Annual ACM Symposium on Principles of Distributed Computing, PODC 1987, pp. 1–12. ACM, New York (1987). https://doi.org/10.1145/41840.41841
3. Dempster, A.P., Laird, N.M., Rubin, D.B.: Maximum likelihood from incomplete data via the em algorithm. J. Roy. Stat. Soc. Ser. B (Methodol.) **39**(1), 1–38 (1977). http://www.jstor.org/stable/2984875
4. Fukunaga, K., Hostetler, L.: The estimation of the gradient of a density function, with applications in pattern recognition. IEEE Trans. Inf. Theory **21**(1), 32–40 (1975)
5. Lloyd, S.: Least squares quantization in PCM. IEEE Trans. Inf. Theory **28**(2), 129–137 (1982)
6. MacQueen, J.: Some methods for classification and analysis of multivariate observations. In: Proceedings of the Fifth Berkeley Symposium on Mathematical Statistics and Probability, Volume 1: Statistics, pp. 281–297. University of California Press, Berkeley (1967)
7. Schubert, E., Sander, J., Ester, M., Kriegel, H.P., Xu, X.: DBSCAN revisited, revisited: why and how you should (still) use DBSCAN. ACM Trans. Database Syst. **42**(3), 19:1–19:21 (2017). https://doi.org/10.1145/3068335

8. Ulm, G., Smith, S., Nilsson, A., Gustavsson, E., Jirstrand, M.: Contraction clustering (RASTER): a very fast big data algorithm for sequential and parallel density-based clustering in linear time, constant memory, and a single pass (2019)
9. Wang, W., Yang, J., Muntz, R.R.: Sting: A statistical information grid approach to spatial data mining. In: Proceedings of the 23rd International Conference on Very Large Data Bases, VLDB 1997, pp. 186–195. Morgan Kaufmann Publishers Inc., San Francisco (1997)
10. Zhang, T., Ramakrishnan, R., Livny, M.: BIRCH: an efficient data clustering method for very large databases. SIGMOD Rec. **25**(2), 103–114 (1996). https://doi.org/10.1145/235968.233324

Surrogate Modeling for Stochastic Assessment of Engineering Structures

David Lehký(✉), Lukáš Novák, and Drahomír Novák

Brno University of Technology, Brno, Czech Republic
lehky.d@fce.vutbr.com

Abstract. In many engineering problems, the response function such as the strain or stress field of the structure, its load-bearing capacity, deflection, etc., comes from a finite element method discretization and is therefore very expensive to evaluate. For this reason, methods that replace the original computationally expensive (high-fidelity) model with a simpler (low-fidelity) model that is fast to evaluate are desirable. This paper is focused on the comparison of two surrogate modeling techniques and their potential for stochastic analysis of engineering structures; polynomial chaos expansion and artificial neural network are compared in two typical engineering applications. The first example represents a typical engineering problem with a known analytical solution, the maximum deflection of a fixed beam loaded with a single force. The second example represents a real-world implicitly defined and computationally demanding engineering problem, an existing bridge made of post-tensioned concrete girders.

Keywords: Surrogate modelling · Artificial neural networks · Polynomial chaos expansion · Stochastic assessment · Uncertainties propagation

1 Introduction

In line with the emphasis on environmental and economic sustainability of the economy, stochastic analysis of structures with respect to uncertainties is becoming increasingly important in the design, assessment, and maintenance of engineering structures, providing valuable information on the variability of response, sensitivity and system reliability. Uncertainties occur at several stages of computational modeling. The modeling of a physical system can be viewed as a mathematical function that depends on input parameters. In computational mechanics, the mathematical function is usually defined using the finite element method (FEM). The input variables characterize geometry, materials or loads and responses can be arbitrary quantities. A probabilistic framework is considered, which means that the input parameters are random variables. Thus, the aim of stochastic analysis is to propagate uncertainties through a mathematical model in order to obtain statistical information about the outputs. The main interest is often reliability analysis, which is performed by calculating the failure probability, i.e., the probability of reaching a certain limit state, e.g., inducing a certain level of stress in the structure, initiating cracks or collapse of the structure.

G. Nicosia et al. (Eds.): LOD 2022, LNCS 13811, pp. 388–401, 2023.
https://doi.org/10.1007/978-3-031-25891-6_29

If the structural response function is discretized using the FEM, its evaluation is very time-consuming and computationally expensive. Therefore, methods that replace the original expensive (high-fidelity) computational model with a simpler (low-fidelity) model that is fast to evaluate are desirable. This new model, called the surrogate model (or metamodel), can be of different types in terms of efficiency.

Among the surrogate models that have gained popularity among researchers in recent decades are polynomial chaos expansion (PCE, Ghanem and Spanos 1991; Blatman and Sudret 2010, 2011; Novák and Novák 2018), artificial neural networks (ANN, Lehký and Šomodíková 2017), support vector machine (SVM, Hurtado 2004, Bourinet et al. 2011) and the Kriging metamodel (Echard et al. 2013). The applications of ANN-based surrogate models in reliability-based optimization are presented in Lehký et al. (2018) and Slowik et al. (2021). Details on ANN-based inverse response surface method used when solving inverse reliability problems can be found in Lehký et al. (2022). This paper focuses on two of above-mentioned surrogate modeling techniques and their potential for stochastic analysis of engineering structures. The first technique is PCE, a representative from the field of uncertainty quantification (UQ). PCE is often employed in applications with limited number of samples since it achieves high accuracy and allows powerful post-processing without additional cost (Sudret 2008). This can yield statistical moments or Sobol sensitivity coefficients, which is a frequent requirement of stochastic analysis. The second technique is ANN, a well-known machine learning model used in many mathematical and engineering fields. ANN is a powerful signal processing model that can efficiently handle large numbers of samples, which determines its common use in big data analysis. Both models have great potential for use in the field of stochastic structural analysis, each with its specific advantages and disadvantages. This paper compares the methods on selected engineering problems and discusses aspects of their application.

2 Surrogate Models

The use of both PCE and ANN methods for surrogate modeling and statistical analysis leads to efficient and accurate stochastic assessment of structures (Lehký and Šomodíková 2017). On the one hand, PCE offers high accuracy using a small number of samples for construction together with powerful analytical post-processing. However, the accuracy of PCE is always limited for highly non-linear functions (or models with discontinuities) due to the polynomial approximation. On the other hand, ANN is a general technique without any assumptions limiting its accuracy for specific functions, although it sometimes needs more samples for training and additional computational cost for post-processing. This paper compares both methods in selected typical engineering applications and shows their synergy for stochastic analysis.

Experimental samples, i.e., evaluation of the FEM model for the optimal set of input parameters, are needed to optimize the parameters of the surrogate models. In our case, due to its efficiency and versatility, the Latin Hypercube Sampling (LHS, McKay et al. 1979; Stein 1987; Novák et al. 2014) is used for the experimental design (ED) of both surrogate models. The LHS belongs to the category of advanced stratified sampling methods, which results in very good estimation of statistical moments of the response using a small number of simulations. More specifically, LHS is considered one of the variance reduction techniques because it provides lower variance of estimates of statistical moments compared to crude Monte Carlo sampling for the same sample size, see e.g., Koehler and Owen (1996). This is why the technique became very attractive in solving computationally intensive problems with a multidimensional space of random variables, resulting in an efficient ED.

2.1 Polynomial Chaos Expansion

Assume a probability space $(\Omega, \mathcal{F}, \mathcal{P})$, where Ω is an event space, $\mathcal{F}$ is a σ-algebra on Ω and $\mathcal{P}$ is a probability measure on $\mathcal{F}$. If the input vector of the mathematical model is random vector $\mathbf{X}(\omega)$, $\omega \in \Omega$, then random model response $Y(\omega)$ is a random variable. Considering $Y = f(\mathbf{X})$ has the finite variance σ^2, i.e. $Y \in L_2(\Omega, \mathcal{F}, \mathcal{P})$. Therefore, the polynomial chaos expansion according to Soize and Ghanem (2004) is in the following form:

$$Y = f(\mathbf{X}) = \sum_{\alpha \in N^M} \beta_\alpha \psi_\alpha(\boldsymbol{\xi}), \tag{1}$$

where M is the number of input random variables, β_α are unknown deterministic coefficients and ψ_α are multivariate basis functions orthonormal with respect to the joint probability density function (PDF) of $\boldsymbol{\xi}$. The basis functions must be selected in dependence to distributions of input random vector $\mathbf{X}$ which must be transformed to associated standardized variables $\boldsymbol{\xi}$. It is common to use Legendre polynomials for Uniform distributions and Hermite polynomials for Normal/Lognormal distributions in engineering applications. Another associated polynomials to specific distributions can be found in the Wiener-Askey scheme (Xiu and Karniadakis 2002).

In practical applications, it is necessary to use PCE truncated to a finite number of terms P. Truncated set of basis functions $\mathcal{A}^{M,p}$ is dependent on given maximal polynomial order p and M as follows:

$$\mathcal{A}^{M,p} = \{\alpha \in N^M : |\alpha| = \sum_{i=1}^{M} \alpha_i \le p\}. \tag{2}$$

Deterministic coefficients β_α can be obtained by non-intrusive approach. Non-intrusive methods utilize the original mathematical model, e.g., FEM model, as a black box, which allows for their easy applications in combination with commercial software. There are generally two types of non-intrusive methods for calculation of deterministic coefficients: spectral projection and linear regression. Regression is typically less expensive than spectral projection and thus it is often employed in engineering applications involving highly computationally expensive models. ED contains sample points in M-dimensional space and corresponding results of the original model. Size of ED must be higher than P (recommended size is typically $3P$). Moreover, sample points should uniformly cover the whole input space. This can be achieved by a Latin hypercube sampling method or any advanced sampling algorithm.

The original mathematical model must be evaluated to obtain a vector of results corresponding to generated sample points. Once the basis functions are created and ED is prepared, PCE coefficients can be estimated by ordinary least square (OLS) regression method. Unfortunately, truncated PCE solved by OLS is not highly efficient and cannot be employed for practical examples with large number of input random variables. The solution is additional reduction of truncated basis set by any model-selection algorithm such as Least Angle Regression (LAR) (Tibshirani et al. 2004). LAR automatically detects the most important basis functions for given ED and create so called sparse PCE (Blatman and Sudret 2010, 2011). Approximation of engineering models by sparse PCE can be justified by the sparsity-of-effects principle, which is often observed in practical applications. For further reduction of computational cost, it is beneficial to employ advanced sampling schemes for sequential enrichment of ED until the desired accuracy is achieved (Novák et al. 2021).

An important part of surrogate modelling is an estimation of the accuracy of the surrogate model. It is common to use the coefficient of determination R^2. However, this can lead to overfitting when model is too complex for the sample data. Therefore, more sophisticated cross-validation techniques, such as Leave-one-out error Q^2, should be employed. This technique helps to assess how well the model fits new observations that weren't used in the model calibration process.

The PCE allows for powerful and efficient post-processing. First of all, the advantage of the explicit form of PCE is its possibility to obtain Q^2 analytically without additional computational demands which can be further used for adaptive construction of PCE approximation. Besides analytical derivation of Q^2 directly from PCE (Blatman and Sudret 2010), it is possible to derive also the first four statistical moments directly from estimated coefficients (Novák 2022). Moreover, PCE allows for analytical derivation of Sobol indices (Sudret 2008).

In this paper, an algorithm for construction of non-intrusive sparse PCE based on LAR is employed for numerical examples (Novák and Novák 2018). The algorithm is depicted in simplified Fig. 1. It consists of 3 virtual parts: pre-processing (construction of stochastic model and ED), processing (adaptive and possibly sequential construction of PCE) and post-processing including analytical part (analysis of PCE function) and numerical part (utilization of PCE as a surrogate model).

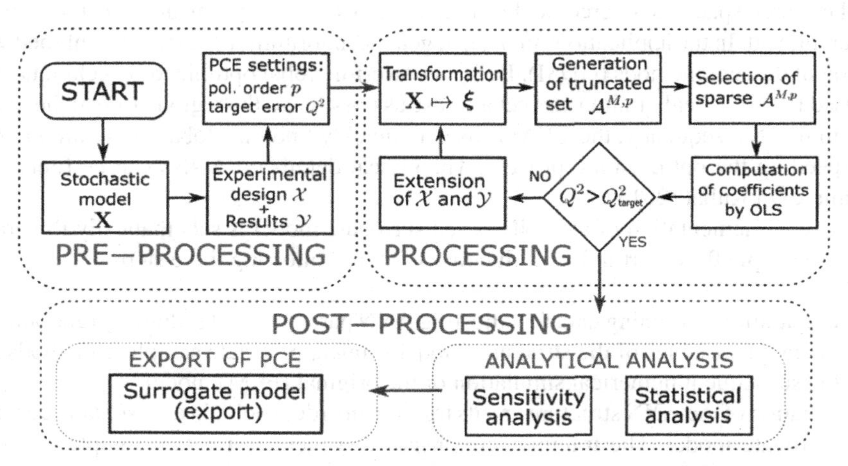

Fig. 1. Simplified algorithm for construction of PCE consisting of three parts: pre-processing, processing and post-processing (including analytical and numerical parts).

2.2 Artificial Neural Network

The second universal and frequently used surrogate model is an artificial neural network-based model belonging to the category of machine learning models. An ANN is a parallel signal processing system composed of simple processing elements, called artificial neurons, which are interconnected by direct links (weighted connections). The aim of such a system is to perform parallel distributed processing in order to solve a desired computational task. One advantage of an ANN is its ability to adapt itself by changing its connection strengths or structure.

Quite a number of neural network types are available; in this paper, a feed-forward multi-layer network type is used. In this type of network, artificial neurons are organized into different layers and the information only moves in the forward direction: data leaves the input nodes and passes through hidden nodes (if there are any) to the output nodes. Such a network is a great mathematical tool for modelling complex relationships between inputs and outputs.

The choice of the number of hidden layers, the number of neurons in them and the type of activation function has a key impact on the complexity and performance of ANNs. Depending on the type of the original model, both linear and nonlinear activation functions can be used. In general, the nonlinear activation function must be employed if one wishes to introduce nonlinearity into the ANN. In that case, we use the hyperbolic tangent activation function.

ANNs must be trained (i.e., the values of synaptic weights and biases must be adjusted) to solve the particular problem for which they are intended. A feed-forward type network is trained using "supervised" learning, where a set of example pairs of inputs and corresponding outputs (p, y), $p \in \mathbf{P}$, $y \in \mathbf{Y}$ is introduced to the network. The aim of the subsequent optimization procedure is to find a neural network function f_{ANN}: $\mathbf{P} \rightarrow \mathbf{Y}$ in the allowed class of functions that matches the examples. ANN training is an optimization task which can sometimes return a local minimum – a point where the function value is smaller than at nearby points, but possibly greater than at a distant point in the search space. To overcome this deficiency, a suitable optimization method should be employed. In the applications in Sect. 4, genetic algorithms (GA) were combined with gradient descent methods (GDM). Being a powerful global optimization technique, GA is used to avoid local minima and get as close as possible to the region close to the global minimum. Subsequently, the GDM is used to quickly find the global minimum in order to speed up the optimization process. Theoretical details on ANNs can be found (for example) in Kubat (2015).

The implementation of an ANN-based surrogate model is schematically illustrated using a simple flowchart in Fig. 2. It consists of the following three steps:

1. Preparation of training data for adjusting ANN parameters. It is done by randomizing the input variables of the structural model using appropriate sampling methods and the subsequent numerical simulation of the original (FEM) model.
2. Creation of the ANN structure and its training in order to obtain a reasonably accurate surrogate model. The training set consists of generated random samples of input variables and the corresponding simulated responses of the structure.

3. Utilization of the ANN surrogate model in sensitivity analysis when the evaluation of structural response for given input data is needed.

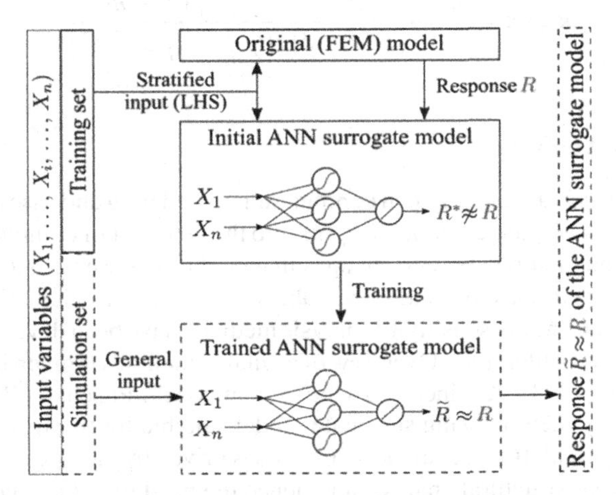

Fig. 2. Flowchart of artificial neural network-based surrogate modeling.

For more details on ANN-based surrogate modeling see Lehký and Šomodíková (2017).

3 Stochastic Structural Analysis

3.1 Statistical Analysis

Statistical analysis is the process of collecting and analyzing data (experimental or numerically simulated) to identify patterns and trends. Descriptive statistics form the first level of statistical analysis and are used to reduce large sets of observations into more compact and interpretable forms. A common objective of the stochastic analysis of engineering structures is the determination of basic descriptive statistics of the structural response related to selected critical limit states. With their help, the so-called design resistance of the structure is subsequently determined using probabilistic or semi-probabilistic methods. The descriptive statistics then serve well for comparing the accuracy of the surrogate model of the analyzed structure. In the application section below, the first four statistical moments are used, which are the mean μ, variance σ^2, skewness γ and kurtosis κ. Their respective estimates from the sample set of variable X containing n samples are determined as:

$$\mu_x \approx m_x = \frac{1}{n}\sum\nolimits_{i=1}^{n} x_i, \tag{3}$$

$$\sigma^2 \approx s^2 = \frac{1}{n-1}\sum\nolimits_{i=1}^{n} (x_i - m_x)^2, \tag{4}$$

$$\gamma \approx g = \frac{n\sqrt{n-1}}{n-2} \frac{\sum_{i=1}^{n} (x_i - m_x)^3}{\left(\sum_{i=1}^{n} (x_i - m_x)^2\right)^{2/3}}, \tag{5}$$

$$\kappa \approx k = \frac{n(n+1)(n-1)}{(n-2)(n-3)} \frac{\sum_{i=1}^{n} (x_i - m_x)^4}{\left(\sum_{i=1}^{n} (x_i - m_x)^2\right)^2}. \tag{6}$$

3.2 Sensitivity Analysis

Sensitivity analysis is a crucial step in computational modeling and assessment. Through sensitivity analysis we gain essential insights into the behavior of computational models, their structure, and their response to changes in model inputs. Sensitivity analysis is also important to reduce a space of random variables for stochastic calculation, building the surrogate model, etc. Many sensitivity analysis methods have been developed, giving rise to a large and growing literature. Overview of available methods is given in review papers, e.g., Novák et al. (1993); Kleijnen (2010); Borgonovo and Plischke (2016). This paper presents sensitivity method using surrogate model suitable for use in combination with the small sample simulation technique. This need is driven by the extreme computational burden of nonlinear structural analysis and hence the need to reduce the computational time to an acceptable level.

The relative effect of each basic random variable on structural response can be measured using the partial correlation coefficient between each basic input variable and the response variable. With respect to the small-sample simulation techniques of the Monte Carlo type utilized for the reliability assessment of time-consuming nonlinear problems, the most straightforward and simplest approach uses non-parametric rank-order statistical correlation (Iman and Conover 1980). This method is based on the assumption that the random variable which influences the response variable most considerably (either in a positive or negative sense) will have a higher correlation coefficient than the other variables. For a detailed discussion of rank-order statistical correlation see Vořechovský (2012). Non-parametric correlation is more robust than linear correlation and more resistant to defects in data. It is also independent of probability distribution. Because the model for the structural response is generally nonlinear, a non-parametric rank-order correlation is used by means of the Spearman correlation coefficient:

$$r_{XY} = \frac{\sum_{i=1}^{n} \left(\hat{x}_i - m_{\hat{x}}\right)\left(\hat{y}_i - m_{\hat{y}}\right)}{\sqrt{\sum_{i=1}^{n} \left(\hat{x}_i - m_{\hat{x}}\right)^2}\sqrt{\sum_{i=1}^{n} \left(\hat{y}_i - m_{\hat{y}}\right)^2}}, \tag{7}$$

where $\hat{x}_i, \hat{y}_i$ are the rank orders of the individual samples and $m_{\hat{x}} = \frac{1}{n}\sum_{i=1}^{n} \hat{x}_i$ is the mean of the rank orders of the variable X samples (for the variable Y the mean value of the rank orders $m_{\hat{y}}$ is determined by analogy). Equation 7 can be used even if the data set of one of the variables contains identical rank values (so-called "tied ranks"). For these values, Eq. 7 works with the average ranks.

The crude Monte Carlo simulation method can be used for the preparation of random samples. However, it is recommended that a proper sampling scheme should be used, e.g., stratified LHS. This method utilizes random permutations of the number of layers of the distribution function of the basic random variables to obtain representative values for the simulation. When using this method, the ranks $\hat{x}_i$, $\hat{y}_i$ in Eq. 7 are directly equivalent to the permutations used in sampling. In addition, the samples used for sensitivity analysis can be further used to build a surrogate model, thus saving a large portion of time.

4 Applications

This section presents two practical applications that demonstrate the applicability of surrogate models in combination with statistical and sensitivity analyses of engineering problems. The first problem was chosen as a benchmark against a known analytical solution. The second application represents a typical more complex structural engineering problem, where the solution is given in implicit form using a FEM computational model.

4.1 Fixed Beam

The first example is represented by a typical engineering problem with a known analytical solution, a maximum deflection of a fixed beam loaded by a single force in mid-span:

$$Y = \frac{1}{16}\frac{FL^3}{Ebh^3}, \tag{8}$$

containing 5 lognormally distributed uncorrelated random variables according to Table 1, where b and h represent the width and height of the rectangular cross-section, E is the modulus of elasticity of concrete, F is the loading force and L is the length of the beam.

Table 1. Stochastic model of mid-span deflection of a fixed beam.

Parameter (unit)	b (m)	h (m)	L (m)	E (GPa)	F (kN)
Mean μ	0.15	0.3	5	30	100
Standard deviation σ	0.0075	0.015	0.05	4.5	20

For this example, it is simple to obtain analytically lognormal distribution of Y, and thus numerical results can be compared to analytical results assumed as a reference solution. The product of lognormally distributed variables is a lognormal variable $Y \sim \mathcal{LN}(\lambda_Y, \zeta_Y)$, where parameters of distribution are obtained as:

$$\lambda_Y = \ln\frac{1}{16} + \lambda_F + 3\lambda_L - \lambda_E - \lambda_b - 3\lambda_h, \tag{9}$$

$$\zeta_Y = \sqrt{\zeta_F^2 + (3\zeta_L)^2 + \zeta_E^2 + \zeta_b^2 + (3\zeta_h)^2}. \tag{10}$$

The sparse PCE is created by the adaptive algorithm depicted in Fig. 1 but generally any algorithm might be employed for efficient construction of PCE. Using the adaptive algorithm, it was possible to build the PCE with experimental design (ED1) containing

100 samples generated by LHS and maximal polynomial order $p = 5$. Since the function is not in a polynomial form, the PCE approximates the original model with an accuracy measured by Leave-one-out cross validation $Q^2 = 0.998$.

The ANN surrogate model consists of five input neurons, one linear output neuron and five nonlinear neurons in a hidden layer. Experimental design for the ANN (ED2) contains once again 100 samples generated by LHS method. Validation of the model was performed using 100 additional test samples with accuracy $Q^2 = 0.998$.

The obtained statistical moments of PCE used for Gram-Charlier (G-C) expansion and from ANN using 1 million Monte Carlo simulations and evaluated using Eqs. 5–8 are listed in Table 2 and compared with analytical reference solution and statistical moments of initial experimental designs ED1 and ED2. As can be seen, 100 samples are enough for the estimation of the first four statistical moments by both PCE and ANN surrogate models, respectively, and both are far more accurate in comparison to simple LHS sampling used to create experimental designs.

Table 2. Comparison of first four statistical moments obtained by surrogate models with analytical solution and experimental designs.

Method	μ	σ^2	γ	κ
Analytical	6.69	4.09	0.91	4.46
ED1 (PCE)	6.65	3.41	0.78	4.10
PCE	6.69	4.08	0.88	4.41
ED2 (ANN)	6.69	4.18	1.09	4.98
ANN	6.69	4.07	0.89	4.25

The results of Spearman non-parametric sensitivity analysis are summarized in Table 3. 1 million samples of the original model are used as a reference solution. Note that in more complex cases such a reference solution is not available for time reasons, see the second application below. The coefficients obtained using 100 samples of ED are in reasonably good agreement with the reference solution. However, the creation of fast to evaluate surrogate models and the subsequent determination of correlation coefficients using a large number of simulations (10 thousand in this case) resulted in a significantly better agreement with the reference solution for both ANN and PCE.

4.2 Post-tension Concrete Bridge

The proposed methodology is applied to the existing post-tensioned concrete bridge with three spans. The superstructure of the middle span is 19.98 m long, has a total width of 16.60 m and is the critical part of the bridge for the assessment. In the transverse direction, each span is constructed of 16 post-tensioned KA-61 bridge girders commonly used in the Czech Republic. The loading is applied according to the Czech National Annex of the Eurocode for the load-bearing capacity of road bridges and corresponds to the exclusive load category (six-axle truck).

Table 3. Spearman correlation coefficients of deflection of a fixed beam.

Variable	Reference	ED	PCE	ANN
b	−0.161	−0.195	−0.165	−0.156
h	−0.489	−0.456	−0.488	−0.484
L	0.097	0.143	0.096	0.094
E	−0.487	−0.506	−0.486	−0.494
F	0.652	0.629	0.653	0.654

The nonlinear FEM model was created using ATENA Science software based on the theory of nonlinear fracture mechanics (Červenka and Papanikolaou 2008). The model of each girder consists of 13,000 hexahedral elements in the bulk of the volume and triangular 'PRISM' elements in the part with complicated geometry, as depicted in Fig. 3 (top left). Reinforcement and tendons are modeled by discrete 1D elements with geometry according to the original bridge documentation. As can be seen from stress field depicted in Fig. 3 (right), girders fail in bending accompanied by significant crack development in midspan region where the tensile stresses reach their highest values (see red color). The numerical model is used to determine the load-bearing capacity when three limit states are reached: (i) the ultimate limit state (ULS, the peak of the load-deformation diagram) to determine the ultimate capacity of the bridge; (ii) the first occurrence of cracks in the superstructure (Cracking), which lead to corrosion of tendons and eventually to failure of the structure; and (iii) the serviceability limit state (SLS), which is represented by the decompression of the structure, i.e., the state when tensile stresses appear in a part of the structure.

The stochastic model contains 4 lognormal random material parameters of a concrete C50/60: Young's modulus E (mean: 36 GPa, coefficient of variation: 0.16); compressive strength of concrete f_c (mean: 56 MPa, CoV: 0.16); tensile strength of concrete f_{ct} (mean: 3.6 MPa, CoV: 0.22) and fracture energy G_f (mean: 195 Jm^2, CoV: 0.22). The last random variable P represents the percentage loss of prestress in the tendons and is normally distributed with a mean of 20 and a CoV of 0.3 Mean values and coefficients of variation were obtained according to Eurocode EN 1992-1-1 (Annex A 2021) for adjustment of partial safety factors for materials and JCSS Probabilistic model code (JCSS 2001). The ED contains 30 numerical simulations generated by LHS. Note that each simulation lasts approximately 24 h and construction of the whole ED took approx. 1 week of computational time.

The PCE is created with maximum polynomial order $p = 5$ using the adaptive algorithm based on LAR presented in Fig. 1. Obtained accuracy measured by Q^2 is higher than 0.98 for all limit states, which is sufficient for statistical and sensitivity analyses. The ANN surrogate models consist of five input neurons, one linear output neuron and five nonlinear neurons in a hidden layer. Validation of the model was performed using 5-fold cross-validation with average accuracy $Q^2 = 0.97$. Although such accuracy is not acceptable for reliability analysis of structures, it is sufficient for statistical or sensitivity analysis (Pan et al. 2021). The obtained first four statistical moments of PCE used for

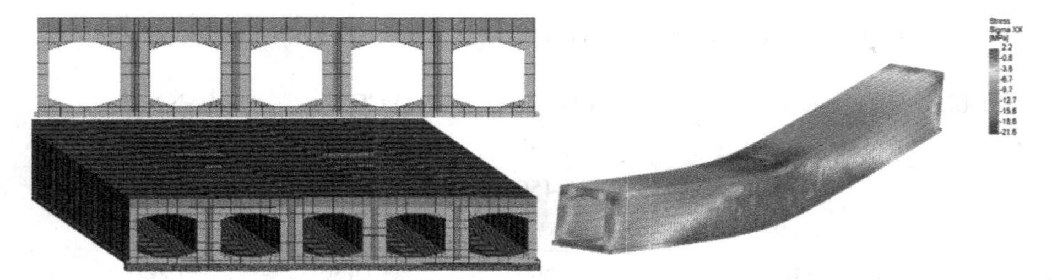

Fig. 3. NLFEM of the analyzed bridge: The cross-section (top left), FEM of the middle 4 girders (bottom left) and stress field of the single girder KA-61 (right).

G-C expansion and from ANN using 1 million Monte Carlo simulations and evaluated using Eqs. 3–6 are listed in Table 4 and compared with the initial ED consisting of 30 samples.

Table 4. Comparison of first four statistical moments obtained by surrogate models with experimental design for three analyzed limit states.

Method	μ	σ^2	γ	κ
ULS				
ED	677	1741	−0.250	3.07
PCE	676	1903	0.116	4.77
ANN	677	1674	−0.216	2.65
Cracking				
ED	555	1703	−0.423	3.39
PCE	556	1637	0.003	3.06
ANN	556	1726	0.124	2.95
SLS				
ED	216	941	−0.106	3.10
PCE	216	908	0.003	3.03
ANN	216	895	−0.085	2.63

The results for all three limit states confirmed a reasonably good agreement of the surrogate models to the ED, especially for the first two statistical moments. For the higher moments (skewness and kurtosis), which are significantly more sensitive, the effect of the low number of ED samples is already evident. Both methods ANN and PCE are in good agreement though there are some differences in higher statistical moments for the SLS and Cracking. However, there is a significant difference between ANN and PCE for the higher statistical moments for the ULS limit state, which exhibits the highest nonlinearity affected by several interactions of the input random parameters. On the one hand, the results obtained by ANN are consistent with a very limited ED, which may be highly affected by the number of samples in the ED. On the other hand, PCE takes into account the whole distribution of the input random variables, and therefore its estimates based on knowledge of complete distribution of the input random vector could be more robust. Conversely, the nonlinearity of the mathematical model might be too challenging for PCE. It is clear from the results that additional simulations of the original mathematical model will be necessary to improve the estimates of the surrogate models, as the two methods lead to significantly different values of the higher statistical moments in the ULS case. Nonetheless, the accuracy in SLS and Cracking is sufficient, and the obtained results can be directly used for further assessment of the structure.

The results of Spearman non-parametric sensitivity analysis for all three analyzed limit states are summarized in Table 5. The results show that both surrogate models lead to almost the same sensitivity coefficients. Compared to the results on ED, the main benefit of using surrogate models is apparent. If surrogate model is a sufficiently accurate approximation of the original model, it allows the sensitivity coefficients to be determined with relatively high accuracy even for a large stochastic model, which is not possible in principle from the limited ED (30 samples). In Table 5, this is particularly obvious for the compressive and tensile strengths of concrete, which, for example, have a negligible effect for ULS, as confirmed by both surrogate models but not by ED. ANN, in contrast to PCE, was also able to detect some degree of influence of fracture energy on crack initiation, which is consistent with the used fracture mechanics-based material model. For the decompression limit state, the value of the prestressing force was clearly identified by both models as the only dominant variable. Here, the tensile strength of the concrete, as incorrectly detected from ED and slightly by ANN, has no effect from a physical point of view.

Table 5. Spearman correlation coefficients obtained by surrogate models and derived from experimental design obtained for three limit states.

Variable	ULS			Cracking			SLS		
	ED	PCE	ANN	ED	PCE	ANN	ED	PCE	ANN
E	0.27	0.20	0.17	0.47	0.41	0.38	0.11	0.00	−0.01
f_c	0.14	0.00	−0.01	0.06	0.00	−0.06	−0.10	0.00	−0.05
f_{ct}	0.23	0.00	0.02	0.31	−0.01	0.10	0.25	0.00	0.09
G_f	0.16	0.27	0.27	0.08	0.00	0.19	−0.08	0.00	0.04
P	−0.94	−0.91	−0.92	−0.87	−0.89	−0.83	−0.99	−0.99	−0.95

5 Conclusions

The paper describes two alternative approaches to surrogate modeling and their potential for stochastic analysis of engineering structures: PCE and ANN. The potential of both methods for routine applications of stochastic assessment of engineering structures is documented by two numerical examples. For the first example, excellent agreement of the stochastic assessment was achieved as there is little nonlinearity and no significant interactions between variables. However, this is not the case for the second numerical example of a bridge structure, which is a highly nonlinear problem with several internal interactions between variables. The differences are significant especially for the higher statistical moments, the first two moments being in good agreement as well as the values of the sensitivity coefficients. Note, however, that from a practical point of view, the higher moments are of little importance. Thus, both alternative approaches can be successfully applied to practical problems of stochastic analysis of engineering structures.

The advantage of PCE is time saving, especially due to efficient post-processing and the inclusion of knowledge of the full distribution of the input random vector. In contrast, ANN is more versatile and scalable. It can easily increase its complexity by adding more hidden elements or learning on additional data. Moreover, in practical problems the correct solution is not usually available, so it is convenient to combine several alternative models. As shown in both applications, the power of surrogate models was strongly demonstrated in the computation of sensitivity coefficients, where qualitatively better results were obtained than by evaluating the limited ED, whose extensibility tends to be difficult in practical structural engineering problems.

Acknowledgement. This work was supported by the project No. 22-00774S, awarded by the Czech Science Foundation (GACR).

References

Borgonovo, E., Plischke, E.: Sensitivity analysis: a review of recent advances. Eur. J. Oper. Res. **248**, 869–887 (2016)

Blatman, G., Sudret, B.: An adaptive algorithm to build up sparse polynomial chaos expansions for stochastic finite element analysis. Probab. Eng. Mech. **25**(2), 183–197 (2010)

Blatman, G., Sudret, B.: Adaptive sparse polynomial chaos expansion based on least angle regression. J. Comput. Phys. **230**(6), 2345–2367 (2011)

Bourinet, J., Deheeger, F., Lemaire, M.: Assessing small failure probabilities by combined subset simulation and support vector machines. Struct. Saf. **33**(6), 343–353 (2011)

Červenka, J., Papanikolaou, V.K.: Three dimensional combined fracture-plastic material model for concrete. Int. J. Plast. **24**(12), 2192–2220 (2008)

Echard, B., Gayton, N., Lemaire, M., Relun, N.: A combined importance sampling and kriging reliability method for small failure probabilities with time-demanding numerical models. Reliab. Eng. Syst. Saf. **111**, 232–240 (2013)

EN 1992-1-1. Eurocode 2: Design of concrete structures – Part 1-1: General rules and rules for buildings, National Annex A. Prague, Czech Republic (2021)

Ghanem, R.G., Spanos, P.D.: Stochastic Finite Elements: a Spectral Approach. Springer, Berlin, Heidelberg (1991). https://doi.org/10.1007/978-1-4612-3094-6

Hurtado, J.E.: An examination of methods for approximating implicit limit state functions from the viewpoint of statistical learning theory. Struct. Saf. **26**(3), 271–293 (2004)

Iman, R.L., Conover, W.J.: Small sample sensitivity analysis techniques for computer models with an application to risk assessment. Commun. Stat. Theory Methods **9**(17), 1749–1842 (1980)

JCSS Probabilistic Model Code. Joint Committee on Structural Safety (2001). www.jcss-lc.org/jcss-probabilistic-model-code/
Kleijnen, J.P.C.: Sensitivity analysis of simulation models: an overview. Procedia Soc. Behav. Sci. **2**, 7585–7586 (2010)
Koehler, J.R., Owen, A.B.: Computer experiments. Handb. Stat. **13**, 261–308 (1996)
Kubat, M.: An Introduction to Machine Learning. Springer International Publishing, Switzerland, Cham (2015). https://doi.org/10.1007/978-3-319-20010-1
Lehký, D., Šomodíková, M.: Reliability calculation of time-consuming problems using a small-sample artificial neural network-based response surface method. Neural Comput. Appl. **28**, 1249–1263 (2017)
Lehký, D., Slowik, O., Novák, D.: Reliability-based design: artificial neural networks and double-loop reliability-based optimization approaches. Adv. Eng. Softw. **117**, 123–135 (2018)
Lehký, D., Šomodíková, M., Lipowczan, M.: An application of the inverse response surface method for the reliability-based design of structures. Neural Computing and Applications (2022). in print
McKay, M.D., Conover, W.J., Beckman, R.J.: A comparison of three methods for selecting values of input variables in the analysis of output from a computer code. Technometrics **21**, 239–245 (1979)
Novák, D., Teplý, B., Shiraishi, N.: Sensitivity analysis of structures: a review. In: International Conference CIVIL COMP 1993, pp. 201–207. Edinburgh, Scotland (1993)
Novák, D., Vořechovský, M., Teplý, B.: FReET: software for the statistical and reliability analysis of engineering problems and FReET-D: degradation module. Adv. Eng. Softw. **72**, 179–192 (2014)
Novák, L.: On distribution-based global sensitivity analysis by polynomial chaos expansion. Comput. Struct. **267**, 106808 (2022)
Novák, L., Novák, D.: Polynomial chaos expansion for surrogate modelling: theory and software. Beton-und Stahlbetonbau **113**, 27–32 (2018)
Novák, L., Vořechovský, M., Sadílek, V., Shields, M.D.: Variance-based adaptive sequential sampling for polynomial chaos expansion. Comput. Methods Appl. Mech. Eng. **386**, 114105 (2021)
Pan, L., Novák, L., Lehký, D., Novák, D., Cao, M.: Neural network ensemble-based sensitivity analysis in structural engineering: Comparison of selected methods and the influence of statistical correlation. Comput. Struct. **242**, 106376 (2021)
Slowik, O., Lehký, D., Novák, D.: Reliability-based optimization of a prestressed concrete roof girder using a surrogate model and the double-loop approach. Struct. Concr. **22**, 2184–2201 (2021)
Sobol, I.: Global sensitivity indices for nonlinear mathematical models and their Monte Carlo estimates. Math. Comput. Simul. **55**, 271–280 (2001)
Soize, C., Ghanem, R.: Physical systems with random uncertainties: chaos representations with arbitrary probability measure. J. Sci. Comput. **26**, 395–410 (2004)
Stein, M.: Large sample properties of simulations using Latin hypercube sampling. Technimetrics **29**(2), 143–151 (1987)
Sudret, B.: Global sensitivity analysis using polynomial chaos expansions. Reliab. Eng. Syst. Saf. **93**(7), 964–979 (2008)
Efron, B., Hastie, T., Johnstone, I., Tibshirani, R.: Least angle regression. Ann. Stat. **32**(2), 407–499 (2004)
Vořechovský, M.: Correlation control in small sample Monte Carlo type simulations II: analysis of estimation formulas, random correlation and perfect uncorrelatedness. Probab. Eng. Mech. **29**, 105–120 (2012)
Xiu, D., Karniadakis, E.G.: The Wiener-Askey polynomial chaos for stochastic differential equations. SIAM J. Sci. Comput. **24**(2), 619–644 (2002)

On Time Series Cross-Validation for Deep Learning Classification Model of Mental Workload Levels Based on EEG Signals

Kunjira Kingphai and Yashar Moshfeghi(✉)

NeuraSearch Laboratory, Department of Computer and Information Sciences, University of Strathclyde, Glasgow, Scotland
{kunjira.kingphai,yashar.moshfeghi}@strath.ac.uk

Abstract. The determination of a subject's mental workload (MWL) from an electroencephalogram (EEG) is a well-studied area in the brain-computer interface (BCI) field. A high MWL level can significantly contribute to mental fatigue, decreased performance, and long-term health problems. Inspired by the success of machine learning in various areas, researchers have investigated the use of deep learning models to classify subjects' MWL levels. A common approach that is used to evaluate such classification models is the cross-validation (CV) technique. However, the CV technique used for such models does not take into account the time series nature of EEG signals. Therefore, in this paper we propose a modification of CV techniques, i.e. a blocked form of CV with rolling window and expanding window strategies, which are more suitable for EEG signals. Then, we investigate the effectiveness of the two strategies and also explore the effects of different block sizes for each strategy. We then apply these models to several state-of-the-art deep learning models used for MWL classification from EEG signals using a publicly available dataset, STEW. There were two classification tasks: Task 1- resting vs testing state, and Task 2- low vs moderate vs high MWL. Our results show that the model evaluated by the expanding window strategy, when it was trained using the 90% of data, provided a better performance than the rolling window strategy and that the BGRU-GRU model outperformed the other models for both tasks.

Keywords: EEG · Time series · Mental workload · Classification · Deep learning

1 Introduction

The concept of the mental workload (MWL) plays an important role in human life, from study design [22,41] to driving fatigue [11,36], pilot performance [15], and performance on other tasks [35]. To measure people's MWL, EEG signals have been thoroughly investigated because these signals are highly correlated with a person's real-time mental workload status [31]. Recently, machine learning

G. Nicosia et al. (Eds.): LOD 2022, LNCS 13811, pp. 402–416, 2023.
https://doi.org/10.1007/978-3-031-25891-6_30

techniques have received much attention and have been developed to capture the variance characteristics of EEG signals to classify MWL levels [13]. Such classification models have often been trained and evaluated using the cross-validation (CV) technique [26].

In a traditional CV technique, an entire dataset is split into K equally sized subsets (also known as folds). The model is trained on $K - 1$ folds (which are called the training set), while the remaining fold is kept apart, unseen by the model, so that it can be used as the test set [27]. The model training process is repeated K times, with a different fold preserved for model evaluation each time. The fundamental idea underlying a CV technique is that a collection of random variables is being drawn from a given probability distribution; these variables are statistically independent of each other, satisfying the independent and identically distributed (i.i.d.) property in probability theory and statistics [9]. However, EEG signals, which are generated over time, represent time series data. Therefore, applying the traditional CV approach (i.e. shuffling and randomly splitting the data into K-folds) can violate the i.i.d. assumption [2] and lead to an unreliable model [6] due to overfitting. This is especially true for a forecasting task in which future information should not be available to the model at the time of training.

Since the standard CV approach is not directly applicable to time series data, researchers have proposed various modifications of CV techniques for this type of data [3,6]. For example, a blocked form of CV with an expanding window strategy is proposed by Bergmeir and Benítez [2]. This CV strategy is implemented by mimicking a real-life scenario, i.e. the test dataset is sequentially moved into the training dataset, and the forecast origin is changed accordingly. Another strategy that can be used the blocked form of CV is the rolling window strategy. Bergmeir and Benítez [2] stated that using this strategy could be beneficial when the characteristics of the previous observation dynamically change throughout time and tend to interrupt model generation. However, discarding some parts of the time series might cause the models to lose some important information, and this could affect their performance. Thus, in this study, we aim to answer two research questions: 1) which time series blocked CV strategies should be applied for the EEG classification task? and 2) how does the size of the block in a time Series cross-Validation (TSCV) influence the effectiveness of the models? To investigate our research questions, we applied our time series blocked CV techniques to the stacked long short-term memory (LSTM), bidirectional LSTM (BLSTM), stacked gated recurrent unit (GRU), bidirectional GRU (BGRU), BGRU-GRU, BLSTM-LSTM and convolutional neural network (CNN) models, which are state-of-the-art and widely used deep learning models, to predict the MWL level [12,20] from EEG signals, using a publicly available mental workload dataset called STEW [17].

The rest of the paper is structured as follows. Section 2 provides background information on the concept of the MWL and describes previous studies related to the prediction of the MWL. Then, in Sect. 3 and Sect. 4, the approach and methodology that are proposed and used in this paper are described. The results and discussion can be found in Sect. 5. Finally, the conclusion is given in Sect. 6.

2 Background

Machine learning models, especially deep learning models have been drawing the attention of researchers in neuroscience, who have used them to classify MWL levels based on EEG signals [13]. CV is a statistical technique for evaluating and comparing machine learning models that involves training the models on a subset of the available input data and evaluating them on the complementary subset. The way that CV is applied is also task-dependent. For example, in a scenario in which we are not attempting to predict the future outcomes, the traditional CV technique, which randomly shuffles data by splitting the data into K-folds, could be used to evaluate the model effectively. However, if we are focusing on predicting a future value, such as the subject's MWL level, we do have to be concerned about the temporal aspect of the data since the time spent on a task typically affects the subjects' performance, which deteriorates over the period of task engagement [16].

2.1 Cross-Validation for Deep Learning Model

Various studies that aim to predict a subject's mental workload have not taken into consideration the temporal aspect of the EEG signal [8]. Additionally, some studies did not provide enough details concerning how they perform CV [5,39]. For example, Ahmadi et al. [1] aimed to detect driver fatigue by using an expert automatic method based on brain region connectivity. They used an EEG dataset that recorded fatigue and alert states. In the model evaluation step, they randomly divided the dataset into five subsets (folds). One of these folds is kept as the validation fold, and the others formed the training set that was used in the feature selection and hyper-parameter tuning stage. Five-time five-fold cross-validation is applied for each subject. However, the way that the authors randomly defined the dataset is implausible in a practical sense. Human fatigue develops over time [34], e.g. a driver's fatigue level at the beginning of a drive may be low, but it increases as time passes. Thus, training a fatigue detection model by using the future fatigue level to detect the previous fatigue level might not result in an accurate model. Five-fold CV was also adopted by Zhang et al. [37], who used a one-dimensional convolutional neural network (1D-CNN) to automatically capture information from different frequency bands and read the subject's mental states. They used an EEG dataset containing 31 recording sessions from 5 subjects and randomly divided independent recording sessions into 5 groups; then they performed CV. As in the previous work, the subjects' mental workload continuously increases with time [28]. Therefore, arbitrarily assigning data from various sessions to training and test datasets is inappropriate. Zeng et al. [36] also aimed to identify the mental states of subjects.

They proposed a light-weight classifier, LightFD, which is based on a gradient boosting framework. In the model evaluation step, they randomly extracted 80% of the EEG signals of each subject to create a training set, and the remaining 20% of the signals were used as a test set. The EEG time series used for MWL classification were also randomly divided into training and test sets to evaluate the proposed deep learning models in [38,39].

As can be seen from the literature, researchers who study EEG signals using the machine learning method have not taken into account the characteristic of time series and the i.i.d. assumption of the elements of time series while performing cross-validation. They applied the traditional strategies by shuffling their EEG signals and then randomly dividing them into training and test sets. This does not take in to account the temporal information of time series data. To the best of our knowledge, there are only two studies that consider the temporal characteristics of EEG data in the CV step, [23,24]. In [23], the authors stated that evaluating the model by using test samples that are chronologically near one of the training samples could introduce an overfitting problem because of EEG signal changes. Therefore, the authors of [24] adopted time-wise cross-validation strategies in their study. In this strategy, the samples of each task in each session are divided into n parts, evenly and continuously. In each fold, the model will be trained using $n - p$ parts of all the tasks and tested on the left-out parts of all the tasks as well. Some parts of the data at the beginning and the end of each task might be cut off to lessen the effect of task transition.

As the temporal aspect of time series has not previously been considered a significant factor in choosing a CV method, it is important to raise awareness of this potential pitfall in the community. To do so, in this work, we aim to present how TSCV can be applied to deep learning models using EEG signals and investigate the effectiveness of applying TSCV to various deep learning models.

3 Approach

We tried to investigate the effect of time series blocked CV strategies and the size of their training and testing detaset on deep learning models for mental workload classification using EEG signals. Background information on time series and time series CV is described in this section.

3.1 Time Series

Time series data consist of a series of data points measured in time order; these data points can be measured every millisecond, every minute, hourly, daily, or annually. The data can be represented by either continuous or discrete datasets. In time series applications, the available past and present data are used to forecast the future values of X. A function $\mathbf{F}$ is calculated to do this. The estimated value $\hat{X}_{t+\tau}$ of X at time $t + \tau$ can be obtained from a function $\mathbf{F}$, where the function $\mathbf{F}$ is computed from a given value of X up to time t (plus additional time-independent variables in multivariate time series analysis):

$$\hat{X}_{t+\tau} = \mathbf{F}(X_t, X_{t-1}, ...) \tag{1}$$

where τ is the lag for prediction. Then, the function of the continuous time series will be mapped onto binary-valued of N classes for a classification approach:

$$\mathbf{F}_c(X_t, X_{t-1}, ...) \rightarrow \hat{c}_i \in C \tag{2}$$

$\mathbf{F}_c(X_t, X_{t-1}, ...) \rightarrow \hat{c}_i \in C$ where C is the set of class labels [7].

The EEG signals that can be used to show human brain activity are measured over a specific time period and are considered time series. The signal from each electrode can be considered a univariate time series, and the MWL levels are considered class labels. Therefore, to perform MWL level classification, we can map the EEG signals from various electrodes into specific classes corresponding to low, moderate, or high MWL levels by using machine learning models. To evaluate such a model and test its performance, CV is commonly used. However, an important distinction in forecasting is that the future is completely unavailable and must be estimated only from what has already happened, so time series cannot be randomly shuffled by the traditional CV method. For example, in an autoregressive (AR(p)) model, there is a parameter p called the order. It can tell us the maximum separation that can exit between events that the outcome variable in the AR process can relate to each other. The model can be written as $\hat{X}_t = (c + \beta_1 X_t - 1 + \beta_2 X_{t-2} + ... + \beta X_{t-p} + \varepsilon)$, where $\beta_1, \beta_2, ..., \beta X$ are the parameters of the model, $X_t - 1, X_{t-2}, ..., X_{t-p}$ are the past time series values, ε represents white noise, and c is a constant. It is clear that, the traditional CV method could cause an overfitting problem [2] because future information could leak into a training dataset. A method for performing CV on a time series dataset is explained in the next section.

3.2 Time Series Cross-Validation (TSCV)

Traditionally, the whole time series dataset is usually used to evaluate machine learning or deep learning models, and this causes a theoretical statistical violation [2]. Therefore, in the time series model evaluation procedure, a final part of the time series needs to be kept for testing, so the corresponding training set consists only of observations that occur before the observations that form the test set. Accordingly, there is no information leaking caused by the use of future observations [2]. Tashmad [29] has suggested four primary strategies for time series forecasting: fixed-origin evaluation; rolling-origin-recalibration evaluation, or expanding window; rolling-origin-update evaluation; and rolling-window evaluation.

In this study, we split the data by ensuring that the test set consists of data that were recorded after the data of the training set; additionally, we employ the expanding window and rolling window strategies in model evaluation. In the expanding window strategy, the test dataset is sequentially moved into the training set, and the forecast origin is changed accordingly. The classification model is recalibrated until all available samples are used. Thus, the test data of the previous fold will be used as a validation set to tune the deep learning model's parameters. Figure 1 shows how we split data into training, validation, and test sets using the expanding window strategy.

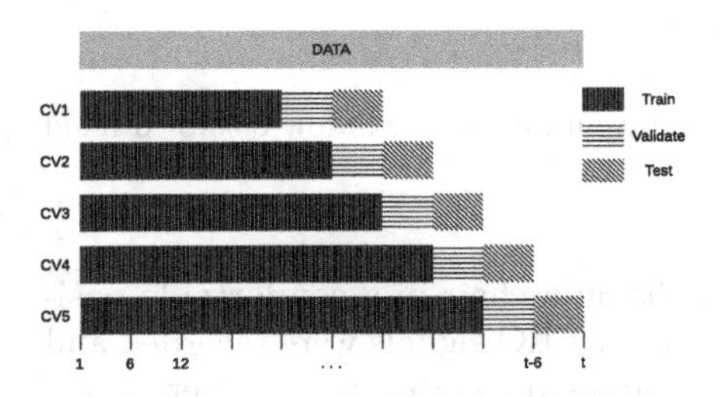

Fig. 1. Expanding window strategy

Fig. 2. Rolling window strategy

In the rolling window strategy, the amount of data used for the training is kept constant. The window is implemented by shifting forward the training and test data at each fold by a constant window size of m seconds (s). Thus, every time that new data enter the series, old data from the beginning of the series are discarded. This method can be applied to time series data, as the model is rebuilt in every window, and the temporal fluctuation of the data does not disturb the creation of the model [29]. Figure 2 shows how we split the data into training, validation, and test sets using the rolling window strategy. In this study, we applied 5-fold time series CV with expanding and rolling window approaches. There are two parameters that must be optimised for TSCV, namely the sizes of the training and testing windows.

We investigate the size of the training window for each TSCV strategy by varying it from 20% to 90% of the data with an incremental step of 10%. Meanwhile, the validation and testing window size of $(10\% * n)/k$ is fixed at the end of the series since this window size provides a good model performance for this dataset [14]. Figure 3 shows how we fixed the validation and test sets at the end of the series. To generalise the model, we feed split data from all subjects into the models simultaneously.

4 Methodology

4.1 Dataset

In this study, the Simultaneous Task EEG Workload (STEW) dataset for multitasking mental workload activity induced by a single-session simultaneous capacity (SIMKAP) experiment taken from [17] was used. The dataset contains the EEG signals of 48 subject collected from 14 electrodes and sampled 128 Hz. The signals have been recorded in two states, the resting state and the testing state.

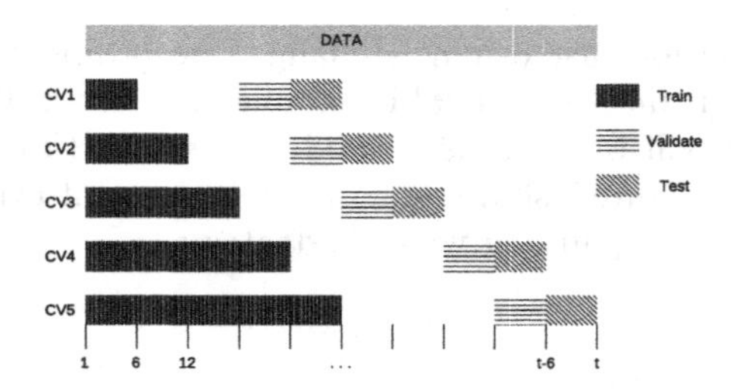

Fig. 3. Expanding window strategy with a validation and test window size of 6 fixed at the end of series

In the resting state, subjects were asked to sit on a chair in a comfortable position and did not perform any task for 3 min; their EEG signals were recorded and considered to represent the resting state. In the testing state, subjects performed the simultaneous capacity (SIMKAP) multitasking activity and their EEG signals were recorded; the final 3 min of the recording are considered the workload task. The first and last 15 s of data from each recording were excluded to reduce the effects of any activity occurring between the tasks. After each segment of the experiment, subjects were also asked to rate their MWL using a rating scale from 1 to 9. In the analysis steps, the 9-point rating scale was considered to indicate three mental workload levels: low (1–3), moderate (4–6), and high (6–9). Therefore, in this study, we performed the following two classification in the two tasks: Task 1 - resting vs testing state, and Task 2 - low vs moderate vs high MWL level.

Data Pre-processing and Artefact Removal: An EEG signal can easily become contaminated by unwanted artefacts; therefore, removing artefacts from EEG signals is usually a prerequisite of signal analysis [32]. In this study, we removed noise from the dataset by using the technique suggested in [13], i.e. an automatic independent component analysis based on ADJUST (ICA-ADJUST) [18], since it has been shown to provide the most improvement in the model performance for this dataset.

4.2 Feature Extraction and Selection

In the machine learning area, feature extraction can save a huge amount of time and resources by preventing the calculation of unnecessary features [19]. Therefore, to capture only relevant EEG signal characteristics, we performed feature extraction by computing a set of features from the frequency domain, time domain, non-linear domain and linear domain. In the frequency domain, we calculated the signal power for each channel at four well-known power spectral density bands by using a fast Fourier transformation (FFT) [21]; delta (0.5–4 Hz), theta (4–8 Hz), alpha (8–12 Hz), beta (12–30 Hz), and gamma (30–100 Hz) bands were used. Each PSD band represents a different state of the human brain

[10]. Meanwhile, the distribution of the signal can be determined from the time-domain features, i.e. the mean, standard deviation, skewness, and kurtosis, and the autoregressive coefficient (AR) with p = 6 [40] was calculated in the linear domain to describe time-varying processes. Finally, the approximate entropy (ApEn) and Hurst exponent (H), which are treated as non-linear features, are used to quantify the unpredictability of fluctuations over time series and to measure the self-similarity of the time series, respectively. In this study, we used a sliding window with a length of 512 sampling points and a shift of 128 sampling points [17]. Due to their extensive usage in a previous study [13], we then chose the PSD alpha, PSD theta, skewness, kurtosis, ApEn and H features as the optimised feature set of the STEW dataset for this scenario. Eventually, this caused us to have 84 (14 × 6) features, which were extracted from all 14 channels. Additionally, we also performed feature standardisation before further analysis. This method will be described in the next subsection.

4.3 Feature Standardisation

Inherent intra- and inter-subject variability are undeniable characteristics of the EEG signal due to time-variant factors and psychological and neurophysiological parameters [33]; they can cause a data distribution shift problem [25]. Consequently, this would cause the extracted features to have poor generalisability. In this paper, a personalised feature standardisation method was applied to alleviate this problem [4,33]. The extracted features were converted so that they all had the same scale across subjects using F_s. Assume that the raw feature value is F. $min(F)$ and $max(F)$ are the upper and lower limits, respectively. The scaled feature value F_s is calculated as follow:

$$F_s = \frac{F - min(F)}{max(F) - min(F)}. \tag{3}$$

4.4 Classification and Model Training

As mentioned in Subsect. 4.1, we performed classification for two tasks, Task 1 and Task 2. We verified our TSCV strategies with the following deep learning models: the stacked LSTM, BLSTM, stacked GRU, BGRU, BGRU-GRU, BLSTM-LSTM, and CNN models. The deep learning model architectures are described in Table 1.

In Table 1, L, G, BL, BG and D refer to LSTM, GRU, BLSTM, BGRU and dense layers, respectively. For example, L128-L64-L40-L32 indicates an LSTM layer with 128 units, followed by a second LSTM layer with 64 units, a third LSTM layer with 40 units, and a dense layer with 32 units. While D1 in the last layer indicates the dense layer with 1 unit used in Task 1, D3 indicates the dense layer with 3 units used in Task 2. In this study, the dropout rate was set to 0.2, and the Adam optimiser with an initial learning rate of 1e-04 was used to train all the deep learning models. Early stopping was also utilised to avoid the overfitting problem. Moreover, we stopped training once the model performance stopped improving for 30 epochs.

Table 1. Deep learning model architectures

Model	Layers
Stacked LSTM	L128-L64-L40-D32-D1(D3)
BLSTM	BL128-D32-D1(D3)
Stacked GRU	G128-G64-G40-D32-D1(D3)
BGRU	BG128-D32-D1(D3)
BLSTM-LSTM	BL256-L128-L64-D32-D1(D3)
BGRU-GRU	BG256-G128-G64-D32-D1(D3)
CNN	1D-CNN(filters = 64, kernel = 3)- MaxPooling-Flatten-D32-D1(D3)

4.5 Metrics

In this study, the model performance is evaluated using six metrics: the sensitivity, specificity, precision, accuracy, false acceptance rate (FAR), and false rejection rate (FRR). In classification, a true positive (TP) is a case that the model correctly predicted to be positive, and a false negative (FN) is a case that the model incorrectly predicted to be negative. A true negatives (TN) is a case that the model correctly predicted to be negative, and a false positive (FP) is a case that the model incorrectly predicted to be positive. Thus, each matrix is calculated as follow:

$$Sensitivity = TP/(TP + FN) \tag{4}$$

$$Specificity = TN/(TN + FP) \tag{5}$$

$$Precision = TP/(TP + FP) \tag{6}$$

$$Accuracy = (TP + TN)/(TP + TN + FP + FN) \tag{7}$$

$$FRR = FN/(TP + FN) \tag{8}$$

$$FAR = FP/(TN + FP). \tag{9}$$

FRR and FAR are usually applied to measure the performance of a biometric system [30]; these measurements are also known as Type I and Type II errors, respectively. That means that the FRR indicates that valid result that should be accepted are rejected, and the FAR indicates that an unauthorised case that should actually be rejected has been accepted.

5 Result and Discussion

5.1 Type of Time Series: Expanding vs Rolling and Training Size

To evaluate the classification models on EEG data and perform a comprehensive study of their effectiveness with several state-of-the-art deep learning models, we adopted a modification of the CV technique called blocked CV with expanding window and rolling window strategies. We also investigated the effect of different block sizes for each strategy.

Task 1: Resting vs Testing State We can observe from the results shown in Table 2 that models evaluated using the expanding window strategy achieved a higher accuracy than those evaluated using the rolling window strategy for different training block sizes. For the expanding window strategy, the model accuracy steadily improves when the training block size is increased. While the highest accuracy score of 93.53% was obtained when 90% of the data were used for model training, the lowest accuracy score of 84.07% was obtained when 20% of the data were used. For the rolling window strategy, we observe the same pattern that we observe for the expanding window strategy. The accuracy scores continuously increase as the block sizes increase; however, the accuracy score in this scenario is slightly lower than the previous accuracy score. While the highest accuracy score is 91.78%, the lowest accuracy score is 78.39%. Nevertheless, we also observed that both the expanding window and rolling window strategies show the same pattern of the standard deviation (S.D.) of model accuracy. The S.D. decreases as the block size of the training data increases. Moreover, we also observe that models trained using the expanding window strategy provide a lower S.D. than those trained using the rolling window strategy.

Task 2: Low vs Moderate vs High MWL Level In this task, we observe a pattern similar to that observed in Task 1, with a few exceptions. These exceptions are related to the expanding window strategy. In this strategy, the highest model accuracy was still obtained when models were trained using 90% of the EEG dataset. However, the pattern of the S.D. score is different. We observe that for models trained using 60% of the data, the S.D. score slightly increases compared to Task 1. For the rolling window strategy, the S.D trend is the same as that in Task 1.

In summary, these results show that the model performance improved as we added more training data to models. Nevertheless, we also observed that the expanding window strategy provides a lower S.D. score than the rolling window strategy. Thus, the way in which the model performance fluctuated as the training dataset changed for the rolling window strategy shows that the model was interfered with by the data in some parts of the time series. Specifically, models trained using the rolling window strategy have more sensitivity to the variance in the data than those trained with the expanding window strategy. Therefore, to mimic a real-life scenario and obtain models with a high accuracy and robustness to the EEG signal, which changes over time, we should use a blocked form of CV with the expanding window strategy in the model evaluation step for the classification of subjects' MWL levels.

5.2 Model Comparison

In the previous subsection, we observed that the model trained using 90% of the data with the expanding window strategy had the best performance for both Task 1 and Task 2. Therefore, we investigate the effectiveness of using a blocked form of CV with the expanding window strategy and performing model training using 90% of the data for several state-of-the-art deep learning models.

Table 2. The average accuracy of deep learning models evaluated using different TSCV strategies and trained using different amounts of data for Task 1 and Task 2

		Task 1		Task 2	
TSCV strategy	Training size	Accuracy	S.D	Accuracy	S.D
Expanding	20%	84.07	2.65	71.92	1.90
	40%	89.07	2.38	76.96	1.60
	60%	91.01	2.50	78.86	1.72
	80%	92.87	2.14	80.72	1.99
	90%	**93.53**	**1.80**	**81.38**	**2.19**
Rolling	20%	78.39	4.67	67.04	4.73
	40%	85.61	4.36	74.27	4.73
	60%	87.58	3.92	76.23	4.59
	80%	90.51	2.27	79.17	3.14
	90%	**91.78**	**1.55**	**80.44**	**1.90**

Table 3. The effect of the ADJUST algorithm, TSCV with the expanding window strategy, and training using of 90% of the data on mental workload classification for Task 1 and Task 2

Task	Model	Sensitivity	Specificity	Precision	Accuracy	FAR	FRR
1: resting vs testing state	Stacked LSTM	90.28	97.08	96.87	93.68	2.92	9.72
	BLSTM	90.60	93.01	91.31	91.81	6.99	9.40
	Stacked GRU	93.33	96.25	96.14	94.79	3.75	6.67
	BGRU	90.67	93.08	92.02	91.88	6.92	9.33
	BLSTM-LSTM	96.30	94.04	95.12	95.12	5.96	3.70
	BGRU-GRU	94.70	97.11	96.19	**95.90**	2.89	5.30
	CNN	89.03	94.03	93.71	91.53	5.97	10.97
2: low vs moderate vs high MWL level	Stacked LSTM	71.17	91.17	81.31	82.34	8.83	28.83
	BLSTM	70.23	90.23	79.44	80.46	9.77	29.77
	Stacked GRU	71.72	91.72	82.42	83.45	8.28	28.28
	BGRU	70.27	90.27	79.51	80.53	9.73	29.73
	BLSTM-LSTM	66.30	84.04	67.50	78.12	15.96	33.70
	BGRU-GRU	72.28	92.28	83.53	**84.56**	7.72	27.72
	CNN	70.09	90.09	79.16	80.19	9.91	29.91

Task 1: Resting vs Testing State As can be seen from the data in Table 3, the top three models in term of the model accuracy are the BGRU-GRU, BLSTM-LSTM, and stacked GRU models, which achieved accuracies of 95.90%, 95.12%, and 94.79%, respectively. The lowest model accuracy of 91.53% was obtaimed by the CNN model. This may imply that GRU-based models perform better than LSTM-based models. Moreover, we observed that the BGRU-GRU model also

obtained the highest specificity score of 97.11% for Task 1. This means that if the subjects were not in a resting state, the BGRU-GRU model was better at identifying the non-resting state than the other models. However, when looking at the sensitivity score, we observed a different pattern; that is, the sensitivity of the BGRU-GRU model is slightly lower than that of the BLSTM-LSTM model. This means that the BGRU-GRU model is less effective at classifying the true level of the subject's MWL compared to the BLSTM-LSTM model. The lowest sensitivity score is also obtained by the CNN model. Nevertheless, we also observed that the GRU-based models perform slightly better than the LSTM-based models, i.e. the stacked GRU model's accuracy is higher than that of the stacked LSTM model, BGRU model's accuracy is higher than that of BLSTM model, and BGRU-GRU model's accuracy is higher than that of BLSTM-LSTM model.

Task 2: Low vs Moderate vs High MWL Level According to the model accuracies shown in Table 3, the best model is still the BGRU-GRU model; however, the second-best model becomes the stacked GRU model, which is followed by the stacked LSTM model. The accuracies of the BGRU-GRU, GRU, and stacked LSTM models are 84.56%, 83.45%, and 82.34%, respectively. Therefore, the GRU-based models still perform better than the LSTM-based models for this task. Nevertheless, the BGRU-GRU model also achieves the highest sensitivity and specificity scores of 72.28% and 92.28%, respectively. This means that the BGRU-GRU model could classify the true level of the subjects' MWL correctly. Furthermore, it was also good at detecting the untrue level of the subject's MWL correctly; for example, when subjects were not at a low MWL level, BGRU-GRU model seemed to correctly indicate medium or high level than other models. In this task, we observed that the least effective model turned to be BLSTM-LSTM model. It provided the lowest accuracy, specificity, and sensitivity scores (78.12%, 84.04%, and 66.30%, respectively). This means that BLSTM-LSTM model could not classify subjects' MWL levels correctly for this dataset. Finally, we still observed that GRU-based models perform better than LSTM-based models in this task.

In summary, these results reveal that the BGRU-GRU model provided the best performance for both tasks and that GRU-based models performed better than LSTM-based models. The key difference between GRU and LSTM is that GRU is less complex than LSTM because it has a smaller number of gates. GRU's bag has two gates that are reset and updated, while LSTM has three gates: input, output and forget. This means that the less sophisticated model, i.e. GRU-based model, can capture relevant information and is preferred for this dataset.

6 Conclusion

Mental workload (MWL) is one of the key concepts that can be used to understand human performance, which can be negatively impacted if the workload is too high or too low. The negative effects of a high MWL can include stress, mood disorders, and illness. To determined subjects' MWL levels, EEG has been used as a tool. To date, machine learning techniques, especially powerful deep learning models, have been employed to capture the variance characteristics of EEG signals to classify the MWL of a subject accurately. One of the main techniques that is used to train and test deep learning models is the CV technique. However, the CV technique used for such models does not take into account the time series nature of EEG signals. Therefore, in this paper, we explored a modification of the CV technique that can be applied to EEG data, i.e. a blocked form of CV with expanding window and rolling window strategies. Moreover, we also investigated the effect of the sizes of the training, validation, and testing window for each TSCV strategy. Then, we verified the TSCV strategies and the window sizes using the following deep learning models: the stacked LSTM, BLSTM, stacked GRU, BGRU, BGRU-GRU, BLSTM-LSTM, and CNN models. We used a publicly available mental workload dataset called STEW [17] to carry out our experiment. Our findings show that the deep learning models evaluated using TSCV with the expanding window strategy provide a better classification performance than those evaluated using TSCV with a rolling window strategy. Moreover, for both classification tasks, every deep learning model achieved the highest accuracy when it was trained using 90% of the data, with a window size of 3 for validation and testing. Moreover, our results also show that the BGRU-GRU model achieved the highest accuracies of 95.90% and 84.56% for Task 1 and Task 2, respectively. Hence, in future work, we should consider applying the BGRU-GRU model and TSCV with the expanding window strategy on other MWL datasets, which are more complex. For example, there are datasets that were collected from different sessions/days, which usually causes a non-stationary or co-variate shift problem. Finally, this proposed CV method can also be applied to other types of time series data, such as stock prices, annual retail sales, or monthly subscribers since all of these data are sequential data that were measured at successive times and have a natural temporal ordering, just like EEG signals.

References

1. Ahmadi, A., Bazregarzadeh, H., Kazemi, K.: Automated detection of driver fatigue from electroencephalography through wavelet-based connectivity. Biocybernetics Biomed. Eng. **41**(1), 316–332 (2021)
2. Bergmeir, C., Benítez, J.M.: On the use of cross-validation for time series predictor evaluation. Inf. Sci. **191**, 192–213 (2012)
3. Bergmeir, C., Costantini, M., Benítez, J.M.: On the usefulness of cross-validation for directional forecast evaluation. Comput. Stat. Data Anal. **76**, 132–143 (2014)
4. Buscher, G., Dengel, A., Biedert, R., Elst, L.V.: Attentive documents: eye tracking as implicit feedback for information retrieval and beyond. ACM Trans. Interact. Intell. Syst. (TiiS) **1**(2), 1–30 (2012)
5. Cao, Z., Yin, Z., Zhang, J.: Recognition of cognitive load with a stacking network ensemble of denoising autoencoders and abstracted neurophysiological features. Cogn. Neurodyn. **15**(3), 425–437 (2021)

6. Cerqueira, V., Torgo, L., Mozetič, I.: Evaluating time series forecasting models: an empirical study on performance estimation methods. Mach. Learn. **109**(11), 1997–2028 (2020)
7. Dorffner, G.: Neural networks for time series processing. In: Neural network world. Citeseer (1996)
8. Hernández, L.G., Mozos, O.M., Ferrández, J.M., Antelis, J.M.: EEG-based detection of braking intention under different car driving conditions. Front. Neuroinform. **12**, 29 (2018)
9. Hoadley, B.: Asymptotic properties of maximum likelihood estimators for the independent not identically distributed case. Annals Math. Stat. 1977–1991 (1971)
10. Islam, M.K., Rastegarnia, A., Yang, Z.: Methods for artifact detection and removal from scalp EEG: a review. Neurophysiol. Clin. Clin. Neurophysiol. **46**(4–5), 287–305 (2016)
11. Islam, M.R., Barua, S., Ahmed, M.U., Begum, S., Di Flumeri, G.: Deep learning for automatic EEG feature extraction: an application in drivers' mental workload classification. In: Longo, L., Leva, M.C. (eds.) H-WORKLOAD 2019. CCIS, vol. 1107, pp. 121–135. Springer, Cham (2019). https://doi.org/10.1007/978-3-030-32423-0_8
12. Jeong, J.H., Yu, B.W., Lee, D.H., Lee, S.W.: Classification of drowsiness levels based on a deep spatio-temporal convolutional bidirectional LSTM network using electroencephalography signals. Brain Sci. **9**(12), 348 (2019)
13. Kingphai, K., Moshfeghi, Y.: On EEG preprocessing role in deep learning effectiveness for mental workload classification. In: Longo, L., Leva, M.C. (eds.) H-WORKLOAD 2021. CCIS, vol. 1493, pp. 81–98. Springer, Cham (2021). https://doi.org/10.1007/978-3-030-91408-0_6
14. Kingphai, K., Moshfeghi, Y.: On time series cross-validation for mental workload classification from EEG signals. In: Neuroergonomics Conference (2021)
15. Lee, D.H., Jeong, J.H., Kim, K., Yu, B.W., Lee, S.W.: Continuous EEG decoding of pilots' mental states using multiple feature block-based convolutional neural network. IEEE Access **8**, 121929–121941 (2020). https://doi.org/10.1109/ACCESS.2020.3006907
16. Lim, J., Wu, W.C., Wang, J., Detre, J.A., Dinges, D.F., Rao, H.: Imaging brain fatigue from sustained mental workload: an ASL perfusion study of the time-on-task effect. Neuroimage **49**(4), 3426–3435 (2010)
17. Lim, W., Sourina, O., Wang, L.: Stew: simultaneous task EEG workload data set. IEEE Trans. Neural Syst. Rehabil. Eng. **26**(11), 2106–2114 (2018)
18. Mognon, A., Jovicich, J., Bruzzone, L., Buiatti, M.: Adjust: an automatic EEG artifact detector based on the joint use of spatial and temporal features. Psychophysiology **48**(2), 229–240 (2011)
19. Murakami, H., Kumar, B.V.: Efficient calculation of primary images from a set of images. IEEE Trans. Pattern Anal. Mach. Intell. **5**, 511–515 (1982)
20. Nagabushanam, P., Thomas George, S., Radha, S.: EEG signal classification using LSTM and improved neural network algorithms. Soft Comput. **24**(13), 9981–10003 (2019). https://doi.org/10.1007/s00500-019-04515-0
21. Nussbaumer, H.J.: The fast fourier transform. In: Fast Fourier Transform and Convolution Algorithms, pp. 80–111. Springer, Berlin (1981). https://doi.org/10.1007/978-3-662-00551-4_4
22. Qayyum, A., Khan, M.A., Mazher, M., Suresh, M.: Classification of EEG learning and resting states using 1d-convolutional neural network for cognitive load assesment. In: 2018 IEEE Student Conference on Research and Development (SCOReD), pp. 1–5. IEEE (2018)

23. Qu, X., Liu, P., Li, Z., Hickey, T.: Multi-class time continuity voting for EEG classification. In: Frasson, C., Bamidis, P., Vlamos, P. (eds.) BFAL 2020. LNCS (LNAI), vol. 12462, pp. 24–33. Springer, Cham (2020). https://doi.org/10.1007/978-3-030-60735-7_3
24. Qu, X., Sun, Y., Sekuler, R., Hickey, T.: EEG markers of stem learning. In: 2018 IEEE Frontiers in Education Conference (FIE), pp. 1–9 (2018). https://doi.org/10.1109/FIE.2018.8659031
25. Saha, S., Baumert, M.: Intra-and inter-subject variability in EEG-based sensorimotor brain computer interface: a review. Front. Comput. Neurosci. **13**, 87 (2020)
26. Schaffer, C.: Selecting a classification method by cross-validation. Mach. Learn. **13**(1), 135–143 (1993)
27. Stone, M.: Cross-validatory choice and assessment of statistical predictions. J. Roy. Stat. Soc.: Ser. B (Methodol.) **36**(2), 111–133 (1974)
28. Szalma, J.L., et al.: Effects of sensory modality and task duration on performance, workload, and stress in sustained attention. Hum. Factors **46**(2), 219–233 (2004)
29. Tashman, L.J.: Out-of-sample tests of forecasting accuracy: an analysis and review. Int. J. Forecast. **16**(4), 437–450 (2000)
30. Taylor, L.P.: Chapter 20 - independent assessor audit guide. In: Taylor, L.P. (ed.) FISMA Compliance Handbook, pp. 239–273. Syngress, Boston (2013). https://doi.org/10.1016/B978-0-12-405871-2.00020-8https://www.sciencedirect.com/science/article/pii/B9780124058712000208
31. Teplan, M., et al.: Fundamentals of EEG measurement. Meas. Sci. Rev. **2**(2), 1–11 (2002)
32. Urigüen, J.A., Garcia-Zapirain, B.: EEG artifact removal—state-of-the-art and guidelines. J. Neural Eng. **12**(3), 031001 (2015). https://doi.org/10.1088/1741-2560/12/3/031001,https://doi.org/10.1088%2F1741-2560%2F12%2F3%2F031001
33. Wang, S., Gwizdka, J., Chaovalitwongse, W.A.: Using wireless EEG signals to assess memory workload in the n-back task. IEEE Trans. Human-Mach. Syst. **46**(3), 424–435 (2015)
34. Wylie, C., Shultz, T., Miller, J., Mitler, M., Mackie, R., et al.: Commercial motor vehicle driver fatigue and alertness study: Technical summary (1996)
35. Yang, S., Yin, Z., Wang, Y., Zhang, W., Wang, Y., Zhang, J.: Assessing cognitive mental workload via EEG signals and an ensemble deep learning classifier based on denoising autoencoders. Comput. Biol. Med. **109**, 159–170 (2019)
36. Zeng, H., et al.: A lightgbm-based EEG analysis method for driver mental states classification. Comput. Intell. Neurosci. **2019** (2019)
37. Zhang, D., Cao, D., Chen, H.: Deep learning decoding of mental state in non-invasive brain computer interface. In: Proceedings of the International Conference on Artificial Intelligence, Information Processing and Cloud Computing, pp. 1–5 (2019)
38. Zhang, J., Li, S.: A deep learning scheme for mental workload classification based on restricted boltzmann machines. Cognit. Technol. Work **19**(4), 607–631 (2017)
39. Zhang, Q., Yuan, Z., Chen, H., Li, X.: Identifying mental workload using EEG and deep learning. In: 2019 Chinese Automation Congress (CAC), pp. 1138–1142. IEEE (2019)
40. Zhang, Y., Liu, B., Ji, X., Huang, D.: Classification of EEG signals based on autoregressive model and wavelet packet decomposition. Neural Process. Lett. **45**(2), 365–378 (2017)
41. Zhou, Y., Xu, T., Li, S., Shi, R.: Beyond engagement: an EEG-based methodology for assessing user's confusion in an educational game. Univ. Access Inf. Soc. **18**(3), 551–563 (2019)

Path Weights Analyses in a Shallow Neural Network to Reach Explainable Artificial Intelligence (XAI) of fMRI Data

José Diogo Marques dos Santos[1,2] and José Paulo Marques dos Santos[3,4,5](✉)

[1] Faculty of Engineering, University of Porto, R. Dr Roberto Frias, 4200-465 Porto, Portugal
[2] Abel Salazar Biomedical Sciences Institute, University of Porto, R. Jorge de Viterbo Ferreira, 4050-313 Porto, Portugal
[3] University of Maia, Av. Carlos de Oliveira Campos, 4475-690 Maia, Portugal
jpsantos@umaia.pt
[4] LIACC - Artificial Intelligence and Computer Science Laboratory, University of Porto, R. Dr Roberto Frias, 4200-465 Porto, Portugal
[5] Unit of Experimental Biology, Faculty of Medicine, University of Porto, Alameda Prof. Hernâni Monteiro, 4200-319 Porto, Portugal

Abstract. A new procedure is proposed to simplify a Shallow Neural Network. With such aim, we introduce the concept of "path weights", which consists of the multiplication of the successive weights along a single path from the input layer to the output layer. This concept is used alongside a direct analysis of the calculations performed by a neural network to simplify a neural network via removing paths determined to be not relevant for decision making.

This study compares the proposed "path weights" method for network lightening with other methods, taken from input ranking, namely the Garson, and Paliwal and Kumar.

All the different methods of network lightening reduce the network complexity, favoring interpretability and keeping the prediction at high levels, well above randomness. However, Garson's and Paliwal and Kumar's methods maintain the totality of the weights between the hidden layer and the output layer, which keeps the number of connections high and reduces the capability of analyzing hidden node importance in the network's decision-making. Yet, the proposed method based on "path weights" provides lighter networks while retaining a high prediction rate, thus providing a high hit per network connection ratio. Moreover, the lightened neural network inputs are coherent with the established literature, identifying the brain regions that participate in motor execution.

Keywords: Explainable Artificial Intelligence (XAI) · Shallow Neural Networks · Backpropagation feedforward artificial neural networks · Human Connectome Project · fMRI

Supplementary Information The online version contains supplementary material available at https://doi.org/10.1007/978-3-031-25891-6_31.

G. Nicosia et al. (Eds.): LOD 2022, LNCS 13811, pp. 417–431, 2023.
https://doi.org/10.1007/978-3-031-25891-6_31

1 Introduction

Artificial Intelligence (AI) and Machine Learning (ML) may be used to model brain functioning to help in its understanding. Decoding the mental states using AI and ML methods has been foreseen for long [1]. One of these methods is Artificial Neural Networks (ANNs), which have been used to locate specialized functional brain areas for categorization [2, 3], or broader approaches like "inferring cognition" [4] or "mind-reading" [5], and even on real-time [6], based on functional magnetic resonance imaging (fMRI) data. ANNs, however, rely on complex calculations and, therefore, their internal processes are challenging to understand, which has been the cause to classify them as black boxes [7]. The developments of Deep Learning in the last decade have been contributing to improved accuracy of the models. Still, due to their even more complicated structure, it has been refraining from the old dream of using ANNs to help decode mental states.

FMRI data is particularly challenging. A typical functional acquisition delivers around 230,000 inputs across about 200–400 timepoints, turning data analysis intractable with ANNs. Therefore, data reduction is necessary. Even so, in the final dataset, one may find task-related data, task-unrelated data (the mind flies), physiology-related data, noise, and also one may expect strongly correlated data. In addition, if one intends to decode mental states, one may not disregard inputs a priori as there are usually no reasons to help the selection process.

The AI and ML final users may not have sufficient confidence in black-box models, which can preclude translating advances into real life. An emerging trend in AI and ML is Explainable Artificial Intelligence (XAI), which aims to construct models that one may understand and help comprehend the process [8, 9]. A critical application is medical and health-related processes, like brain functioning [10].

Although XAI is a recent framework, some methods have been proposed and compared to extract useful knowledge from ANNs' models. This is the case of neural network pruning, which is a set of techniques that reduce "the size of a network by removing parameters" [11]. Usually, pruning is applied to reduce the size of deep convolutional neural networks aiming to decrease computational infrastructure costs, which, however, is not the purpose here. Pruning may be mainly exerted over filters, nodes, or connections and, commonly, the techniques search for and merge redundant parameters, thus lightening the network [12, 13]. A critical stage in network pruning is ranking the inputs by sequencing those that contribute the most to the least [7, 14–19]. Furthermore, some of the ranking methods post-hoc act on trained networks [14, 15], while others work during the training stage [16, 18].

The procedure we propose here relies on a simpler version of ANNs, Shallow Neural Networks (SNNs), which have only one layer with a few nodes. This lighter structure, however, must keep an interesting accuracy. Besides identifying redundant nodes or weights and ranking the inputs per importance, the procedure aims to identify and deplete the least necessary weights in the SNN structure, as fMRI data usually includes task-related information, task-unrelated, physiology, and noise. Thus, by making it lighter, we expect the structure becomes more interpretable.

Although it was found that pruning nodes are preferable [20], because the base network is an SNN, we will prune weights. There are already known methods for weight pruning [21], but we introduce the concept of "path weight", expecting each hidden node to concentrate coherent rationales. That is why we preserve all the input and hidden nodes unless the method depletes all the afferent and efferent connections. The path weight considers all the weights found in one path from an input node until a node in the output layer. Because SNNs have one hidden layer only, the paths explored have two legs, one from the node in the input into a node in the hidden layer and one from the node in the hidden layer into a node in the output. Therefore, the path weight is calculated according to:

$$path\ weight_{ijk} = \left| w_{I_iH_j} \times w_{H_jO_k} \right| \quad (1)$$

where $w_{I_iH_j}$ is the weight between the input node I_i and the hidden node H_j, and $w_{H_jO_k}$ is the weight between the hidden node H_j and the output node O_k. Therefore, $path\ weight_{ijk}$ is the modulus of the product of the weights found in the path from input I_i to output O_k, passing by the hidden node H_j. The analysis of the path weights aims to identify which magnitudes are further from zero. These "heavier" path weights contribute more to the perceptron equation than underweighted paths, close to zero, although signal magnitude also has a role here. Therefore, such path weights may identify the inputs that hold important information for correct predictions, like other methods mentioned above, and, in addition, identify the nodes in the hidden layer and the connections from the hidden layer to the output that support partial processes. Considering all the path weights, one may establish a threshold and prune the network from the path weights that did not survive thresholding. The result is a lightened network, which maintains accuracy at an acceptable level with increased interpretability.

Although artificial data is preferred for testing methods because one only knows "the true importance of the variables" [14] in such cases, which, in the perspective of these authors, means that artificial data is the desirable way for generalizations, it was found that methods that cope well with artificial data drop performance when applied in real data [19]. Thus, other than using artificial data, our option is to test with real fMRI data. However, we chose a simple motor paradigm because neuroscience literature may be used to contrast the findings [22, 23], i.e., instead of using artificial data with known goals, we expect the method screens off the neural networks which are known to participate in the motor tasks, compare with established literature, and thus explain the SNN with the task-related natural process (pertinent brain functioning).

2 Method

2.1 Human Connectome Project Data

The data is a subset of 30 subjects from the HCP (Human Connectome Project) Young Adult database[1], subjects 100307 to 124422, from the 100 Unrelated Subjects subset [24–26]. HCP's motor paradigm is adapted from [23, 27]. It consists of two runs per subject. Each run encompasses a sequence of five stimuli where subjects are asked to:

- squeeze their left foot (LF);

[1] https://www.humanconnectome.org/study/hcp-young-adult.

- tap their left-hand fingers (LH);
- squeeze their right foot (RF);
- tap their right-hand fingers (RH);
- move their tongue (T).

A 3-s cue precedes a 12-s stimulus-response time. Both are visually projected. There are also three periods with a fixation cross (FIX), each lasting 15 s. Besides the stimuli sequences themselves, the difference between both runs is the fMRI scanner's phase encoding: one is acquired with right-to-left phase encoding (RL), and the other run with left-to-right (LR):

- RL sequence: FIX-RH-LF-T-RF-LH-FIX-T-LF-RH-FIX-LH-RF-FIX;
- LR sequence: FIX-LH-RF-FIX-T-LF-RH-FIX-LH-T-RF-RH-LF-FIX.

Subjects make no responses in this task. The repetition time (TR) is set to 0.72, and the run duration is 3:34. Each subject originates 284 volumes in each fMRI session, as the first fixation cross is cut to synchronize the data files with the other run for the same subject and between subjects. Data files are already subjected to brain extraction and registered to a standard image.

2.2 Data Processing

The data originated in the fMRI scanning sessions is analyzed by the mean of a back-propagation feedforward shallow neural network, but it is previously pre-processed using ICA (Independent Component Analysis). The purpose of the ICA pre-processing is to reduce data dimensionality.

The 30 subjects are randomly split into two groups: 20 are assigned to the train group, and the remaining 10 to the test group. Because the stimuli sequence is different in each session and to concatenate the data files, each session is interpreted as a separate subject, i.e., for the data reduction with ICA and pre-processing, each data file is considered one subject, and, therefore, there are 40 data files in the train group, and 20 in the test group. It is important to stress that each subject keeps in the respective group, i.e., there is no data of the same subject in both groups.

The 40 fMRI data files of the train group enter the ICA analysis as implemented in MELODIC (Multivariate Exploratory Linear Optimized Decomposition into Independent Components) v. 3.15 [28], part of FSL[2]. Importantly, to overcome the limitations imposed by using MIGP (MELODIC's Incremental Group-PCA) in v. 3.15, which precludes the output of ICs' complete time-courses, this step is run in line command including the option --disableMigp. The following data pre-processing is applied to the input data: masking of non-brain voxels; voxel-wise de-meaning of the data; normalization of the voxel-wise variance. Pre-processed data is whitened and projected into a 46-dimensional subspace using probabilistic Principal Component Analysis. The number of dimensions is estimated using the Laplace approximation to the Bayesian evidence of the model order [28, 29]. The whitened observations are decomposed into

[2] https://fsl.fmrib.ox.ac.uk/fsl/fslwiki/FSL.

sets of vectors that describe signal variation across the temporal domain (time-courses), the session/subject domain and across the spatial domain (maps) by optimizing for non-Gaussian spatial source distributions using a fixed-point iteration technique [30]. Estimated component maps are divided by the standard deviation of the residual noise and thresholded by fitting a mixture model to the histogram of intensity values [28].

Features are then extracted from each of the 46-time courses of the ICs (independent components). The strategy adopted is to average the seventh, eighth and ninth signals after the stimulus onset. The mean time difference to the stimulus onset is 5.285 s, proximal to the maximum of the canonical hemodynamic response in the brain [31], i.e., this feature maximizes the difference between task activation and the baseline. The data is standardized. At the end of this stage, the result is a matrix with 400 rows (20 subjects × 2 sessions × 5 × 2 stimulus/session), each corresponding to an epoch and 46 columns corresponding to one IC. This matrix is the training set input.

The test data is obtained with a different procedure. The 46 brain activation maps obtained with the train group are used as masks to average the individual time courses in the raw NIfTI files of each subject pertaining to the test group. The same procedure for feature calculation is adopted. The seventh, eighth and ninth acquisitions after stimulus onset are averaged. The data is standardized. Finally, a similar matrix with 200 rows (10 subjects × 2 sessions × 5 × 2 stimulus/session) and 46 columns is obtained. This matrix is the test set input.

2.3 Shallow Neural Network Design

The AMORE package v. 0.2–15 [32] implemented in R[3] v. 4.1.2 [33] and RStudio[4] v. 2021.09.01 Build 372 is used to design and perform the necessary calculations of the backpropagation feedforward shallow neural network.

Exploratory analysis searched for the best tunning parameters. Firstly, the global learning rate varied from 0.05 to 0.50 in 0.05 steps and then, more finely, from 0.03 to 0.10 in 0.01 steps, and the momentum ranged from 0.3 to 0.9 in 0.1 steps. The best combination yields a global learning rate of 0.10 and a global momentum of 0.8. Because the purpose of the study is to deliver interpretable shallow neural networks, it is considered a single hidden layer with 10 nodes fully connected with the inputs (46), each one corresponding to an IC, and outputs (5), each one corresponding to a stimulus category (LF, LH, RF, RH, and T).

The selected activation function for the hidden nodes is "tansig", while "sigmoid" is for output nodes. Tansig is chosen for the hidden layer nodes because it *a priori* outputs in the range [–1, 1], i.e., it may yield in the negative spectrum, then representing inhibitory components, besides higher learning speed [34]. Its canonical representation is:

$$f(v) = a \cdot \tanh(bv) \quad (2)$$

[3] https://www.r-project.org/.

[4] https://www.rstudio.com/.

where suitable values for the parameters a and b are [34, 35]:

$$a = \frac{1}{\tanh(b)} = \frac{1}{\tanh\left(^{2}/_{3}\right)} \approx 1.7159 \tag{3}$$

$$b = arc\tanh\left(\sqrt{^{1}/_{3}}\right) \approx {}^{2}/_{3} \tag{4}$$

Thus, the activation function tansig outputs in the range [–1.7159, 1.7159]. The activation function sigmoid in the output layer outputs in the range [0, 1].

The script is run 50,000 times, and the network with the higher hit rate is chosen (“best network”). Its performance is characterized in Table 1. The global hit rate is 166, corresponding to a global accuracy of 83.0%. As is expected, this data is intrinsically correlated (cf. Section S1 and Table S1 of the Supplementary Information; Section S2.1 and Table S2 list all the network biases and weights).

Table 1. Confusion matrix of the predictions of the neural network based on the test data, including the partial and global accuracies and precisions (LF: left foot; LH: left hand; RF: right foot; RH: right hand; T: tongue).

Stimulus		Prediction					Total
		LF	LH	RF	RH	T	
Input	LF	27	1	6	5	1	40
	LH	3	36	0	1	0	40
	RF	8	0	31	1	0	40
	RH	0	2	1	37	0	40
	T	4	0	1	0	35	40
Total		42	39	39	44	36	200
Accuracy (%)		67.5	90.0	77.5	92.5	87.5	83.0
Precision (%)		64.3	92.3	79.5	84.1	97.2	

2.4 Shallow Neural Network Lightening

Two lightening processes are conducted in parallel. The first, which is the purpose of this study, relies on Eq. (1). The second lightening process analyzes the summations’ summands (biases excluded) to identify those contributing to the activation function more. In addition, the second analysis aims to have a contrasting background during the network lightening, which may signal excessive performance drops.

The first lightening process uses Eq. (1) and is thresholded in two levels, so one has a perspective on the impacts of the lightening process:

- once, keeping the top 10 path weights per stimulus (output), “top” meaning higher absolute value;

- the second analysis keeps the top 10% path weights per stimulus (top 10% means the top 46 path weights because it is 10% × 46 connections input-hidden × 10 connections hidden-output).

The procedure of the second lightening process, summand analysis, is:

1. calculate the feedforward equations with the 40 cases per stimulus (output) according to:

$$v_i = bias_i + \sum_{j=1}^{n} s_{ij} w_{ij} \tag{5}$$

where v_i is the result to input in the activation function in node *i*, $bias_i$ is the bias in node *i*, *n* is the number of connections afferent to the node, s_{ij} is the signal from node *j* to node *i*, and w_{ij} is the weight of the connection that links node *j* to node *i*;
2. calculate the ratio between each summand in the summation and v_i, i.e., yields the fraction contribution of each summand in the overall result, i.e., yields the fraction contribution of each connection:

$$ratio_i = \frac{s_{ij} w_{ij}}{v_i} \tag{6}$$

3. for each stimulus (output) retains the $ratio_i$ that correspond to correct hits (i.e., discards the $ratio_i$ that correspond to incorrect predictions);
4. for each stimulus (output), and each connection averages the $ratio_i$ retained in the previous step;
5. the averages are thresholded; the threshold level is arbitrary; in this case, we thresholded the averages in the output layer with >0.25 (i.e., we retained the averages of the connections whose contribution to the activation function is higher than 25%), and thresholded the averages in the hidden layer with an absolute value >0.25 (i.e., we retained the averages of the connections whose contribution to the activation function is higher than 25% or lesser than –25%).

Thus, the best network is lightened:

- considering the top 10 path weighs only;
- considering the top 10% (46) path weighs only;
- considering the connections that survived to the summand analysis.

2.5 Comparison with Other Methods

Two methods are used as "gold standard": Garson [15] and Paliwal and Kumar [16]. The former ranks the inputs of an established network (post-hoc analysis), and the latter does a statistical calculus considering the data generated during the training stage.

The method proposed by Garson [15] has been widely used. The formula for the rank was published elsewhere [19]. Besides ranking the inputs, and for lightening purposes, we here retain the top 10 inputs (ICs) per stimulus (output) and their connections to the hidden layer. However, the network remains fully connected between the hidden layer and the output. We test this lightened network with the test dataset.

Paliwal and Kumar [16] proposed a method to interpret the inputs' importance in an ANN-based on the network connections' weights. The method relies on statistical analysis over a large set of training examples, yielding an ordered list of the input nodes as per their relative importance. The relative importance of each input node results from averaging the interquartile range of all afferent connections of that input node across all the training examples. In this case, 50,000 examples are averaged for each input node. Similarly to Garson's method, and to make these methods comparable both with the path weight and summand analyses, the top 10 most important input nodes are considered, i.e., the weights of the remaining connections that connect the input and the hidden layer are discarded. However, the second part of the SNN, which links the hidden and the output layers, remains fully connected.

The total sum of connections measures the complexity of each network in the final. More connections negatively impact interpretability, and less interpretability may mean lesser explainability. The total sum of connections is reported in Table 2.

Table 2. Total number of connections in each kind of network

Network	Connections
Best network	510
Path weights (top 10)	36
Path weights (top 10%; 46)	126
Summand analysis	203
Garson (top 10)	170
Paliwal and Kumar (top 10)	150

3 Results

3.1 Path Weights and Summand Analyses

Figure S1, in Supplementary Information, depicts the 460 path weights of each stimulus (46 inputs $\times$ 10 hidden nodes) ordered from the highest to the smallest. The initial values are large, but the path weight value drops quickly for every case. The path weights (top 10) analysis considers each stimulus's first 10 path weights, while the path weights (top 10%; 46) analysis considers the first 46. Graphically, the curves flatten after then.

The top 10 and the top 10% (46) networks are characterized in the Supplementary Information (cf. Table S3 containing the highest path weights per output, and Fig. S2 and Fig. S3 depict both networks). The top 10% (46) network delivers 148 hits, and Table 3 reports its accuracies and precisions, while the top 10 network provides 128 hits, and Table 4 reports its accuracies and precisions.

The results of the inverted networks are reported in Supplementary Information (cf. Table S6 and Table S7). In the inverted networks, while the hidden and the output layers are fully connected, the connections between the input and the hidden layer are all those that did not survive the top 46 or the top 10 selection. The inverted top 10% (46) network yielded 70 hits, and the inverted top 10 returned 65 hits. As a result, the global accuracies dropped to 35.0% and 32.5%, respectively.

Table 3. Confusion matrix of the predictions of the top 10% (46) path weights per stimulus (output) of the neural network, including the partial and global accuracies and precisions (LF: left foot; LH: left hand; RF: right foot; RH: right hand; T: tongue).

Stimulus		Prediction					Total
		LF	LH	RF	RH	T	
Input	LF	27	2	6	2	3	40
	LH	4	28	3	4	1	40
	RF	9	0	28	0	3	40
	RH	2	2	2	33	1	40
	T	3	0	4	1	32	40
Total		45	32	43	40	40	200
Accuracy (%)		67.8	70.0	70.0	82.5	80.0	74.0
Precision (%)		60.0	87.5	65.1	82.5	80.0	

Table 4. Confusion matrix of the predictions of the top 10 path weights per stimulus (output) of the neural network, including the partial and global accuracies and precisions (LF: left foot; LH: left hand; RF: right foot; RH: right hand; T: tongue).

Stimulus		Prediction					Total
		LF	LH	RF	RH	T	
Input	LF	19	1	9	2	9	40
	LH	1	31	4	4	0	40
	RF	5	0	17	8	10	40
	RH	0	1	2	36	1	40
	T	7	1	3	4	25	40
Total		32	34	35	54	45	200
Accuracy (%)		47.5	77.5	42.5	90.0	62.5	64.0
Precision (%)		59.4	91.2	48.6	66.7	55.6	

The network constructed with the connections that survived the summand analysis is reported in Table S8. The test returned 150 hits, with accuracies and precisions reported in the same table. Table S9 in Supplementary Information reports the ratios of the connections between the hidden and the output layers, highlighting those above the threshold (>0.25).

3.2 Analyses Based on Garson's and Paliwal and Kumar's Methods

The application of the method proposed by Garson [15] ranked the 46 inputs (ICs) according to their importance measured by the respective weights. The top 10 higher values per stimulus (output) were selected from this list. The remaining were removed from the network (cf. Supplementary Information section S2.4, Table S10, and Fig. S4 for a graphical representation). Testing this network yielded 154 correct hits with the partials, accuracies, and precision reported in Table S11.

The network lightened considering the top 10 input nodes selected using Paliwal and Kumar [16] method (note that this network maintains its full connection between the hidden and output layers) delivers 129 hits. Supplementary Information section S2.5 details the procedure. The values are listed in Table S12 and Fig. S5 depicts the network. Fig. S6 plots the ICs' interquartile range averages ordered by decreasing magnitude and Table S13 reports its accuracies and precisions. Similarly to Fig. S1, the initial values are large, but they drop quickly after then.

3.3 Networks Comparison

Table 5 compares side by side the partial and global accuracies of the six types of networks considered in the present study. Garson's method performs best, close to the original network (Best network). Its hits rate is higher (+25 hits) than the network depleted considering the inputs ranking method proposed by Paliwal and Kumar, although it encompasses 20 more connections. The summand analysis network has more connections but performs worse than the network lightened using the inputs ranking method proposed by Garson. Both networks lightened using the path weights procedures described above have the higher hits per connection ratio. The performance of the Path weights (top 10%) network gets close to the Garson's network (148 versus 154 hits, respectively) but with considerably fewer connections (126 versus 170, respectively). Path weights (top 10) network has the worst performance but, because it encompasses 36 connections only, is the network with the highest hits per connection ratio. This network keeps 77.1% of the hits' ratio but has a much lighter structure, making it suitable for interpretation.

Table 5. Accuracies, partial and global, and structures' complexity (number of connections) of the six types of networks.

Network	Conn.	Hits	Hits per conn.	Accuracies (%)					
				LF	LH	RF	RH	T	Global
Best network	510	166	0.325	67.5	90.0	77.5	92.5	87.5	83.0

(*continued*)

Table 5. *(continued)*

Network	Conn.	Hits	Hits per conn.	Accuracies (%)					
				LF	LH	RF	RH	T	Global
Path weights (top 10)	36	128	3.556	47.5	77.5	42.5	90.0	62.5	64.0
Path weights (top 10%; 46)	126	148	1.175	67.8	70.0	70.0	82.5	80.0	74.1
Summand analysis	203	150	0.739	50.0	80.0	77.5	90.0	77.5	75.0
Garson (top 10)	170	154	0.906	65.0	80.0	77.5	75.0	87.5	77.0
Paliwal and Kumar (top 10)	150	129	0.860	52.5	55.0	60.0	70.0	85.0	64.5

3.4 Identification of the Main Input (ICs) and Hidden Nodes

The input nodes (ICs) that contributed more to the top 10 network and their respective path weight are identified in Table S14 in Supplementary Information. ICs 5, 7, 11, and 12 are depicted in Fig. S7. Most of the activations of these ICs are located in the precentral and postcentral gyri, which means motor and primary somatosensorial cortices. Such areas border the central sulci (all four cases) and the interhemispheric fissure (ICs 7, 11, and 12, mainly the latter).

Table 6 identifies the connections between the hidden layer and the output that survived the lightening process and their respective magnitudes. The positive ones have a red background, while blue in the negative. The hidden nodes h4 and h9 are present in all outputs, although with different patterns. While, in h4, it is negative for the hand, h9 is negative for the left. The combination of h4, h9, and h10 is sufficient to categorize each type of stimulus.

Table 6. Connections' weights between the hidden node and the output for the top 10 network. "---" signify the connection was depleted during the lightening process.

		Hidden nodes									
		h1	h2	h3	h4	h5	h6	h7	h8	h9	h10
Output nodes	LF	---	---	2.874	4.661	---	---	---	-2.823	-3.339	3.238
	LH	---	---	---	-3.304	-1.705	---	---	---	-2.399	---
	RF	---	---	---	5.811	---	---	---	---	2.159	-3.813
	RH	---	-1.473	---	-4.198	---	---	---	---	2.665	-1.906
	T	---	---	-2.606	0.882	---	---	---	---	1.704	2.716

4 Discussion

The first interesting finding is that a simple network architecture, as the SNN, is able to model fMRI data and make predictions well above the chance level, signifying that it extracts pertinent information from the raw data and organizes it into a useful model. The procedure's success includes extracting features considering the canonical representation of the hemodynamic response of brain processes, data dimensionality reduction with ICA, and masking brain activity for the test dataset construction.

The summand analysis network serves as a reference for comparison. In its lightening process, the connections are depleted one by one, depending on each contribution to the activation function. As summarized in Table S8, the prediction performance of the summand analysis network just drops 9.6% (166 to 150) but reduces the connections by 60.2% (510 to 203). Their performance differed when comparing the two networks lightened with input ranking methods, Garson, and Paliwal and Kumar. While the Paliwal and Kumar (top 10) network prediction drops 22.2%, the network is reduced by 70.6%, which means a ratio hits per connection of 0.860, the Garson (top 10) network performance just drops 7.2%, i.e. less than the summand analysis network. The total connections reduce by 66.7%., meaning a hits-per-connection ratio of 0.906. Therefore, Garson's method is better. The inconvenience of these methods is that, although they effectively select the most important inputs, they keep fully connected between the hidden and the output layers, which precludes the hidden nodes' interpretability. In any case, they serve as a reference.

Considering the two path weights-based lightening, logically, the top 10% (46) network prediction drops 10.8%, versus 22.9% in the top 10 network. In what it concerns the network lightening, the top 10% (46) reduces by 75.3% (ratio hits per connection 1.175), and the top 10 network reduces by 92.9% (ratio 3.556). All these lightened networks keep predictions well above the chance level. They are, however, much lighter than the original, which favors interpretability.

In what respects interpretability, the top 10 network exhibits the best balance between prediction performance and lightness, which is translated by the hits-per-connections ratio of 3.556. Such a lighter network permits both:

- exploring decisions in the hidden nodes;
- identify the biological networks that contribute more in the input layer.

Some nodes in the hidden layer contribute to all, or almost all, outputs, such as h4, h9, and h10. The interpretation of the connection weights allows identifying those that distinguish between feet and tongue from hand (h4), right and center from the left (h9), or left and center from the right (h10). Considering the weights' signals of these nodes is sufficient for correctly classifying the five types of stimuli, which contributes to explaining the decision process. In addition, node h3 distinguishes between the left foot and tongue, and the nodes h2, h5, and h8 have an inhibitory contribution respectively for the right hand, the left hand, and the left foot.

The method retained ICs 5, 7, 11, and 12 in the input layer, and these inputs have the highest weights in the connections between the input and the hidden layer. As reported in Sect. 3.4, all these ICs represent biological neural networks that are known to be involved in motor tasks, mainly in movements of the feet, hands, and tongue. The method was able to select and retain the sources of biological information that would have to be selected and retained for corrected modeling purposes, at least taking into account the actual state of knowledge [22, 23], matching the "gold standard". Thus, the high hit rate in the pruned SNN may be explained by the ability of the method to select, retain, and organize the pertinent sources of information in the brain. Therefore, it may be explored to decode brain function.

5 Conclusion

The present work explores the suitability of SNNs to model fMRI data, aiming to reach high correct predictions in the test stage. Afterwards, the neural network is pruned in order to favor its interpretability. The lightened neural network's performance is explained either by the arrangement of the connection weights between the hidden layer nodes and the output layer, and the selected and retained input nodes. The latter represent biological neural networks known to participate in motor tasks that involve the feet, hands, and tongue. Thus, the method achieves what it would have to achieve, i.e., it selects, retains, and organizes the pertinent sources of biological information for modeling the brain processes.

Further work may be developed from this stage. For example, the two pruned SNNs may be extra trained, aiming to recover the initial network's hit rate, accuracies and precisions. In addition, the process may be iterated, targeting even more frugal networks, but maintaining the higher hit rates. Another aspect that should deserve attention is the data dimension reduction, which is a necessary stage that highly impacts granularity in the findings. In parallel, the network structure may be improved from the simple SNN, seeking an architecture more adequate to model the canonical hemodynamic response, all of which help in decoding the brain processes.

Acknowledgements. This work was partially financially supported by Base Funding - UIDB/00027/2020 of the Artificial Intelligence and Computer Science Laboratory – LIACC - funded by national funds through the FCT/MCTES (PIDDAC).

References

1. Haynes, J.-D., Rees, G.: Decoding mental states from brain activity in humans. Nat. Rev. Neurosci. **7**, 523–534 (2006). https://doi.org/10.1038/nrn1931
2. Hanson, S.J., Matsuka, T., Haxby, J.V.: Combinatorial codes in ventral temporal lobe for object recognition: Haxby (2001) revisited: is there a "face" area? Neuroimage **23**, 156–166 (2004). https://doi.org/10.1016/j.neuroimage.2004.05.020
3. Onut, I.-V., Ghorbani, A.A.: Classifying cognitive states from fMRI data using neural networks. In: Proceedings. 2004 IEEE International Joint Conference on Neural Networks, pp. 2871–2875 (2004). https://doi.org/10.1109/IJCNN.2004.1381114
4. Sona, D., Veeramachaneni, S., Olivetti, E., Avesani, P.: Inferring cognition from fMRI brain images. In: de Sá, J.M., Alexandre, L.A., Duch, W., Mandic, D. (eds.) ICANN 2007. LNCS, vol. 4669, pp. 869–878. Springer, Heidelberg (2007). https://doi.org/10.1007/978-3-540-74695-9_89
5. Marques dos Santos, J.P., Moutinho, L., Castelo-Branco, M.: 'Mind reading': hitting cognition by using ANNs to analyze fMRI data in a paradigm exempted from motor responses. In: International Workshop on Artificial Neural Networks and Intelligent Information Processing (ANNIIP 2014), pp. 45–52. Scitepress (Science and Technology Publications, Lda.), Vienna, Austria (2014). https://doi.org/10.5220/0005126400450052

6. Weygandt, M., Stark, R., Blecker, C., Walter, B., Vaitl, D.: Real-time fMRI pattern-classification using artificial neural networks. Clin. Neurophysiol. **118**, e114–e114 (2007). https://doi.org/10.1016/j.clinph.2006.11.265
7. de Oña, J., Garrido, C.: Extracting the contribution of independent variables in neural network models: a new approach to handle instability. Neural Comput. Appl. **25**(3–4), 859–869 (2014). https://doi.org/10.1007/s00521-014-1573-5
8. Adadi, A., Berrada, M.: Peeking inside the black-box: a survey on explainable artificial intelligence (XAI). IEEE Access **6**, 52138–52160 (2018). https://doi.org/10.1109/ACCESS.2018.2870052
9. Samek, W., Müller, K.-R.: Towards explainable artificial intelligence. In: Samek, W., Montavon, G., Vedaldi, A., Hansen, L.K., Müller, K.-R. (eds.) Explainable AI: Interpreting, Explaining and Visualizing Deep Learning. LNCS (LNAI), vol. 11700, pp. 5–22. Springer, Cham (2019). https://doi.org/10.1007/978-3-030-28954-6_1
10. Tjoa, E., Guan, C.: A survey on explainable artificial intelligence (XAI): toward medical XAI. IEEE Trans. Neural Netw. Learn. Syst. **32**, 4793–4813 (2021). https://doi.org/10.1109/TNNLS.2020.3027314
11. Blalock, D., Gonzalez Ortiz, J.J., Frankle, J., Guttag, J.: What is the state of neural network pruning? In: Dhillon, I., Papailiopoulos, D., Sze, V. (eds.) 3rd Conference on Machine Learning and Systems, MLSys 2020, vol. 2, pp. 129-146, Austin (TX), USA (2020)
12. Zhao, F., Zeng, Y.: Dynamically optimizing network structure based on synaptic pruning in the brain. Frontiers in Systems Neuroscience 15, 620558 (2021). https://doi.org/10.3389/fnsys.2021.620558
13. Mirkes, E.M.: Artificial neural network pruning to extract knowledge. In: 2020 International Joint Conference on Neural Networks (IJCNN), pp. 1–8. IEEE (2020). https://doi.org/10.1109/IJCNN48605.2020.9206861
14. Olden, J.D., Joy, M.K., Death, R.G.: An accurate comparison of methods for quantifying variable importance in artificial neural networks using simulated data. Ecol. Model. **178**, 389–397 (2004). https://doi.org/10.1016/j.ecolmodel.2004.03.013
15. Garson, D.G.: Interpreting neural network connection weights. AI Expert **6**, 46–51 (1991)
16. Paliwal, M., Kumar, U.A.: Assessing the contribution of variables in feed forward neural network. Appl. Soft Comput. **11**, 3690–3696 (2011). https://doi.org/10.1016/j.asoc.2011.01.040
17. Fischer, A.: How to determine the unique contributions of input-variables to the nonlinear regression function of a multilayer perceptron. Ecol. Model. **309–310**, 60–63 (2015). https://doi.org/10.1016/j.ecolmodel.2015.04.015
18. de Sá, C.R.: Variance-based feature importance in neural networks. In: Kralj Novak, P., Šmuc, T., Džeroski, S. (eds.) Discovery Science. Lecture Notes in Computer Science (Lecture Notes in Artificial Intelligence), vol. 11828, pp. 306–315. Springer, Cham (2019). https://doi.org/10.1007/978-3-030-33778-0_24
19. Luíza da Costa, N., Dias de Lima, M., Barbosa, R.: Evaluation of feature selection methods based on artificial neural network weights. Expert Syst. Appl. **168**, 114312 (2021). https://doi.org/10.1016/j.eswa.2020.114312
20. Bondarenko, A., Borisov, A., Alekseeva, L.: Neurons vs weights pruning in artificial neural networks. In: 10th International Scientific and Practical Conference on Environment. Technologies. Resources, vol. 3, pp. 22–28. Rēzekne Academy of Technologies, Rēzekne (2015)
21. Karnin, E.D.: A simple procedure for pruning back-propagation trained neural networks. IEEE Trans. Neural Netw. **1**, 239–242 (1990). https://doi.org/10.1109/72.80236
22. Penfield, W., Boldrey, E.: Somatic motor and sensory representation in the cerbral cortex of man as studied by electrical stimulation. Brain **60**, 389–443 (1937). https://doi.org/10.1093/brain/60.4.389

23. Yeo, B.T.T., et al.: The organization of the human cerebral cortex estimated by intrinsic functional connectivity. J. Neurophysiol. **106**, 1125–1165 (2011). https://doi.org/10.1152/jn.00338.2011
24. Van Essen, D.C., Glasser, M.F.: The human connectome project: progress and prospects. In: Cerebrum 2016, cer-10–16 (2016)
25. Van Essen, D.C., Smith, S.M., Barch, D.M., Behrens, T.E.J., Yacoub, E., Ugurbil, K.: The WU-Minn human connectome project: an overview. Neuroimage **80**, 62–79 (2013). https://doi.org/10.1016/j.neuroimage.2013.05.041
26. Elam, J.S., et al.: The human connectome project: a retrospective. Neuroimage **244**, 118543 (2021). https://doi.org/10.1016/j.neuroimage.2021.118543
27. Buckner, R.L., Krienen, F.M., Castellanos, A., Diaz, J.C., Yeo, B.T.T.: The organization of the human cerebellum estimated by intrinsic functional connectivity. J. Neurophysiol. **106**, 2322–2345 (2011). https://doi.org/10.1152/jn.00339.2011
28. Beckmann, C.F., Smith, S.M.: Probabilistic independent component analysis for functional magnetic resonance imaging. IEEE Trans. Med. Imaging **23**, 137–152 (2004). https://doi.org/10.1109/TMI.2003.822821
29. Minka, T.P.: Automatic choice of dimensionality for PCA. Technical Report 514, MIT Media Lab Vision and Modeling Group. MIT (2000)
30. Hyvärinen, A.: Fast and robust fixed-point algorithms for independent component analysis. IEEE Trans. Neural Netw. **10**, 626–634 (1999). https://doi.org/10.1109/72.761722
31. Buckner, R.L.: Event-related fMRI and the hemodynamic response. Hum. Brain Mapp. **6**, 373–377 (1998). https://doi.org/10.1002/(SICI)1097-0193(1998)6:5/6%3c373::AID-HBM8%3e3.0.CO;2-P
32. Limas, M.C., Meré, J.B.O., Marcos, A.G., de Pisón Ascacibar, F., Espinoza, A.P., Elías, F.A.: A MORE flexible neural network package (0.2–12). León (2010)
33. R Development Core Team: R: A Language and Environment for Statistical Computing. R Foundation for Statistical Computing, Vienna (2010)
34. Haykin, S.: Neural Networks and Learning Machines. Prentice Hall, New Jersey (2009)
35. Le Cun, Y.: Efficient learning and second-order methods. Tutorial NIPS **93**, 61 (1993)

Audio Visual Association Test in Non Synesthetic Subjects: Technological Tailoring of the Methods

Costanza Cenerini[1](✉), Luca Vollero[1], Giorgio Pennazza[1], Marco Santonico[2], and Flavio Keller[3]

[1] Department of Engineering, Campus Bio-Medico University of Rome, Rome, Italy
costanza.cenerini@unicampus.it
[2] Department of Science and Technology for Humans and the Environment, Campus Bio-Medico University of Rome, Rome, Italy
[3] Department of Medicine and Surgery, Campus Bio-Medico University of Rome, Rome, Italy

Abstract. Synesthesia is a relatively rare condition for which a specific stimulus activates one or more senses different from the ones usually designed for its evaluation; this can happen in any combination of senses or cognitive paths. In this work, we focus on audio-visual synesthesia, one of the most common. Many investigators have studied the connection between music and colours in synesthetic and non-synesthetic subjects in the past years, exploring the possibility of an emotional gate between the perception of these stimuli. In this study, an experiment was designed to verify whether this connection can be extended to music and images, considering not only colour but also shape and their spatial distribution. In the study, participants' musical and visual abilities will also be tested to explore any correlation with their music-images association.

Keywords: Synesthesia · Music and color association · Generative art

1 Introduction

According to Rothen [1], synesthesia has neural bases: it is thought that the synesthetic experience of colour activates brain regions responsible for normal colour perception, contrary to what non-synesthetic subjects experience when remembering or imagining colours. There are many theories on this atypical neural activation. The most credited are cross-activation, in which the cause could be an extraneuronal connection in contiguous areas of the brain responsible for senses processing, and disinhibited feedback theory, in which it is caused by some differences in subjects' neuronal transmission. Ward states that these differences could be present at a cortical level, showing the complex nature of the inducer/concurrent process [2]. All the models agree in explaining synesthesia as caused by an atypical pattern of connections between the processing brain centres, which could be functional or structural [3].

G. Nicosia et al. (Eds.): LOD 2022, LNCS 13811, pp. 432–437, 2023.
https://doi.org/10.1007/978-3-031-25891-6_32

The first study on music-colour association was conducted by Sabaneev in 1929 [4], in which participants were asked to choose a colour that best described notes and harmonies that they were listening to; this type of experiment has been reproduced many times over the years with the addition of other factors: in Howell's study subjects were also administered Ishihara and Seashore test [5], Marks asked subjects to associate volume and lightness volume [6], and finally, Odbert introduced emotion in the experiment [7]. Since it turned out to be an important factor, many studies replicated the experiment, asking participants to associate music and emotion, colour and emotion, and music and colour [8,9]. Our study aims to investigate the connection between music, emotion, and images, adding shape and spatial distribution of elements in pictures as other features. As a second goal, it aims to evaluate the influence of one's visual and musical abilities in the image-music-emotion association. In this paper, we present the test and preliminary results obtained on a small population of subjects.

2 Material and Methods

This study received the approval of the university's ethical committee on February 16 2022 with number of clinical studies register 2021.236. All tests are delivered through a computer, and the interaction is achieved through a graphical interface.

2.1 Experiment Design

Once enrolled, participants will be asked to complete a form, stating their basic information, their musical and artistic background, their likes and dislikes of musical genres, the level of emotion evoked from certain colours and shapes and if they have ever experienced synesthesia. Subjects will be sent the link to access the experiment via email and they will be given the possibility to complete the test in one or more sessions, depending on their fatigue after each session. Before the beginning of the test, participants are informed of each test's duration, they are asked to be alone, to use a PC with a high resolution screen, to deactivate night shift on their monitor and to wear headphones. The test consists of a pretest and two main phases.

Pretest: This part consists of 5 different tests: (i) Ishihara Test for Colour Blindness detection; (ii) Perfect Pitch Test; (iii) Melodic discrimination test [10]; (iv) Mistuning Perception test [11]; (v) Beat alignment test [12]. These tests are simplified versions of the cited tests. In particular:

- Ishihara Test: The subjects are shown 24 tables and asked to identify numbers or lines in the picture and write it in the input box underneath the image.
- Perfect Pitch Test: The subjects can play a tone only once and choose the corresponding note on a 2 octave keyboard. Each time a key is pressed, users are given the possibility to listen to the next tone.

- Melodic Discrimination test: The participants can play only once three melodies. Two of them have the same melodic structure, the last has a different note. The second and third melodies are higher in pitch than the first one by a semitone and a tone, respectively. They have to detect the different melody. They can play the melodies of an example before starting the trial.
- Mistuning Perception Test: The subjects can play the same song twice and have to choose in which one of them the singer is in tune.
- Beat Alignment Test: The subjects can play the same song twice and have to choose in which one of them the triangle is on time.

The pretest assesses the user's visual abilities and then its musical ones from the most to the least challenging.

Phase One: In this phase the participants compose an image configuring a set of parameters while listening to a song and are then asked about their feelings on the song. The subjects can see the image they are generating, while setting the image's parameters. The parameters are: number of objects, objects dimensions, objects dispersion, objects shape, hue, saturation and brightness. Objects shape ranges from an angular to a rounded figure similar to the Maluma/Takete [13]. They can play and listen to the song once, and are only able to submit their image after the song is over. Once they submit their image, they can move 9 emotion sliders and, if they want, they can listen to the song a second time. Emotions are expressed with the 9-term version of GEMS with the replacements made in Aljanaki's study [14]. Once they are finished, they can move to the next song. Before starting the test, they can play with a trial image's parameters to understand how image composition works. All songs are composed by Andrea Sorbo, a composer hired specifically for the study, to avoid any influence of memories generated by song recognition. Moreover, the songs belong to different musical genres.

Phase Two: In this phase the subjects are shown twenty one images in which all parameters listed in phase one are set on neutral values and only one parameter is exaggerated (e.g. a very bright image). The respondent are asked to describe their emotions upon seeing the image by modifying the same emotion sliders of phase one. Figure 1 shows examples of images composed by the subjects or presented to the participants in our tests.

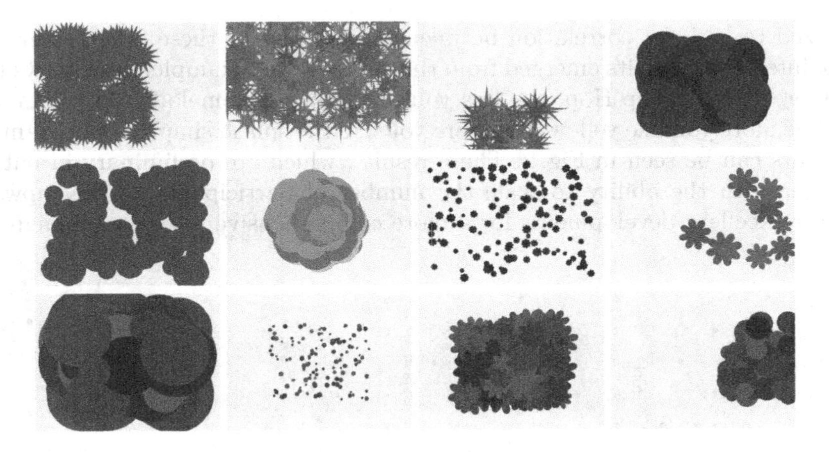

Fig. 1. Rows 1–2: examples of images generated in phase one. **Row 3**: examples of images shown in phase two.

Test Development: The test is administered remotely via a website. Both front-end and back-end development of the website was realized using `javascript` [15], in the first case using `p5.js`[1] and in the second `node.js`[2]. The users' answers and their actions on the elements on the screen are saved on a `firebase`[3] database at the end of each test.

2.2 Pilot Trial Data Analysis

Before conducting the synesthesia trial on a 100-subject recruited cohort, a pilot study was conducted aimed at verifying the whole testing platform and exploring some of the outcomes that could be generated from the tests. The pilot study was conducted on 14 subjects, and the findings are briefly discussed below. In order to have a better understanding of the information collected in the pilot trial, the form data and the results from the pretest were analyzed using scatter plot matrix and linear regression in MATLAB R2020a. Data from phase one and two was not analyzed at this stage due to its complexity compared to the small number of the subjects.

3 Results and Discussion

The platform collected the information entered by participants in the tests correctly, with no reported errors. The satisfaction in using the system was high, all participants found it accessible and easy to use. Collected data was split in 3 groups: musical preferences, artistic preferences and pretest results and was then

[1] https://p5js.org/.
[2] https://nodejs.org/.
[3] https://firebase.google.com/.

analyzed to look for correlation between the features of these different groups. Some interesting results emerged from this analysis: for example, data shows that the more you like Hip-Hop, the less you perform in the melodic discrimination test, the more you like yellow, the more you like the splash shape. Some examples of results can be seen in Fig. 2. These results, which are preliminary in nature, together with the ability to scale the number of participants quickly upward, suggest excellent developments for a more comprehensive testing campaign.

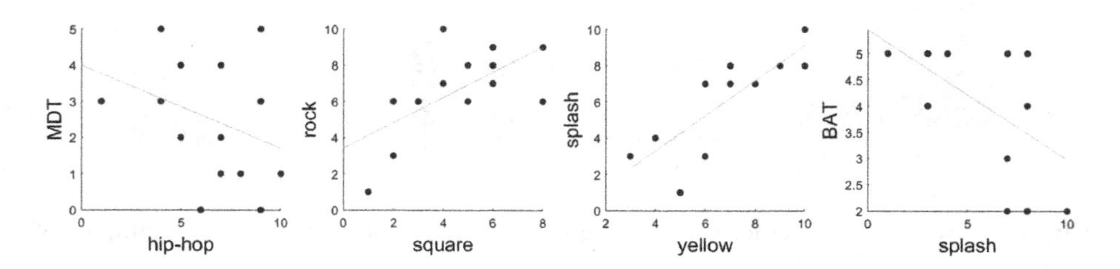

Fig. 2. Examples of data analysis results. MDT = Melodic Discrimination test, BAT = Beat Alignment Test

4 Conclusion and Future Developments

In this article, we presented a web-based platform designed and developed for the assessment of audio-visual association. The platform includes several standard tests and some custom-developed tests, integrated into a web-based delivery system. Some preliminary results supporting the functionality of the system have been presented. Future developments concern the use of the platform in an extended experimentation from which actual results of interdependence between stimuli can be deduced. Moreover, the developed approach builds the foundation for the development of new tests aimed at revealing the interdependence between the different artistic qualities that a subject can possess. These studies represent basic research towards the exploitation of the emerging interdependence between senses to develop new inclusive smart systems for sensory integration and augmentation in subjects with disabilities.

References

1. Rothen, N., Meier, B., Ward, J.: Enhanced memory ability: insights from synaesthesia. Neurosci. Biobehav. Rev. **36**(8), 1952–1963 (2012)
2. Jamie, W.: Synesthesia. Annu. Rev. Psychol. **64**(1), 49–75 (2013)
3. Blake, R., Palmeri, T.J., Marois, R., Kim, C.-Y.: On the perceptual reality of synesthetic color. Synesthesia: Perspect. Cognit. Neurosci. 47–73 (2005)
4. Sabaneev, L., Pring, S.W.: The relation between sound and colour. Music Lett. **10**(3), 266–277 (1929)

5. Howells, T.H.: The experimental development of color-tone synesthesia. J. Exp. Psychol. **34**(2), 87 (1944)
6. Marks, L.E.: On associations of light and sound: the mediation of brightness, pitch, and loudness. Am. J. Psychol. 173–188 (1974)
7. Odbert, H.S., Karwoski, T.F., Eckerson, A.B.: Studies in synesthetic thinking: I. musical and verbal associations of color and mood. J. Gen. Psychol. **26**(1), 153–173 (1942)
8. Palmer, S.E., Schloss, K.B., Xu, Z., Prado-León, L.R.: Music-color associations are mediated by emotion. In: Proceedings of the National Academy of Sciences, vol. 110, no. 22, pp. 8836–8841 (2013)
9. Whiteford, K.L., Schloss, K.B., Helwig, N.E., Palmer, S.E.: Color, music, and emotion: bach to the blues. i-Perception **9**(6), 2041669518808535 (2018)
10. Harrison, P.M.C., Musil, J.J., Müllensiefen, D.: Modelling melodic discrimination tests: descriptive and explanatory approaches. J. New Music Res. **45**(3), 265–280 (2016)
11. Larrouy-Maestri, P., Harrison, P.M.C., Müllensiefen, D.: The mistuning perception test: a new measurement instrument. Behav. Res. Methods **51**(2), 663–675 (2019). https://doi.org/10.3758/s13428-019-01225-1
12. Harrison, P., Müllensiefen, D.: Development and validation of the computerised adaptive beat alignment test (ca-bat). Sci. Rep. **8**(1), 1–19 (2018)
13. Holland, M.K., Wertheimer, M.: Some physiognomic aspects of naming, or, maluma and takete revisited. Percept. Motor Skills **19**(1), 111–117 (1964)
14. Aljanaki, A., Wiering, F., Veltkamp, R.: Computational modeling of induced emotion using gems. In: Proceedings of the 15th Conference of the International Society for Music Information Retrieval (ISMIR 2014), pp. 373–378 (2014)
15. ECMA Ecma. 262: Ecmascript language specification. ECMA (European Association for Standardizing Information and Communication Systems), pub-ECMA: adr (1999)

Flow Control: Local Spectral Radius Regulation

Fabian Schubert[1] and Claudius Gros[2(✉)]

[1] Department of Informatics, University of Sussex, Brighton, UK
f.schubert@sussex.ac.uk
[2] Institute for Theoretical Physics, Goethe University Frankfurt a.M., Frankfurt, Germany
gros07@itp.uni-frankfurt.de
https://profiles.sussex.ac.uk/p579198-fabian-schubert,
https://itp.uni-frankfurt.de/~gros/

Abstract. The spectral radius R_a of a synaptic weight matrix determines whether neuronal activities grow exponentially, for $R_a > 1$, when not limited by non-linearities. For $R_a < 1$, the autonomous activity dies out. The spectral radius is a global property, which can however be regulated using only locally available information. Regulating the flow of activities, neurons can homeostatically regulate R_a online, even in the presence of a continuous flow of external inputs. The resulting adaptation rule, flow control, is shown to be robust, leading to highly performant echo-state networks with the correct size-scaling for the synaptic weights.

Keywords: Spectral radius · Neural networks · Flow control

1 Introduction

Neurons actively adapt their internal parameters in order to achieve a preset working regime [11]. On an individual level, intrinsic plasticity regulates the type and the frequency of output activities [10]. Globally, on a network level, neuronal homeostasis has been shown to support network self-organization by mitigating synaptic plasticity [3,15] and to improve network performance in neural reservoirs [2,13]. While the latter works acknowledged the possible computational advantages of homeostatically controlling higher moments of the distribution of neuronal activities, they still identified the spectral radius of the recurrent weight matrix as a quantity that remains a crucial determinant for the information processing capabilities of recurrent networks.

The underlying dynamics allowing networks to solve time-dependent tasks are a collective result of neuronal activity, where dynamics are embedded in low-dimensional manifolds as a result of network-wide activity [9,14]. Therefore, homeostatic adaptation that adjusts intrinsic parameters based on local information processing objectives [2,10,13] necessarily omit the global dynamics. In contrast, flow control relates locally available information to the spectral radius of the recurrent weights as a global quantity.

G. Nicosia et al. (Eds.): LOD 2022, LNCS 13811, pp. 438–442, 2023.
https://doi.org/10.1007/978-3-031-25891-6_33

We propose a homeostatic mechanism that allows the network to approximately control the spectral radius using only locally available information [8]. The spectral radius is given by the largest eigenvalue (in magnitude) of the synaptic connection matrix. As such, it controls whether the activity increases or decreases on average, with a spectral radius of one denoting the transition between an inactive absorbing, and a chaotic state [5], viz the locus of the absorbing phase transition [4]. Here, we examine the scaling of recurrent weights with respect to the number of afferent weights and the performance that echo-state networks achieve when adapted using flow control.

2 Model and Methods

A repository containing the network model can be found under https://github.com/FabianSchubert/ESN_Frontiers. The membrane potential $\mathbf{x}$ of the discrete-time rate-encoding model studied here is given in vector notation by $\boldsymbol{x}(t) = \boldsymbol{x}_r(t) + \boldsymbol{I}(t)$, with $\boldsymbol{x}_r(t) = \widehat{W}_{\mathrm{a}}\boldsymbol{y}(t-1)$ denoting the recurrent contribution and $\boldsymbol{I}$ the external input. Flow control is based on rescaling the bare synaptic weights W_{ij} by a local factor $a_i = a_i(t)$, which defines the rescaled matrix $\widehat{W}_{\mathrm{a}}$ via $W_{\mathrm{a},ij}(t) = a_i(t)W_{ij}$. The firing rate is $\boldsymbol{y}(t) = \tanh(\boldsymbol{x}(t) - \boldsymbol{b}(t-1))$, with $\mathbf{b}(t)$ being an adaptable threshold. The connection probability is set to $p = 0.1$, and we draw the non-zero bare synaptic weights W_{ij} from a Gaussian distribution with zero mean and variance σ_w^2. Flow control locally adapts the scaling factors $a_i(t)$ via

$$a_i(t) = a_i(t-1)\big[1 + \epsilon_{\mathrm{a}}\Delta R_i(t)\big], \qquad \Delta R_i(t) \equiv R_{\mathrm{t}}^2 y_i^2(t-1) - x_{\mathrm{r},i}^2(t), \tag{1}$$

which regulates the spectral radius R_{a} of the effective weight matrix $\widehat{W}_{\mathrm{a}}$ to the desired target R_{t} [8]. A small ϵ_{a} results in slow adaptation. By means of (1), the flow of activity through a neuron is regulated online, i.e. in the presence of a time-dependent external input $I_i(t)$.

In contrast to flow control, which controls activity fluctuations via an overall rescaling of the afferent synaptic weights, the thresholds b_i serve to regulate mean neuronal activities. Given a target average activity level μ_{t}, we use $b_i(t) = b_i(t-1) + \epsilon_{\mathrm{b}}[y_i(t) - \mu_{\mathrm{t}}]$, where ϵ_{b} is a suitable adaptation rate.

For the proposed synaptic scaling rule given by (1), we assumed that presynaptic activities are independent of the corresponding random weights. As (1) enforces $\langle x_{\mathrm{r},i}^2\rangle_t = R_{\mathrm{t}}^2\langle y_i^2\rangle_t$, we can therefore state

$$\langle x_{\mathrm{r},i}^2\rangle_t = R_{\mathrm{t}}^2\langle y_i^2\rangle_t \approx \frac{1}{N}\sum_{k=1}^{N}\langle y_k^2\rangle_t \sum_{j=1}^{N} a_i^2 W_{ij}^2 \,. \tag{2}$$

By summing over i, one finds $R_{\mathrm{t}}^2 \approx \sum_{j=1}^{N} a_i^2 W_{ij}^2 \approx R_{\mathrm{a}}^2$, where the latter approximate equivalence applies to large random matrices with zero mean [7].

3 Results

Spectral radius adaptation using (1) for a network with $N = 500$ neurons is shown in Fig. 1A . For the first input type, the values of the inputs $I_i(t)$ are

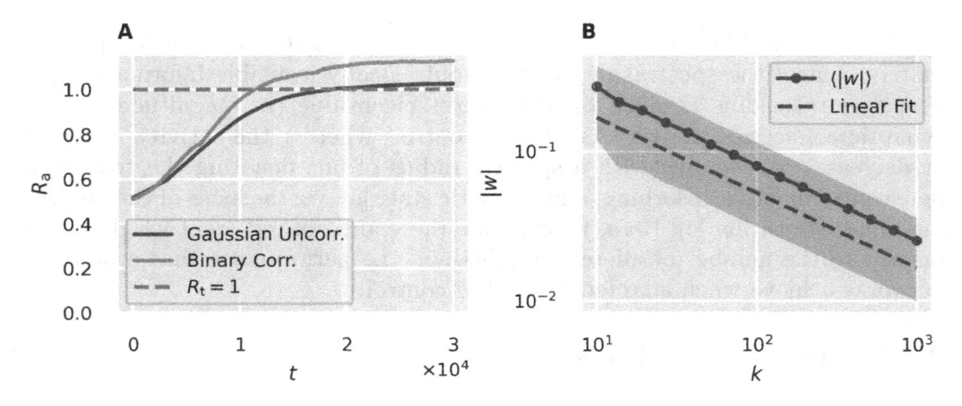

Fig. 1. A: Dynamics of the effective spectral radius R_{a} for Gaussian uncorrelated and binary correlated input. B: The absolute values of the effective weights $a_i W_{ij}$ after adaptation as a function of the number k of afferent connections. The solid line shows mean values based on an equidistant binning of k in the log space. The 25% to 75% percentiles are enclosed by the blue transparent area. A linear fit in log-log space (dashed, shifted) yields a slope of -0.502.

normal distributed and heterogeneous, i.e. $I_i(t) \sim \mathcal{N}\left(\mu = 0, \sigma^2 = \sigma^2_{\mathrm{ext},i}\right)$, where the local variances $\sigma^2_{\mathrm{ext},i}$ are drawn from a half-Gaussian distribution with a mean of 0.5. In contrast, the input streams $I_i(t) = u(t)w_i$ for the second example are correlated, where $u(t) \in \{-1, 1\}$ is a randomly generated binary sequence and $w_i \sim \mathcal{N}\left(\mu = 0, \sigma^2 = 0.5\right)$.

For both types of input protocols, flow control settles into a stationary state for the effective spectral radius R_{a}. However, while the adaptation settles at the target ($R_{\mathrm{t}} = 1$) with very high precision for uncorrelated Gaussian inputs, correlated binary stimulation leads to a certain deviation between R_{a} and R_{t} in the stationary state. As shown in Fig. 1B, flow control also leads to a scaling of weights with the number of afferent connections k by $1/\sqrt{k}$. This relation is in line with experimental evidence [1] and matches the scaling proposed in the theory of balanced state networks [12].

3.1 Performance of Flow-Controlled Networks

We also investigated the effect of flow control on the network's sequential information processing capabilities. We trained the weights of a linear readout unit $y_{\mathrm{out}}(t)$ on two tasks, after adaptation with flow control: performing a time-delayed XOR operation on a random binary input sequence, and time series prediction on a chaotic Mackey-Glass system [6].

The output target $f(t)$ for the XOR-task is defined as $f_\tau(t) = \mathrm{XOR}(u(t) - \tau, u(t) - \tau - 1)$, with $u(t) \in \{-1, 1\}$ being the random binary input sequence, and $\tau \in \mathbb{N}$ a time delay. The performance of the network is then measured by the (non-linear) memory capacity $\mathrm{MC}_{\mathrm{XOR}} = \sum_{\tau=1}^{\infty} \rho^2(f_\tau, y_{\mathrm{out}})$, where $\rho^2(f_\tau, y_{\mathrm{out}})$ is the squared Pearson correlation between the target for a given delay τ, and the

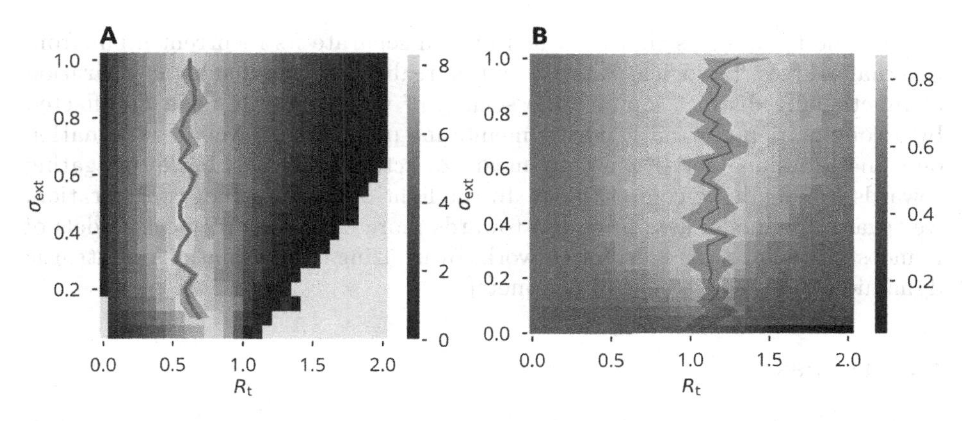

Fig. 2. A: Memory capacity for the time-delayed XOR task of a network after adaptation using flow control. B: Correlation coefficient of the time series generated from a chaotic Mackey-Glass system (MGS) and the prediction of an adapted network. The time series was generated with a constant time step of $\Delta t = 0.1$ and the readout weights were trained after adaptation using flow control to predict 50 time steps into the future. The results in A and B are averages over 5 samples. The orange line shows optimal performance for a given σ_{ext}. (Color figure online)

network output. The Mackey-Glass system is defined by the delayed differential equation $\dot{x} = \alpha x(t-\tau)/\left[1 + x(t-\tau)^{\beta}\right] - \gamma x(t)$. We chose the parameters of the system such that the solution is chaotic. The numerical solution of the equation was sampled into a discrete input time series $u(t)$ using $\Delta t = 0.1$. The learning target was then set to $f(t) = u(t+50)$, i.e. to predict 50 time steps into the future. The performance was quantified by the Pearson correlation $\rho(u, f)$.

The network ($N = 500$) was adapted using flow control under the same type of external input that would thereafter be used to train the output. The performances as a function of the target spectral radius R_{t} and the standard deviation of the average external input, σ_{ext} are shown in Fig. 2. While the target value R_{t} for achieving optimal performance varies across the experiments, it does not significantly change as a function of the strength of the external driving given by σ_{ext}. This implies that the homeostatic target does not need to be fine-tuned under changing external input statistics.

4 Conclusions

Flow control as a homeostatic control of the spectral radius was found to be dynamically stable under different input protocols. However, strong correlations in the external input led to significant deviations from the homeostatic target. Yet, these deviations had no detrimental effect on reliably tuning the network for optimal task performance: Achieving optimal tuning for a given task was only weakly dependent on the strength of the external driving.

Our model assumes that each neuron can separate its recurrent input from external sources. Biologically, this could be realized by the dendritic separation of functionally distinct inputs. The scaling of weights would then be affected by a combination of local dendritic membrane potentials as well as information on somatic spiking activity, e.g. by means of action potentials backpropagating towards the dendritic compartment. In the light of these further consideration, we regard our model as a first step towards more detailed biological models of homeostatic control in recurrent networks by utilizing the fluctuations of intrinsic dynamic variables as a control parameter.

References

1. Barral, J., D Reyes, A.: Synaptic scaling rule preserves excitatory-inhibitory balance and salient neuronal network dynamics. Nature Neurosci. **19**(12), 1690–1696 (2016)
2. Boedecker, J., Obst, O., Mayer, N.M., Asada, M.: Initialization and self-organized optimization of recurrent neural network connectivity. HFSP J. **3**(5), 340–349 (2009)
3. Effenberger, F., Jost, J., Levina, A.: Self-organization in balanced state networks by STDP and homeostatic plasticity. PLoS Comput. Biol. **11**(9), 1–30 (2015)
4. Gros, C.: Complex and Adaptive Dynamical Systems: A Primer. Springer, Cham (2015)
5. Gros, C.: A devil's advocate view on 'self-organized'brain criticality. J. Phys. Complex. **2**(3), 031001 (2021)
6. Mackey, M.C., Glass, L.: Oscillation and chaos in physiological control systems. Science **197**(4300), 287–289 (1977)
7. Rajan, K., Abbott, L.F.: Eigenvalue spectra of random matrices for neural networks. Phys. Rev. Lett. **97**, 188104 (2006)
8. Schubert, F., Gros, C.: Local homeostatic regulation of the spectral radius of echo-state networks. Front. Comput. Neurosci. **15**, 12 (2021)
9. Sussillo, D., Churchland, M.M., Kaufman, M.T., Shenoy, K.V.: A neural network that finds a naturalistic solution for the production of muscle activity. Nat. Neurosci. **18**(7), 1025–1033 (2015)
10. Triesch, J.: Synergies between intrinsic and synaptic plasticity in individual model neurons. In: Advances in Neural Information Processing Systems, vol. 17 (2004)
11. Turrigiano, G., Abbott, L., Marder, E.: Activity-dependent changes in the intrinsic properties of cultured neurons. Science **264**(5161), 974–977 (1994)
12. van Vreeswijk, C., Sompolinsky, H.: Chaos in neuronal networks with balanced excitatory and inhibitory activity. Science **274**(5293), 1724–1726 (1996)
13. Wang, X., Jin, Y., Hao, K.: Echo state networks regulated by local intrinsic plasticity rules for regression. Neurocomputing **351**, 111–122 (2019)
14. Yu, B.M., Cunningham, J.P., Santhanam, G., Ryu, S., Shenoy, K.V., Sahani, M.: Gaussian-process factor analysis for low-dimensional single-trial analysis of neural population activity. In: Koller, D., Schuurmans, D., Bengio, Y., Bottou, L. (eds.) Advances in Neural Information Processing Systems, vol. 21. Curran Associates, Inc. (2008)
15. Zenke, F., Hennequin, G., Gerstner, W.: Synaptic plasticity in neural networks needs homeostasis with a fast rate detector. PLoS Comput. Biol. **9**(11), 1–14 (2013)

Neural Correlates of Satisfaction of an Information Need

Sakrapee Paisalnan[1], Frank Pollick[1], and Yashar Moshfeghi[2](✉)

[1] University of Glasgow, Glasgow, UK
s.paisalnan.1@research.gla.ac.uk, frank.pollick@gla.ac.uk
[2] NeuraSearch Laboratory, University of Strathclyde, Glasgow, UK
yashar.moshfeghi@strath.ac.uk

Abstract. Knowing when a searcher's information needs (IN) are satisfied is one of the ultimate goals in information retrieval. However, the psycho-physiological manifestation of the phenomenon remains unclear. In this study, we investigate brain manifestations of the moment when an IN is satisfied compared to when an IN is not satisfied. To achieve this, we used functional Magnetic Resonance Imaging (fMRI) to measure brain activity during an experimental task. The task was purposefully designed to simulate the information-seeking process and suit the fMRI experimental procedure. Twenty-nine participants engaged in an experimental task designed to represent a search process while being scanned. Our results indicated that both affective and cognitive processes are involved when an information need was being satisfied. These results are in distinction to when satisfaction was not obtained. These results provide insight into features of brain activity that can ultimately be developed to detect satisfaction in search systems with more portable brain imaging devices.

Keywords: Satisfaction · fMRI · Information need · Neural correlates · Search · Cognition · Affect · Brain

1 Introduction

Information seeking refers the behaviour of how an individual looks for pieces of information in order to satisfy their needs [5]. From a searcher's perspective, the satisfaction of an information (SAT) can be seen to combine both the accuracy and the completeness of the task that a searcher performs [1]. It can also be thought to be a part of the information seeking process [20,23]. A searcher who is performing a search will continue their search and examine multiple sources of information repeatedly to gather pieces of relevant information [23]. The search will be successfully terminated when a searcher believes that they no longer have an information need. This is similar to an image puzzle game where a player finds puzzle pieces and the game will end when the puzzle can be solved, i.e., the need for information has been satisfied. In the real world, there are many different

G. Nicosia et al. (Eds.): LOD 2022, LNCS 13811, pp. 443–457, 2023.
https://doi.org/10.1007/978-3-031-25891-6_34

ways that information can be gathered and internally represented, which makes the satisfaction of an IN a complex and dynamic process to understand.

Several behavioural studies have been conducted to understand the satisfaction of an information need by observing human search behaviour and conducting surveys [15,17]. However, these studies are limited in their ability to inform what is happening to the internal states of searchers when their IN is being satisfied. In order to illuminate the internal states of users during search, recent research has increasingly used neuroscience to measure brain activity, using techniques such as electroencephalography (EEG) and functional Magnetic Resonance Imaging (fMRI). This neuroscience research has focused on the topics of relevance [2,10,26,34] and IN [28–30], while SAT has been addressed only in the context of the transition of brain activity between different periods of the search process [27]. More detailed examination of SAT has not yet been performed. Understanding the brain regions involved during the satisfaction of an information need will fundamentally advance our theories of search and add to a growing body of evidence that supports the application of neuroscience to search systems. Thus, in this paper, we aim to investigate the following research questions (RQ) related to SAT in Information Retrieval:

- **RQ1:** What brain regions correspond to the moment of obtaining information that satisfies an information need?
- **RQ2:** What brain regions correspond to the moment of obtaining information that does not satisfy an information need?
- **RQ3:** What differs in brain activity between the moment an IN is satisfied and the moment when it is not satisfied?

To answer our research questions, we used an fMRI technique, which allows us to acquire brain signals of participants during search. The technique of fMRI was chosen as it provides superior spatial resolution of brain signal, particularly in deep regions of the brain that are not readily accessible by other popular neurotechnologies such as EEG and functional Near Infrared Spectroscopy (fNIRS). This is important as previous investigations of IN and satisfaction have revealed that deep brain regions such as the insula and cingulate cortex play an important role [27,30].

2 Related Work

Information Seeking describes the process of how an individual seeks for information to resolve an information need [5]. The process can be thought of as finding missing pieces of a puzzle in order to solve the puzzle [7,36]. The behavioural pattern is explained as a searcher purposefully seeking pieces of information as a result of a need of information to fulfil a gap of knowledge or an information need in order to resolve a task [5].

Once the information need is fulfilled by pieces of relevant information, the searcher themselves will express their satisfaction of information need by terminating a search and make use of the information to complete their task [20].

So, the satisfaction of an information need can be considered as an ultimate goal of the information seeking process [20,23], which could be influenced by both cognitive and affective factors [1]. Recently, many studies in the field of information science have put great effort in developing models of predicting user satisfaction using feedback collected during a web search, for example dwell time and click-though [11,15,21].

A new emerging field in Information Science integrating neuroscience techniques in order to understand concepts of information science is called "Neurasearch" [25]. The imaging techniques include fMRI, EEG, and magnetoencephalography (MEG). Previous works in the field have made use of these techniques to explore brain regions associated with several concepts in Information Science, including the realisation of an information need, relevance judgement, and other search behaviours [22,26–30,32]. Other related works also investigated graded relevance using EEG [34] as well as attempts to build a prediction model that can predict relevance judgements using EEG [10]. These works have provided clear evidence of neural correlates of the concepts in Information Retrieval driven by the measurement of brain activity of a searcher. However, evidence of the neural correlates of satisfaction and how it differs from the ongoing search still remain unclear.

3 Experimental Procedure

Design: This study used a within-subject design. Participants were first given a caption and then presented with separate image tiles with the task of determining if the caption matched to the image. The independent variable was satisfaction of the information need and had two levels. One level, labelled as **SAT**, corresponded to when participants could determine that the caption shown matched to the image. The other level, labelled as **UNSAT**, corresponded to trials where, despite seeing every tile making up the image, participants could not determine that the caption matched to the image, i.e. an established and ongoing search that is not satisfied. The time-point of tile presentation used to represent the UNSAT condition for the fMRI analysis was obtained based on behavioural task performance. A final condition, labelled as **FIX**, corresponded to the time of seeing a fixation cross at the beginning of a trial. The dependent variable was the brain activity revealed by the fMRI Blood Oxygen Level Dependent (BOLD) signal.

Stimuli: First, we carefully selected a list of images and converted them into gray-scale images and divided into nine tiles based on a 3×3 matrix. Next, we conducted two preliminary studies to generate a proper set of stimuli for the experiment. In the first study, we aimed to distinguish the difficulty levels of the images. In this study, the participants were recruited to view a caption explaining details of a particular image. The participants then viewed each tile of the image individually presented in random order. They were asked to press the button anytime that they recognised that the caption was proper to the image. As a result, the set of images was categorised into three groups based on difficulty including the levels of easy, moderate, and hard. Next, we asked three assessors to provide a rating of relevance on the moderate group of images in order to generate a pre-defined sequence for each image by sorting the tiles from the least relevant tile to the most relevant tile. Consequently, we obtained a total of 24 stimuli for the fMRI experiment.

Task: We applied the idea of an image puzzle to mimic the information seeking scenario when participants were being scanned. This could allow us to overcome some practical limitations of using fMRI. With this idea, participants have to gather pieces information by viewing a sequence of tiles and integrate the pieces of information in order to answer a particular question.

In the task, participants encounter an image puzzle task, where they have to determine whether a presented caption matches to an image or not. In the task, the image was divided into nine pieces. The participants viewed only one piece of the image at a time. Relating to a web search, each piece of the image could be thought of as a document in a web search that might satisfy an information need of a searcher. A piece of the image could contain relevant/irrelevant information to what the searcher is looking for. This experimental design is an imitation of the information search paradigm in the information seeking process, where a searcher has to gather pieces of information to fulfill their information need. Each piece of an image in this experiment can be thought of as a relevant/irrelevant piece of information in the search process. The searcher will terminate their search when they realise that they have enough information to determine that the image matches the caption.

For each trial, participants were first presented with a caption and given the task of deciding whether the caption was an appropriate description of an image, which was presented to them, one-by-one, as a sequence of nine, image tiles. They then viewed the sequence of image tiles, as described in the Stimuli section. Participants were instructed to view the individual tiles as they were presented and to press the response button as soon as they determined that the caption was an appropriate image description (i.e., realised they satisfied their task). After pressing the button, the whole image was shown to participants, and they were asked to validate their response by pressing the button again if the response they had provided was correct.

Procedure: Ethical permission was obtained from the Ethics Committee of the College of Science and Engineering, University of Glasgow. Participants were instructed that the duration of the experiment was around one hour, approximately fifty minutes to perform the task and ten minutes to obtain the anatomical structure of the brain. Participants were informed that they could leave at any point during the experiment and would still receive payment (payment rate of £6/hr). They were then asked to sign a consent form. Before beginning the experiment participants underwent a safety check to guarantee that they did not possess any metal items inside or outside of their body, or had any contraindications for scanning, such as certain tattoo inks. They were then provided with gear (similar to a training suit) to wear for the duration of the experiment to avoid potential interference with the fMRI signal from any metal objects in their clothes.

All participants were then asked to complete an entry questionnaire, which assessed the participants' demographics and online search experience. Following this, participants were given a corresponding set of example trials in order to familiarise themselves with the procedure. Next, they performed the main task, as shown in Fig. 1 where each participant encountered the experimental conditions related to information search while being scanned, i.e. information gathering and response validation for each of the 24 stimuli. At the completion of the study an exit questionnaire was given to assess participants' experience of the experiment.

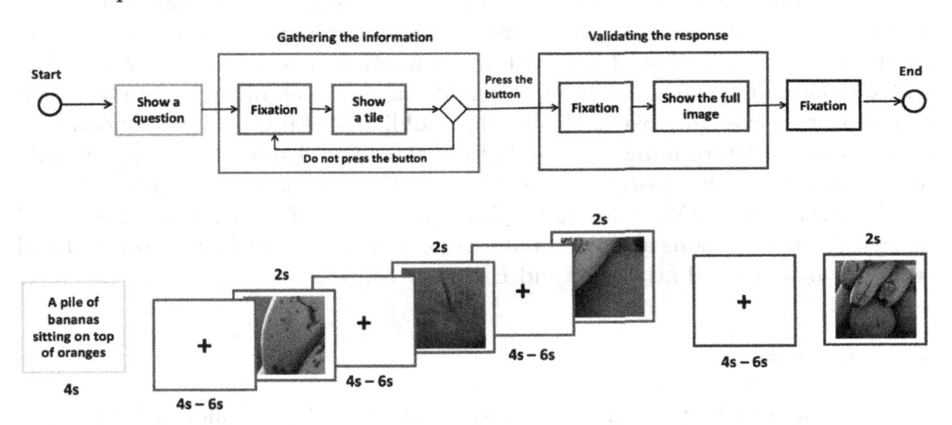

Fig. 1. Experimental procedure shown by both a flowchart and a timeline. Participants were asked to evaluate whether the caption matched to the image. The participants were presented one piece of the image at a time. They would press the button as soon as they were able to make their decision. Then the participants were shown all the tiles together as a single image and asked to press the button again if their judgement was correct.

Image Acquisition: Imaging was performed using a 3T Siemens TIM Trio MRI scanner at the Centre for Cognitive Neuroimaging, University of Glasgow. Functional volumes were acquired using a T2*-weighted gradient echo, echo-planar imaging sequence (68 interleaved slices, TR: 2000 ms, TE: 30 ms, voxel size: $3 \times 3 \times 3$ mm, FOV: 210 mm, matrix size: 70×70). Also, a high-resolution anatomical volume was acquired at the end of the scanning using a T1-weighted sequence (192 slices, TR: 1900 ms, TE: 2.52 ms, voxel size: $1 \times 1 \times 1$ mm, FOV: 210mm, matrix size: 256×256).

fMRI Preprocessing: The fMRI data were analysed using Brain Voyager 20. A standard pipeline of preprocessing of the data was performed for each participant [12]. This included slice scan time correction using trilinear interpolation. Three-dimensional motion correction was performed to detect and correct for small head movements by spatial alignment of all the volumes to the first volume by rigid body transformation. In addition, linear trends in the data were removed, and high pass filtering with a cutoff of 0.0025 Hz performed to reduce artefact from low-frequency physiological noise. The functional data were then co-registered with the anatomic data and spatially normalised into the Montreal Neurological Institute (MNI) space. Finally, the functional data of each individual underwent spatial smoothing using a Gaussian kernel of 6mm to facilitate analysis of group data.

General Linear Model (GLM) Analysis: Analysis began with first-level modeling of the data of individual participants using multiple linear regression of the BOLD-signal response in every voxel. Predictor time courses were adjusted by convolution with a hemodynamic response function. Group data were tested with a second-level analysis using a random effects analysis of variance, using search condition as a within-participants factor. Three contrasts of brain activity included examination of satisfaction vs fixation, no-satisfaction vs fixation and satisfaction vs no-satisfaction. To address the issue of multiple statistical comparisons across all voxels, cluster thresholding was used. This included first using a cluster determining threshold of $p < 0.0005$ and submitting these results to a cluster threshold algorithm that determined chance level for voxels to form a contiguous cluster. MNI coordinates of the peak value of the contrasts were used to identify brain regions and Brodmann areas using Harvard-Oxford cortical and subcortical structural atlases[1,2] and Bioimage Suite[3].

4 Results

Twenty-nine healthy participants joined in the study (16 females and 13 male). All participants were right handed and had normal or corrected-to-normal vision. Participants were under the age of 44, with the largest group between the ages of 18–23 (51.72%) followed by a group between the ages of 24–29 (31.03%). The occupation of most participants was student (89.66%), followed by unemployed (6.9%) and researcher (3.45%). A majority of participants were native English speakers (58.62%) followed by participants with advanced English proficiency (37.93%). On average, participants had around 14 years experience in online searching, often using search services for website browsing, followed by videos and images.

Out of 29 participants, seven participants had head motion exceeding 3 mm or 3° while being scanned and therefore we had to exclude them from the analysis due to data artefacts in the Blood Oxygen Level Dependent (BOLD) signal that arises from head movement. Three participants did not correctly follow the instructions. Thus, the fMRI data captured from nineteen participants were used for the final analysis.

Task Perception: To assess participants task perception two questions were asked in the exit questionnaire: (i) "The task we asked you to perform was: [easy/stressful/simple/satisfying]" (with the answer being a 5-point Likert scale (1. Strongly agree to 5. Strongly disagree") and (ii) "I believe I have succeeded in

[1] https://people.cas.sc.edu/rorden/mricron/index.html.
[2] https://fsl.fmrib.ox.ac.uk/fsl/fslwiki/Atlases.
[3] https://bioimagesuiteweb.github.io/webapp/.

my performance of this task" (with the answer being from a 5-point Likert scale (1-Never to 5-Always). The responses of the participants to this questionnaire are shown in Fig. 2. The results show that the participants found the tasks moderately easy, simple, satisfying, and not stressful. They also reported that they were very successful in performing the task.

Behavioural Performance: On average, the 19 participants responded correctly on 20 (SD 2.2) of a possible 24 trials to report that the caption matched the image. The distribution of these satisfaction responses are shown in Fig. 3. Trials where participants were able to match the caption to the image were subsequently used in the fMRI analysis to represent the SAT condition. There were 3 participants who reported that the caption was matched to the image on all possible trials leaving only 16 participants with UNSAT trials. The UNSAT condition included trials where participants could not match the caption to the image, even after seeing all the tiles.

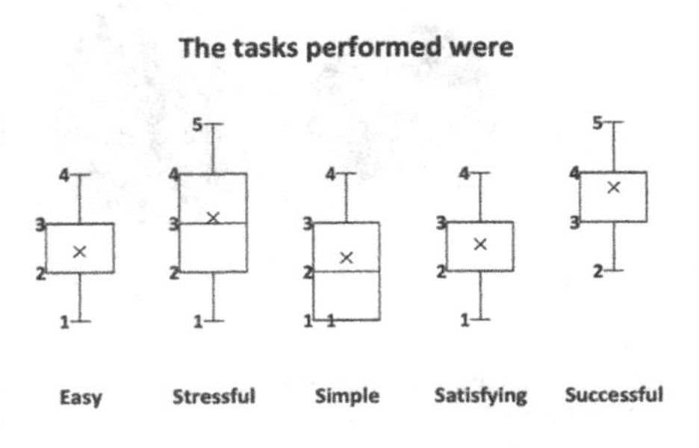

Fig. 2. Task perception results gathered from the exit questionnaire of the participants.

Based on the behavioural results for SAT judgments during the information gathering phase we wished to choose a step of information gathering for the UNSAT condition that would be a suitable for comparison to SAT. For the UNSAT condition we wanted a representative time for an ongoing and established search process that is not satisfied. We choose the seventh step of information gathering (tile 7) to represent the UNSAT condition for several reasons. These included that it had the largest number of responses and corresponded to the time of roughly 75% of the cumulative SAT responses. This choice reduced the difference in time spent searching between SAT and UNSAT. It also reduced fatigue effects that might arise if we used the final step.

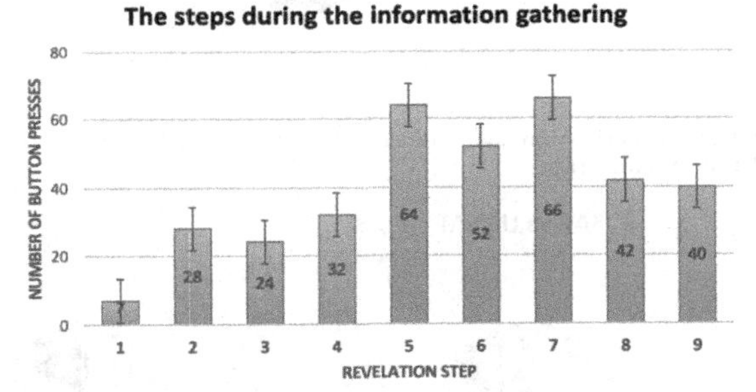

Fig. 3. Results of the nineteen participants during information gathering shown as a bar chart to display the number of button presses indicating satisfaction for each step (termed revelation step) for trials at which satisfaction occurred.

Analysis of the Contrast of SAT vs FIX: In this analysis, we compared brain activity between the SAT and FIX conditions. We hypothesised that this contrast would include brain regions associated with the moment of satisfaction

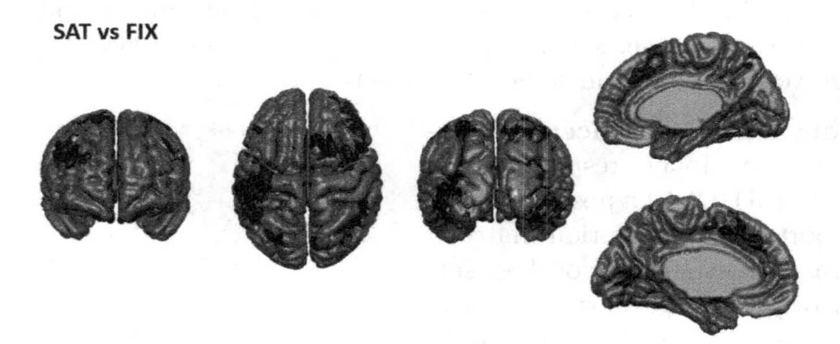

(a) Brain activation maps obtained from contrasting SAT vs FIX, projected on to the surface of the human brain

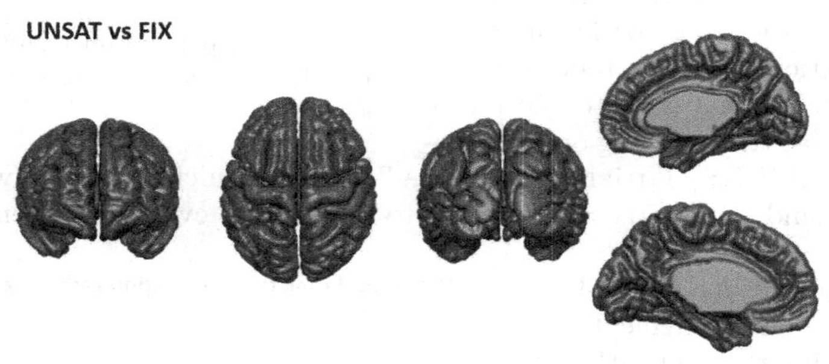

(b) Brain activation maps obtained from contrasting UNSAT vs FIX, projected on to the surface of the human brain

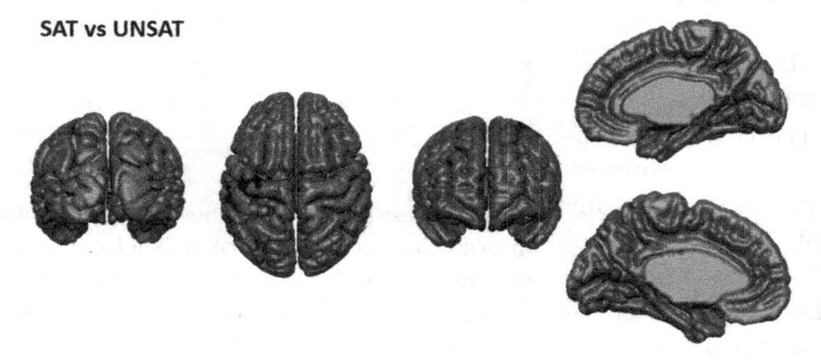

(c) Brain activation maps obtained from contrasting SAT vs UNSAT projected on to the surface of the human brain

Fig. 4. The figures illustrate three different contrasts (including SAT vs FIX, UNSAT vs FIX, and SAT vs UNSAT) projected on to the three-dimensional surface of a human brain. Each figure also shows four different perspectives of the brain including front view, top view, back view, and medial view.

of an information need. In addition, given that SAT contained a button press and complex visual processing, it was expected that the contrast would also reveal brain activity related to the planning and execution of the button press as well as possibly increased visual processing.

The results revealed 13 clusters and their activity was higher during SAT than FIX (shown in Fig. 4a and Table 1). These regions included postcentral gyrus, insula, superior frontal gyrus, middle frontal gyrus, inferior frontal gyrus, superior temporal gyrus, and basal ganglia. Activity centred in the postcentral gyrus is consistent with our prediction of increased motor activity in the SAT condition. The postcentral gyrus is associated with the somatosensory system and this large cluster in the left hemisphere extended into related motor cortex, which would have been active in the planning and execution of the button press. Similarly, the middle temporal gyrus is known for visual information processing as well as regulating semantic information retrieval processes [6,42]. Thus, finding a cluster in this region is consistent with increased visual processing due to visually processing a tile rather than a fixation cross.

The remaining brain areas found in the frontal cortex and insula can be associated with higher order cognition and evaluation. For example, the inferior frontal gyrus is known to play an important role in memory retrieval and decision making [3,9,24]. It is also implicated with the integrative processing of the low-level sensory information in decision making [16,33]. The superior frontal gyrus is known to be involved in a network of high-level processing areas for working memory and higher order cognitive functions [4]. This region has also been found to be activated during judgments of relevance [30]. The insula has been shown to play a role in affective processing [35,40]. It has also been implicated in the salience network, which is involved in the identification of stimulus properties of behavioural relevance [39]. As such, the results of this contrast reveal SAT to be a multilayered response suggesting processing in both cognitive and affective domains.

Table 1. Results for the contrast of SAT vs FIX, including the anatomic label, location (MNI coordinates), Brodmann Area (BA), effect size as indicated by F statistic and p-value, and volume for the different brain regions.

	Brain area	H	MNI coordinates				Effect size		Voxels
			X	Y	Z	BA	F(1, 21)	p-value	mm^3
SAT > FIX	Postcentral gyrus	R	63.0	−16.0	28.0	3	30.07	0.000033	701
	Postcentral gyrus	L	−42.0	−25.0	46.0	2	102.89	<0.000001	133206
	Superior frontal gyrus	R	36.0	47.0	34.0	9	63.17	<0.000001	38645
	Insula	R	33.0	26.0	−5.0	13	72.12	<0.000001	4537
	Superior temporal gyrus	R	54.0	−25.0	−11.0	21	40.30	0.000006	503
	Lentiform nucleus	R	18.0	−1.0	4.0	*	51.48	0.000001	994
	Lentiform nucleus	R	21.0	8.0	1.0	*	32.07	0.000023	767
	Lentiform nucleus	L	−18.0	−1.0	4.0	*	38.59	0.000007	4195
	Claustrum	L	−30.0	20.0	4.0	48	53.16	0.000001	3836
	Middle frontal gyrus	L	−33.0	50.0	16.0	10	29.44	0.000037	575
	Insula	L	−36.0	−1.0	13.0	13	32.34	0.000022	391
	Middle frontal gyrus	L	−42.0	29.0	43.0	8	36.68	0.00001	1581
	Inferior frontal gyrus	L	−51.0	8.0	25.0	9	63.40	<0.000001	4260

Analysis of the Contrast of UNSAT vs FIX: In this analysis we compared brain activity in the UNSAT and FIX conditions. We hypothesised that this contrast would reveal brain regions associated with cognition related to the ongoing process of search. In addition, given that UNSAT included complex visual processing, it was expected that the contrast would reveal evidence of increased visual processing.

Results obtained six clusters, shown in Fig. 4b and Table 2 and all were found to have higher activation during the UNSAT condition than FIX. The results included middle occipital gyrus, inferior frontal gyrus, medial frontal gyrus, and cuneus. Consistent with our expectation that the UNSAT condition would include higher activation in visual areas we found activation in the region of middle occipital gyrus which is typically known for visual perception and object recognition [41,44]. As well, the activation in the cuneus can be related to visual processing. The remaining brain areas, which were all located in frontal cortex can be associated more with higher order cognition than with affective processing. For example, the inferior frontal gyrus region has been reported to be responsible for processes of memory retrieval and integrating information from sensory regions in the decision making process [3,9,24]. The left medial frontal gyrus is known for executive functions and decision-related processes [37].

Table 2. Results for the contrast of UNSAT vs FIX, including the anatomic label, location (MNI coordinates), Brodmann Area (BA), effect size as indicated by F statistic and p-value, and volume for the different brain regions.

	Brain area	H	MNI coordinates				Effect size		Voxels
			X	Y	Z	BA	F(1,21)	p-value	mm^3
UNSAT > FIX	Middle occipital gyrus	R	27.0	−97.0	−2.0	18	43.63	0.000008	396
	Cuneus	R	9.0	−85.0	4.0	17	55.78	0.000002	1119
	Cuneus	L	−46.0	−73.0	4.0	30	36.60	0.000022	900
	Medial frontal gyrus	L	−3.0	20.0	49.0	8	44.50	0.000007	388
	Medial frontal gyrus	L	−3.0	29.0	40.0	6	41.26	0.000011	203
	Inferior frontal gyrus	L	−39.0	20.0	−11.0	47	37.81	0.000019	502

Analysis of the Contrast of SAT vs UNSAT: This analysis examined differences in brain activity between SAT and UNSAT conditions. We hypothesised that this contrast would reveal brain regions that differ between these two cognitive states, as well as differences in motor activity due to the presence of the button press in the SAT condition.

Results, based on contrast of the nineteen participants in the SAT condition versus the sixteen participants in the UNSAT condition revealed six clusters. The results are shown in Fig. 4c and Table 3. All six clusters had higher activation during SAT than UNSAT, which included postcentral gyrus, precentral gyrus, inferior frontal gyrus, superior parietal lobule, declive, and middle occipital gyrus. The postcentral and precentral regions are associated with primary somatosensory functions and motor planning. Thus, these regions were likely involved in the motor planning of the button press during the SAT condition.

The remaining brain areas included the right inferior frontal gyrus, which is known to be involved in working memory and attentional control [14,38]. It also included the superior parietal lobule, a region implicated in manipulating information in working memory [19]. It is also known for regulating visuospatial attention as part of frontoparietal attention network [43].

Table 3. Results for the contrast of SAT vs UNSAT, including the anatomic label, location (MNI coordinates), Brodmann Area (BA), effect size as indicated by F statistic and p-value, and volume for the different brain regions.

	Brain area	H	MNI coordinates				Effect size		Voxels
			X	Y	Z	BA	F(1,21)	p-value	mm^3
SAT > UNSAT	Postcentral gyrus	R	51.0	−25.0	52.0	2	31.04	0.000053	527
	Inferior frontal gyrus	R	45.0	8.0	25.0	9	42.85	0.000009	654
	Declive	R	33.0	−64.0	−17.0	*	45.89	0.000006	457
	Superior parietal lobule	L	−30.0	−61.0	46.0	7	33.04	0.000039	908
	Precentral gyrus	L	−42.0	−16.0	61.0	4	43.24	0.000009	6384
	Middle occipital gyrus	L	−42.0	−73.0	-2.0	19	39.42	0.000015	1965

5 Discussion and Conclusion

In this paper, we aimed to understand how the satisfaction of information need physically emerges inside the human brain. Satisfaction of information need is a topic of crucial importance in the information retrieval and information science communities. The field has made great progress in developing systems that can deliver a piece of relevant information to a searcher. However, understanding the internal mechanisms of how searchers determine that they have enough information for their task remains unclear. To advance our understanding of the satisfaction of information need, we conducted an experiment to measure the brain activity of nineteen participants while they performed an experimental task scenario that simulated information seeking.

The contrast of brain activity between the moment of satisfaction and the baseline measurement of fixation helps us to answer **RQ1**. Results of this comparison of brain activity, SAT>FIX, revealed activity in 13 brain regions. Of particular interest are two regions of activity that include bilateral regions of both the frontal cortex and the insula. The regions of the frontal cortex include the left inferior frontal gyrus (BA9). This region is associated with a variety of cognitive functions and would likely have been included in the large cluster previously reported in a study of information need [30] that was centred in the left dorsolateral prefrontal cortex. This previous region was found to be more active when a question could be answered than when it could not. This provides converging evidence that the left inferior frontal gyrus plays an important role when an information need is satisfied. This region has been widely suggested to be involved in selection demands when performing semantic tasks [13,31] along with the controlled retrieval of semantic information [31]. Interestingly, the previous result [30] was obtained with a text-based task, while the current task was image-based, raising the possibility that the involvement of the left inferior frontal gyrus might be independent of modality. The finding of insula involvement is interesting from the standpoint of its general role in affective and other evaluative aspects of brain response. In particular, the insula is part of the salience network, which is activated when items in the world are determined to have behavioural relevance. The salience network has also been found to be active in the contrast of brain activity between satisfaction judgments and relevance judgments [27]. Thus, converging evidence points towards involvement of the insula, possibly in its affective capacity of determining behavioural relevance, in the satisfaction of information need.

The contrast of brain activity between a moment of ongoing information search that does not satisfy the information need and a baseline measurement of fixation helps us to answer **RQ2**. This contrast of UNSAT>FIX revealed activity in 6 brain regions. Of particular interest are three regions in the left frontal cortex that have been implicated in complex cognition. Two of these regions were in the medial frontal gyrus, which is a known heterogeneous region associated with many cognitive functions [8]. The other region, the inferior frontal gyrus (BA 47) has been reported to be involved in semantics of language, though it is not clear how language would be involved in the current task. Overall, the results addressing **RQ2** point to areas of the frontal cortex involved in cognition. Though it is possibly difficult to infer more than the involvement of high level cognition, it is notable that the results do not provide evidence to support an affective response.

Brain activity at the moment of satisfaction compared to a moment of ongoing not satisfied information search helps us to answer **RQ3**. In the SAT>UNSAT condition, we found 6 brain regions, which are mainly located in the left superior parietal lobule and the right inferior frontal gyrus (BA9). Potentially, the parietal cortex activation could result from more robust recruitment of attention to items in an image tile that satisfies an IN, compared to an image tile which does not. Finally, the larger activity in the right inferior frontal gyrus is found on the other side of the brain to the inferior frontal gyrus finding from the contrast of SAT vs FIX, making it harder to directly relate the two. The region has been ascribed to higher order cognition and thus we can take the result as indicating a difference in cognitive processing in trials where the information need is satisfied.

Our results addressing **RQ1** provide converging evidence of the importance of the inferior frontal gyrus and the insula in understanding the moment of satisfaction. They further suggest that both cognitive and affective processes are at work during the moment of satisfaction. However, while the data supporting **RQ2** and **RQ3** clearly implicate regions of the frontal cortex, they provide less clear an interpretation of the nature of these cognitive functions. Future research should explore in more detail the ongoing brain activity preceding the moment of satisfaction in order to more fully understand the particular brain processes at work. In the current work we examined this preceding brain activity at a single point in time and future work could benefit from exploring how brain activity might reflect an accumulation of information that leads to satisfaction. One limitation of the current research was that for some of the contrasts, the brain activity found could possibly be related to the button press. Although it is possible to discount these activities, future research might design a task where the effect of the button press is less of an issue.

In conclusion, the results from our study revealed cognitive and affective brain regions involved in satisfying an information need, suggesting differences in brain activation associated with cognitive processing involved in an information search. By investigating SAT alone and the contrast SAT to UNSAT the results go past previous reports [27] that only contrasted relevance to satisfaction. We believe that this study will provide a great contribution to the community. The study revealed the brain regions underlying both the moments of satisfaction and no-satisfaction. These results, particularly for deep brain regions such as the insula, might be helpful in the future with the use of other technologies such as EEG, because knowledge of the activity of deep brain structures can be used to shape interpretation of EEG results [18]. Thus, these results not only advance our theoretical understanding of affective and cognitive processes during satisfaction of search, but they also could be used to advance the effectiveness of more portable brain imaging technology such as EEG and fNIRS.

References

1. Al-Maskari, A., Sanderson, M.: A review of factors influencing user satisfaction in information retrieval. J. Am. Soc. Inform. Sci. Technol. **61**(5), 859–868 (2010)
2. Allegretti, M., Moshfeghi, Y., Hadjigeorgieva, M., Pollick, F.E., Jose, J.M., Pasi, G.: When relevance judgement is happening?: an EEG-based study. In: Proceedings of the 38th International ACM SIGIR Conference on Research and Development in Information Retrieval, pp. 719–722. ACM (2015)
3. Barbey, A.K., Koenigs, M., Grafman, J.: Dorsolateral prefrontal contributions to human working memory. Cortex **49**(5), 1195–1205 (2013)
4. Boisgueheneuc, F.D., et al.: Functions of the left superior frontal gyrus in humans: a lesion study. Brain **129**(12), 3315–3328 (2006)
5. Case, D.O., Given, L.M.: Looking for Information: A Survey of Research on Information Seeking, Needs, and Behavior (2016)
6. Davey, J., et al.: Exploring the role of the posterior middle temporal gyrus in semantic cognition: integration of anterior temporal lobe with executive processes. Neuroimage **137**, 165–177 (2016)
7. Davies, R.: The creation of new knowledge by information retrieval and classification. J. Doc. (1989)
8. De La Vega, A., Chang, L.J., Banich, M.T., Wager, T.D., Yarkoni, T.: Large-scale meta-analysis of human medial frontal cortex reveals tripartite functional organization. J. Neurosci. **36**(24), 6553–6562 (2016)
9. Ding, L., Gold, J.I.: Caudate encodes multiple computations for perceptual decisions. J. Neurosci. **30**(47), 15747–15759 (2010)
10. Eugster, M.J., et al.: Predicting term-relevance from brain signals. In: Proceedings of the 37th International ACM SIGIR Conference on Research & Development in Information Retrieval, pp. 425–434. ACM (2014)

11. Fox, S., Karnawat, K., Mydland, M., Dumais, S., White, T.: Evaluating implicit measures to improve web search. ACM Trans. Inf. Syst. (TOIS) **23**(2), 147–168 (2005)
12. Goebel, R.: Brainvoyager qx, vers. 2.1. Brain Innovation BV, Maastricht (2017)
13. Gold, B.T., Buckner, R.L.: Common prefrontal regions coactivate with dissociable posterior regions during controlled semantic and phonological tasks. Neuron **35**(4), 803–812 (2002)
14. Hampshire, A., Chamberlain, S.R., Monti, M.M., Duncan, J., Owen, A.M.: The role of the right inferior frontal gyrus: inhibition and attentional control. Neuroimage **50**(3), 1313–1319 (2010)
15. Hassan, A., Shi, X., Craswell, N., Ramsey, B.: Beyond clicks: query reformulation as a predictor of search satisfaction. In: Proceedings of the 22nd ACM International Conference on Information & Knowledge Management, pp. 2019–2028. ACM (2013)
16. Heekeren, H.R., Marrett, S., Bandettini, P.A., Ungerleider, L.G.: A general mechanism for perceptual decision-making in the human brain. Nature **431**(7010), 859 (2004)
17. Jiang, J., Hassan Awadallah, A., Shi, X., White, R.W.: Understanding and predicting graded search satisfaction. In: Proceedings of the Eighth ACM International Conference on Web Search and Data Mining, pp. 57–66. ACM (2015)
18. Keynan, J.N., et al.: Limbic activity modulation guided by functional magnetic resonance imaging-inspired electroencephalography improves implicit emotion regulation. Biol. Psychiat. **80**(6), 490–496 (2016)
19. Koenigs, M., Barbey, A.K., Postle, B.R., Grafman, J.: Superior parietal cortex is critical for the manipulation of information in working memory. J. Neurosci. **29**(47), 14980–14986 (2009)
20. Kuhlthau, C.C.: Inside the search process: information seeking from the user's perspective. J. Am. Soc. Inf. Sci. **42**(5), 361–371 (1991)
21. Liu, Y., et al.: Different users, different opinions: predicting search satisfaction with mouse movement information. In: Proceedings of the 38th International ACM SIGIR Conference on Research and Development in Information Retrieval, pp. 493–502. ACM (2015)
22. Mackie, S., Azzopardi, L., Moshfeghi, Y.: An initial analysis of e-procurement search behaviour. In: Smits, M. (ed.) iConference 2022. LNCS, vol. 13193, pp. 3–12. Springer, Cham (2021). https://doi.org/10.1007/978-3-030-96960-8_1
23. Marchionini, G., White, R.: Find what you need, understand what you find. Int. J. Hum. Comput. Interact. **23**(3), 205–237 (2007)
24. Mitchell, A.S., Chakraborty, S.: What does the mediodorsal thalamus do? Front. Syst. Neurosci. **7**, 37 (2013)
25. Moshfeghi, Y.: NeuraSearch: neuroscience and information retrieval. In: Proceedings of the Second International Conference on Design of Experimental Search & Information REtrieval Systems, Padova, Italy, 15–18 September 2021. CEUR Workshop Proceedings, vol. 2950, pp. 193–194. CEUR-WS.org (2021)
26. Moshfeghi, Y., Pinto, L.R., Pollick, F.E., Jose, J.M.: Understanding relevance: an fMRI study. In: Serdyukov, P., Braslavski, P., Kuznetsov, S.O., Kamps, J., Rüger, S., Agichtein, E., Segalovich, I., Yilmaz, E. (eds.) ECIR 2013. LNCS, vol. 7814, pp. 14–25. Springer, Heidelberg (2013). https://doi.org/10.1007/978-3-642-36973-5_2
27. Moshfeghi, Y., Pollick, F.E.: Search process as transitions between neural states (2018)

28. Moshfeghi, Y., Pollick, F.E.: Neuropsychological model of the realization of information need. J. Am. Soc. Inf. Sci. **70**(9), 954–967 (2019)
29. Moshfeghi, Y., Triantafillou, P., Pollick, F.: Towards predicting a realisation of an information need based on brain signals. In: The World Wide Web Conference, pp. 1300–1309 (2019)
30. Moshfeghi, Y., Triantafillou, P., Pollick, F.E.: Understanding information need: an fMRI study. In: Proceedings of the 39th International ACM SIGIR Conference on Research and Development in Information Retrieval, pp. 335–344. ACM (2016)
31. Moss, H., et al.: Selecting among competing alternatives: selection and retrieval in the left inferior frontal gyrus. Cereb. Cortex **15**(11), 1723–1735 (2005)
32. Paisalnan, S., Pollick, F., Moshfeghi, Y.: Towards understanding neuroscience of realisation of information need in light of relevance and satisfaction judgement. In: Nicosia, G., et al. (eds.) LOD 2021. LNCS, vol. 13163, pp. 41–56. Springer, Cham (2021). https://doi.org/10.1007/978-3-030-95467-3_3
33. Philiastides, M.G., Sajda, P.: EEG-informed fMRI reveals spatiotemporal characteristics of perceptual decision making. J. Neurosci. **27**(48), 13082–13091 (2007)
34. Pinkosova, Z., McGeown, W.J., Moshfeghi, Y.: The cortical activity of graded relevance. In: Proceedings of the 43rd International ACM SIGIR Conference on Research and Development in Information Retrieval, pp. 299–308 (2020)
35. Quarto, T., et al.: Association between ability emotional intelligence and left insula during social judgment of facial emotions. PLoS One **11**(2), e0148621 (2016)
36. Swanson, D.R., et al.: Historical note: information retrieval and the future of an illusion. J. Am. Soc. Inf. Sci. **39**(2), 92–98 (1988)
37. Talati, A., Hirsch, J.: Functional specialization within the medial frontal gyrus for perceptual go/no-go decisions based on "what","when", and "where" related information: an fMRI study. J. Cogn. Neurosci. **17**(7), 981–993 (2005)
38. Tops, M., Boksem, M.A.: A potential role of the inferior frontal gyrus and anterior insula in cognitive control, brain rhythms, and event-related potentials. Front. Psychol. **2**, 330 (2011)
39. Uddin, L.Q.: Salience processing and insular cortical function and dysfunction. Nat. Rev. Neurosci. **16**(1), 55–61 (2015)
40. Uddin, L.Q., Nomi, J.S., Hébert-Seropian, B., Ghaziri, J., Boucher, O.: Structure and function of the human insula. J. Clin. Neurophysiol. **34**(4), 300–306 (2017)
41. Vandenberghe, R., Price, C., Wise, R., Josephs, O., Frackowiak, R.S.: Functional anatomy of a common semantic system for words and pictures. Nature **383**(6597), 254–256 (1996)
42. Visser, M., Jefferies, E., Embleton, K.V., Lambon Ralph, M.A.: Both the middle temporal gyrus and the ventral anterior temporal area are crucial for multimodal semantic processing: distortion-corrected fMRI evidence for a double gradient of information convergence in the temporal lobes. J. Cogn. Neurosci. **24**(8), 1766–1778 (2012)
43. Wu, Y., et al.: The neuroanatomical basis for posterior superior parietal lobule control lateralization of visuospatial attention. Front. Neuroanat. **10**, 32 (2016)
44. Yonelinas, A.P., Hopfinger, J.B., Buonocore, M., Kroll, N.E.A., Baynes, K.: Hippocampal, parahippocampal and occipital-temporal contributions to associative and item recognition memory: an fMRI study. NeuroReport **12**(2), 359–363 (2001)

Role of Punctuation in Semantic Mapping Between Brain and Transformer Models

Zenon Lamprou[1], Frank Pollick[2], and Yashar Moshfeghi[1](✉)

[1] NeuraSearch Laboratory, University of Strathclyde, Glasgow, UK
{zenon.lamprou,yashar.moshfeghi}@strath.ac.uk
[2] University of Glasgow, Glasgow, UK
frank.pollick@gla.ac.uk

Abstract. Modern neural networks specialised in natural language processing (NLP) are not implemented with any explicit rules regarding language. It has been hypothesised that they might learn something generic about language. Because of this property much research has been conducted on interpreting their inner representations. A novel approach has utilised an experimental procedure that uses human brain recordings to investigate if a mapping from brain to neural network representations can be learned. Since this novel approach has been introduced, more advanced models in NLP have been introduced. In this research we are using this novel approach to test four new NLP models to try and find the most brain aligned model. Moreover, in our effort to unravel important information on how the brain processes text semantically, we modify the text in the hope of getting a better mapping out of the models. We remove punctuation using four different scenarios to determine the effect of punctuation on semantic understanding by the human brain. Our results show that the RoBERTa model is most brain aligned. RoBERTa achieves a higher accuracy score on our evaluation than BERT. Our results also show for BERT that when punctuation was removed a higher accuracy was achieved and that as the context length increased the accuracy did not decrease as much as the original results that include punctuation.

Keywords: fMRI · Transformers · Explainable AI · Punctuation symbols

1 Introduction

Continuous advancements in the area of natural language processing (NLP) have produced advanced deep learning models. How these models represent language internally is currently not very well understood because these models often tend to be large and complex. However, what is clear is that these deep learning models don't learn from any explicit rules regarding language, rather they learn generic information about language itself. Some of these models are trained on

G. Nicosia et al. (Eds.): LOD 2022, LNCS 13811, pp. 458–472, 2023.
https://doi.org/10.1007/978-3-031-25891-6_35

a large generic corpus and then fine-tuned in a downstream task. These models tend to perform better than models that are trained on just one specific NLP task [12,20,30].

There has been much effort to explain the inner representations of such models. Some of these efforts have focused on designing specific NLP tasks that target specific linguistic information [10,26,41].

Other research has tried to offer a more theoretical explanation regarding what word embeddings can represent [7,29,40]. Toneva and Whebe [37] in their effort to explain the inner representation of four NLP models proposed a novel approach that used brain recordings obtained from functional magnetic resonance (fMRI) and magnetoencephalography (MEG). They hypothesised that by aligning brain data with the extracted features of a model, the mapping from brain to model representation can be learned. This mapping primarily can then offer insights on how these models represent language. They refer to this approach as aligning neural network representation with brain activity.

One of the models tested by [37] was BERT. BERT, since its introduction [12] has generated a great deal of related research. Much of this work has been done to either improve its performance or to fix particular aspects of BERT or to fine-tune BERT to produce a version of BERT expert in a particular subject domain [2,9,19,23,24,33,42]. Toneva and Whebe [37] were able to learn a mapping from the representations of BERT, ELMo [30], USE [6] and T-XL [11] to brain data. However, this alignment was not tested against more sophisticated versions of BERT. We use the same novel approach as [37] to examine four of these more sophisticated versions of the BERT model: RoBERTa [42], DistiliBERT [33], ELECTRA [9] and ALBERT [24].

Moreover, because this novel approach is a proof of concept to show alignment of brain and neural networks we wanted to explore whether removing different punctuation symbols would influence results, potentially to produce a better alignment between brain and neural networks. Research investigating punctuation [27] has shown effects of punctuation on reading behaviour. Punctuation symbols can help the reader skim through text faster, but less is known about the semantic role of punctuation symbols and how they are processed in the brain. There is a substantial amount of research that examines how the brain processes text semantically and syntactically [5,31,34]. In addition there is research [1] that examined which brain areas are responsible for different operations, e.g. a brain area that is responsible for understanding the meaning of complex words. However, even if the goals of these research are different, their approaches share a common fMRI experimental approach, which involves participants reading a set of sentences. When sentences appear to the participants in these experiments, a choice is made whether or not to include punctuation in the pre-processed text. Though there appears to be no general agreement if the punctuation should be included or not.

Contributions: The contributions of this research are divided into two parts that we present in the subsequent sections. Firstly, we examined the degree to which the four different, BERT-related transformer models aligned to brain activity. Secondly, by using the concept of learning a mapping from brain to neural network introduced by [37] we try to use the neural network to obtain an hypothesis about the brain rather than using the brain to make observations about the network. In particular by using altered sequences, in terms of punctuation, we hypothesise that if we can get better alignment then punctuation might be processed differently for the human brain to comprehend the semantics of language.

2 Related Work

A frequent design used in fMRI experiments is to contrast brain activity in two conditions [16]. Often one condition is a specifically controlled baseline and the other condition reflects a process of interest. For example, to capture how the brain semantically processes information, the conditions are often a sentence vs a list of words [16]. However, some research [3,4,21,25,38] has used a narrative to capture how the brain semantically processes information. The work done by Toneva and Wehbe [37] was done in a non-controlled environment and the condition they were trying to capture was how semantically the brain processes text sequences. They used these text sequences to learn a mapping from neural model features to brain features. However this approach didn't take into account how punctuation influences semantic processing but rather how the brain processes the text sequence as a whole.

Previous work [38] managed to align MEG brain activity with a Recurrent Neural Network (RNN) and its word embedding, and that informed sentence comprehension by looking at each word. Other research demonstrated that one could align a Long Short-Term Memory (LSTM) [22] network with fMRI recordings to measure the amount of context in different brain regions. These results provided a proof of concept that one could meaningfully relate brain activity to machine learning models. Other approaches combining brain data and neural networks have tried to identify brain processing effort by using neural networks to learn a mapping from brain activity to neural network representations [15,18].

Previous research has explored the role of punctuation in natural language processing. From a natural language processing view, one of the standard text pre-processing steps is to remove punctuation symbols from the text to reduce the noise in the data. Some researchers have removed punctuation to achieve better results in spam detection [14], or plagiarism detection [8]. Such results have provided mixed results in the sense that in some cases removing punctuation can lead to better results, but not every time. Recent research [13] explored punctuation and differences in performance of MultiGenre Natural Language Inference (MNLI) using BERT and other RNN models. Their results showed that no model is capable of taking into account cases where punctuation is meaningful, and that punctuation is generally insignificant semantically.

From a neuroscience point of view there is limited research that assesses the role of punctuation in semantics and its purpose in brain processing of text. Related to the role of punctuation in the syntax of language, some research has tried to focus on creating multidimensional-features that are specific to syntactical parts of language [31]. Then using these features they tried to model the syntactical representation of punctuation symbols in the text. There is variability in the literature in how punctuation is treated. For example, using the same fMRI data but with different research goals, one study preprocessed the text seen by the participants with punctuation [5] and the other without [34]. Furthermore, other research [1] tried to show that the combination of words can create more complex meaning and they tried to find which brain regions are responsible for representing the meaning of these words. In their pre-processing step they decided to remove the punctuation from text before presenting to their participants. This research further illustrates how there is no consensus on how punctuation is processed from the brain semantically and syntactically, and if it is needed to be included in a corpus to achieve better results.

Overall, it can be seen that there is not enough research that evaluates or improves NLP models through brain recordings as proposed by Toneva and Wehbe [37]. Though there is research examining cognition that evaluates whether word embeddings contain relevant semantics [35]. Furthermore, other research [17] tried to create new embeddings that align with brain recordings to asses if these new embeddings aligned better with behavioural measures of semantics. We believe that by using new state-of-the-art models to assess how their representations align with brain activity we can help expand this important and growing research area. We can unravel what training choices might lead to better alignment with brain data by assessing how the training choices of particular models improve or worsen this alignment.

3 Methodology

3.1 Data

The fMRI data were obtained originally as part of previously published research [38] and were made publicly available to use.[1] Their experimental setup involved 8 subjects reading Chapter 9 from Harry Potter and the Sorcerer's stone [32] in its entirety. The data acquisition was done in 4 runs. Every word in the corpus appeared for 0.5 s on the screen and a volume of brain activity represented by the BOLD signal would be captured every 2 s (1 TR). Because of this timing arrangement, in each TR 4 words would be contained in each brain volume. These 4 words are not in random order, they are presented in the order they appear in the Harry Potter book. The data for every subject were preprocessed and smoothed by the original authors and the preprocessed version is the one available online. In their paper they note the use of MEG data as well but in this research we only focus using the fMRI since only these data have been made publicly available.

[1] The data and the original code can be found at this link.

3.2 Transformer Models

In our effort to find a more advanced transformer model that might produce better brain alignment we tested four different models against BERT, which we used as our baseline. Each model was selected based on its new characteristics compared to BERT.

- **RoBERTa.** RoBERTa was one of the first variations of BERT that came after its initial introduction. [42] stated that training neural networks is often expensive and its performance really relies on hyper-parameter choices. By comparing different techniques on the two main tasks that BERT was pre-trained on, Masked Language Modelling (MLM) and Next Sentence Prediction (NSP) they were able to produce a better performing version of BERT. The interest in using RoBERTa to try and understand its inner representation is to understand if choosing different hyper-parameters in a natural language experiment can lead to better representations that are closer to how the brain represents contextual information.
- **DistiliBERT.** Transformer model architectures using NLP tend to be large and slow to fine-tune and pre-train. [33] in their pursuit to combat those problems proposed a BERT architecture that is reduced in size by 40% and is 60% percent faster to train compared to the original BERT model. Moreover during their experiments they managed to retain 97% of its language understanding capabilities. We thought that this design choice makes DistiliBERT interesting to test and show that indeed its internal representations are not affected by the reduction in size.
- **ALBERT.** ALBERT was firstly introduced by [24] it tried to address the factor that modern day natural language processes are too big and we can easily reach memory limits when using a GPU or TPU. To combat that issue they proposed a two parameter reduction technique. The first technique they used is called factorised embedding parameterization. They hypothesised that by exposing the embedding space to a lower embedding space before exposing it to the hidden space of the model they can reduce the embedding space dimension without compromising the performance. The second technique that ALBERT implements is cross-layer parameter sharing. All the parameters in ALBERT are shared across all its layers. The features of the model in the original research [37] are extracted for each layer independently. The prediction and evaluation on the brain data is done layer by layer using the extracted features of each layer. We hypothesise that if ALBERT's features yield a better accuracy when used on predicting brain data, then this would suggest that the brain has a uniform weighting mechanism across its own layers.
- **ELECTRA.** ELECTRA was developed by [9]. Their purpose for this model was to redefine the Masked Language Modelling (MLM) task, which was one of the two tasks that the original BERT model was pre-trained on. Because modern natural language models generally require a large amount of computational power to train, they proposed an alternative to the MLM task where instead of using the [MASK] token to mask a token in the sentence

and have the model predict a possible value for that token they replace it with a proposed solution generated from a small generation network. Then the main model is pre-trained by trying to predict if the generated token is indeed a correct one or if is not correct, similarly to a binary classification task. They hypothesised that this process results in a model with better contextual representations than BERT. This hypothesis makes this model a good candidate to assess its performance when aligned with brain data and determine if its inner representations are closer to how the brain represents contextual information.

3.3 Experimental Procedure

Our experimental procedure can be divided into three main operations. First we extracted the features from the desired models using different sequence lengths. Then we trained a ridge regression model using the extracted features and used the ridge regression to predict the brain data. Finally, we evaluated the predictions using searchlight classification. In the rest of this section we present each of these operations in detail. The code used to run the procedure that follows was made publicly available[2].

Extracting Features from Models. The first step of the experimental procedure was to extract features from different sequences of length S. In the original research [37] the analysis was done using sequence lengths of 4, 5, 10, 15, 20, 25, 30, 35, 40 so we followed the same pattern. The sequences are formed from Chap. 9 from Harry Potter and the Sorcerer' s Stone book. A dictionary was created that contains an entry for every layer in the model, plus the embedding layer if it exists. The representations of the embeddings layer were extracted first for every word in the chapter. The next step was to get the first sequence of S length and extract the representation of every layer for that sequence S times. Then the representation of every sequence of S length was extracted using a sliding window the same way as the first sequence. The first sequence was extracted S times to ensure that the number of sequences was the same as the number of words in the dataset. A separate script was used for every model we chose. In order to use the pre-trained checkpoints of the models the hugging face library was used so we could download this checkpoints.

Making Predictions Using the fMRI Recordings and the Extracted Features. The second step was to train a ridge regression model to learn a mapping from brain representation to neural network representation. This approach followed some previous work [21,28,36,38,39] that learned a linear function with a ridge penalty to learn the mapping from brain to neural network representation. Although a ridge regression is a relatively simple model we chose to use it since it previously has been demonstrated to be useful [37]. However as a

[2] https://github.com/NeuraSearch/Brain-Transformer-Mapping-Punctuation.git.

future work, new more complex models can be used with the hope of producing a better mapping. We used this ridge regression model to make predictions using the brain data and the features extracted from the previous step. The predictions were done for every layer, for every subject and for every model in a cross validation setting. We used a 4-fold cross validation setting for the predictions which followed the same procedure followed by Toneva and Wehbe [37]. Because the fMRI data were gathered in four runs for every fold the fMRI data gathered in the corresponding run were treated as test data for that fold. First the extracted features from the models were loaded and using PCA were reduced to ten dimensions. The next step was to time align the corresponding fMRI data with the corresponding features extracted from the model so we knew the mapping between the features and brain data for the ridge regression to be trained on. A total of 1351 images were recorded for every subject. By removing the edges of every run, to combat the edge effect, the 1351 were reduced to 1211. In order to align the features with the fMRI data we find out in which TR each word was presented to the user. Using that mapping because the sequences are the same number as the words then the sequence in the same index as a word corresponds to the same TR. After the mapping from TR to sequence is complete the t–4, t–3, t–2, t–1 at any given time t were concatenated to t, because by concatenating representation in the previous four time points can give better representation to the model and yield better accuracy at the end [37]. After the previous steps were done we ended up with an array of 1351 values which are the sequences time-stamped. The last part was to remove the edges in every run, and that resulted in 1211 time-stamped sequences. Moving on, time stamping was done and the ridge regression model was trained. Different weights were obtained for every voxel using a set of different lambda values where

$$\lambda = 10^x where - 9 \geq x \leq 9$$

for the ridge regression model. Then for every voxel the lambda value that produced the lowest error score was used to construct the weights of the final model. Then the model generated from this procedure was used to make predictions on the test data.

Evaluating the Predictions. The final step was to evaluate predictions from the ridge regression model. To do that [37] used a searchlight classification algorithm to binary classify neighbours for every voxel, for every subject. The neighbourhoods have been pre-computed and were provided from the original research. We used the same approach and the same pre-computed neighbourhood for our evaluation method. We used the recorded brain data and a random chunk of 20 TRs was chosen and this corresponds to the correct predictions. From the predicted data the same chunk was chosen and then another random chunk of 20 TRs was chosen from the predicted data to be the wrong chunk. Then using the neighbourhoods the euclidean distance between the 2 chunks from the predicted chunk was calculated. If the distance from the correct chunk was less than the wrong chunk the prediction was marked as correct. This process was done 1000 times for every voxel and at the end the average accuracy for every voxel for every fold was obtained for every subject.

Removing Punctuation. To be able to assess how the human brain processes punctuation symbols semantically we used the same experimental procedure described above but altering slightly the first part. We used four different scenarios to remove the punctuation characters before inputting the sequences to the models. Then we extracted the features in the same manner as described above but as an input we used the altered text sequences. After the sequences were extracted then the same approach was followed. The four scenarios we tested were:

1. Replace fixation ("+") symbol with the ["UNK"] token (Removing fixation). This scenario was used to assess if using the fixation symbol in the sequences as text can influence the quality of the extracted features and how brain aligned they are. The ["UNK"] token indicated in this scenario that this is an unknown token for the model.
2. Replace fixation symbol ("+") with the ["PAD"] token (Padding fixation). This scenario was used to assess if using the fixation symbol in the sequences as text can influence the quality of the extracted features and how brain aligned they are. The ["PAD"] token indicated to the model that this symbol should be ignored when extracting the sequences.
3. Replace ["–","...","—"] with the ["PAD"] token (Padding all). When looking at the text from the Harry Potter chapter we identified in the text some special characters. We wanted to assess if these special characters can influence the performance when learning the brain-network mapping.
4. Replace ["–","...","—",".","?"] with the ["PAD"]token (Padding everything). Motivated from the previous scenario we extended the list of the special characters by adding "." and "?" to the list of tokens to be ignored since they are one of the most commonly used punctuation symbols. Again we wanted to test how this alteration in sequences can affect the mapping.

4 Results and Discussion

Comparing the Models: We compared the performance of BERT against the models presented in Sect. 3.2. In this section we discuss the differences of these networks and what the results might show in terms of how the human brain functions. A common observation that can be seen in all the neural networks including BERT is that as the sequence length increases the accuracy of the ridge regression model decreases. All the results discussed below can be seen in Fig. 1.

BERT. Following the experimental procedure described in Section 3.3, and using the same data, we were not able to exactly reproduce the same original

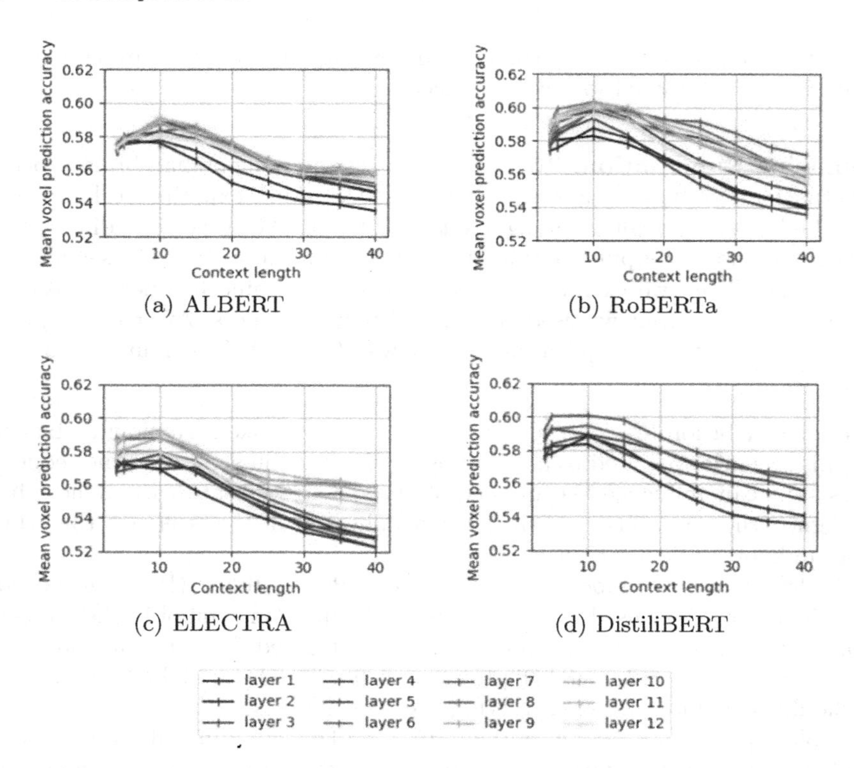

(a) ALBERT (b) RoBERTa

(c) ELECTRA (d) DistiliBERT

Fig. 1. This figure shows the overall accuracy across all subjects for the four different models we investigated.

results as [37]. Thus, our reproduced baseline, used to compare with the other models was obtained by our own code and is presented in Fig. 2 along with the original results. We believe we did not have the same results for three reasons. Firstly, when averaging across subjects and across folds for every layer the exact procedure of the averaging did not appear to be completely described in the original paper. Moreover when PCA is used the random state is not provided so we believe that this might be another reason for not replicating the exact same results. Finally our hardware is not the same as used by the original authors so this might also contribute to getting different results. However, we believe the results are accurate. By examining the results in Fig. 2 which show the original results of [37] and our reproduced results we can see that qualitatively the overall shape of the graphs are the same but quantitatively there are some differences in the exact quantities. Similarities include that our reproduced results reaches its peak at sequence length 10, the same as the original results. Furthermore, as in the original results the accuracy decreases as the sequence length increases.

ALBERT. The first model tested was ALBERT. The interesting characteristic to assess on ALBERT was that all the layers shared the same representations.

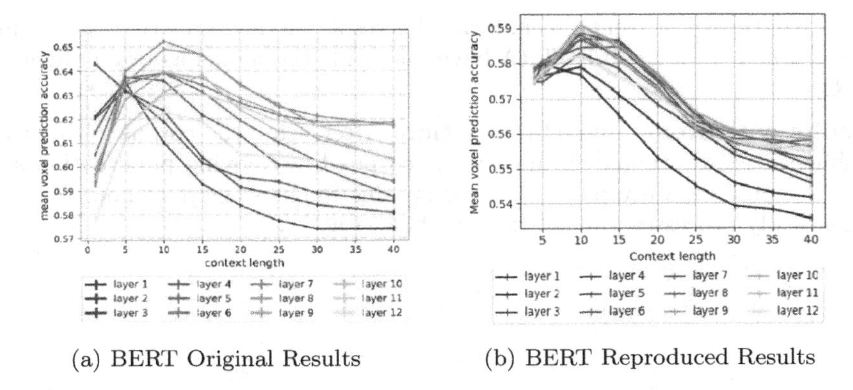

(a) BERT Original Results (b) BERT Reproduced Results

Fig. 2. On the left hand side are the original results reported in [37]. On the right hand side are the reproduced results we obtained when running our code for BERT. Note that the y-axis has a different minimum and maximum value for the two panels, original and reproduced

By looking closer it can be seen that there are not any major differences in the accuracy. This would suggest that the changes made on ALBERT to distinguish it from BERT have not made its representations more brain-aligned than obtained with BERT. The fact that shared weights across the layers does not improve the alignment between brain and neural networks might suggest that the brain has a shared "weights" mechanism but this is something that needs to be investigated further.

RoBERTa. The next model tested was RoBERTa. The same trend in the graphs that is present in BERT seems to be present when using RoBERTa. An immediate observation is that RoBERTa, in almost every layer, at its peak achieves a slightly better accuracy than BERT. Taking that into consideration indicates that the decision to choose different hyper parameters when training and choosing the best performing ones leads to inner representations that are slightly better aligned with those of the human brain.

DistiliBERT. Moving on to examine the results with DistiliBERT shows that even though DistiliBERT has fewer layer, and is smaller in size than BERT the performance of its layers is better for all the layers compared to the layers of BERT. The increase is not so substantial but it indicates that the inner representation of the DistiliBERT can capture semantic information as good as BERT even though it is smaller in size. It also shows that layers of DistiliBERT are as brain aligned as the ones in BERT.

ELECTRA. Looking at the results for ELECTRA we can see that ELECTRA doesn't surpass the baseline results obtained with BERT, even though the same trend in both graphs can be seen. From this result we can argue that the training choices when training ELECTRA have not made the model as brain aligned as BERT.

Results with Removing Punctuation: After using the corpus without any modifications, we wanted to modify the corpus by removing punctuation symbols. In doing so, we wanted to see what this might suggest of how the brain semantically processes punctuation symbols (Fig. 3).

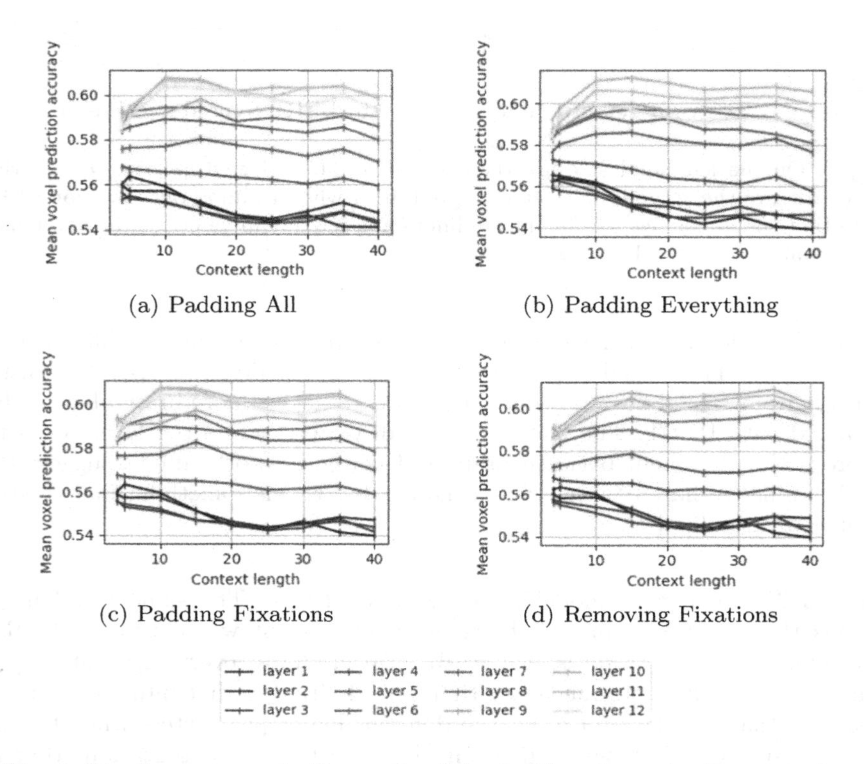

(a) Padding All

(b) Padding Everything

(c) Padding Fixations

(d) Removing Fixations

Fig. 3. The figures presents the results of the 4 different punctuation scenarios.

Our results showed that in all four punctuation-modification scenarios the accuracy increases only on layers 7–12. The increase is not always substantial, and the biggest boost in performance is almost 1.5%. The most interesting thing that can be seen is that removing the punctuation as padding tokens makes the model not lose as much accuracy as the length of the sequence increases. Moreover, layer 6 seems to be acting as a divisor between the starting and ending layers. This resonates with the observation of [37] that the first 6 layers are not as brain aligned as the last 6 layers and that removing the attention mechanism from the 6 first layers might result in better representations.

Taking into account the observations mentioned above and the observation of [37] that some layers of BERT are brain aligned, one can hypothesise that the brain might have limited use of punctuation to understand a sentence semantically. It can also be seen that as the sequence length becomes longer, which means more information is presented to the model and brain, the performance does not drop as much, revealing the limited contribution of punctuation.

5 Conclusion

We have investigated four different models: ELECTRA, RoBERTa, ALBERT and DistiliBERT to determine which of the models, compared to a baseline obtained with BERT, can output language representations that are more closely related to brain representations of language as indicated by the alignment between model and fMRI brain data. The most aligned models appear to be RoBERTa and DistiliBERT as they outperform our baseline. In addition, using the four different scenarios we wanted to test how the semantic processing of punctuation symbols might vary. By looking at the work done by Toneva and Wehbe [37] and using their proof of concept that a mapping from the neural network features to the brain features can be learned, we tested if we can get a better mapping by removing these punctuation symbols. Indeed in all four cases the performance of the model increased. Also, the drop in accuracy when predicting brain data was not as big as originally obtained when the sequence length increased. These results support a view that the brain might have limited use for punctuation symbols to understand semantically a sentence and that as the context text length gets longer then there is less need for punctuation symbols. We believe that our research can be extended in the future for more transformer models to be analysed with the goal of finding a more brain-aligned model. We also believe that the experimental procedure created by Toneva and Wehbe [37] can be used to evaluate new models and produce models that are more brain aligned. Last, but not least we believe that the matter of how the brain processes punctuation symbols semantically should be investigated more in depth, using different models, in the quest of unravelling the capabilities of the human brain.

References

1. Acunzo, D.J., Low, D.M., Fairhall, S.L.: Deep neural networks reveal topic-level representations of sentences in medial prefrontal cortex, lateral anterior temporal lobe, precuneus, and angular gyrus. Neuroimage **251**, 119005 (2022). https://doi.org/10.1016/j.neuroimage.2022.119005
2. Beltagy, I., Lo, K., Cohan, A.: SciBERT: A Pretrained Language Model for Scientific Text. In: Proceedings of the 2019 Conference on Empirical Methods in Natural Language Processing and the 9th International Joint Conference on Natural Language Processing (EMNLP-IJCNLP), pp. 3615–3620. Association for Computational Linguistics, Hong Kong, November 2019. https://doi.org/10.18653/v1/D19-1371

3. Blank, I.A., Fedorenko, E.: Domain-general brain regions do not track linguistic input as closely as language-selective regions. J. Neurosci. Official J. Soc. Neurosci. **37**(41), 9999–10011 (2017). https://doi.org/10.1523/JNEUROSCI.3642-16.2017
4. Brennan, J., Nir, Y., Hasson, U., Malach, R., Heeger, D.J., Pylkkänen, L.: Syntactic structure building in the anterior temporal lobe during natural story listening. Brain Lang. **120**(2), 163–173 (2012). https://doi.org/10.1016/j.bandl.2010.04.002
5. Caucheteux, C., King, J.R.: Language processing in brains and deep neural networks: computational convergence and its limits. Technical report, BioRxiv, January 2021. https://doi.org/10.1101/2020.07.03.186288. Section: New Results Type: article
6. Cer, D., et al.: Universal sentence encoder for English. In: Proceedings of the 2018 Conference on Empirical Methods in Natural Language Processing: System Demonstrations, pp. 169–174. Association for Computational Linguistics, Brussels, November 2018. https://doi.org/10.18653/v1/D18-2029
7. Chen, Y., Gilroy, S., Maletti, A., May, J., Knight, K.: Recurrent neural networks as weighted language recognizers. In: Proceedings of the 2018 Conference of the North American Chapter of the Association for Computational Linguistics: Human Language Technologies, Volume 1 (Long Papers), pp. 2261–2271. Association for Computational Linguistics, New Orleans, June 2018. https://doi.org/10.18653/v1/N18-1205
8. Chong, M., Specia, L., Mitkov, R.: Using natural language processing for automatic detection of plagiarism. In: Proceedings of the 4th International Plagiarism Conference (IPC-2010) (2010)
9. Clark, K., Luong, M.T., Le, Q.V., Manning, C.D.: ELECTRA: pre-training text encoders as discriminators rather than generators (2020)
10. Conneau, A., Kruszewski, G., Lample, G., Barrault, L., Baroni, M.: What you can cram into a single $&!#* vector: probing sentence embeddings for linguistic properties. In: Proceedings of the 56th Annual Meeting of the Association for Computational Linguistics (Volume 1: Long Papers), pp. 2126–2136. Association for Computational Linguistics, Melbourne, July 2018. https://doi.org/10.18653/v1/P18-1198
11. Dai, Z., Yang, Z., Yang, Y., Carbonell, J., Le, Q.V., Salakhutdinov, R.: Transformer-XL: attentive language models beyond a fixed-length context. arXiv:1901.02860 [cs, stat], June 2019. arXiv: 1901.02860
12. Devlin, J., Chang, M.W., Lee, K., Toutanova, K.: BERT: pre-training of deep bidirectional transformers for language understanding. In: Proceedings of the 2019 Conference of the North American Chapter of the Association for Computational Linguistics: Human Language Technologies, Volume 1 (Long and Short Papers), pp. 4171–4186. Association for Computational Linguistics, Minneapolis, June 2019. https://doi.org/10.18653/v1/N19-1423
13. Ek, A., Bernardy, J.P., Chatzikyriakidis, S.: How does punctuation affect neural models in natural language inference. In: Proceedings of the Probability and Meaning Conference (PaM 2020), pp. 109–116. Association for Computational Linguistics, Gothenburg, June 2020
14. Etaiwi, W., Naymat, G.: The impact of applying different preprocessing steps on review spam detection. Procedia Comput. Sci. **113**, 273–279 (2017). https://doi.org/10.1016/j.procs.2017.08.368
15. Frank, S.L., Otten, L.J., Galli, G., Vigliocco, G.: The ERP response to the amount of information conveyed by words in sentences. Brain Lang. **140**, 1–11 (2015). https://doi.org/10.1016/j.bandl.2014.10.006

16. Friederici, A.D.: The brain basis of language processing: from structure to function. Physiol. Rev. **91**(4), 1357–1392 (2011). https://doi.org/10.1152/physrev.00006.2011
17. Fyshe, A., Talukdar, P.P., Murphy, B., Mitchell, T.M.: Interpretable semantic vectors from a joint model of brain- and text- based meaning. In: Proceedings of the 52nd Annual Meeting of the Association for Computational Linguistics (Volume 1: Long Papers), pp. 489–499. Association for Computational Linguistics, Baltimore, June 2014. https://doi.org/10.3115/v1/P14-1046
18. Hale, J., Dyer, C., Kuncoro, A., Brennan, J.R.: Finding syntax in human encephalography with beam search. arXiv:1806.04127, June 2018
19. Hong, W., Ji, K., Liu, J., Wang, J., Chen, J., Chu, W.: GilBERT: generative vision-language pre-training for image-text retrieval. In: Proceedings of the 44th International ACM SIGIR Conference on Research and Development in Information Retrieval, pp. 1379–1388. SIGIR 2021, Association for Computing Machinery, New York, July 2021. https://doi.org/10.1145/3404835.3462838
20. Howard, J., Ruder, S.: Universal language model fine-tuning for text classification. In: Proceedings of the 56th Annual Meeting of the Association for Computational Linguistics (Volume 1: Long Papers), pp. 328–339. Association for Computational Linguistics, Melbourne, July 2018. https://doi.org/10.18653/v1/P18-1031
21. Huth, A.G., de Heer, W.A., Griffiths, T.L., Theunissen, F.E., Gallant, J.L.: Natural speech reveals the semantic maps that tile human cerebral cortex. Nature **532**(7600), 453–458 (2016). https://doi.org/10.1038/nature17637
22. Jain, S., Huth, A.G.: Incorporating context into language encoding models for fMRI. Technical report, BioRxiv, November 2018. https://doi.org/10.1101/327601. Section: New Results Type: Article
23. Jia, Q., Li, J., Zhang, Q., He, X., Zhu, J.: RMBERT: news recommendation via recurrent reasoning memory network over BERT. In: Proceedings of the 44th International ACM SIGIR Conference on Research and Development in Information Retrieval, pp. 1773–1777. SIGIR 2021, Association for Computing Machinery, New York, July 2021. https://doi.org/10.1145/3404835.3463234
24. Lan, Z., Chen, M., Goodman, S., Gimpel, K., Sharma, P., Soricut, R.: ALBERT: a lite bert for self-supervised learning of language representations, April 2020
25. Lerner, Y., Honey, C.J., Silbert, L.J., Hasson, U.: Topographic mapping of a hierarchy of temporal receptive windows using a narrated story. J. Neurosci. **31**(8), 2906–2915 (2011). https://doi.org/10.1523/JNEUROSCI.3684-10.2011. publisher: Society for Neuroscience Section: Articles
26. Linzen, T., Dupoux, E., Goldberg, Y.: Assessing the ability of LSTMs to learn syntax-sensitive dependencies. arXiv:1611.01368, November 2016
27. Moore, N.: What's the point? The role of punctuation in realising information structure in written English. Funct. Linguist. **3**(1), 1–23 (2016). https://doi.org/10.1186/s40554-016-0029-x
28. Nishimoto, S., Vu, A.T., Naselaris, T., Benjamini, Y., Yu, B., Gallant, J.L.: Reconstructing visual experiences from brain activity evoked by natural movies. Curr. Biol. **21**(19), 1641–1646 (2011). https://doi.org/10.1016/j.cub.2011.08.031
29. Peng, H., Schwartz, R., Thomson, S., Smith, N.A.: Rational recurrences. arXiv:1808.09357, August 2018
30. Peters, M.E., et al.: Deep contextualized word representations. In: Proceedings of the 2018 Conference of the North American Chapter of the Association for Computational Linguistics: Human Language Technologies, Volume 1 (Long Papers), pp. 2227–2237. Association for Computational Linguistics, New Orleans, June 2018. https://doi.org/10.18653/v1/N18-1202

31. Reddy, A.J., Wehbe, L.: Can fMRI reveal the representation of syntactic structure in the brain? In: Advances in Neural Information Processing Systems, vol. 34, pp. 9843–9856. Curran Associates, Inc. (2021)
32. Rowling, J.K.: Harry Potter and the Philosopher's Stone, vol. 1, 1st edn. Bloomsbury Publishing, London (1997)
33. Sanh, V., Debut, L., Chaumond, J., Wolf, T.: DistilBERT, a distilled version of BERT: smaller, faster, cheaper and lighter
34. Shain, C., et al.: 'Constituent length' effects in fMRI do not provide evidence for abstract syntactic processing. Preprint, Neuroscience, November 2021. https://doi.org/10.1101/2021.11.12.467812
35. Søgaard, A.: Evaluating word embeddings with fMRI and eye-tracking. In: Proceedings of the 1st Workshop on Evaluating Vector-Space Representations for NLP, pp. 116–121. Association for Computational Linguistics, Berlin, August 2016. https://doi.org/10.18653/v1/W16-2521
36. Sudre, G., et al.: Tracking neural coding of perceptual and semantic features of concrete nouns. Neuroimage **62**(1), 451–463 (2012). https://doi.org/10.1016/j.neuroimage.2012.04.048
37. Toneva, M., Wehbe, L.: Interpreting and improving natural-language processing (in machines) with natural language-processing (in the brain). In: Advances in Neural Information Processing Systems, vol. 32. Curran Associates, Inc. (2019)
38. Wehbe, L., Murphy, B., Talukdar, P., Fyshe, A., Ramdas, A., Mitchell, T.: Simultaneously uncovering the patterns of brain regions involved in different story reading subprocesses. PLoS ONE **9**(11), e112575 (2014). https://doi.org/10.1371/journal.pone.0112575
39. Wehbe, L., Vaswani, A., Knight, K., Mitchell, T.: Aligning context-based statistical models of language with brain activity during reading. In: Proceedings of the 2014 Conference on Empirical Methods in Natural Language Processing (EMNLP), pp. 233–243. Association for Computational Linguistics, Doha, October 2014. https://doi.org/10.3115/v1/D14-1030
40. Weiss, G., Goldberg, Y., Yahav, E.: On the practical computational power of finite precision RNNs for language recognition. In: Proceedings of the 56th Annual Meeting of the Association for Computational Linguistics (Volume 2: Short Papers), pp. 740–745. Association for Computational Linguistics, Melbourne, July 2018. https://doi.org/10.18653/v1/P18-2117
41. Zhu, X., Li, T., de Melo, G.: Exploring semantic properties of sentence embeddings. In: Proceedings of the 56th Annual Meeting of the Association for Computational Linguistics (Volume 2: Short Papers), pp. 632–637. Association for Computational Linguistics, Melbourne, July 2018. https://doi.org/10.18653/v1/P18-2100
42. Zhuang, L., Wayne, L., Ya, S., Jun, Z.: A robustly optimized BERT pre-training approach with post-training. In: Proceedings of the 20th Chinese National Conference on Computational Linguistics, pp. 1218–1227. Chinese Information Processing Society of China, Huhhot, August 2021

A Bayesian-Optimized Convolutional Neural Network to Decode Reach-to-Grasp from Macaque Dorsomedial Visual Stream

Davide Borra[1(✉)], Matteo Filippini[2], Mauro Ursino[1,3], Patrizia Fattori[2,3], and Elisa Magosso[1,3]

[1] Department of Electrical, Electronic and Information Engineering "Guglielmo Marconi" (DEI), University of Bologna, Cesena Campus, Cesena, Italy
davide.borra2@unibo.it

[2] Department of Biomedical and Neuromotor Sciences (DIBINEM), University of Bologna, Bologna, Italy

[3] Alma Mater Research Institute for Human-Centered Artificial Intelligence, University of Bologna, Bologna, Italy

Abstract. Neural decoding is crucial to translate the neural activity for Brain-Computer Interfaces (BCIs) and provides information on how external variables (e.g., movement) are represented and encoded in the neural system. Convolutional neural networks (CNNs) are emerging as neural decoders for their high predictive power and are largely applied with electroencephalographic signals; these algorithms, by automatically learning the more relevant class-discriminative features, improve decoding performance over classic decoders based on handcrafted features. However, applications of CNNs for single-neuron decoding are still scarce and require further validation. In this study, a CNN architecture was designed via Bayesian optimization and was applied to decode different grip types from the activity of single neurons of the posterior parietal cortex of macaque (area V6A). The Bayesian-optimized CNN significantly outperformed a naïve Bayes classifier, commonly used for neural decoding, and proved to be robust to a reduction of the number of cells and of training trials. Adopting a sliding window decoding approach with a high time resolution (5 ms), the CNN was able to capture grip-discriminant features early after cuing the animal, i.e., when the animal was only attending the object to grasp, further supporting that grip-related neural signatures are strongly encoded in V6A already during movement preparation. The proposed approach may have practical implications in invasive BCIs to realize accurate and robust decoders, and may be used together with explanation techniques to design a general tool for neural decoding and analysis, boosting our comprehension of neural encoding.

Keywords: Neural decoding · Convolutional neural networks · Dorsomedial visual stream · V6A · Bayesian optimization

G. Nicosia et al. (Eds.): LOD 2022, LNCS 13811, pp. 473–487, 2023.
https://doi.org/10.1007/978-3-031-25891-6_36

1 Introduction

Neural decoding (i.e., prediction of observable output variables, such as movements or stimuli, from neural time series) is a central aim in neuroscience and in neural engineering, for its practical and theoretical implications. Indeed, neural decoding is at the core of brain-computer interfaces (BCIs) where neural activity is translated into output commands for assistive and therapeutic purposes [1]. Moreover, neural decoding may advance our understanding of how information is represented and encoded in the neural system, as the accuracy of neural decoding reveals the amount of information the neural signals contain about an external variable (e.g., movement or sensation) and this can be evaluated across different brain regions and/or across different time intervals over the course of the sensory or sensorimotor task [2–8].

Currently, recent advances in machine learning, such as deep learning algorithms, are receiving growing attention for their predictive power in neural decoding applications [9]. Among deep learning architectures, Convolutional Neural Networks (CNNs) are preferentially adopted over other architectures (such as recurrent neural networks or fully-connected neural networks) for the classification of electroencephalographic signals (EEG) [10]. CNNs are feed-forward neural networks that learn convolutional filters to identify the features in the input signal that better discriminate among the predicted conditions. Attractive characteristics of these networks are a lower number of trainable parameters (e.g., vs. recurrent architectures), without hampering decoding performance, and an easier interpretability of the learned features in specific domains (e.g., temporal, spatial, and frequency domains) by using explanation techniques [6, 8]. CNNs have been successfully applied to EEG in a large spectrum of classification problems, such as emotion classification [11], event related potential detection and analysis [7, 8, 12, 13], motor execution/imagery classification [6, 14]. While CNN-based decoding of non-invasive neural recordings (EEG) have been widely investigated, CNNs are scarcely applied to single-neuron recordings acquired invasively in non-human mammals (in particular, non-human primates) and human patients. To overcome this limitation and to explore the potentialities of single-neuron decoding via CNNs, a recent study by Filippini et al. [2] designed and applied a CNN to decode the activity of neurons recorded from the posterior parietal cortex (PPC) of macaque (areas V6A, PEc, PE) while the animal performed a delayed reaching task towards 9 positions in the 3D space. PPC host areas integrating sensorimotor stimuli to dynamically guide the interaction with the surrounding environment and neurons in these areas are known to encode information regarding reaching endpoints, goals and trajectories [15–17]. In that study, Filippini et al. [2], employed a CNN whose configuration was optimized in its hyper-parameters (i.e., the parameters defining the functional form of the learning system, e.g., the number of convolutional kernels) via Bayesian optimization (BO), an efficient automatic hyper-parameter search algorithm. Results proved that the CNN was able to accurately decode the position of the reached points over the entire time course of the task, from target presentation to the end of reaching movement, with modulation across task phases and recording areas. Furthermore, the CNN outperformed a linear Naïve Bayes (NB) classifier suggesting that the CNN may represent a better framework to analyze how sensorimotor information are temporally encoded in neural representations. However, despite these promising results, the design and application of CNNs to motor decoding require a further validation on different motor tasks.

In this study, we aim to further validate the design and the application of CNNs to single-cell recordings, using a similar decoding workflow to the one adopted in [2] (CNN architecture and BO for hyper-parameter search), while addressing a different motor decoding problem. Specifically, a CNN was used here to decode different grip types from the activity of V6A neurons, recorded while monkeys performed a delayed reach-to-grasp task towards objects of different shapes. V6A is a visuomotor area involved in the transformation of sensory information to guide prehension movements [18]. Its neurons are not only modulated by reaching targets located in different spatial positions but also by grasping information (e.g., grip type) and here we derived a Bayesian-optimized CNN configuration to decode such information and tested its decoding accuracy over the entire time course of the task, with a high time resolution (5 ms). Furthermore, we also evaluated the CNN performance while progressively using a reduced number of cells or a reduced number of training trials inside the recorded dataset, to test the robustness of the decoder by simulating practical scenarios where smaller datasets are available.

2 Methods

2.1 Dataset and Pre-processing

This study reanalyzes the data obtained in [3], where single-cell activity was recorded from V6A area in two male Macaca fascicularis monkeys, using invasive electrode penetrations. The activity of 93 and 75 cells was recorded from the two monkeys (more details about the recording procedure can be found in [3]). Action potentials (spikes) were isolated and sampled at 100 kHz. Monkey sat on a primate chair with its head fixed in front of a rotating panel containing 5 different objects. The objects were chosen to evoke reach-to-grasp with different hand configurations and are illustrated in Fig. 1. These were: a ball (l_0: whole-hand prehension), ring (l_1: hook), plate (l_2: primitive precision grip), stick-in-groove (l_3: advanced precision grip), handle (l_4: finger prehension). Objects were presented to the monkey one at a time, in a randomized order. For each object, 10 trials were recorded, overall resulting in 50 trials per monkey. Each trial consisted of different phases, hereafter referred as 'epochs'. The trial started when the monkey pressed a 'home button' in complete darkness and then, the animal waited for instructions in darkness for 1 s (*free epoch*, epoch 0). Subsequently, the fixation LED turned green, and the monkey had to wait for the LED to change its color to red, without performing any movement. After a fixation period of 0.5–1.0 s, LEDs surrounding the object to grasp turned on, illuminating the object. The monkey maintained the fixation on the LED without releasing the home button for a period of 1.5–2.0 s. The first 0.5 s portion of this period (corresponding to a first object visualization interval [3]) was excluded from the analysis. The remaining 1.0–1.5 s portion of this period formed the *delay epoch*, subdivided into the *early delay* (epoch 1) and *late delay* (epoch 2), by extracting 1 s epoch after the start of the delay epoch and before the end of the delay epoch, respectively. Then, the LED turned red, representing the go-signal for the reach-to-grasp movement (*reaction time epoch* and *reach-to-grasp epoch,* epochs 3,4, respectively). Once performed the movement, the monkey had to keep holding (*hold epoch*, epoch 5) the grasped object

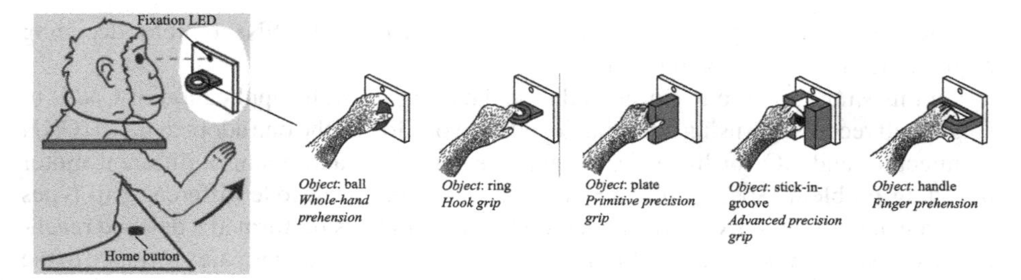

Fig. 1. Schematic representations of grip types and objects.

until the LED switched off (0.8–1.2 s). The LED switch-off cued the monkey to release the object and press the home button again, starting a new trial with a different object to reach and grasp.

For each neuron and each trial, spikes were binned at 5 ms. As trials and epochs may have a different duration across different neurons and trials, to obtain the same number of bins across neurons and trials, the average number of bins for each epoch was computed, then, the activity of each neuron and trial was re-binned using that number of average bins per epoch (thus, slightly changing respect to the original 5 ms binning). Then, firing rates were computed; thus, in this study the multi-variate neural activity was described by means of neuron firing rate. Firing rates recorded during the early delay, late delay, reaction time, reach-to-grasp, and hold epochs were collected and used in this study. Ten-fold stratified cross-validation was applied to partition the dataset of each monkey. Therefore, within each fold, 5 trials (one for each grip) were used as test set; then, 5 (one for each grip) and 40 of the remaining trials were used as validation and training sets, respectively.

2.2 Sliding Window Neural Decoding

To analyze the temporal dynamics of reach-to-grasp encoding inV6A, a sliding window decoding approach was applied, as performed in previous studies [2–5] that analyzed the neural activity (neuron firing rates or EEG) by using the prediction of a machine learning algorithm as measure of neural encoding of motor- or cognitive-related brain states. This approach consists in decoding small portions of neurons' signals (hereafter referred as 'chunks') within each single recorded trial, enabling the study of the time course of neural encoding across all the phases of the task. To this aim, the neurons' firing rates were processed as follows, for each monkey (see upper panel of Fig. 2).

Let X_t be the t-th trial of shape (N, T), representing the multi-variate neural activity, where N is the number of recorded neurons (different across animals, here $N = 93$ and $N = 75$) and T is the number of time samples in the trial ($T = 812$ and $T = 816$ for the two monkeys, having 5 ms resolution). Overlapped chunks $X_{t,i}$ of shape (N, T_z) were extracted with a stride of T_s, where T_z is the number of time samples of each chunk. Each sampled chunk ($X_{t,i}$) was fed as input to the neural decoder.

$$X_{t,i} = X_t[:, iT_s : iT_s + T_z - 1], 0 \leq i \leq M - 1, \tag{1}$$

where i is the chunk index and M is the total number of chunks that can be extracted using T_z and T_s as chunk size and stride, respectively, i.e., $M = (T - T_z)/T_s + 1$. T_z and T_s are hyper-parameters of the decoding approach and were set to $T_z = 60$ (= 300 ms) and $T_s = 10$ (= 50 ms) during the training phase of the decoder (as in [2]), while $T_s = 1$ (= 5 ms) during the testing phase. That is, during training a higher stride was used to speed up the computation, while during testing chunks were extracted with the maximum overlap, producing an inference with a high time resolution of 5 ms-step.

The addressed decoding problem was the classification of 5 different grip types from neuron firing rates (5-way classification). Each trial X_t was associated to a single label corresponding to the specific shape of the object the monkey had to grasp in that trial, i.e., $y_t \in L = \{l_k\}, 0 \leq k \leq 4$ (see Sect. 2.1 for the association label ID-grip type). Therefore, while performing sliding window decoding, the label associated to each sampled chunk ($y_{t,i}$) was the one associated to the trial the chunk was extracted from.

$$y_{t,i} = y_t, 0 \leq i \leq M - 1. \tag{2}$$

The CNN can be described by a probabilistic model $f(X_{t,i}; \vartheta, h) : \mathbb{R}^{N \times T_z} \rightarrow L$ parametrized in the trainable parameters and hyper-parameters contained in the arrays ϑ, h, respectively. In this study, monkey-specific decoders were designed, by using monkey-specific datasets to tune trainable parameters and hyper-parameters of CNNs. Hyper-parameters must be set before the model training starts; these parameters are optimized on the validation set via the hyper-parameter search procedure. Trainable parameters are the collection of weights and biases that model connections across the artificial neurons included in the network; these are learned on the training set during the network training.

2.3 Architecture and Parameter Tuning of the CNN

Architecture. The adopted CNN topology is inspired from the architecture recently proposed for the decoding of reaching targets from V6A neuron activity [2]. The CNN is composed by two modules and a schematization is reported in Fig. 2 (lower panel).

The first module (*convolutional module*) includes only trainable sparse connections across artificial neurons. This is composed by N_b blocks; each block in turn is composed by the sequence of N_c 2-D convolutional layers. The very first convolutional layer of the architecture performed convolutions in both space and time domains (mixed spatio-temporal convolutions) using kernels of size (N, F), while all other convolutional layer performed convolutions in the time domain using kernels of size $(1, F)$. Each layer learned K kernels using unitary stride and a padding of $(0, F//2)$, where // is the floor division operator. After each layer, batch normalization [19] is optionally included and, then, Exponential Linear Unit (ELU) non-linearity [20], i.e., $f(x) = x, x > 0$ and $f(x) = \exp(x) - 1, x \leq 0$, is applied. Between each block, average pooling with a pool size of $(0, 2)$ is performed, halving the temporal dimension, and then dropout [21] with dropout rate p_{drop} is included. Over the network, batch normalization and dropout act as regularizers to improve model generalization. The main hyper-parameters of this module were searched via BO and are detailed in Table 1.

The second module (*fully-connected module*) includes only trainable dense connections across neurons. In particular, feature maps provided by the convolutional module are flattened and given as input to a fully-connected layer with 5 neurons, corresponding to the output layer. This layer is activated with a softmax activation function to convert neuron outputs into the conditional probabilities $p(l_k|X_{t,i}), 0 \leq k \leq 4$, and then, the predicted class is the one with the highest probability.

Table 1. Hyper-parameters of the convolutional module searched with Bayesian optimization: distributions and admitted values.

Hyper-parameter	Distribution	Values
No. of blocks (N_b)	Uniform	[1, 2]
No. of conv. Layers (N_c)	Uniform	[1, 2, 3]
No. of kernels (K)	Uniform	[4, 8, 16, 32]
Kernel size (F)	Uniform	[(N, 11), (N, 21), (N, 31), (N, 41)]
Dropout rate (p_{drop})	Uniform	[0, 0.25, 0.5]
Use batch norm	Uniform	[False, True]
Learning rate (lr)	Log-uniform	[1e−4, 5·1e−4, 1e−3, 5·1e−3, 1e−2]

Hyper-parameter Optimization. Hyper-parameter optimization is devoted to find the optimal hyper-parameters of a learning system on a validation set (different from the training and test sets). Denoting with h the array containing the hyper-parameters to search, hyper-parameter optimization finds the h^* that minimizes an objective function $k(h)$ on the validation set, i.e., $h^* = argmin_h k(h)$. In this study, we used $k(h) = 1-acc(h)$ as objective function, where $acc(h)$ is the accuracy (averaged across all time samples and across epochs) obtained with a specific hyper-parameter configuration; for each configuration, a new training (as specified in Section Trainable parameter optimization) and evaluation stages must be performed, increasing the computational cost.

Hyper-parameter search algorithms (e.g., grid search and random search) generally search h^* without exploiting results from past iterations to select the array h to be evaluated in the next iteration (uninformed algorithms), often wasting time on unpromising h values. BO [22] overcomes this limitation, by suggesting, in an informed way, the next hyper-parameters h to be evaluated using a selection criterion, and thus investigating only hyper-parameters that seem promising based on past evaluations. In particular, a Bayesian statistical model $p(k|h)$ of the objective function (surrogate model) is used and it is updated after each iteration by keeping track of past evaluation results (i.e., each pair h, $k(h)$). Crucially, this surrogate model is easier to optimize than the objective function $k(h)$ [22].Thus, the next set of hyper-parameters to be evaluated on the actual objective function is chosen by selecting the hyperparameters that perform best on the current surrogate model. The criterion used to optimize the surrogate is called 'selection function'. BO was performed for 100 iterations by using tree-structured Parzen estimator as surrogate model and expected improvement as selection function. A more complete description of BO used to tune hyper-parameters of a CNN for neural decoding can be

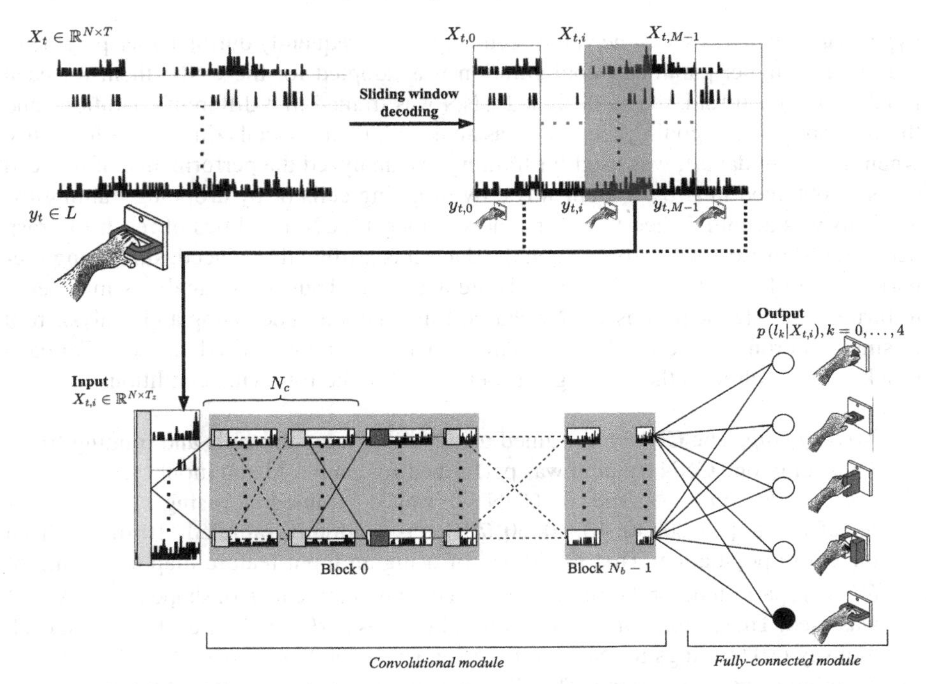

Fig. 2. Sliding window decoding approach and CNN architecture. In the CNN structure, only the main layers are displayed (i.e., only convolutional, pooling and fully-connected layers) with their output feature maps. Convolutional kernels are displayed with light-blue boxes, while pooling kernels are displayed with red boxes. Convolutional blocks are displayed as grey boxes; note that for brevity, only the first convolutional block (block 0) is detailed in its layers, while in the last block (block $N_b - 1$) only the first and last feature maps are displayed.

found in [2]. BO was performed for each cross-validation fold (10 in total) and each monkey (2 in total), leading to 20 optimal hyper-parameter configurations.

Trainable Parameter Optimization. The cross-entropy between the predicted probability distribution (provided by the probabilistic learning system) and the empirical distribution (provided by the labelled dataset) was used as loss function ($j(\vartheta)$) while learning the trainable parameters. Adam [23] was used as optimizer, searching for $\vartheta^* = argmin_{\vartheta} j(\vartheta)$. The learning rate was selected via BO (see Table 1) together with the other searched hyper-parameters. Furthermore, the mini-batch size was set to 64 and the maximum number of training epochs to 250. The optimization stopped when the validation accuracy did not decrease after 50 consecutive epochs (early stopping).

2.4 Analysis of Decoding Performance

For each hyper-parameter, the most frequent value across the configurations selected via BO was derived; thus, we identified a single hyper-parameter configuration where each hyper-parameter was set to the value occurring more frequently during hyper-parameter search. This hyper-parameter configuration was adopted for the CNN; then, for each monkey and each fold, the so designed CNN was trained in 3 different conditions and the corresponding decoding accuracy was analyzed. Besides analyzing the performance when the entire dataset was used for training, we analyzed the performance of reduced datasets obtained in two different ways, by dropping cells or by dropping training trials. This was accomplished to understand whether the CNN abilities of reach-to-grasp decoding from V6A still persist when the dataset is artificially reduced, simulating scenarios where less cells or training trials are available. Thus, these analyses may serve to further validate the proposed CNN-based framework as a decoding and analysis tool of single-neuron time series, by artificially generating variable-sized datasets. For each monkey and each fold, the training was performed in the following conditions:

a) *No dropping*. The CNN was trained using all the recorded cells and training trials. Thus, only one CNN training was performed for each fold and monkey.
b) *Cell dropping*. The CNN was trained using a subset of $N^{'} \in \{10, 20, 30, 40, 50, 60, 70\}$ cells randomly sampled (10 times) from the entire population. That is, instead of using an input feature map consisting of (N, T_z) spatio-temporal samples, a reduced input feature map of shape $(N^{'} < N, T_z)$ was used. Thus, a total of 10·7 CNN trainings was performed for each fold and each monkey (10 trainings for each of the seven values of $N^{'}$).
c) *Training example dropping*. The CNN was trained using a subset of training examples corresponding to the 12.5%, 25%, 37.5%, 50%, 62.5%, 75%, 87.5% randomly sampled (10 times) from the entire training set. Thus, in this case too, a total of 10·7 CNN trainings was performed for each fold and each monkey (10 trainings for each of the seven percentages).

For each CNN training in the previous conditions, the trained CNN was tested within each fold and monkey, computing the accuracy chunk by chunk. Note that in this way, for each fold, we obtained a temporal pattern of decoding accuracy thanks to the sliding window decoding approach, that enabled to highlight the dynamics of the reach-to-grasp task encoded in V6A with a high time resolution (5 ms). Furthermore, in condition b) and c), the accuracy was averaged, chunk by chunk, across the 10 random extractions for each dropping value. Therefore, one temporal pattern of accuracy per fold and monkey was obtained in the condition a), while 7 averaged temporal patterns of accuracy were obtained per fold and monkey in conditions b) and c), each pattern corresponding to a different dropping value.

In each condition, the performance of the proposed approach was compared with the one obtained with a NB classifier (the same used in [2–4]). This decoder takes as input the multi-variate neural activity and it linearly combines inputs assuming independences between time samples. Permutation cluster tests (1000 iterations) with threshold-free cluster enhancement (TFCE) [24] were performed to test for differences between the accuracy temporal dynamics obtained with the CNN and with NB for each analyzed condition (conditions a–c).

3 Results

Figure 3 reports the hyper-parameter distributions resulting from BO. The most frequent configuration was a shallow CNN with one convolutional layer ($N_b = 1$, $N_c = 1$), learning the maximum number of allowed feature maps, i.e., kernels ($K = 32$) thus resulting in a shallow but wide network, with a kernel size of $(N, 21)$, corresponding to learning temporal features within approximately 100 ms of the input multi-variate neural activity. Furthermore, BO selected more frequently dropout ($p_{drop} = 0.5$) as regularizer, instead of batch normalization.

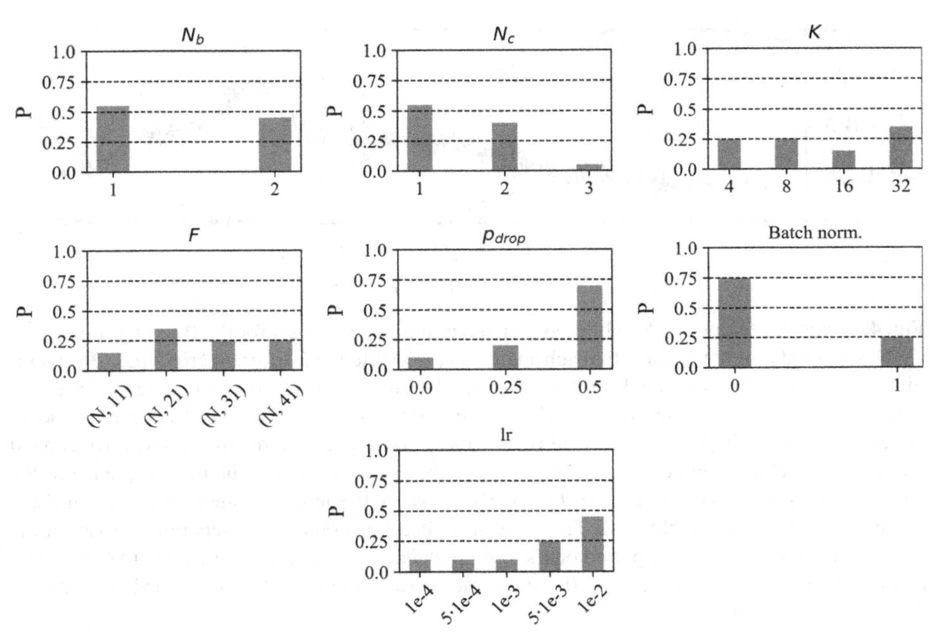

Fig. 3. Hyper-parameter probability distributions. Each bar plot shows the frequency of occurrence of a specific hyper-parameter value among the admitted ones (see Table 1) across the 20 BO configurations (one per fold and monkey).

Figure 4 reports the dynamic of the prediction (panel A) and of the accuracy (panel B) over the trial course of the proposed Bayesian-optimized CNN, when no dropping was applied (see Sect. 2.4). Specifically, Fig. 4A displays the output probabilities for the correct class, separately for each object to be reached and grasped, while Fig. 4B reports the decoding accuracy compared with the NB algorithm inspired from [2–4]. The CNN significantly outperformed ($p < 0.05$) the NB algorithm, across the entire time course, already from the last portion of the early delay epoch (epoch 1) up to the hold epoch (epoch 5). Notably, in Fig. 4B the accuracy is reported also in the free interval (epoch 0), in which the animal was not engaged in the motor task. As expected, here both algorithms performed at the chance level (20%), thus, algorithms proved to be able also to detect the absence of motor-related signatures from the neural activity.

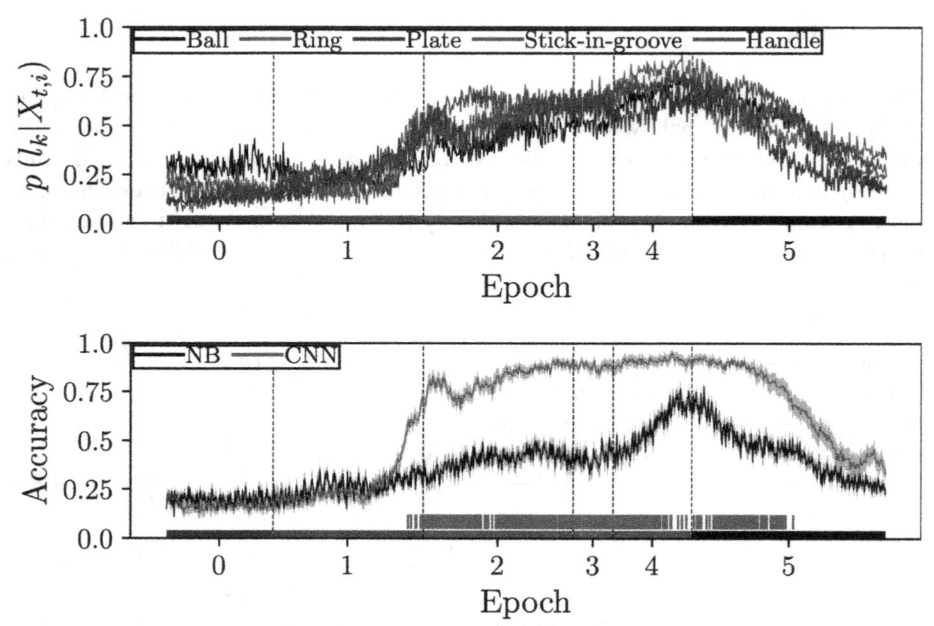

Fig. 4. Panel A - Output probabilities associated to the correct labels for the Bayesian-optimized CNN, separately for the 5 object to reach and grasp (associated to 5 different grip types). Probabilities were averaged across monkeys and cross-validation folds. Panel B - Decoding performance of the proposed Bayesian-optimized CNN (red) and of the NB algorithm (black). Decoding accuracies are reported in their mean value (thick lines) and standard error of the mean (overlayed area) across monkeys and cross-validation folds. The epochs outlining the time sequence of the task are color-coded as: purple-free; blue-early delay; red: late delay; magenta: reaction time; green: reach-to-grasp; black: hold. The vertical dashed lines denote the separation between the epochs. Grey vertical strips reported on the bottom denote time intervals where the two decoding algorithms are significantly ($p < 0.05$) different, as resulting from the performed permutation cluster test. (Color figure online)

Finally, Fig. 5 reports the time course of the decoding accuracy scored with NB and the Bayesian-optimized CNN when reduced-size datasets were simulated by sampling a variable number of cells (Fig. 5A) and training examples (Fig. 5B) from the entire dataset (see Sect. 2.4). Some considerations can be drawn. First, the proposed CNN significantly outperformed the NB baseline (even up to 50% of accuracy) across the entire ranges of cell and training example dropped out, especially in the second part of the delay (epoch 2 and last portion of epoch 1), during the reaction time and the first part of reach-to-grasp and hold epochs. Second, the CNN accuracy was more susceptible to deterioration when few cells were available, in particular below 40 cells, while a reduction in the number of training examples affected less the CNN performance, as only the lowest percentage of training examples produced an evident performance decrease.

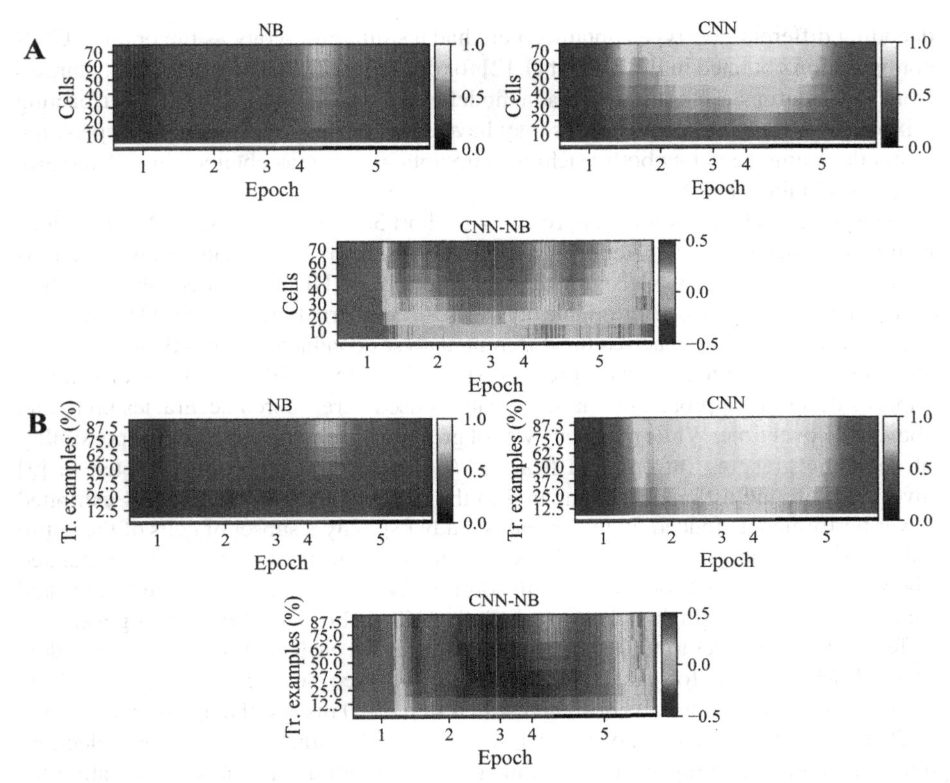

Fig. 5. Time course of the decoding accuracy scored by NB and Bayesian-optimized CNN while randomly dropping out cells (Fig. A) and training trials (Fig. B) from the dataset. In both Fig. A and B, the decoding accuracy of each algorithm (NB, CNN) was averaged across folds and monkeys for each dropping value and displayed as an heatmap (upper panels). In addition, the difference between the average accuracy scored with the Bayesian-optimized CNN and with NB is reported in the bottom panel (CNN-NB), where the significant ($p < 0.05$) differences resulting from the performed permutation cluster test were colored (leaving in grey the unsignificant ones). The epochs outlining the time sequence of the task are color-coded as: blue-early delay; red: late delay; magenta: reaction time; green: reach-to-grasp; black: hold. (Color figure online)

4 Discussion and Conclusions

In this study, a Bayesian-optimized CNN was designed and applied to decode reach-to-grasp from the activity of neurons in V6A, a pivotal parietal area of the dorsomedial visual stream of macaque brain. The decoding capabilities of the learning system were analyzed under different conditions, i.e., while using the entire datasets available and while using reduced datasets, both in terms of number of cells and of training trials.

By using a similar methodology as the one adopted in [2] (similar CNN structure coupled with BO), the present study serves to further validate a CNN-based decoding workflow for general motor decoding, instead of proposing novel CNN workflows, often tailored to one specific application. Notably, the optimal CNN configuration for decoding different grip types obtained here had a similar topology as the optimal CNN configuration obtained in Filippini et al. [2] for decoding the location of reaching targets from PPC neurons. This suggests that a model based on a single-layer CNN performing mixed spatio-temporal convolutions may have enough capacity for general upper-limb motor decoding, decoding both reaching endpoints in space (as obtained in [2]) and grip types (as obtained here).

Despite the adopted sliding approach with short 300 ms windows, CNN predictions resulted consistent through the trial course, especially from the late delay epoch, as shown in Fig. 4A. The CNN significantly outperformed NB, thus, was able to capture more relevant features to discriminate between different grip types compared to NB, over the entire task course and across the different analyses conducted (Fig. 4B and Fig. 5). It is important and fair to note that a previous study [3], using a NB classifier with a sliding window decoding approach on the same dataset used here, scored accuracies up to and above 80% over time. While that study was of great importance to evidence the possibility of decoding grasping information from V6A neurons, the procedure adopted in [3] involves profound differences compared to the one adopted here, where NB exhibited lower performance. Indeed, in the previous study [3], only a subset of cells of the entire dataset (79 cells across the two monkeys), previously identified as being grip-modulated via ANOVA, were used for decoding. In addition, NB was separately trained and tested for each step of the sliding window approach (i.e., $\forall i$, see Eq. 1 and Eq. 2), thus generating different decoders over time, i.e., a different decoder for each analyzed 300 ms-window (overall, M decoders for each cross-validation fold, see Sect. 2.1). Lastly, in [3] the sliding window decoding was designed with a step of 20 ms and the dataset was binned at 20 ms. Here, all the cells of the dataset were used, without any a priori selection, thus, leaving the learning system the ability to explore all the available information for decoding and to learn to discard task-unrelated information. Furthermore, in this study, the decoder was trained and tested considering all chunks at once, a procedure more parsimonious in terms of number of trained decoders (one decoder per fold, instead of M decoders per fold), but more challenging, as being time-aspecific. In addition, here, signals were binned within a shorter window (5 ms vs 20 ms), enabling to take advantage also of high frequency components (e.g., high gamma band) for decoding, which may play a key role during movement [25]. Furthermore, the decoding and analysis of reach-to-grasp was performed with a finer time resolution of 5 ms. All these points represent more challenging conditions for the neural decoder and the CNN proved to outperform NB with these settings.

Besides the performance difference between the CNN and NB decoders, they showed different time patterns of decoding accuracy (Fig. 4B). NB exhibited only a small increase in performance from early to late delay epoch, increasing rapidly only during the reach-to-grasp epoch, and peaking at the end of movement. Conversely, the CNN showed a strong increase in performance already from the last portion of early delay, then slowly increased over time and peaked between the reach-to-grasp and hold epochs as NB. That is, the CNN was able to capture grip-discriminant features already while the animal was attending the object to grasp and waiting for the go-signal, further supporting that grip-related neural signatures are progressively encoded in V6A during movement preparation, as known in the literature [3], reflecting a visuomotor processing of the information. These differences between the CNN and NB may be associated to the ability of the CNN to better capture temporal dynamics and non-linearities encoded in neural activity [2], better extracting relevant features from the input signal and discarding task-unrelated information. Lastly, as expected, accuracies gradually deteriorated as the number of cells and training example was artificially reduced. However, the CNN exhibited good robustness, as its accuracy markedly deteriorated only in an extreme low-data regime, suggesting its usefulness also when less cells are recorded or few training examples are available. This result support the potentialities of CNNs as decoders in BCIs, as while transposing decoders from the monkeys to humans, performing a BCI calibration as short as possible and designing a decoder robust to signal degradation over time (thus, reducing the number of recorded cells) are desired aspects.

In conclusion, this study further validates the design and application of CNNs for motor decoding from neurons' recordings. A single-layer wide CNN resulted the optimal design for reach-to-grasp decoding, matching the findings of [2] where a similar structure resulted optimal for reach decoding, and was able to accurately decode grip type early from the start of the task, i.e., from the beginning of movement preparation. However, the present study is affected by the following limitations that will be addressed in the future. First, a more complete performance comparison with also other deep neural networks is needed. Second, selecting a model design by taking the most frequent optimal value for each hyper-parameter may have neglected potential correlations between hyper-parameters. Lastly, the comparison with traditional decoders (e.g., NB) should be further validated as a function of the selection of modulated neurons performed before decoding (as adopted in [3]). Despite the previous limitations, the results obtained in this study, together with the ones obtained in [2], may have some relevant perspectives. Indeed, CNNs compared to other deep learning approaches are more easily interpretable in their learned features. Thus, future studies, by using explanation techniques (e.g., saliency maps [26]) to interpret the features, can exploit CNNs to analyze neural signatures in spatial, temporal and frequency domains [6, 8], identifying differences in neural motor encoding not only across time samples but also, for example, across subpopulations of neurons (at different spatial locations) within area V6A. Finally, the proposed CNN resulted an accurate and light non-linear decoder that, in prospective, may find applicability in BCIs.

Fundings. This study was supported by PRIN 2017 – Prot. 2017KZNZLN and MAIA project. MAIA project has received funding from the European Union's Horizon 2020 research and innovation program under grant agreement No 951910. This article reflects only the author's view, and the Agency is not responsible for any use that may be made of the information it contains.

References

1. Wolpaw, J., Wolpaw, E.W.: Brain-Computer Interfaces: Principles and Practice. Oxford University Press, USA (2012)
2. Filippini, M., Borra, D., Ursino, M., Magosso, E., Fattori, P.: Decoding sensorimotor information from superior parietal lobule of macaque via Convolutional Neural Networks. Neural Netw. **151**, 276–294 (2022). https://doi.org/10.1016/j.neunet.2022.03.044

3. Filippini, M., Breveglieri, R., Akhras, M.A., Bosco, A., Chinellato, E., Fattori, P.: Decoding information for grasping from the macaque dorsomedial visual stream. J. Neurosci. **37**, 4311–4322 (2017). https://doi.org/10.1523/JNEUROSCI.3077-16.2017
4. Filippini, M., Breveglieri, R., Hadjidimitrakis, K., Bosco, A., Fattori, P.: Prediction of reach goals in depth and direction from the parietal cortex. Cell Rep. **23**, 725–732 (2018). https://doi.org/10.1016/j.celrep.2018.03.090
5. Solon, A.J., Lawhern, V.J., Touryan, J., McDaniel, J.R., Ries, A.J., Gordon, S.M.: Decoding P300 variability using convolutional neural networks. Front. Hum. Neurosci. **13**, 201 (2019). https://doi.org/10.3389/fnhum.2019.00201
6. Borra, D., Fantozzi, S., Magosso, E.: Interpretable and lightweight convolutional neural network for EEG decoding: application to movement execution and imagination. Neural Netw. **129**, 55–74 (2020). https://doi.org/10.1016/j.neunet.2020.05.032
7. Borra, D., Fantozzi, S., Magosso, E.: A lightweight multi-scale convolutional neural network for p300 decoding: analysis of training strategies and uncovering of network decision. Front. Hum. Neurosci. **15**, 655840 (2021). https://doi.org/10.3389/fnhum.2021.655840
8. Borra, D., Magosso, E.: Deep learning-based EEG analysis: investigating P3 ERP components. J. Integr. Neurosci. **20**, 791–811 (2021). https://doi.org/10.31083/j.jin2004083
9. Livezey, J.A., Glaser, J.I.: Deep learning approaches for neural decoding across architectures and recording modalities. Brief. Bioinform. **22**, 1577–1591 (2021). https://doi.org/10.1093/bib/bbaa355
10. Craik, A., He, Y., Contreras-Vidal, J.L.: Deep learning for electroencephalogram (EEG) classification tasks: a review. J. Neural Eng. **16**, 031001 (2019). https://doi.org/10.1088/1741-2552/ab0ab5
11. Suhaimi, N.S., Mountstephens, J., Teo, J.: EEG-based emotion recognition: a state-of-the-art review of current trends and opportunities. Comput. Intell. Neurosci. **2020**, 1–19 (2020). https://doi.org/10.1155/2020/8875426
12. Simões, M., et al.: BCIAUT-P300: a multi-session and multi-subject benchmark dataset on autism for p300-based brain-computer-interfaces. Front. Neurosci. **14**, 568104 (2020). https://doi.org/10.3389/fnins.2020.568104
13. Borra, D., Magosso, E., Castelo-Branco, M., Simoes, M.: A Bayesian-optimized design for an interpretable convolutional neural network to decode and analyze the P300 response in autism. J. Neural Eng. **19** (2022). https://doi.org/10.1088/1741-2552/ac7908
14. Schirrmeister, R.T., et al.: Deep learning with convolutional neural networks for EEG decoding and visualization. Hum. Brain Mapp. **38**, 5391–5420 (2017)
15. Mulliken, G.H., Musallam, S., Andersen, R.A.: Decoding trajectories from posterior parietal cortex ensembles. J. Neurosci. **28**, 12913–12926 (2008). https://doi.org/10.1523/JNEUROSCI.1463-08.2008
16. Aflalo, T., et al.: Decoding motor imagery from the posterior parietal cortex of a tetraplegic human. Science **348**, 906–910 (2015). https://doi.org/10.1126/science.aaa5417
17. Chinellato, E., Grzyb, B.J., Marzocchi, N., Bosco, A., Fattori, P., del Pobil, A.P.: The Dorso-medial visual stream: from neural activation to sensorimotor interaction. Neurocomputing **74**, 1203–1212 (2011). https://doi.org/10.1016/j.neucom.2010.07.029
18. Fattori, P., Breveglieri, R., Bosco, A., Gamberini, M., Galletti, C.: Vision for prehension in the medial parietal cortex. Cereb. Cortex. bhv302 (2015). https://doi.org/10.1093/cercor/bhv302
19. Ioffe, S., Szegedy, C.: Batch normalization: accelerating deep network training by reducing internal covariate shift. In: Bach, F. and Blei, D. (eds.) Proceedings of the 32nd International Conference on Machine Learning. pp. 448–456. PMLR, Lille (2015)
20. Clevert, D.-A., Unterthiner, T., Hochreiter, S.: Fast and accurate deep network learning by exponential linear units (ELUs). arXiv preprint (2015)
21. Srivastava, N., Hinton, G., Krizhevsky, A., Sutskever, I., Salakhutdinov, R.: Dropout: a simple way to prevent neural networks from overfitting. J. Mach. Learn. Res. **15**, 1929–1958 (2014)

22. Frazier, P.I.: A tutorial on Bayesian optimization (2018). http://arxiv.org/abs/1807.02811
23. Kingma, D.P., Ba, J.: Adam: a method for stochastic optimization. arXiv:1412.6980 [cs] (2017)
24. Smith, S., Nichols, T.: Threshold-free cluster enhancement: Addressing problems of smoothing, threshold dependence and localisation in cluster inference. Neuroimage **44**, 83–98 (2009). https://doi.org/10.1016/j.neuroimage.2008.03.061
25. Nowak, M., Zich, C., Stagg, C.J.: Motor cortical gamma oscillations: what have we learnt and where are we headed? Curr. Behav. Neurosci. Rep. **5**(2), 136–142 (2018). https://doi.org/10.1007/s40473-018-0151-z
26. Simonyan, K., Vedaldi, A., Zisserman, A.: Deep inside convolutional networks: visualising image classification models and saliency maps. arXiv:1312.6034 [cs] (2014)

Brain-like Combination of Feedforward and Recurrent Network Components Achieves Prototype Extraction and Robust Pattern Recognition

Naresh Balaji Ravichandran[1(✉)], Anders Lansner[1,2], and Pawel Herman[1,3]

[1] Division of Computational Science and Technology, School of Electrical Engineering and Computer Science, KTH Royal Institute of Technology, Stockholm, Sweden
{nbrav,ala,paherman}@kth.se

[2] Department of Mathematics, Stockholm University, Stockholm, Sweden

[3] Digital Futures, KTH Royal Institute of Technology, Stockholm, Sweden

Abstract. Associative memory has been a prominent candidate for the computation performed by the massively recurrent neocortical networks. Attractor networks implementing associative memory have offered mechanistic explanation for many cognitive phenomena. However, attractor memory models are typically trained using orthogonal or random patterns to avoid interference between memories, which makes them unfeasible for naturally occurring complex correlated stimuli like images. We approach this problem by combining a recurrent attractor network with a feedforward network that learns distributed representations using an unsupervised Hebbian-Bayesian learning rule. The resulting network model incorporates many known biological properties: unsupervised learning, Hebbian plasticity, sparse distributed activations, sparse connectivity, columnar and laminar cortical architecture, etc. We evaluate the synergistic effects of the feedforward and recurrent network components in complex pattern recognition tasks on the MNIST handwritten digits dataset. We demonstrate that the recurrent attractor component implements associative memory when trained on the feedforward-driven internal (hidden) representations. The associative memory is also shown to perform prototype extraction from the training data and make the representations robust to severely distorted input. We argue that several aspects of the proposed integration of feedforward and recurrent computations are particularly attractive from a machine learning perspective.

Keywords: Attractor · Associative memory · Unsupervised learning · Hebbian learning · Recurrent networks · Feedforward networks · Brain-like computing

Funding for the work is received from the Swedish e-Science Research Centre (SeRC), European Commission H2020 program. The authors gratefully acknowledge the HPC RIVR consortium (www.hpc-rivr.si) and EuroHPC JU (eurohpcju.europa.eu) for funding this research by providing computing resources of the HPC system Vega at the Institute of Information Science (www.izum.si).

G. Nicosia et al. (Eds.): LOD 2022, LNCS 13811, pp. 488–501, 2023.
https://doi.org/10.1007/978-3-031-25891-6_37

1 Introduction

Recurrent networks are ubiquitous in the brain and constitute a particularly prominent feature of the neocortex [1–3]. Yet, it is not clear what function such recurrence lends to cortical information processing. One popular hypothesis is that recurrent cortical networks perform associative memory, where assemblies of coactive neurons (cell assembly) act as internal mental representation of memory objects [4–6]. Several theoretical and computational studies have shown that recurrently connected neuron-like binary units with symmetric connectivity can implement associative attractor memory: the network is guaranteed to converge to attractor states corresponding to local energy minima [4]. Learning memories in such networks typically follows Hebbian synaptic plasticity, i.e., the synaptic connections between units are strengthened when they are coactive (and weakened otherwise). Attractor networks provide a rich source of network dynamics and have produced several important network models of the neocortex that explain complex cognitive functions [7–9].

Typically, attractor networks are trained with artificially generated orthogonal or random patterns, since overlapping (non-orthogonal) inputs cause interference between memories, so-called crosstalk, and lead to the emergence of spurious memories [10]. Consequently, attractor networks have not really been combined with high-dimensional correlated input and the problem of extracting suitable representations from real-world data has not received much attention in the context of associative memory. It is hypothesized that desirable neural representations in the brain are extracted from sensory input by feedforward cortical pathways. This biological inspiration has been loosely adopted in deep neural networks (DNN) developed for pattern recognition on complex datasets, e.g., natural images, videos, audio, natural languages. The main focus in the DNN community is on learning mechanisms to obtain high-dimensional distributed representations by extracting the underlying factors from the training data [11]. The success of DNNs has been predominantly attributed to the efficient use of the backprop algorithm that adjusts weights in the network to minimize a global error in supervised learning scenarios. Since it is hard to justify the biological plausibility of backprop, there has been growing interest in recent years on more brain plausible local plasticity rules for learning representations [12–17].

The work we present here demonstrates a step towards a brain-like integration of recurrent attractors networks with feedforward networks implementing representation learning. In our model, the recurrent attractors are trained on feedforward-driven representations and the network exhibits multiple associative memory capabilities such as prototype extraction, pattern completion and robustness to pattern distortions. The model incorporates several core biological details that are critical for computational functionality [18]: unsupervised learning, Hebbian plasticity, sparse connectivity, sparse distributed neural activations, columnar and laminar cortical organization. We show results from evaluating our new model on the MNIST handwritten digit dataset, a popular machine learning benchmark.

2 Model Description

2.1 Model Overview

Our network consists of an input layer (representing the data) and a hidden layer (Fig. 1A). The feedforward network component connects the input layer to the hidden layer and the recurrent component connects the hidden layer to itself. The organization of within the hidden layer can be described in terms of vertical and horizontal organization as discussed below.

The vertical organization of the hidden layer comprises minicolumns and hypercolumns, which derives from the discrete columnar organization of the neocortex of large mammalian brains [19, 20]. The cortical minicolumn (shown as cylinders in Fig. 1A) comprises around 80–100 tightly interconnected neurons having functionally similar response properties [19–21]. The minicolumns are arranged in larger hypercolumn modules (shown as squares enclosing the cylinders in Fig. 1B), defined by local competition between minicolumns through shared lateral inhibition [20, 22].

The hidden layer is also organized horizontally in term of laminae as derived from the laminar structures found through the depths of the cortical sheet. The basis for such laminar organization in the cortex is mainly neuroanatomical, differing in their cell types, cell densities, afferent, and efferent connectivity, etc. In our network, we model units in layer 4 (L4; granular layer) and layers 2/3 (L2/3; supragranular) (we do not model L1, L5 and L6). The L4 units receive the feedforward connections from the input layer (sparse connectivity). The L2/3 units receive recurrent connections from other L2/3 units in the hidden layer (full connectivity) and the L4 units within the same minicolumn. The extensive recurrent connectivity within the L2/3 units in the hidden layer implements associative memory function [5, 23].

The computations performed by the units are based on the Bayesian Confidence Propagation Neural Network model (BCPNN) [24–27]. BCPNN converts probabilistic inference (naïve Bayes) into neural and synaptic computation [24, 25, 27] and the BCPNN computations are applicable to both the recurrent and feedforward components.

2.2 Activation Rule

When source units (indexed by i) send connections to a target unit (index by j), the activity propagation rule is:

$$s_j = b_j + \sum\nolimits_i \pi_i w_{ij}, \tag{1}$$

$$\pi_j = \frac{e^{s_j}}{\sum_k e^{s_k}}, \tag{2}$$

where s_j is the total input received by the j-th target unit, b_j and w_{ij}, are the bias and weight parameters, respectively (Fig. 1B). The activation π_j is calculated by a softmax nonlinear activation function that implements a soft-winner-takes-all competition between the units within each hypercolumn from the same lamina [22].

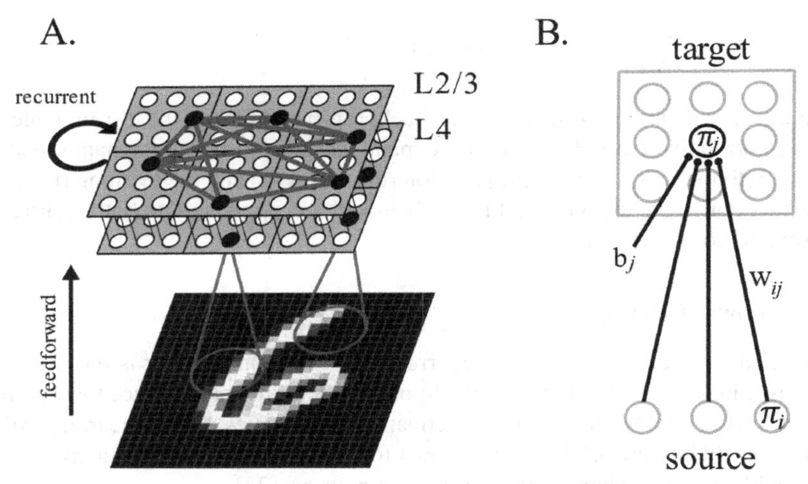

Fig. 1. Network schematic. A. The feedforward component connects the input layer to the hidden layer and the recurrent component connects the hidden layer to itself. The hidden layer is organized vertically in terms of minicolumns and hypercolumns, shown as cylinders and gray squares enclosing the cylinders respectively, and horizontally in terms of laminar layers L4 and L2/3, shown as gray sheet at the bottom and on top respectively. L4 units receive feedforward connections from the input layer and the L2/3 units receive recurrent connections from other L2/3 units (and from the L4 unit within the same minicolumn). In the schematic, we show $H_h = 6$ hypercolumns, each with $M_h = 9$ minicolumns. The activities of the units are indicated by black (1) and white (0) circles, and the red connection link indicate the cell assembly dynamically formed within the recurrent network component. **B.** In the BCPNN model, a target unit's activation is driven by source activations and determined by local competition within the target hypercolumn (gray square box). The BCPNN computations are universal and apply to both the feedforward and recurrent components.

2.3 Learning Rule

The synaptic plasticity in our model is based on the incremental form of Hebbian-Bayesian learning [25, 27], where local traces of co-activation of pre- and post- synaptic units are used to compute Bayesian weights. The learning step involves incrementally updating three p-traces: probability of pre-synaptic activity, p_i, probability of post-synaptic activity, p_j, and joint probability of pre-synaptic and post-synaptic activities, p_{ij}, as follows:

$$p_i := (1 - \alpha)p_i + \alpha\pi_i, \tag{3}$$

$$p_j := (1 - \alpha)p_j + \alpha\pi_j, \tag{4}$$

$$p_{ij} := (1 - \alpha)p_{ij} + \alpha\pi_i\pi_j, \tag{5}$$

where α is the learning rate. The bias and weight parameters are computed from the p-traces as follows:

$$b_j = \log p_j, \tag{6}$$

$$w_{ij} = \log \frac{p_{ij}}{p_i p_j}. \quad (7)$$

As a crucial departure from traditional backprop-based DNNs, the learning rule above is local, correlative, and Hebbian, i.e., dependent only on the pre-synaptic and post-synaptic activities. The activity propagation rule (Eq. 1–2) and the learning rule (Eq. 3–7) together constitute the universal BCPNN model, i.e. applicable to both recurrent and feedforward components [24–27].

2.4 Network Model

The BCPNN model organized as a recurrent network (source layer is the same as the target layer in Eq. 1–7) implements fixed-point attractor dynamics since the weights are symmetric (Eq. 7). Due to the sparse activations and Bayes optimal learning rule, the recurrent-BCPNN network has been shown to have higher storage capacity compared to Hopfield networks when trained on random patterns [28].

Recent work has demonstrated that a feedforward-BCPNN model (source layer encodes data and the target layer is the hidden layer in Eq. 1–7) can perform unsupervised representation learning [16, 17]. This is possible when the model is augmented with structural plasticity mechanisms that learns a sparse set of connections from the input layer to hidden layer. The representations learnt by the model were evaluated on linear classification on various machine learning benchmark datasets and the feedforward BCPNN network was found to favorably compare with many unsupervised brain-like neural network models as well as with multi-layer perceptrons [16, 17].

Table 1. List of simulation parameters.

Symbol	Value	Description
H_i	784	Num. of hypercolumns in input layer
M_i	2	Num. of minicolumns per hypercolumn in input layer
H_h	100	Num. of hypercolumns in hidden layer
M_h	100	Num. of minicolumns per hypercolumn in hidden layer
p_{conn}	10%	Connectivity from input layer to hidden layer
α	0.0001	Synaptic learning time constant
T	20	Number of timesteps for iterating recurrent attractor

We incorporated both recurrent attractor and feedforward components in our network model (Fig. 1A). For this, we modelled a L4 + L2/3 hidden layer: L4 units receive feedforward connections from the input layer, and L2/3 units receive input from other L2/3 units (and from the L4 unit within the same minicolumn). We employed the feedforward BCPNN augmented with structural plasticity and used a hidden layer with 100 hypercolumns ($H_h = 100$), each with 100 minicolumns ($M_h = 100$). The recurrent component in the hidden layer was implemented with full connectivity. The network was trained by first learning the feedforward BCPNN component [17], then the feedforward-driven hidden activations were used to train the recurrent weights with the same Hebbian-Bayesian learning. We trained on the 60000 samples of MNIST hand-written digits dataset. For evaluation purposes we used the 10000 samples from the test set and the recurrent attractor was run for $T = 20$ timesteps. Table 1 summarizes all the parameters used in the simulations.

3 Results

3.1 Feedforward-Driven Representations Are Suitable for Associative Memory

We trained the network on the MNIST dataset and simulated the attractors over the test dataset. To visualize the attractor representation, we back projected (top down) the hidden representations to the input layer using the same weights as feedforward connectivity. For comparison, we also trained a recurrent attractor on the input data. We observed that the recurrent attractors trained on the feedforward-driven representations converged in a few steps, and their reconstruction correspond to prototypical digit images (Fig. 2B). This was in contrast with the severe crosstalk reflected in the behavior of the recurrent attractor network trained directly on the input data (Fig. 2A). The results suggest that the recurrent attractor network operating on the hidden representations enables better associative memory. This is likely because the feedforward-driven hidden activations are orthogonalized, i.e., the representations from the same class are similar and overlapping, while representations from different classes are more distinct. In that case, the activation patterns used for recurrent network learning tend towards "orthogonal" patterns, an ideal condition for associative memory storage to avoid interference between attractor states.

To test this more systematically, we computed the pair-wise similarity matrix on *i*) the input data (Fig. 2C), *ii*) activations after running the recurrent attractor on the input data (Fig. 2D), *iii*) feedforward-driven hidden representations (Fig. 2E), and iv) recurrent attractor hidden representations (Fig. 2F). We used the cosine similarity score as the metric and sorted the rows and columns by the sample class (label). Ideally, the similarity for samples from the same class should be close to one (high overlap between representations), and close to zero for samples from different classes (minimal overlap between representations). We quantified the overall degree of orthogonality by calculating a ratio of mean similarity of samples within the same class and mean similarity across all samples (higher ratio implies more orthogonal representations). The similarity matrix for the input data showed class ambiguity (the orthogonality ratio was 1.37; Fig. 2C) as samples were similar to other samples within and outside the class, and the cosine score was roughly around 0.5 for most sample pairs indicating considerable overlap. The similarity matrix of attractors when run on the input data was largely disordered (the orthogonality ratio was 1.47; Fig. 2D), indicating heavy crosstalk from learning the input data. The similarity matrix for the feedforward-driven hidden representations were markedly different (the orthogonality ratio was 3.79; Fig. 2E), with most of the cosine score close to zero (because the activations are sparse), and the few non-zero scores occurred within the same class (small square patches along the diagonal in Fig. 2E). The feedforward network can be said to untangle the representations, i.e., those from the same class are made similar and overlapping, while representations from different classes are pushed far-apart. When the recurrent attractor layer was trained on these feedforward representations, the attractors learnt were even more orthogonal (the orthogonality ratio was 7.54; Fig. 2F) since many samples from the same class settled on the same prototype attractor. Hence, the feedforward-driven hidden activations of the BCPNN model are sparse, distributed, high-dimensional and orthogonal – in other words, having all the attributes ideal for associative memory.

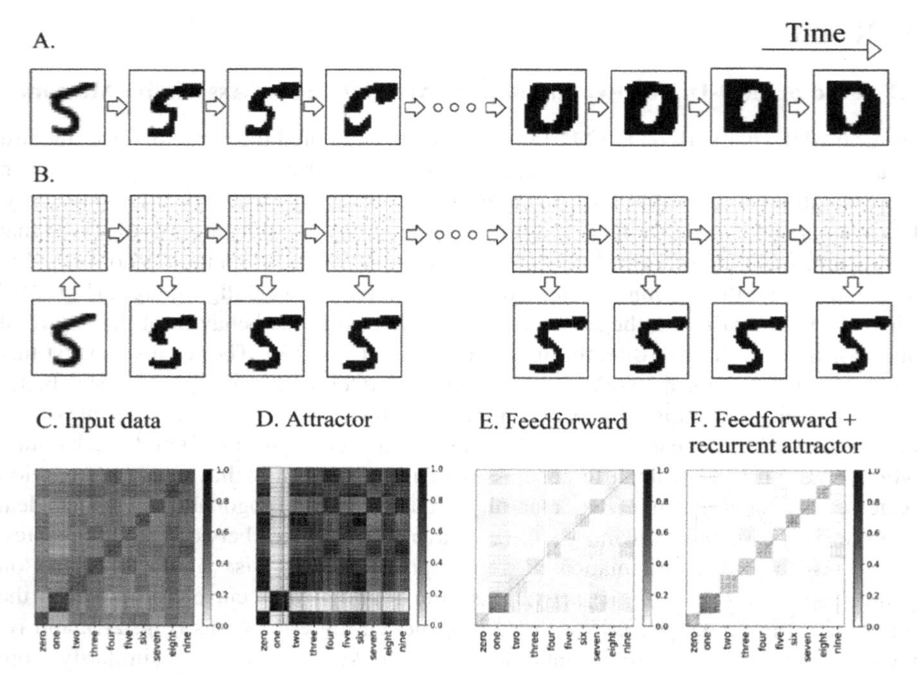

Fig. 2. Feedforward-driven representations are suitable for associative memory. A. Recurrent attractor activations of the network trained on the input data and tested on the image of digit "5" show convergence to a spurious pattern because of severe crosstalk between stored memories. **B.** Recurrent attractors trained on the feedforward-driven representations (boxes in the top row with dots representing active minicolumns) and tested on the same digit "5" converged in a few steps to one of the attractor prototypes, as reflected by the input reconstruction (boxes in the bottom row). **C, D, E, F.** Pair-wise similarity matrix for the input data (C), attractor activations when trained on the input data (D), feedforward-driven representations (E), attractor activations when trained on the feedforward-driven representations (F).

3.2 Associative Memory Extracts Prototypes from Training Data

Prototype extraction facilitates identification of representative patterns for subsets of data samples. Here we examine if the attractors learnt by the recurrent network component can serve as prototypes even if the network is not explicitly trained on any prototypical data. We expect that the attractor prototypes account for semantically similar data, for example, data samples sharing many common features or images from the same class.

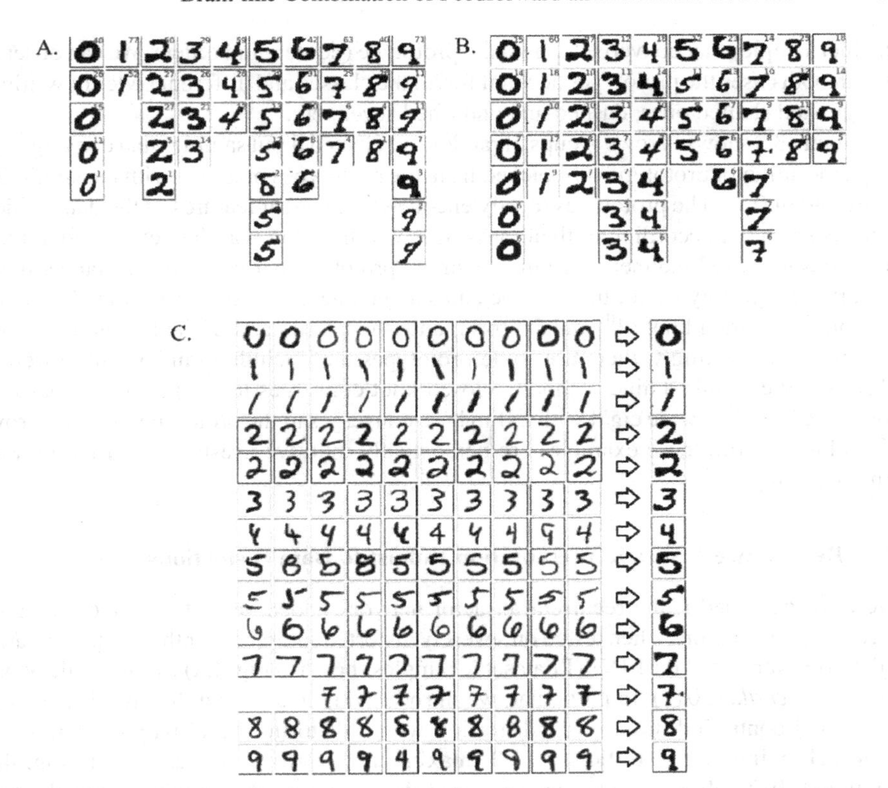

Fig. 3. Prototypes extraction. A, B: Top-down reconstructions of the unique prototypes extracted by the recurrent attractor network for 1000 test samples. Prototype patterns were considered unique if the cosine similarity was more than a threshold value set to 0.5 (A) or 0.9 (B). The red number shown in the top right corner of each image indicates the number of input samples (out of 1000) that converged to the prototype. **C.** Randomly selected input samples that converged to the same prototype attractor (input reconstruction in the last column).

To test this, following the training process described earlier we ran our full network model (feedforward propagation followed by recurrent attractor for T steps) with 1000 samples randomly selected from the unseen test subset of the MNIST dataset to obtain the final attractor pattern for each sample. We examined these attractor patterns to assess if they represented unique prototypes. Since we used real-valued activation patterns (from Eq. 2), we could not directly check if patterns are identical to one another. Instead, we computed the cosine similarity between attractor patterns (same as Fig. 2D) and used a threshold to determine if they were similar enough to be considered the same prototype. As a result, the number of "unique" prototypes depends on the threshold. For illustrative purposes we used two different threshold values and computed the prototypes as well as their corresponding reconstructions. For the threshold of 0.5 we found 46 prototypes (shown in Fig. 3A) and for 0.9−248 prototypes (prototypes with more than 5 input samples are shown in Fig. 3B). In Figs. 3A-B the prototype reconstructions are manually grouped into columns by their class and sorted in rows by the number of distinct samples that converge to the same prototype (descending order). We can observe that in both cases the prototypes account for all the classes and different styles of writing: upright and slanted ones, dashed and undashed seven, etc.

Figure 3C shows in rows groups of randomly selected input samples that converged to the same attractor prototypes (14 picked from the prototypes obtained with the similarity threshold of 0.9). The prototypes mostly encode some salient features of the data, which reflects grouping according to their class. It is worth noting that the network identified in an unsupervised manner more than a single prototype per class. For instance, there are different prototypes for upright one and slanted one (2nd and 3rd row, Fig. 3C), twos without and with a knot (4th and 5th row), undashed seven and dashed seven (11th and 12th row). Interestingly, there are some prototypes that include samples with different classes, for example a nine without a closed knot converges to the prototype four (7th row, 9th column), a wide eight with a slim base converges to the prototype nine (15th row, 8th column). Still, these examples are perceptually intuitive irrespective of the ground truth labelling.

3.3 Recurrence Makes Representation Robust to Data Distortions

We next examined if the recurrent attractor network added value to the feedforward-driven representations when tested on severely distorted samples. For this we first created a distorted version of the MNIST dataset (examples shown in Fig. 4A) following the work of George *et al.*, (2017). In particular, we introduced nine different distortions (Fig. 4A rows) and controlled the level of distortion with a distortion level parameter ranging from 0.1 (minimum distortion) to 1.0 (maximum distortion) in step of 0.1 (Fig. 4A columns). In total, we created the distorted dataset with 900 samples (from the test set). Then we ran the network, trained beforehand on the original undistorted MNIST training dataset as before, with these distorted samples as inputs to comparatively study the recurrent network attractor activations and their corresponding input reconstructions. Except for high distortion levels (0.5 or higher), the attractor network was very robust to the distortions, and the reconstructed images showed that most of the distortions were removed upon attractor convergence. Figure 4B shows examples of two digits under all distortion types (distortion level of 0.3) and the reconstructions of the corresponding attractor activations.

To quantify the network's robustness to input distortions in the pattern recognition context, we used a 10-class linear classifier (one-vs-rest), which was trained on the feedforward-driven activations (on the original undistorted MNIST training set). We then compared the classification accuracy from five random trials obtained for recurrent attractor representations and the feedforward-driven hidden representations on the distorted MNIST dataset (Fig. 4C and 4D). We found that the recurrent attractor representations performed better compared to feedforward-driven representations on all distortion levels greater than 0.1 (Fig. 4C). We also examined the performance across the different distortion types and the recurrent attractor representations turned out more resilient in most cases (Fig. 4D). Further investigation is needed to understand why feedforward-driven representations scored higher on three distortion types: "Grid", "Deletion", and "Open". Both feedforward-driven and recurrent attractor representations result from unsupervised learning, so the distortion resistance provided by the recurrent component is achieved without any access to data labels.

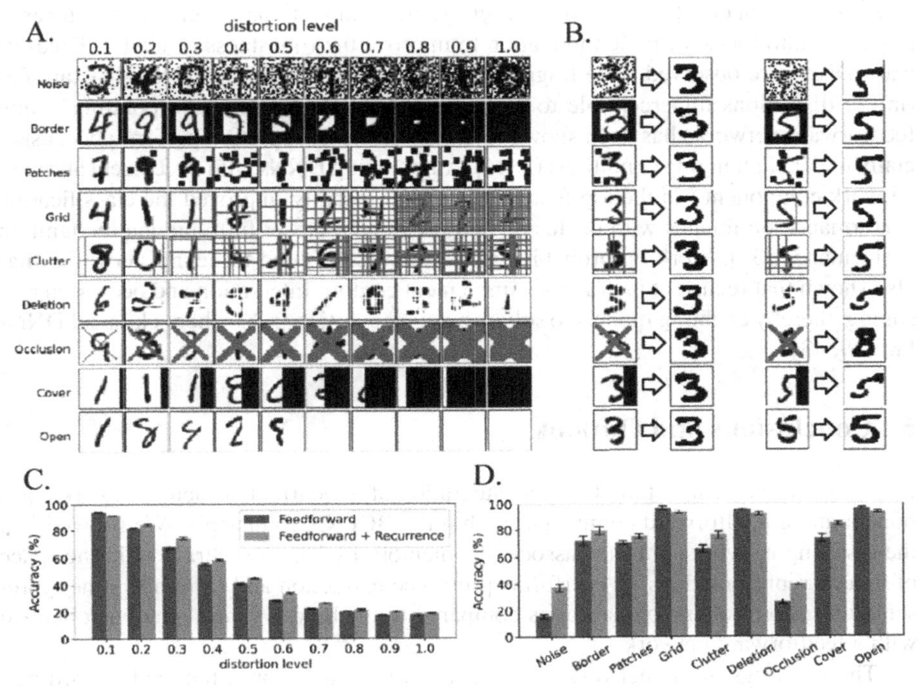

Fig. 4. Robustness to distortion A. Samples from the distorted MNIST dataset, with 9 distortion types (rows) and varying distortion levels (columns). **B.** Examples of digits "3" (left) and "5" (right) under all distortion types (distortion level = 0.3), and the input reconstruction after convergence of the recurrent layer. For most cases, the attractor reaches the prototype pattern while removing all distortions. **C.** Classification performance for different distortion levels comparing feedforward-driven and recurrent attractor representations. **D.** Classification performance for all distortion types with distortion level at 0.3.

4 Related Work

Feedforward networks are currently the dominant approach for building perceptual systems [11]. Although feedforward networks make use of some salient aspects of the brain, such as the hierarchical processing stages found in the cortical sensory pathway [30, 31], there are reasons to believe this is not sufficient. Given the abundance of recurrent structures in the brain, their function in perception and biological vision has been studied extensively over the years. There is considerable evidence that recurrence is responsible for many important visual computations such as pattern completion [32], figure-ground segmentation [33], occlusion reasoning [34, 35] and contour integration [36, 37]. Specifically, object recognition tasks under challenging conditions (e.g. occlusions, distortions) are thought to engage recurrent processes [38].

Early computational models incorporated recurrent interactions between units in the hidden layer as a mechanism to encourage competition. Such competition creates mutually inhibiting neurons, and allows feedforward learning of localist representations that clusters the input data [39–41]. More recently, the computational potential of recurrence has come into focus with the motivation to improve the robustness of DNNs. Feedforward DNNs are observed to be fragile to distortions as well as "adversarial" examples (minor distortions imperceptible to humans) [29, 42, 43]. Thus, incorporating it into feedforward networks has been shown to enhance their recognition performance, especially under challenging conditions [2, 44]. Augmenting DNNs with a recurrent attractor network component (on the top feature layer) significantly improved the classification performance on images with occlusion and provided a closer fit to human recognition performance [32]. Work on probabilistic models (as opposed to neural networks) has also shown that recurrent interactions implement contour integration and occlusion reasoning, thereby enabling models to solve object recognition tasks where classical DNNs fail [29, 43].

5 Conclusions and Outlook

In this work, we demonstrated a tight integration of a recurrent attractor network component and a feedforward component within the BCPNN framework. We showed that the resulting network (1) forms associative memory by learning attractors in the space of hidden representations, (2) performs prototype extraction and (3) renders the representations more robust to distortions compared to the hidden representations obtained with a feedforward network.

The workings of the network can be summarized as follows: the feedforward network component learns in an unsupervised manner a set of salient features underlying the data and forms sparse distributed representations, while the recurrent attractor network component acts on the feedforward-driven hidden representation and performs pattern reconstruction by integrating information across the distributed representation and forming a coherent whole in the spirit of Gestalt perception. This design principle comes directly from the circuitry found in the neocortex: within each cortical minicolumn feedforward connections arrive at excitatory neurons located in the granular layer 4, L4 neurons project to neurons in the superficial layer 2/3, L2/3 neurons make extensive recurrent connectivity with L2/3 neurons in other minicolumns before projecting to L4 neurons of higher cortical areas. Such circuit motifs are considered canonical in the neocortex [19, 20] and offer powerful computational primitives that, we believe, need to be understood for building intelligent computer perception systems as well as for exploring the perceptual and cognitive functions of the cortex.

Prior work has shown that DNNs perform poorly on distorted images (such as those used in this work) despite having high accuracy on clean images [29]. As the next step, we plan to extensively compare our model with other state-of-the-art methods [29, 32] in such challenging pattern recognition scenarios. Other outstanding problems include among others a seamless functional integration of top-down back-projections in the spirit of predictive coding (which we utilized in this work only for visualization purposes) [45, 46]. Future work will also focus on how multilayer deep network formed by stacking such L4-L2/3 modules can be built and trained most efficiently. Extending from the static data domain, the synaptic traces at the core of the BCPNN model can also be configured to perform temporal association and sequence learning and generation [47, 48].

The research direction we pursued here integrates various architectural details from neurobiology, especially from the mammalian neocortex, into an abstract cortical model, which is then operationalized as a system of functionally powerful unsupervised learning algorithms. The resulting information processing machinery holds promise for creating novel intelligent brain-like computing systems with a broad spectrum of technical applications.

References

1. Douglas, R.J., Martin, K.A.C.: Recurrent neuronal circuits in the neocortex. Curr. Biol. **17**, R496–R500 (2007). https://doi.org/10.1016/J.CUB.2007.04.024
2. van Bergen, R.S., Kriegeskorte, N.: Going in circles is the way forward: the role of recurrence in visual inference. Curr. Opin. Neurobiol. **65**, 176–193 (2020). https://doi.org/10.1016/J.CONB.2020.11.009
3. Stepanyants, A., Martinez, L.M., Ferecskó, A.S., Kisvárday, Z.F.: The fractions of short- and long-range connections in the visual cortex. Proc. Natl. Acad. Sci. U. S. A. **106**, 3555–3560 (2009). https://doi.org/10.1073/pnas.0810390106
4. Hopfield, J.J.: Neural networks and physical systems with emergent collective computational abilities (associative memory/parallel processing/categorization/content-addressable memory/fail-soft devices). Proc. Natl. Acad. Sci. U. S. A. **79**, 2554–2558 (1982)
5. Lansner, A.: Associative memory models: from the cell-assembly theory to biophysically detailed cortex simulations. Trends Neurosci. **32**, 178–186 (2009). https://doi.org/10.1016/j.tins.2008.12.002
6. Hebb, D.O.: The Organization of Behavior. Psychology Press (1949). https://doi.org/10.4324/9781410612403
7. Lundqvist, M., Herman, P., Lansner, A.: Theta and gamma power increases and alpha/beta power decreases with memory load in an attractor network model. J. Cogn. Neurosci. **23**, 3008–3020 (2011). https://doi.org/10.1162/jocn_a_00029
8. Silverstein, D.N., Lansner, A.: Is attentional blink a byproduct of neocortical attractors? Front. Comput. Neurosci. **5**, 13 (2011). https://doi.org/10.3389/FNCOM.2011.00013/BIBTEX
9. Fiebig, F., Lansner, A.: A spiking working memory model based on Hebbian short-term potentiation. J. Neurosci. **37**, 83–96 (2017). https://doi.org/10.1523/JNEUROSCI.1989-16.2016
10. MacGregor, R.J., Gerstein, G.L.: Cross-talk theory of memory capacity in neural networks. Biol. Cybern. **65**, 351–355 (1991). https://doi.org/10.1007/BF00216968
11. LeCun, Y., Bengio, Y., Hinton, G.: Deep learning. Nature **521**(7553), 436–444 (2015). https://doi.org/10.1038/nature14539
12. Mattar, M.G., Daw, N.D.: Prioritized memory access explains planning and hippocampal replay. Nat. Neurosci. **21**, 1609–1617 (2018). https://doi.org/10.1038/S41593-018-0232-Z
13. Krotov, D., Hopfield, J.J.: Unsupervised learning by competing hidden units. Proc. Natl. Acad. Sci. U. S. A. **116**, 7723–7731 (2019). https://doi.org/10.1073/pnas.1820458116
14. Bartunov, S., Santoro, A., Hinton, G.E., Richards, B.A., Marris, L., Lillicrap, T.P.: Assessing the scalability of biologically-motivated deep learning algorithms and architectures. In: Advances in Neural Information Processing Systems, pp. 9368–9378 (2018)

15. Illing, B., Gerstner, W., Brea, J.: Biologically plausible deep learning—but how far can we go with shallow networks? Neural Netw. **118**, 90–101 (2019). https://doi.org/10.1016/j.neunet.2019.06.001
16. Ravichandran, N.B., Lansner, A., Herman, P.: Learning representations in Bayesian confidence propagation neural networks. In: Proceedings of the International Joint Conference on Neural Networks. (2020). https://doi.org/10.1109/IJCNN48605.2020.9207061
17. Ravichandran, N.B., Lansner, A., Herman, P.: Brain-like approaches to unsupervised learning of hidden representations - a comparative study. In: Farkaš, I., Masulli, P., Otte, S., Wermter, S. (eds.) ICANN 2021. LNCS, vol. 12895, pp. 162–173. Springer, Cham (2021). https://doi.org/10.1007/978-3-030-86383-8_13
18. Pulvermüller, F., Tomasello, R., Henningsen-Schomers, M.R., Wennekers, T.: Biological constraints on neural network models of cognitive function. Nat. Rev. Neurosci. **228**(22), 488–502 (2021). https://doi.org/10.1038/s41583-021-00473-5
19. Mountcastle, V.B.: The columnar organization of the neocortex (1997). https://academic.oup.com/brain/article/120/4/701/372118. https://doi.org/10.1093/brain/120.4.701
20. Douglas, R.J., Martin, K.A.C.: Neuronal circuits of the neocortex (2004). www.annualreviews.org. https://doi.org/10.1146/annurev.neuro.27.070203.144152
21. Buxhoeveden, D.P., Casanova, M.F.: The minicolumn hypothesis in neuroscience. Brain **125**, 935–951 (2002). https://doi.org/10.1093/BRAIN/AWF110
22. Carandini, M., Heeger, D.J.: Normalization as a canonical neural computation. Nat. Rev. Neurosci. **131**(13), 51–62 (2011). https://doi.org/10.1038/nrn3136
23. Fransen, E., Lansner, A.: A model of cortical associative memory based on a horizontal network of connected columns. Netw. Comput. Neural Syst. **9**, 235–264 (1998). https://doi.org/10.1088/0954-898X_9_2_006
24. Lansner, A., Ekeberg, Ö.: A one-layer feedback artificial neural network with a Bayesian learning rule. Int. J. Neural Syst. **01**, 77–87 (1989). https://doi.org/10.1142/S0129065789000499
25. Sandberg, A., Lansner, A., Petersson, K.M., Ekeberg, Ö.: A Bayesian attractor network with incremental learning. Netw. Comput. Neural Syst. **13**, 179–194 (2002). https://doi.org/10.1080/net.13.2.179.194
26. Lansner, A., Holst, A.: A higher order Bayesian neural network with spiking units (1996). https://doi.org/10.1142/S0129065796000816
27. Tully, P.J., Hennig, M.H., Lansner, A.: Synaptic and nonsynaptic plasticity approximating probabilistic inference. Front. Synaptic Neurosci. **6**, 8 (2014). https://doi.org/10.3389/FNSYN.2014.00008/ABSTRACT
28. Johansson, C., Sandberg, A., Lansner, A.: A capacity study of a Bayesian neural network with hypercolumns. Rep. Stud. Artif. Neural Syst. (2001)
29. George, D., et al.: A generative vision model that trains with high data efficiency and breaks text-based CAPTCHAs. Science (80), 358 (2017). https://doi.org/10.1126/SCIENCE.AAG2612
30. Yamins, D.L.K., DiCarlo, J.J.: Using goal-driven deep learning models to understand sensory cortex. Nat. Neurosci. **193**(19), 356–365 (2016). https://doi.org/10.1038/nn.4244
31. Felleman, D.J., Van Essen, D.C.: Distributed hierarchical processing in the primate cerebral cortex. Cereb. Cortex. **1**, 1–47 (1991). https://doi.org/10.1093/cercor/1.1.1
32. Tang, H., et al.: Recurrent computations for visual pattern completion. Proc. Natl. Acad. Sci. U. S. A. **115**, 8835–8840 (2018). https://doi.org/10.1073/PNAS.1719397115/SUPPL_FILE/PNAS.1719397115.SAPP.PDF
33. Roelfsema, P.R.: Cortical algorithms for perceptual grouping (2006). https://doi.org/10.1146/annurev.neuro.29.051605.112939

34. Wyatte, D., Curran, T., O'Reilly, R.: The limits of feedforward vision: recurrent processing promotes robust object recognition when objects are degraded. J. Cogn. Neurosci. **24**, 2248–2261 (2012). https://doi.org/10.1162/jocn_a_00282
35. Fyall, A.M., El-Shamayleh, Y., Choi, H., Shea-Brown, E., Pasupathy, A.: Dynamic representation of partially occluded objects in primate prefrontal and visual cortex. Elife **6**, (2017). https://doi.org/10.7554/eLife.25784
36. Li, W., Piëch, V., Gilbert, C.D.: Learning to link visual contours. Neuron **57**, 442–451 (2008). https://doi.org/10.1016/J.NEURON.2007.12.011
37. Li, W., Gilbert, C.D.: Global contour saliency and local colinear interactions. J. Neurophysiol. **88**, 2846–2856 (2002). https://doi.org/10.1152/JN.00289.2002
38. Lamme, V.A.F., Roelfsema, P.R.: The distinct modes of vision offered by feedforward and recurrent processing. Trends Neurosci. **23**, 571–579 (2000). https://doi.org/10.1016/S0166-2236(00)01657-X
39. Grossberg, S.: Competitive learning: from interactive activation to adaptive resonance. Cogn. Sci. **11**, 23–63 (1987). https://doi.org/10.1016/S0364-0213(87)80025-3
40. Rumelhart, D.E., Zipser, D.: Feature discovery by competitive learning. Cogn. Sci. **9**, 75–112 (1985). https://doi.org/10.1016/S0364-0213(85)80010-0
41. Földiák, P.: Forming sparse representations by local anti-Hebbian learning. Biol. Cybern. **64**, 165–170 (1990). https://doi.org/10.1007/BF02331346
42. Szegedy, C., et al.: Intriguing properties of neural networks. In: International Conference on Learning Representations, ICLR (2014)
43. Lake, B.M., Ullman, T.D., Tenenbaum, J.B., Gershman, S.J.: Building machines that learn and think like people. Behav. Brain Sci. **40** (2017). https://doi.org/10.1017/S0140525X16001837
44. Kietzmann, T.C., Spoerer, C.J., Sörensen, L.K.A., Cichy, R.M., Hauk, O., Kriegeskorte, N.: Recurrence is required to capture the representational dynamics of the human visual system. **116** (2019). https://doi.org/10.1073/pnas.1905544116
45. Rao, R.P.N., Ballard, D.H.: Predictive coding in the visual cortex: a functional interpretation of some extra-classical receptive-field effects. Nat. Neurosci. **21**(2), 79–87 (1999). https://doi.org/10.1038/4580
46. Bastos, A.M., Usrey, W.M., Adams, R.A., Mangun, G.R., Fries, P., Friston, K.J.: Canonical microcircuits for predictive coding. Neuron **76**, 695–711 (2012). https://doi.org/10.1016/J.NEURON.2012.10.038
47. Tully, P.J., Lindén, H., Hennig, M.H., Lansner, A.: Spike-based Bayesian-Hebbian learning of temporal sequences. PLOS Comput. Biol. **12**, e1004954 (2016). https://doi.org/10.1371/JOURNAL.PCBI.1004954
48. Martinez, R.H., Lansner, A., Herman, P.: Probabilistic associative learning suffices for learning the temporal structure of multiple sequences. PLoS One **14**, e0220161 (2019). https://doi.org/10.1371/JOURNAL.PONE.0220161

Thinking Fast and Slow in AI: The Role of Metacognition

M. Bergamaschi Ganapini[1(✉)], Murray Campbell[2], Francesco Fabiano[3], Lior Horesh[2], Jon Lenchner[2], Andrea Loreggia[4], Nicholas Mattei[5], Francesca Rossi[2], Biplav Srivastava[6], and Kristen Brent Venable[7]

[1] Union College, Schenectady, USA
bergamam@union.edu
[2] IBM Research, New York, USA
{mcam,lhoresh,lenchner}@us.ibm.com, francesca.rossi2@ibm.com
[3] University of Parma, Parma, Italy
francesco.fabiano@unipr.it
[4] University of Brescia, Brescia, Italy
[5] Tulane University, New Orleans, USA
[6] University of South Carolina, Columbia, USA
[7] IHMC, Pensacola, USA
bvenable@uwf.edu

Abstract. Artificial intelligence (AI) still lacks human capabilities, like adaptability, generalizability, self-control, consistency, common sense, and causal reasoning. Humans achieve some of these capabilities by carefully combining their thinking "fast" and "slow". In this work we define an AI architecture that embeds these two modalities, and we study the role of a "meta-cognitive" component, with the role of coordinating and combining them, in achieving higher quality decisions.

1 Overview

AI systems have seen dramatic advancement in recent years, bringing many applications that pervade our everyday life. However, we are still mostly seeing instances of narrow AI: many of these recent developments are typically focused on a very limited set of competencies and goals, e.g., image interpretation, natural language processing, classification, prediction, and many others. Moreover, while these successes can be accredited to improved algorithms and techniques, they are also tightly linked to the availability of huge datasets and computational power [23]. State-of-the-art AI still lacks many capabilities that would naturally be included in a notion of (human) intelligence. Examples of these capabilities are generalizability, adaptability, robustness, explainability, causal analysis, abstraction, common sense reasoning, ethical reasoning [22,30], as well as a complex and seamless integration of learning and reasoning supported by both implicit and explicit knowledge [21].

We argue that a better study of the mechanisms that allow humans to have these capabilities can help us understand how to imbue AI systems with these

G. Nicosia et al. (Eds.): LOD 2022, LNCS 13811, pp. 502–509, 2023.
https://doi.org/10.1007/978-3-031-25891-6_38

competencies [5,29]. We focus especially on D. Kahneman's theory of thinking fast and slow [17], and we propose a multi-agent AI architecture (called SOFAI, for SlOw and Fast AI) where incoming problems are solved by either system 1 (or "fast") agents (also called "solvers"), that react by exploiting only past experience, or by system 2 (or "slow") agents, that are deliberately activated when there is the need to reason and search for optimal solutions beyond what is expected from the system 1 agent. Both kinds of agents are supported by a model of the world, containing domain knowledge about the environment, and a model of "self", containing information about past actions of the system and solvers' skills. Given the need to choose between these two kinds of solvers, a meta-cognitive agent is employed, performing introspection and arbitration roles, and assessing the need to employ system 2 solvers by considering resource constraints, abilities of the solvers, past experience, and expected reward for a correct solution of the given problem [31,32]. To do this balancing in a resource-conscious way, the meta-cognitive agent includes two successive phases, the first one faster and more approximate, and the second one (if needed) more careful and deliberate. Different approaches to the design of AI systems inspired by the dual-system theory have also been published recently [2,4,7,15,16,24,26].

Many real-life settings present sequential decision problems. Depending on the availability of system 1 and/or system 2 solvers that can tackle single decisions or a sequence of them, the SOFAI architecture employs the meta-cognitive agent at each decision, or only once for a whole sequence [3,14,26]. The first modality provides additional flexibility, since each call of the meta-cognitive module may choose a different solver to make the next decision, while the second one allows to exploit additional domain knowledge in the solvers.

We hope that the SOFAI architecture will support more flexibility, faster performance, and higher decision quality than a single-modality system where meta-cognition is not employed, and/or where there is no distinction between system 1 and system 2 agents. In this short paper, we describe the overall architecture and the role of the meta-cognitive agent, for a more extensive discussion and preliminary results refer to this contribution [12]. We provide motivation for the adopted design choices and we describe ongoing work to test instances of the architecture on sequential decision problems such as (epistemic) planning and path finding in constrained environments, against criteria intended to measure the quality of the decisions.

2 Thinking Fast and Slow in Humans

According to Kahneman's theory, described in his book "Thinking, Fast and Slow" [17], human decisions are supported and guided by the cooperation of two kinds of capabilities, that, for sake of simplicity are called *systems*: system 1 provides tools for intuitive, imprecise, fast, and often unconscious decisions ("thinking fast"), while system 2 handles more complex situations where logical and rational thinking is needed to reach a complex decision ("thinking slow").

System 1 is guided mainly by intuition rather than deliberation. It gives fast answers to simple questions. Such answers are sometimes wrong, mainly because of unconscious bias or because they rely on heuristics or other short cuts [13], and usually do not have explanations. However, system 1 is able to build models of the world that, although inaccurate and imprecise, can fill knowledge gaps through causal inference, allowing us to respond reasonably to the many stimuli of our everyday life.

When the problem is too complex for system 1, system 2 kicks in and solves it with access to additional computational resources, full attention, and sophisticated logical reasoning. A typical example of a problem handled by system 2 is solving a complex arithmetic calculation, or a multi-criteria optimization problem. To do this, humans need to be able to recognize that a problem goes beyond a threshold of cognitive ease and therefore see the need to activate a more global and accurate reasoning machinery [17]. Hence, introspection and meta-cognition is essential in this process.

When a problem is new and difficult to solve, it is handled by system 2 [19]. However, certain problems over time pass on to system 1. The reason is that the procedures used by system 2 to find solutions to such problems are used to accumulate experience that system 1 can later use with little effort. Thus, over time, some problems, initially solvable only by resorting to system 2 reasoning tools, can become manageable by system 1. A typical example is reading text in our own native language. However, this does not happen with all tasks. An example of a problem that never passes to system 1 is finding the correct solution to complex arithmetic questions.

3 Thinking Fast and Slow in AI: A Multi-agent Approach

As shown in Fig. 1, we are working on a multi-agent architecture which is inspired by the "Thinking Fast and Slow" theory of human decisions. In the SOFAI architectures, incoming problems/tasks are initially handled by system 1 (S1) solvers that have the required skills to tackle them, analogous to what is done by humans who unconsciously react to an external stimulus via their system 1 [17].

We assume S1 solvers act in constant time (that is, their running time is not a function of the size of the input problem) by relying on the past experience of the system, which is maintained in the model of self. The model of the world maintains the knowledge accumulated by the system over the external environment and the expected tasks, while the model of others maintains knowledge and beliefs over other agents that may act in the same environment. The model updater agent acts in the background to keep all models updated as new knowledge of the world, of other agents, or new decisions are generated and evaluated.

Once an S1 solver has solved the problem (for sake of simplicity, let's assume it is just one S1 solver), the proposed solution and the associated confidence level are available to the meta-cognitive (MC) module. At this point the MC agent starts its operations, with the task of choosing between adopting the S1 solver's

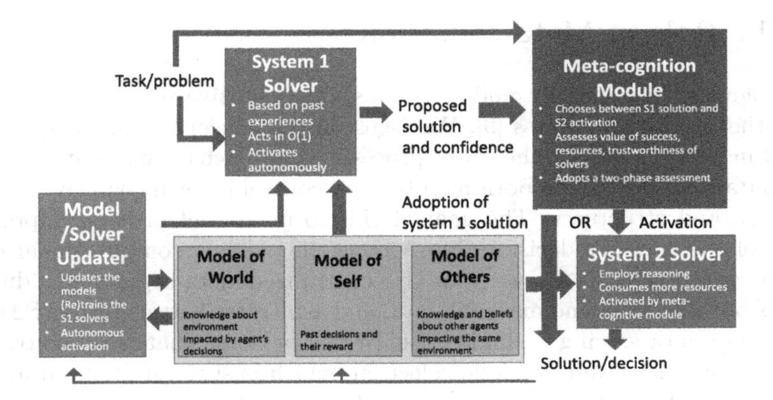

Fig. 1. The SOFAI architecture, supporting system 1/system 2 agents and meta-cognition.

solution or activating a system 2 (S2) solver. S2 agents use some form of reasoning over the current problem and usually consume more resources (especially time) than S1 agents. Also, they never work on a problem unless they are explicitly invoked by the MC module.

To make its decision, the MC agent assesses the current resource availability, the expected resource consumption of the S2 solver, the expected reward for a correct solution and for each of the available solvers, as well as the solution and confidence evaluations coming from the S1 solver. In order to not waste resources at the meta-cognitive level, the MC agent includes two successive assessment phases, the first one faster and more approximate, related to rapid unconscious assessment in humans [1,28], and the second one (to be used only if needed) more careful and resource-costly, analogous to the conscious introspective process in humans [6]. The next section will provide more details about the internal steps of the MC agent.

This is clearly an S1-by-default architecture, analogous to what happens in humans: whenever a new problem is presented, an S1 solver with the necessary skills to solve the problem starts working on it, generating a solution and a confidence level. The MC agent will then decide if there is the need to activate an S2 solver. This allows for minimizing time to action (since S1 solvers act in constant time) when there is no need for S2 processing. It also allows the MC agent to exploit the proposed action and confidence of S1 when deciding whether to activate S2, which leads to more informed and hopefully better decisions by MC.

Notice that we do not assume that S2 solvers are always better than S1 solvers, analogously to what happens in humans [13]. Take for example complex arithmetic, which usually requires humans to employ system 2, vs perception tasks, which are typically handled by our system 1. Similarly, in the SOFAI architecture we allow for tasks that are better handled by S1 solvers, especially once the system has acquired enough experience on those tasks.

4 The Role of Meta-cognition

Meta-cognition is generally understood as any cognitive process that is about some other cognitive process [9]. We focus on the concept of meta-cognition as defined in [11,25], that is, the set of processes and mechanisms that could allow a computational system to both monitor and control its own cognitive activities, processes, and structures. The goal of this form of control is to improve the quality of the system's decisions [9]. Among the existing computational models of meta-cognition [8,20,27], we propose a centralized meta-cognitive module that exploits both internal and external data, and arbitrates between S1 vs S2 solvers in the process of solving a single task. Notice however that this arbitration is different than algorithm portfolio selection, which is successfully used for many problems [18], because of the characterization of S1 and S2 solvers and the way the MC agent controls them.

The MC module exploits information coming from two main sources:

1. The system's internal models, which are periodically updated by the model updater agent:
   - Model of self, which includes information about:
     - Solvers' past decisions in specific tasks.
     - Resource consumption, e.g., memory and time of using solvers for specific tasks.
     - Rewards of solvers' decisions.
     - Available system's resources, e.g., memory, time.
     - Expected reward for solving a task.
     - Past resource consumption of the MC agent.
   - Model of the world, which contains information about:
     - Tasks, e.g., their input, goal, skills needed to solve it, etc.
     - Decision environment.
     - Actions available to the solvers in each state and for each task.
   - Model of others, which maintains knowledge and beliefs over other agents that may act in the same decision environment.
2. S1 solver(s):
   - Proposed decision for a task.
   - Confidence in the proposed decision.

The first meta-cognitive phase (MC1) activates automatically as a new task arrives and a solution for the problem is provided by an S1 solver. MC1 decides between accepting the solution proposed by the S1 solver or activating the second meta-cognitive phase (MC2). MC1 makes sure that there are enough resources, specifically time and memory, for completing both MC1 and MC2. If not, MC1 adopts the S1 solver's proposed solution. MC1 also compares the confidence provided by the S1 solver with the expected reward for that task: if the confidence is high enough compared to the expected reward, MC1 adopts the S1 solver's solution. Otherwise, it activates the next assessment phase (MC2) to make a more careful decision. The rationale for this phase of the decision process is that we envision that often the system will adopt the solution proposed by the S1 solver, because it is good enough given the expected reward for solving the task, or because there are not enough resources to invoke more complex reasoning.

Contrarily to MC1, MC2 decides between accepting the solution proposed by the S1 solver or activating an S2 solver to solve the task. To do this, MC2 evaluates the expected reward of using the S2 solver in the current state to solve the given task, using information contained in the model of self about past actions taken by this or other solvers to solve the same task, and the expected cost of running this solver. MC2 then compares the expected reward for the S2 solver with the expected reward of the action proposed by the S1 solver: if the expected additional reward of running the S2 solver, as compared to using the S1 solution, is greater than the expected cost of running the S2 solver, then MC2 activates the S2 solver. Otherwise, it adopts the S1 solver's solution.

To evaluate the expected reward of an action (here used to evaluate the S1 solver's proposed action), MC2 retrieves from the model of self the expected immediate and future reward for the action in the current state (approximating the forward analysis to avoid a too costly computation), and combines this information with the confidence the S1 solver has in the action. The rationale for the behavior of MC2 is based on the design decision to avoid costly reasoning processes unless the additional cost is compensated by an even greater additional expected value for the solution that the S2 solver will identify for this task, analogous to what happens in humans [31].

For the sake of simplicity and lack of space, in this paper we did not consider the possibility of having several S1 and/or S2 solvers for the same task, which would bring additional steps for both MC1 and MC2 to identify the most suitable solver. We also did not consider issues related to problem similarity, and we assumed we have enough past experience about exactly the same problem to be able to evaluate the expected reward and utility of solvers and actions. Finally, we did not discuss how the expected value of an action and the associated confidence are combined: currently we are working with a risk-adverse approach where these two quantities are multiplied, but we also plan to explore other approaches.

5 Instances of the SOFAI Architecture

As mentioned, we believe that many real-life settings present sequential decision problems. Available solvers could be able to choose one single decision of a whole sequence, or they may be able to build a whole decision sequence (or everything in between). In the first scenario, the MC agent is used to select the appropriate solver for each decision. In the second setting, the MC agent is used only once, so its influence on the performance of the system and the quality of its decisions could be more limited. However, usually the second setting occurs when we have solvers with a deep knowledge of the decision domain, and can adapt their capabilities to the available resources and expected reward for optimality and correctness of the decision sequences. We envision that these parameters and domain knowledge can be exploited by the meta-cognitive agent both in the MC1 and the MC2 phase.

We are currently working on implementing two main instances of the SOFAI architecture, that cover both extremes of this spectrum. In the first instance, we consider a decision environment in the form of a constrained grid (where constraints are over states, moves, and state features), where solvers are only able to decide on the next move, but entire trajectories (from an initial state to a goal state) need to be built to solve the given task [14]. In the second instance, we consider epistemic planning problems [10] where solvers are indeed planners, so they are able to build an entire plan and not just a single step in the plan. In this case, the meta-cognitive module includes also steps regarding problem and solver simplifications, as well as problem similarity assessment.

Acknowledgements. Nicholas Mattei was supported by NSF Awards IIS-RI-2007955, IIS-III-2107505, and IIS-RI-2134857, as well as an IBM Faculty Award and a Google Research Scholar Award. K. Brent Venable are supported by NSF Award IIS-2008011.

References

1. Ackerman, R., Thompson, V.A.: Meta-reasoning: Monitoring and control of thinking and reasoning. Trends Cogn. Sci. **21**(8), 607–617 (2017)
2. Anthony, T., Tian, Z., Barber, D.: Thinking fast and slow with deep learning and tree search. In: Advances in Neural Information Processing Systems, pp. 5360–5370 (2017)
3. Balakrishnan, A., Bouneffouf, D., Mattei, N., Rossi, F.: Incorporating behavioral constraints in online AI systems. In: Proceedings of the 33rd AAAI Conference on Artificial Intelligence (AAAI) (2019)
4. Bengio, Y.: The consciousness prior. arXiv preprint arXiv:1709.08568 (2017)
5. Booch, G., et al.: Thinking fast and slow in AI. In: Proceedings of the AAAI Conference on Artificial Intelligence, vol. 35, pp. 15042–15046 (2021)
6. Carruthers, P.: Explicit nonconceptual metacognition. Philos. Stud. **178**(7), 2337–2356 (2021)
7. Chen, D., Bai, Y., Zhao, W., Ament, S., Gregoire, J.M., Gomes, C.P.: Deep reasoning networks: Thinking fast and slow. arXiv preprint arXiv:1906.00855 (2019)
8. Cox, M.T.: Metacognition in computation: a selected research review. Artif. Intell. **169**(2), 104–141 (2005)
9. Cox, M.T., Raja, A.: Metareasoning: Thinking About Thinking. MIT Press, Cambridge (2011)
10. Fagin, R., Moses, Y., Halpern, J.Y., Vardi, M.Y.: Reasoning About Knowledge. MIT press, Cambridge (2003)
11. Flavell, J.H.: Metacognition and cognitive monitoring: a new area of cognitive-developmental inquiry. Am. Psychol. **34**(10), 906 (1979)
12. Ganapini, M.B.: Combining fast and slow thinking for human-like and efficient decisions in constrained environments. In: Proceedings of the 16th International Workshop on Neural-Symbolic Learning and Reasoning (NeSy 2022) Co-located with (IJCLR 2022), vol. 3212 of CEUR Workshop Proceedings, pp. 171–185 (2022). CEUR-WS.org
13. Gigerenzer, G., Brighton, H.: Homo heuristicus: why biased minds make better inferences. Top. Cogn. Sci. **1**(1), 107–143 (2009)

14. Glazier, A., Loreggia, A., Mattei, N., Rahgooy, T., Rossi, F., Venable, K.B.: Making human-like trade-offs in constrained environments by learning from demonstrations. arXiv preprint arXiv:2109.11018 (2021)
15. Goel, G., Chen, N., Wierman, A.: Thinking fast and slow: optimization decomposition across timescales. In: IEEE 56th Conference on Decision and Control (CDC), pp. 1291–1298. IEEE (2017)
16. Gulati, A., Soni, S., Rao, S.: Interleaving fast and slow decision making. arXiv preprint arXiv:2010.16244 (2020)
17. Kahneman, D.: Thinking, Fast and Slow. Macmillan, New York (2011)
18. Kerschke, P., Hoos, H.H., Neumann, F., Trautmann, H.: Automated algorithm selection: survey and perspectives. Evol. Comput. **27**(1), 3–45 (2019)
19. Kim, D., Park, G.Y., John, P., Lee, S.W., et al.: Task complexity interacts with state-space uncertainty in the arbitration between model-based and model-free learning. Nat. Commun. **10**(1), 1–14 (2019)
20. Kralik, J.D., et al.: Metacognition for a common model of cognition. Procedia Comput. Sci. **145**, 730–739 (2018)
21. Littman, M.L., et al.: gathering strength, gathering storms: The one hundred year study on artificial intelligence (AI100) 2021 study panel report. Stanford University (2021)
22. Loreggia, A., Mattei, N., Rahgooy, T., Rossi, F., Srivastava, B., Venable, K.B.: Making human-like moral decisions. In: Proceedings of the 2022 AAAI/ACM Conference on AI, Ethics, and Society, AIES'22, pp. 447–454. Association for Computing Machinery, New York, NY, USA (2022)
23. Marcus, G.: The next decade in AI: four steps towards robust artificial intelligence. arXiv preprint arXiv:2002.06177 (2020)
24. Mittal, S., Joshi, A., Finin, T.: Thinking, fast and slow: combining vector spaces and knowledge graphs. arXiv preprint arXiv:1708.03310 (2017)
25. Nelson, T.O.: Metamemory: a theoretical framework and new findings. In: Psychology of Learning and Motivation, vol. 26, pp. 125–173. Elsevier (1990)
26. Noothigattu, R., et al.: Teaching AI agents ethical values using reinforcement learning and policy orchestration. IBM J. Res. Dev. **63**(4/5), 2:1-2:9 (2019)
27. Posner, I.: Robots thinking fast and slow: on dual process theory and metacognition in embodied AI (2020)
28. Proust, J.: The Philosophy of Metacognition: Mental Agency and Self-awareness. OUP Oxford, Oxford (2013)
29. Rossi, F., Loreggia, A.: Preferences and ethical priorities: thinking fast and slow in AI. In: Proceedings of the 18th International Conference on Autonomous Agents and Multiagent Systems, pp. 3–4 (2019)
30. Rossi, F., Mattei, N.: Building ethically bounded AI. In: Proceedings of the 33rd AAAI Conference on Artificial Intelligence (AAAI) (2019)
31. Shenhav, A., Botvinick, M.M., Cohen, J.D.: The expected value of control: an integrative theory of anterior cingulate cortex function. Neuron **79**(2), 217–240 (2013)
32. Thompson, V.A., Turner, J.A.P., Pennycook, G.: Intuition, reason, and metacognition. Cogn. Psychol. **63**(3), 107–140 (2011)

Confidence as Part of Searcher's Cognitive Context

Dominika Michalkova[1], Mario Parra Rodriguez[2], and Yashar Moshfeghi[1](✉)

[1] NeuraSearch Laboratory, Computer and Information Sciences, University of Strathclyde, Glasgow, UK
{dominika.michalkova,yashar.moshfeghi}@strath.ac.uk

[2] School of Psychological Sciences and Health, University of Strathclyde, Glasgow, UK
mario.parra-rodriguez@strath.ac.uk

Abstract. A growing body of interdisciplinary *NeuraSearch* research in the Information Retrieval (IR) domain, built on the investigations evaluating searchers' neurophysiological activity captured during the information search interactions, advances the understanding of the searchers' cognitive context. Regarding the searchers' information needs, the cognitive context represents the surroundings of a knowledge anomaly perceived in their state of knowledge. Memory retrieval is a fundamental mechanism that drives the users' informativeness about their knowledge and knowledge gaps. Moreover, the confidence perceptions manifest the quality and attribute of the users' memories and could be, thus, used as a sign of the quality of memories aiding the user to appraise their knowledge abilities. We used the methodology of NeuraSearch to reduce the cognitive burden commonly in traditional IR scenarios posed to the users to understand and interpret their subjective perceptions and feelings. We investigated the patterns of spatio-temporal brain activity (captured by EEG) in 24 neurologically-healthy volunteers engaged in textual general knowledge Question Answering (Q/A) Task. We looked for i) the evidence of functional processes leading to descriptive (factual) knowledge memory retrieval and ii) their interaction effects incorporating retrospective confidence judgments. Our investigation raises further questions informing research in IR and the area of user information seeking.

Keywords: Information retrieval · Information search · Proactive information retrieval · NeuraSearch · Anomalous state of knowledge · Information need · Confidence · Memory retrieval · EEG · ERP

1 Introduction

Searcher's cognitive context defines a state of knowledge concerning a problematic situation the searcher is facing [26]. A state is characterised as anomalous, i.e. ASK [6], reflecting the user's recognition of a knowledge gap in this context.

G. Nicosia et al. (Eds.): LOD 2022, LNCS 13811, pp. 510–524, 2023.
https://doi.org/10.1007/978-3-031-25891-6_39

Information needs (IN) are produced representing what the users think they need to know to fill these gaps and are, thus, the triggers for search engagement and IR system interaction. The output of Information Retrieval (IR) should fit into what the IR user already knows in order to transform the ASK into a coherent state of knowledge. Searcher's "personal frame of reference" [29] refers to the memory as a repository of searchers' internal knowledge, which supports the user in a given context. There is an analogy between the information retrievals on both i) the searcher side, i.e. searcher accesses the memory to determine the knowledge anomaly and formulates IN, and ii) the IR side, i.e. to provide the ultimate goal of relevant documents retrievals as a response to users' INs. Whereas the IR is a well-modelled process with numerous studies done in the user-centric domain [11,38,43], the representation of the user's cognitive context in IR is left behind.

As the research has uncovered [29], central to the search process is the feelings of uncertainty and confusion, even feelings of anxiety, both affective and cognitive, exceptionally high at the beginning phase, when the user engages with the system and faces their feelings of unease [10] and feeling of dissatisfaction [41]. Generally, confidence is regarded as a signal of memory strength [12]. Traditional methods of user-based investigations in IR commonly rely on users' subjective interpretations of their perceptions, feelings and experiences either on their own or via a moderator. In contrast, interdisciplinary *NeuraSearch* research within IR [32] is driven by the study of unbiased neurological measures to explore the underlying cognitive manifestations of the users' information search process to improve the understanding of the user's cognitive context in the user-centric IR models. To confirm the cognitive IR theories [10,24,25], the memory retrieval component has been identified in a user study [33] to actively participate in the process of knowledge gap realisation.

Memory is a complex process proved by the existing taxonomy of memory [39]. To specify, the emphasis of this study is on declarative memory, which holds the facts and events and has been highlighted by the findings in the study by Arguello et al. [2]. The study discovered that IR users tend to express their INs using knowledge retrieved from the declarative memory, both factual and episodic.

We developed an interactive Question-Answering (Q/A) study by capturing the brain signals of 24 participants via EEG. We examined the persons' information processing with the focus on memory during the sequential presentation of stimuli, i.e. question of general knowledge followed by judgments of confidence. The brain signal was encoded into one of the three defined Memory Retrieval (MR) levels that aligned with the participant's responses to a given stimulus: 1) Correct (MR-C), 2) Incorrect (memory falsely endorsed as correct, MR-I), 3) Not Know (no recollection, MR-N). Such segmented data were further split into two levels depicting searchers' perceived Confidence: A) High (HIGH) and B) Low (LOW). We utilised a data-driven analytical framework focusing on the contrast of ERP components to reveal the significantly different activities and answer the following research questions: **RQ1:** What spatio-temporal differences of neural correlates are exhibited for the spectrum of our three defined MR levels, i.e. MR-C, MR-I, MR-N? **RQ2:** How do two confidence levels, i.e. HIGH and LOW, modulate these neural correlates?

The approach reflects our intention to obtain insight into the cognitive underpinnings of the memory attribute of confidence that has not yet been subjected to a NeuraSearch study.

2 Background

Memory in IR. The Information Search Process model (ISP) [29] drew attention to the search process as a process of sense-making for the user who actively searches for new information to fit into what they already know on a particular topic. The dynamic character was highlighted in [24] that referred to IR as an interactive problem-solving process where individual knowledge structures play a vital part in solving these problems. Knowledge plays a central element in the information exchange between the user (subjective internal knowledge) and the system (objective external knowledge). Hence, giving the idea to the concept of "transactive memory systems" forming between the system and the user [37]. The illustration of information processing follows human perceptual system that firstly filters and then passes sensory stimuli data (e.g., a problematic event the user faces) to process by short and long-term memory [24]. Expanding on this approach, Cole [10] framed the user's internal mechanism of knowledge usage and generation according to i) Minsky's Frame Theory [31], an internal representation of the world based on knowledge frames, and ii) the person's perceptual-cognitive system. Information needs set off the cognitive processes by employing memory search in long-term memory that extracts the relevant knowledge frames. These are inputted into working memory, where multiple retrieval structures work in parallel to evaluate them and contribute, thus, to informed decisions and generating actions.

Referring to the outcomes of NeuraSearch research, Moshfeghi and Pollick [33] proposed "Neuropsychological model of the realisation of IN". On the basis of findings from an fMRI-based user study, they conceptualised two components playing a vital part in IN realisation: Memory Retrieval and Information Flow to create a link between internal (memory) and external search. A study by Arguello et al. [2] performed a qualitative study of IN requests where it was revealed that users use recollections from declarative memory, both semantic and episodic [39] to articulate their INs.

Modelling Uncertainty and Confidence in IR. Naturally evoked searcher's uncertainty, both as an affective and cognitive feeling, can be seen as an early perceived manifestation of INs [29]. Both Taylor [41] and Belkin et al. [6] approached in their studies the notion of uncertainty as a user-specific aspect of IN realisation. According to Taylor, the searcher's inquiry is driven by the "area of doubt", demonstrating an absence of confidence and certainty. The term "area" intuitively evokes a broader term. Not only does it refer to the information space, it also suggests the employment of the user's cognitive and exploratory capabilities necessary to understand own INs. These requirements conform to ASK's the realisation of anomaly accompanied by doubts, uncertainty which is hard to express [6], but in paradox, seen as a trigger of one's engagement in information seeking and search scenario. Moreover, ISP Model by Kuhlthau [29] established that uncertainty causes discomfort and anxiety, affecting articulation of the problem and information relevancy judgements. The intensity of the user's confidence varies depending on the stage of the search process. The early manifestations of confusion, apprehension, frustration, and doubt are shifted in the later stages towards the feelings of satisfaction, sure and relief. Confidence correlates with the levels of information clarity and focused thinking, which corresponds to the evidence of ISP as a sense-making constructive process for the searcher. The source of uncertainty is the limited understanding and clarity about the problem and INs, i.e. ill-defined INs [10].

3 Methodology

Design. A "within-subjects" design was used in the present study, in which the participants performed a Q/A task of general knowledge (see Sect. 3.1). Addressing our RQs, we created a study with the involvement of simultaneous data capturing using EEG in order to evaluate the participants' neurophysiological measures associated with information processing in order to evoke i) memory retrieval (MR) based on the recognition task asking the participant to select the correct answer among a set of mnemonic cues and ii) followed by judgments of confidence. In general, tests of memory performance is studied either 1) at the time of information encoding into memory, with the focus on memorability where the researchers have control over the information input, or 2) as a function of events that takes place at the time of retrieval [42], similarly as our study aimed. Based on our data structure we created two models that aligned with our research goals to uncover: i) unbiased neurophysiological evidence triggering different MR outcomes (Model MR) and ii) how sensitive this brain activity is to different perceptions of confidence (Model MR+Conf).

Model MR. The independent variables were linked to the participants' responses which were controlled by responding to the questions viewed on the computer screen: (1) "Correct" (MR-C), (2) "Incorrect" (MR-I) and (3) "I do not know", meaning acknowledged state of not knowing (MR-N).

Model MR+Conf. The participants judged retrospectively their confidence on two levels: High (HIGH) or Low (LOW). They were not asked to rate their confidence for all three MR levels. They provided these judgments only if their MR actually resulted in factual information outcome (i.e., MR-C or MR-I levels) resulting thus in levels MR-C+HIGH, MR-C+LOW, MR-I+HIGH, MR-I+LOW. By limiting the response space to three (Model MR) and two levels, respectively (Model MR+Conf), we had control over the categorisation of participants and the information processing associated with these levels. For each level, we extracted the relevant ERP activity. The dependent variable in each model was the mean amplitude of the relevant ERP components (associated with the responses) drawn from the EEG signals synchronised with the Q/A task.

Participants. Twenty-four healthy university students volunteered an age range between 18 and 39 years and the mean age of 24 years (sd 6). Initial insight into data showed that data were unbalanced across our Scenarios. As the within-subject methods of the analytical framework we devised (see Sect. 3.2) rely on individual ERP averages, we needed sufficient samples in each scenario. We, thus, filtered[1] the data where only participants satisfying a threshold of a minimum number of responses sampling stable ERP waveforms in each level will be moved to final sample and undergo the following statistical analyses. For (a) Model MR: we obtained 14 participants and (b) the Model MR+Conf: 9 participants. Participants completed the task (without the breaks) on average in 44 min (sd 4.62, med 43.40).

Q/A Dataset. We constructed a dataset of 180 general knowledge questions taken from the following sources (1) TREC-8 and TREC-2001[2] (widely applied in IR and NeuraSearch studies [34,36]) and (2) B-KNorms Database[3] (used in cognition and learning studies [9]). Two independent assessors evaluated the questions' difficulty (Cohen's Kappa: 0.61). We then selected a subset of 120 questions where both annotators agreed upon their judgments and had an equal distribution of easy and difficult questions. Questions covered a diverse range of topics and were of open domain with closed-ended answers. The question length (measured by the number of words the question consisted of) resulted from 3 words to 13 words. None of the question attributes (difficulty, topic, length) were considered as independent variables in our models as we did not investigate their effects on the cortical activity.

3.1 Procedures

Ethical permission to carry out this study was obtained from the Ethics Committee at the University of Strathclyde. All participants fulfilled the inclusion criteria to participate in the study, i.e. healthy people between 18–55 years, fluent in English, and without any prior or current psychiatric or neurological scenarios that could influence EEG signals. Participants were informed about the task and were requested to sign an informed consent. To ensure that all the participants understood the task well, they underwent a practice session. The practice session consisted of five questions not used in the main task. In the main session, the question order and the answer options on the screen were randomised across the participants. There was no time limit to provide responses, which were entered by the participants via a button press using three keys previously allocated to each option. Two breaks were provided during the task to avoid fatigue. Participants were asked to remain still and minimise body movements, particularly

[1] The details of the procedure are due to page limit omitted.

[2] https://trec.nist.gov/data/qamain.html.

[3] http://www.mangelslab.org/bknorms.

blinking. Before data collection, a pilot study with two volunteers was conducted to ensure that the experimental procedure ran smoothly. Feedback from the participants in the pilot study was used towards the procedure improvements, and the data collected in the pilot study are not used in the present analysis.

Procedure of the Main Experimental Task. Figure 1 illustrates the sequence of the trial with an example of a 5-word question stimulus.

Trial. Each trial started with viewing a fixation cross in the middle of the screen for 2000 ms indicating the location of the stimuli on the screen as a way to minimise eye movements on the screen. Next, the participants viewed a sequential presentation of a question randomly selected from the dataset. Each word was displayed for 800 ms. After the last word of the question was presented, ("Wn" in Fig. 1), participants moved to Step 3. Here, they were presented with a fully-displayed question and three on-screen answer choices associated with the question. They were requested to select the correct answer or choose the option "I do not know". If the participants answered MR-C or MR-I, they were asked to judge their confidence that their answer was correct (Step 4.1). Alternatively, if the participants answered MR-N, they were presented with the correct answer (Step 4.2), which they terminated by a button press and started a new trial. The process was repeated for all 120 questions.

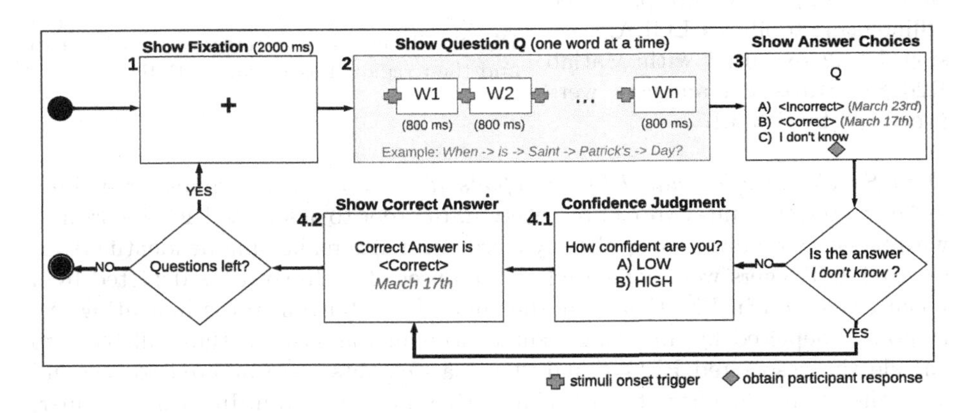

Fig. 1. Diagram of the task flow.

Experimental Approach. **Sequential presentation** of a stimulus has been applied in studies examining neurological correlates of reading [16], as well as in IR-related studies of relevance [36] or query construction [27]. It aims to control free-viewing and minimise the presence of any confounding artefacts (i.e., saccades). As a result, the ERPs were time-locked to the word onset presentation, i.e. we captured the temporal processing of each stimulus. The words on the screen were presented using a **Fixed Rapid Serial Visual Presentation** (fixed-RSVP) of 800 ms for each word. As previous studies of sentence processing [17,22] pointed out, a ratio above 700 ms enables engagement of higher cognitive abilities. These are important for us as we want to capture the human information processing and extract EEG signals associated with particular words, i.e. events. The outcomes pilot study determined the final ratio of 800 ms. It was found sufficient for fluent reading [17] and to avoid the overlapping effect of two consecutive words on the ERPs [18].

Apparatus. An interactive Q/A system running the experimental task was developed in a behavioural research software e-Prime2 and synchronised with a 40-channel NeuroScan Ltd. system with a 10/20 configuration cap used for EEG data acquisition. EEG was recorded with a sampling frequency of 500 Hz. Impedances were kept below 10 kΩ, and signals were filtered online within the band of 0.1–80 Hz. EEG recordings were subsequently pre-processed offline using toolbox EEGLAB version 14.1.2 executed with Matlab R2018a. Statistical analyses were processed by R 3.6.1.

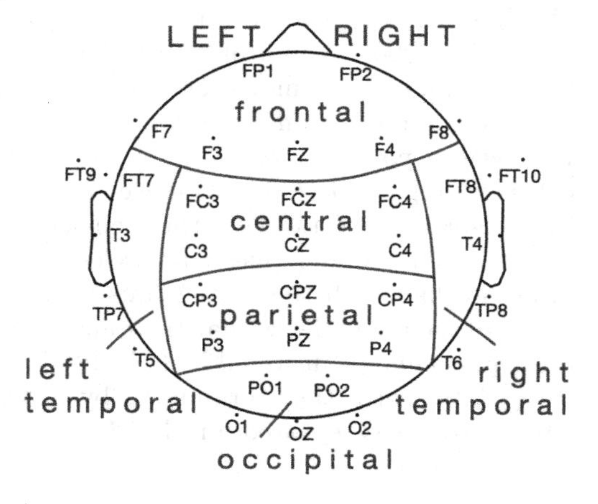

Fig. 2. Placement of electrodes on EEG cap and their regional assignment (ROI).

Data Synchronisation and EEG Artefacts Reduction. EEG signals were time-locked to the stimulus (word) presentation. In order to associate EEG recordings with behavioural data, we applied synchronous triggers, i.e. unique identifiers for each trial. Triggers were generated at the onset of each word (as depicted by a green cross icon in Fig. 1) and at the time of the button, press indicating the response (depicted by an orange square icon in Fig. 1) and, thus, allowing to encode the associated EEG signal. The participants were allowed to respond after the entire sequence of words completing a question ran. In such a manner, we eliminated the EEG signal contamination by motor response artefacts (i.e., a hand movement to make a button click) which generally count as a response-preparation component affecting the underlying brain signal [28].

3.2 ERP Analytical Framework

Initial inspection of EEG data showed a clear manifestation of ERP components, which resulted in our decision to apply a combination of exploratory and component-driven approaches, focusing on the evaluation of the ERP deflection within specific time windows where such ERP deflection occurs. Spatio-temporal investigation of neural activity requires unbiased selection of i) relevant spatial regions where the activity is significant and ii) splitting of the overall timeline into smaller time windows. We framed a procedure that allows us to achieve unbiased results. Its details are provided later in the text.

Data Pre-processing Pipeline. The individual participant data were pre-processed using the pipeline constructed according to guidelines for the standardisation of processing steps for large-scale EEG data [7] we now describe. Before passing the data through a high-pass filter at 0.5 Hz and then through a low-pass filter at 30 Hz, we removed the power line noise at 50 Hz. Next, we down-sampled data to 250 Hz. We then proceeded to reconstruct a low-quality EEG signal of selected electrode/-s whose recordings were found to be very noisy or their recording was interrupted. After that, we used average re-reference, suggested as an appropriate approach [14] in a case of data-driven exploratory analyses of ERP components and ROIs. In the next step, we performed Independent Component Analysis (ICA), a data cleaning technique used to separate the noise-introducing artefacts from the genuine brain signal, i.e. the sole brain effect on stimuli. Completion of ICA resulted in an artefact-free signal which was then epoched in every channel. Each data epoch contained the brain signal from 200ms pre-stimulus presentation to 800 ms post-stimulus presentation (i.e., word of the particular question). All epochs associated with a single question were baseline-corrected using the −200 to 0 ms window using the baseline activity from prior to the onset of the question's first word (represented by "W1" in Fig. 1). After that, for every participant, we averaged all epochs that belonged to the particular level determined by our independent variables. Finally, these average participant data entered the further stages of the statistical analysis.

Identifying Regions of Interest (ROI). Literature [28] suggested dealing with both the selection of time windows and ROIs by separating statistical tests for each time point at each electrode, combined with correction for multiple comparisons. Following this approach, we searched for significant differences across the levels of independent variables with a combination of the 2-sample paired Monte Carlo permutation test and non-parametric bootstrapping running 10,000 permutations[4]. The outcome of each pairwise comparison was a set of significant (p-value < 0.001) electrodes and their assigned time point where the activity significantly differed. We assigned these electrodes into clusters (ROI) based on their spatio-temporal properties, i.e. local proximity (according to Fig. 2) and the individual significance exhibiting in the same time window.

Time Windows. To ensure that setting the boundaries of time frames for capturing ERP components is not arbitrary, we used an unbiased data-driven procedure as follows. For each model, we averaged all epochs of the levels of independent variables for every ROI across all the participants. The output value reflects the overall brain responses regardless of task scenarios and the topological distributions with baseline set $y = 0$. Next, we applied this value to

[4] A solution for multiple comparison problems and does not depend on multiple comparisons correction or Gaussian assumptions about the probability distribution of the data.

the corresponding grand average waveform ROI to adjust the baseline. The time point where the waveform is reaching the new baseline represents a point where EEG deflection departs from the baseline respectively returns to a baseline. We then calculated the mean signal within the intervals for each relevant ROI.

Statistical Methods. Our main comparative measure is the mean signal, calculated as the mean of the ERP activity, precisely the amplitude that describes the ERP activity, which occurred within a particular time window and significant ROI. We created a mixed linear model for each time window with the parameters: 3 MR levels (Model MR), 2 MR levels × 2 confidence judgements (Model MR+Conf) as independent variables, "Participant" as the random variable, "Mean signal within ROI" as the dependent variable. To test if and how much each participant's neurocognitive response (amount of potential elicited) varied across the Scenarios, we applied a two-way ANOVA repeated measures test with a 3-level factorial design (Model MR) and 2 × 2 factorial design (Model MR+Conf) for each time window (Model MR) and for each significant ROI. Post-hoc pairwise tests applied Bonferroni corrections.

Table 1. Model MR: Significant differences in ERP amplitudes and the pairwise contrasts (p-value adjusted using Bonferroni corrections <0.05 *, <0.01 **, <0.001 ***)

Time window	ERP	ROI*	F value	M_{diff}	p-value
90–150 ms	N1	RF/RFC	F[2, 26] = 11.80	$\bar{x}$(MR-I) = −0.02 µV, $\bar{x}$(MR-N) = −0.39 µV	***
				$\bar{x}$(MR-C) = −0.08 µV, $\bar{x}$(MR-N) = −0.39 µV	**
		LTP	F[2, 26] = 4.95	$\bar{x}$(MR-I) = −0.07 µV, $\bar{x}$(MR-N) = 0.27 µV	*
		PO/O	F[2, 26] = 5.22	$\bar{x}$(MR-C) = 0.35 µV, $\bar{x}$(MR-N) = 0.72 µV	*
				$\bar{x}$(MR-I) = 0.35 µV, $\bar{x}$(MR-N) = 0.72 µV	*
150–270 ms	P2	TP	F[2, 26] = 3.97	$\bar{x}$(MR-I) = −1.64 µV, $\bar{x}$(MR-N) = −1.39 µV	*
270–430 ms	Onset N400	RFT	F[2, 26] = 5.03	$\bar{x}$(MR-C) = −0.39 µV, $\bar{x}$(MR-I) = −0.73 µV	*
		C/CP	F[2, 26] = 4.30	$\bar{x}$(MR-I) = 0.04 µV, $\bar{x}$(MR-N) = −0.15 µV	*
430–570 ms	Offset N400				
570–800 ms	P6	RFC/RC	F[2, 26] = 3.97	$\bar{x}$(MR-C) = −0.004 µV, $\bar{x}$(MR-N) = 0.19 µV	*
		LTP	F[2, 26] = 4.51	$\bar{x}$(MR-I) = −0.05 µV, $\bar{x}$(MR-N) = −0.28 µV	*

* For spatial reference see Fig. 2: L - Left, R - Right, F - Frontal, C - Central, T - Temporal, P - Parietal, O - Occipital.

4 Results and Discussion

Model MR (Addressing RQ1). The results from the statistical analysis are summarised in Table 1. The EEG waveforms showed a separation of the brain activity into four ERP components: N1, P2, N400 and P6. In general, the early emergence of N1 is considered to be triggered independently of the task demands [23] and used to measure early perceptual processing [3]. The early significant differences in elicited potentials across MR levels might be evidence of the early availability of knowledge cues driving these differences. The lowered N1 amplitude for both MR-C and MR-I indicates that when people think they know the answer, fewer neural resources are deployed [30] than when they think they do not (MR-N). Therefore, taking the highest neural activity elicited for MR-N level early in the time might predict the absence of knowledge (and potentially stand as the earliest sign of IN realisation). Next, the occurrence of P2 component indicated further support for input processing, such as registration and input classification [42]. Significantly higher P2 amplitudes for MR-I in contrast to MR-N pronounced in TP region might suggest differences in the cognitive effort linked to memory recall [4,13,19] with the differences driven by the memory strength. The manner of the involved bilateral TP regions, generally considered memory-oriented, implies the engagement of memory search knowledge and deployment of neural resources to link attention and memory. As was found, P2 could also be linked to an increased attention [35] to an item which has a certain degree of familiarity, possibly initialising the retrieval from contextual memory [21] and recall of past experiences [13]. The facilitation of prior knowledge could be reflected in the outcome of the higher negativity of P2 measured in the P/O region for MR-I level and support the integration of words to the context [15].

The subsequent ERP deflection attributed to the emergence of the later N400 seems to indicate the link between early and late processes [15]. First, the P2 emergence confirms the availability of memory and, thus, supports the decision by memory search and verification, correlated with the modulation of N400. There is a lack of consensus on what N400 component indexes. As Diana et al. [13] suggests, the later ERP effects might index co-occurring memory phenomena or the initiation of memory search and, thus, not necessarily index just one process, but parallel processes of memory [10].

Furthermore, the spatial differences in the mean amplitude of N400 for different pairs of MR levels suggest spatially-dependent support for different MR levels. Specifically, MR-I elicits the highest activity over the RFT region, whilst the CP/P region was found to support the MR-N level. A frontal difference between the pair of MR-C and MR-I pronounced in the RFT area signalises memory-driven processes and supports with increased resources (i.e., increased N400 amplitude) false impressions of knowing a correct answer (MR-I). The amplified MR-N over the MR-I level with CP/P distribution might indicate the decision of an external search, i.e. a realisation that the internal knowledge abilities are insufficient. This finding is strengthened by the Information Flow component in the Model of IN realisation by Moshfeghi and Pollick [33] which regulates a broader memory search when a person realises unavailability of information internally (i.e., a potential realisation of IN), the attention must be shifted towards external search. Our findings complete this model with the temporal signature of the Information Flow component activity.

The late sustained activity shows the significantly highest positive potentials of P6 over RFC and LTP driven by MR-N level. The pattern of P6 across MR levels indicates that these potentials scale with the success and strength of declarative MR [42]. This notion could reflect the highest averages for MR-N level, indicating the low success of MR and high strength, in contrast to reduced amplitudes for MR-C and MR-I. For P6, as the final stage of information processing, more neural resources were, similarly to in the early processing, found to be recruited for the MR-N level. The late processing might index final verification checks and maintenance of the previously triggered processes of awareness, memory search and the flow of information between significant parts of the brain resulting in conscious response to stimuli, as suggested in [40].

Recruitment of more resources might indicate that MR-I utilises a deeper memory search. The source of MR-I can be either a false memory already encoded in the user's memory [8] or a wrong interpretation of the memories and the distortion of the memory-retrieved output. From the IR search perspective, the question arises of how to prevent the users from misattribution of their knowledge, rectify it, or help with the correct interpretation of the information. Employing the interdisciplinary research, such as NeuraSearch, can expand on the components indexing memory-oriented processes, such as N400 suggested by our study, in relation to the user-information interaction. Further research surrounding the information veracity concerning the areas such as information use and knowledge generation could be a further intervention in this field.

Model MR+Conf (Addressing RQ2). We found statistically significant differences during the evaluation of Model MR+Conf in two time windows:

1. 90–150 ms. We found a significant main effect of Confidence ($F[1, 8] = 5.0$, p-value < 0.05) in the left frontal/front-temporal ROI (LF/LFT). The post-hoc tests specified significantly larger N1 negativity measured for low confidence ($-0.45\ \mu V$) than the N1 amplitude of high confidence ($-0.25\ \mu V$). In addition to the main effect of Confidence, we found in the same ROI a statistically significant effect of the interaction of MR and Confidence ($F[1, 1, 8] = 7.84$, p-value < 0.01). Pairwise contrasts with Bonferroni corrections revealed a significant difference between MR-I+HIGH and MR-I+LOW. The negativity of N1 amplitude of MR-I+LOW ($-0.62\ \mu V$) was significantly higher ($p < 0.001$) than the mean amplitude of MR-I followed by high confidence, MR-I+HIGH ($-0.16\ \mu V$).
2. 270–570 ms. In the time window spanning the N400 component, we found a significant main effect of Confidence ($F[1, 8] = 5.15$, p-value < 0.03) distributed over the previously found LF/LFT ROI. Here, the N400 amplitude of high confident responses ($-0.35\ \mu V$) resulted in a significantly larger mean negativity of N400 ($p < 0.05$) in contrast to the mean negativity of low confidence ($-0.25\ \mu V$).

We analysed the qualitative judgements of memory confidence to further subset the brain signal to obtain a more granular insight into data and investigate factual MR attributes. We found that the latency of the early N1 component and late N400 was significantly modulated consistently over LF/LFT ROI by the Confidence condition (main effect of Confidence). In addition to the main effects of Confidence, we found the interaction between MR-I and Confidence during the time window depicting the N1 component. The confidence levels vary with the MR-I level, indicating more in-depth processing.

The pattern of activity distributed over F/FT ROI depicting N400 as well as the largest frontal negative deflection of N400 for high confident responses showed a similarity with the findings of frontal N400 evoked during the MR with subjective confidence judgments [44]. The literature usually relates the negative frontal deflection, peaking between 300–500 ms, to the effect of familiarity (i.e., no specifics are retrieved) or recollection (i.e., specific information is retrieved). Our results could support both as our study did not subjectively distinguish between these two. In summary, our results suggest that early amplified activity associated with low confidence levels might index false memory. Later stages saw the opposite pattern with high confidence eliciting greater activation but in an undifferentiated manner (regardless of MR levels). Continuing in the present scenario would benefit from larger sample size to test the robustness and scalability of the present results.

5 Conclusions

The present application of the interdisciplinary framework shows the potential for IR system design to objectively and proactively detect the surroundings of the user's cognitive context [26]. The recognition of the states when the users need or do not need (any more of) information affects the efficacy of the fundamental aim of the IR system, i.e., to deliver the information to the users to satisfy their INs. The effectiveness of this performance and user-system interaction could be further increased by reducing the overload of its users, both cognitive and information [5] and the behaviours the users often tend to apply when feeling the overload (e.g., satisficing behaviour [20]). The more accurate information the system possesses about the user and their context, the better the results of IR in terms of improved user satisfaction, increased confidence and the reduction of searchers' effort. Our study, built on the evaluation of the EEG data captured synchronously during the user-system interactions, offers a novel insight into the neural mechanisms constituting the user's awareness of their state of knowledge context. A future evaluation can be applied in a more realistic setting, allowing to test a broader range of user interactions and information-seeking procedures, including monitoring the manifestations of the user's prior- and post-retrieval confidence indicative of the subjective IR's performance. The diversification of the data analysis methods and the expansion of contrast measures (e.g., power spectrum, EEG topomaps) would help to produce a multidimensional insight into neural correlates linked with information searching. The direct identification of the spatio-temporal features of the user's cognitive context is relevant to advancing the development of a framework for a class of proactive interactive IR systems [1] built on the intelligence learnt from the searcher's neurophysiological data.

References

1. Andolina, S., et al.: Investigating proactive search support in conversations. In: Proceedings of the 2018 Designing Interactive Systems Conference, DIS 2018, pp.

1295–1307. Association for Computing Machinery, New York (2018). https://doi.org/10.1145/3196709.3196734
2. Arguello, J., Ferguson, A., Fine, E., Mitra, B., Zamani, H., Diaz, F.: Tip of the tongue known-item retrieval: a case study in movie identification. In: Proceedings of the 2021 Conference on Human Information Interaction and Retrieval, CHIIR 2021, pp. 5–14. Association for Computing Machinery, New York (2021). https://doi.org/10.1145/3406522.3446021
3. Bauer, E., Wilson, K., MacNamara, A.: Cognitive and affective psychophysiology (2020). https://doi.org/10.1016/B978-0-12-818697-8.00013-3
4. Bauer, P., Jackson, F.: Semantic elaboration: ERPs reveal rapid transition from novel to known. J. Exp. Psychol. Learn. Mem. Cogn. **41**, 271 (2014). https://doi.org/10.1037/a0037405
5. Belabbes, M., Ruthven, I., Moshfeghi, Y., Pennington, D.: Information overload: a concept analysis. J. Doc. **79**, 144–159 (2022)
6. Belkin, N., Oddy, R., Brooks, H.: ASK for information retrieval: part I. Background and theory. J. Doc. **38**, 61–71 (1982). https://doi.org/10.1108/eb026722
7. Bigdely-Shamlo, N., Mullen, T., Kothe, C., Su, K.M., Robbins, K.A.: The prep pipeline: standardized preprocessing for large-scale EEG analysis. Front. Neuroinform. **9**, 16 (2015). https://doi.org/10.3389/fninf.2015.00016
8. Brainerd, C., Reyna, V.: The Science of False Memory. Oxford University Press, Oxford (2005)
9. Chua, E.F., Ahmed, R., Garcia, S.M.: Effects of HD-tDCS on memory and metamemory for general knowledge questions that vary by difficulty. Brain Stimul. **10**(2), 231–241 (2017). https://doi.org/10.1016/j.brs.2016.10.013
10. Cole, C.: A theory of information need for information retrieval that connects information to knowledge. Information Today Inc. (2012)
11. Cooper, W.: A definition of relevance for information retrieval. Inf. Storage Retrieval **7**(1), 19–37 (1971). https://doi.org/10.1016/0020-0271(71)90024-6, https://www.sciencedirect.com/science/article/pii/0020027171900246
12. Diana, R., Ranganath, C.: Recollection, familiarity and memory strength: confusion about confounds. Trends Cogn. Sci. **15**, 337–8 (2011). https://doi.org/10.1016/j.tics.2011.06.001
13. Diana, R.A., Vilberg, K.L., Reder, L.M.: Identifying the ERP correlate of a recognition memory search attempt. Cogn. Brain Res. **24**(3), 674–684 (2005). https://doi.org/10.1016/j.cogbrainres.2005.04.001
14. Dien, J.: Issues in the application of the average reference: review, critiques, and recommendations. Behav. Res. Methods **30**, 34–43 (1998). https://doi.org/10.3758/BF03209414
15. Dien, J., Michelson, C., Franklin, M.: Separating the visual sentence n400 effect from the p400 sequential expectancy effect: cognitive and neuroanatomical implications. Brain Res. **1355**, 126–40 (2010). https://doi.org/10.1016/j.brainres.2010.07.099
16. Dimigen, O., Sommer, W., Hohlfeld, A., Jacobs, A., Kliegl, R.: Coregistration of eye movements and EEG in natural reading: analyses and review. J. Exp. Psychol. Gener. **140**, 552–72 (2011). https://doi.org/10.1037/a0023885
17. Ditman, T., Holcomb, P.J., Kuperberg, G.R.: An investigation of concurrent ERP and self-paced reading methodologies. Psychophysiology **44**(6), 927–935 (2007). https://doi.org/10.1111/j.1469-8986.2007.00593.x
18. Eugster, M.J., et al.: Natural brain-information interfaces: Recommending information by relevance inferred from human brain signals. Sci. Rep. (2016). https://doi.org/10.1038/srep38580

19. Franke, N., Radach, R., Jacobs, A., Hofmann, M.: No one way ticket from orthography to semantics in recognition memory: N400 and p200 effects of associations. Brain Res. **1639**, 88–98 (2016). https://doi.org/10.1016/j.brainres.2016.02.029
20. Gould, S.J., Cox, A.L., Brumby, D.P.: Task lockouts induce crowdworkers to switch to other activities. In: Proceedings of the 33rd Annual ACM Conference Extended Abstracts on Human Factors in Computing Systems, CHI EA 2015, pp. 1785–1790. Association for Computing Machinery, New York (2015). https://doi.org/10.1145/2702613.2732709
21. Guo, C., Duan, L., Li, W., Paller, K.A.: Distinguishing source memory and item memory: brain potentials at encoding and retrieval. Brain Res. **1118**, 142–154 (2006)
22. Hagoort, P.: Interplay between syntax and semantics during sentence comprehension: ERP effects of combining syntactic and semantic violations. J. Cogn. Neurosci. **15**(6), 883–899 (2003). https://doi.org/10.1162/089892903322370807
23. Hillyard, S., Hink, R., Schwent, V., Picton, T.: Electrical signs of selective attention in the human brain. Science (New York, N.Y.) **182**, 177–80 (1973). https://doi.org/10.1126/science.182.4108.177
24. Ingwersen, P.: Psychological aspects of information retrieval. Soc. Sci. Inf. Stud. **4**(2), 83–95 (1984). https://doi.org/10.1016/0143-6236(84)90068-1. Special Issue Seminar on the Psychological Aspects of Information Searching
25. Ingwersen, P.: Cognitive perspectives of information retrieval interaction: Elements of a cognitive IR theory. J. Doc. **52**, 3–50 (1996). https://doi.org/10.1108/eb026960
26. Ingwersen, P., Järvelin, K.: The Turn: Integration of Information Seeking and Retrieval in Context. Springer, Germany (2005). https://doi.org/10.1007/1-4020-3851-8
27. Kangassalo, L., Spapé, M., Jacucci, G., Ruotsalo, T.: Why do users issue good queries?: neural correlates of term specificity. In: Proceedings of the 42nd International ACM SIGIR Conference on Research and Development in Information Retrieval, SIGIR 2019, pp. 375–384. ACM, Association for Computing Machinery, USA (2019). https://doi.org/10.1145/3331184.3331243
28. Kappenman, E., Luck, S.: Best practices for event-related potential research in clinical populations. Biol. Psychiatry Cogn. Neurosci. Neuroimaging **1**, 110–115 (2015). https://doi.org/10.1016/j.bpsc.2015.11.007
29. Kuhlthau, C.C.: Inside the search process: information seeking from the user's perspective. J. Am. Soc. Inf. Sci. **42**(5), 361–371 (1991)
30. Luck, S.: An Introduction to the Event-Related Potential Technique (2005)
31. Minsky, M.: A Framework for Representing Knowledge. MIT-AI Laboratory Memo (306) (1974)
32. Moshfeghi, Y.: NeuraSearch: neuroscience and information retrieval. In: Alonso, O., Marchesin, S., Najork, M., Silvello, G. (eds.) Proceedings of the Second International Conference on Design of Experimental Search & Information REtrieval Systems, Padova, Italy, 15–18 September 2021. CEUR Workshop Proceedings, vol. 2950, pp. 193–194. CEUR-WS.org (2021). https://ceur-ws.org/Vol-2950/paper-27.pdf
33. Moshfeghi, Y., Pollick, F.: Neuropsychological model of the realization of information need. J. Assoc. Inf. Sci. Technol. **70**, 954–967 (2019). https://doi.org/10.1002/asi.24242
34. Moshfeghi, Y., Triantafillou, P., Pollick, F.E.: Understanding information need. In: Proceedings of the 39th International ACM SIGIR conference on Research and Development in Information Retrieval - SIGIR 2016 (2016). https://doi.org/10.1145/2911451.2911534

35. Mueller, V., Brehmer, Y., von Oertzen, T., Li, S.C., Lindenberger, U.: Electrophysiological correlates of selective attention: a lifespan comparison. BMC Neurosci. **9**, 18 (2008). https://doi.org/10.1186/1471-2202-9-18
36. Pinkosova, Z., McGeown, W.J., Moshfeghi, Y.: The cortical activity of graded relevance. In: Proceedings of the 43rd International ACM SIGIR Conference on Research and Development in Information Retrieval, SIGIR 2020, pp. 299–308. Association for Computing Machinery, New York (2020). https://doi.org/10.1145/3397271.3401106
37. Risko, E.F., Ferguson, A.M., McLean, D.: On retrieving information from external knowledge stores: feeling-of-findability, feeling-of-knowing and internet search. Comput. Hum. Behav. **65**(C), 534–543 (2016). https://doi.org/10.1016/j.chb.2016.08.046
38. Savolainen, R.: Information need as trigger and driver of information seeking: a conceptual analysis. Aslib J. Inf. Manag. **69**(1), 2–21 (2017). https://doi.org/10.1108/AJIM-08-2016-0139
39. Squire, L.R., Zola, S.M.: Structure and function of declarative and nondeclarative memory systems. Proc. Natl. Acad. Sci. **93**(24), 13515–13522 (1996). https://doi.org/10.1073/pnas.93.24.13515
40. Stróżak, P., Bird, C., Corby, K., Frishkoff, G., Curran, T.: Fn400 and LPC memory effects for concrete and abstract words. Psychophysiology **53**, 669–1678 (2016). https://doi.org/10.1111/psyp.12730
41. Taylor, R.S.: Question-negotiation and information seeking in libraries. Coll. Res. Libr. **29**(3), 178–194 (1968). https://doi.org/10.5860/crl_29_03_178
42. Voss, J.L., Paller, K.A.: pp. 81–98. No. September 2016, 3rd edn. Elsevier (2016). https://doi.org/10.1016/B978-0-12-809324-5.21070-5
43. Wilson, T.: On user studies and information needs. J. Doc. **37**(1), 3–15 (1981). https://doi.org/10.1108/eb026702
44. Wynn, S.C., Daselaar, S.M., Kessels, R.P., Schutter, D.J.: The electrophysiology of subjectively perceived memory confidence in relation to recollection and familiarity. Brain Cogn. **130**, 20–27 (2019). https://doi.org/10.1016/j.bandc.2018.07.003

Brain Structural Saliency over the Ages

Daniel Taylor[1(✉)], Jonathan Shock[1,2,3], Deshendran Moodley[1], Jonathan Ipser[1], and Matthias Treder[4]

[1] University of Cape Town, Rondebosch, Cape Town, South Africa
tyldan008@myuct.ac.za

[2] The National Institute for Theoretical and Computational Sciences, Stellenbosch, South Africa

[3] Institut National de la Recherche Scientifique, Montreal, Canada

[4] Cardiff University, Laval, Wales

Abstract. Brain Age (BA) estimation via Deep Learning has become a strong and reliable bio-marker for brain health, but the black-box nature of Neural Networks does not easily allow insight into the features of brain ageing. We trained a ResNet model as a BA regressor on T1 structural MRI volumes from a small cross-sectional cohort of 524 individuals. Using Layer-wise Relevance Propagation (LRP) and DeepLIFT saliency mapping techniques, we analysed the trained model to determine the most relevant structures for brain ageing for the network, and compare these between the saliency mapping techniques. We show the change in attribution of relevance to different brain regions through the course of ageing. A tripartite pattern of relevance attribution to brain regions emerges. Some regions increase in relevance with age (e.g. the right Transverse Temporal Gyrus); some decrease in relevance with age (e.g. the right Fourth Ventricle); and others are consistently relevant across ages. We also examine the effect of the Brain Age Gap (BAG) on the distribution of relevance within the brain volume. It is hoped that these findings will provide clinically relevant region-wise trajectories for normal brain ageing, and a baseline against which to compare brain ageing trajectories.

1 Introduction

Brain Age (BA) prediction by Deep Learning (DL) methods using structural Magnetic Resonance Imaging (MRI) data has shown to be an accurate and reliable bio-marker for brain health [5]. The Brain Age Gap (BAG) of an individual – the difference between the predicted BA and the chronological age – is an increasingly popular and predictive bio-marker for brain health, and has been shown to predict accelerated or slowed brain ageing [7,27,38].

DL frameworks have shown great efficacy in the BA regression task [4,19, 22,27]. An advantage of DL models is that they are able to analyse whole-brain structural images with minimal pre-processing. Other summary statistics such as measures of cortical thickness, volume and surface area [45] treat the brain as being modular, and may therefore lack details contained in minimally-processed

G. Nicosia et al. (Eds.): LOD 2022, LNCS 13811, pp. 525–548, 2023.
https://doi.org/10.1007/978-3-031-25891-6_40

MRI volumes (which represent the entire structure with all its integrated substructures). Convolutional Neural Networks (CNNs) have been shown to provide extremely small Mean Absolute Error (MAE) between true age and BA. The state-of-the-art in BA regression at the time of writing is around 2.5y MAE [7,22,27]. Such accuracies have thus far only been achieved on large datasets, and some of these datasets are limited in their age ranges, such as the UK BIOBANK [40] (45y–80y).

The black-box nature of Neural Networks makes it difficult to attribute BAG to specific brain regions. Many methods of Explainable Artificial Intelligence (XAI) have aimed at alleviating this issue. These attempt to explain why the models make their predictions. A popular post-hoc method of reasoning for DL models is saliency mapping. This is a group of methods by which areas of input are assigned relevance scores proportional to their saliency to the model decision. We compare the results of two widely used saliency mapping methods. We explore Layer-wise Relevance Propagation (LRP) [3] and DeepLIFT [32] as saliency mapping methods for the BA task. These methods were chosen due to their past performance in the literature in DL-based brain imaging tasks [9,12,14,16,23]. In this work, we use the terms 'salience' and 'relevance' interchangeably.

Many factors can contribute to accelerated BA (positive BAG), such as the presence of neurodegenerative disease, type 2 diabetes or HIV, and past physical activity [5]. If we wish to create a model that accurately predicts BA, and therefore accurately assesses BAG, we must ensure that the model accurately captures a path of 'normal' brain ageing. It must thus be trained on data in which the subjects do not have any underlying pathologies that can affect BA independent of actual age (to the extent that such pathologies are able to be detected). Even in a 'healthy' cohort, however, there will be non-negligible deviation from an average ageing trajectory. Smith et al. [38] note that some meaningful BAG should exist between the predicted and chronological age of an individual, as long as the model does not badly over-fit.

In this work we aim to shed light on the contributions of specific brain regions to BA and BAG through the course of ageing. Specifically we apply saliency mapping techniques to a BA regression model and compare the results to known characteristics of brain ageing. We then analyse the differences between saliency mapping techniques' distributions of relevance throughout the brain volume. We examine the link between region-specific saliency and accelerated brain ageing both in older and younger individuals. Finally, we create region-specific trajectories of BA saliency across age groups from a population study.

The primary contribution of this research is the development of methods to determine trajectories of BA saliency over the course of age; allowing us not only to determine the saliency of the region towards age and the change of this through time, but also to create baseline saliency trajectories against which to compare and assess BA of individuals at a region-specific level. Another contribution of the work is the analysis of regional saliency in individuals with accelerated BA (large BAG), which allows us to determine key contributing regions to accelerated BA in younger and older adults. Our final contribution is the comparison of different saliency mapping techniques, allowing us not only to compare the similarities and differences between them, but also to shed light on different aspects of BA through these differences.

From our findings in the current literature, we expected that regions that are deemed highly relevant in general towards BA would increase in their relevance with age. Areas that are generally less salient would then necessarily decrease in assigned relevance.

2 Related Work

Levakov et al. [23] used SmoothGRAD [35] to create population-level saliency maps for a BA regression model. The authors examined which brain regions corresponded with the highest attribution of saliency by thresholding the top-1% of voxels in saliency maps and examining clusters with > 100 voxels. Most salience was attributed to the cisterns and ventricles.

Hofmann et al. [16] used LRP to create saliency maps for two multi-ensemble BA regression models. They note that the SmoothGRAD method employed by [23] is not directional, while LRP highlights areas of input that both agree with and contradict the output. They utilised the LRP_{CMP} saliency method with $\alpha = 1$ (see Appendix (C)). The authors verified the method on a simulated brain ageing model and found that in the regression task, regions of higher relevance argue for greater age, while regions of lower relevance argue for lower age. After verifying the method, the authors created the LRP saliency maps for the real BA regression task. Like [23] they found that relevance was greatest in and around the ventricles. They also found a significant attribution to the grey-matter-dense regions at the cortical surface. The authors also contrasted the saliency maps of individuals in a younger and an older cohort to determine the difference in attribution of relevance.

There have been no previous attempts in the literature to compare the results of different saliency mapping techniques on the BA regression task. While [16] compared a younger and older cohort in their study, no previous work has focused on the change in relevance attributions to brain regions as a function of subject age. The authors also only examined statistical significance toward BAG in an older cohort, and not age-related contributors to BAG. Furthermore, the concern raised by Geirhos et al. [11] and Sixt et al. [34] about modified backpropagation algorithms like LRP has not been addressed. Their concern is that such methods attempt to recreate the input to the model, as opposed to focusing solely on areas of relevance.

3 Methods

3.1 BA Regression Model

The model used in the BA regression task was a ResNet [15] with filter number sequence [32, 32, 64, 64, 128] in the main branches of the 5 residual blocks.

This configuration of residual blocks is borrowed from [19]. We used a softplus activation function for all the nonlinearities in the model, as recommended by Dombrowski et al. [8], for the robustness of the saliency maps.

We used the cross-sectional Cam-CAN T1-weighted MRI dataset [31,43] for our experiments. The volumes are from 656 healthy individuals ranged 18–89y, with 49.4% or 324 male participants.

We trained our ResNet model on a random 80% training split of the dataset (524 subjects). 80 of these were randomly held out for validation (15% validation split from the beginning of training). The remaining 20% of the dataset (132 subjects) was used in testing. We used grid-search hyper-parameter tuning to find the best optimiser for the model (Adam [20] vs RMSProp), as well as the best loss function (Mean Squared Error vs Mean Absolute Error) and the best starting learning rate (5×10^{-4} vs 10^{-3} vs 5×10^{-3}). Taking the average of three trials for each hyper-parameter configuration, it was found that the best configuration was using the Mean Squared Error (MSE) loss function with the Adam optimiser and an initial learning rate of 5×10^{-3}. For a set of n predictions $[f(x_i)]$ and the corresponding set of n true ages $[y_i]$, the MSE is defined as $\mathrm{MSE}\left([f(x_i)], [y_i]\right) = \frac{\sum_i (f(x_i) - y_i)^2}{n}$. We did not perform hyper-parameter tuning on any other hyper-parameters as the rest of the model configuration was based off of the model of [19]. The learning rate was halved every 20 epochs using a learning rate schedule. We also utilised an early stopping callback in training such that if there was no validation improvement for 20 epochs, the training would cease. Due to the large size of the data ($233 \times 189 \times 197$ voxels per scan at $1\,\mathrm{mm}^3$ resolution), the batch size was limited to 4. The metrics used to evaluate the model were the MAE, MSE and Mean Average Percentage Error (MAPE). Appendix (A) details the pre-processing of the data.

3.2 Saliency Mapping

We performed saliency mapping to analyse the trained model using LRP [3,26] and DeepLIFT [32,33]. We used three variations of $\mathrm{LRP}_{\mathrm{CMP}}$ (as this method is considered to be best-practice [21]), with the parameters $\alpha \in \{1, 2, 3\}$ (the same values implemented by Grigorescu et al. [12]). We refer to these by the shorthand LRP_1, LRP_2 and LRP_3. In all three cases, we used the *Epsilon Rule* for dense layers and the $z^{\mathcal{B}}$*-Rule* for the input layer [26]. We used two variations of the DeepLIFT method. In the first case ($_{\mathrm{bg}}$), we used a reference input of all zeros (the MRI background). In the second case ($_{\mathrm{comp}}$) we used as reference a composite MRI volume, which was the aggregate of all the subjects' volumes from the test set. For the nonlinear softplus activation layers, we used the Reveal-Cancel Rule [32], which was found to produce more consistent and interpretable saliency maps than the Rescale Rule. Both DeepLIFT methods allow for positive and negative relevance assignment.

For all methods, we used permutation-based t-tests to determine statistically significant differences from the input volumes. This is in order to address the concern of [11] that the LRP methods simply recreate the input.

Regional Relevance Attribution Between Methods. To compare the saliency mapping methods, we examined the attribution of top-1% relevance (T1R) to different brain areas across the test set, per method. T1R refers to the voxels with the first percentile of activations in the brain volume for a saliency map. This is similar to the methodology used in [23]. We count the number of such voxels per region, and normalise by the region volume to get the proportion of the region assigned T1R. We assessed the similarities and differences in regional relevance attribution between the methods, to determine how each of the methods explains BA.

To determine which brain regions saliency voxels were attributed to, we used the CerebrA 2009c atlas [6,10] and a corresponding standard MNI brain volume [25] to which all of the dataset volumes were aligned before training. This allowed us to map regions directly onto each individual's brain volume (and thus the corresponding saliency map) to determine the proportion of T1R per region. It is by this method that we are able to compare the regional attributions of T1R among participants.

From the findings of [16], we know that regions of high saliency argue for higher age and those of lower saliency argue for lower age in the context of BA regression, and so the directionality of relevance is important. We are most interested in regions which contribute to accelerated brain ageing, and so we consider only positive relevance in T1R (see the discussion of directionality by [16]).

Regional Relevance Attribution Based on BAG. We used the δ_2 definition of Smith et al. [38], which uses an orthogonality matrix multiplication on the traditional chronological-predicted BA difference to remove dependence of the BAG on chronological age.

To examine the effect of large BAG on the distribution of relevance in the brain, we thresholded BAG in the test set and compared the T1R distributions of those lying above and below the threshold δ^*. This has not been done before in the literature, and so we made the simple choice of a threshold for BAG of $\delta^* = \sigma$, where σ is the standard deviation of the BAG values for the test set. We compared the distribution of T1R within the brains of those with BAG δ lying above the threshold ($\delta \geq \delta^*$) to those lying within the threshold ($|\delta| < \delta^*$). To calculate BAG, we used the age-orthogonal value δ_2 of [38]. We compare the T1R distributions of three groups for each method: $\delta_2 \geq \delta^*$ for younger individuals (< 50y), $|\delta_2| < \delta^*$, and $\delta_2 \geq \delta^*$ for older individuals (> 50y).

Regional Relevance Attribution Across Age Brackets. To assess the change in the distribution of relevance as a function of age, we examined how the T1R assignment for each method changes between equally spaced age brackets. We did this for each region individually, by calculating the average proportion of T1R assigned to the region across the test set members who lie within each age bracket.

Of greatest interest were highly-relevant regions whose relevance changes most or least across ages. To quantify the greatest change, we examined those with the highest standard deviation (SD) in the proportion that is assigned

T1R. To quantify the least change, we examined those with lowest coefficient of variation (CoV – standard deviation normalised by mean).

We did not focus on those regions with lowest SD since those tend to be the regions with lowest mean relevance attribution. Similarly we did not focus on those with highest CoV, since these also tend to be regions with low mean attributions, where a small change in relevance attribution can lead to a very large CoV. Our aim was to focus on the changes in relevance for highly-relevant structures. More is given on this choice of metrics in Appendix (D).

4 Results

4.1 BA Regression Model

The training was stopped by the early stopping callback at the end of the 143rd epoch, at which pint it had reached a test MAE of 6.55y, with an MSE of 72.55y^2 and an MAPE of 13.53. The performance on the evaluation metrics is given in Table (1), and the regression plot is shown in Fig. (5); both in Appendix (B).

4.2 Saliency Mapping

We report here the findings of the saliency mapping tasks. Figure (1) shows sections of the composite volume overlaid with aggregate maps for each method. The composite volume was created as the aggregate of all the brain volumes

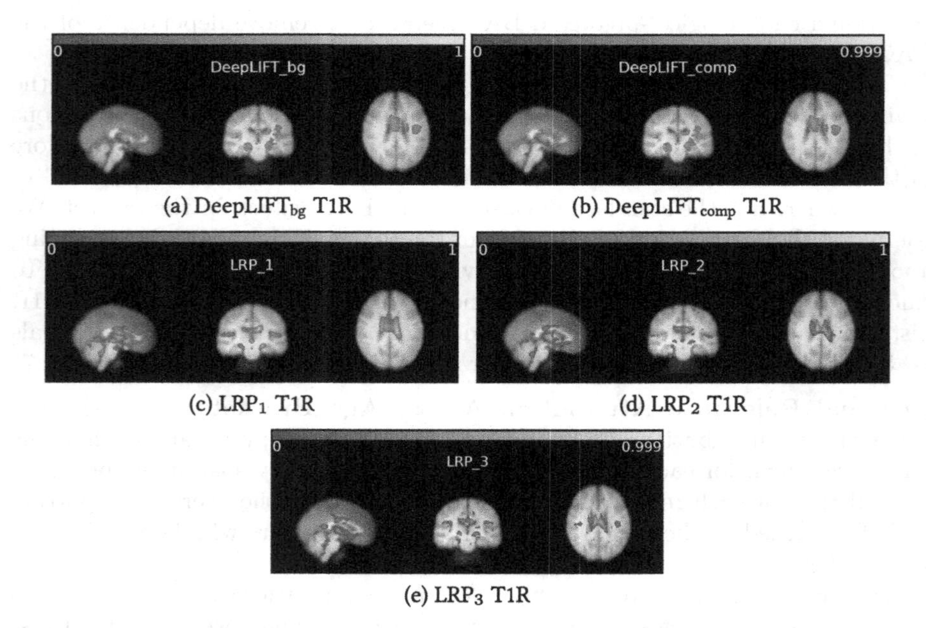

(a) DeepLIFT$_{bg}$ T1R (b) DeepLIFT$_{comp}$ T1R (c) LRP$_1$ T1R (d) LRP$_2$ T1R (e) LRP$_3$ T1R

Fig. 1. Sections of T1R for each method, overlaid on the composite MRI volume.

from the test set, and each overlaid saliency map is similarly the aggregate of all the saliency maps from the test set for that method. The aggregate map was thresholded to the top-1% of voxel values to show the average distribution of T1R over the test set. We also show in Fig. 6 in Appendix (C) an example of overlaid LRP salience on sections of the volume of a 60-year-old male subject.

The permutation-based t-tests showed that there were statistically significant ($p < 0.001$) differences between the saliency maps and the input volumes for each method in almost the entire brain volume.

Regional Relevance Attribution Between Methods. Here we show the results of the assignment of relevance to brain regions by the five saliency mapping techniques. We compare the results to previous findings as well as to some of the medical BA literature, and examine differences between the methods.

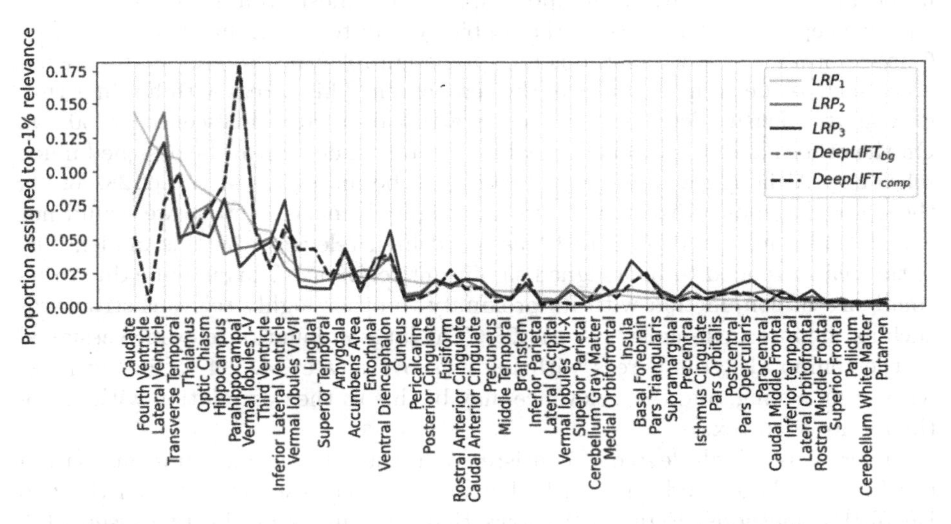

Fig. 2. Proportion of each region assigned Top-1% Relevance per method. Regions are ordered in descending proportion of T1R as determined by LRP_1.

Figure (2) shows the proportion of each region that is assigned T1R, for each method. The relevance assignment in this case was averaged between the two hemispheres, since there was a large degree of symmetry in the distribution between hemispheres (see Appendix (C)). The ordering of the regions along the x-axis is in descending order of proportion for LRP_1.

It is clear that the LRP methods assign high T1R to the ventricles (particularly the lateral and fourth), which agrees not only with the medical literature [2,30], but also with the findings of Levakov et al. [23] and Hofmann et al. [16]. The DeepLIFT methods also assign high relevance to the lateral and inferior lateral ventricles, but less so than the LRP methods. Upon inspection of the DeepLIFT multipliers, it was clear that the same trend was present regardless of multiplication (masking) by the input. It appears that instead of assigning relevance to the ventricles themselves like the LRP methods, DeepLIFT tends to assign relevance to the immediate surrounds of the ventricles. DeepLIFT appears to highlight the areas which contract with age, whereas the LRP methods highlight the ventricles themselves, which dilate over time. This trend is displayed in the saliency sections of Fig. (1).

The transverse temporal gyrus was consistently assigned significant relevance by all methods, as were many limbic structures, such as the caudate nucleus, hippocampus, thalamus, parahippocampal gyrus (most notably by DeepLIFT), and diencephalon. Relevance of the limbic system to ageing has been shown by Gunbey et al. [13]. Both methods tend to assign relevance to grey-matter-dense areas such as the vermal lobules of the cerebellum. This agrees with findings that grey matter density decreases with age, starting from late adolescence [28,30,39]. On the other hand, the cerebellum white matter tended not to be assigned much relevance. Although white matter lesions can be indicative of BA [2,28,30,39], these are not shown distinctively in T1-weighted images, and so we would not expect white matter regions to be assigned high relevance by this model. The optic chiasm tended to be assigned significant relevance. This may be due to its small size in conjunction with its proximity to other highly relevant structures such as the interpeduncular cistern [23]. The large amount of relevance assigned to the temporal gyrus agrees with findings by Lutz et al. [24] of involvement with brain ageing, and this relevance may be due to the degradation with age of the auditory cortex.

There was a high degree of consistency in the relevance assignments within the LRP methods, and an even higher degree of consistency between the two DeepLIFT methods. Figure (2) shows that the curves of the two DeepLIFT methods almost perfectly overlap. There was some similarity in the trend of relevance assignment between DeepLIFT and LRP. The biggest difference was that DeepLIFT assigns higher relevance to limbic structures like the parahippocampal gyrus, while LRP assigns more relevance to the ventricles.

Regional Relevance Attribution Based on BAG. Here we compare the distribution of relevance in individuals with small BAG to those with large BAG, both older and younger. We examine how regional relevance associates with BAG for older and younger individuals.

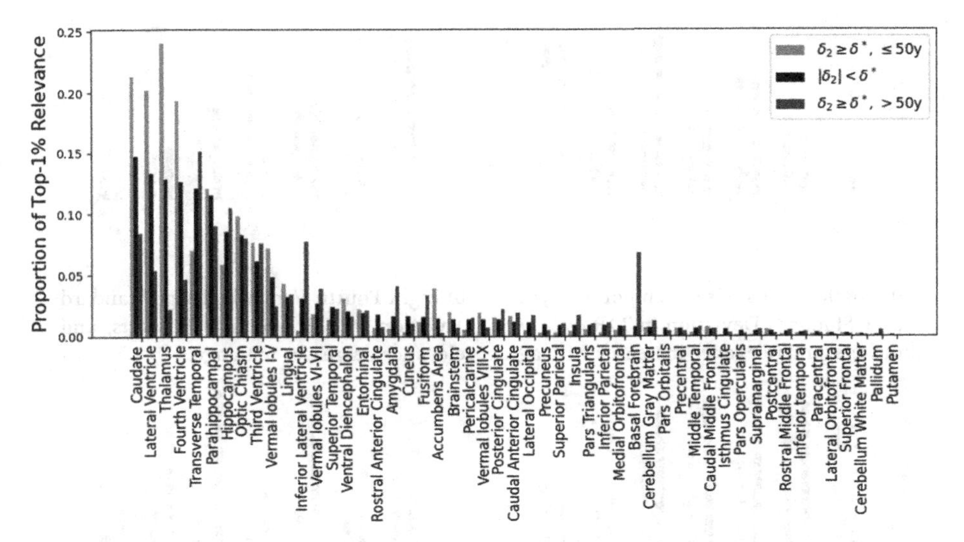

Fig. 3. Distributions of Top-1% Relevance, via LRP_1, in young ($\leq$ 50y) individuals with high BAG ($\delta_2 > \delta^*$), elderly individuals (> 50y) with high BAG, and individuals with small-to-moderate BAG ($|\delta_2| < \delta^*$). Regions are ordered by descending proportion of T1R in the small-to-moderate BAG group.

Figure 8 in Appendix (D) shows the distributions of the quantities δ_1 and δ_2, as defined in [38], for the test set. The correction from orthogonalisation to age shifted the distribution to become approximately Gaussian. This allowed us to threshold BAG reliably to compare BAG-salient regions in younger and older groups. On the test set, our threshold value was $\delta^* = \sigma = 11.58$y.

Figures 9, 10, 11 and 12 in Appendix (D) compare the distributions of T1R for these groups according to each method. We have included in Fig. (3) this distribution according to LRP_1. We see that several regions displayed significant changes in T1R assignment according to grouping by BAG and age. By way of example from Fig. (3), for large BAG values, DeepLIFT assigned to the fourth ventricle a significantly greater relevance in younger individuals than in older individuals. The opposite was true for the transverse temporal gyrus. Since increased relevance is predictive of higher BAG, these results suggest that markers of BAG change through the course of ageing. It appears that regions can be more indicative of BAG at some ages than others.

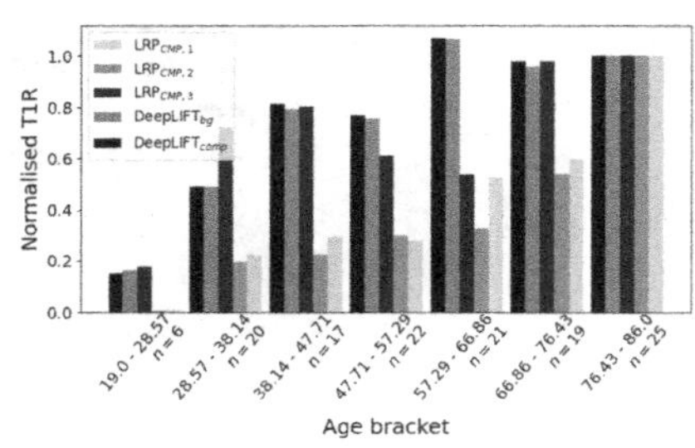

(a) Right Transverse Temporal Gyrus. Large Standard Deviation in T1R over age brackets, and relevance increases with age.

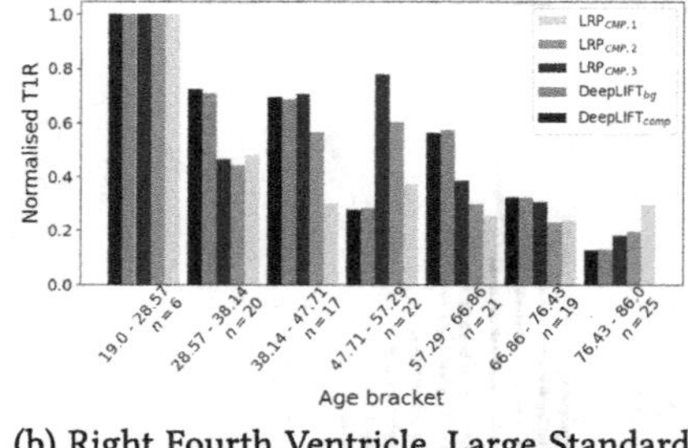

(b) Right Fourth Ventricle. Large Standard Deviation in T1R over age brackets, and relevance decreases with age.

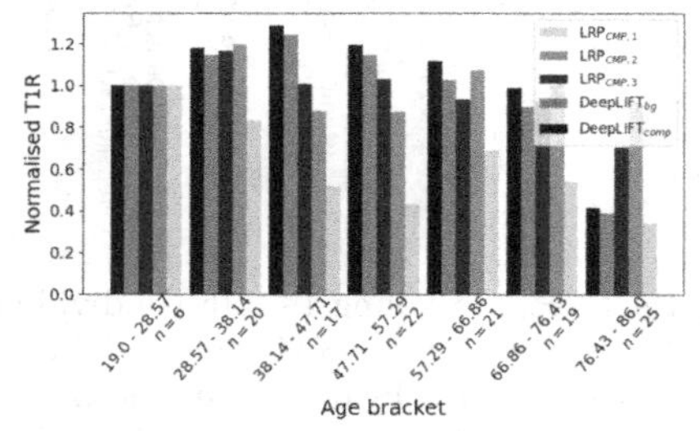

(c) Right Lateral Ventricle. Low CoV in T1R over age brackets.

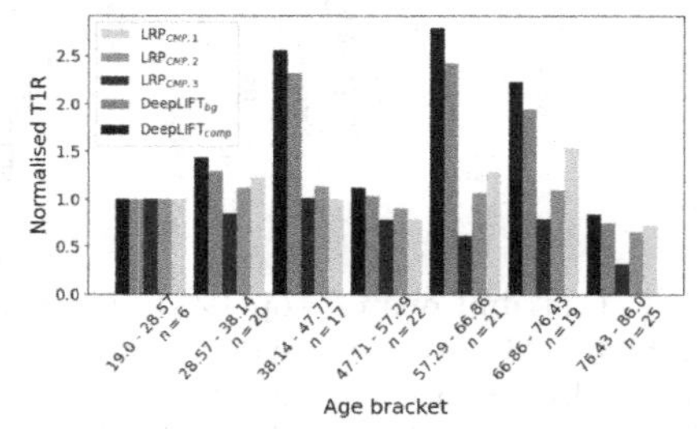

(d) Right Caudate Nucleus. Low CoV in T1R over age brackets.

Fig. 4. 'Proportion' of Top-1% Relevance per method over the age brackets. The proportion of T1R is normalised in each sub-figure such that either the youngest or oldest group have T1R assignment of 1. Brackets all have an age range of 9.57y. Four example regions are shown.

Regional Relevance Attribution Across Age Brackets. Here we examine the change in relevance across age brackets for brain regions according to each saliency mapping method. In Fig. 13 in Appendix (E) we show the standard deviations and coefficients of variation in the proportion of regions assigned T1R over all age brackets.

One region which has a large SD over ages for all methods is the transverse temporal gyrus. We show the distribution across age brackets for the right hemispheric structure in Fig. (4a) for each method. For uniformity in scale we normalise the proportions per method such that either the oldest or youngest age bracket is 1 for all methods. All methods, especially LRP, show an increase in relevance of this region with greater age.

Greater attribution of relevance in older subjects than in younger subjects suggests that the structure bears more information about BA and BAG in older individuals. This agrees with our findings about relevance distribution as a function of BAG. The transverse temporal gyrus for example is a more salient indicator of accelerated brain ageing in older subjects than in younger subjects. On the other hand, we see in Fig. (4b) the distribution across age brackets of the same quantity for the right fourth ventricle. This region too is shown to have significant SD across age brackets. There is a marked decline in assignment of T1R from younger to older ages. This suggests that while the fourth ventricle is universally relevant to BA and BAG, it is most telling of ageing in younger individuals.

Figure (4c) shows the distribution across age brackets of the same quantity, now for the right lateral ventricle. This was shown to have uniformly low CoV across the methods. We see that there is a decrease in relevance in the oldest age bracket for DeepLIFT and LRP_3, but otherwise the relevance assignment is uniform.

Figure (4d) shows the age bracket distribution of T1R for the right caudate nucleus. The region has a low CoV for each method. Although there is some disparity in assigned relevance over the age brackets (especially within DeepLIFT relative to the others), there is no overarching upward or downward trend.

5 Discussion and Conclusions

While the test set MAE achieved by the model of 6.55y is not SOTA, no BA model in the literature has achieved such accuracy on a dataset of comparable size. This would indicate that while very large datasets may be necessary to achieve SOTA performance, they may not be necessary simply to train a BA regression model of reasonable accuracy (with a coefficient of correlation to chronological age of $r = 0.89$ – see Fig. (5) of Appendix (B)). Since it can be difficult to access very large datasets of healthy individuals across the full adult lifespan, it may be preferable for some studies to use smaller datasets such as Cam-CAN. Table (2) of Appendix (B) compares the MAE performance of several high-performing BA regression models and shows the dataset size used in each study.

The three LRP methods are relatively consistent in their distribution of T1R among the brain regions, and the two DeepLIFT functions even more so. By way of comparison, there is a notable similarity between LRP and DeepLIFT in BA relevance assignment. This is to be expected, since the methods are very closely related [33], and DeepLIFT can in fact represent a generalisation of some LRP methods [1]. The LRP methods assigned T1R to the ventricles in large proportion, agreeing with the findings of Levakov et al. [23] and Hofmann et al. [16]. DeepLIFT, however, tends to assign relevance to the immediately surrounding regions, and more heavily to limbic areas like the parahippocampal gyrus. There is no precedent for such differences in explanations in the literature, past suggestions for what such differences may mean, or where they may come from. We speculate that LRP and DeepLIFT may highlight the importance of the same phenomenon from two different perspectives, in that the dilation of the ventricles and the atrophy of the surrounding regions are causally coupled.

Both LRP and DeepLIFT assign high relevance to the ventricles and to limbic structures, which agrees both with our expectations and with literature on BA [2,13,24,30]. High proportions of relevance are assigned to grey-matter-dense areas, which also agrees with some medical literature [2,13,30].

There are non-trivial differences in the distributions of relevance between LRP and DeepLIFT, but both methods highlight known areas of importance to brain ageing. The utility of one method over the other may depend on the user's main regions of interest. It may be best to use a combination of LRP and DeepLIFT to form BA explanations. LRP_1 is the most commonly used LRP method, and is the least noisy of those we implemented, since higher values of α create greater contrast in the saliency maps [3,12]. This is illustrated in Figs. (1c)–(1e), where higher values of α decrease the smoothness of T1R clusters. As we have seen, the two DeepLIFT methods perform extremely similarly, and both are simple to implement; thus we cannot conclude that one is more useful for BA explanation than the other.

We found that large BAG values are associated with distributions in relevance among brain regions that can vary to a large degree depending on age. This suggests that the degree of saliency of a region towards BAG is age-dependent. Upon inspection of age-bracket distribution of relevance to brain regions, we showed that a tripartite pattern emerges. Some regions increase in relevance with age, some decrease with age and others retain relatively uniform relevance with age. From our findings with BAG, decreases with age would suggest that the region is more informative of BAG in younger individuals than in older individuals (for example, the fourth ventricle). Similarly increases with age would suggest that the region is more informative of BAG in older individuals than in younger ones (for example, the transverse temporal gyrus). Uniformity across ages in relevance would suggest that the region is consistently informative of BAG (for example, the caudate nucleus).

This tripartite pattern of trajectories was unexpected, since previous studies have only reported on increased relevance with age [16]. Indeed, many regions decrease in the proportion assigned T1R with age. This can be explained for a structure like the right fourth ventricle, by the fact that although the structure is informative of BA at all ages, it is expected to dilate with age. A younger individual with dilated ventricles is clearly experiencing accelerated brain ageing.

These findings may be used as baseline regional trajectories for BA salience in a clinical setting. As part of a DL pipeline for BA regression, clinicians can create saliency maps for patients' BA predictions and compare the regional distributions of T1R to these baselines. Large individual BAG can be assessed regionally by these methods. Saliency mapping can be performed post-hoc on the regression model in close to real-time, so as to have BA predictions accompanied by an explanation in a clinical setting.

Without access to this model and its accompanying results, clinicians will still be able to use the underlying methods – perhaps with more powerful models trained on larger datasets – to create and compare relevance trajectories for clinical use.

In addition to the BA regression model and explanation pipeline, and the baseline regional saliency trajectories, we have also contributed a regional analysis of BA saliency in individuals with accelerated BA (large BAG), showing that some regions show increased BA in younger individuals and others show increased BA in older individuals. Finally, we have also contributed a comparison of several saliency mapping techniques from two classes of technique, showing that there are some significant differences in explanations between the classes of explanation techniques.

A limitation of this work is the use of single-fold cross-validation in the analysis of the BA regression model. A 5-fold cross-validation applied to larger datasets would ensure robust model accuracy on unseen data. In the future it would also be of great interest to examine how the distribution of relevance within the brain may change according to different model architectures and across different datasets. Another potential avenue for future work is to see how the size of a dataset may influence the distribution of relevance through the brain volume. It would also be of interest to compare the saliency maps of other, possibly more computationally expensive techniques, such as Integrated Gradients [41,42].

Acknowledgements. Funding for this work was provided by the University of Cape Town Postgraduate Funding Office, the University of Cape Town Computer Science Department, and the ETDP-SETA. Training and saliency mapping was performed on the Lambda Cloud GPU platform, using their A6000 instances. Preprocessing utilised FSL's [37] BET and FLIRT tools [17,18,36]. The randomise tool was used as well [44], for the permutation-based t-tests. Brain regions were determined using the CerebrA 2009c atlas [6,10,25].

A Appendix 1

The standard pre-processing of the Cam-CAN dataset is detailed by Taylor et al. [43]. We used FSL's Brain Extraction Tool (BET) [36] to perform skull-stripping on the T1-weighted volumes, using the recommended fractional intensity of 0.5. We then aligned all volumes to MNI space by registration to a standard MNI volume [25], using the FLIRT tool [17,18]. This was done primarily for spatial correspondence with the region atlas [6,10]. Voxel values were normalised using Global Contrast Normalisation (GCN) to have unit variance within each volume.

B Appendix 2

Figure (5) shows the regression plot for our ResNet model on the test set, and Table (1) shows the performance metrics. We note that the MAE of 6.5y is far from state-of-the-art, but also that the size of the dataset is far less than those of previous works. Table (2) shows the performances of some key studies for BA regression, with the number of individuals in the datasets used.

Our ResNet had filter sizes [32, 32, 64, 64, 128] in the main branches of the five residual modules. This configuration was adapted from the lightweight model of Jonsson et al. [19].

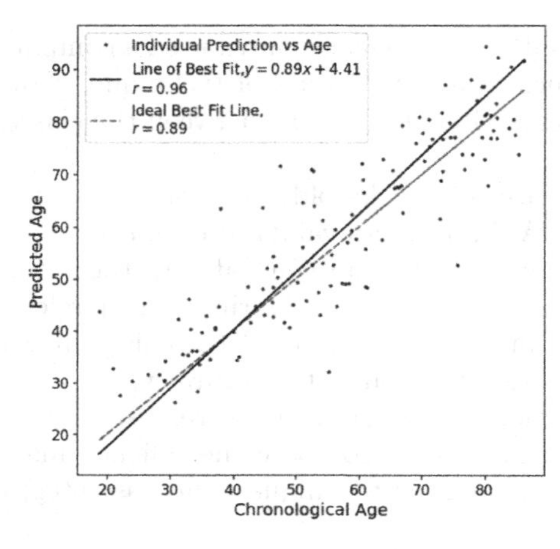

Fig. 5. Test set regression plot

Table 1. Metrics

Metric	Result
MAE	6.5457
MSE	72.5481
MAPE	13.5305

Table 2. Test MAE performances of several key BA regression studies, as compared to ours. Our dataset is at least three times smaller than any other used here.

Author	Ours	Cole et al. (2017)	Jónsson et al. (2019)	Kossaifi et al. (2020)	Levakov et al. (2020)	Hofmann et al. (2021)	Peng et al. (2021)
Dataset Size	656	2001	12378	19100	10176	2016	14503
Test MAE (y) on T1 Volumes	6.55	4.65	4.00	2.69	3.02	3.95	2.14

C Appendix 3

LRP and DeepLIFT are relevance decomposition methods which assign scores to the elements of a model's input space according to their saliency to the model's decision. Both methods take into account the learned parameters of the model and the activations at each layer of the model specific to the inference of a given input. Relevance decomposition refers to the backward propagation of relevance from the output layer of a model to its input. We chose LRP and DeepLIFT because these are both fast and computationally inexpensive saliency mapping methods. Both methods have also been used before in reasoning for brain imaging [9,12,14,16,29], and offer directionality in their explanations, unlike simpler method such as SmoothGRAD [35]).

The decomposition rules for LRP_{CMP} [21] from a layer $(L+1)$ to a layer L are as follows:

For fully-connected layers, the ε-rule [3] is used:

$$R_i^{(L)} = \sum_j \frac{z_{ij}}{\sum_{i'} z_{i'j} + \varepsilon} R_j^{(L+1)} \tag{1}$$

where $R_i^{(L)}$ is the relevance assigned to neuron i in layer L, $z_{ij} = x_i w_{ij}$ is the product of previous layer activations and the learned weights between the layers, and ε is a small constant for numerical stability.

For convolutional, BatchNorm and pooling layers (apart from the input layer), the LRP$_{\alpha\beta}$ rule [3] is used:

$$R_i^{(L)} = \sum_j \left(\alpha \frac{(x_i w_{ij})^+}{\sum_{i'} (x_i w_{ij})^+ + b_j^+} - \beta \frac{(x_i w_{ij})^-}{\sum_{i'} (x_i w_{ij})^- + b_j^-} \right) R_j^{(L+1)} \tag{2}$$

with the constraint that $\alpha - \beta = 1$. w_{ij}^+ refers to the positive parts only of the weights and b_j^+ to the positive parts only of the biases.

For the input layer, the $z^{\mathcal{B}}$-rule [26] is used:

$$R_i^{(L)} = \sum_j \frac{z_{ij} - l_i w_{ij}^+ - h_i w_{ij}^-}{\sum_{i'} z_{i'j} - l_{i'} w_{i'j}^+ - h_{i'} w_{i'j}^-} R_j^{(L+1)} \tag{3}$$

where h_i and l_i are the highest and lowest input values respectively.

The parameters α and β of LRP act as contrast parameters, with higher values of α giving higher contrast in the saliency maps [3].

The DeepLIFT decomposition rules from a neuron y in a layer to a neuron x in a layer above are given by contribution scores [32], allocated according to layer-specific rules. The *Linear Rule* for all linear layers, allocates contribution scores as follows:

$$\begin{aligned}
C_{\Delta x_i^+ \Delta y^+} &= 1\{w_i \Delta x_i > 0\}\, w_i \Delta x_i^+, \\
C_{\Delta x_i^- \Delta y^+} &= 1\{w_i \Delta x_i > 0\}\, w_i \Delta x_i^-, \\
C_{\Delta x_i^+ \Delta y^-} &= 1\{w_i \Delta x_i < 0\}\, w_i \Delta x_i^+, \\
C_{\Delta x_i^- \Delta y^-} &= 1\{w_i \Delta x_i < 0\}\, w_i \Delta x_i^-
\end{aligned}$$

where $C_{x_i y}$ is the contribution of x to y, Δx refers to the difference-from-reference of the neuron x, and Δx^+ is the positive-only part of that quantity. The difference-from-reference for a neuron x is calculated as the difference between the activations of that neuron for the given input and a reference input, $\Delta x = x^0 - x$.

The *Reveal-Cancel Rule* is used for non-linear layers and allocates contribution scores as follows:

$$\begin{aligned} C_{\Delta x^+ \Delta y^+} =& \frac{1}{2}\left(f(x^0 + \Delta x^+) - f(x^0)\right) \\ &+ \frac{1}{2}\left(f(x^0 + \Delta x^- + \Delta x^+) - f(x^0 + \Delta x^-)\right) \\ C_{\Delta x^- \Delta y^-} =& \frac{1}{2}\left(f(x^0 + \Delta x^-) - f(x^0)\right) \\ &+ \frac{1}{2}\left(f(x^0 + \Delta x^+ + \Delta x^-) - f(x^0 + \Delta x^+)\right) \end{aligned}$$

where f refers to the nonlinearity function – in our case the Softplus function.

The reference input is a choice of the user for DeepLIFT. The authors [32] state that there can be multiple 'good' choices for reference inputs, and that the choice is dependent on the task. This is our reason for having chosen two different reference inputs to compare. Indeed, the results were almost identical for distribution of T1R. Ancona et al. [1] show that LRP and DeepLIFT are similar enough that DeepLIFT relevance decomposition can be rewritten to look like LRP, with difference-from-reference values. The authors note that the significant difference between the methods is the use of reference inputs in DeepLIFT.

We show in Fig. 6 the overlay of T1R of a single subject onto their specific brain volume, as an example.

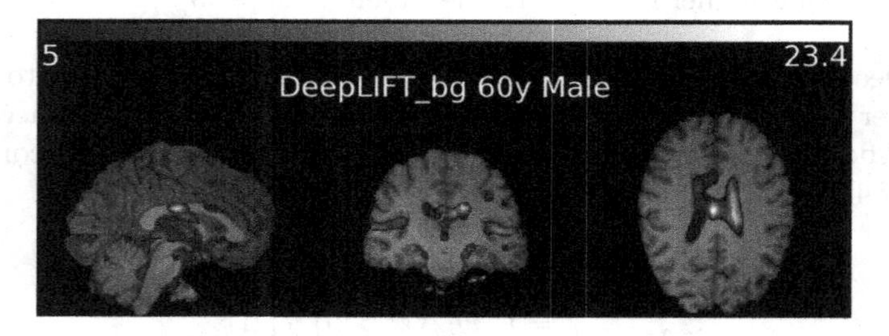

Fig. 6. Individual saliency map for a 60-year-old male subject, thresholded to T1R and overlaid onto the subject's MRI volume

Figure 7 shows the distributions of the coefficients of correlation for relevance attribution to the same region in opposite hemispheres, for each method. We see that the majority of regions have high coefficients of correlation between hemispheres. A few do not have high correlations, and these tend to be regions which are assigned low proportions of T1R. This correlation between hemispheres implies a symmetry across the hemispheres, and therefore allows us to average relevance between hemispheres in our analysis.

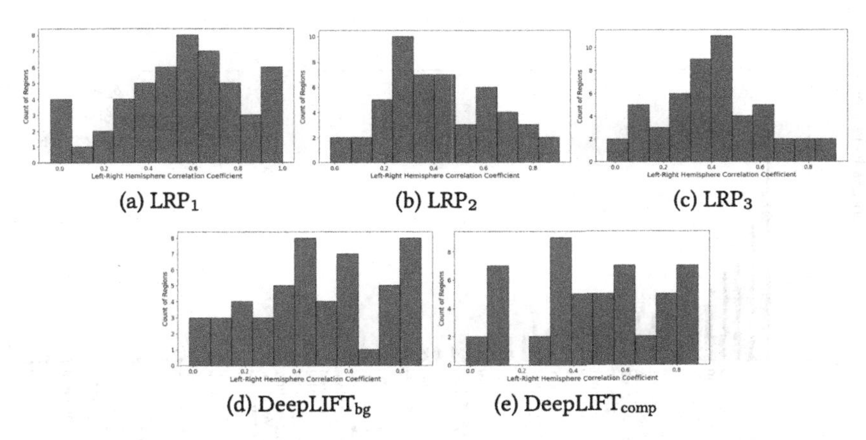

(a) LRP_1 (b) LRP_2 (c) LRP_3

(d) $DeepLIFT_{bg}$ (e) $DeepLIFT_{comp}$

Fig. 7. Histograms of correlation coefficients of T1R between left- and right-hemispheric structures.

D Appendix 4

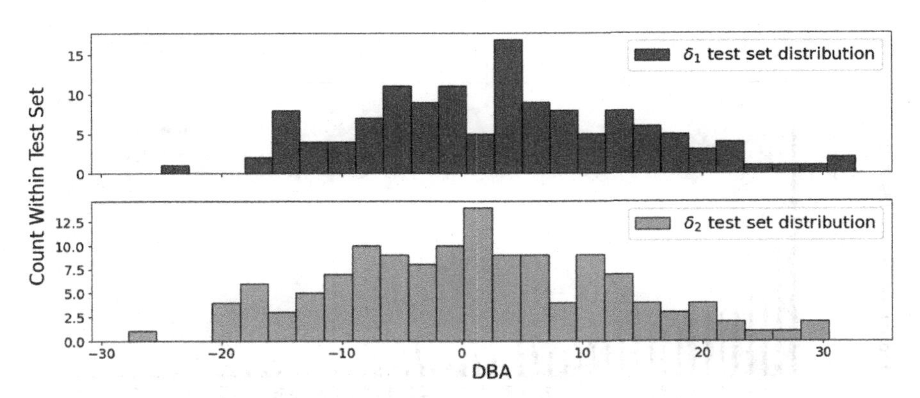

Fig. 8. Distributions of BAG (sometimes called DBA) by δ_1 and δ_2 (as defined by [38]) on the test set. The age-orthogonality correction of δ_2 shifts this distribution to be centered approximately at 0 (mean moves from 3.30y to 0.98y), and distributed more evenly to either side, becoming approximately Normal (range shifts from $(-24.98, 32.70)$ to $(-27.79, 30.47)$).

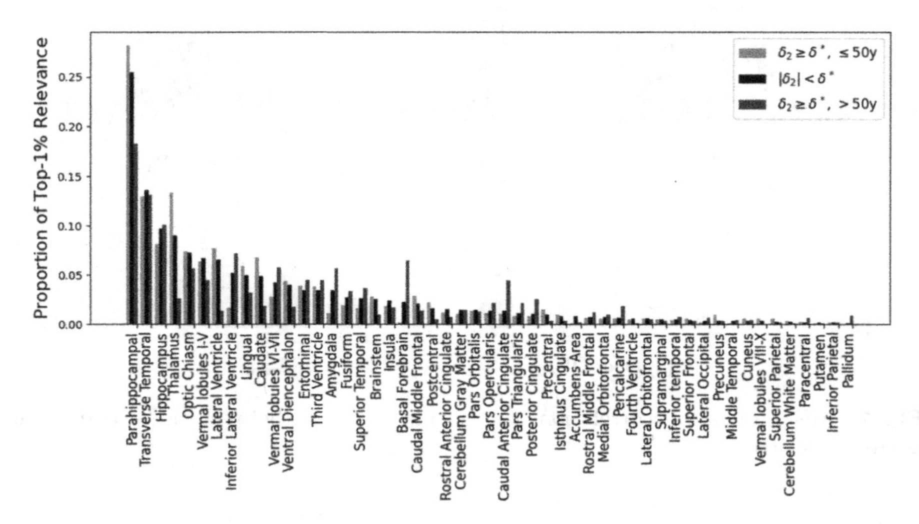

Fig. 9. Distributions of Top-1% Relevance, via $\text{DeepLIFT}_{\text{bg}}$, in young ($\leq$ 50y) individuals with high BAG ($\delta_2 > \delta^*$), elderly individuals ($>$ 50y) with high BAG, and individuals with small-to-moderate BAG ($|\delta_2| < \delta^*$). Regions are ordered by descending proportion of T1R in the small-to-moderate BAG group.

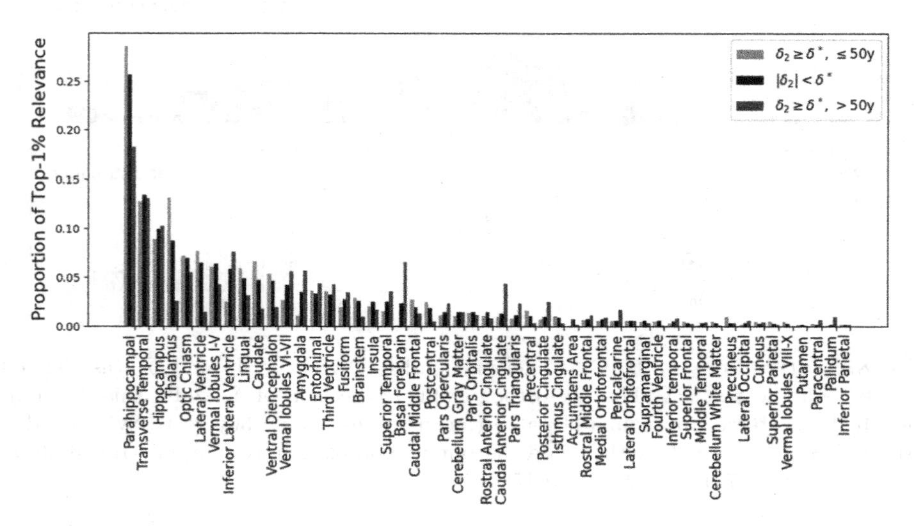

Fig. 10. Distributions of Top-1% Relevance, via $\text{DeepLIFT}_{\text{comp}}$, in young ($\leq$ 50y) individuals with high BAG ($\delta_2 > \delta^*$), elderly individuals ($>$ 50y) with high BAG, and individuals with small-to-moderate BAG ($|\delta_2| < \delta^*$). Regions are ordered by descending proportion of T1R in the small-to-moderate BAG group.

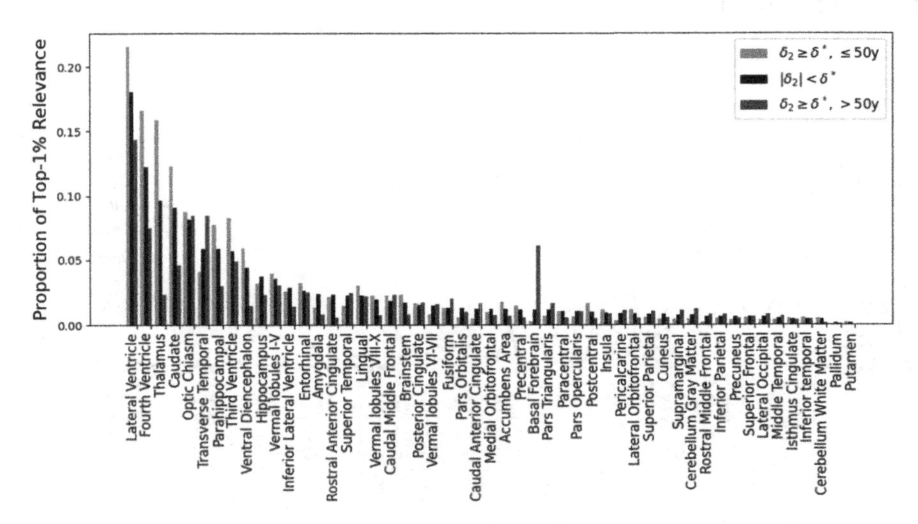

Fig. 11. Distributions of Top-1% Relevance, via LRP_2, in young ($\leq$ 50y) individuals with high BAG ($\delta_2 > \delta^*$), elderly individuals (> 50y) with high BAG, and individuals with small-to-moderate BAG ($|\delta_2| < \delta^*$). Regions are ordered by descending proportion of T1R in the small-to-moderate BAG group.

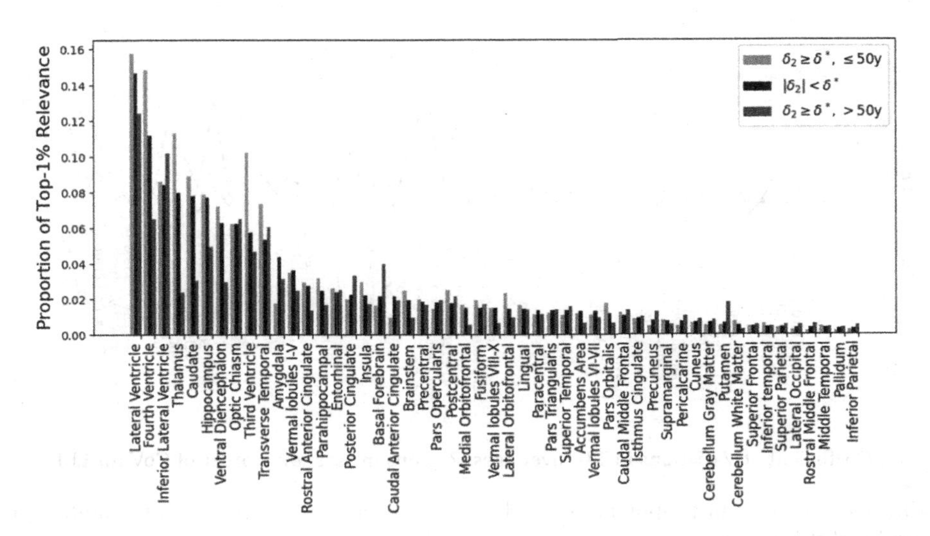

Fig. 12. Distributions of Top-1% Relevance, via LRP_3, in young ($\leq$ 50y) individuals with high BAG ($\delta_2 > \delta^*$), elderly individuals (> 50y) with high BAG, and individuals with small-to-moderate BAG ($|\delta_2| < \delta^*$). Regions are ordered by descending proportion of T1R in the small-to-moderate BAG group.

E Appendix 5

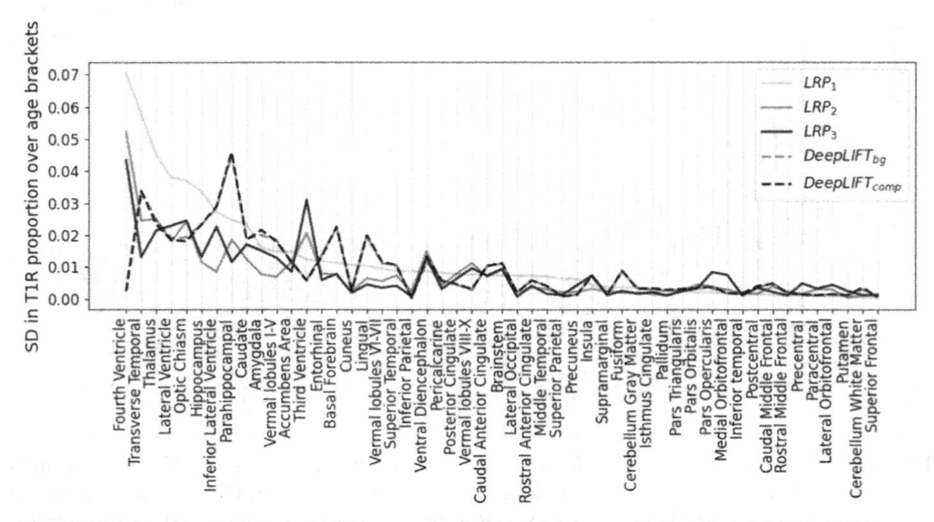

(a) Standard Deviations in T1R over ages. Regions in descending order of SD for LRP_1.

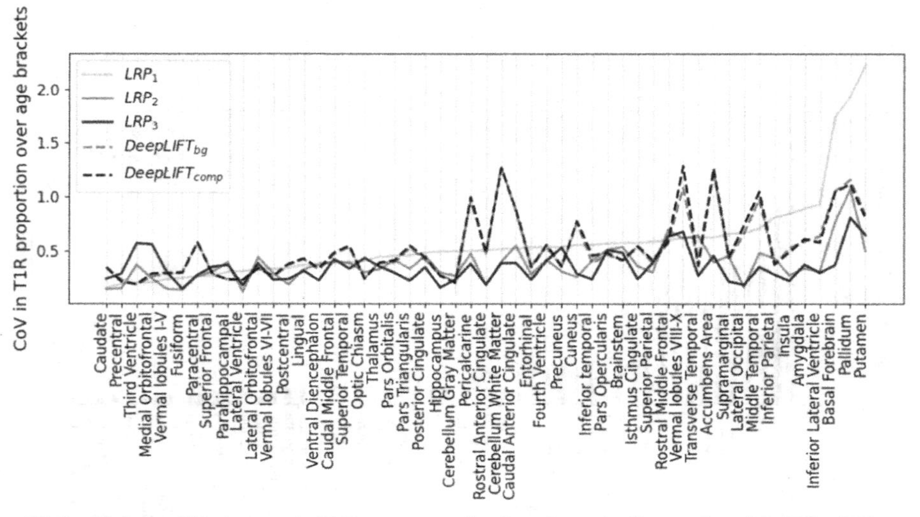

(b) Coefficient of Variations in T1R over ages. Regions in ascending order of CoV for LRP_1.

Fig. 13. Changes in proportion of each region allocated T1R for each method, over age brackets

Figures (13a) and (13b) show the Standard Deviations and Coefficients of Variation respectively of the proportion of T1R in each brain region across age brackets for each method.

Regions with largest SD are those salient areas which change most in their assignment of relevance over ages. Regions with low SD often tend to be regions that are simply assigned very low proportions of T1R, such as the Cerebellum White Matter.

We chose to use the CoV as a metric to find those salient regions which change least in their relevance assignment over ages. This is due to the fact that normalising the SD by the mean allows us to measure which regions have the least change in T1R relative to their mean T1R. The caudate nucleus, for example, has a large proportion of assigned T1R, and changes very little in this assignment across age brackets. On the other hand, regions with high CoV tend to be regions that are assigned low relevance overall. These regions can have small fluctuations in relevance assignment over age brackets, which leads to large CoVs. An example is the Putamen, which has very large CoV and very low proportions of assigned T1R (also, very small SD).

References

1. Ancona, M., Ceolini, E., Öztireli, C., Gross, M.: Towards better understanding of gradient-based attribution methods for deep neural networks. In: 6th International Conference on Learning Representations, ICLR 2018 - Conference Track Proceedings, pp. 1–16 (2018)
2. Anderton, B.H.: Ageing of the brain. Mech. Ageing Dev. **123**, 811–817 (2002). https://doi.org/10.1016/S0047-6374(01)00426-2
3. Bach, S., Binder, A., Montavon, G., Klauschen, F., Müller, K.R., Samek, W.: On pixel-wise explanations for non-linear classifier decisions by layer-wise relevance propagation. PLoS ONE **10**(7), 1–46 (2015). https://doi.org/10.1371/journal.pone.0130140
4. Cole, J.H., et al.: Brain age predicts mortality. Mol. Psychiatry (2018). https://doi.org/10.1038/mp.2017.62
5. Cole, J.H., et al.: Predicting brain age with deep learning from raw imaging data results in a reliable and heritable biomarker. Neuroimage **163**, 115–124 (2017). https://doi.org/10.1016/j.neuroimage.2017.07.059
6. Collins, D., Zijdenbos, A., Baare, W., Evans, A.: ANIMAL+INSECT: improved cortical structure segmentation. In: Kuba, A., Saamal, M., Todd-Pokropek, A. (eds.) IPMI. LNCS, vol. 1613, pp. 210–223. Springer, Heidelberg (1999). https://doi.org/10.1007/3-540-48714-X_16
7. Dinsdale, N.K., et al.: Learning patterns of the ageing brain in MRI using deep convolutional networks. Neuroimage **224**, 117401 (2021). https://doi.org/10.1016/J.NEUROIMAGE.2020.117401
8. Dombrowski, A.K., Alber, M., Anders, C.J., Ackermann, M., Müller, K.R., Kessel, P.: Explanations can be manipulated and geometry is to blame. In: Advances in Neural Information Processing Systems (2019)
9. Eitel, F., et al.: Uncovering convolutional neural network decisions for diagnosing multiple sclerosis on conventional MRI using layer-wise relevance propagation. NeuroImage: Clin. **24**, 102003 (2019). https://doi.org/10.1016/j.nicl.2019.102003

10. Fonov, V., Evans, A., Almli, C., McKinstry, R., Collins, D.: Unbiased nonlinear average age-appropriate brain templates from birth to adulthood. Neuroimage **47**, s102 (2009). https://doi.org/10.1016/S1053-8119(09)70884-5
11. Geirhos, R., Michaelis, C., Wichmann, F.A., Rubisch, P., Bethge, M., Brendel, W.: ImageNet-trained CNNs are biased towards texture; increasing shape bias improves accuracy and robustness. In: 7th International Conference on Learning Representations, ICLR 2019 (c), pp. 1–22 (2019)
12. Grigorescu, I., Cordero-Grande, L., Edwards, A.D., Hajnal, J., Modat, M., Deprez, M.: Interpretable convolutional neural networks for preterm birth classification, pp. 1–4 (2019). https://arxiv.org/abs/1910.00071
13. Gunbey, H.P., Ercan, K., Fjndjkoglu, A.S., Bulut, H.T., Karaoglanoglu, M., Arslan, H.: The limbic degradation of aging brain: a quantitative analysis with diffusion tensor. Imaging (2014) https://doi.org/10.1155/2014/196513
14. Gupta, S., Chan, Y.H., Rajapakse, J.C.: Decoding brain functional connectivity implicated in AD and MCI. In: Shen, D., et al. (eds.) MICCAI 2019. LNCS, vol. 11766, pp. 781–789. Springer, Cham (2019). https://doi.org/10.1007/978-3-030-32248-9_87
15. He, K., Zhang, X., Ren, S., Sun, J.: Deep residual learning for image recognition. In: Proceedings of the IEEE Computer Society Conference on Computer Vision and Pattern Recognition (2016). https://doi.org/10.1109/CVPR.2016.90
16. Hofmann, S.M., et al.: Towards the interpretability of deep learning models for human neuroimaging. BioRxiv (2021). https://doi.org/10.1101/2021.06.25.449906,https://biorxiv.org/content/early/2021/08/26/2021.06.25.449906.abstract
17. Jenkinson, M., Bannister, P., Brady, J., Smith, S.: Improved optimisation for the robust and accurate linear registration and motion correction of brain images. Neuroimage **17**(2), 825–841 (2002)
18. Jenkinson, M., Smith, S.: A global optimisation method for robust affine registration of brain images. Med. Image Anal. **5**(2), 143–156 (2001)
19. Jonsson, B.A., et al.: Brain age prediction using deep learning uncovers associated sequence variants. Nature Commun. **10**(1), 5409 (2019). https://doi.org/10.1038/s41467-019-13163-9
20. Kingma, D.P., Ba, J.L.: Adam: a method for stochastic optimization. In: 3rd International Conference on Learning Representations, ICLR 2015 - Conference Track Proceedings (2014). https://arxiv.org/abs/1412.6980v9
21. Kohlbrenner, M., Bauer, A., Nakajima, S., Binder, A., Samek, W., Lapuschkin, S.: Towards best practice in explaining neural network decisions with LRP, pp. 1–5 (2019). https://arxiv.org/abs/1910.09840
22. Kossaifi, J., Kolbeinsson, A., Khanna, A., Furlanello, T., Anandkumar, A.: Tensor regression networks. J. Mach. Learn. Res. **21**, 1–21 (2020). https://jmlr.org/papers/v21/18-503.html
23. Levakov, G., Rosenthal, G., Shelef, I., Raviv, T.R., Avidan, G.: From a deep learning model back to the brain-Identifying regional predictors and their relation to aging. Hum. Brain Mapp. **41**(12), 3235–3252 (2020). https://doi.org/10.1002/HBM.25011/FORMAT/PDF
24. Lutz, J., et al.: Evidence of subcortical and cortical aging of the acoustic pathway: a diffusion tensor imaging (DTI) study. Acad. Radiol. **14**(6), 692–700 (2007). https://doi.org/10.1016/J.ACRA.2007.02.014
25. Manera, A., Dadar, M., Fonov, V., Collins, D.: CerebrA, registration and manual label correction of Mindboggle-101 atlas for MNI-ICBM152 template. Sci. Data **7**(1), 1–9 (2020)

26. Montavon, G., Lapuschkin, S., Binder, A., Samek, W., Müller, K.R.: Explaining nonlinear classification decisions with deep Taylor decomposition. Pattern Recogn. **65**(2016), 211–222 (2017). https://doi.org/10.1016/j.patcog.2016.11.008
27. Peng, H., Gong, W., Beckmann, C.F., Vedaldi, A., Smith, S.M.: Accurate brain age prediction with lightweight deep neural networks. Med. Image Anal. **68**, 101871 (2021). https://doi.org/10.1016/J.MEDIA.2020.101871
28. Peters, R.: Ageing and the brain (2006). https://doi.org/10.1136/pgmj.2005.036665
29. Pianpanit, T., et al.: Interpreting deep learning prediction of the Parkinson's disease diagnosis from SPECT imaging (2019). https://arxiv.org/abs/1908.11199https://arxiv.org/abs/1908.11199
30. Raz, N., Rodrigue, K.M.: Differential aging of the brain: patterns, cognitive correlates and modifiers (2006). https://doi.org/10.1016/j.neubiorev.2006.07.001
31. Shafto, M.A., et al.: The Cambridge centre for ageing and neuroscience (Cam-CAN) study protocol: a cross-sectional, lifespan, multidisciplinary examination of healthy cognitive ageing. BMC Neurol. (2014). https://doi.org/10.1186/s12883-014-0204-1
32. Shrikumar, A., Greenside, P., Kundaje, A.: Learning important features through propagating activation differences. In: 34th International Conference on Machine Learning, ICML, July 2017, pp. 4844–4866 (2017)
33. Shrikumar, A., Greenside, P., Shcherbina, A.Y., Kundaje, A.: Not just a black box : learning important features through propagating activation differences. In: Proceedings of the 33rd International Conference on MachineLearning (2016)
34. Sixt, L., Granz, M., Landgraf, T.: When explanations lie: why many modified BP attributions fail, December 2019 (2019). https://arxiv.org/abs/1912.09818
35. Smilkov, D., et al.: SmoothGrad: removing noise by adding noise. arXiv:1706.03825, June 2017. https://ui.adsabs.harvard.edu/abs/2017arXiv170603825S/abstract
36. Smith, S.: Fast robust automated brain extraction. Hum. Brain Mapp. **17**(3), 143–155 (2002)
37. Smith, S., et al.: Advances in functional and structural MR image analysis and implementation as FSL. Neuroimage **23**(S1), 208–19 (2004)
38. Smith, S.M., Vidaurre, D., Alfaro-Almagro, F., Nichols, T.E., Miller, K.L.: Estimation of brain age delta from brain imaging. Neuroimage **200**, 528–539 (2019). https://doi.org/10.1016/J.NEUROIMAGE.2019.06.017
39. Sowell, E.R., Peterson, B.S., Thompson, P.M., Welcome, S.E., Henkenius, A.L., Toga, A.W.: Mapping cortical change across the human life span. Nat. Neurosci. (2003). https://doi.org/10.1038/nn1008
40. Sudlow, C., et al.: UK biobank: an open access resource for identifying the causes of a wide range of complex diseases of middle and old age. PLOS Med. **12**(3), e1001779 (2015). https://doi.org/10.1371/JOURNAL.PMED.1001779, https://journals.plos.org/plosmedicine/article?id=10.1371/journal.pmed.1001779
41. Sundararajan, M., Taly, A., Yan, Q.: Gradients of counterfactuals, November 2016. https://arxiv.org/abs/1611.02639
42. Sundararajan, M., Taly, A., Yan, Q.: Axiomatic attribution for deep networks. In: 34th International Conference on Machine Learning, ICML, July 2017, pp. 5109–5118 (2017)
43. Taylor, J.R., et al.: The Cambridge centre for ageing and neuroscience (Cam-CAN) data repository: structural and functional MRI, MEG, and cognitive data from a cross-sectional adult lifespan sample. Neuroimage **144**, 262–269 (2017). https://doi.org/10.1016/J.NEUROIMAGE.2015.09.018

44. Winkler, A., Ridgway, G., Webster, M., Smith, S., Nichols, T.: Permutation inference for the general linear model. Neuroimage **92**, 381–397 (2014)
45. Zhao, L., Matloff, W., Ning, K., Kim, H., Dinov, I.D., Toga, A.W.: Age-related differences in brain morphology and the modifiers in middle-aged and older adults (2019). https://doi.org/10.1093/cercor/bhy300, https://biobank.ctsu.ox

Revisiting Neurological Aspects of Relevance: An EEG Study

Zuzana Pinkosova[1(✉)], William J. McGeown[2], and Yashar Moshfeghi[1]

[1] NeuraSearch Laboratory, University of Strathclyde, Glasgow, UK
{zuzana.pinkosova,yashar.moshfeghi}@strath.ac.uk
[2] Neuroanalytics Laboratory, University of Strathclyde, Glasgow, UK
william.mcgeown@strath.ac.uk

Abstract. Relevance is a key topic in Information Retrieval (IR). It indicates how well the information retrieved by the search engine meets the user's information need (IN). Despite research advances in the past decades, the use of brain imaging techniques to investigate complex cognitive processes underpinning relevance is relatively recent, yet has provided valuable insight to better understanding this complex human notion. However, past electrophysiological studies have mainly employed an event-related potential (ERP) component-driven approach. While this approach is effective in exploring known phenomena, it might overlook the key cognitive aspects that significantly contribute to unexplored and complex cognitive processes such as relevance assessment formation. This paper, therefore, aims to study the relevance assessment phenomena using a data-driven approach. To do so, we measured the neural activity of twenty-five participants using electroencephalography (EEG). In particular, the neural activity was recorded in response to participants' binary relevance assessment (relevant vs. non-relevant) within the context of a Question Answering (Q/A) Task. We found significant variation associated with the user's subjective assessment of relevant and non-relevant information within the EEG signals associated with P300/CPP, N400 and, LPC components, which confirms the findings of previous studies. Additionally, the data-driven approach revealed neural differences associated with the previously not reported P100 component, which might play important role in early selective attention and working memory modulation. Our findings are an important step towards a better understanding of the cognitive mechanisms involved in relevance assessment and more effective IR systems.

Keywords: Information retrieval · Relevance assessment · Binary relevance · Brain signals · EEG · ERPs · Cognitive processes

1 Introduction

Relevance assessment plays a central role in Information Retrieval (IR), denoting how well the document retrieved by an IR system meets the searcher's information need (IN) submitted as a query to the system (see Fig. 1). Although IR covers documents containing different modalities (e.g. videos or images) the most

G. Nicosia et al. (Eds.): LOD 2022, LNCS 13811, pp. 549–563, 2023.
https://doi.org/10.1007/978-3-031-25891-6_41

information consumption happens in textual format [3]. Assessing relevance of textual documents, given IN, involves several cognitive processes including reading comprehension. Therefore, it is one of the most complex cognitive activities in IR [13]. In addition, despite recent findings supporting the idea of categorical thinking [38], relevance assessment has been primarily investigated in binary terms (i.e. content judged as 'relevant' vs 'non-relevant') [42]. Therefore, this work aims to focus on textual binary relevance assessment, which would enable us to compare experimental results obtained using a data-driven approach with previously reported results associated with textual binary relevance assessment formation.

Fig. 1. The interactive IR components.

Recently, NeuraSearch research, which is user-centred research bridging neuroscience and IR [29], has significantly contributed to our understanding of relevance assessment and associated cognitive processes. However, past brain-imaging studies investigating relevance assessment have either misplaced relevance within the context of word-relatedness (i.e. IN of the user was not considered) [14] or (when considering IN) predominantly focused on ERP component-driven analysis [38]. Although investigating relevance assessment employing the component-driven approach provides invaluable insights, its ability to detect and quantify other previously not reported potential components that could arise from this phenomenon is limited. This work, in contrast, employs a data-driven approach to avoid the potential analytical bias introduced by the restriction to distinct ERPs reported by previous studies [44]. This is the first NeuraSearch EEG study investigating binary relevance assessment using a data-driven approach to gain a holistic view of ERP components underpinning complex cognitive processes associated with relevance assessment.

2 Related Work

Relevance Assessment. Relevance has been considered in IR as a multidimensional [42], dynamic and complex phenomenon [8]. Previous studies have shown that relevance is difficult to quantify and depends on the users' perception of information relating to the specific IN situation at a certain time point [42]. Therefore, relevance assessment strongly depends on the user's internal cognitive states. In terms of relevance granularity, binary division has been the prevalent approach in the field, keeping the assessment cost low while maximising the number of relevant documents per topic, guaranteeing measure stability [42,46][1]. In this work, thus, we capture the users' binary relevance assessment in real-time as they engage in the question-answering task.

Neurasearch Research. Neuroscience has significantly contributed to the understanding of IN [34], query formulation [17], search [31] and relevance assessment (e.g. [20,38]) leading to the potential development of novel information search models [31] that can incorporate user's neurophysiological responses to the presented information. The most frequently used neuroimaging methods in the IR have been functional magnetic resonance (fMRI) [30–33,36,37], magnetoencephalography (MEG) [20] and electroencephalography (EEG) [2,17,22]. Within the relevance assessment context, the above-mentioned brain imaging techniques have been employed to investigate different aspects of neurological relevance assessment, such as the brain's functional connectivity [31], underlying cognitive processes and their timing [2].

NeuraSearch and Relevance Assessment. The neuroscientific approach might be categorised in two ways based on the experimental design used to investigate relevance assessment. The first line of brain-imaging research (NeuraSearch) has considered users' IN and positioned relevance assessment within the IR task. Moshfeghi and colleagues [30] used fMRI to localise cortical activity differences during the processing of relevant vs non-relevant images that were related to visuospatial working memory [31,34]. Relevance assessment has also been studied within the context of the IR task for stimuli of different modalities, such as text [17], images [2] and videos [22]. Allegretti et al. examined the processing of relevant vs non-relevant images, finding the most significant differences to occur between 500–800 ms, reaching the peak in the central scalp areas [2]. Kim and Kim [22] explored the topical relevance of video skims and classified the data based on two specific ERP components (N400 and P600), which are indicators of relevant and non-relevant assessments. Furthermore, recent findings have shown that relevance assessment can be automatically predicted using EEG data while the user engages with the IR task [17]. The results of these studies suggest that human mental experience during the relevance assessment can be understood and accurately decoded.

Another approach employing EEG has placed relevance assessments in the word associations context [14,15]. Within the task, participants did not

[1] Recent studies have discussed higher relevance granularity [38,52], however, it is out of the scope of this paper.

experience IN, but they judged associations between words and topics. The study findings have shown that neurological signals differ when subjects process relevant vs non-relevant words [15]. Later, Eugster et al. [14] introduced a brain-relevance paradigm enabling information recommendation to users without any explicit user interaction, based on EEG signals alone evoked by users' text engagement.

Data-Driven Approach. Relevance assessment is a holistic cognitive response with underlying neuropsychological mechanisms that form more basic perceptional and cognitive features of some sort. While past feature classification and theory-driven approaches have contributed to the understanding of relevance assessment formation, these approaches can constrain knowledge by potentially overlooking all the possible features or dimensions that synthesise complex phenomena such as relevance assessment. On the other hand, a data-driven approach is a useful tool when it comes to making sense of complex behavioural responses during complex tasks. Despite its advantages, the data-driven analysis might be challenging because the EEG data frequently exhibit spatial heterogeneity, have non-stationary and multiscale dynamics, and are typified by substantial individual variability [12]. Many scientific observations of brain activity may be scope limited, constrained by opportunistic participant sampling, and have reduced reproducible controls. Relying on varied background assumptions while employing a data-driven exploratory approach might help to overcome the above-mentioned limitations and provide significant benefits associated with the potential discovery of novel, previously not reported cognitive phenomena.

3 Methodology

Participants. The study was carried out with a sample of twenty-five individuals who reported themselves to be neurologically and physically healthy. Seven participants were excluded from the analysis due to the high number of physiological artefacts present in the EEG data. The eighteen remaining participants (11 females and 7 males) were between 19 to 39 years old and with the mean age of 24.5 and the standard deviation (SD) 4.91 years. Participants were either native English speakers (8) or had high English proficiency. On average, participants indicated using search engines on average several times a day.

Design. This study followed a within-subject design in which participants engaged in the Q/A task. The independent variable was the user's binary relevance assessment (with two levels: "Non-Relevant" ('nr') and "Relevant" ('rel')). The dependent variable was the EEG signal gathered during the answer presentation. We controlled the number of relevant and non-relevant answers presented to the participant, but we did not control the number of words each participant saw. This allowed us to simulate an information search and retrieval, as participants were not required to read through the whole answer. Instead, they were able to terminate the answer presentation once the relevance assessment was made.

Stimuli Presentation. The stimuli were presented on a 22-inch colour Mitsubishi Diamond Pro 2040u CRT monitor using E-Prime 2.0. Participants were seated approximately 60 cm from the monitor, and response keys were located on a QWERTY keyboard. Text events were presented in Arial font, size 16.

Questionnaires. Throughout the experiment, participants filled in the Entry and Post-Task Questionnaires. The Entry questionnaire was administered at the beginning of the experiment to gather participants' demographic and background information. All participants were screened for eligibility to take part in the study. Any participants with a pre-existing neurological condition that might impact the EEG recordings were excluded. After completing the task, participants filled in the Post-Task questionnaire, assessing their task perception.

Question-Answering Data Sets. The data set employed in the study was developed and used by Moshfeghi, Triantafillou, and Pollick [34]. We have chosen this data set as it has been proven effective in investigating IR phenomena from a neuroscience standpoint. The data set was further adapted to address our research questions and expanded through the additional question and answer selection from TREC-8 and TREC-2001. These two Tracks were selected as they (i) cover a wide discipline range, (ii) they are independent of one another, and (iii) they provide a correct question answer. We ensured that the selected questions were accurate and not time-dependent or ambiguous. The data set contained 128 questions, answers, and relevance assessments in total. To reduce the fatigue, 64 questions were carefully selected for each participant to be balanced based on relevance (i.e. 50% relevant and 50% non-relevant), length (i.e. 50% long vs 50% short) and difficulty (i.e. 50% difficult vs 50% easy)[2]. This was done to reduce any possible bias that could result from the focus on one form of Q/A. An example of an easy question presented to the participants was "What is epilepsy?", which was followed by the short, relevant answer "Epilepsy is a brain disorder characterised by seizures". The order of the presented questions was randomised also for each participant.

EEG Recordings. Brain signals were acquired using 128-channel HydroCel Geodesic Sensor Nets (Electrical Geodesic Inc) and recorded within the standard EGI package Net Station 4.3.1. A Net Amps 200 amplifier was used for the recording and to facilitate the synchronisation between the behavioural response of the participant and their brain signals. To set the system for recording, we followed Electrical Geodesic Inc guidelines. We aimed to keep the electrode impedances below 50 kΩ, according to the recommended system value. Raw EEG data were recorded at a sampling rate of 1000 Hz Hz and referenced to the vertex electrode (Cz).

Procedure. The experiment was carried out in accordance with the University of Strathclyde Ethics Committee guidelines. All participants were briefed and confirmed their willingness to voluntarily participate in the experiment.

[2] To assess the difficulty level, two annotators separately judged question difficulty (i.e. difficult vs easy). The overall inter-annotator agreement was reasonably high (Cohen's kappa, $\kappa = 0.72$).

After that, they filled in an Entry Questionnaire. Prior to the main experimental trials, participants underwent a number of training trials, which resembled the main experimental task. Participants were able to repeat the training until they confirmed to have a good understanding of the procedure. In total, every participant completed 64 trials. To avoid fatigue, the trials were split into two equally long blocks separated by a break. On average, participants were presented with 810.06 words ($\pm$134.77) and the main experimental task lasted approximately 53.69 min $\pm$ 9.74). After completing the main experimental task, participants were instructed to fill out the Post-Task Questionnaire.

Task. The schematic task representation is depicted in Fig. 2. At the beginning of the task, participants were presented with the instructions. Next, they viewed a randomly selected question from the data set, followed by the fixation cross, which indicated the location of the answer presentation. To control free-viewing and minimise the presence of any confounding artefacts (i.e. saccades), the answer was presented in the middle of the screen word by word. Each word was presented for 950ms, which has been deemed to be a sufficient duration to model fluent reading and to avoid the overlapping effect of two consecutive words on the ERPs [14]. The ERP components were, therefore, time-locked to the word presentation. This approach has been commonly applied to examine neurological signatures of reading in the ERP studies (e.g. [11]).

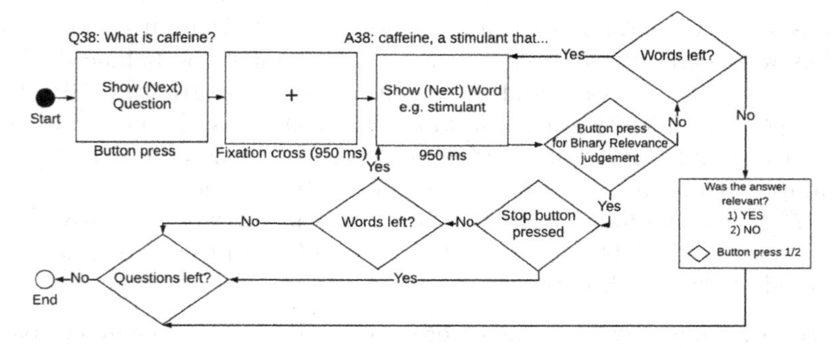

Fig. 2. The flow diagram of the task.

Participants were instructed to carefully read individual words that would form either relevant or non-relevant answers. Once participants gathered enough information, they had an option to terminate the word presentation sequence (and to continue to the next step), or to view the sequence in full. As brain activity was recorded during the reading, to avoid the possibility of confounding hemispheric effects (due to motor planning or execution), counter-balancing was used, and participants were instructed to interact with the keyboard using either their left or right hand. The interpretation of binary relevance categories depended on each participant's subjective understanding, which enabled capturing the subjective nature of relevance assessment [42].

Pilot Studies. Before commencing the main user study, we performed a pilot study with four participants whose data were not included in the final analysis.

Based on participants' feedback, it was determined that the participants were able to complete the study without problems.

Pre-processing Steps. The brain activity was recorded from participants as they engaged with relevant and non-relevant content, up to the point where the participant stopped the answer presentation. To prepare data for analysis, an automated pre-processing pipeline was built using EEGLAB and its associated toolboxes [10]. The EEG data pre-processing steps were based on Makoto's Pre-processing Pipeline[3]. All EEG data were first visually inspected. Then a low-pass filter 30 Hz was applied. We down-sampled the data from 1000 Hz 250 Hz. Down-sampling, a commonly applied procedure, is used to reduce file size for easier data manipulation. Then a high pass filter of 0.3 Hz was applied. Filtering is another common procedure used to attenuate frequencies associated with noise rather than a signal of interest. We then automatically rejected bad channels (EEG sensors that were not functioning properly during the data acquisition and that were high in noise throughout the task. On average, we removed 13.94 bad channels (±7.67). The re-referencing to average (across all electrodes) was subsequently performed (to provide an approximation of zero microvolts for the reference at each timepoint). The CleanLine EEGLAB plugin was used to filter line noise. All epochs (the time windows of interest) were then extracted from 200ms before stimulus presentation to 950ms afterwards. To detect and remove components associated with ocular, cardiac and muscular artefacts based on their power spectrum and time-course, we performed Independent Component Analysis and then used ADJUST [28] to automatically reject components. A mean number of 18.17 (±9.17) components were removed. Bad channels were interpolated (reconstructed) using a spherical interpolation method. Next, we removed the two outermost belts of electrodes of the sensor net[4] which are prone to show muscular artefacts, following the approaches of Bian et al. [4] and Calbi et al. [6]. Epochs were then extracted again from 100ms before stimulus presentation to 950ms afterwards based on the stimulus labels for every condition of interest (i.e. 'rel' and 'nr'). We have used automatic epoch rejection based on thresholding (i.e. rejecting epochs by detecting outlier values greater than $\pm 100 \mu V$). The mean number and SD of accepted and rejected epochs are displayed in Table 1. All epochs were baseline corrected. After pre-processing the data, we calculated the grand average for epochs of interest.

Table 1. The Mean number and SD of rejected epochs for 'rel' and 'nr' conditions

Condition	Accepted epochs		Rejected epochs	
	Mean	SD	Mean	SD
rel	399.11	88.01	69.39	89.55
nr	410.94	95.99	76.56	90.76

[3] https://sccn.ucsd.edu/wiki/Makoto's_preprocessing_pipeline.

[4] We removed 38 peripheral channels: E1, E8, E14, E17, E21, E25, E32, E38, E43, E44, E48, E49, E56, E57, E63, E64, E68, E69, E73, E74, E81, E82, E88, E89, E94, E95, E99, E100, E107, E113, E114, E119, E120, E121, E125, E126, E127, E128.

Statistical Analysis of EEG data. Participants' brain activity was recorded while they engaged in relevant and non-relevant assessments (i.e. 'rel' vs 'nr') within the Q/A task. After data pre-processing, 49.65% of accepted trials were marked as 'rel' and 50.35% as 'nr'. To test whether there are statistically significant differences in the neurological processing associated with the judgement of 'rel' vs 'nr' information, we employed a data-driven approach, which is particularly effective in whole-brain analysis of complex mental phenomena as it minimises the upfront assumptions and allows for the contribution of many distinct areas at different time points [7,27]. To identify significant cortical differences, we compared the values for 109 electrode pairs at every time point (every 4 ms, 237-time points in total) over the 100–950 ms time window. The initial time interval (0–100 ms) was excluded from the main analysis as we were not interested in the initial sensory processing of stimulus features [25]. The data-driven approach applied a non-parametric permutation-based paired t-test (1000 permutations) using the *statcond* function implemented in the EEGLAB toolbox [10]. Differences were considered significant at a threshold of $p < 0.05$.

ROIs. As this study uses a data-driven approach, for optimal effect detection, the Regions of Interest (ROIs)[5] were determined based on statistically significant differences between compared conditions of interest. Therefore, we have used features of the data under analysis to position the ROIs. We were not interested in isolated electrodes where a test statistic might happen to be large. Instead, we applied the method used by Laganaro and colleagues [23]. To identify potential ROIs, we have only considered clusters with at least five electrodes next to each other extending over at least 20 ms and retained with an alpha criterion of 0.05 [23].

4 Results

Questionnaire Results. Before analysing the main results, we measured participants' task perception using the Post-task Questionnaire using a 7-point Likert Scale (answers: 1: "Strongly Disagree", 2: "Disagree", 3: "Somewhat Disagree", 4: "Neither Agree or Disagree", 5: "Somewhat Agree", 6: "Agree", 7: "Strongly Agree"). Overall, the Post-task Questionnaire results suggest that participants did not perceive difficulties associated with the experimental design that might have caused them discomfort and impacted their engagement (Fig. 3).

Binary Relevance. As described above, we used a data-driven analysis to investigate the brain activity differences during the 'rel' vs 'nr' content processing and this section presents the main experimental results.

100–200 ms. The earliest differences in neural activity for the comparison 'rel' vs 'nr' conditions emerged in the 100–200 ms interval. The 'rel' condition was associated with a significantly greater positivity in the right centro-parietal

[5] Region of Interest refers to a selected region of neighbouring electrodes that jointly and significantly contribute towards neurophysiological phenomena of interest.

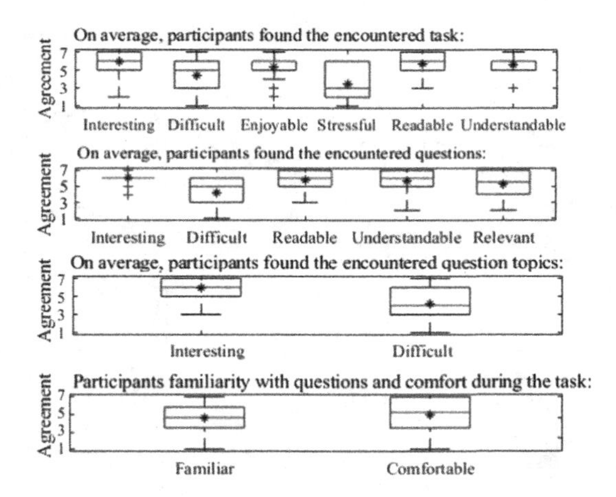

Fig. 3. Post-task questionnaire results: The asterisk (*) denotes the mean values and the dot (·) denotes the outlier values.

region and significantly greater negativity in the left fronto-central region compared to the 'nr' condition. Significant electrode clusters, time intervals, and ERP waveforms, as well as topographic plots, are displayed in Fig. 4, row I. Given the topographies and waveform peaks, the differences are likely to reflect variability in the P100 ERP component (similar distributions reported, e.g., by [24]). The P100 ERP component reflects initial visual field activation and enhanced P100 amplitude observed during the processing of relevant information might suggest early selective attention allocation, with greater early attention allocated to relevant stimuli [26]. This early P100 selective stimulus encoding might affect later ERP components associated with working-memory [41], such as the P300 and LPC commonly reported in the relevance assessment studies [38].

250–300 ms. The comparison of 'rel' and 'nr' conditions was associated with statistically significant differences in the right centro-parietal region within the 250–300 ms interval (see Fig. 4, row II). Closer inspection of the topographies revealed that the above effects were driven by greater positivity of the P300/Centro-parietal positivity (CPP) component across the 'rel' condition compared to the 'nr' condition. Obtained topographies with centro-parietal positivity are very similar to the P300 and CPP signal, peaking at around 300ms post-stimulus [48]. Both the CPP and P300 ERPs share many characteristics, such as dynamics and topography. While the P300 amplitude reflects attentional allocation [19] and task relevance [16], the CPP amplitude modulations are related to evidence accumulation during decision formation [48]. Significant differences might signal difference in effort for participants to process 'rel' and 'nr' information, which may induce differences in working memory load [50].

400–600 ms. The processing of 'nr' compared to 'rel' content was associated with lower amplitude in an electrode cluster that bridged the right centro-parieto-temporal negativity within the 400–600 ms time interval, as displayed

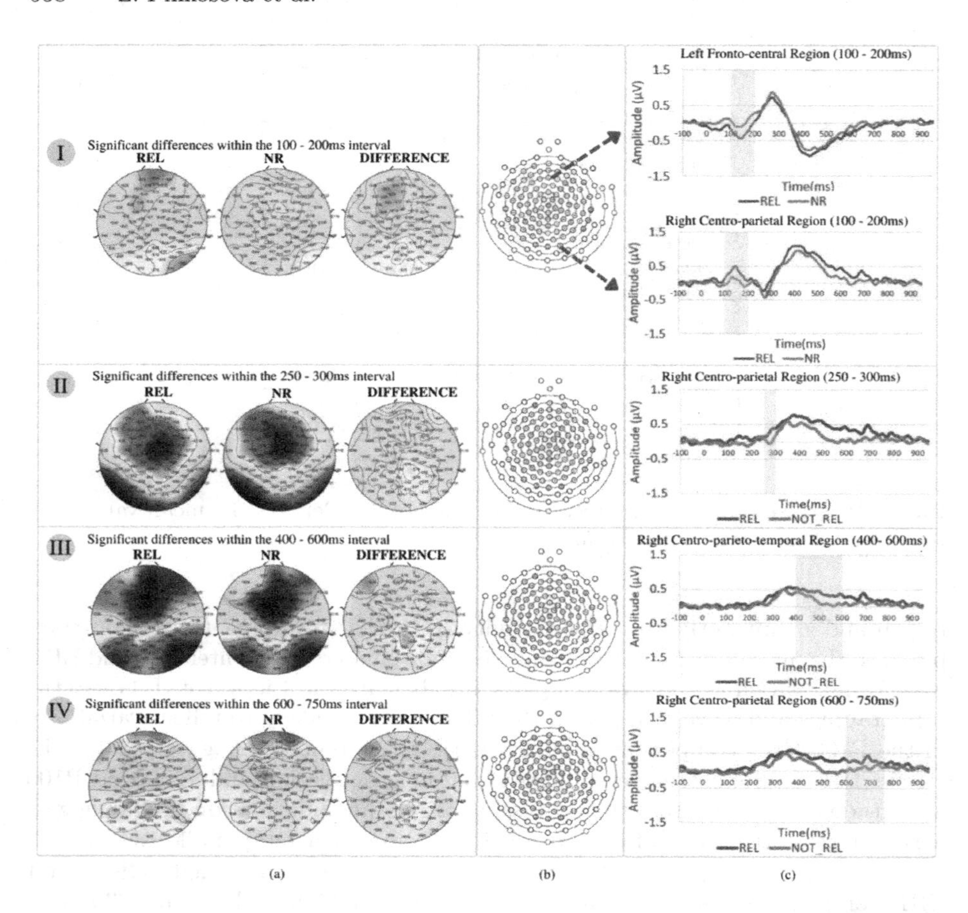

Fig. 4. (a) Topographic plots for 'rel' vs 'nr' conditions, including a mean difference plot for the 100–200 ms (I), 250–300 ms (II), 400–600 ms (III) and 600–750 ms (IV) time windows. Reddish colours of the scalp topography indicate positive ERP values, whereas bluish colours indicate negative ERP values. (b) The 128-channel net graph with highlighted statistically significant electrode sites for each significant time interval. (c) The comparison of grand averaged ERP waveforms for 'rel' (blue) vs 'nr' (orange) condition. Significant time intervals are highlighted in grey for each significant time period.

in Fig. 4, row III. The significant differences were driven by the higher centro-parieto-temporal negativity associated with 'nr' compared to the 'rel' condition. Observed anterior negativity and co-occurring posterior positivity reflect the N400 ERP component, with similar topographic distributions reported by previous studies [43,47]. Lower N400 amplitudes in response to 'nr' content might be linked to the higher semantic incongruity between the question context and the given answer. That 'nr' content is linked with less positivity in this time window bears similarity to the results of e.g. [22], who found that irrelevant content produced more negative N400 responses.

600–750 ms. The posterior positivity observed in the topographic plots within the 600–750 ms time frame (see Fig. 4, row IV) can be attributed to the LPC (Late Positive Component)[6] (e.g. [14,15]). The LPC component is a positive-going deflection, emerging around 600 ms post-stimulus, usually largest over the posterio-medial brain areas [9]. The positivity was significantly greater during the processing of 'rel' compared to the 'nr' content. Higher LPC amplitudes are associated with memory processing and decision-making, which might reflect effort invested in retaining relevant information during cumulative information exposure [22,51]. Furthermore, higher LPC amplitude deflection in response to task-related stimuli have previously been reported [5], which is consistent with our findings.

5 Discussion and Conclusion

The current experiment was carried out to investigate binary relevance assessment using a data-driven approach to gain a better understanding of complex cognitive processes underpinning relevance assessment formation. In particular, the user's neural signals associated with 'rel' and 'nr' relevance assessments were recorded in real-time during a Q/A relevance assessment task. Using a data-driven approach we compared information judged as relevant to that which was not relevant. We found significant differences in neural activity associated with the user's perceived relevance in distinct significant time intervals linked to differences in topographies and ERP waveform distributions. Along with previously reported P300/CPP, N400 and LPC ERP components in studies investigating binary relevance (e.g., [22]), we have also observed significant differences in neural activity in the early time interval associated with the P100 component. Differences in the P100 component might signal early stages of selective attention allocation, indicating enhanced attention to 'rel' information during early sensory facilitation, which is then passed along to higher levels of cognitive processing.

P100. The 100–200 ms time interval seems to be an early time point associated with the P100 ERP component, reflecting the participant's selective attention modulation of task-relevant information. Past studies have provided a direct correlative link between the P100 and working memory performance, which might suggest that the P100 reflects participants' initial ability and processing effort to recognise relevant stimuli during relevance assessment formation [41]. Further studies could explore the relationship between the P100, P300 and LPC ERP components, related to attention during relevance assessment formation.

P300/CPP. Observed P300/CPP topographies and ERP waveforms might indicate the allocation of greater attentional resources to information perceived

[6] The terms LPC and P600 are commonly interchanged. Relevance assessment has frequently been linked to the P600 ERP component (e.g. [14]. However, the P600 component is mainly associated with 'syntactic re-analyses' in language studies. Therefore, the label LPC might be more appropriate to use while focusing on relevance assessment, as the LPC has been linked to memory and recognition processes.

as relevant. Relevant information might also be easier to process in terms of memory access and retrieval, leading to reduced cognitive load [1,39]. The CPP component is equivalent to the P300 and previous research indicates that both components represent decision formation, through the information accumulation process, determining behaviour via a boundary-crossing criterion [35]. In addition, the P300/CPP amplitude is proportional to attentional engagement [18,21], quality of useful information [19,21], and the quality of information transmitted [21,40]. Our results are consistent with previous studies suggesting that the P300 amplitude is higher for the target ('rel') than for non-target ('nr') stimuli [49].

N400. Significantly greater amplitude was observed for the 'rel' compared to the 'nr' condition, as indicated through the differences observed within the N400 topographies and time-window (see Fig. 4, line III). The N400 component has been extensively researched in the sense of semantic processing. Past studies suggest that the N400 represents aspects of semantic information retrieval and integration [11]. N400 negativity is amplified by the processing of semantic mismatch (see e.g., [45]). Less negative deflections (indicated through the higher amplitude) for the 'rel' condition might, therefore, reflect higher answer relatedness to the question, which has been previously reported by [38], who found a reduced N400 for highly relevant words.

LPC. The amplitudes associated with the LPC component are significantly greater for content assessed as 'rel' compared to 'nr', as indicated through the differences observed within the LPC topographies (see Fig. 4, line IV). The LPC is commonly reported to follow the N400 component and it is related to cumulative evidence exposure during decision-dependent tasks [51]. Therefore, the LPC amplitudes appear to be modulated by the participant's response to a stimulus, when the memory judgement at hand requires consideration of the relevant dimension in search tasks [51]. Furthermore, higher amplitudes recorded during the 'rel' condition might suggest the ease of the word categorisation process.

To conclude, our findings further our understanding of the concept of relevance and provide the evidence needed to strengthen its theoretical foundations. Overall, our results confirmed the empirical findings of previous studies examining textual relevance processing associated with the P300/CPP, N400 and LPC components. Additionally, the data-driven approach revealed neural differences in an early P100 component, which is a novel, previously not reported finding. Finally, we believe our conclusions constitute an important step in unravelling the nature of relevance assessment in terms of its electrophysiological modulation and operationalising it for the IR process.

Acknowledgement. This work was supported by the Engineering and Physical Sciences Research Council [grant number EP/R513349/1].

References

1. Ahmed, L., de Fockert, J.W.: Working memory load can both improve and impair selective attention: evidence from the navon paradigm. Attention Percept. Psychophysics **74**(7), 1397–1405 (2012)
2. Allegretti, M., Moshfeghi, Y., Hadjigeorgieva, M., Pollick, F.E., Jose, J.M., Pasi, G.: When relevance judgement is happening? an EEG-based study. In: SIGIR'15, pp. 719–722. ACM, NY, USA (2015)
3. Barral, O., et al.: Extracting relevance and affect information from physiological text annotation. User Model. User-Adap. Inter. **26**(5), 493–520 (2016). https://doi.org/10.1007/s11257-016-9184-8
4. Bian, Z., Li, Q., Wang, L., Lu, C., Yin, S., Li, X.: Relative power and coherence of EEG series are related to amnestic mild cognitive impairment in diabetes. Front. Aging Neurosci. **6**, 11 (2014)
5. Bouaffre, S., Faita-Ainseba, F.: Hemispheric differences in the time-course of semantic priming processes: evidence from event-related potentials (ERPS). Brain Cogn. **63**(2), 123–135 (2007)
6. Calbi, M., et al.: How context influences the interpretation of facial expressions: a source localization high-density EEG study on the kuleshov effect. Sci. Rep. **9**(1), 1–16 (2019)
7. Calhoun, V.: Data-driven approaches for identifying links between brain structure and function in health and disease. Dialogues Clin. Neurosci. **20**(2), 87 (2018)
8. Cool, C., Frieder, O., Kantor, P.: Characteristics of text affecting relevance judgments. In: Proceedings of the 14th National Online Meeting 14 (1993)
9. Curran, T.: Brain potentials of recollection and familiarity. Memory Cognition **28**(6), 923–938 (2000)
10. Delorme, A., Makeig, S.: Eeglab: an open source toolbox for analysis of single-trial EEG dynamics including independent component analysis. J. Neurosci. Methods **134**(1), 9–21 (2004)
11. Dien, J., Michelson, C.A., Franklin, M.S.: Separating the visual sentence n400 effect from the p400 sequential expectancy effect: cognitive and neuroanatomical implications. Brain Res. **1355**, 126–140 (2010)
12. Dimitriadis, S.I., Salis, C., Tarnanas, I., Linden, D.E.: Topological filtering of dynamic functional brain networks unfolds informative chronnectomics: a novel data-driven thresholding scheme based on orthogonal minimal spanning trees (OMSTS). Front. Neuroinform. **11**, 28 (2017)
13. Elleman, A.M., Oslund, E.L.: Reading comprehension research: implications for practice and policy. Policy Insights Behav. Brain Sci. **6**(1), 3–11 (2019)
14. Eugster, M.J.: Natural brain-information interfaces: recommending information by relevance inferred from human brain signals. Sci. Rep. **6**, 38580 (2016)
15. Eugster, M.J., et al.: Predicting term-relevance from brain signals. In: SIGIR'14, pp. 425–434. ACM, NY, USA (2014)
16. Farwell, L.A., Donchin, E.: The truth will out: Interrogative polygraphy (lie detection) with event-related brain potentials. Psychophysiology **28**(5), 531–547 (1991)
17. Jacucci, G., et al.: Integrating neurophysiologic relevance feedback in intent modeling for information retrieval. JASIST **70**, 917–930 (2019)
18. Johnson, R., Jr.: The amplitude of the p300 component of the event-related potential: review and synthesis. Adv. Psychophysiol. **3**, 69–137 (1988)
19. Johnson, R., Jr., Donchin, E.: On how p300 amplitude varies with the utility of the eliciting stimuli. Electroencephalogr. Clin. Neurophysiol. **44**(4), 424–437 (1978)

20. Kauppi, J.P.: Towards brain-activity-controlled information retrieval: decoding image relevance from meg signals. Neuroimage **112**, 288–298 (2015)
21. Kelly, S.P., O'Connell, R.G.: Internal and external influences on the rate of sensory evidence accumulation in the human brain. J. Neurosci. **33**(50), 19434–19441 (2013)
22. Kim, H.H., Kim, Y.H.: ERP/MMR algorithm for classifying topic-relevant and topic-irrelevant visual shots of documentary videos. JASIST **70**(9), 931–941 (2019)
23. Laganaro, M., Perret, C.: Comparing electrophysiological correlates of word production in immediate and delayed naming through the analysis of word age of acquisition effects. Brain Topogr. **24**(1), 19–29 (2011)
24. LePendu, P., Dou, D., Frishkoff, G.A., Rong, J.: Ontology database: a new method for semantic modeling and an application to brainwave data. In: Ludäscher, B., Mamoulis, N. (eds.) SSDBM 2008. LNCS, vol. 5069, pp. 313–330. Springer, Heidelberg (2008). https://doi.org/10.1007/978-3-540-69497-7_21
25. Liu, Y., et al.: Early top-down modulation in visual word form processing: Evidence from an intracranial SEEG study. J. Neurosci. **41**(28), 6102–6115 (2021)
26. Luck, S.J.: An Introduction to the Event-related Potential Technique. MIT Press, Cambridge (2014)
27. Meghdadi, A.H., Karić, M., Berka, C.: EEG analytics: benefits and challenges of data driven eeg biomarkers for neurodegenerative diseases. In: 2019 IEEE International Conference on Systems, Man and Cybernetics (SMC), pp. 1280–1285 (2019)
28. Mognon, A., Jovicich, J., Bruzzone, L., Buiatti, M.: Adjust: an automatic EEG artifact detector based on the joint use of spatial and temporal features. Psychophysiology **48**(2), 229–240 (2011)
29. Moshfeghi, Y.: Neurasearch: neuroscience and information retrieval. In: CEUR Workshop Proceedings, vol. 2950, pp. 193–194 (2021)
30. Moshfeghi, Y., Pinto, L.R., Pollick, F.E., Jose, J.M.: Understanding relevance: an fMRI study. In: Serdyukov, P., et al. (eds.) ECIR 2013. LNCS, vol. 7814, pp. 14–25. Springer, Heidelberg (2013). https://doi.org/10.1007/978-3-642-36973-5_2
31. Moshfeghi, Y., Pollick, F.E.: Search process as transitions between neural states. In: International World Wide Web Conferences Steering Committee, WWW'18, Republic and Canton of Geneva, CHE, pp. 1683–1692 (2018)
32. Moshfeghi, Y., Pollick, F.E.: Neuropsychological model of the realization of information need. JASIST **70**(9), 954–967 (2019)
33. Moshfeghi, Y., Triantafillou, P., Pollick, F.: Towards predicting a realisation of an information need based on brain signals. In: WWW'19, pp. 1300–1309. ACM, NY, USA (2019)
34. Moshfeghi, Y., Triantafillou, P., Pollick, F.E.: Understanding information need: an FMRI study. In: SIGIR'16, pp. 335–344. ACM, NY, USA (2016)
35. O'connell, R.G., Dockree, P.M., Kelly, S.P.: A supramodal accumulation-to-bound signal that determines perceptual decisions in humans. Nat. Neurosci. **15**(12), 1729 (2012)
36. Paisalnan, S., Moshfeghi, Y., Pollick, F.: Neural correlates of realisation of satisfaction in a successful search process. In: Proceedings of the Association for Information Science and Technology, vol. 58, no. 1, pp. 282–291 (2021)
37. Paisalnan, S., Pollick, F., Moshfeghi, Y.: Towards understanding neuroscience of realisation of information need in light of relevance and satisfaction judgement. In: International Conference on Machine Learning, Optimization, and Data Science, pp. 41–56. Springer, Cham (2021). https://doi.org/10.1007/978-3-030-95467-3_3
38. Pinkosova, Z., McGeown, W.J., Moshfeghi, Y.: The cortical activity of graded relevance. In: SIGIR'20, pp. 299–308. ACM, NY, USA (2020)

39. Polich, J.: Updating p300: an integrative theory of p3a and p3b. Clin. Neurophysiol. **118**(10), 2128–2148 (2007)
40. Ruchkin, D., Sutton, S.: Emitted p300 potentials and temporal unvertainty. Electroencephalogr. Clin. Neurophysiol. **45**(2), 268–277 (1978)
41. Rutman, A.M., Clapp, W.C., Chadick, J.Z., Gazzaley, A.: Early top-down control of visual processing predicts working memory performance. J. Cogn. Neurosci. **22**(6), 1224–1234 (2010)
42. Saracevic, T.: Relevance: a review of the literature and a framework for thinking on the notion in information science. part iii: behavior and effects of relevance. JASIST 58(13), 2126–2144 (2007)
43. Savostyanov, A., Bocharov, A., Astakhova, T., Tamozhnikov, S., Saprygin, A., Knyazev, G.: The behavioral and ERP responses to self-and other-referenced adjectives. Brain Sci. **10**(11), 782 (2020)
44. Schmüser, L., Sebastian, A., Mobascher, A., Lieb, K., Tüscher, O., Feige, B.: Data-driven analysis of simultaneous EEG/FMRI using an ICA approach. Front. Neurosci. **8**, 175 (2014)
45. Sitnikova, T., Salisbury, D.F., Kuperberg, G., Holcomb, P.J.: Electrophysiological insights into language processing in schizophrenia. Psychophysiology **39**(6), 851–860 (2002)
46. Sormunen, E.: Liberal relevance criteria of TREC-: Counting on negligible documents? In: SIGIR'02, pp. 324–330. ACM (2002)
47. Spironelli, C., Angrilli, A.: Complex time-dependent ERP hemispheric asymmetries during word matching in phonological, semantic and orthographical matching judgment tasks. Symmetry **13**(1), 74 (2021)
48. Tagliabue, C.F., Veniero, D., Benwell, C.S., Cecere, R., Savazzi, S., Thut, G.: The EEG signature of sensory evidence accumulation during decision formation closely tracks subjective perceptual experience. Sci. Rep. **9**(1), 1–12 (2019)
49. Wang, F., et al.: A novel audiovisual brain-computer interface and its application in awareness detection. Sci. Rep. **5**(1), 1–12 (2015)
50. Wang, L., Zheng, J., Huang, S., Sun, H.: P300 and decision making under risk and ambiguity. Comput. Intell. Neurosci. **2015** (2015)
51. Yang, H., Laforge, G., Stojanoski, B., Nichols, E.S., McRae, K., Köhler, S.: Late positive complex in event-related potentials tracks memory signals when they are decision relevant. Sci. Rep. **9**(1), 1–15 (2019)
52. Zhitomirsky-Geffet, M., Bar-Ilan, J., Levene, M.: How and why do users change their assessment of search results over time? ASIST **52**(1), 1–4 (2015)

BioLCNet: Reward-Modulated Locally Connected Spiking Neural Networks

Hafez Ghaemi[1], Erfan Mirzaei[2], Mahbod Nouri[3], and Saeed Reza Kheradpisheh[4](✉)

[1] Politecnico di Torino, Torino, Italy
hafez.ghaemi@studenti.polito.it
[2] University of Tehran, Tehran, Iran
erfunmirzaei@ut.ac.ir
[3] The University of Edinburgh, Edinburgh, UK
m.nouri@sms.ed.ac.uk
[4] Shahid Beheshti University, Tehran, Iran
s_kheradpisheh@sbu.ac.ir

Abstract. Brain-inspired computation and information processing alongside compatibility with neuromorphic hardware have made spiking neural networks (SNN) a promising method for solving learning tasks in machine learning (ML). Spiking neurons are only one of the requirements for building a bio-plausible learning model. Network architecture and learning rules are other important factors to consider when developing such artificial agents. In this work, inspired by the human visual pathway and the role of dopamine in learning, we propose a reward-modulated locally connected spiking neural network, BioLCNet, for visual learning tasks. To extract visual features from Poisson-distributed spike trains, we used local filters that are more analogous to the biological visual system compared to convolutional filters with weight sharing. In the decoding layer, we applied a spike population-based voting scheme to determine the decision of the network. We employed Spike-timing-dependent plasticity (STDP) for learning the visual features, and its reward-modulated variant (R-STDP) for training the decoder based on the reward or punishment feedback signal. For evaluation, we first assessed the robustness of our rewarding mechanism to varying target responses in a classical conditioning experiment. Afterwards, we evaluated the performance of our network on image classification tasks of MNIST and XOR MNIST datasets.

Keywords: Spiking neural networks (SNN) · Bio-plausible learning · Spike-timing dependent plasticity (STDP) · Image classification

H. Ghaemi, E. Mirzaei, M. Nouri—Equal contribution

Supplementary Information The online version contains supplementary material available at https://doi.org/10.1007/978-3-031-25891-6_42.

G. Nicosia et al. (Eds.): LOD 2022, LNCS 13811, pp. 564–578, 2023.
https://doi.org/10.1007/978-3-031-25891-6_42

1 Introduction

Deep convolutional neural network (DCNN) has been one of the best-performing architectures in the field of computer vision and object recognition [12,22]. Despite the emergence of novel methods and architectures, such as visual transformers [6], DCNNs are still ubiquitous, and one of the most popular architectures employed for solving machine learning tasks. In addition, the representations formed in convolutional networks are similar to those found in the primate visual cortex [36]. Nevertheless, traditional DCNN has fundamental differences with biological visual system.

First of all, neuron activations in an artificial neural network (ANN) are static real-numbered values that are modeled by differentiable, non-linear activation functions. This is in contrast to biological neurons that use discrete, and mostly sparse spike trains to transmit information between each other, and in addition to the rate of spikes (spatial encoding), they also use spike timing to encode information temporally [40]. Therefore, a spiking neural network (SNN) is more akin to the neural networks in the brain. Spiking neural networks are more energy-efficient and compatible with neuromorphic hardware [5].

Secondly, the brain is incapable of error backpropagation, as done in traditional ANNs. One issue with error backpropagation in ANNs is the weight transport problem, i.e., the fact that weight connectivity in feedforward and feedback directions is symmetric [2,24]. Additionally, error feedback propagation, which does not affect neural activity, is not compliant with the feedback mechanisms that biological neurons use for communication [25].

Furthermore, although convolutional neural network has shown great potential in solving any translation-invariant task, its use of weight sharing is biologically implausible. There is no empirical support for explicit weight sharing in the brain [31]. However, local connections between neurons are biologically plausible because neurons in the biological visual system exploit them to have local visual receptive fields [13]. Also, Poggio et al. [30] showed that DCNN can avoid the curse of dimensionality for compositional functions not due to its weight sharing, but its depth.

One of the main challenges in designing SNNs is learning from labeled data in a bio-plausible manner. To address this challenge and to enable the development of networks with deeper architectures, we decompose feature extraction from classification. We consider the classification task as a decision making problem, and propose a dopamine-based learning scheme.

Considering the above-mentioned arguments, and to move towards more bio-plausible architectures and learning methods in SNNs, in this work, we are proposing BioLCNet, a reward-modulated locally connected spiking neural network. Local filters enable the extension of the famous Diehl and Cook's architecture [8] into a more realistic model of the biological receptive fields in the visual pathway of the brain. We extended this locally-connected structure with a fully-connected layer to mimic the hierarchical structure of the visual pathway and to decompose the tasks of feature extraction and recognition. Our network is trained using the unsupervised spike-timing-dependent plasticity and its semi-supervised variant, reward-modulated STDP (R-STDP). The input images are encoded proportional to the pixels intensity using Poisson spike-coding that converts intensity to average neuron firing rate in Hertz. The hidden layer extracts features using unsupervised STDP learning. In the output layer, there are neuronal groups for each class label, trained using R-STDP, and decision making is based on aggregated number of spikes during the decision period. In this layer, in addition to normal R-STDP, we employed TD-STDP in SNNs [10] that is inspired by the dopamine hypothesis in the brain [37] and incorporates the element of surprise in STDP learning. After conducting a classical conditioning experiment to prove the effectiveness of our decoding scheme and rewarding mechanism, we evaluate the classification performance of our network with different sets of hyperparameters on MNIST [21] and XOR MNIST datasets.

2 Related Work

Neuroscientists and deep learning researchers have long been searching for more biologically plausible deep learning approaches in terms of neuronal characteristics, learning rules, and connection types. Regarding neuronal characteristics, researchers have turned to biological neuronal models and spiking neural networks. The vanishing performance gap between deep neural networks (DNNs) and SNNs, and the compatibility of SNNs with neuromorphic hardware and online on-chip training [35] has piqued the interest of researchers [27].

Spiking neurons are activated by discrete input spike trains. This differs from artificial neurons used in an ANN that have differentiable activation functions and can easily employ backpropagation and gradient-based optimization. There are works that use gradient-based methods with SNNs and some of them have achieved great performances [3,20]. On the other hand, many works in this area use derivations of the Hebbian learning rule where changes in connection weights depend on the activities of the pre and post-synaptic neurons. Spike-timing-dependent plasticity (STDP) and its variants, apply asymmetric weight updates based on the temporal activities of neurons. Normal STDP requires an external read-out for classification [28], and have been applied to image reconstruction and classification tasks by many researchers [1,18,19]. Reward-modulated STDP (R-STDP) uses a reward (or punishment) signal to directly modulate the STDP weight change, and can be used to decode the output without an external cue. Izhikevich [17] tried to address the distal reward problem in reinforcement learning by using a version of R-STDP with decaying eligibility traces that gives recent spiking activity more importance. Around the same time, Florian [9] showed that R-STDP can be employed to solve a simple XOR task with both Poisson and temporal encoding of the output. Historically, R-STDP was first adopted with temporal (rank-order) encoding for image classification [28]. In this work, Mozafari et al. employed a convolutional architecture that uses a time-to-first-spike decoding scheme. An extended architecture was later developed which had multiple hidden layers [27]. The use of R-STDP with Poisson-distributed spike-coding has been mostly limited to fully-connected architectures for solving reinforcement learning robot navigation tasks [4]. In a recent work, Weidel et al. [41] proposed a spiking network with clustered connectivity and R-STDP learning to solve multiple tasks including classification of a small subset of MNIST with three classes.

Despite the biological nature of local connections, they mostly underperform convolution-based methods with weight sharing in the visual domain, especially on large-scale datasets [2]. This weaker performance may be mainly attributed to the smaller number of parameters and better generalization in CNNs. Fewer parameters in CNNs would also require less memory and computational cost, and would lead to faster training [30]. Studies are being done to bridge the performance gap between convolutional and locally-connected networks [2,25].

The most prevalent architectures used for image classification in deep learning with both DNNs and SNNs are based on convolutional layers and weight sharing. However, as mentioned before, there are arguments against the biological plausibility of these approaches [2,31]. Locally connected (LC) networks are an alternative to the convolutional ones. Illing et al. [16] show that shallow networks with localized connectivity and receptive fields perform much better than fully-connected networks on the MNIST benchmark. However, Bartunov et al. [2] showed that the lower generalization of LC networks compared to CNNs results in their underperforming CNNs in most image classification tasks, and prevents their scalability to larger datasets such as ImageNet. Recently, Pogodin et al. [31] proposed bio-inspired dynamic weight sharing and adding lateral connections to locally-connected layers to achieve the same regularization goals of weight sharing and normal convolutional filters. Saunders et al. [32] and Illing et al. [16] used local filters for learning visual feature representations in SNNs, and achieved a good performance on the MNIST benchmark with supervised decoding mechanisms.

3 Theory

In this section, we will outline the theoretical foundations underlying our proposed method. Specifically, the dynamics of the spiking neuronal model, the learning rules used, and the connection type employed in our network will be described.

3.1 Adaptive LIF Neuron Model

The famous leaky and integrate fire neuronal model is governed by the following differential equation [11],

$$\tau_m \frac{du}{dt} = -[u(t) - u_{rest}] + RI(t), \tag{1}$$

where $u(t)$ denotes the neuron membrane potential and is a function of time, R is the membrane resistance, $I(t)$ is any arbitrary input current, and τ_m is the membrane time constant. Equation (1) dictates that the neuron potential exponentially decays to a constant value u_{rest} over time. When a pre-synaptic neuron fires (spikes), it generates a current that reaches its post-synaptic neurons. In the simple leaky integrate and fire (LIF) model, a neuron fires when its potential surpasses a **constant** threshold u_{thr}. After firing, the neuron's potential resets to a constant u_{reset} and will not be affected by any input current for a period of time known as the refractory period (Δt_{ref}).

A variant of the LIF model uses adaptive firing thresholds. In this model, u_{thr} can change over time based on the neuron's rate of activity [8]. When a neuron fires, its tolerance to the input stimuli and consequently its firing threshold increases by a constant amount, g_0, otherwise the threshold decays exponentially with a time constant τ_g to the default threshold u_{thr_0}. Equations (2) to (4) explain the dynamics of the adaptive LIF model,

$$u_{thr}(t) = u_{thr_0} + g(t), \tag{2}$$

where,

$$\tau_g d_g/d_t = -g(t), \tag{3}$$

and

$$spike \Rightarrow g(t) = g(t-1) + g_0, \tag{4}$$

3.2 Reward-Modulated STDP

Spike-timing-dependent plasticity is a type of biological Hebbian learning rule that is also aligned with human intuition ("Neurons that fire together wire together." [26]). The normal STDP is characterized by two asymmetric update rules. The synaptic weights are updated based on the temporal activities of pre and post-synaptic neurons. When a pre-synaptic neuron fires shortly **before** its post-synaptic neuron, the causal connection between the first and the second neuron temporal activity is acknowledged, and the connection weight is increased. On the other hand, if the post-synaptic neuron fires shortly **after** the pre-synaptic neuron, the causality is undermined and the synaptic strength will decrease [15]. These weight updates, called long-term potentiation (LTP) and long-term depression (LTD), can be performed with asymmetric learning rates to adapt the learning rule to the excitatory-to-inhibitory neuron ratio or the connection patterns of a specific neural network. A popular variant of STDP that integrates reinforcement learning into the learning mechanism of spiking neural networks is reward-modulated STDP (also known as R-STDP or MSTDP [9]). In R-STDP, a global reward or punishment signal, which can be a function of time, is generated as the result of the network's activity or task performance. Using a notation similar to Florian [9], to mathematically formulate both STDP and R-STDP, we can define the spike train of a pre-synaptic neuron as the sum of Dirac functions over the spikes of the post-synaptic neurons,

$$\Phi(t) = \sum\nolimits_{\mathscr{F}_i} \delta(t - t_i^f). \tag{5}$$

where t_i^f is the firing time of the ith post-syanptic neuron. Now, we can define the variables P_{ij}^+ and P_{ij}^- to respectively track the influence of pre or post-synaptic spikes on weight updates. Now, the spike trace ξ for a given spike from neuron i to j can be defined as below,

$$\xi_{ij} = P_{ij}^+ \Phi_i(t) + P_{ij}^- \Phi_j(t), \tag{6}$$

where,

$$dP_j^+/dt = -P_j^+/\tau_+ + \eta_{post}\Phi_j(t), \tag{7}$$

$$dP_i^-/dt = -P_i^-/\tau_- - \eta_{pre}\Phi_i(t), \tag{8}$$

where we assumed that $P_{ij} = P_j$ for all pre-synaptic connections related to neuron j, and $P_{ij} = P_i$ for all post-synaptic connections related to neuron i.

The variables $\tau_\pm$ are the time constants determining the time window in which a spike can affect the weight updates. Using larger time constants will cause spikes that are further apart to also trigger weight updates. The variables η_{post} and η_{pre} determine the learning rate for LTP and LTD updates respectively. We denote the reward or punishment signal with $r(t)$. The R-STDP update rules for positive and negative rewards can be written as,

$$\frac{dw_{ij}(t)}{dt} = \gamma r(t)\xi_{ij}(t), \tag{9}$$

where γ is a scaling factor. The update rule for normal STDP can also be written as,

$$\frac{dw_{ij}(t)}{dt} = \gamma \xi_{ij}(t). \tag{10}$$

Based on Eq. (9), we note that R-STDP updates only take effect when a non-zero modulation signal is received at time step t. However, STDP updates do not depend on the modulation signal, and are applied at every time step. In other words, STDP can be considered a special case of R-STDP where the reward function is equal to 1 in every time step. This causes STDP to respond to the most frequent patterns regardless of their desirability.

3.3 Local Connections

A local connection in a neural network is similar to a convolutional connection but with distinct filters for each receptive field. As seen in Fig. 1, in normal convolutional connections, there is one filter for each channel that is convolved with all receptive fields as it moves along the layer's input. This filter has one set of weights that are updated using the network's update rule. However, In local connection (LC), after taking each stride, a new set of parameters characterize a whole new filter for the next receptive field. This type of connectivity between the input and the LC layer resembles the physical structure of retinal Ganglion cells. Because there are more filters in an LC, the number of distinct synapses in a local connection is greater than a convolutional connection, yet much lower than a dense connection. Similar to a convolutional connection, assuming square filters, and equal horizontal and vertical strides, we can specify a local connection by the number of channels (ch_{lc}), the kernel size (k), and the stride (s).

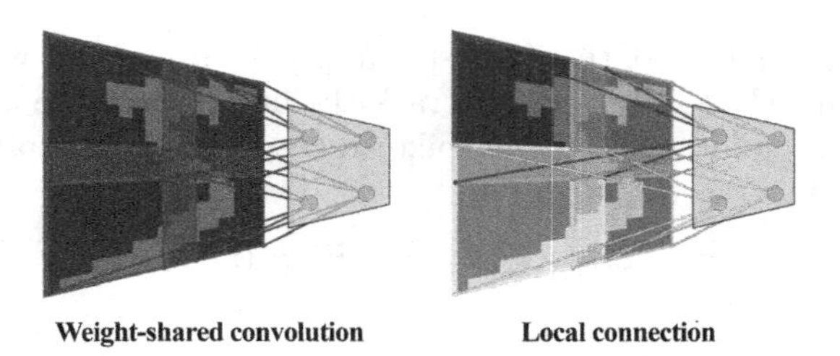

Fig. 1. Visual comparison of convolutional and local connections for a given channel; in convolutional connections, the weights are shared between all receptive fields. However, in a local connections, each receptive field has its own set of weights.

4 Architecture and Methods

BioLCNet consists of an input layer, a locally connected hidden layer, and a decoding layer. Each layer structure and its properties alongside the training and rewarding procedure will be delineated in this section. A graphical representation of our network is presented in Fig. 2. During our experiments, the simulation time T is divided into three phases, adaptation period (T_{adapt}), decision period (T_{dec}), and learning period (T_{learn}). The details of each phase will be specified in the remainder of this section.

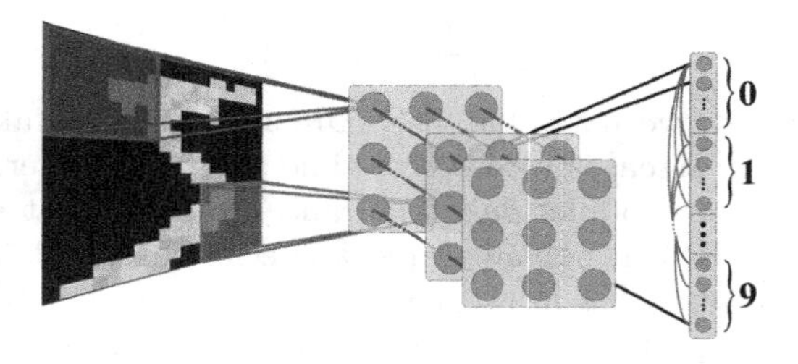

Fig. 2. Graphical representation of the proposed network; locally connected filters will be applied to the Poisson-distributed input image. Based on a winner-take-all inhibition mechanism, the most relevant features from each receptive field transmit their spikes to the decoding layer, which selects the most active neuronal group as the predicted label exploiting lateral inhibitory connections. The red lines indicate inhibitory connections. (Color figure online)

4.1 Encoding Layer

The input of the network is an image of dimensions (ch_{in}, h_{in}, w_{in}). For a grayscale image dataset such as MNIST, ch_{in} equals to one. Each input channel is encoded using a Poisson-distributed scheme, i.e., the spiking neuron corresponding to each pixel has an average firing rate proportional to the intensity of that pixel. By choosing the maximum firing rate f_{max}, the spike trains average firing rates will be distributed in the interval $[0, f_{max}]$ Hertz based on the pixel values.

4.2 Feature Extraction Layer (Local Connections)

The encoded input at each simulation time step passes through local connections with ch_{out} distinct filters for each receptive field. Therefore, the output of this layer will have dimensions $(ch_{out}, h_{out}, w_{out})$, where the output size depends on the size of the kernel and the stride. There are generally two approaches in the SNN literature for training a feature extraction layer with Poisson-distributed inputs using STDP to attain a rich feature representation and also prevent the weights from growing too large. One is allowing the weights to have negative values, which corresponds to having inhibitory neurons, as done in the convolutional layers used by Lee et al. [23]. The other is to use a combination of recurrent inhibitory connections and adaptive thresholds [8,32,33]. In this work, we used the latter approach for our feature extraction LC layer. We use adaptive LIF neurons and inhibitory connections between neurons that share the same receptive field. This is equivalent to the winner-take-all inhibition mechanism which causes a competition between neurons to select the most relevant features. The inhibitory connections are non-plastic and they all have a static negative weight w_{inh} with a large absolute value. As hypothesized by Diehl and Cook [8], the adaptive threshold of the neurons in this layer is a measure that may counterbalance the large number of inhibitory connections to each neuron, that is not compliant with the biological 4:1 excitatory to inhibitory ratio [7].

In normal STDP, the LTP learning rate (η_{post}) is usually chosen larger than the LTD rate (η_{pre}) to suppress the random firing of neurons that triggers many LTD updates during the early stages of training. However, this may become problematic in the later stages, and the weights may grow too large. Therefore, in practice, different mechanisms, such as weight clipping and normalization are used to prevent the weights running amok. In this work, we clipped the weights to stay in the range $[0, 1]$. We also employed the normalization technique [8,32] and normalized the pre-synaptic weights of each neuron in the LC layer to have a constant mean of c_{norm} at the end of each time step.

4.3 Decoding Layer and Rewarding Mechanisms

The final layer of our network is a fully connected layer for reward-based decoding. The layer is divided into n_c neuronal groups where n_c is the number of classes related to the task. Consequently, the n_{out} neurons in this layer are divided equally into n_c neuronal groups. The predicted label for a given test sample is the class whose group has the most number of spikes aggregated over the decision period (T_{dec}). This decoding layer is trained using reinforcement learning and R-STDP during the learning period (T_{learn}) based on the modulation signal generated by the rewarding mechanism. We exploited two different rewarding mechanisms, normal R-STDP [9], and TD-STDP [10]. In normal R-STDP, we use a fixed reward or punishment signal for the whole learning period (T_{learn}) based on the prediction of the network for the ith training sample,

$$r_i = \begin{cases} 1 : predicted\ label = target\ label \\ -1 : \qquad otherwise \end{cases} \tag{11}$$

The second mechanism, TD-STDP is based on the reward prediction error theory in neuroscience and reinforcement learning. According to this theory, the dopaminergic neurons in the brain release dopamine proportional to the difference between the actual reward and the expected reward, not solely based on the actual reward [37,39]. This mechanism involves the element of surprise in learning; the agent receives an amplified reward signal when it has a correct prediction after a sequence of wrong ones. Similarly, it receives an amplified punishment signal when a wrong prediction comes after a sequence of correct ones. We can formulate TD-STDP with exponential moving average as below,

$$M_i = \eta_{rpe}(r_i - \text{EMA}_r) \tag{12}$$

where M_i is the scalar TD-STDP modulation signal used during the whole learning period (T_{learn}) of the ith training sample, r_i is the reward signal received based on the prediction in (11), and EMA_r is the exponential moving average of the reward signals with a smoothing factor α.

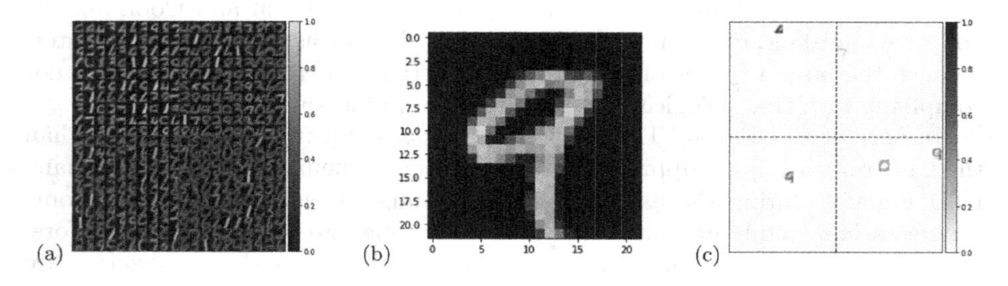

Fig. 3. Input and LC layer visualizations. (a) LC layer learned filters; the red lines separate filters corresponding to each receptive field. (b) A sample input image. (c) The LC layer activation map corresponding to the sample input image shown. (Color figure online)

4.4 Training Procedure

The network is trained in a layer-wise fashion. After initializing the weights uniformly random between $[0, 1]$, we train the feature extraction LC layer in a completely unsupervised manner using STDP. Simulation time for training the feature extraction layer is T_{learn} time steps. After this layer is trained, the weights are freezed, and we train the decoding FC layer in a semi-supervised manner using R-STDP and the selected rewarding mechanism. Training this layer requires all three simulation phases. The input image is first presented to the network for T_{adapt} time steps to let the LC layer neurons adapt to the input image and relevant features are selected based on the WTA mechanism. During T_{dec} time steps, the decoding layer accumulates the number of spikes received by each neuronal group to determine the predicted label. It was observed that a small T_{adapt} highly affected the quality of the network's decision. In other words, it is important to let the WTA mechanism take effect before entering the decision phase. Afterwards, the modulation signal is generated and the decoding layer weights are updated using R-STDP for a duration of T_{learn} time steps. It should be emphasized that we do not use the ground truth label in any of the training steps, and the feedback signal is generated only based on the validity of the network's decision.

When training the LC layer, we observed that after a specific number of iterations (training samples), the weights of this layer converge and remain constant. Figure 3a visualizes the filters learned after 2000 iterations for 100 channels with a filter (kernel) size of 15 with a stride of 4 applied to the input images of the MNIST dataset. This fast convergence is an evidence showing the strength of STDP learning. Considering these observations, and to save computation time, we limit the number of training sample of the LC layer to 2000 for all of the hyperparameter configurations for the two classification tasks. Given a sample MNIST input image (Fig. 3b), we plotted the activation map of the LC layer (Fig. 3c). This map shows that the post-synaptic neurons corresponding to the relevant features activate, and suppress the other neurons in accordance with the WTA inhibition mechanism.

The network is implemented using PyTorch [29], and mostly on top of the BindsNet framework [14]. We reimplemented the local connection topology to make it compatible with multi-channel inputs and a possible deep extension of our network. For transparency and to foster reproducibility, the code of the experiments are available publicly[1].

5 Experiments and Discussion

5.1 Classical Conditioning

In order to show the effectiveness of our rewarding mechanism, we perform a classical (Pavlovian) conditioning experiment. This type of conditioning pairs up a neutral stimulus with an automatic conditioned response by the agent. In this experiment, we present the network with images belonging to one class of the MNIST dataset as the neutral stimuli. We used the pre-trained feature extraction layer of a model with 25 channels with filter size of 13 and stride of 3, following by a decoding layer with 20 neurons for a two-class prediction task. In the first half of the experiment (task 1), the target response is class 1, and the network receives a constant reward of 1 if it predicts this class regardless of the input. A punishment signal of –1 is received if the agent predicts class 0. We monitor the rate of the reward and punishment received during the experiment. After the convergence in about 50 iterations, Fig. 4 shows that the agent has become completely conditioned on the rewarding response. After 200 iterations, we swap the rewarding and punishing classes, and continue running the network. In task 2, the network should predict the input images as class 0. The RL agent (the network) adapts to the change notably fast, and completely changes its behavior

[1] https://github.com/Singular-Brain/BioLCNet.

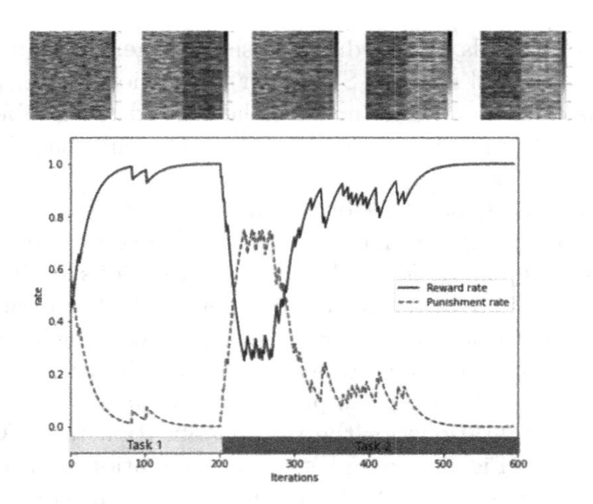

Fig. 4. Classical conditioning experiment; in this experiment, we tested the adaptability of the network to varying target responses. The plot shows the rate of receiving reward and punishment averaged over 20 runs, and the decoding layer weight maps at iterations 0, 200, 300, 400, and 600. The right half of the weight heat maps correspond to the task 1 target response neurons, and the left half corresponds to the task 2 target response neurons. The weights adapt to the varying target response during the experiment.

after about 100 iterations. The heat maps in Fig. 4 visualize the weights of the output layer through the training.

The reward adaptability of an RL agent is critical because in many real-world problems the environment is non-stationary. Integration of reward adaptation into spiking neural networks, as done in this work, can pave the path for models that simulate human behaviour with the same spike-based computation as done in the human brain.

5.2 MNIST

To evaluate our network's classification performance, we trained our model on the MNIST benchmark [21]. Some of the hyperparameters were fixed and others were subject to grid search. The full list of hyperparameters for this experiment are given in Table 1 in supplementary materials.

Considering the hyperparameters mentioned in Table 1 of the supplementary materials, we report in Table 1, the classification accuracy on the whole MNIST test set (10000 samples) for four hyperparameter configurations chosen based on the highest test accuracy obtained after conducting a grid search. The number of neurons and synapses for each model are also reported in this table. The final models were all trained using 10000 training samples from the MNIST training set. Using more training samples did not improve the classification performance as can be observed from Fig. 5. The mean and standard deviations reported are estimated from ten independent runs. In addition to the RL-based models, another classification approach was employed. In this approach, for each training sample, we create a feature vector containing the number of spikes aggregated over T_{learn} time steps for every filter in the LC layer. We use these feature vectors to train a support vector machine (SVM) classifier. The SVM results are also obtained by training on 10000 training samples, and testing on the whole MNIST test set. The SVM test results for two different hyperparameter configurations are reported in Table 1 and are compared to the RL-based results. The best performance of SVM and RL-based classification are 87.50, and 76.40 respectively.

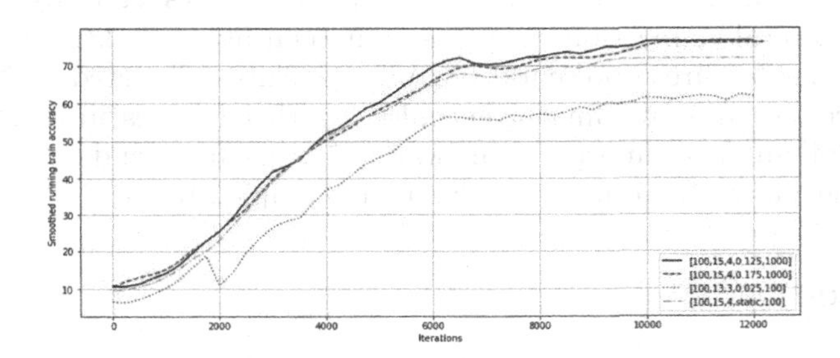

Fig. 5. Smoothed running accuracy over the training set for four sets of hyperparameters using the R-STDP classifier

Table 1. MNIST test dataset accuracies obtained by four different sets of hyperparameters; the test accuracies are averaged over ten independent runs

Parameters $[k,\ s,\ \eta_{rpe},\ n_{out}]$	$n_{neurons}$	$n_{synapses}$	Test accuracy	SVM test accuracy
[13, 3, 0.025, 100]	1700	430400	61.30 ± 3.14	87.5 ± 1.32
[15, 4, 0.175, 1000]	1884	490000	75.00 ± 2.68	83.3 ± 1.74
[15, 4, 0.125, 1000]	1884	490000	76.40 ± 2.43	83.3 ± 1.74
[15, 4, (static), 100]	984	130000	68.8 ± 2.87	83.3 ± 1.74

We observed that the features that are activated in the hidden layer and transferred to the decoding layer may overlap in samples with different labels but similar form, which may lower the classification performance of the RL-based and SVM classifiers. This claim is supported by the low SVM performance compared to its usual performance on the MNIST benchmark. Two important observations can be made from Table 1. First, the classification accuracy has a positive correlation with the filter size, and the number of neurons in the decoding layer. Consequently, training a larger network with higher number of channels and decoding neurons may be a possible way to improve the classification performance. This solution was neglected due to an exponential growth of the training time when increasing the number of channels. Secondly, as shown in Table 1, using TD-STDP instead of normal (static) R-STDP improved the classification performance. Indeed, in TD-STDP, the agent exploits the element of surprise to modulate its learning signal, which in turn, would result in a more stable learning. The results of another experiment on the XOR MNIST dataset is provided in the supplementary materials.

6 Conclusions and Future Work

In this work, we examined the capabilities of a neural network with three-fold biological plausibility; spiking neurons, local visual receptive fields, and a reward-modulated learning rule. In the future, by bringing ideas such as dynamic weight sharing and lateral connections [31] to spiking neural networks, we may be able to obtain richer feature representations using locally connected SNNs. Exploiting structures, such as pooling layers, along with advances in SNN minibatch processing [34] and neuromorphic hardware [35], we can extend our network with larger and deeper architectures to solve more complex tasks.

References

1. Allred, J.M., Roy, K.: Unsupervised incremental stdp learning using forced firing of dormant or idle neurons. In: 2016 International Joint Conference on Neural Networks (IJCNN), pp. 2492–2499. IEEE (2016)
2. Bartunov, S., Santoro, A., Richards, B., Marris, L., Hinton, G.E., Lillicrap, T.: Assessing the scalability of biologically-motivated deep learning algorithms and architectures. In: Bengio, S., Wallach, H., Larochelle, H., Grauman, K., Cesa-Bianchi, N., Garnett, R. (eds.) Advances in Neural Information Processing Systems, vol. 31. Curran Associates, Inc. (2018)
3. Bellec, G., et al.: A solution to the learning dilemma for recurrent networks of spiking neurons. Nat. Commun. **11**(1), 1–15 (2020)
4. Bing, Z., Jiang, Z., Cheng, L., Cai, C., Huang, K., Knoll, A.: End to end learning of a multi-layered snn based on r-stdp for a target tracking snake-like robot. In: 2019 International Conference on Robotics and Automation (ICRA), pp. 9645–9651. IEEE (2019)
5. Cao, Y., Chen, Y., Khosla, D.: Spiking deep convolutional neural networks for energy-efficient object recognition. Int. J. Comput. Vis. **113**(1), 54–66 (2015)
6. Carion, N., Massa, F., Synnaeve, G., Usunier, N., Kirillov, A., Zagoruyko, S.: End-to-end object detection with transformers. In: Vedaldi, A., Bischof, H., Brox, T., Frahm, J.-M. (eds.) ECCV 2020. LNCS, vol. 12346, pp. 213–229. Springer, Cham (2020). https://doi.org/10.1007/978-3-030-58452-8_13
7. Connors, B.W., Gutnick, M.J.: Intrinsic firing patterns of diverse neocortical neurons. Trends Neurosci. **13**(3), 99–104 (1990)
8. Diehl, P.U., Cook, M.: Unsupervised learning of digit recognition using spike-timing-dependent plasticity. Front. Comput. Neurosci. **9**, 99 (2015)
9. Florian, R.V.: Reinforcement learning through modulation of spike-timing-dependent synaptic plasticity. Neural Comput. **19**(6), 1468–1502 (2007)
10. Frémaux, N., Gerstner, W.: Neuromodulated spike-timing-dependent plasticity, and theory of three-factor learning rules. Front. Neural Circ. **9**, 85 (2016)

11. Gerstner, W., Kistler, W.M., Naud, R., Paninski, L.: Neuronal Dynamics: From Single Neurons to Networks and Models of Cognition. Cambridge University Press, Cambridge (2014)
12. Goodfellow, I., Bengio, Y., Courville, A.: Deep Learning. MIT Press (2016). https://www.deeplearningbook.org
13. Gregor, K., LeCun, Y.: Emergence of complex-like cells in a temporal product network with local receptive fields (2010)
14. Hazan, H.: Bindsnet: a machine learning-oriented spiking neural networks library in python. Front. Neuroinform. **12**, 89 (2018)
15. Hebb, D.O.: The Organisation of Behaviour: A Neuropsychological Theory. Science Editions New York, New York (1949)
16. Illing, B., Gerstner, W., Brea, J.: Biologically plausible deep learning-but how far can we go with shallow networks? Neural Netw. **118**, 90–101 (2019)
17. Izhikevich, E.M.: Solving the distal reward problem through linkage of stdp and dopamine signaling. Cereb. Cortex **17**(10), 2443–2452 (2007)
18. Kheradpisheh, S.R., Ganjtabesh, M., Masquelier, T.: Bio-inspired unsupervised learning of visual features leads to robust invariant object recognition. Neurocomputing **205**, 382–392 (2016)
19. Kheradpisheh, S.R., Ganjtabesh, M., Thorpe, S.J., Masquelier, T.: Stdp-based spiking deep convolutional neural networks for object recognition. Neural Netw. **99**, 56–67 (2018)
20. Kheradpisheh, S.R., Masquelier, T.: Temporal backpropagation for spiking neural networks with one spike per neuron. Int. J. Neural Syst. **30**(06), 2050027 (2020)
21. LeCun, Y., Cortes, C., Burges, C.: The mnist dataset of handwritten digits (images). NYU: New York, NY, USA (1999)
22. LeCun, Y., Bengio, Y., Hinton, G.: Deep learning. Nature **521**(7553), 436–444 (2015)
23. Lee, C., Srinivasan, G., Panda, P., Roy, K.: Deep spiking convolutional neural network trained with unsupervised spike-timing-dependent plasticity. IEEE Trans. Cogn. Dev. Syst. **11**(3), 384–394 (2018)
24. Liao, Q., Leibo, J., Poggio, T.: How important is weight symmetry in backpropagation? In: Proceedings of the AAAI Conference on Artificial Intelligence, vol. 30 (2016)
25. Lillicrap, T.P., Santoro, A., Marris, L., Akerman, C.J., Hinton, G.: Backpropagation and the brain. Nat. Rev. Neurosci. **21**(6), 335–346 (2020)
26. Lowel, S., Singer, W.: Selection of intrinsic horizontal connections in the visual cortex by correlated neuronal activity. Science **255**(5041), 209–212 (1992)
27. Mozafari, M., Ganjtabesh, M., Nowzari-Dalini, A., Thorpe, S.J., Masquelier, T.: Bio-inspired digit recognition using reward-modulated spike-timing-dependent plasticity in deep convolutional networks. Pattern Recogn. **94**, 87–95 (2019)
28. Mozafari, M., Kheradpisheh, S.R., Masquelier, T., Nowzari-Dalini, A., Ganjtabesh, M.: First-spike-based visual categorization using reward-modulated stdp. IEEE Trans. Neural Netw. Learn. Syst. **29**(12), 6178–6190 (2018)
29. Paszke, A., et al.: Pytorch: an imperative style, high-performance deep learning library. In: Wallach, H., Larochelle, H., Beygelzimer, A., d'Alché-Buc, F., Fox, E., Garnett, R. (eds.) Advances in Neural Information Processing Systems, vol. 32, pp. 8024–8035. Curran Associates, Inc. (2019). https://papers.neurips.cc/paper/9015-pytorch-an-imperative-style-high-performance-deep-learning-library.pdf
30. Poggio, T., Mhaskar, H., Rosasco, L., Miranda, B., Liao, Q.: Why and when can deep-but not shallow-networks avoid the curse of dimensionality: a review. Int. J. Autom. Comput. **14**(5), 503–519 (2017)

31. Pogodin, R., Mehta, Y., Lillicrap, T., Latham, P.E.: Towards biologically plausible convolutional networks. Adv. Neural. Inf. Process. Syst. **34**, 13924–13936 (2021)
32. Saunders, D.J., Patel, D., Hazan, H., Siegelmann, H.T., Kozma, R.: Locally connected spiking neural networks for unsupervised feature learning. Neural Netw. **119**, 332–340 (2019)
33. Saunders, D.J., Siegelmann, H.T., Kozma, R., et al.: Stdp learning of image patches with convolutional spiking neural networks. In: 2018 International Joint Conference on Neural Networks (IJCNN), pp. 1–7. IEEE (2018)
34. Saunders, D.J., Sigrist, C., Chaney, K., Kozma, R., Siegelmann, H.T.: Minibatch processing for speed-up and scalability of spiking neural network simulation. In: 2020 International Joint Conference on Neural Networks (IJCNN), pp. 1–8. IEEE (2020)
35. Schemmel, J., Brüderle, D., Grübl, A., Hock, M., Meier, K., Millner, S.: A wafer-scale neuromorphic hardware system for large-scale neural modeling. In: 2010 IEEE International Symposium on Circuits and Systems (ISCAS), pp. 1947–1950. IEEE (2010)
36. Schrimpf, M., et al.: Brain-score: which artificial neural network for object recognition is most brain-like? BioRxiv, p. 407007 (2020)
37. Schultz, W., Dayan, P., Montague, P.R.: A neural substrate of prediction and reward. Science **275**(5306), 1593–1599 (1997)
38. Sun, S.H.: Multi-digit mnist for few-shot learning (2019). https://github.com/shaohua0116/MultiDigitMNIST
39. Sutton, R.S., Barto, A.G.: Reinforcement Learning: An Introduction. MIT Press, Cambridge (2018)
40. Tavanaei, A., Ghodrati, M., Kheradpisheh, S.R., Masquelier, T., Maida, A.: Deep learning in spiking neural networks. Neural Netw. **111**, 47–63 (2019)
41. Weidel, P., Duarte, R., Morrison, A.: Unsupervised learning and clustered connectivity enhance reinforcement learning in spiking neural networks. Front. Comput. Neurosci. **15**, 18 (2021)

Author Index

G. Nicosia et al. (Eds.): LOD 2022, LNCS 13811, pp. 579–582, 2023.
https://doi.org/10.1007/978-3-031-25891-6